AF807364

DIRECT IMAGING OF EXOPLANETS:
SCIENCE AND TECHNIQUES

IAU COLLOQUIUM No. 200

COVER ILLUSTRATION: Watercolor of *Villefranche-sur-Mer*

Dealing with such distant worlds as exoplanets should not make us forget of nice historical places on Earth like the city of *Villefranche-sur-Mer*, founded by *Charles II D'Anjou* (1285-1309), as a city (*ville*) free of taxes (*franche*).

The colloquium was held in the *Citadelle Saint-Elme*, a fortification build in 1566 by *André Provana de Leyni*, for the *Duc de Savoie Emmanuel-Philibert*.

Since 1388, *Nice* and its *Conté* were part of the *Duché de Savoie*.

The *Citadelle* was erected to protect *Nice* from the invaders, after the disaster of 1543 in which the Turkish flotilla of *Frédéric Barberousse*, allied to the King of France *François* 1er, landed in the protected harbor of *Villefranche-sur-Mer* to invade *Nice*. A local washer woman and heroine of *Nice*, *Catherine Ségurane*, saved the city from its Turkish assailants by inventing a new technique of war.

The watercolor is by *Yves Rabbia*, an astronomer at the *Observatoire de la Côte d'Azur* and an amateur painter when not working with coronagraphs and nulling interferometers.

IAU SYMPOSIUM AND COLLOQUIUM PROCEEDINGS SERIES

2005 EDITORIAL BOARD

Chairman

K.A. VAN DER HUCHT, IAU Assistant General Secretary
SRON National Institute for Space Research
Sorbonnelaan 2, NL-3584 CA Utrecht, The Netherlands
K.A.van.der.Hucht@sron.nl

Advisors

O. ENGVOLD, IAU General Secretary, *Institute of Theoretical Astrophysics, University of Oslo, Norway*
E.J. DE GEUS, *Netherlands Foundation for Research in Astronomy, Dwingeloo, The Netherlands*
M.C. STOREY, *Australia Telescope National Facility, Australia*
P.A. WHITELOCK, *South African Astronomical Observatory, South Africa*

Members

IAUS227
R. CESARONI, *Osservatorio Astrofisico di Arcetri, I.N.A.F., Firenze, Italia*
IAUS228
V. HILL, *Observatoire de Meudon, Meudon, France*
IAUS229
D. LAZZARO, *Observatorio Nacional, Rio de Janeiro, Brasil*
IAUS230
E.J.A. MEURS, *Dunsink Observatory, Dublin, Ireland*
IAUS231
D.C. LIS, *Downs Laboratory of Physics, Caltech, Pasadena, CA, USA*
IAUS232
M. DENNEFELD, *Institut d'Astropysique de Paris, Paris, France*
IAUC198
H. JERJEN, *Mt Stromlo Observatory, Australian National University, Canberra, Australia*
IAUC199
P.R. WILLIAMS, *Shanghai Astronomical Observatory, Chinese Academy of Sciences, Shanghai, China*
IAUC200
C. AIME, *Laboratoire Universitaire d'Astrophysique de Nice, Nice, France*

INTERNATIONAL ASTRONOMICAL UNION

UNION ASTRONOMIQUE INTERNATIONALE

International Astronomical Union

DIRECT IMAGING OF EXOPLANETS: SCIENCE AND TECHNIQUES

PROCEEDINGS OF THE 200th COLLOQUIUM OF THE INTERNATIONAL ASTRONOMICAL UNION HELD IN VILLEFRANCHE SUR MER, FRANCE OCTOBER 3–7, 2005

Edited by

CLAUDE AIME and FARROKH VAKILI

Laboratoire Universitaire d'Astrophysique de Nice

UNIVERSITE DE NICE SOPHIA ANTIPOLIS

06108 NICE Cedex 2

CAMBRIDGE

UNIVERSITY PRESS

CAMBRIDGE UNIVERSITY PRESS
The Edinburgh Building, Cambridge CB2 2RU, UK
40 West 20th Street, New York, NY 10011 4211, USA
477 Williamstown Road, Port Melbourne, VIC 3207, Australia
Ruiz de Alarcón 13, 28014 Madrid, Spain
Dock House, The Waterfront, Cape Town 8001, South Africa

© International Astronomical Union 2006

This book is in copyright. Subject to statutory exception
and to the provisions of relevant collective licensing agreements,
no reproduction of any part may take place without
the written permission of the International Astronomical Union.

First published 2006

Printed in the United Kingdom at the University Press, Cambridge

Typeset in System LaTeX 2_ε

A catalogue record for this book is available from the British Library

Library of Congress Cataloguing in Publication data

ISBN 978 0 85607 2
ISBN 0 521 85607 8 hardback
ISSN 1743-9213

Table of Contents

Part 1. Science: theoretical and observational aspects of exoplanetary systems

Part 2. Instrumentations and Techniques

Section A. Classical and nulling interferometry of exoplanets

Section B. Coronagraphy, adaptive optics and related instrumentations

Section C. Data processing for high dynamic imagery

Preface

The topic of the IAU Colloquium No 200 "Direct Imaging of Exoplanets: Science and Techniques" was the detection and analysis of photons from terrestrial and giant exoplanets. The Colloquium was held in the citadel of Villefranche-sur-Mer, a 16th century castle located at sea level in the beautiful bay of Villefranche, near Nice.

The duration of the Colloquium was five days, starting Monday October 3rd, 2005, with a break on Wednesday afternoon. The two first days of the Conference were devoted to science goals of exoplanets detection: interest of their direct observation, physics of terrestrial and giant planets, instrumental requirements in the visible and the infrared and reports of the first observations of very massive exoplanets. The three last days of the Conference were on ground and space instrumentation to achieve these goals. There was half a day for interferometric nulling techniques, one day for coronagraphy and the rest of the communications were on innovative techniques and data processing.

We had 66 oral presentations, including 7 invited talks. We deliberately avoided having SOC members as invited speakers. The 16 sessions of the conference have been successively chaired by Michael Shao, Anne-Marie Lagrange, Sara Heap, Victoria Meadows, Farrokh Vakili, Neville Woolf, Malcolm Fridlund, Yves Rabbia, Wesley Traub, Motohide Tamura, France Allard, Virginia Ford, Andreas Quirrenbach and Daniel Rouan. The oral presentations were complemented by about 70 Poster presentations split in two sessions of 2 hours each, chaired by Claude Aime and André Ferrari. Except for a few rare papers not received in time, these Proceedings contain all the contributions regrouped on a thematic basis. They described some of the latest theoretical developments in understanding the formation and evolution of planetary systems. The state of the art for both the instrumental concepts and data analysis methods were largely presented and debated. All world-wide top experts were present. We also favored the participation of a large number of students, post-docs and young scientists developing their own observing science concepts and instruments relevant to the selected topics of this colloquium.

The context of the meeting was the preparation of both ground-based and space-borne ambitious projects for the detection of exoplanets by imaging, especially terrestrial planets. Key-projects of the major world agencies such as ESA, ESO, NASA, NASDA were outlined. The impact of the conference can also be measured by the participation of scientists in charge of the major European, Japanese and US projects in exoplanet imaging.

The meeting was not only useful for delivering new information on several instrumental topics and scientific perspectives. It has also underlined several conclusions for the present status and future directions for the actions of the exoplanet community and agencies scientific policy. First, there is a remarkable agreement among the exoplanets community on the goal of characterizing them by imaging. Some important questions remain, that are guidelines for future actions: Should we consider different definitions of "a planet"? We should prepare ourselves scientifically about unexpected characteristics of planets. Do Earth-like planets exist nearby? How do we prepare for the public response to their discovery? What if the result is "No Earthlike planet out there"? Are precursor ground and space missions needed? The relative merits of ground-based ELT and space missions should be strongly investigated further. How can we stimulate international co-operation, necessary for large projects? What could be the benefits for other branches of astrophysics?

In addition to the usual scientific debates and discussions between participants, semi-formal contacts were initiated in view of establishing future international cooperation on the most ambitious ground and space projects for the coming decade and the long term of research in exoplanetary astrophysics.

The major part of the presentations given during the Colloquium is available in PDF Format at the web site of the *Laboratoire Universitaire d'Astrophysique de Nice* (http://www-luan.unice.fr).

Claude Aime and Jean Schneider, co-chairs SOC,
Farrokh Vakili, chair LOC
Nice, France, December 21, 2005

THE ORGANIZING COMMITTEE

Scientific

C. Aime (chair, France)
T. Henning (Germany)
A. Quirrenbach (the Netherlands)
D. Rouan (France)
S. Seager (USA)
M. Tamura (Japan)
M. Fridlund (Sweden)
A-M. Lagrange (France)
R. Ragazzoni (Italy)
J. Schneider(co-chair, France)
M. Shao (USA)
W. Traub (USA)

Local

F. Vakili (chair)
V. Cheron
A. Ferrari
C. Aime
M. Carbillet
Y. Rabbia

Acknowledgements

The organizers acknowledge support of the International Astronomical Union via Division III (Planetary Systems Sciences), Commission No.9 (Instrumentation and Techniques), Commission No.51 (Bioastronomy: Search for Extraterrestrial Life), Working Group on Extrasolar Planets and Working Group on Future Large Scale Facilities.

The Local Organizing Committee acknowledge funding from:

Centre National d'Etudes Spatiales
Région Provence Alpes Côte d'Azur
Université de Nice Sophia Antipolis
Observatoire de Paris
Laboratoire Gemini de l'Observatoire de la Côte d'Azur
Laboratoire Universitaire d'Astrophysique de Nice
Ecole Doctorale Sciences Fondamentales et Appliques
Action Spécifique Haute Résolution angulaire du
Centre National de la Recherche Scientifique
Laboratoire d'Etudes Spatiales et d'Instrumentation en Astrophysique

We heartily acknowledge Ms Anne-Marie Lagrange for her personal contribution to the expenses of the Colloquium with her *Grand Prix de l'Académie des Sciences.*

Photograph by
Laurent Thareau
[l.thareau@free.fr].

Participants

Lyu **Abe**, National Astronomical Observatory of Japan, Mitaka, Japan abe@optik.mtk.nao.ac.jp
Cristina **Afonso**, Max-Planck fr Astronomie, Heidelberg, Germany afonso@mpia.de
Claude **Aime**, Laboratoire Universitaire d'Astrophysique de Nice, France Claude.Aime@unice.fr
France **Allard**, CRAL-ENS Lyon cédex, France fallard@ens-lyon.fr
J. Roger P. **Angel**, University of Arizona, Tucson, USA rangel@as.arizona.edu
Jonathan **Arenberg**, Northrop Grumman Redondo Beach,USA jon.arenberg@ngc.com
Luc **Arnold**, Observatoire de Haute-Provence, Saint-Michel-l'Observatoire, France arnold@obs-hp.fr
Kunjithapatham **Balasubramanian**, Jet Propulsion Lab, Pasadena, USA kbala@jpl.nasa.gov
Marc **Barillot**, ALCATEL SPACE, Cannes la Bocca, France mark.barillot@alcatelaleniaspace.com
Richard **Barry**, NASA, Greenbelt, USA richard.k.barry@nasa.gov
Pierre **Baudoz**, Observatoire de Meudon, France pierre.baudoz@obspm.fr
Charles **beichman**, Caltech, Pasadena, CA, USA chas@pop.jpl.nasa.gov
Ruslan **Belikov**, Princeton University, Princeton, NJ, USA rbelikov@princeton.edu
Adrian **Belu**, Laboratoire Universitaire d'Astrophysique de Nice, France adrian.belu@unice.fr
Philippe **Bendjoya**, Laboratoire Universitaire d'Astrophysique de Nice, France bendjoya@unice.fr
Slimane **Bensammar**, Observatoire de Paris, France slimane.bensammar@obspm.fr
Willy **Benz**, University of Bern, Bern, Switzerland wbenz@phim.unibe.ch
Alessandro **Berton**, MPIA - Heidelberg, Germany berton@mpia.de
Jean-Luc **Beuzit**, Laboratoire d'Astrophysique de Grenoble, France Jean-Luc.Beuzit@obs.ujf-grenoble.fr
Gabriel **Bihain**, Instituto de Astrofsica de Canarias,La Laguna, Spain gbihain@ll.iac.es
Beth **Biller**, University of Arizona, Tucson, Arizona, USA bbiller@as.arizona.edu
Anthony **Boccaletti**, Obs Paris-Meudon, Meudon France anthony.boccaletti@obspm.fr
Daniel **Bonneau**, Observatoire de la Côte d'Azur, Département GEMINI, Grasse, France Daniel.Bonneau@obs-azur.fr
Pascal **Bordé**, Harvard-Smithsonian Center for Astrophysics, Pasadena, USA pborde@cfa.harvard.edu
Julien **Borgnino**, Laboratoire Universitaire d'Astrophysique de Nice, France borgnino@unice.fr
Virginie **Borkowski**, LISE - Collège de France, St Michel l'Observatoire France borkowski@obs-hp.fr
Pierre **bourget**, CNRS - Service d'Aéronomie PARIS, France pierre@aerov.jussieu.fr
James **Breckinridge**, California Institute of Technology, Pasadena, USA jbreckin@jpl.nasa.gov
Robert **Brown**, Space Telescope Science Institute, Silver Spring, MD, USA rbrown@stsci.edu
Christophe **Buisset**, O.C.A. Grasse, France christophe.buisset@obs-azur.fr
Juan **Cabrera**, Observatoire de Paris, Meudon, France juan.cabrera@obspm.fr
John **Campbell**, James Cook University, Cairns, Australia John.Campbell@jcu.edu.au
Marcel **Carbillet**, Laboratoire Universitaire d'Astrophysique de Nice, France marcel.carbillet@unice.fr
Joseph **Carson**, Jet Propulsion Laboratory, Pasadena, USA Joseph.C.Carson@jpl.nasa.gov
Céline **Cavarroc**, Observatoire de Paris-Meudon, Meudon, France celine.cavarroc@obspm.fr
Bruno **Chazelas**, IAS Orsay, France bruno.chazelas@laposte.net
Alain **Chelli**, Laboratoire d'Astrophysique de Grenoble (LAOG), Grenoble, Cedex 9, France Alain.Chelli@obs.ujf-grenoble.fr
Jessie **Christiansen**, University of New South Wales, Sydney, Australia jessiec@phys.unsw.edu.au
Laird **Close**, University of Arizona, Tucson, USA lclose@as.arizona.edu
Martin **Cohen**, University of California, Berkeley, USA mcohen@astro.berkeley.edu
Vincent **Coudé du Foresto**, LESIA - Paris Observatory, Meudon, France vincent.foresto@obspm.fr
Daniel **Coulter**, Jet Propulsion Laboratory, Pasadena, USA Daniel.R.Coulter@jpl.nasa.gov
Athena **Coustenis**, Paris-Meudon Observatory, 92195 MEUDON Cedex, France athena.coustenis@obspm.fr
Justin **Crepp**, University of Florida, Gainesville, USA jcrepp@astro.ufl.edu
Jean-Baptiste **Daban**, Laboratoire Universitaire d'Astrophysique de Nice, France jean-baptiste.daban@unice.fr
William **Danchi**, NASA GSFC, Greenbelt, USA wcd@iri1.gsfc.nasa.gov
Christian **De Duve**, Institute of Cellular Pathology Christian de Duve, Brussels, BELGIUM deduve_private@icp.ulc.ac.be
Cor **de Vries**, SRON, Utrecht, Netherlands C.P.de.Vries@sron.nl
Roland **den Hartog**, ESA / ESTEC, Noordwijk, Netherlands rdhartog@rssd.esa.int
Kjetil **Dohlen**, LAM-OAMP Marseille Cedex 4, France kjetil.dohlen@oamp.fr
Youssef **El Azhari**, ENS-Marrakech / UCAM Marrakech, Morocco elazhari@ucam.ac.ma
Keigo **Enya**, ISAS/JAXA, Kanagawa, Japan enya@ir.isas.jaxa.jp
Nicolas **Epchtein**, Laboratoire Universitaire d'Astrophysique de Nice, France epchtein@unice.fr
James **Fanson**, Jet Propulsion Laboratory, Pasadena, USA James.L.Fanson@jpl.nasa.gov
Markus **Feldt**, Max Planck Institute for Astronomy, Neckargemnd, Germany mfeldt@mpia.de
Marc **Ferrari**, LAM Marseille cedex 4, France marc.ferrari@oamp.fr
André **Ferrari**, Laboratoire Universitaire d'Astrophysique de Nice, France ferrari@unice.fr
Virginia **Ford**, Jet Propulsion Laboratory California Institute of Technology, MS 301-451, Pasadena, USA virginia.g.ford@jpl.nasa.gov
Eric **Ford**, Eidetic Optical Systems La Canada, CA 91011, USA eidetic@dslextreme.com
Jonathan **Fortney**, NASA Ames Research Center Moffett Field, CA USA jonathan-fortney@yahoo.com
Eric **Fossat**, Laboratoire Universitaire d'Astrophysique de Nice, France eric.fossat@unice.fr
Franois **Fressin**, OCA - Dpt. Gemini NICE Cédex 4, France fressin@obs-nice.fr
Malcolm **Fridlund**, ESA ESTEC Noordwijk, Netherlands malcolm.fridlund@esa.int
Thierry **Fusco**, ONERA, Chatillon, France thierry.fusco@onera.fr
Eric **Gaidos**, University of Hawaii, Honolulu, USA gaidos@hawaii.edu
Raphal **Galicher**, ENS Ulm, Paris, France galicher@clipper.ens.fr
Jean **Gay**, OCA - Dpt. Gemini, NICE Cédex 4, France gay@obs-nice.fr
Kerstin **Geissler**, MPIA, Heidelberg, Germany geissler@mpia.de
Amir **Give'on**, Princeton University Princeton, USA agiveon@princeton.edu
Pierre-Marie **Gori**, Université de Corse Pascal Paoli, Corte, Corse Pierre-Marie.gori@unice.fr
Nick **Gorkavyi**, Crimean Astrophysical Observatory, Haymarket, USA simeiz@aol.com
Raffaele **Gratton**, INAF, Osservatorio Astronomico di Padova, Padova, Italy gratton@pd.astro.it
Catherine **Grec**, Laboratoire Universitaire d'Astrophysique de Nice, France catherine.grec@unice.fr
Géraldine **Guerri**, Laboratoire Universitaire d'Astrophysique de Nice, France guerri@unice.fr
Olivier **Guyon**, Subaru Telescope Hilo, HI, USA guyon@naoj.org
Slim **Hamdani**, OHP, St Michel l'Observatoire, France hamdani@obs-hp.fr
Sara **Heap**, NASA/Goddard, Greenbelt, MD, USA Sally.Heap@nasa.gov
Thomas **Henning**, Max-Planck-Institut für Astronomie, D-69117 Heidelberg, Germany henning@mpia.de
Cédric **Jacob**, Laboratoire Universitaire d'Astrophysique de Nice, France cjacob@unice.fr
Slobodan **Jankov**, Laboratoire Universitaire d'Astrophysique de Nice, France Slobodan.Jankov@unice.fr
James **Jenkins**, University of Hertfordshire, Hatfield, United Kingdom jsj@star.herts.ac.uk
Franco **Joos**, Institute of Astronomy, ETH Zurich, Zürich, Switzerland fjoos@astro.phys.ethz.ch
Lisa **Kaltenegger**, CfA, Harvard Smithsonian Center for Astrophysics, Cambridge, USA lkaltenegger@cfa.harvard.edu
Markus **Kasper**, ESO Garching, Germany mkasper@eso.org
Kacper **Kornet**, MPIA Heidelberg, Germany kornet@mpia.de
Nadiya **Kostogryz**, Main astronomical Observatory of NAS of Ukraine, Kiev, Ukraine kosn@mao.kiev.ua
Rodolphe **Krawczyk**, Alcatel Space, Cannes la Bocca, France rodolphe.krawczyk@alcatelaleniaspace.com
John **Krist**, Jet Propulsion Laboratory, Pasadena, USA john.e.krist@jpl.nasa.gov
Lia **La Piana**, NASA Headquarters Washington, DC, USA Lia.S.LaPiana@nasa.gov
Antoine **Labeyrie**, LISE - Collge de France, St Michel l'Observatoire, France labeyrie@obs-hp.fr
Anne-Marie **Lagrange**, L.A.O.G. , Grenoble Cedex 09, France Anne-Marie.Lagrange@obs.ujf-grenoble.fr
Maud **Langlois**, OAMP, Allauch, France maud.langlois@oamp.fr
Olivier **Lardiere**, Collge de France, St-Michel-l'observatoire, France lardiere@obs-hp.fr
Jean-Michel **Le Duigou**, CNES, Toulouse, Cedex 09, France Jean-Michel.LeDuigou@cnes.fr
Brice **Le Roux**, Osservatorio Astrofisico di ARCETRI, firenze, Italy leroux@arcetri.astro.it
Alain **Leger**, IAS, Orsay, France Alain.Leger@ias.fr
B. Martin **Levine**, California Inst. of Tech. NASA Jet Propulsion Laboratory, Pasadena, USA bmlevine@s383.jpl.nasa.gov
Charles **Lillie**, Northrop Grumman Space Technology, Redondo Beach, USA chuck.lillie@ngc.com

Michael **Liu**, IfA/Hawaii, Honolulu, USA mliu@ifa.hawaii.edu
James Lloyd **Cornell**, University Ithaca, USA jpl@astro.cornell.edu
Bruno **Lopez**, Observatoire de la Côte d'Azur Nice Cédex 4 France bruno.lopez@obs-nice.fr
Richard **Lyon**, NASA/Goddard Space Flight Center Greenbelt, MD, USA Richard.G.Lyon@nasa.gov
Russell **Makidon**, Space Telescope Science Institute, Baltimore, USA makidon@stsci.edu
Salvatore **Mancuso**, INAF Osservatorio Astronomico di Torino, Pino Torinese (To), Italy mancuso@to.astro.it
Stefan **Martin**, Jet Propulsion Laboratory, Pasadena, USA stefan.r.martin@jpl.nasa.gov
Neil **Martin**, NASA Greenbelt, MARYLAND,USA Neil.F.Martin@nasa.gov
Patrice **Martinez**, LESIA Epinay, France patrice.martinez@hotmail.com
Anthony **Martino**, NASA Greenbelt, MD, USA anthony.j.martino@nasa.gov
Elena **Masciadri**, Max-Planck Institut für Astronomie, Heidelberg, Germany masciadr@mpia.de
Dimitri **Mawet**, University of Liège, Liège, Belgium mawet@astro.ulg.ac.be
Victoria **Meadows**, Spitzer Science Center, Pasadena, USA vsm@ipac.caltech.edu
Bertrand **Mennesson**, Jet Propulsion Laboratory, Pasadena, USA bmenness@jpl.nasa.gov
Jean-Luc **Menut**, Observatoire de la Côte d'Azur, Nice Cédex 4 France menut@obs-nice.fr
Florentin **Millour**, LUAN / LAOG Nice Cédex 2, France florentin.millour@online.fr
David **Mouillet**, Observatoire Midi-Pyrénées, Tarbes France mouillet@bagn.obs-mip.fr
Denis **Mourard**, Observatoire de la Côte d'Azur, Grasse France Denis.Mourard@obs-azur.fr
Claire **Moutou**, LAM Les 3 Lucs - Marseille, France claire.moutou@oamp.fr
Laurent **Mugnier**, ONERA Châtillon Cédex, France Laurent.Mugnier@onera.fr
Ramon **Navarro**, ASTRON, Dwingeloo, Netherlands navarro@astron.nl
Ralph **Neuhaeuser**, AIU Univ. Jena, Germany rne@astro.uni-jena.de
Andrzej **Niedzielski**, Torun Center for Astronomy, Torun, Poland aniedzi@astri.uni.torun.pl
Eric **Nielsen**, Steward Observatory University of Arizona, Tucson, USA enielsen@as.arizona.edu
Jun **Nishikawa**, National Astronomical Observatory of Japan, Tokyo, Japan jun.nishikawa@nao.ac.jp
Raymond **Ohl**, NASA/Goddard Space Flight Center, Greenbelt, MD, USA raymond.g.ohl@nasa.gov
Marc **Ollivier**, Institut d'Astrophysique Spatiale d'ORSAY, ORSAY, France marc.ollivier@ias.u-psud.fr
Jose E. **Oti**, Gonzalez Universidad de Cantabria, Santander, Spain otije@unican.es
Furdui **Ovidiu**, Zentrum für Astronomie, D-69120 Heidelberg, Germany ovidiu@ari.uni-heidelberg.de
Fabien **Patru**, Observatoire de la Côte d'Azur, Grasse, France fabien.patru@obs-azur.fr
Romain **Petrov**, Laboratoire Universitaire d'Astrophysique de Nice, France petrov@unice.fr
Bettina **Posselt**, Max-Planck-Institut für extraterrestrische Physik, Garching, Germany posselt@mpe.mpg.de
Florence **Puech**, ESO, Garching, Germany fpuech@eso.org
Laurent **Pueyo**, Princeton University, princeton, USA lpueyo@princeton.edu
Andreas **Quirrenbach**, Leiden University, Leiden, Netherlands quirrenb@strw.leidenuniv.nl
Yves **Rabbia**, Observatoire Côte d'Azur, Grasse, France rabbia@obs-azur.fr
Jayadev **Rajagopal**, NASA Goddard Space Flight Center, Greenbelt, MD, USA jayadev@iri1.gsfc.nasa.gov
Shanti **Rao**, Jet Propulsion Laboratory Pasadena, CA 91109-8099, USA shanti.r.rao@jpl.nasa.gov
Xavier **Rejeaunier**, Alcatel Alenia Space, Cannes La Bocca, France xavier.rejeaunier@alcatelaleniaspace.com
Jeremy **Richardson**, NRC/NASA Goddard Greenbelt, MD, USA jeremy@iri1.gsfc.nasa.gov
Stephen **Ridgway**, NASA/NOAO, Washington, USA ridgway@noao.edu
Jean-Pierre **Rivet**, OCA Nice, Cédex 04, France rivet@obs-nice.fr
Clélia **Robert**, ONERA, Châtillon, France clelia.robert@onera.fr
Katherine **Roth**, Gemini Observatory, Hilo, HI, USA roth@gemini.edu
Daniel **Rouan**, LESIA - Observatoire de Paris, Meudon, France daniel.rouan@obspm.fr
Jean-Francois **Sauvage**, ONERA - DOTA, Chatillon, France Jean-Francois.Sauvage@onera.fr
Hans Martin **Schmid**, ETH Zurich, Switzerland schmid@astro.phys.ethz.ch
Franois-Xavier **Schmider**, Laboratoire Universitaire d'Astrophysique de Nice, France schmider@unice.fr
Edouard **Schmidtlin**, Jet Propulsion Laboratory, Pasadena, USA edouard@huey.jpl.nasa.gov
Jean **Schneider**, Paris Observatory Meudon, France jean.schneider@obspm.fr
Eugene **Serabyn**, Jet Propulsion Laboratory Pasadena, USA Eugene.Serabyn@jpl.nasa.gov
Denis **Serre**, LATT - OMP Toulouse, France denis.serre@ast.obs-mip.fr
Stuart **Shaklan**, Jet Propulsion Laboratory, Pasadena, USA stuart.shaklan@jpl.nasa.gov
Michael **Shao**, Jet Propulsion Laboratory, Pasadena, USA mshao@huey.jpl.nasa.gov
Shiri **Sharam**, NASA/GSFC Greenbelt, MD, USA rshiri@mirage.gsfc.nasa.gov
Anand **Sivaramakrishnan**, American Museum of Natural History New York, NY, USA anand@stsci.edu
Tomasz **Soltysinski**, Nicolaus Copernicus Astronomical Center, Warsaw, Poland tomasz-s@camk.edu.pl
Remi **Soummer**, American Museum of Natural History, Baltimore, USA rsoummer@amnh.org
Frank **Spaan**, Heriot-Watt University, Edinburgh, United Kingdom F.Spaan@hw.ac.uk
Julien **Spronck**, Delft University of Technology Delft, Netherlands J.Spronck@tnw.tudelft.nl
Daphne **Stam**, University of Amsterdam, Amsterdam, Netherlands dstam@science.uva.nl
John **Storey**, University of New South Wales, Sydney, Australia j.storey@unsw.edu.au
Mark **Swain**, Max Planck Institute for Astronomy, Heidelberg, Germany swain@mpia-hd.mpg.de
Motohide **Tamura**, NAOJ, Mitaka, Japan hide@optik.mtk.nao.ac.jp
Shinichiro **Tanaka**, ISAS/JAXA, Kanagawa, Japan stanaka@ir.isas.jaxa.jp
Alexander **Tavrov**, Tokyo University of Agriculture and Technology, Tokyo, Japan tavrov@cc.tuat.ac.jp
Eric **Thiebaut**, Observatoire de Lyon, St-Genis Laval Cédex France thiebaut@univ-lyon1.fr
Volker **Tolls**, Harvard-Smithsonian Center for Astrophysics, Cambridge, USA vtolls@cfa.harvard.edu
Wesley **Traub**, Jet Propulsion Laboratory, Pasadena, USA Wesley.A.Traub@jpl.nasa.gov
John **Trauger**, Jet Propulsion Laboratory, Pasadena, USA John.T.Trauger@jpl.nasa.gov
Tony **Travouillon** Caltech-TMT, Pasadena, MC102-08, USA tonyt@caltech.edu
Zlatan **Tsvetanov**, NASA Headquarters, Washington, DC, USA zlatan.tsvetanov@nasa.gov
Farrokh **Vakili**, Laboratoire Universitaire d'Astrophysique de Nice, France Farrokh.Vakili@unice.fr
Bruno **Valat**, Laboratoire Universitaire d'Astrophysique de Nice, France valat@unice.fr
Andrew **Vanden Heuvel**, University of Florida Gainesville - Archer, FL 32618, USA avh@astro.ufl.edu
Thangasamy **Velusamy**, JPL, Pasadena, USA velusamy@jpl.nasa.gov
Christophe **Verinaud**, ESO Garching-bei-München, Germany cverinau@eso.org
J. Kent **Wallace**,Jet Propulsion Laboratory, Pasadena, USA James.K.Wallace@jpl.nasa.gov
Debra **Wallace**, NASA/GSFFC, Greenbelt, USA wallace@exo1.gsfc.nasa.gov
Sebastian **Wolf**, MPI for Astronomy, Heidelberg, Germany swolf@mpia.de
Robert **Woodruff**, LOCKHEED Martin CO, Boulder, USA robert.a.woodruff@lmco.com
Neville **Woolf**, U. Arizona Tucson, AZ, USA nwoolf@as.arizona.edu
Hans **Zinnecker**, Astrophysikalisches Institut, Potsdam, BRD hzinnecker@aip.de

Address by the Local Organizing Committee

Dear colleagues,

The idea of organizing the present meeting emerged in early 2004, presented by professor Claude Aime to and soon accepted by IAU in Spring 2004. This left indeed little time to the local organizers to make happen the IAU colloquium 200 on extrasolar planets on the French Riviera. The small local organizing team from Laboratoire Universitaire d'Astrophysique de Nice (LUAN) and Observatoire de la Côte d'Azur: Yves Rabbia, Cathy Blanc, Claude Aime, Farrokh Vakili, André Ferrari, Marcel Carbillet was joined by the colloquium executive secretary Valérie Chéron and her colleagues Marie-Claude Pophillat and Jocelyne Bettini and a number of devoted students from LUAN namely Géraldine Guerri, Florentin Millour, Adrian Belu, Pierre-Marie Gori and Catherine Grec. Recording and organizing almost 200 participants arrivals, lodging and attendance to IAU 200 colloquium, the last of colloquia series, was both a harvest and a reward for our team and a surprise for the Citadelle de Villefranche and the CNRS DR20 local administration who have always offered us their generous support.

The LOC wishes to warmly acknowledge professor Albert Marouani, President of the University of Nice and professor Nicole Tournois the Vice-President of UNSA and the Administration Council as well as the Mayor of Villefranche, to having so nicely sponsored and hosted the present IAU colloquium. Finally the editors are indebted to Gilbert Ricort, André Ferrari, Marcel Carbillet and Cédric Jacob for their careful reading of the author's manuscripts and precious help to finalize the proceedings.

It must be pointed out that the organizers have been lucky enough to offer a partial solar eclipse occurring on the very first day of the meeting, although hampered by a cloudy weather. On the perspective of search for habitable zones and life by astronomers the colloquium was highlighted by the open-session conference given by the Nobel Prize of Medecine Professor Christian De Duve on "Chance and Necessity Revisited" on Tuesday 4th. May we be fortunate enough to attend a next IAU Symposium in a few decades from now in this nice Citadelle to discuss the impact of astrobiological discoveries on astrophysics and extra-solar planetary science made by us and guiding the work of the youngest scientists of this conference to whom we wish the best luck.

Farrokh Vakili, chair LOC
Villefranche, 3 October 2005

Address by the Scientific Organizing Committee

Dear colleagues,

These Proceedings of the 200th Colloquium of the International Astronomical Union deal with one of the oldest well-formulated scientific questions, the search for planets around other stars. The quest started with the Greeks and has led, until the first scientific approach by Huygens in 1698, only to speculations by thinkers like Epicurus, Bruno, Descartes ... A profound question behind these other worlds is the existence of life outside the Solar System. It occupies a remarkable place in human research: it is in the forefront of science and at the same time it is one of the easiest to understand by the general public. Its philosophical and psychological foundations provide strong motivations to researchers.

Until the recent times, astronomical instrumentation was powerless to detect the other planetary systems. Starting in the late '30s, the search for extrasolar planets has first begun with astrometry. After some unfortunate false alarms (*e.g.* the case of the Barnard star), no exoplanet could be detected until the late '80s. The first successes came from the radial velocity and the pulsar timing techniques. These discoveries by the way present an interesting case of the non-linearity of the development of science. On the one hand, what was found was a series of complete surprises. On the other hand, with the interplay of announcements, retractations, rediscoveries of planets and the ambiguities of the mass value for the planets γ Cephei b, ϵ Eridani b and HD 114762 b, it is not easy to tell what was the first planet discovered. All these discoveries, making use of the parent star's wobble, detect the planets only indirectly, they do not "see" it. The amount of information on the planet is rather poor and the stellar wobble alone does not help for the goal behind all these efforts: the detection of extrasolar life.

The problem is difficult, but accessible. High contrast is mandatory because the planets are so faint compared to their parent stars. Since they are so close to the star, they also require a sufficient angular resolution. These two areas of astronomical instruments are experiencing spectacular developments with coronagraphy and interferometry. They lead to ambitious instrumental developments at the major ground-based telescopes (such as the Very Large Telescope operated by the European Southern Observatory Organization) and space mission projects such as the European Darwin project and the numerous NASA projects. Major technological efforts on subsystems and components such as adaptive optics and coronagraphic imagers are necessary. Of great importance is also the analysis of the astrophysical aspects of the science goals, such as reflected light vs thermal emission approaches.

The world-wide astronomical community is very active in both the scientific and the instrumental aspects of the search for planets by imaging. Many of its members have met for one week in Villefranche-sur-Mer, near Nice, France, and these Proceedings present an up-to-date account of this promising field of research.

Claude Aime and Jean Schneider, co-chairs SOC
Villefranche sur Mer, October, 3rd, 2005

Direct Imaging of Exoplanets: Science & Techniques
Proceedings IAU Colloquium No. 200, 2005
C. Aime and F. Vakili, eds.
© 2006 International Astronomical Union
doi:10.1017/S1743921306009008

Exoplanets: The tip of the iceberg?

W. Benz[1], Y. Alibert[1], C. Mordasini[1] and D. Naef[2]
[1]Physikalisches Institut, University of Bern, Switzerland
[2]European Southern Observatory, Alonso de Cordova 3107, Casilla 19001 Santiago 19, Chile

Abstract. As the number of large scale ground- and space-bound planet detection and imaging projects is growing, the need for theoretical guidance in order to optimize instrumental design is rapidly mounting. In an effort to provide this required framework, we present the results of Monte Carlo simulations of the formation of giant planets and compare them with the current population of exoplanets. Our models show that due to the severe current observational detection bias only a small percentage (3.6 %) of the potentially existing planets can be detected. Indeed, a large number of planetary embryos never grow enough to become giant planets giving raise to a large populations of bodies with masses smaller than $\simeq 5\ M_{jup}$. In addition, this observational bias, coupled with the fact that systems enriched in heavy elements tend to form more massive planets, explains the currently observed correlation between stellar metallicity and likelihood to host planets. Finally, we show that the disk gas delivery rate during the late stages of formation actually determines the maximum planetary mass.

Keywords. Planet formation, core accretion, planetary mass distribution, metallicity.

1. Introduction

At the time of this Colloquium, roughly ten years after the first announcement by Mayor and Queloz (1995), 168 exoplanets have been discovered. From the beginning, they have shattered the understanding of the formation of planetary systems that had been patiently constructed based upon the study of a single example: our own solar system. The diversity amongst the new systems by baffling the earlier concepts has taught us how dangerous it can be to build theories on too small a data set. At a time when searching for life *as we know it* becomes possible, this comes as a serious reminder to keep our approaches as unbiased and open as possible.

The lack of good theoretical understanding creates two major problems. First, we cannot explain what is observed and second, we cannot predict what should be observed. The latter is especially severe at a time when major new ground- and space-based instrumentation is being discussed to detect and image smaller planets and possibly search for life. Progress in understanding has been difficult for many reasons. First and foremost, the formation and evolution of planet is a difficult subject involving large changes of scales, long timescales and many non-linearities and feed-back mechanisms. Furthermore, the topic is multi-disciplinary involving astronomy, cosmo-chemistry, material sciences, planetary sciences, climate physics, and, if one is looking for life, biology. Unfortunately, a tradition of exchange and collaboration between these fields is still lacking and more often than none, studies are carried out in one field ignoring the constraints from stemming from the others. Progress therefore relies on a better pooling of ressources, better multi-dsciplinary collaborations and, of course, more data and new bright ideas.

While over 168 exoplanets may seem a lot, the constraints they impose on formation theories are not easily usable in a quantitative fashion. Indeed, in most cases, these planets having been detected by indirect methods (radial velocity techniques essentially), we only know the orbital parameters and a minimum mass. In the few cases where the planet is actually transiting the star, more constraints can be derived to check structure models but since these planets may have been at least partially evaporated, their current characteristics are not necessarily directly related to their formation epoch. Short of direct imaging and the corresponding spectroscopic analysis, the only possibility of extracting constraints on formation models is a statistical approach in which the "observed" population is compared to synthetic populations computed under various assumptions. Differences can be quantified statistically and model parameters adjusted as to minimize differences.

We present here some first results of such a statistical approach and show that already promising results can be obtained. We present our model in section 2. Particular attention is given to the choice of initial conditions (section 2.2) which we chose based as much as possible on observations. We also compute the expected observational detection biases (section 2.3) and correct our synthetic planet populations before comparing them to the actual data. Finally, we present some of our results in section 3 with a particular focus on the differences between the underlying planet population and the fraction actually detectable for those are especially interesting for the design of the new generation of instruments. In particular, we show that the currently detected population of exoplanets may only be the very tip of the iceberg of a much larger populations of planets whose masses are too small to be detectable with the current radial velocity search techniques.

2. Theoretical approach

In this section we briefly explain how we generate various populations of synthetic planets which we then compare to actual observations. We discuss the formation model itself (section 2.1), the way in which we generate initial conditions (section 2.2) and finally the corrections involved in order to take into account the observational detection biases (section 2.3).

2.1. *Giant planet formation model*

We compute a population of synthetic planets within the framework of the core accretion scenario as it is currently the only formation model allowing quantitative comparisons to be made with observations. In this scenario a solid core surrounded by a tenuous gaseous envelope is first formed by the accretion of solid planetesimals which themselves were formed by sedimentation and coagulation of small dust grains (Wetherill & Steward 1989, Lissauer 1993). As the core grows beyond a critical mass (of the order of 15 $M_\oplus$ at 5 AU, but depending on different physical parameters), radiative losses can no longer be offset by planetesimal accretion and gas accretion runs away leading to the very rapid build up of a massive gaseous envelope (Pollack *et al.* 1996).

Recently, we have extended this model (Alibert *et al.* 2005a) to take into account both the evolution of the proto-planetary disk (gas and solid phase) and the migration of the forming planet (due to the momentum transfer between the planet and the gaseous disk). We showed that an important consequence of including these effects is to drastically reduce the formation timescale (Alibert *et al.* 2004) bringing it to an excellent agreement with the observationally derived lifetime of proto-planetary disks (Haisch *et al.* 2001). The numerical tools we have developed, as well as the tests we have performed to validate them can be found in Alibert *et al.* (2005a).

We have also used our approach to study the formation of the giant planets in our own solar system. The wealth of detailed *in situ* and remote measurements existing for Jupiter and Saturn makes them mandatory benchmarks for all formation models.

We have been able to show that our approach is capable of satisfying all the available relevant constraints for both planets within the same self-consistent formation model (Alibert, Mousis, & Benz 2005; Alibert *et al.* 2005b). To the best of our knowledge, this is the first time that a single formation model is capable of accounting quantitatively for all these characteristics.

2.2. *Initial conditions*

To allow a meaningful statistical comparison between our synthetic planet populations and the actually observed population, it is necessary to specify not only the initial conditions but also the probability of occurrence of each one. To keep the calculations tractable, we have limited the number of variables we vary to five: the proto-planetary disk mass, the metallicity (dust to gas ratio), the initial location of the planetary embryo ($0.6\,M_\oplus$ in mass), the rate of photo-evaporation and the age of the disk when the planetary embryo is created. In addition, we have also varied the gas mass accretion rate allowed through the gap. In one set of calculation we limit this accretion rate to the value derived by Veras & Armitage (2004) and in the other case, we allow for the maximum rate provided by the viscosity (α parameter) of the disk. As we shall see below, this gas accretion rate through the gap directly determines the maximum mass a planet can reach.

All other parameters (see Alibert *et al.* 2005a for a detailed description of our model) are kept constant, in particular the stellar mass ($1M_\odot$), the initial surface density distribution of the gas and solids (power law of index $-3/2$), the dissipation parameter ($\alpha = 0.002$), the reduction factor for type I migration ($f_I = 0.001$). Because the computation of a single model can take up to 2 to 3 hours of CPU time, we build a large 5 dimensional master table from $60'480$ simulations ($9 \times 4 \times 20 \times 21 \times 4$ respectively in the independent variables mentioned above). We then use a Monte Carlo approach to generate initial conditions as described below and interpolate the corresponding results linearly in our master table. Notice that in our calculations we always assume that one and only one planetary embryo forms in orbit around a given star.

Given a fixed initial surface density distribution $\Sigma(r, t = 0) = \Sigma_0(a/a_0)^{-3/2}$, the mass of the initial gaseous disk in terms of Σ_0 is easily obtained once a_0, the inner (a_{in}) and outer (a_{out}) edge of the disk are specified. We chose $a_0 = 5$ AU, $a_{in} = 0.25$ AU, and $a_{out} = 50$ AU and generate disk masses at random varying Σ_0 from 100 g/cm^2 to 900 g/cm^2 with a probability of occurrence given by the mass distribution of circumstellar disks derived from observations (Beckwith & Sargent 1996). The disks obtained span a mass range going from about $0.008\,M_\odot$ to about $0.07\,M_\odot$. Following Murray *et al.* (2001), we relate the dust to gas ratio f_{dg} to the metallicity Fe/H of the disk (it is assumed that initially the star and the disk have the same composition) by the relation $Fe/H = log(f_{dg}/f_{dg\odot})$, where $f_{dg\odot} = 0.0167$ is the dust to gas ratio corresponding to solar composition. We vary f_{dg} from 0.00667 to 0.0333 using the metallicity distribution of stars in the solar neighborhood (Nordström *et al.* 2004) to define the probability of occurrence of a given disk. The resulting disks span a range of metallicity going from about $Fe/H = -0.4$ to $Fe/H = 0.3$.

The distribution of the starting position of planetary embryos is not constrained by observations and only theoretical arguments can be used. Here we follow Ida & Lin (2004a) and assume a uniform distribution in $log(a)$ between 1 AU and 20 AU for the starting positions. For a lack of better constraints, the distribution of the photo-evaporation rates is taken to be uniformly distributed between $3 \times 10^{-9}\,M_\odot$/y and $1.5 \times 10^{-8}\,M_\odot$/y (see Clarke *et al.* 2001; Armitage *et al.* 2003; Veras & Armitage 2004). These values together with our choice of the dissipation parameter α yield disk ages almost linearly decreasing between 0 and 8 Myr in good agreement with the observed age distribution (Haisch *et al.*

2001). Finally, the starting times of our embryos are taken uniformly distributed between 0 and 1 Myr after the start of disk evolution. Allowing embryos to start later results in lower mass planets thus further reducing the fraction of detectable planets. For example, starting times taken uniformly between 0 and 4 Myr, results in only 2 % of detectable planets, a number probably too small compared to actual detection rates.

2.3. *Detection biases*

A working planet formation model is not enough to statistically compare the computed population of synthetic planets with the one actually observed. To compare both populations it is also necessary to understand the various detection biases entering in the observational process and correct for them before comparison. This is not an easy task as these biases are technique and instrument (and probably also observer) dependent. Since the overwhelming majority of planets has been detected using radial velocity techniques, we shall focus here on biases affecting this detection technique.

To first order, in radial velocity searches the detection probability goes down with decreasing planetary mass and/or increasing distance to the star following Kepler's law and the instrumental accuracy determines whether the body can be detected or not. In fact, a large number of other quantities also play a role such as the magnitude of the star, its metallicity and rotation rate, orbital parameters of the planet, actual measurement schedule, jitter, etc. Using the method developed by Naef (2004, 2005) for the spectrograph Elodie, we computed for each planetary mass (in a grid ranging from 0.025 to 40 M_{jup}) and period (in a grid ranging from 35 to 38′000 days) 50′000 randomly chosen planetary orbits and determined from a χ^2 analysis the fraction of this number that can actually be detected. This fraction represents the detection probability corresponding to this planetary mass and orbital period.

We illustrate the results of these calculations in Fig. 1 in which we plot the detection probability assuming a instrumental accuracy of 5 m/s of a number of planets with masses ranging from 10 to 5000 $M_\oplus$. Note that with this instrumental accuracy we can expect that nearly all planets with masses larger than 1000 $M_\oplus$can be detected out to 4 AU while the discovery probability of a Jupiter mass planet drops below 50% beyond 3 AU.

We use these probabilities to determine which planets from a given synthetic populations are actually detectable and compare only those with the observed ones.

3. Results

In this section we present some results from our planet population simulations. In all cases, we have generated an ensemble of 250′000 planets. Note that our simulations always start with one and only one embryo with a mass of 0.6 $M_\oplus$. Hence, our simulations do not address multiple systems nor the question of whether a given system can grow embryos of this size. Here we simply assume that they can grow regardless of the mass, age or metallicity of the disk. We show the importance of the detection biases and compare the fraction of synthetic planets actually detectable to the one observed (section 3.1). We show that this detectable fraction is actually a function of the content in heavy elements. This dependance turns out to explain the correlation between planet detection probability and host star metallicity (section 3.2). Finally, we compare the mass distribution obtained with observation and illustrate the effect of limiting the gas accretion rate that flows through the gap on the characteristics of the final planets (section 3.3).

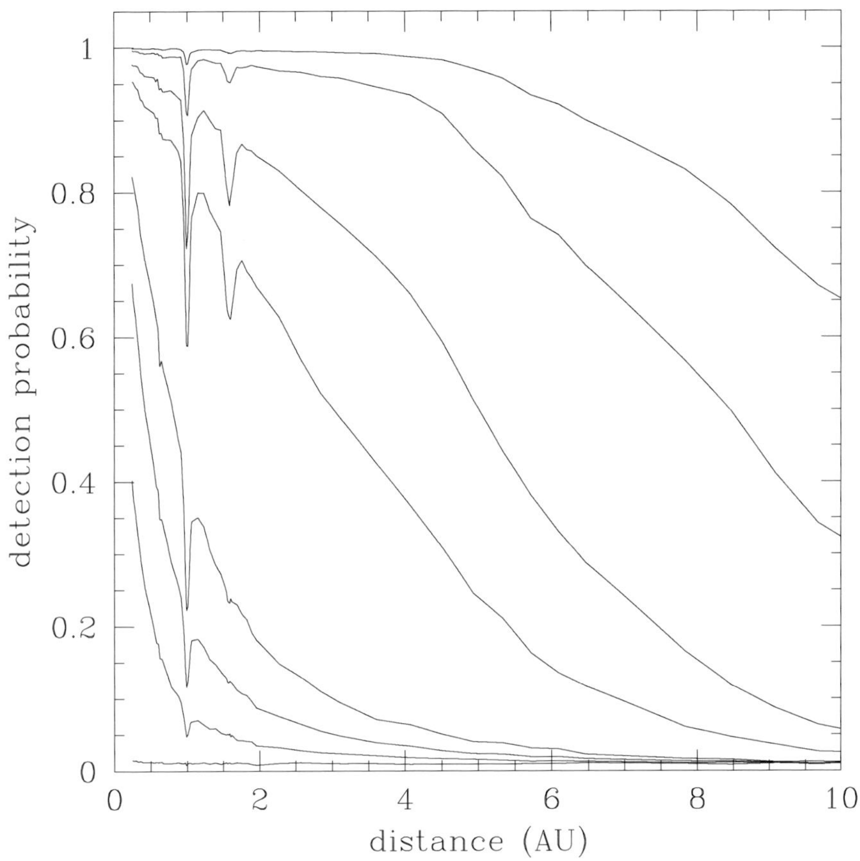

Figure 1. Detection probability as a function of distance to the star assuming an instrumental accuracy of 5 m/s. The various curves are for different planet masses: $10, 50, 100, 200, 300, 500, 1000, and 5000\ M_{\oplus}$ from bottom to top.

3.1. *Existing and detectable planets*

Using the method described in section 2.2, we have generated a population of $250'000$ planets using the maximum gas delivery rate provided by the disk's viscosity (no limitations due to the gap). We have then determined which of these planets are potentially detectable using the probabilities calculated in section 2.2. The resulting distributions are show in Fig. 2.

The most striking feature in this figure is clearly that the detectable planets represent only a small fraction of the underlying planet population since of the $250'000$ bodies actually only $8'948$ or 3.6% are detectable. Note that this fraction depends on the starting time distribution of the embryos. Those which are starting late generally remain small and are therefore, on average, not observable thus reducing the fraction of detectable planets. The numbers here correspond to embryos starting within the first million year of the disk. Allowing for embryos to start within the first 4 Myr of disk evolution leads to a fraction of detectable planets of the order of 2%.

A further inspection of Fig. 2 reveals that although our models reproduce reasonably well the overall planetary mass distribution (see section 3.3 below), they seem unable to reproduce the proper radial distribution. The paucity of synthetic planets between

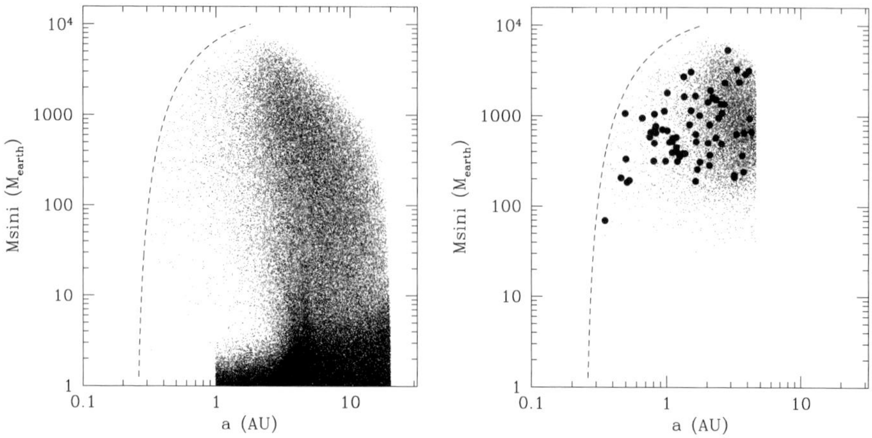

Figure 2. Actual population of synthetic planets (left panel) compared to the observable population assuming a 5 m/s instrumental accuracy (right panel). The actually observed population is shown as big dots. The sharp cutoff at about 5 AU corresponds to about a 10 year period, the actual observation baseline. To avoid inner boundary problems, we only considered planets with feeding zones (four times the Hill radius on either side) not reaching the inner edge of the disk located at 0.25 AU).The dashed line indicates the loci where the planet's feeding zone reaches this inner boundary.

0.5 – 2 AU can have different origins. First, it is possible that the detection biases are worse than what has been computed here and that the probability of detecting planets of several Jupiter masses located at 2 – 5 AU from their parent star is actually significantly lower than the 20 – 40% used here (for example, some of the early instruments did not have 5 m/s accuracy). The second possibility is that the orbital migration of the growing planets is underestimated in our models. This hypothesis which is currently being examined can be used to illustrate the potential power of such a Monte Carlo approach in discussing the different processes governing planet formation.

3.2. *Correlation with metallicity*

We have seen in the previous section that the overall percentage of detectable planet is quite small. Here we show that this fraction is actually a strong function of the assumed heavy elements content (metallicity) in the system at the beginning of planet formation. We recall that our simulations were started with a metallicity distribution corresponding to the one for F and G stars in the solar neighborhood. Hence, we can compute the detection probability as a function of metallicity by simply computing the ratio between detectable to existing planets in fixed metallicity bins and compare the results directly with observations. The results of this computation are show in Fig. 3 in which we also plotted the observationally determined probability as given by Marcy *et al.* (2005).

The explanation of this correlation is straightforward and is due to two effects. First, we find that, within the framework of our models, metal rich systems favor the formation of massive planets. This is quite intuitive as for these systems the core reaches the critical mass earlier therefore allowing more gas to be accreted before the disappearance of the disk. The second effect is simply related to the fact that the detection of massive planets is actually easier than the detection of smaller ones. We recall that in our models we assumed that all systems regardless of their metallicity can form embryos of 0.6 $M_\oplus$. It is therefore the subsequent evolution of these embryos coupled to the detection biases that is at the origin of the correlation. We cannot exclude that the formation of these

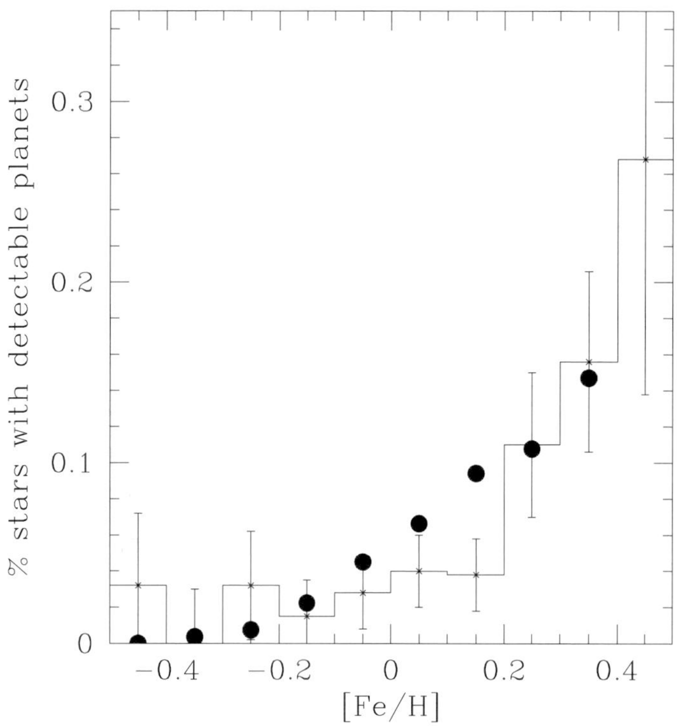

Figure 3. Percentage of stars with detectable planets as a function of the metallicity of the host star. Histogram: Observations (from Marcy *et al.* 2005); Black dots: Present models.

embryos is in itself a function of metallicity. However, we note that if this were the case, the correlation should even be stronger than the one observed. The future detection of smaller and smaller mass objects will allow us to address this issue more quantitatively.

3.3. *Mass distribution of planets*

The actual mass distribution of the planets shown in Fig. 2 is displayed in Fig 4. We again separate the existing planet mass distribution (left panel) from the detectable planet distribution (right panel). In the latter case, we also plotted the distribution of the "observed" planets to allow for comparison. In addition we also computed the mass distribution obtained using a different formation model in which the gas accretion rate is limited in the presence of a gap to a maximum value which is a function of the planet's mass (Veras & Armitage 2004).

We notice again that the population of planet that are actually detectable represents only a small fraction of all the planets forming in our models. We point out that these models start with seed embryos of 0.6 $M_{\oplus}$ and therefore cannot be considered as reliable Earth-like planet formation models. However, the presence of low mass planets can be taken as an indication that our formation model predicts that a large fraction of these embryos will not be able to grow substantially even less become giant planets. Hence, it is an indication that terrestrial like planets should be in fact quite more frequent than giant planets.

Fig. 4 also shows that the rate at which the disk can feed the planet during the late stages of formation determines directly the maximum mass that can be reached. The maximum mass is reached using the gas delivery rate set by the disk's internal dissipation

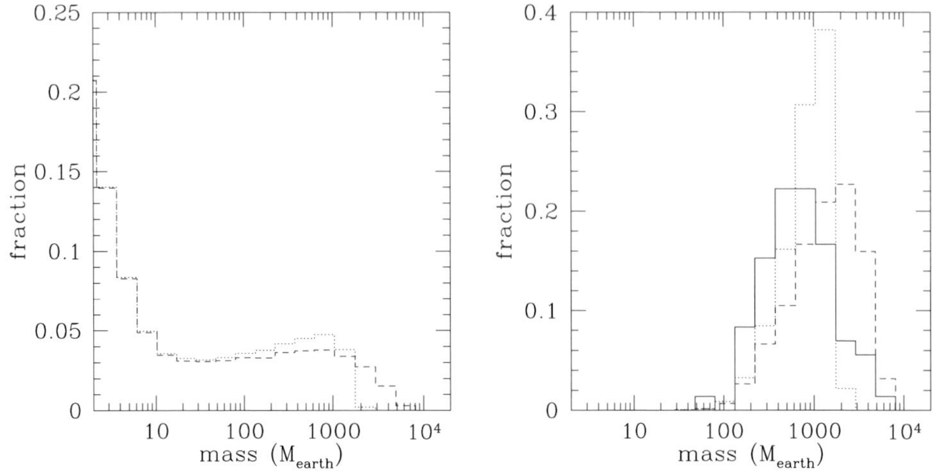

Figure 4. Mass distribution of planets. The mass distribution of the actual planets is shown on the left panel while the one of detectable planets is shown on the right (with a different normalization). In both panels, the dashed line represents the models in which the maximum gas delivery rate to the planet is given by the disk while models where this rate is limited by the gap (Veras & Armitage 2004) are shown by a dotted line. Finally, the solid line in the right panel shows the distribution of the actually detected panels that fall in the domain in which models have been calculated.

mechanism (dashed line). If the formation of a gap is limiting the amount of gas the disk can deliver to the planet, the maximum planetary mass will be correspondingly reduced. To show this, we computed also models in which the gas delivery rate is reduced due to the presence of a gap (dotted line) as given by Veras & Armitage (2004). We notice that there is a difference in the maximum mass reached of almost a factor 5.

Finally we note that when comparing our mass distribution to the "observed" one, the agreement is not too bad even though our models seem to predict systematically more massive planets. More work will be required to pinpoint the origin of this discrepancy.

4. Conclusions

Extended core accretion models allow for a quantitative comparison with observations. While there are still important discrepancies between computed and observed planet populations, this statistical approach is promising and allows already to explain some of the observed trends. The correlation between stellar metallicity and likelihood to host planets can be explained in terms of an observational bias connected to the detection method (radial velocity techniques are most sensitive to large mass objects) and the fact that systems with more heavy elements tend to form more massive planets. Finally, our models show that the maximum mass a giant planet can reach is determined by the amount of gas the disk can deliver to the planet during the runaway growth phase. This delivery rate is set by the flow of gas through the gap as well as by possible disk instabilities occurring during these late stages.

Our calculations also indicate that a large fraction of planetary embryos do not grow and become giant planets detectable by the current radial velocity searches. Hence, for every currently detected exoplanet there should exist of order 20 − 30 times more "failed" giant planet cores with masses smaller than $\simeq 5\ M_\oplus$. The currently known population of exoplanets appears therefore to be truly only the tip of the iceberg.

Acknowledgements

The authors gratefully acknowledge partial support from the Swiss National Science Foundation.

References

Alibert Y., Mordasini, C. & Benz, W. 2004, *A&A* 417, L25

Alibert, Y., Mordasini, C., Benz, W., & Winisdoerffer, C. 2005a, *A&A* 434, 343

Alibert, Y., Mousis, O., & Benz, W. 2005, *ApJ* 622, L145

Alibert, Y., Mousis, O., Mordasini, C., & Benz, W. 2005b, *ApJ* 626, 57

Armitage, P. J., Clarke, C. J., & Palla, F. 2003, *MNRAS* 342, 1139

Beckwith, S. V. W. & Sargent, A. I. 1996, *Nature* 383, 139

Clarke, C. J., Gendrin, A., & Sotomayor, M. 2001, *MNRAS* 328, 485

Haisch, K. E. Lada, E. A., & Lada, C. J. 2001, *ApJ* 553, L153

Ida, S., & Lin, D. N. C. 2004a, *ApJ* 604, 388

Lissauer, J. J.1993, *Ann. rev. Astron. Astrophys.* 31, 129

Marcy, G., Butler, R. P., Fischer, D., Vogt, S., Wright, J. T., Tinney, C. G., & Jones, H. R. A. 2005, *Prog. Theo. Phys. Supp.* 158, 24

Mayor, M. & Queloz, D. 1995, *Nature* 378, 355

Murray, N., Chaboyer, B., Arras, P., Hansen, B., & Noyes, R. W. 2001, *ApJ* 555, 801

Naef, D. 2004, PhD thesis, Geneva University

Naef, D., Mayor, M., Beuzit, J.-L., Perrier, C., Queloz, D., Sivan, J.-P., & Udry, S. 2005, *Proceedings of the 13th Cambridge Workshop on Cool Stars, Stellar Systems and the Sun* ESA SP-560, p. 833

Nordström, B., Mayor, M., Andersen, J., Holmberg, J., Pont, F., Jorgensen, B. R., Olsen, E. H., Udry, S., & Mowlavi, N. 2004, *A&A* 418, 989

Pollack, J. B., Hubickyj, O., Bodenheimer, P., Lissauer, J. J., Podolak, M., & Greenzweig, Y. 1996, *Icarus* 124, 62

Veras, D. & Armitage, P. J. 2004, *MNRAS* 347, 613

Wetherill, G.W. & Sterwart, G. R. 1989, *Icarus* 77, 330

All photographs: Laurent Thareau [`l.thareau@free.fr`].

Direct Imaging of Exoplanets: Science & Techniques
Proceedings IAU Colloquium No. 200, 2005
C. Aime and F. Vakili, eds.
© 2006 International Astronomical Union
doi:10.1017/S174392130600901X

On the Age, Spectral Type, Orbit, and Comparison to Evolutionary Models of AB Dor C

L. M. Close[1], E. L. Nielsen[1], J. C. Guirado[2], B. A. Biller[1], R. Lenzen[3], W. Brandner[3], M. Hartung[4] and C. Lidman[4]

[1]Steward Observatory, University of Arizona, 933 N. Cherry Ave, Tucson, AZ 85721 USA,
email: lclose@as.arizona.edu
[2]Departament d'Astronomia i Astrofisica, Universitat de Valencia, E-46100 Burjassot, Valencia, Spain
[3]Max-Plank-Institut für Astronomie, Königstuhl 17, D-69117 Heidelberg, Germany
[4]European Southern Observatory, Alonso de Cordova 3107, Santiago 19, Chile

Abstract. We expand upon the results of Close *et al.* 2005 regarding the young, low-mass object AB Dor C and its role as a calibration point for theoretical tracks. We argue for a new 70±30 Myr age estimate and present two additional detections of C with the Simultaneous Differential Imaging (SDI) camera. Our improved analysis (Nielsen *et al.* 2005) confirms our spectral type of M8 (±1) and mass of 0.090 ±0.003 M_{sun} for AB Dor C. However, Luhman & Potter (2006) argue for a hotter spectral type (M6±1). Here we adopt the consistent spectral range of M7±1 in this paper as a final spectral type. Plotting AB Dor C (and all other young (<0.12 Myr), low-mass ($0.3 - 0.05 M_{sun}$) objects with accurate dynamical masses) on the HR diagram suggests a trend where current evolutionary models tend to over-predict the temperature (or under-predict the mass) for young low-mass stars and high-mass brown dwarfs. With our uncertainties, there is a ∼90% chance that the mass of AB Dor C is underestimated by the DUSTY tracks in the HR diagram.

Keywords. Infrared: Stars, Stars: Formation, Stars: Low-Mass, Brown Dwarfs, Stars: Pre-Main-Sequence, Stars: Individual: AB Dor.

---

1. Introduction

The study of young, low-mass objects has been yielding increasingly fruitful science, yet the field remains dependent on evolutionary models to properly interpret the data that are collected from these objects. In particular, mass, while a fundamental property, is very rarely measured directly, and instead must be inferred from theoretical tracks (e.g., Burrows, Sudarsky, and Lunine 2003, Chabrier *et al.* 2000). It is thus of great interest to find calibrating objects that can link a dynamically measured mass with observables such as NIR (1-2 μm) fluxes and spectral types.

In our previous work (Close *et al.* 2005), we reported the first direct detection of the low-mass companion to the young star AB Dor A (separation 0.15″, ΔH=5.2 mag), along with measurements of the JHKs fluxes, spectral type, and dynamically determined mass of AB Dor C. Upon comparing these results with the predictions of Chabrier *et al.* 2000, we found the models to be systematically over-predicting the fluxes and temperature of AB Dor C, given an age of the system of 50 Myr. Put another way, the model masses seem to be underestimating the mass of a low-mass object given its age, NIR fluxes, and spectral type. Since the publication of these results, another calibrating object has been reported by Reiners, Basri, and Mohanty 2005: USco CTIO 5. While this equal-mass binary is younger (∼8 Myr) and more massive (total $\text{mass}_A \geqslant 0.32\,M_{sun}$) than AB Dor

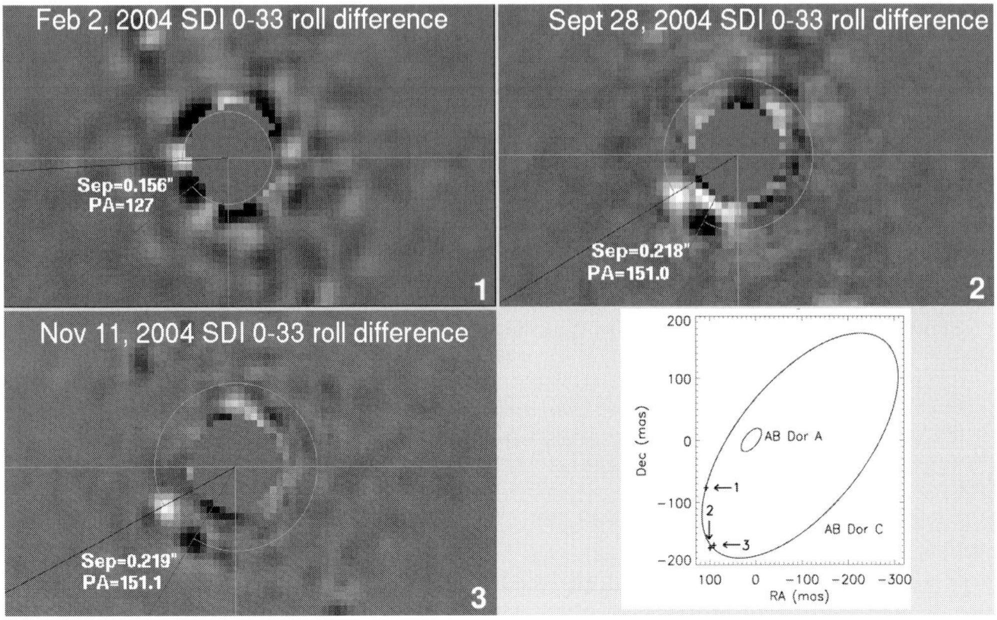

Figure 1. The different epochs of AB Dor C measured during 2004 with the VLT SDI device (from Nielsen *et al.* 2005). Each of the numbered panels shows an individual epoch of SDI observations, with images at position angles of 0° and 33° subtracted from each other, showing a positive and negative signal from AB Dor C. The inner pixels of AB Dor A have been intentionally saturated, and have been removed from the image. The orbital motion of the companion can clearly be seen over this span of time. The bottom-right panel shows these locations against a plot of the full orbit of AB Dor C. The 11 additional VLBI/Hipparcos measurements of the reflex motion of AB Dor A (which went into finding the orbital solution) are not shown on this plot (See Close *et al.* 2005).

C, Reiners *et al.* find the same trend of models under-predicting masses based simply on photometric and spectroscopic data applied to the HR diagram. A similar trend for such masses was previously noted by Hillenbrand and White 2004. Moreover, this trend has been theoretically predicted for higher masses by Mohanty, Jayawardhana, and Basri 2004, and by Marley *et al.* 2005 for planetary masses.

In this paper, we seek to expand on our earlier results from Close *et al.* 2005 and Nielsen *et al.* 2005, using an improved spectral reduction. We also present an improved orbital fit based on additional astrometric data from Nielsen *et al.* (2005), as well we address concerns raised by Luhman, Stauffer, and Mamajek 2005 regarding the age of the AB Dor system. In addition we incorporate the new spectra type determination of Luhman & Potter (2006) into our analysis.

2. An Improved Spectral Reduction

As described in Close *et al.* 2005, in February 2004 we obtained 20 minutes of K-band spectra using the Very Large Telescope (VLT), following our initial detection of AB Dor C.

Improved spectra from that of Close *et al.* were re-reduced by Nielsen *et al.* 2005. The spectra for AB Dor C is shown in Figure 2 from Nielsen *et al.* 2005. All of the templates used in Figure 2 are significantly younger than AB Dor C. To properly bound the spectral

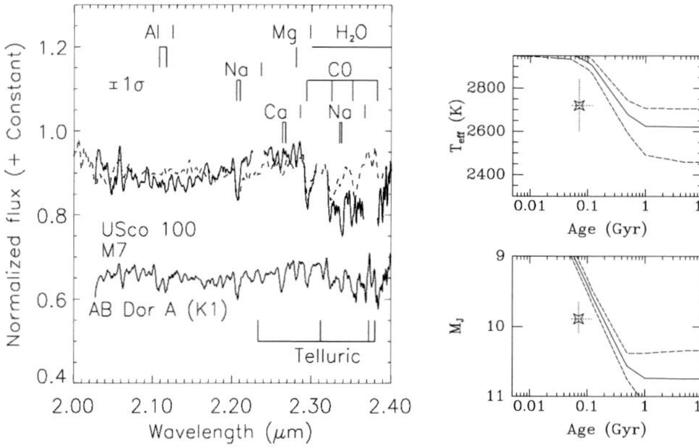

Figure 2. LEFT: The spectrum of AB Dor C shown against a number of young (~10 Myr), low-surface gravity objects. RIGHT: old field objects. The features seem to be bound between the M7 and M9.5 templates. From Nielsen *et al.* 2005.

type, we also compare to field objects, as we did in Nelsen *et al.* 2005. The spectrum seems to fit best in the sequence between an M7 and an M9.

Recently (after the IAUC 200 conference) an independent reduction of our dataset by Luhman & Potter (2006) finds an earlier spectral type M6±1 for AB Dor C when comparing to old M dwarfs. A spectral type of M7 is consistent with all AB Dor C spectral and photometric studies. Since these studies used independent approaches to the data reduction it is reasonable to adopt a final spectral type of M7±1 which is consistent with all the colors and spectral types of Close *et al.* (2005), Nielsen *et al.* (2005), and Luhman & Potter (2006) measured for AB Dor C to date.

3. Discussion

4. Improved Orbit

Our earlier paper (Close *et al.* 2005) was based on observations conducted at the VLT in February of 2004. Since this work was published, we have reduced additional Simultaneous Differential Imaging (SDI, see Close *et al.* 2003; Lenzen *et al.* 2004) data from September and November of 2004. While data through the narrow-band SDI filters do not provide us with any improved photometric information (beyond confirmation that between AB Dor A and C, $\Delta H = 5.20$), these images do give us additional astrometric data points, allowing us to refine the orbit. Figure 1 shows each epoch of measuring the position of AB Dor C with respect to A.

The most immediate consequence of the fit of Nielsen *et al.* 2005 is that the mass of AB Dor C remains at 0.090 M_{sun}. As reported in Guirado *et al.* 2005, we notice that the error bars shrink from 0.005 M_{sun} to 0.003 M_{sun}. This confidence with which we know the mass of AB Dor C makes it an ideal object for calibrating theoretical evolutionary tracks.

4.1. *The Age of AB Dor*

In Close *et al.* 2005 it was argued that due to the excess luminosity of AB Dor A and C compared to the Pleiades, and A's large ~300mA Li equivalent width and very fast (12.3h) rotation, that the age of the system was 30-100 Myr. An age of 50 (-20, +50) Myr was adopted, which was consistent with the 50 Myr published age of the AB Dor moving group (Zuckerman, Song, and Bessell 2004). Recently Luhman *et al.* 2005 has

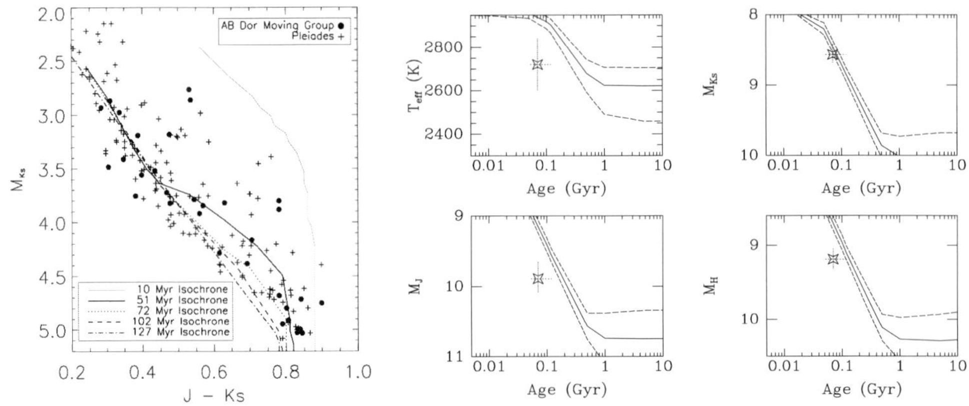

Figure 3. LEFT: A NIR color-magnitude diagram of medium-mass members of the Pleiades (compiled from the literature) and the AB Dor moving group (Zuckerman *et al.* 2004), with the theoretical isochrones of Baraffe *et al.* 1998. The offset between the single star locus of the two groups redward of J-Ks ~0.4 suggests a younger age for the AB Dor moving group, closer to 70 Myr. RIGHT: A comparison between our Close *et al.* (2005) photometry and effective temperature to the dusty models for the 0.085-0.095 mass tracks appropriate of AB Dor C. Ages of 70±30 Myr are adopted for AB Dor C. Note how the models overestimate the values observed. The J, H and Ks fluxes of Luhman & Potter (2006) agree with these values; however, their uncertainties are higher and they adopt an older age (100-150 Myr) for AB Dor C –leading to closer agreement with the models than we find here.

suggested a slightly older age of 70-150 Myr. However, as Luhman *et al.* 2005 notes, the AB Dor moving group is systematically over-luminous compared to the Pleiades (age 100-120 Myr) by 0-0.1 magnitudes in M_{Ks} vs. V-Ks plots (see their Figure 1).

We have found in Nielsen *et al.* 2005 similar results with a near-infrared color magnitude diagram, as seen in Figure 3. While the two groups of stars appear similar for the early-type members (where the isochrones overlap), beyond a color of J-Ks ~0.4, the lower main sequence of the AB Dor Moving group appears to be above that of the Pleiades by about 0.15 magnitudes. We have run a series of simulations that suggest that only ~10% of the time would a group of Pleiades aged stars appear 0.15 magnitudes above the single-star locus as is observed for all the AB Dor group members.

We note that a 0.15 magnitude offset from the Pleiades single star locus suggests a group age of ~70 Myr from the Lyon group's models (Baraffe *et al.* 1998; see fig 3). We adopt an average age of 70 ± 30 Myr with 1σ error bars. Hence, there is a ~10% chance that the AB Dor moving group is as old as the Pleiades based on these color-magnitude diagrams.

Luhman *et al.* 2005 conclude that their increase in the age of the system (from 50 to 120 Myr) implies that the luminosity is correctly predicted by the models. But, as we will see in Section 4.2, even if the luminosity is close to (but less) than the predicted value at an age of 120 Myr and 0.09 M_{sun}, there is still an error in the temperature which leads to a significant underestimation of the mass.

4.2. *HR Diagram and Evolutionary Models*

In order to further compare our observations of AB Dor C with the theoretical models, we consider an HR diagram with our measured values and the DUSTY models. Using the adopted spectral type of M7 and the absolute Ks from Close *et al.* 2005, we can derive an effective temperature and bolometric luminosity for AB Dor C. We plot AB Dor C in such an HR diagram in Figure 4, along with the DUSTY tracks, AB Dor Ba/Bb,

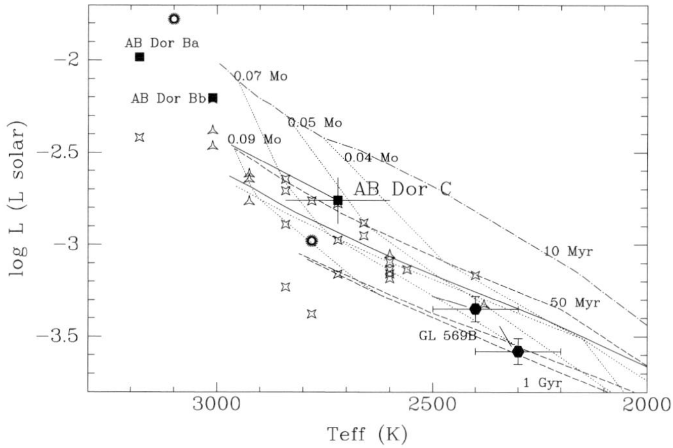

Figure 4. HR diagram showing low-mass Pleiades objects from Martín *et al.* 2000 (open stars), other low-mass members of the Pleiades taken from the literature (open triangles), and AB Dor Ba/Bb (filled boxes). Both AB Dor Ba/Bb and PPL 15 A/B are shown both as individual objects and as a single, blended source (rings). The dotted vertical lines are iso-mass contours for the DUSTY models (from left to right, 0.09, 0.07,0.05, and 0.04 M_{sun}), while the more horizontal, dashed lines are the DUSTY isochrones (top to bottom, 10, 50, 100, 120, 500, 1000 Myr). Note that the DUSTY models predict a 70-100 Myr object of 0.09 M_{sun} should be ~300 K hotter than observed. *From the location of AB Dor C on the HR diagram, one would derive a mass of 0.053* M_{sun}*, a factor of* ~*1.7 underestimate in mass.* As the temperature and luminosities of the Pleiades objects in this plot were determined in the same manner used for AB Dor C, and these Pleiades points mostly fall along the appropriate 120 Myr DUSTY isochrone, we are assured that our temperature scale and bolometric correction are reasonable (so these results cannot be improved by adopting another temperature scale since that would offset the Pleiades points from the 120 Myr isochrone). With our 1σ error bars, there is a ~90% chance that the DUSTY models underestimate the mass of AB Dor C from the HR diagram.

and low-mass members of the Pleiades. We compile Pleiades members from Martín *et al.* 2000, as well as from other sources in the literature (Cluster identifications and spectral types from Briggs and Pye 2004, Pinfield *et al.* 2003, Terndrup *et al.* 1999, Festin 1998, Martín, Rebolo, and Zapatero-Osorio 1996, and Ks-band fluxes from the 2MASS catalog) The bolometric luminosities and temperatures for all these objects (AB Dor Ba/Bb, C, and the Pleiades members) are derived using Allen *et al.* 2003 and Luhman 1999 (dwarf scale), respectively.

As is seen in Figure 4, AB Dor C is overluminous, above the Pleiades sequence (as expected from a younger object, ~70 Myr). We also show an thick line to its position in the HR diagram predicted by the DUSTY models appropriate to its age and 0.09 M_{sun} mass.

It has been suggested that this overluminosity could be explained if AB Dor C were a close, unresolved binary (Martín private communication; Marois *et al.* 2005). Were this the case, AB Dor C would split into two points in Figure 4, and move downward (as AB Dor Ba/Bb and PPL 15 do when deblended), appearing consistent with the Pleiades locus. While this interpretation cannot be currently ruled out it is unlikely. To be stable against the pull of AB Dor A (~2 AU distant) the suggested binary brown dwarf would have to be <0.1 AU in separation to remain bound. Based on the survey of Reid 2003 we estimate the likelihood to be <5% that such a tightly bound brown dwarf would be found. Moreover, as we will see in Figure 5 that this trend is seen for other low mass objects

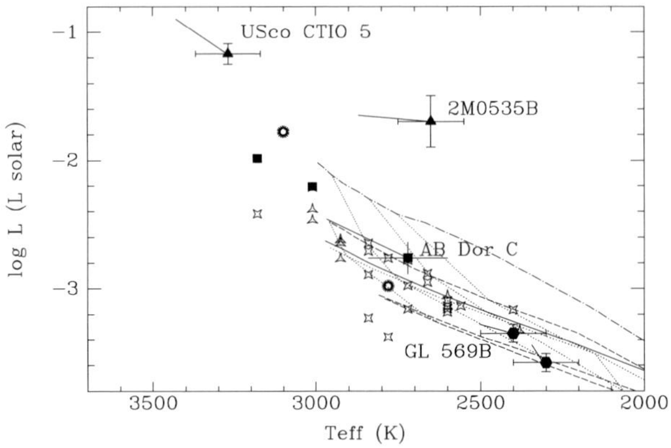

Figure 5. An expanded HR diagram with the Pleiades and the known low-mass (0.3-0.05 M_{sun}) with dynamical masses. The points with error bars mark objects with dynamical masses, with the diagonal lines representing the displacement from the measured luminosity and temperature to the values predicted by the DUSTY and Next-Gen models. Upper Sco CTIO 5 (Reiners *et al.* 2005; with the hottest "intermediate" temperature scale of Luhman *et al.* 1999), AB Dor C, 2MASS0535B (Stassun *et al.* 2005), and Gl 569 Ba/Bb (Zapatero Osorio *et al.* 2004) all show a systematic trend where the measured HR diagram location is cooler and fainter than the models' predictions. Seen another way, the masses predicted by the models are underestimates of the actual masses. *This appears to be a general trend for all young low mass stars and high mass brown dwarfs.* In the case of the older (300 Myr) Gl 569 B system, the offset is within the 1σ uncertainties.

with dynamical masses as well – and they all cannot be spectroscopic binaries. Perhaps future very high spatial and spectral resolution observations will be able to investigate this possibility further.

Finally, we note an overall trend for young, low-mass objects where dynamic masses have been measured, as shown in Figure 5. There is a global offset to higher luminosities and temperatures for AB Dor C, USco CTIO 5, and Gl 569 Ba/Bb. These results suggest that there is a systematical trend of theoretical evolutionary tracks being somewhat too hot and slightly too luminous for measured dynamical masses in the range 0.3-0.05 M_{sun} and ages ~0.01-0.12 Myrs.

Acknowledgements

We thank Gael Chauvin for providing an electronic copy of the spectrum of GSC 8047-0232, and Nadja Gorlova for providing spectra of many young, low-mass objects. We also thank the organizers of the IAUC 200 conference for the chance to present this work. LMC acknowledges support from the NASA Orgins and NSF CAREER programs that supported this work.

This publication makes use of data products from the Two micron All Sky Survey, which is a joint project of the University of Massachusetts and the Infrared Processing and Analysis Center/California Institute of Technology, funded by the National Aeronautics and Space Administration and the National Science Foundation.

References

Allen, P.R., Trilling, D.E., Korner, D.W., & Reid, I.N.: 2003 ApJ 595, 1222
Baraffe, I., Chabrier, G., Allard, F., & Hauschildt, P.H.: 1998 A&A 337 403
Burrows, A., Sudarsky, D., & Lunine, J.I.: 2003 ApJ 569, 587

Briggs, K.R. & Pye, J.P.: 2004 MNRAS 353, 673
Chabrier, G., Baraffe, I., Allard, F., & Hauschildt, P.H.: 2000 ApJ 542, 464
Chauvin, G., Lagrange, A.M., Lacombe, F., *et al.*: 2005 A&A 430, 1027
Close, L.M., Lenzen, R., Guirado, J.C., *et al.*: 2005 Nature 433, 286
Cushing, M.C., Rayner, J.T., & Vacca, W.D.: 2005 ApJ 623, 1115
Nielsen E.L., Close, L.M., Guirado, J.C., *et al.* 2005, Astron. Nachrichten, in press, Astroph/0509400
Festin, L.: 1998 MNRAS 2981, 34
Gorlova, N.I., Meyer, M.R., Rieke, G.H., & Liebert, J.: 2003 ApJ 593, 1074
Guirado, J.C., Martí-Vidal, I., Marcaide, J.M., *et al.*: 2005 A&A in press, astroph/0510199
Lenzen, R., Close, L., Brandner, W., Biller, B., & Hartung, M.: 2004 SPIE 5492, 970
Lenzen, R., Hartung, M., Brandner, W., *et al.*: 2003 SPIE 4841, 944
Luhman, K.L.: 1999 ApJ 525, 466
Luhman, K.L., Stauffer, J.R. & Mamajek, E.E.: 2005 ApJ 628, 69L
Luhman, K.L., & Potter, D. 2006 ApJ, in press, astroph/0510666
Hillenbrand, L.A. & White, R.J.: 2004 ApJ 604, 741
Maiolino, R., Rieke, G.H. & Rieke, M.J.: 1996 AJ 111, 537
Marley, M.S., Fortney, J., Hubickyj, O., Saumon, D., *et al.* 2005 Presentation, 2005 Winter Conference on Astrophysics
Marois, C., Macintosh, B., Song, I., & Barman, T.: 2005 astro-ph/0502382
Martín, E.L.: 2005 private communication
Martín, E.L., Brandner, W., Bouvier, *et al.*: 2000 ApJ 543, 299M
Martín, E.L., Rebolo, R., & Zapatero-Osorio, M.R.: 1996 ApJ 469, 706M
Mohanty, S., Jayawardhana, R., & Basri, G.: 2004 ApJ 609, 885M
Pinfield, D.J., Dobbie, P.D., Jameson, R.F., Steele, I.A., Jones, H.R.A. & Katsiyannis, A.C.: 2003 MNRAS 342, 1241
Reiners, A., Basri, G., & Mohanty, S.: 2005 accepted to ApJ, astro-ph/0506501
Stassun K. G. *et al.* 2005, Protostars & Planets V conf. abstract
Terndrup, D.M., Krishnamurthi, A., Pinsonneault, M.H., & Stauffer, J.R.: 1999 AJ 118, 1814
Zapatero Osorio, M. R., Lane, B.F., Pavlenko, Ya., Martín, E.L., Britton, M., & Kulkarni, S.R.: 2004 ApJ 615, 958
Zuckerman, B., Song, I., & Bessell, M.S.: 2004 ApJ 613L, 65

Discussion

Eric Gaidos: Then is your conclusion that the [dusty] evolutionary models are too hot for a given mass?

Laird Close: Yes, we find that in the case of AB Dor C the observed effective temperature appears lower than would be predicted by the models for a young 0.09 M_{sun} companion. Or in other words the HR diagram would have predicted a lower mass from the dusty models given our observed spectral type and luminosity. A similar trend has been recently reported for the low mass binary star Upper Sco CTIO 5 system by Reiners *et al.*

Ralph Neuhauser: The gravity analysis of Monhanty et al. agrees with you that low-mass stars and high mass brown dwarfs will be underestimated in mass by these models

Laird Close: Yes

Ralph Neuhauser: But they also suggest that the masses for *very low mass* brown dwarfs maybe *over-estimated* by the same models.

Laird Close: Indeed, but we cannot confirm this prediction for exoplanets with our data since AB Dor C is too massive. It is interesting to note that the eclipsing binary brown dwarf in Orion (2MASS0535; Stassun *et al.* PPV abs.) may show just such a trend comparing its ~0.035 and ~0.55 M_{sun} components to the models.

Direct Imaging of Exoplanets: Science & Techniques
Proceedings IAU Colloquium No. 200, 2005
C. Aime and F. Vakili, eds.
© 2006 International Astronomical Union
doi:10.1017/S1743921306009021

Search for giant extrasolar planets around white dwarfs: direct imaging with NICMOS/HST and NACO/VLT

Hans Zinnecker[1], Serge Correia[1], Wolfgang Brandner[2], Susanne Friedrich[3], and Mark McCaughrean[4]

[1]Astrophysikalisches Institut Potsdam, An der Sternwarte 16, D-14482 Potsdam, Germany
email: hzinnecker@aip.de
email: scorreia@aip.de

[2]Max-Planck-Institut für Astronomie, Auf dem Königsstuhl, D-69117 Heidelberg, Germany
email: brandner@mpia.de

[3]Max-Planck-Institut für Extraterrestrische Physik, Giessenbachstr., D-85748 Garching, Germany
email: sfriedr@mpe.mpg.de

[4]Astrophysics Group, School of Physics, University of Exeter, Stocker Road, Exeter EX4 4QL, UK
email:mjm@astro.ex.ac.uk

Abstract. We report preliminary results from our search for massive giant planets (6 – 12 Jupiter masses) around the known seven single white dwarfs in the Hyades cluster at sub-arcsec separations. At an age of 625 Myr, the white dwarfs had progenitor masses of about 3 solar masses, and massive gaseous giant planets should have formed in the massive circumstellar disks around these ex-Herbig A0 stars, probably at orbital separations similar or slightly larger than that of Jupiter. Such planets would have survived the post-Main-Sequence mass loss of the parent star and would have migrated outward adiabatically to orbital separations of about 25 AU. At the distance of the Hyades (45 pc) this corresponds to angular separations of about 0.5 arcsec which can be resolved with NICMOS/HST; the expected contrast in the J and H bands amount to 7.5 ± 1.5 mag.

Evaluation of our NICMOS data set did not reveal any evidence for planetary mass companions with masses down to about 10 Jupiter masses nor brown dwarfs around any of the seven white dwarfs for separations larger than 0.5 arcsec. However, we detected a low-mass, probably stellar, companion to a field white dwarf (WD1847-223J, distance ~50 pc, age ~1 Gyr; separation ~0.5 arcsec, $\Delta H \sim 2.5$ mag), using the NACO adaptive optics system at the VLT.

Keywords. extrasolar planets, white dwarfs, imaging, HST, VLT.

1. Planets around White Dwarfs

Since the discovery of the first extrasolar planet by Mayor & Queloz (1995) the number of known extrasolar planets has grown to more than 150. Most discoveries are based on measurements of radial velocities of stellar lines. Due to the huge contrast in brightness between the main sequence star and its planet(s) no planet has so far been seen directly with the possible exception of a planetary-mass companion to a young brown dwarf (Chauvin *et al.* 2005). The contrast problem can be reduced if one searches for planets around white dwarfs, which are about a factor of ten smaller than giant planets ($F \sim R^2$) and typically 10^3 to 10^4 times less luminous than main sequence stars of spectral type G or K.

For our search we have chosen the only young cluster in the near vicinity with a large enough sample of white dwarfs: the Hyades. Compared to isolated white dwarfs the Hyades cluster white dwarfs have the advantage that their age is known from the turn-off age of the cluster (625 Myr, Perryman *et al.* 1998). This also implies that any planet will be of the same age. A Hyades main sequence star with $3\,M_\odot$ (the turn-off mass) will eventually become a $0.66\,M_\odot$ white dwarf (Weidemann 2000). All single stars with masses above the turn-off mass and below about $8\,M_\odot$ have already left the main sequence and will develop into or have already become white dwarfs. While planets closer than 3 – 5 AU will probably not survive this post-main-sequence evolution (and will migrate inwards and merge) planets farther away will survive (Duncan & Lissauer 1998; Burleigh *et al.* 2002; Rasio *et al.* 1996), and their semimajor axis will increase by a factor $M_{\rm initial}/M_{\rm final}$ as the central star loses mass. Thus any giant planet in a Jupiter-like orbit ($\geqslant 5$ AU) will migrate outwards to a new equilibrium radius of at least:

$$R = 5\,\mathrm{AU} \times (3M_\odot\,/\,0.66M_\odot) \approx 22.5\,\mathrm{AU}$$

At a distance of the Hyades (45 pc) 22.5 AU correspond to 0.5 arcsec. Furthermore, a giant planet with mass somewhat below 13 Jupiter masses (which is one of the definition for "planet") orbiting a white dwarf in the Hyades has **a near-infrared magnitude difference of only about 7.5 mag**. Resolving such a favorable contrast at such a favorable angular separation is feasible with NICMOS in the F110W and F160W filters with spatial resolutions of 0.1 arcsec and 0.15 arcsec, respectively, as our simulations have shown. The same is true for near-infrared direct imaging with the NACO adaptive optics system at the VLT.

2. Observations and Data Reduction

The seven single white dwarfs (HZ4, LB227, VR7, VR16, HZ7, HZ14, LP475-242) have been observed with HST/NICMOS at two roll angles separated by 20 degrees through the F110W and F160W filters (program GO 9737). Two and four dithered frames, respectively, of 320 s integration time each were recorded for each roll angle and each target in these filters, which corresponds to a total of one and two orbits of HST time per target, respectively. The single dithered frames were reduced using the HST pipeline, then oversampled by a factor two using cubic spline before registration and a FFT-shift based combination for each roll angle. Subsequent bad-pixel masking was further applied in order to compensate for the non-optimal bad-pixel mask available through the pipeline. For that purpose, remaining not corrected bad-pixels were identified using a criterion based on the deviation from the median of all frames available for each target. In each filter, the resulting frames corresponding to different roll angles were subtracted from each other. This step was done after the use of a 2D cross-correlation for the registration of these frames, which provided a sub-pixel accuracy, and a FFT-shift based combination. The result obtained for HZ7 is presented in Fig. 1 as an example.

The NACO/VLT observations were obtained on 2004, May 3, in service mode in relatively good seeing (0.5 – 0.7 arcsec). The resulting ESO pipeline H-band image is presented in Fig. 4.

3. Method Validation and Contrast Ratio

We validated the method used in the data reduction through simulations. For this purpose a fake companion of given brightness was introduced in each frame as a scaled version of the target, adding the appropriate noise, and checking that we could detect the signature of the companion after all reduction steps described above. Fig. 2 shows the

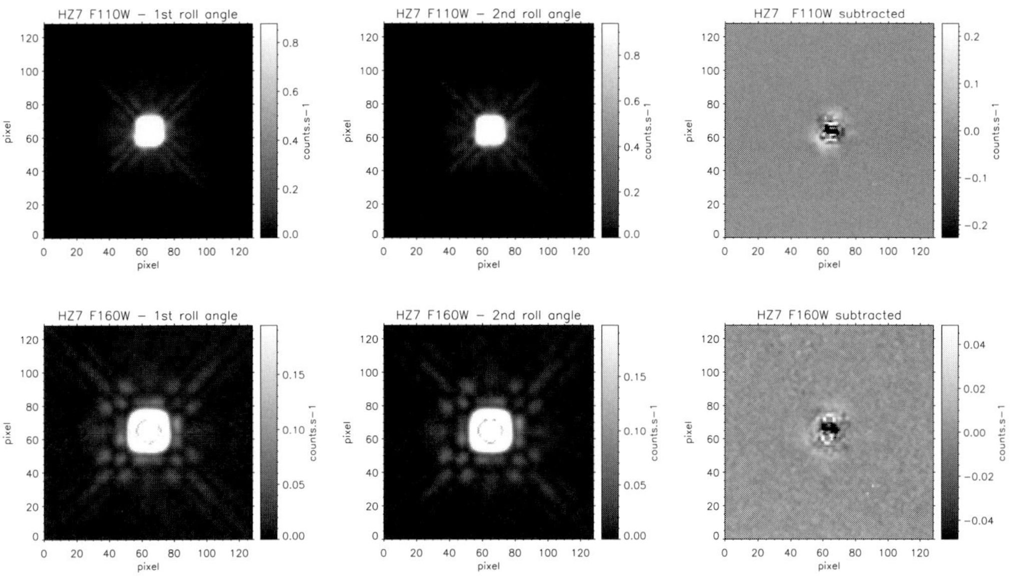

Figure 1. The observations of HZ7 at two different roll angles and the result of their subtractions in the F110W filter (upper panel) and the F160W filter (lower panel). The field of view is about 2.75 arcsec $\times$ 2.75 arcsec (0.0215 arcsec pixel size). Maximum cuts are at 2 % of the peak intensity for the observations, and $\pm$ 0.5 % for the subtracted frames.

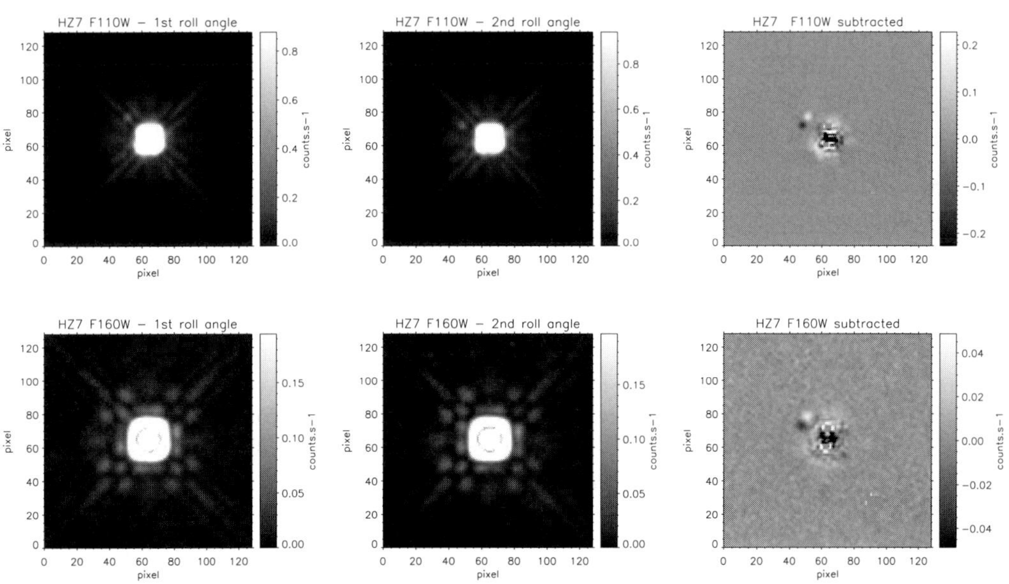

Figure 2. Same as Fig. 1 with an artificial planetary mass companion added (6 mag brightness difference, 0.4 arcsec separation).

results of such a simulation for both filters and the same target as in Fig. 1, with a planet at 0.4 arcsec separation and 6 magnitudes difference in brightness. One can easily notice the pair of negative and positive images of the companion at the given position from the star in the subtracted frame. Such a simulation allows to estimate the limiting difference in brightness as a function of separation and given SNR of the detection. Fig. 3 shows the result for the same target as in Fig. 1 in each filter and a detection threshold of $5\,\sigma$. The uncertainties are estimated from the scatter obtained using four different position angles of the companion. According to the COND models of Baraffe *et al.* (2003), the limiting mass of the planetary mass companion that we are able to detect is typically of about 10 Jupiter masses at 0.5 arcsec separation (Friedrich *et al.* 2005).

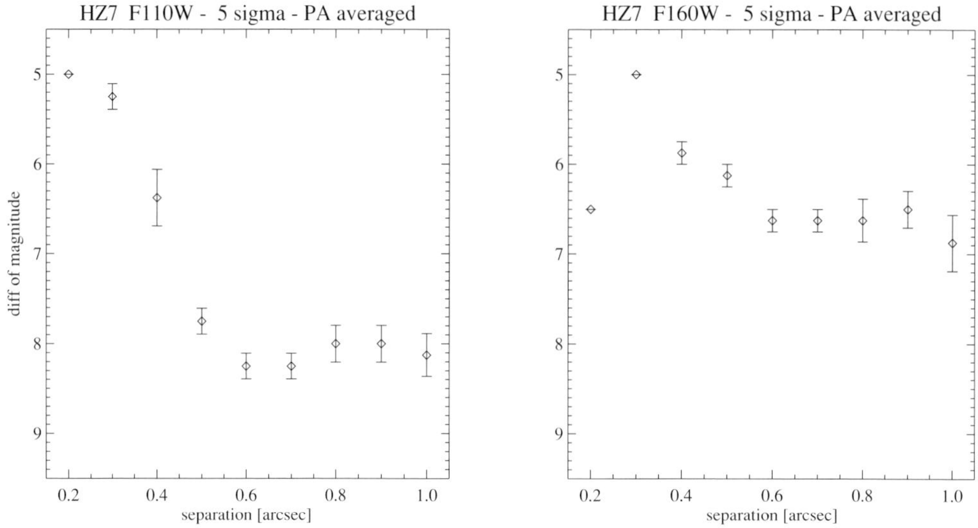

Figure 3. Estimated sensitivity limits as a function of separation for HZ7. Uncertainties are estimated from the scatter obtained using four different position angles of the companion.

4. Conclusion and Discussion

We did not find any evidence for neither planetary mass companions with masses down to about 10 Jupiter masses nor brown dwarfs around any of the seven white dwarfs in the Hyades for separations larger than 0.5 arcsec. Possible reasons for the non-detection include

- small number statistics (giant $\sim$10 M_{Jup} planets are rare, $< 10\,\%$)
- limited sensitivity (can't detect $\sim$1 M_{Jup} planets yet, too faint)
- hot planetary nebula (X-rays) may have evaporated the giant planet
- planet-planet perturbations during mass loss unbinds giant planets
- original giant planet has spiralled in (gas drag) and was swallowed

At present, we cannot decide whether our null result is due to the non-existence of giant planets in orbit around the progenitor stars or due to them not surviving post-Main Sequence evolution. The latter seems more likely (Villaver & Livio 2005, private communication; Debes & Sigurdsson 2002, Duncan & Lissauer 1998), especially if the giant planets have small orbital radii. In this context, we note that a giant planet of $M \sin i = 6.5\,M_{Jup}$ with semi-major axis of some 2 AU and eccentricity $e = 0.4$ has been found by the radial velocity technique around a G5 red giant star (HD 11977) of about $2\,M_{\odot}$, i.e. an A-star Main Sequence progenitor (Setiawan *et al.* 2005); thus giant planets appear to exist around early-type progenitors like our Hyades stars (good news), but it is clear that this giant planet is in danger to spiral into the red giant envelope and get lost (bad news). Indeed, we made a simple estimate, based on the Bondi-Hoyle formalism, how long it takes for a $5\,M_{Jup}$ planet on a circular orbit of radius 1 AU to spiral into a red giant of $3\,M_{\odot}$, given that the envelope mass density at 1 AU is $10^{-7} gcm^{-3}$ (Herwig, private communication). The answer is about 100 yr (this is actually the timescale on which the giant planet is going to double its own mass by envelope gas accretion, and the corresponding accretional drag will make the orbit decay on the same timescale).

5. Outlook

Despite the NICMOS/HST non-detection, the direct imaging search for giant planets around white dwarfs is a promising avenue and should be further pursued in the future, e.g. with JWST. Meanwhile, the next step is to search for mid-infrared excess on white dwarfs caused by thermal emission of the planets, using the Spitzer Space Telescope camera (IRAC). Although its angular resolution is only 1.5 arcsec, implying that any detection would likely be spatially unresolved, its sensitivity is such that we could see giant planets at 4.5 microns with masses of 2 to $3\,M_{Jup}$ around nearby young hot white dwarfs, such as WD 1134+300 at a distance of ~15 pc and an age of ~0.2 Gyr (25 – 50 % excess emission). Spitzer observing programmes are underway, including one by our team (no results yet).

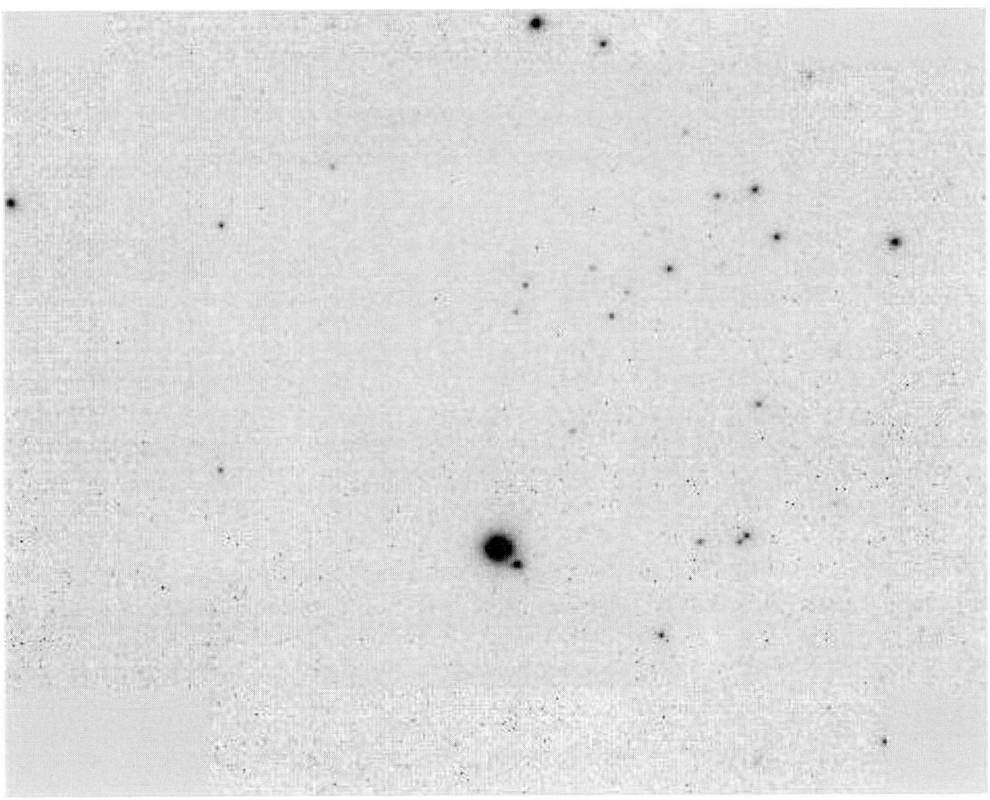

Figure 4. NACO/VLT H-band image of the field white dwarf WD1847-223J showing a ~10 times fainter companion towards the south-west (separation ~0.5 arcsec).

Acknowledgements

We thank Matt Burleigh for sharing with us his calculations of the detectability of giant planets around white dwarfs through mid-infrared Spitzer observations. We also thank Eva Villaver for pointing out the possibility of X-ray evaporation of giant planets during the hot planetary nebula phase. Furthermore, thanks are due to Falk Herwig for making his red giant structural models available. Last but not least, congratulations to the organizers for a wonderful last IAU Colloquium!

References

Baraffe, I., Chabrier, G., Barman, T.S., Allard, F., & Hauschildt, P.H. 2003, *A&A* 402, 701

Burleigh, M.R., Clarke, F.J., & Hodgkin, S.T. 2002, *MNRAS* 331, L41

Chauvin, G., Lagrange, A.-M., Dumas, C., Zuckerman, B., Mouillet, D., Song, I., Beuzit, J.-L., & Lowrance, P. 2005, *A&A* (Letters) 438, L25

Debes, J.H. & Sigurdsson, S. 2002, *ApJ* 572, 556

Duncan, M.J. & Lissauer, J.J. 1998, *Icarus* 134, 303

Friedrich, S., Zinnecker, H., Brandner, W., Correia, S., & McCaughrean, M. 2005, astro-ph/0501148

Mayor, M. & Queloz, D. 1995, *Nature* 378, 355

Neuhäuser, R., Guenther, E.W., Wuchterl, G., Mugrauer, M., Bedalov, A., & Hauschildt, P.H. 2005, *A&A* (Letters) 435, L13

Perryman, M.A.C., Brown, A.G.A., Lebreton, Y. Gomez, A., Turon, C., de Strobel, G.C., Mermilliod, J.C., Robichon, N., Kovalevsky, J., & Crifo, F. 1998, *A&A* 331, 81

Rasio, F.A., Tout, C.A., Lubow, S.H., Livio, M. 1996, *ApJ* 470, 1187

Setiawan, J., Rodmann, J., da Silva, L. *et al.* 2005 *A&A* 437, L31

Weidemann, V. 2000, *A&A* 363, 64

Direct Imaging of Exoplanets: Science & Techniques
Proceedings IAU Colloquium No. 200, 2005
C. Aime and F. Vakili, eds.
© 2006 International Astronomical Union
doi:10.1017/S1743921306009033

Modelling the Diversity of Extrasolar Terrestrial Planets

Victoria S. Meadows[1]†

[1] *Spitzer* Science Center, California Institute of Technology, 1200 E California Blvd, Pasadena, CA 91125, USA
email: vsm@ipac.caltech.edu

Abstract. As we prepare to undertake the observational search for extrasolar terrestrial planets, theoretical modeling studies can help to prepare us for the likely diversity of the extrasolar terrestrial planets. This diversity may arise as a function of planetary system architecture and formation history, which results in a variety of initial planetary properties, as well as stellar, planetary and biological evolutionary processes. Modeling of the physical and chemical processes of the planetary environment, and their interaction with the parent star, allows us to understand the nature of the planetary characteristics that indicate habitability and life, and how these manifest in the planetary spectra. Here we present disk-averaged spectra of planets in our own solar system, and models of the Earth through several eons to understand the types of planetary characteristics that are likely to be observed by planned planet detection and characterization missions.

Keywords. astrobiology, techniques: spectroscopic telescopes, instrumentation: spectrographs, Earth, planets and satellites: general, stars: planetary systems.

---

1. Introduction

1.1. *The Search For Habitable Worlds*

The search for habitable or Earth-like planets beyond our Solar System will be conducted with space-based observatories and advanced observing techniques. Yet, the planetary characteristics sought, the analysis tools used, and the ultimate interpretation of what we find must rely on an interdisciplinary synthesis of theory and existing observation.

The NASA Terrestrial Planet Finder Coronagraph (TPF-C), and ESAs Darwin and NASAs Terrestrial Planet Finder Interferometer (TPF-I) missions will have the capability to directly detect and characterize terrestrial planets around several hundred nearby stars, and both will have low-resolution spectral capability (($R \sim 70$ in the visible, and $R \sim 20$ in the mid-infrared (MIR)). TPF-C's visible wavelength observations will be sensitive to surface and atmospheric composition, while TPF-I and Darwin's MIR capability is most sensitive to surface and atmospheric temperatures and atmospheric trace gases.

Extrasolar terrestrial planets will appear as unresolved point sources, precluding spatial resolution of continents, oceans or clouds. The planet's characteristics, including life, must be inferred from a spectrum averaged over the visible disk. Self-consistent models of extrasolar terrestrial planet environments must therefore by combined with advanced remote-sensing retrieval methods to identify signs of habitability and life in the context of the planetary characteristics that affect this disk-averaged spectrum, especially as the spectra will also likely significantly time-averaged as well. We also cannot expect that the planets found will be similar to known Solar System bodies, due to likely differences in planetary architecture, initial composition, evolutionary development, and age.

† member of the NASA Astrobiology Institute

Our search strategy, detection, and analysis techniques for extrasolar terrestrial planets must be flexible enough to determine the characteristics of worlds completely unlike our own. Self-consistent models of planet formation, climate, and atmospheric compositions can inform these search strategies by using a specific planetary context (stellar type, mass, orbital geometry, bulk composition) to help define the plausible range of planetary environmental conditions and their resulting disk-averaged spectral signatures.

1.2. *The Potential Diversity of Other Worlds*

The Earth is the only known habitable, living world, and yet the complexity of terrestrial planet structure, composition and processes, with myriad possibilities for origins and evolution, suggest that an Earth twin is unlikely to be the only other habitable planet found. Recent ground-based observations show an unexpected diversity of extrasolar planetary system architectures (Marcy *et al.* 2005), that modeling indicates translate into differences in the formation, composition, evolution and characteristics of terrestrial planets. In particular, initial planetary mass, the initial volatile inventory, and the subsequent evolution of volatile delivery may differ significantly, and can be affected by stochastic variations during formation (Wetherill, 1996; Morbidelli *et al.*, 2000), or systematic variations due to the initial disk mass and metallicity, and the location and orbits of Jovian planets (Chambers and Cassen, 2002; Levison and Agnor, 2003; Raymond *et al.*, 2004).

While terrestrial planets more massive than the Earth are not represented in our Solar System, they are seen in planet formation simulations (c.f. Raymond *et al.*, 2004). The recent discovery of a 7.5 Earth mass planet provides the first evidence of a likely large-mass terrestrial planet (Rivera *et al.*, 2005). The environmental characteristics of these larger planets are unknown, but are important to model, as they are likely to be the first terrestrial planets detected via astrometry and characterized via direct detection.

Initial volatile inventory is also strongly affected by the planetary system architecture. For planetary systems with a hot Jupiter, or an eccentric outer Jovian, the formation of water-rich worlds in the habitable zone can become increasingly more difficult (Raymond *et al.*, 2004; Raymond *et al.*, 2005a) favoring the formation of water-poor worlds. Yet, for Solar Systems similar to our own, recent simulations suggest that the majority of terrestrial planets are formed with a higher water inventory than our current Earth (Raymond *et al.* 2006). Retention of volatiles over the course of evolution of the planet may then be a significant factor in continued habitability and planetary diversity.

Evolutionary processes therefore provide an additional mechanism for planetary diversity, as the divergent evolution of Venus and Earth, near "twins" in size, mass and bulk composition, shows. The planetary state is dictated by both the planets characteristics and its evolving interaction with its parent star, including possible planetary migration, atmospheric mass loss, large impacts and other exogenic processes. The interaction of life with our evolving planet has resulted in a diversity of planetary states at different points in our history. The more stable and long-lived the planetary environmental state, the more likely it is to be seen in astronomical observations. As our neighboring stars have ages from newborn to 10Gy old (Nordstrom *et al.*, 2004) the planned planet detection and characterization missions will have the exciting opportunity to observe planets at different stages of planetary and biological evolution.

2. Characterizing Terrestrial Planets

2.1. *Planetary System Environments*

For the planned planet detection and characterization missions, the target stars to be searched for planets will likely be well characterized prior to mission launch. With regard to planetary habitability, the star's most important characteristics are its full-wavelength spectrum and its activity level (e.g. flares, luminosity variations). Other aspects of the planetary system environment that will affect habitability include the presence of small bodies and the presence, nature and orbits of any giant planets in the system.

2.2. *The Signs of Habitability*

For extrasolar terrestrial planets, habitability is defined as the environmental conditions required to maintain liquid water on the surface of the planet. Photometric variability may provide clues to planetary surface inhomogeneity from continents and oceans (Ford *et al.*, 2001), although to first order, cloud-cover will produce the majority of this variability (Tinetti *et al.*, 2006) and repeated observations at relatively high temporal resolution will be needed to discriminate surface from cloud effects.

Once a terrestrial planet is identified, we will use both models and observations to attempt to determine if its surface temperature and pressure are consistent with liquid water. From Earth and planetary studies, we know that surface temperature (and possibly pressure) are the product of the planetary global energy balance, which depends on the planets orbit(which gives planetary distance, and therefore solar insolation, as a function of time), the properties of the parent star, the atmospheric mass and composition, and the planetary surface properties (albedo, thermal inertia). Direct observation of surface temperature at optical and infrared wavelengths can be thwarted by the presence of clouds, or a thick, absorbing atmosphere. Direct measurements of atmospheric pressure are also difficult, especially in the presence of clouds. For terrestrial planets, clues to atmospheric pressure may be provided from the slope of Rayleigh scattering in the 0.4-0.7μm region (Arnold *et al.*, 2002, Woolfe *et al.*, 2002). However, this spectral region is also sensitive to surface absorption from iron oxides (as seen on Mars) and absorption by ozone and other trace gases (SO_2, NO_2) in both a clear and cloudy atmosphere.

Spectra of Earth provide clues to habitability. In the MIR, strong CO_2 absorption at 15μm indicates the presence of an atmosphere and provides clues about its thermal structure. For Earth, the structure of this broad absorption feature indicates a warm emitting surface that is overlain by a cooler troposphere, and a hotter stratosphere. A hotter stratosphere may be due to the presence of a UV absorber, like the Earth's O_3, which can act as a UV shield for the surface below. Water vapor is also a spectral indicator of potential habitability, and can be seen at both MIR and visible wavelengths (Figure 1). There may, however, be other indicators of habitability, or conversely, other atmospheric and surface constituents that, if present, would indicate that liquid water was unlikely to exist on the planetary surface. For Venus, although both CO_2 and small amounts of H_2O can be seen in the spectrum, the simultaneous presence of SO_2 and H_2SO_4, both unstable in the presence of liquid water, indicates that surface water is unlikely.

2.3. *Biosignatures: The Signs of Life*

The concept of an astronomical biosignature was introduced as early as 1965 when Lovelock (1965) proposed that the life on a planet would change the atmospheric composition by many orders of magnitude. Such changes could be recognized even at astronomical distances. Initial identification and estimates of biosignature detectability for TPF and Darwin are described in Des Marais *et al.*, 2002 and Selsis *et al.*, 2003. Here we define three different types of biosignatures; atmospheric, surface, and temporal.

Atmospheric Biosignatures: are constituents in the planetary atmosphere that are significantly out of chemical equilibrium, or likely byproducts of life processes. The abundant O_2 in our atmosphere is considered a good biosignature (Lovelock & Margulis, 1974; Des Marais *et al.*, 2002), but both CH_4 and N_2O, especially when seen in the presence

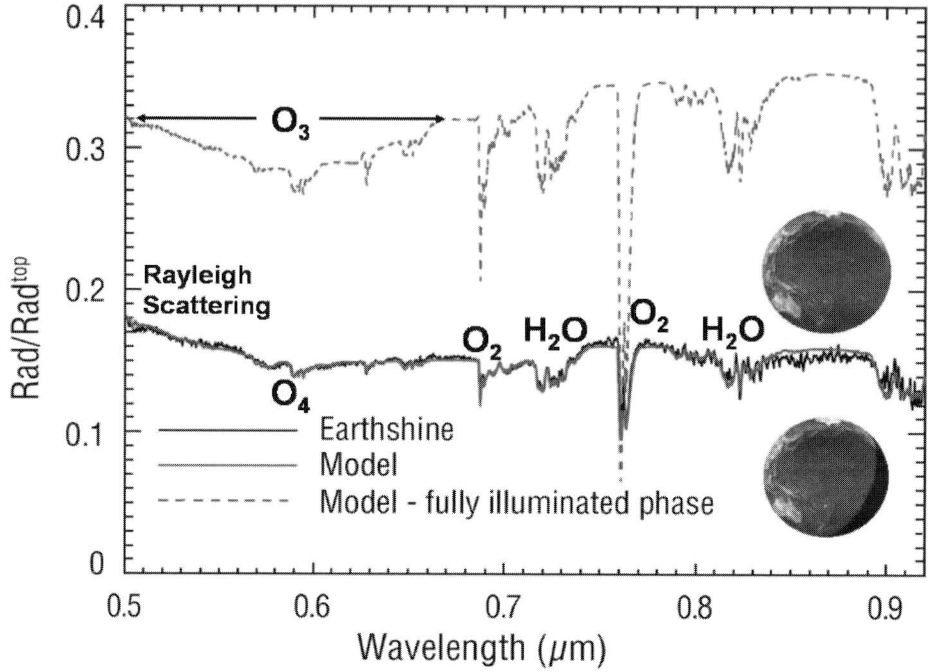

Figure 1. The observed disk-averaged Earthshine spectrum (from Woolf *et al.*, 2002), fit with synthetic spectra generated by the VPL 3-D Earth model for the observed gibbous phase, and the same model extrapolated to full-phase (after Fig 8(c), Tinetti *et al.*, 2006). Water vapor can be clearly seen, especially at the longer wavelengths, and the biosignatures O_2, O_3 and O_4 are also indicated. The pressure sensitive O_2 dimer (O_4) may become more detectable in higher mass atmospheres.

of O_2, provide a much stronger potential biosignature because they indicate a chemical disequilibrium (Lovelock & Margulis, 1974).

Surface Biosignatures: are spectral features in the surface reflectance that are associated with bio-pigments (e.g. chlorophyll), or are due to the physical structure of life, such as the *red edge* reflectivity of leaves (Gates *et al.* 1965; Seager *et al.*, 2005). Although the red edge is seen near 0.7μm on Earth, it could be shifted for extrasolar planets, due to different planetary atmospheres and host star spectra(Wolstencroft & Raven, 2002)

Temporal Biosignatures: are atmospheric or surface characteristics that change as a function of time due to seasonal variations in biomass or planet-wide life processes. On the Earth, seasonal periodicities in both atmospheric CO_2 and CH_4 are associated with respiration in the land biosphere.

False Positives: occur when abiotic planetary processes mimic a biosignature. Examples include the abiotic production of CH_4, O_2 production during loss of an ocean, surface mineral reflectivities, and seasonal changes in planetary reflectivity or composition induced by dust storms or photochemistry. Once modeling has identified the nature and likelihood of false positives, observations can be planned to ensure sufficient wavelength coverage and sensitivity to help to rule them out.

3. Modelling: The Virtual Exploration of New Worlds

While the planned NASA and ESA mission suite will obtain the first data on the range of extrasolar terrestrial planet environments, these technically-challenging missions have decades-long lead times. The current field of extrasolar terrestrial planet characterization consequently relies on modeling investigations which allow us to explore a larger range of planetary environments and spectra, including planets of different masses and ages. Models can also be used to visualize even the more familiar Solar System planets as disk-averages, over larger wavelength ranges and from different viewing geometries than are currently readily available.

Modeling planetary spectra for extrasolar terrestrial planets involves three major steps, input data collection, generation of a planetary environment (surface and atmospheric temperature structure and composition), and spectral generation. The input data includes stellar spectra and the optical and spectral properties of the constituent gases and aerosols. As planetary habitability is affected by both the UV portion of the spectrum (photochemistry, surface incident radiation), and the longer wavelengths (global energy balance and climate), the full wavelength range for the input stellar spectra is required, via both data gathering and modeling. For the planet models shown here, we have used stellar spectra generated by M. Cohen (http://vpl.ipac.caltech.edu/spectra/stellar) and molecular linelists from the HITRAN database (Rothman *et al.*, 2005)

To generate the planetary environment, coupled-climate-chemistry models are used. These models produce vertical profiles for atmospheric temperature, pressure and constituents. Fluxes of gases at the atmospheric boundaries (surface and exosphere) can be specified, or generated by models that are coupled to the climate-chemistry model. Coupled climate-chemistry models are more likely to produce temperature and constituent profiles that are self-consistent with each other and the incoming stellar radiation. This is particularly important for determining the detectability of planetary spectral features in the MIR, where the atmospheric temperature structure can enhance or reduce the strength of constituent absorption bands. Here we have principally used a 1-D (vertical) coupled climate-chemical Radiative-Convective Equilibrium model developed by J. Kasting, and additional models as specified below.

The planetary temperature, pressure and constituent profiles, and stellar spectra are input to radiative transfer models which generate synthetic spectra of the planet. Here we have used a radiative transfer code based on the Spectral Mapping Radiative Transfer or SMART code developed by D. Crisp (Meadows & Crisp, 1996; Crisp, 1997).

4. Synthetic Spectra of Terrestrial Planets

To understand the spectra that might be seen by extrasolar terrestrial planet characterization missions, observations, data and models can be combined to progressively move from studying objects we know relatively well, such as the Earth and Solar System planets, to planetary environments that are less well known, such as the Earth's environment during its 4.6Gy history, and planets around stars of different spectral type.

4.1. *Synthetic Spectra of Solar System Planets*

Figure 2 shows full resolution synthetic spectra for terrestrial planets in our Solar System from 0.5 to 1.7μm. The Mars and Earth spectra are fully 3-dimensional radiative transfer simulations of the disk-averaged spectrum using atmospheric GCM or observational data as input. The spectra have been validated to spacecraft and Earthshine observations (Tinetti *et al.*, 2005, 2006). The Venus spectrum is a simulated disk-average for an average solar illumination and viewing angle. Also included is a 3-D model disk-average of the Earth as it would appear if completely covered by high cirrus clouds.

This figure shows the remarkable diversity of terrestrial planet spectra, even in our own Solar System. The Venus spectrum is dominated principally by strong CO_2

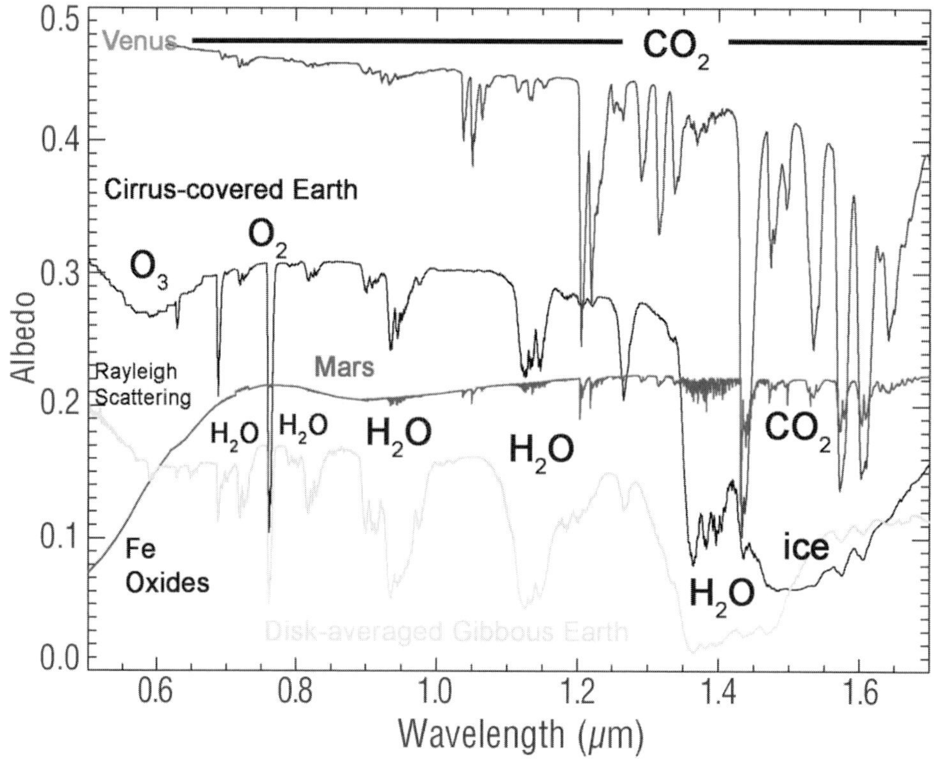

Figure 2. Synthetic albedo spectra of Solar System terrestrial planets generated by the Virtual Planetary Laboratory (VPL). The Mars and Earth spectra are from fully-spatially resolved models that have been degraded to the disk average (Tinetti *et al.*, 2005, 2006). Also shown is a model of a cirrus cloud covered Earth, which exhibits enhancements in the O_3 absorption from 0.5-0.7μm and water ice bands near 1.55μm.

absorption, although extremely weak water bands can also be seen. As Venus represents the likely end state for terrestrial planet evolution, this type of environment and spectrum may be relatively common for extrasolar terrestrial planets. The Mars spectrum is also dominated by CO_2, but these relatively weak features are only likely to be detected longward of 1.4μm. The Earth spectra are dominated by water vapor, and Earth's O_2 biosignatures are clearly seen. In Earth's spectrum from 0.5-0.7μm, ozone absorption and Rayleigh scattering are both present. However, in the presence of high clouds, the observable atmospheric column is reduced, and the Earth's Rayleigh scattering slope is correspondingly diminished. However, due to multiple scattering within the clouds, the O_3 absorption is enhanced. Attempting to use Rayleigh scattering as an indicator of atmospheric density is also difficult for Venus-like and Mars-like planets. In the case of Venus, the planetwide cloud-deck truncates the atmospheric column, and the unknown UV absorber reduces the albedo of the planet shortward of 0.55μm. For Mars, strong iron oxide absorption from the surface dramatically reduces the albedo of the planet shortward of 0.7μm.

Figure 3 shows a comparison of Earth and Venus with the smaller, CH_4-dominated bodies, Titan and Neptune. The planetary spectra were convolved with a triangular slit function to achieve a constant wavenumber resolution corresponding to R$\sim$70 at 0.76μm. The Venus and Earth models are as for Figure 2 above, and the Neptune and Titan input albedos were taken from observations (Karkoschka, 1994).

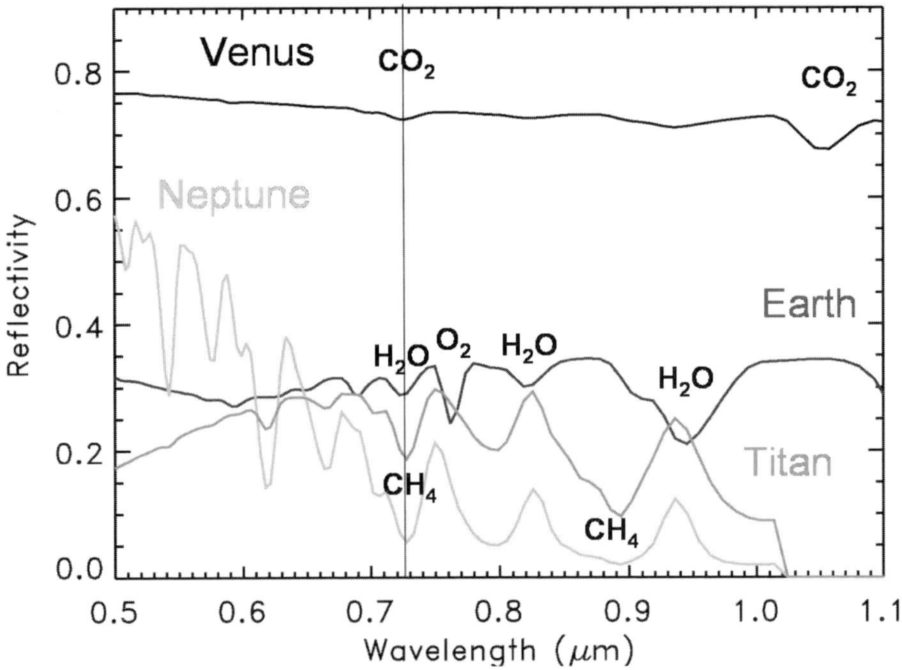

Figure 3. Synthetic spectra at R∼70 of the Earth (Tinetti et al., 2006) and Venus compared with measured albedos for Neptune and Titan (Karkoshka,1994).

Titan also shows a downturn due to absorption in the Rayleigh scattering region. Note also that at this resolution, absorption at 0.725μm is variously due to CO_2 (Venus), H_2O (Earth), or CH_4 (Titan and Neptune), depending on the different planetary composition. This redundancy however can be quickly overcome by searching for corroborating bands of each candidate molecule elsewhere in the spectrum, and argues for making characterization measurements over as large a wavelength range as possible.

4.2. *The Earth Through Time*

The Earth's history provides a suite of environments different to the modern Earth, which serve as analogs for habitable extrasolar planets. Prior to 2.3 billion years ago (Gya) the Earth's atmosphere likely had little O_2, and higher concentrations of CH_4 and CO_2. During the Proterozoic period, O_2 levels rose, but CH_4 levels may have remained high. To simulate spectra for the Earth's history, we used our radiative transfer model to generate synthetic spectra from a climate-chemistry model of the early Proterozoic environment (Pavlov et al., 2003; Segura et al., 2004) and a coupled atmosphere-ecosystem model for the Archean (Karecha *et al.*, 2005). For the Proterozoic, atmospheric O_2 was at 10% of present levels, with 100ppm of CH_4. For the Archean, the atmosphere was assumed to be 99.8% N_2, with 2000ppm of CO_2, 1000ppm of CH_4 and 100ppm of H_2. The resulting spectra for cloud-free cases over an ocean surface are shown in Figure 4 (vis-NIR) and in Figure 5 (MIR) (see also Kaltenegger *et al.*, this proceedings).

In the visible/NIR, one of the largest differences between the spectra of the three eons can be seen in the 0.5-0.7μm range, where the decreasing column depth of ozone from the modern to the Archean eon causes the slope of the spectrum to change significantly. In the modern and Proterozoic eons, O_2 can be seen at 0.76μm, although significantly diminished in the Proterozoic, but is not present in the Archean. Similar behaviour is

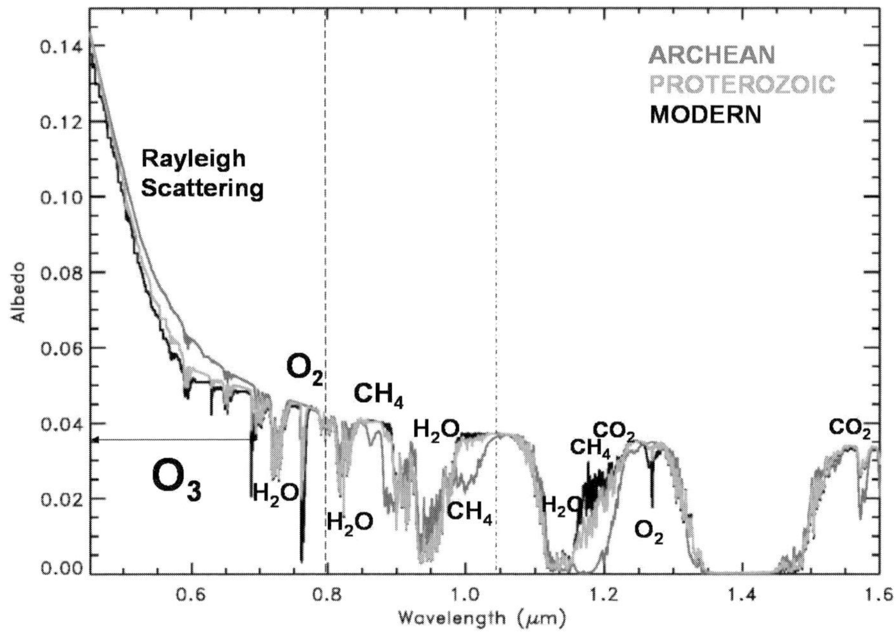

Figure 4. Synthetic visible wavelength spectra of the cloudless Earth during the late Archean (3.2-2.3Gya), Proterozoic (2.3-0.8 Gya) and Modern (present day) eons.

seen for the 1.27μm O_2 band. CH_4 absorption changes markedly throughout the Earth's history, with the strongest absorption seen in the Archean. For a TPF-C wavelength range, changes in atmospheric CH_4 abundance are most readily seen near 1.18μm, although strong bands also occur near 1.0μm and 0.88μm. Changes in CO_2 abundance are also seen in the spectra, although relatively weak. The strongest band is at 1.57μm with a weaker effect seen near 1.24μm. Note also that although CO_2 is visible at 1.05μm in the Venus spectrum (Figure 2) that band is typically extremely weak at Earth CO_2 concentrations and likely to be undetectable in Earth twins.

The MIR spectra are dominated by the bands of CO_2, CH_4, O_3 and H_2O. As we move back in time, the CO_2 band strengthens as the atmospheric CO_2 concentration increases. Similar behavior is also seen in the CH_4 absorption near 7.7μm. Interestingly, the O_3 absorption does not decrease in strength in the Proterozoic, even though the oxygen abundance of the atmosphere has been decreased ten-fold. This is due to both the non-linear relationship between oxygen concentration and ozone production, and the coupled effects of reduced ozone and reduced stratospheric temperature, which enhances absorption of the hotter surface radiation. Consequently, detection of the classic atmospheric biosignature, the simultaneous presence of CH_4 and O_2, or its proxy O_3, may be easier in a Proterozoic type environment than for modern-day Earth.

4.3. *Earth-like Planets Around Other Stars*

Planetary modeling also allows us to explore the interaction between terrestrial planets and the spectral energy distributions of their host stars. Segura *et al.* (2003, 2005) used coupled-climate chemistry models to explore the equilibrium atmospheric states of an Earth-like planet around F, G, K and active M stars. The planets modeled had an atmospheric composition and similar surface fluxes of biogenic gases to modern-day Earth. However, the modeled interaction of the planet with the spectral energy distribution of

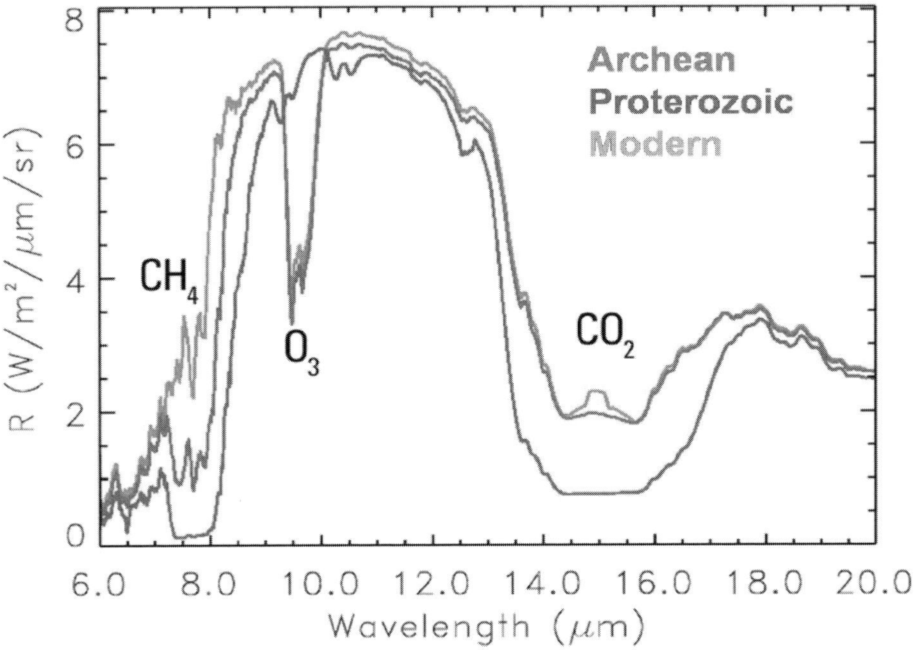

Figure 5. Synthetic mid-infrared wavelength spectra of the cloudless Earth during the late Archean (3.2-2.3Gya) Proterozoic (2.3-0.8 Gya) and Modern (present day) eons.

the parent star resulted in photochemistry that preferentially enhanced the atmospheric lifetimes of some biogenic gases (CH_4, and CH_3Cl) over others (N_2O) around K and M stars. This was shown to increase the detectability of atmospheric biosignatures that are only present trace quantities in the Earths present atmosphere for planets around K and M stars (see, for example, Figures 5-8 from Segura *et al.*, 2005)

5. Conclusions

Although our Solar System provides an excellent starting point from which to understand terrestrial planet spectra, extrasolar terrestrial planets may have different masses, orbits, ages and compositions, and be found in planetary systems unlike our own. Until the TPF and Darwin missions launch, modeling will allow us to explore a wider diversity of planets, and provide a basis to interpret and constrain the first order characterization data obtained by these missions. Modeling of the Earth's history and planets around other stars has already shown us that for Proterozoic-like periods and for planets around late type stars, life may be easier to detect than for the present day Earth. Finally, the diversity of planetary systems found so far indicates that we need to understand the effects of metallicity on planet formation, the processes and outcomes of planetary migration, and be flexible enough to characterize planets in planetary systems with architectures, orbits and evolutionary histories that may be quite unlike those currently known.

Acknowledgements

This material is based upon work performed by the NASA Astrobiology Institutes Virtual Planetary Laboratory Lead Team, supported by NASA through the NASA Astrobiology Institute, under Cooperative Agreement No. CAN-00-OSS-01.

References

Arnold, L., Gillet, S., Lardire, O., Riaud, P. & Schneider, J. 2002, *A&A*, 392
Chambers, J.E., & Cassen, P. 2002, *Meteor. & Plan. Sci.*, 37, 1523.
Crisp, D. 1997, *Geophys. Res. Lett.* 24(5), 571
Des Marais, D.J., Harwit, M., Jucks, J., Kasting, J.F., Lunine, J.I., Lin, D., Seager, S., Schneider, J., Traub, W., & Woolf, N. 2002, *Astrobiology*,2(2), 153
Ford, E.B., Seager, S., & Turner, E.L. 2001, *Nature*, 415, 885
Gates, D.M., Keegan, H.J., *et al.* 1965. *Appl. Opt.*, 4(1), 11
Karecha, P., Kasting, J., & Siefert, J. 2005, *Geobiology*, 3, 53
Kaltenegger, L., Jucks, K., & Traub, W. 2006, *this proceedings*
Karkoschka, E. 1994, *Icarus*, 111, 174
Levison, H.F. & Agnor, C. 2003, *ApJ*, 125, 2692
Lovelock, J.E. 1965, *Nature* 207, 568
Lovelock, J.E. & Margulis, L. 1974, *Icarus*, 21, 471
Marcy, G.W., Butler, R.P., Fischer, D.A., Vogt, S.S., Wright, J.T., Tinney, C.G., & Jones, H.R.A. 2005, *Progress of Theor. Phys. Supp.* 158, 24
Meadows, V.S. & Crisp, D. 1996, *JGR-Planets*, 101(E2), 4595
Morbidelli, A., Chambers, J., Lunine, J.I.,Petit, J.M., Robert,F., Valsecchi, G.B., & Cyr, K.E. 2000, *Meteor. & Plan. Sci.*, 35, 1309
Nordstrom, *et al.*, 2004, *A&A*418, 989
Pavlov, A.A., Hurtgen, M.T., Kasting, J.F., & Arthur, M.A. 2003, *Geology* 31, 87
Raymond, S.N., Quinn, T., & Lunine J.I. 2004, *Icarus*, 168, 1
Raymond, S.N., Quinn, T., & Lunine J.I. 2005a, *Icarus*, 177, 256
Raymond, S.N., Quinn, T., & Lunine J.I. 2006, *Icarus*, submitted
Rivera, E., Lissauer, J., Butler, R.P., Marcy, G.W., Vogt, S.S., Fischer, D.A., Brown, T., & Laughlin, G. 2005, *ApJ*, accepted
Rothman, L.S. *et al.* 2005, *J. Quant. Spect. & Rad. Trans.*, 96, 139
Seager, S., Turner, E.L., Schafer, J., & Ford, E.B. 2005, *Astrobiology*, 5(3), 372
Segura, A., Krelove, K., Kasting, J.F., Sommerlatt, D., Meadows, V., Crisp, D., Cohen, M., & Mlawer, E. 2003, *Astrobiology*, 3, 689
Segura, A., Kasting, J.F., Meadows, V., Cohen, M., Scalo, J., Crisp, D., Butler, R.A.H., & Tinetti, G. 2005, *Astrobiology*, 5(6), 706
Selsis, F., Despois, D., & Parisot, J.-P. 2002, *A&A*, 388, 985
Tinetti, G., Meadows, V.S., Crisp, D., Fong, W., Velusamy, T., & Snively, H. 2005, *Astrobiology*, 5(4)
Tinetti, G., Meadows, V.S., Crisp, D., Fishbein, E., Turnbull, M., & Bibring, J.-P. 2006, *Astrobiology*, accepted, in press
Wetherill, G.W. 1996, *Icarus*, 119, 219
Wolstencroft, R.D., & Raven, J.A. 2002, *Icarus*, 157, 535
Woolf, N.J., Smith, P.S., Traub, W.A., & Jucks, K.W. 2002, *ApJ*, 574(1), 430

Discussion

HEAP: Have you modeled planets around stars of different metallicity?

MEADOWS: Not yet. Although some of the planetary spectra I showed are modeled around stars of different spectral type, the planets themselves are Earth-like, and a direct connection between planetary composition and the formation history of the planet has not been made. However, VPL team members Raymond and Kress are collaborating to couple planet formation and disk chemistry models to generate planetary bulk-compositions consistent with the stellar metallicity. These planets will then serve as the starting point for our climate-chemistry, planetary evolution and spectral models.

Direct Imaging of Exoplanets: Science & Techniques
Proceedings IAU Colloquium No. 200, 2005
C. Aime and F. Vakili, eds.
© 2006 International Astronomical Union
doi:10.1017/S1743921306009045

Direct imaging search for substellar companions around neutron stars

B. Posselt[1,2], R. Neuhäuser[1] and F. Haberl[2]

[1]Astrophysikalisches Institut und Universitätssternwarte, FSU Jena, Schillergässchen 3-5, 07745 Jena, Germany

[2]Max-Planck-Institut fuer extraterrestrische Physik, Giessenbachstrasse, Garching, Germany
email: posselt@mpe.mpg.de

Abstract. Currently two planetary systems around pulsars are known - discovered by the pulse timing technique, an indirect method. These planets were a surprise and gave rise to diverse planet formation scenarios, some of them very different to the common planet formation models around solar type stars and thus physically very interesting. Furthermore, neutron star planets are not only interesting themselves but also to study properties of the poorly understood neutron stars. After a summary about the current state of pulsar planets and the theoretical formation models, we present our own direct-imaging search for thermal emission of neutron star planets using the VLT. The project sample includes the fascinating radio-quiet isolated neutron stars, which are some of the closest and probably youngest neutron stars we know. Companions around them can only be found by direct imaging. Detecting planets around neutron stars by direct imaging differs significantly from using this technique for other, e.g. solar type, stars. As great advantage there is no need to reject the starlight of the primary.

Keywords. Stars: neutron, Infrared.

1. Introduction

The first earth-mass planets discovered have been neutron star planets. Meanwhile four planets are known around the old millisecond(ms)-pulsar PSR 1257+12 (Wolszczan & Frail (1992), Konacki & Wolszczan (2003), Wolszczan (2005)) . They have up to four earth masses and are very close to the primary – 0.2 AU to around 3 AU. Interestingly, for three of them nearly circular orbits are reported (Konacki & Wolszczan (2003)). Another planetary system around a neutron star is the binary system of the ms-pulsar PSR 1620+10 and its white dwarf companion (Sigurdsson (2003), Sigurdsson & Thorsett (2004)). Actually, a 2.5 Jupiter mass object is orbiting the white dwarf at 23 AU.

These unxepected dicoveries rised several questions about the formation of these planets. Did they formed before the supernova and how they did survive this event ? If they formed after the supernova, how did they form ? In the case of PSR 1257+12 it is extremely unlikely that all four planets are captured from other stars, while this is the likely explanation of PSR 1620+10 where probabaly an exchange of the white dwarf companion took place (Sigurdsson & Thorsett (2004)). There are several models for planet formation after the supernova or survival, well described by Podsiadlowski (1995) and references therein. If the planets around a neutron star survived the supernova, an initial orbital separation larger than 4 AU (to survive the giant stage) and asymmetric supernovae are needed. The nearly circular orbits of three planets make this survival model unlikely for PSR 1257+12. For formation after the supernova, the supernova fallback model, the circumbinary disk model and the disrupted companion model are currently discussed (e.g. Podsiadlowski (1995)). The first formed disks have different properties. e.g. in masses, sizes and composition. The following planet formation could differ from those around solar-like stars. Thus, the study of neutron star planetary systems in comparison to those around solar-like stars can be very fruitful for the understanding of planet formation in gerneral. For PSR 1257+12 the circumbinary disk model and the disrupted companion model are currently favoured (Podsiadlowski (1995)).

But substellar companions of neutron stars are not only interesting for planet formation theory. They provide also insights into the supernova event and maybe the progenitor star(s). Furthermore, the study of low-mass companions around neutron stars, their masses and orbital characteristics, can give masses of the neutron stars. Together with radius determinations these neutron star masses constrain the neutron stars (degenerate) equation-of-state. Currently no radius and mass of a neutron star are known at the same time and the equation of state of ultradense matter in the Universe is one of the big questions in astrophysics. While masses of neutron stars can be derived for binary systems (e.g. Jonker *et al.* (2003)), the best radius estimation comes from the radio-quiet X-ray thermal neutron stars, which are isolated without stellar companions (e.g. Trümper *et al.* (2004)).

2. How to find substellar companions around neutron stars?

By radio pulse arrival time variations, one can detect only relatively close companions (like in radial velocity planet searches among normal solar-like stars), so that the question of wide companions remains open. Furthermore only radio pulsars with very stable pulses, like ms-pulsars can be investigated. Radio-quiet neutron stars are not covered. As the neutron stars are very faint (fainter than 25 mag in the optical), the commonly used radial velocity method or the search for astrometric wobbling should preferably be done in X-rays. Unfortunately the spectral and spatial resolution in X-ray astronomy is currently not enough for these methods. Furthermore the neutron stars have active region, pulsar beams and even larger companions than the primary, which also rule out the transit method as good choice. Substellar companions of all possible separations can be detected only by direct imaging, especially in the near infrared if they are young (They are several mags fainter in the optical.). The companions are brighter than the neutron stars in the NIR, even though lower in mass, because they are larger in size and still contracting when young. Direct detection of very close and faint substellar companions is currently possible only around neutron stars.

2.1. *Direct Imaging Search*

2.1.1. *Sample selection*

The magnitudes of substellar objects (brown dwarfs and giant planets) can be predicted following the non-gray theory of substellar objects by Burrows *et al.* (1997), who have computed IR magnitudes of substellar objects depending on age and mass etc. At 100 Myrs and 300 pc, a companion at the Deuterium burning mass limit ($\sim$13 M_{jup}) would have an H-band magnitude of 22 mag (Burrows *et al.* (1997)). In case that possible companions to the neutron stars formed before the supernova, their age is the sum of the age of the progenitor star and the time elapsed since the neutron stars was born as such (in the supernova). The former is negligible compared to the latter. We have therefore chosen to select all neutron stars younger than 100 Myrs and closer than 300 pc.

Within 300 pc there are seven radio plusars younger than 100 Myrs, selected from the pulsar catalog of Taylor *et al.* (1993) in its latest online revision. To that sample, we added the seven radio-quiet neutron stars discovered recently as very bright and soft X-ray sources with ROSAT. These neutron stars are also called X-ray thermal neutron stars, because they are thermal X-ray emitters, showing no non-thermal component, especially no radio emission; for reviews, see e.g. Neuhäuser & Trümper (1999), and Haberl (2004). Their blackbody-like spectra results in the best radius estimation currently avaiable (Trümper *et al.* (2004)), making them especially interesting for the determination of the equation of state of neutron stars. The X-ray brightest dim thermal neutron star (RX J1856.5-3754) was identified as a V=25.7 mag star using both the HST (Walter & Matthews (1997)) and the NTT (Neuhäuser *et al.* (1998)). For two dim thermal neutron stars only upper limits (R >23 mag, V >24 mag) are avaiable, but the limits presumably exclude everything except an isolated neutron star (Haberl (2004) and references therein). The remaining four have magnitudes up to $B = 27.2$ mag (Motch *et al.* (2005), Haberl (2004), Haberl *et al.* (2004)). These deep optical observation exclude all companions down to substellar masses, hence these neutron stars are isolated.

The ages of the radio pulsars in our sample are known to be $\leqslant$100 Myrs from the spin-down time. The additional radio-quiet neutron stars found as bright and soft ROSAT sources are most probably also young neutron stars, indicated by large proper motion and high magnetic fields. They are probably several Myrs old, with cooling (but still hot) surfaces which emit the X-ray emission. For all seven radio-quiet X-ray thermal neutron stars, it appears unlikely that they are old with X-ray emission due to accretion from the interstellar medium, because their X-ray emission is constant, and also for statistical reasons (Neuhäuser & Trümper (1999)). Three radio-quiet X-ray thermal neutron stars have a known proper motion, which is so high, that it implies too low Bondi-Hoyle accretion efficiencies to explain these stars as old neutron stars re-heated by accretion from the ISM. Thus, 100 Myrs can be savely used as an upper age limit for all 14 neutron stars.

Most of 14 neutron stars are at high galactic latitude and for all fields, we know that the absorbing column density is quite low. For the radio pulsars, distance estimates are available from the dispersion measure (Taylor *et al.* (1993)). For the radio-quiet X-ray thermal neutron stars, a direct optical detection of the neutron star itself and the measurement of its parallax is available only in one case, namely RX J1856.5-3754 (the brightest in X-ray and, hence, probably also the brightest in the optical, V=25.7 mag), for which Walter & Lattimer (2002) determined $\sim$117 pc. The other X-ray thermal neutron stars appear to be very similar. The measured column densities from the X-ray spectra are smaller than the galactic values (towards the respective direction), which means that these neutron stars are also quite nearby. One can assume the same order for the distance as for RX J1856.5-3754, i.e. roughly 100 pc.

2.1.2. *First epoch observations*

Of the 14 young nearby neutron stars twelve were observed in H or K-band using the VLT with the instrument ISAAC, the Keck telescope (NIRCSPEC) or the 2.2 m telescope of the University of Hawaii (QUIRC). We considered possible orbital separation up to 5000 AU, taking into account recently found wide substellar companions at around 4000 AU and possible increase of orbits due to the mass loss of the primary due to the supernova. Using deep optical observation (e.g. down to 26 mag in B or V) avaiable for some of the neutron stars we can derive colors and roughly estimate spectral types of faint NIR-objects. Comparing these color to the know ones for M-dwarfs (e.g. Kenyon & Hartmann (1995)) we can already exclude many probable stars from further considerations. In nine out of the twelve observed fields there are several candidates up to 5000 AU. As an example the field of RX J0720.4-3125 is shown in Fig. 1. The deep

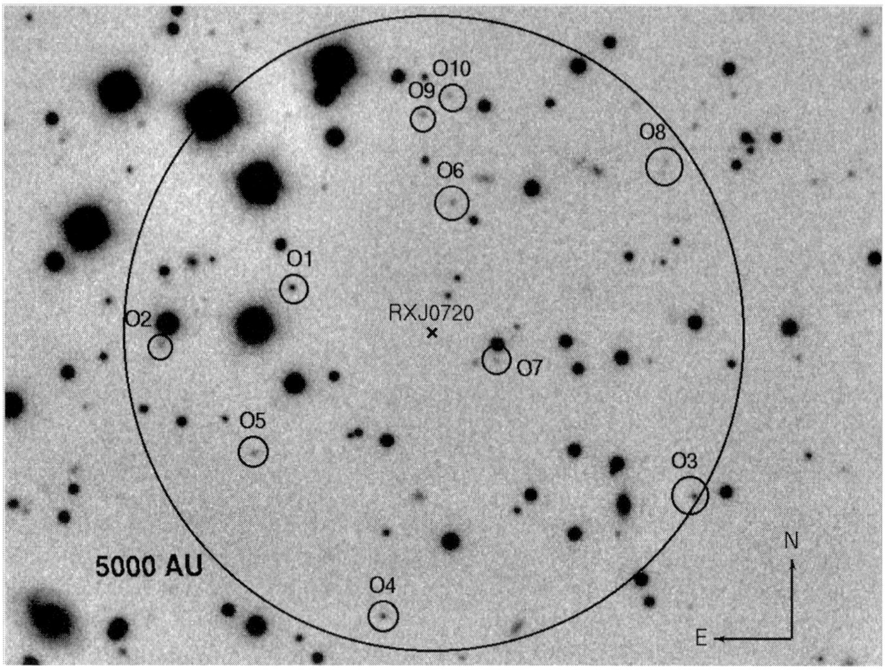

Figure 1. This 1st epoch ISAAC H-band image of RX J0720.4-3125 shows interesting substellar candidates considered for 2nd epoch observation (box size: 75 arcsec × 55 arcsec). The neutron star in the center (RA: 07:20:24.92, DEC: -31:25:50.90) is surrounded by a circle of 5000 AU radius, which was searched for possible substellar companion candidates having at least a SNR of three. The indicated sources O1 to O10 are faint enough to be substellar companion candidates. The magnitudes range from H=20.9 mag (O1) to H=22.3 mag (O10). O5 to O10 are faint enough (H $\gtrsim$ 21.5 mag) to be even planet candidates following the estimations by Burrows *et al.* (1997). The B-H color of O4 excludes a star earlier than M6V as counterpart. For all other candidate objects the FORS1 B-band limit (28 mag) results in B-H colors larger than 5.5, excluding stars earlier than M2V as counterparts.

optical observations are VLT FORS1 B-band observations by Motch *et al.* (2005) of the same year.

2.1.3. *Outlook/ Future observations*

The faint substellar companions candidates found around the nine neutron stars can not be confirmed using only their JHK-colors. From JHK, one could indeed simultaneously determine reddening and spectral type (i.e. temperature). However, even if the color is consistent with a substellar companion (of spectral type L or T) and if the background possibility would be very unlikely (from galactic latitude, angular separation, and faintness), the object could still be an unrelated reddened background object. Thus, proper motion confirmation is necessary. The known proper motion for the investigated neutron stars range from around 40 mas/yr to 300 mas/yr. For the fastest neutron stars the second epoch observations will be carried out soon. Together with JHK-colors and – if bright enough – even spectra the nature of the found substellar companions can be clearified.

Future extremely large telescopes like the planned 30 m Telescope (TMT webpage (2005)), the 50 m EURO50 (EURO50 webpage (2005)) or the 100 m OWL Telescope (ESO's OWL webpage (2005)) would increase the search possibilities significantly. Even the lowest-mass gas giant planets would be seen within our search criteria of young and closeby neutron stars. Alternatively, one can also expand the search to older objects (e.g. 1Gyr) or more far away neutron stars. According to the recent pulsar catalogue (ATNF Pulsar Database (2005)) one could cover around 40 young pulsars (up to 1.1 kpc), 80 (up to 1.9 kpc) or 120 young pulsars (up to 3.8 kpc) by the use of 30 m, 50 m or 100 m telescope respectively. Hopefully by the time of the first light of these telescopes also more than the seven radio-quiet X-ray thermal neutron stars have been found, especially interesting to get constraints on the equation of state of ultadense matter in the universe.

References

The ATNF Pulsar Database September 2005, *atnf.csino.au/research/pulsar/psrcat*
Burrows, A. *et al.* 1997, *ApJ* 491, 865
ESO OWL webpage 2005, *www.eso.org/projects/owl*
EURO50 webpage 2005, *www.astro.lu.de/% 7Etorben/euro50/*
Haberl, F. 2004, *AdSpR* 33, 638
Haberl, F. *et al.* 2004, *A&A* 424, 635
Jonker, P.G., Van der Klis, M., & Groot, P.J. 2003, *MNRAS* 339, 663
Kenyon, S.J. & Hartmann L. 1995, *ApJS* 101, 117
Konacki, M. & Wolszczan, A. 2003, *ApJ* 591, 147
Motch, C. 2005, *A&A* 429, 257
Neuhäuser, R. *et al.* 1998, *Msngr* 92, 27
Neuhäuser, R. & Trümper, J. 1999, *A&A* 143, 151
Podsiadlowski, P. 1995, *ASP Conference Series*, 72
Sigurdsson, S. *et al.* 2003, *Science* 301, 193
Sigurdsson, S. & Thorsett, S.E. 2004, *ASP Conference Series* TBD
Taylor, J.H., Manchester, R.N., & Lyne, A.G. 1993, *ApJS* 88, 529
Trümper, J. *et al.* 2004, *Nucl. Phys.* 132C, 560
TMT webpage 2005, *tmt.ucolick.org*
Walter, F.M. & Matthews, L.D. 1997, *Nature* 389, 358
Walter, F.M. & Lattimer, A.G. 2002, *ApJ* 576, 145
Wolszczan, A. & Frail, D.A. 1992, *Nature* 355, 145
Wolszczan, A. 7. February 2005, *www.science.psu.edu/alert/wolszczan2-2005.htm*

All photographs: Laurent Thareau [`l.thareau@free.fr`].

Direct Imaging of Exoplanets: Science & Techniques
Proceedings IAU Colloquium No. 200, 2005
C. Aime and F. Vakili, eds.
© 2006 International Astronomical Union
doi:10.1017/S1743921306009057

Direct imaging and spectroscopy of planets around young stars: GQ Lupi

Ralph Neuhäuser[1], Christopher Broeg[1], Markus Mugrauer[1] and Eike Guenther[2]

[1]Astrophysikalisches Institut und Universitäts-Sternwarte, Schillergäßchen 2, D-07745 Jena, Germany
emails: rne, broeg, markus@astro.uni-jena.de

[2]Thüringer Landessternwarte Tautenburg, Sternwarte 5, D-07778 Tautenburg, Germany
email: guenther@tls-tautenburg.de

Abstract. Neuhäuser *et al.* (2005a) presented evidence for a sub-stellar, common proper motion companion to GQ Lup, which is almost certainly a planet imaged directly, based on two independent mass estimates by means of theoretical models. The magnitude of GQ Lup b was determined by relative photometry compared to GQ Lup A. During a one-month monitoring of the primary star, we found is to be a variable with K= 6.9 to 7.3 mag (absolute photometry). Then, we re-observed GQ Lup A+b in May 2005, again with VLT/NACO. Due to the monitoring in April 2005, we could predict the exact K-band magnitude range of the primary during the new AO observation. Hence, we could determine a more precise error of the K-band magnitude of GQ Lup b by relative photometry compared to the primary. The new value is consistent with the previous estimate within the error margin, so that also the conclusions in Neuhäuser *et al.* (2005a,b) about its very low mass remain valid.

Keywords. GQ Lup b, direct imaging

1. Introduction

Neuhäuser *et al.* (2005a) presented direct evidence for a companion of the $\leqslant$ 2 Myr young classical T Tauri star GQ Lup in the Lupus star forming region at 140 ± 50 pc from imaging, astrometry, and spectroscopy. With direct K-band imaging using VLT/NACO, we detected an object 6 mag fainter than GQ Lup located $0.7''$ west of it. Compared to images obtained 2 to 5 years earlier with Subaru/CIAO and HST/PC, this object shares the proper motion of GQ Lup by 5 and 7σ, respectively, hence it is a co-moving companion. Its $K - L'$ color is consistent with a spectral type early to mid L. Our NACO K-band spectrum yields spectral type M9-L4 with H_2O and CO absorption, consistent with the new GAIA-Dusty template spectrum for log $g \simeq 2$ to 3 and $T_{\rm eff} \simeq 2000$ K with $\sim$1 to 2 $R_{\rm jup}$ radius at $\sim$140 pc, hence few Jupiter masses.

Using the core accretion formation model from Wuchterl *et al.* (2000) for planets, with some new estimates added, the mass is again around 1 to 3 Jupiter masses. Following the formation model for stars and brown dwarfs by direct gravitational collapse from Wuchterl & Tscharnuter (2003), the mass is below 13 Jupiter masses. See fig. 4 in Neuhäuser *et al.* (2005a).

A comparison of our NACO K-band spectrum with the GAIA-dusty model (Brott & Hauschildt, in preparation), an update of the AMES-dusty model, yielded the radius to be $1.2 \pm 0.5\, R_{\rm Jup}$ and the gravity to be $\log g = 2.0$ to 3.3 (g in cgs units). For M9-L4 spectral type, the temperature is 2050 ± 450 K. At the given flux of the GQ Lup companion ($K_s = 13.10 \pm 0.15$ mag at 140 ± 50 pc), this yields a mass of $\leqslant 2\, M_{\rm Jup}$ (for $\log g \simeq 4$ and $2\, R_{\rm Jup}$, it is $\sim 6\, M_{\rm Jup}$).

Mohanty *et al.* (2004a) measured gravities for isolated young brown dwarfs and free-floating planetary mass objects. Their coolest objects have spectral type M7.5 and gravities as low as $\log g = 3.125$ (GG Tau Bb). This lead Mohanty *et al.* (2004b) to mass estimations as low as $\sim 10\,M_{Jup}$. GQ Lup A is younger than the Mohanty *et al.* Upper Sco objects. Its companion is at least as late in spectral type, probably even cooler. An object younger and cooler must be lower in mass. The GQ Lup companion is fainter than the faintest Mohanty *et al.* object (USco 128, $\sim 9\,M_{Jup}$), so that the mass estimate for the GQ Lup companion would be $\leqslant 8\,M_{Jup}$ (Neuhäuser *et al.* 2005b). However, Reiners (2005) argue that some TiO oscillator strength were wrong in some inputs used in Mohanty *et al.* (2004a,b), so that all temperatures have to be increased by 150 to 200 K, and, hence, also mass estimates. Therefore, the mass estimate for USco 128 and, hence, the upper mass limit for GQ Lup b is now $\sim 15\,M_{Jup}$.

Mugrauer & Neuhäuser (2005) then performed a more detailed astrometric analysis of the GQ Lup system, using all the (different) proper motions published for the primary. The common proper motion is significant in all cases, also when taking into account the error in parallax or distance (140 ± 50 pc). When using the weighted mean, the significance for common proper motion of GQ Lup and its companion is $6\,\sigma + 4\,\sigma$ for no change in separation plus $7\,\sigma$ for no change in position angle. They also discuss the question, whether GQ Lup and its common-proper motion companion are not bound, but share the same or similar proper motion as two independent members of the Lupus T association, which is a moving group, where most members should have the same motion anyway. This hypothesis can be rejected by several σ: The probability to find by chance an L-dwarf fainter than $K_s = 14$ mag within $0.7325''$ with (almost) the same proper motion of GQ Lup is only $\leqslant 3 \cdot 10^{-10}$. The orbital motion of the system is not yet detected ($1.4 \pm 2.2''$ mas/yr), but is probably smaller than the escape velocity (5.3 ± 2.1 mas/yr), so that the system may well be gravitationally bound and stable. This is different for the companion of 2MASSWJ 1207334-393254 (2M1207), which is not yet shown to be bound and, if formed together, will not be stable. Also, for AB Pic A and its common proper motion companion, it is not yet shown that the remaining motion between them is smaller than the expected ejection velocity. However, if bound, then it is long-term stable, because the total mass or bounding energy is sufficient given the high mass of the primary.

2. The variability of GQ Lup A and consequences for GQ Lup b

In Neuhäuser *et al.* (2005a), the magnitude of GQ Lup b was determined by relative photometry compared to GQ Lup A, a brightness difference of 6.00 ± 0.15 mag in K. They did not take into account the possible variability of star A, but used its 2MASS value ($K = 7.096 \pm 0.020$ mag). The variability of star A would introduce an additional uncertainty (or rather a small shift when knowing the phase and correcting for this effect) not taken into account in Neuhäuser *et al.* (2005a).

Now, we have taken a new image of GQ Lup A with b on 27 May 2005 between 03:30h and 04:17h UT, again using NACO at the VLT/Yepun, as in Neuhäuser *et al.* (2005a). We have reduced the data as in 2004.

From an extensive photometric monitoring campaign of GQ Lup A during April 2005 with ANDICAM at CTIO in the filters R, I, J, and K, as well as quasi-simultaneous radial velocity monitoring with HARPS at the ESO 3.6m on La Silla, we could obtain the amplitude of its K-band variation, few tenth of mag, probably due to rotational modulation by a spotted surface, typical for classical T Tauri stars.

Hence, we can predict the K-band magnitude of GQ Lup A of K = 6.9 to 7.3 mag for May 2005, i.e. shortly after the monitoring in April 2005, so that no long-term variability is expected. The absolute value (or rather range) is obtained from data of two photometric nights during our April 2005 monitoring by using three stars from 2MASS, which appeared to constant with respect to eachother, so that they can be assumed to long-term constant.

Now, with a the magnitude difference between GQ Lup A and b on the image from 27 May 2005 being 6.10 ± 0.15 mag, and with K=6.9 to 7.3 mag for star A, we can obtain the magnitude of GQ Lup b to be K=13.20 ± 0.25 mag (on 27 May 2005). Then, with $B.C._K = 3.3 \pm 0.1$ mag and 140 ± 50 pc, its luminosity will be $\log L/L_\odot = -2.41 \pm 0.43$ (instead of -2.37 ± 0.41 previously). Given the large error on the distance and, hence, luminosity, this does not change the mass estimate according to Figure 4 in Neuhäuser *et al.* (2005a), which is still 1 to 3 Jupiter masses.

The problem of variability of the primary (GQ Lup A) probably also affects the L-band magnitude (of GQ Lup b), however, the variability amplitude in L is expected to be smaller than in K, if due to spots.

Liu (this conference) presented his own observations of GQ Lup A and b. He determined narrow-band colors for GQ Lup b within the J-band with HST/Nicmos and broad-band colors from the ground; he then compared the GQ Lup b colors in a color-color diagram with M5 to L0 dwarfs (with spectral types known from ground-based observations). The position of GQ Lup b in that color-color diagram was consistent with those dwarfs, so that Liu derived the temperature of GQ Lup b to be 2200 to 2650 K. However, because Liu did not obtain the same narrow-band colors for L1-4 dwarfs, this temperature estimate has to be considered as upper limit (consistent with T=1600 to 2500 K from Neuhäuser *et al.* 2005a).

Liu also obtained some of his magnitudes for GQ Lup b by relative photometry compared to A, but did not take into account the variability of GQ Lup A, so that it is not surprising, that some magnitudes varry from observation to observation. Liu then compared GQ Lup b with Burrows *et al.* (1997) and Baraffe *et al.* (2002) models, where the GQ Lup b error ellipse crosses the D burning mass limit. He also compared its location with young brown dwarf companions, concluding that they are all quite similar in lumonisity and temperature. However, one also needs to take into account that GQ Lup b is much younger than all brown dwarf companion candidates used in that comparison, hence lower in mass.

We would like to note that the same problem of possible variability of the primary can have occured in the determination of the magnitude and, hence, luminosity and mass, of companions of other young stars such as, most recently, 2M1207 (Chauvin *et al.* 2005a), AB Pic (Chauvin *et al.* 2005b), and AB Dor (Close *et al.* 2005), the latter known to be one of the most variable stars on the sky.

3. Following the orbital motion of GQ Lup b around A

With the new AO observation of GQ Lup A+b in May 2005, we can also check again for common proper motion and possibly orbital motion. See figure 1.

The separation between GQ Lup A and b is still constant within small error bars, so that common proper motion is highly significant.

Orbital motion is not yet detected. Orbital motion would be detectable as follows: A new data point would need to be significantly different from the full line (constant separation), but within the dotted lines (i.e. orbital motion detected); as long as a data point is consistent with the full line, orbital motion would not yet be detected, and one

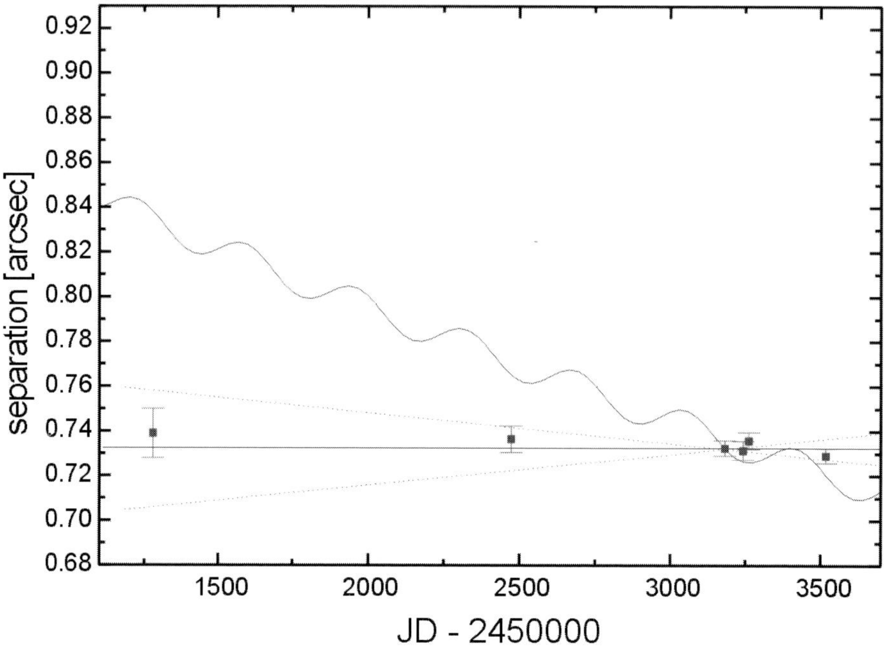

Figure 1. Observed separation between primary star GQ Lup A and its companion b with HST/PC (left), Subaru/CIAO (middle) and VLT/NACO (right, first the June 2004 image, then the Aug and Sept 2004 aquisition images, and finally our new data point from May 2005) compared to the expectation when bound (no change in separation as straight full line few mas/yr for orbital motion as straight dotted lines) or when background (motion of GQ Lup A alone to the SW due to its proper motion ± its error with wobble due to expected parallactic motion of 7.1 mas). We can reject the background hypothesis by many σ.

would have to wait longer; if a data point would be on the curved upper lines, the fainter object would be a non-moving background object; if a data point would be anywhere else in the diagram, outside the dotted line cones, then it would not be bound.

The remaining possible differential motion between A and b is 1.4 ± 2.2 mas/yr. Based on the mass of GQ Lup A (0.7 solar masses) and GQ Lup b (negligible), we can expect orbital motion of 3.7 ± 1.5 mas/yr, which is hence detectable within a few years. In the other hand, if the system would not be bound, we would expect an ejection velocity of b on the order of 5.2 ± 2.1 mas/yr. The fact that the remaining motion is smaller than this expected escape velocity shows that GQ Lupi A+b is most certainly bound.

4. Conclusions

All the available evidence regarding GQ Lup b is consistent with it being a planet imaged directly (Neuhäuser *et al.* 2005a). Two other similar claims were published since then: 2M1207 (Chauvin *et al.* 2005a) and AB Pic (Chauvin *et al.* 2005b). In table 1, we compare all the available data for these three planets or planet candidates (adapted from Neuhäuser 2005, where the parameters were derived in a homogeneous way, but updated for GQ Lup b).

Mass determination of directly detected planets is obtained thru theoretical models. If the objects involved are young, then their formation is to be taken into account. This is not yet implemented in Tucson or Lyon models.

Table 1. Parameters of three planet candidates

Parameter	Objects GQ Lup b	2M1207 b	AB Pic b
distance [pc]	140 ± 50	70 ± 20:	47.3 ± 1.8
membership	Lupus I	TWA (?)	TucHorA
age [Myr]	$\leqslant 2$	5-12:	30-40
epoch difference [yr]	5	1	1.5
separation	0.7", 100 AU	0.8", 54 AU	0.5", 258 AU
significance for CPM (1) [σ]	$6+4+7$	$2+2+4+4$	$3+5$
remaining motion A/b [mas/yr]	1.4 ± 2.2	4.1 ± 8.2	6.9 ± 13.2
orbital motion expected [mas/yr]	3.7 ± 1.5	1.9 ± 0.6	6.9 ± 0.4
escape velocity expected [mas/yr]	5.2 ± 2.1	2.7 ± 0.9	9.8 ± 0.6
long-term stable ? (2)	yes	no	yes
Spectral type	M9-L4	L5-L9.5	L0-3
spectrum resolution	700	< 700	700
spectrum S/N ratio	45	low	high
T_{eff} [K]	2520-1600	2000-1100	2400-1600
gravity $\log g$ [cgs]	2.0-3.3 (3)	unknown	unknown
radius [R_{jup}]	1.2 ± 0.6 (4)	unknown	unknown
M_K [mag]	7.47 ± 0.94 (*)	12.70 ± 0.75	10.77 ± 0.14
$\log L_{bol}/L_{\odot}$	-2.41 ± 0.43 (*)	-4.49 ± 0.34	-3.730 ± 0.039
mass [M_{jup}] Lyon/Tucson	3-50	2-70	11-70
mass [M_{jup}] Wuchterl	1-3	n/a (5)	n/a (5)

Remarks: (*) this paper; (1) significance for common proper motion in Gaussian σ; (2) according to criteria in Weinberg *et al.* (1987) and Close *et al.* (2003); (3) from fit to theoretical GAIA-dusty template spectrum; (4) from fit to spectrum with flux and temperature known; (5) not applicable, because outside of plotted or calculated range.
Ref.: Neuhäuser *et al.* 2005a, Mugrauer & Neuhäuser 2005, Chauvin *et al.* 2005a, Chauvin *et al.* 2005b, Neuhäuser *et al.* 2005c.

The model by Wuchterl & Tscharnuter (2003) for stars and brown dwarfs does take into account their formation, so that it can be valid for very young objects. The tracks for planets shown in Fig. 4 in Neuhäuser *et al.* (2005a) are calculated based on the nucleated instability hypothesis (Wuchterl *et al.* 2000).

For higher confidence in the mass determination and, hence, classification as planets, one would need to test and possibly calibrate the theoretical models used, both the formation models (such as Wuchterl & Tscharnuter 2003) as well as the atmospheric models (such as Peter Hauschildt's GAIA-dusty). Such tests can be done with objects, where the masses have been determined dynamically, e.g. transiting planets with direct or indirect (secondary eclipse) magnitude estimations, or resolved very low-mass binaries with solved orbits.

Alternatively, higher resolution higher S/N spectra of the companions would also be helpfull to constrain gravity, temperature, radius, and, hence, their masses.

Acknowledgements

We would like to acknowledge many fruitful discussions with Guenther Wuchterl and Andreas Seifahrt.

References

Chauvin G., Lagrange A.M., Dumas C., Zuckerman B., Mouillet D., Song I., Beuzit J.L., & Lowrance P. 2005a, *A&A* 438, L25

Chauvin G., Lagrange A.M., Zuckerman B., *et al.* 2005b, *A&A* 438, L29

Close L.M., Siegler N., Freed M., & Biller B. 2003, *ApJ* 587, 407

Close L.M., Lenzen R., & Guirado J.C. 2005, *Nature* 433, 286

Mohanty S., Basri G., Jayawardhana R., Allard F., Hauschildt P., & Ardila D. 2004a, *ApJ* 609, 854

Mohanty S., Jayawardhana R., & Basri G. 2004b, *ApJ* 609, 885

Mugrauer M. & Neuhäuser R. 2005, *AN* 326, 701 (astro-ph/0509162)

Neuhäuser R., Guenther E.W., Wuchterl G., Mugrauer M., Bedalov A., & Hauschildt P. 2005a, *A&A* 435, L13

Neuhäuser R., Guenther E.W., & Hauschildt P.H. 2005b, in: A. Richichi & F. Paresce (ed.), *ESO Workshop The power of optical/IR interferometry*, in press (astro-ph/0506011)

Neuhäuser R. 2005, in: M. Petr-Goetzens (ed.), *ESO Workshop Multiple Stars accross the H-R Diagram*, in press

Reiners A. 2005, *AN* in press (ULMSF proceedings, astro-ph/0510502)

Weinberg M.D., Shapiro S.L., & Wasserman I. 1987, *ApJ* 312, 367

Wuchterl G. & Tscharnuter W.M. 2003, *A&A* 398, 1081

Wuchterl G. & Guillot T., Lissauer J. 2000, in: V. Mannings *et al.* (eds.), *in: Protostars and Planets IV*, 1081

Discussion

CLOSE: The magnitude of AB Dor C was obtained by relative photometry compared to AB Dor A and B, whose magnitudes where known from 2MASS, so that I believe our magnitude of AB Dor C is correct, even though of the large and fast variability of AB Dor A.

NEUHÄUSER: AB Dor B, as companion to the young and very variable AB Dor A, should also be young and variable. For relative photometry compared to young and variable stars, one needs to know the exact rotational period and magnitude amplitude, and the rotational phase of one's own and the 2MASS observations.

LIU: In previous comparisons of observed spectra of sub-stellar objects to AMES-dusty models, one has often underestimated the gravity. Why should this be different in your comparison of the GQ Lup b spectrum with the new GAIA-dusty model ?

NEUHÄUSER: The GAIA-dusty model (Brott & Hauschildt, in prep.) is an update of the AMES-dusty model (Allard *et al.* 2001) with improved molecular dissociation constants, more dust species with opacities, spherical symmetry, and a mixing length parameter $1.5 \cdot H_p$. In particular, the new model uses spherical symmetry, which is most important for young and sub-stellar objects as GQ Lup A and b with low gravities. Hence, we expect less problems with the gravities than in previous models. The gravity estimate from the GAIA-dusty model for GQ Lup b is consistent with both the estimate just from the CO-lines (using the Gorlova *et al.* (2003) calibration) and with the gravity obtained for GQ Lup b from the Wuchterl model using the observed luminosity and temperature (see figure 4 in Neuhäuser *et al.* 2005a). In September 2005, we have taken a higher resolution spectrum, R=4000 in the K-band, with Sinfoni at the VLT, being reduced.

Direct Imaging of Exoplanets: Science & Techniques
Proceedings IAU Colloquium No. 200, 2005
C. Aime and F. Vakili, eds.
© 2006 International Astronomical Union
doi:10.1017/S1743921306009069

Outermost planets of Beta Pictoris, Vega and Epsilon Eridani: goals for direct imaging

Nick Gorkavyi[1]† and Tanya Taidakova[2]

[1]Crimean Astrophysical Observatory, Crimea, Ukraine
email: gorkavyi@gist.us

[2]Greenwich Institute for Science and Technology,
P.O.Box 797, Haymarket, VA
email: taidakova@gist.us

Abstract. We discuss our numerical approach to the high-resolution modeling of the 3D structure and infrared emission of circumstellar dust disks. We examine the resonant structures of a dusty disk induced by the presence of giant planet, that outermost from a star. These features can serve as indicators of outermost planets embedded in the circumstellar dust disk and, moreover, can be used to determine its position, major orbital parameters and even the mass of the planet. Such planets are attractive goals for direct imaging. Our simulations indicate that Vega may have a massive planet ~2 Jupiter mass at a distance >50 AU, and other giant planet(s) at a smaller distance, and Epsilon Eri may have a less massive planet ~0.2 Jovian mass at a distance of 55–60 AU. Theoretical models and non-direct observations show that Beta Pictoris system can be a multiplanetary system with set of giant planets. Our dynamical model of the origin of the warping of the Beta Pictoris disk includes the gravitational influence of a planet with a mass of about 10 masses of Earth, at a distance of 70 AU, and a small inclination (2.5 deg) of the planetary orbit to the main dust disk. The direct signatures of this planet were discovered on 2002 by Keck observations.

Keywords. Beta Pictoris, Vega, Epsilon Eridani, exoplanets, resonant structures.

---

1. Introduction

Basic elements of planetary systems are: inner and outer planets; cometary-asteroidal belts; dust disk and Jackson-Zook resonant rings. Origin of planets is the result of fast accretion in dense areas of a planetary system. Outer cometary-asteroidal belt (TNO in the Solar system) is a common element of planetary systems as the result of low peripheral surface density and large period of orbital revolution. These two factors lead to time of accretion larger than life time of planetary system. Origin of inner asteroidal belt (as Main Asteroidal Belt in the Solar system) is the result of interaction between next dynamical mechanisms: 1. Collisions between asteroids and large dust particles lead to origin of micron-size dust. 2. Solar radiation pressure push of small particles to Jupiter zone. 3. 50% of small dust escape in interstellar space by scattering on Jupiter gravitional field (Gorkavyi *et al.*, 2000). As result of this efficient mechanism, asteroidal belt lost most of initial mass and accretion time drastically increased. Additional factor of growing of accretion time is reasonant influence of the Jupiter to relative velocities of asteroids. Dust disk and Jackson-Zook rings are the result of re-distribution and resonant accumulation of dust from asteroids and comets. Any planetary system can demonstrate complex dynamical interaction between all basic components: planets, minor bodies, dust and gas on early stage. Important that dust is the most visible part of planetary

† Present address: Greenwich Institute, P.O.Box 797, Haymarket, VA.

systems. Link between dust disk features, JZ-rings and invisible planets can be important for search and direct imaging of planets.

2. The multiplanetary systems

Outermost (~50 AU) giant (~10–1000 M_{Earth}, where M_{Earth} is the mass of the Earth) planets are attractive goals for direct imaging. ¿From our point of view, planetary systems of Vega, Beta Pictoris and Epsilon Eridani is good sample of multiplanetary systems with outer giants. The results of our study reveal a remarkable similarity of theoretical models with various types of highly asymmetric circumstellar disks around Beta Pictoris, Epsilon Eridani and Vega (Gorkavyi *et al.* 2000, 2004).

According our modeling, Epsilon Eridani may have outermost giant planet with ~0.2 Jovian mass at a distance of 55–60 AU. Inner panet of Epsilon Eridani (3.4 AU, ~$1M_J$, e~0.6, M_J is the mass of Jupiter) was discovered by Doppler effect (Hatzes etal, 2000) and we can decide from general theoretical arguments, that Epsilon Eridani can have another massive ~0.1–$1M_J$ planets between 5–50 AU.

Beta Pictoris is the planetary system with specific spectral activity (Lagrange 1995), that is the clear signature of inner giant planets, for example, in (Gorkavyi and Taidakova 1995) was considered system from two planets at radii 3.2 and 5.9 AU with masses 0.0019 and 0.00095 of mass of star. Our dynamical model of the origin of the warping of the Beta Pictoris disk (Gorkavyi *et al.* 2004) includes the gravitational influence of a planet with a mass of about 10 masses of Earth, at a distance of 70 AU, and a small inclination (2.5 deg) of the planetary orbit to the main dust disk. Jackson-Zook ring with similar radii and inclination were discovered on 2002 by Keck observations (Wahhaj *et al.* 2002) and we can consider this JZ-ring as the direct signatures of outermost theoretically predicted planet. Totally four JZ rings was discovered by (Wahhaj *et al.* 2002) with radii of 14, 28, 52 and 82 AU and we believe that Beta Pictoris is a multiplanetary system with at least 5–6 giant planets between 3 and 80 AU. Our simulations indicate that Vega may have a massive planet ~2 Jupiter mass at a distance >50 AU, and other giant planet(s) at a smaller distance. This planetary system will be considered below in more details.

3. The planetary system around Vega

We applied our numerical method (Taidakova and Gorkavyi 1999) to compute the thermal emission of a dusty resonant structure around Vega (Ozernoi *et al.* 2000, Gorkavyi and Taidakova 2002). Both the Poynting–Robertson and stellar wind drags tend to induce cometary dust inflow toward the star. The modeling has shown that a planet produces an asymmetric resonant dust belt via resonances and gravitational scattering. Our simulations have three main steps:

1. simulation of distribution of cometary population;
2. calculation of distribution of cometary dust (see Fig. 1);
3. determination of thermal emission of dust (see Fig. 2).

Our high resolution (2–3 AU) simulation is in good agreement with the new observation for a ring arc at 95 AU near Vega (Koerner *et al.* 2001). Our modeling indicate that Vega may have a outermost planet at a distance of 90–100 AU with coordinates near 18 35'16.25″ (on image from (Koerner *et al.* 2001), right ascension, B1950) and 38 44'15″ (declination). Another symmetrical position, 18 35'14.75″ and 38 44'32.5″, of the planet is also possible. We can decide between these two positions after measuring the revolution direction of the resonant pattern, which must have an angular velocity of 0.6 deg/yr. Radii of clumps (and planet's orbital radius) may be smaller: 60–75 AU (according to

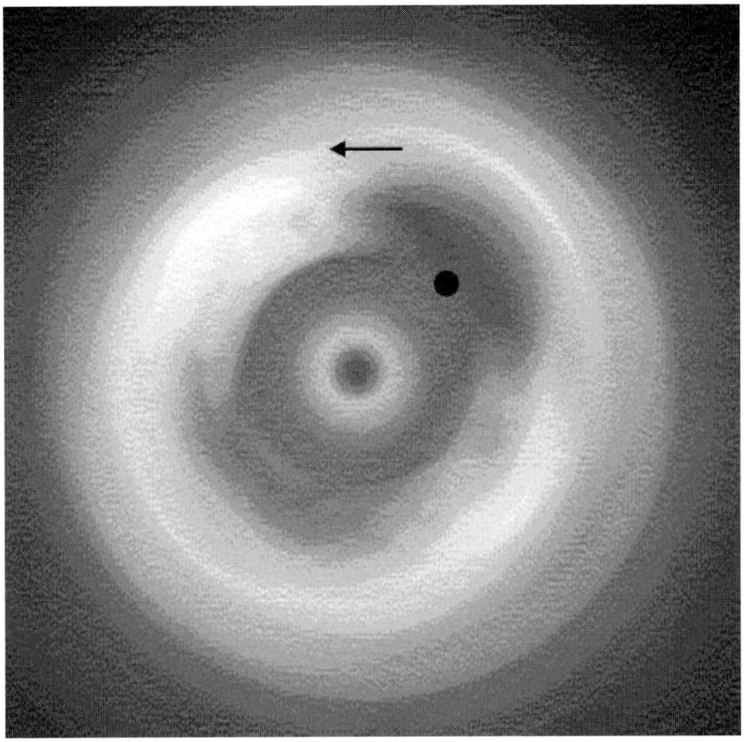

Figure 1. Simulated surface density of circumstellar dust disk with one Jupiter-like planet (dark circle in right top area inside the cavity). Orbital radius is 30 pixels of image (1 pixel = 2.5−3 AU for the Vega). Counterclockwise rotation of disk.

Wilner *et al.* 2002). In this case the planet stay approximately on the line between the two possible points marked above and the angular velocities of the planet and dust pattern will be up to 1.2 deg/year. Preliminary considerations give the mass of outer planet less than ∼2 Jupiter masses. Our simulations in the one planet approximation show that it must have bright emission from inner quasisymmetrical dust disk with radius <30–40 AU. Observations by Koerner *et al.* (2001) don't show emission from such inner dust disk and we can predict existence of giant planet(s) at distance <50–60 AU, which destroyed inner circumstellar dust disk by gravitational scattering. If weak dusty inner clumps (Koerner *et al.* 2001) have resonant link to inner planets, they must revolve around Vega much faster than the outermost arcs.

An example of the results that are obtained from our method is given in Fig. 1 for a surface density. Fig. 2 shows emission for this surface density.

4. Conclusions

1. Predicted in 2000 the outermost (∼70 AU) planet in the Beta Pictoris system at inclined (2.5 deg) orbit was confirmed by observation in 2002 (Wahhaj *et al.* 2002).

2. Our modeling indicates that the Vega has a multiplanetary system. Orbital radius of the outermost planet is approximately the same as the orbital radius of dust clumps (90–100 AU from observation by Koerner *et al.* (2001) or 60–75 AU from observation by Wilner *et al.* (2002).

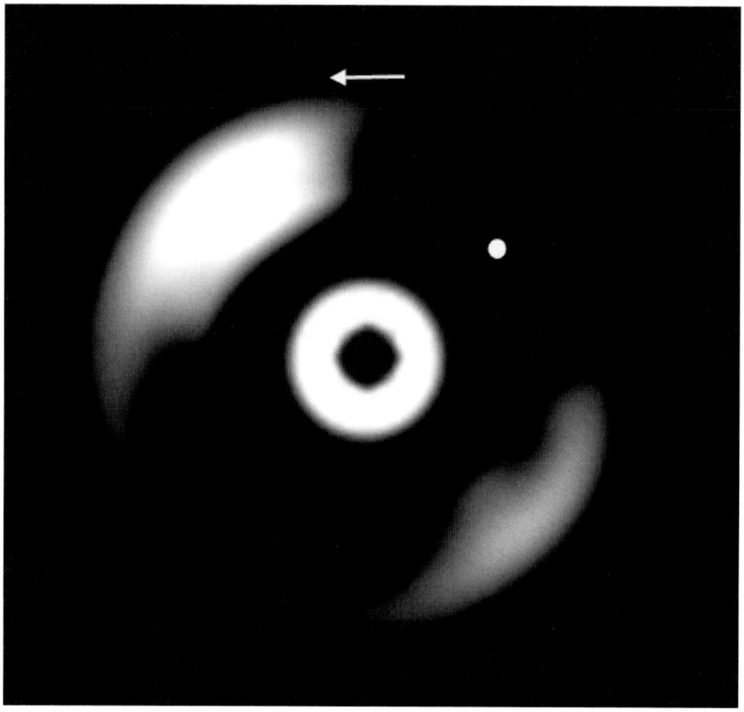

Figure 2. Thermal emission from circumstellar disk (model from Fig. 1). 5 pixel' averaging was used for comparison of our simulation with observation by Koerner *et al.* 2001.

3. Dust clumps near Vega are made of resonant particles (2:1; 3:1 etc); they rotate around Vega exactly with the planet's orbital speed (precession is very small): 0.6–1.2 deg/year depending on orbital radius of parent planet (see point 2).

4. The most dense clump is ahead of the planet in the direction of the orbital motion.

5. The planet stays in the empty area between the clumps.

6. The orbit of the planet may be circular or with small eccentricity.

7. Mass of the outermost Vega's planet is near or less than 2 Jupiter masses.

8. An outer cometary belt (like TNO' belt) there exists beyond the outermost planet.

9. Transparency of the inner part of Vega's planetary system is a signature of inner giant planet(s) with orbital radii $<$50–60 AU.

Acknowledgements

N. G. was supported by IAU travel grant.

References

Gorkavyi, N. & Taidakova, T. 1995, in: R.Ferlet & A.Vidal-Madjar (eds.), *Circumstellar Dust Disks and Planet Formation*, Editions Frontieres, Gif sur Yvette Cedex - France, p. 99

Gorkavyi, N., Ozernoy, L., Mather, J. & Taidakova, T. 2000, in: E.P. Smith & K.S. Long (eds.), *NGST Science and Technology Exposition*, ASP, v.207, p. 462

Gorkavyi, N. & Taidakova, T. 2002, *Planetary system around Vega* 2002, Washington DC, AAS; *BAAS* 2002, 33, N4

Gorkavyi, N., Heap, S., Ozernoy, L., Taidakova, T. & Mather, J. 2004, in: A. Penny, P. Artymowicz, A.-M. Lagrange & S. Russell (eds.), *Planetary Systems in the Universe: Observation, Formation and Evolution*, IAU/ASP, Symp. 202, p.331; *astro-ph/0012470*, 21 Dec 2000

Koerner, D.W., Sargent, A.I., & Ostroff, N.A. 2001, *astro-ph/0109424*, 24 Sept 2001

Lagrange, A.-M. 1995, in: R.Ferlet & A.Vidal-Madjar (eds.), *Circumstellar Dust Disks and Planet Formation*, Editions Frontieres, Gif sur Yvette Cedex - France, p. 19

Ozernoy, L., Gorkavyi, N., Mather, J., & Taidakova, T. 2000, *ApJ* (Letters) 537, L147

Taidakova, T. & Gorkavyi, N. 1999, in: B.A.Steves & B.A.Roy (eds.), *The Dynamics of Small Bodies in the Solar Systems: A Major Key to Solar Systems Studies*, Kluwer Academic Publishers, p. 393

Wahhaj, Z., Koerner, D.W., Ressler, M.E., Werner, M.W., Backman, D.E., & Sargent, A.I. 2002, *ApJ* (Letters) 584, L27

Wilner, D.J., Holman, M.J., Ho, P.T.P., & Kuchner, M. 2002, *astro-ph/0203264*, 17 Mar 2002

All photographs: Laurent Thareau [l.thareau@free.fr].

Direct Imaging of Exoplanets: Science & Techniques
Proceedings IAU Colloquium No. 200, 2005
C. Aime and F. Vakili, eds.
© 2006 International Astronomical Union
doi:10.1017/S1743921306009070

A Survey of Close, Young Stars with SDI at the VLT and MMT

Beth A. Biller[1], **Laird M. Close**[1], **Elena Masciadri**[2], **Rainer Lenzen**[2], **Wolfgang Brandner**[2], **Donald McCarthy**[1], **Thomas Henning**[2], **Eric Nielsen**[1], **and Markus Hartung**[3]

[1]Steward Observatory, University of Arizona, Tucson, AZ 85721
email: bbiller@as.arizona.edu

[2]Max-Planck-Institut für Astronomie, Königstuhl 17, 69117 Heidelberg, Germany
[3]European Southern Observatory, Alonso de Cordova 3107, Santiago 19, Chile

Abstract. We discuss the preliminary results of a survey of young (<300 Myr), close (<50 pc) stars with the Simultaneous Differential Extrasolar Planet Imager (SDI) implemented at the VLT and the MMT. SDI uses a quad filter to take images simultaneously at 3 wavelengths surrounding the 1.62 μm methane bandhead found in the spectrum of cool brown dwarfs and gas giants. By performing a difference of images in these filters, speckle noise from the primary can be significantly attenuated, resulting in photon noise limited data. In our survey data, we achieved H band contrasts >25000 (5σ ΔF1(1.575μm)>10 mag, ΔH>11.5 mag for a T6 spectral type) at a separation of 0.5" from the primary star. With this degree of attenuation, we should be able to image (5σ detection) a 2-4 Jupiter mass planet at 5 AU around a 30 Myr star at 10 pc. We are currently completing our survey of young, nearby stars, with complete datasets for 35 stars in the southern sky (VLT) and 7 stars in the northern sky (MMT). We believe that our SDI images are the highest contrast astronomical images ever made from ground or space for methane rich companions.

Keywords. (stars:) planetary systems, instrumentation: adaptive optics.

---

1. Introduction

Direct detection of extrasolar giant planets is extremely difficult. Giant gas planets seen in reflected light are >20 magnitudes fainter than their primary stars and likely lie within ~1" of their primary stars. The problem is slightly easier with younger, hotter planets – 100 Myr old extra-solar planets are 10^{4-7} times more self-luminous than old (5 Gyr) extra-solar planets, whereas their primary stars are only slightly (2-5 times) brighter at early ages. In theory, adaptive optics (AO) systems that are "photon noise limited" can detect an object up to 10^5 times fainter than its primary at separations >1". However, numerous surveys for extrasolar planets using large telescopes with AO systems have yielded useful limits but few confirmed giant planet candidates (Kaisler *et al.* 2003, Masciadri *et al.* 2005, Chauvin *et al.* 2005, Neuhäuser *et al.* 2005).

AO surveys for young extrasolar planets only address half of the difficulty of direct detection – the contrast limit problem. Beyond the contrast limit problem, all AO systems suffer from a limiting "speckle noise" floor (Racine *et al.* 1999). Within 1" of the primary star, the field is filled with speckles left over from instrumental features and residual atmospheric turbulence after adaptive optics correction. These speckles vary as a function of time and color. For photon noise limited data, the signal to noise S/N is proportional to $t^{0.5}$, where t is the exposure time. For speckle-noise limited data, the S/N does not increase with time past a specific speckle-noise floor (limiting contrasts to ~10^3 at 0.5").

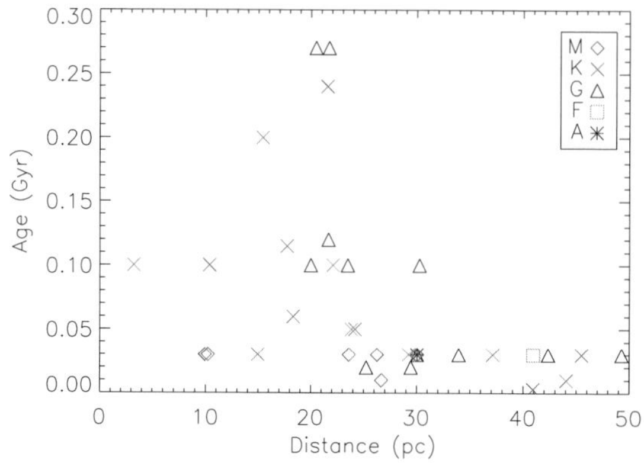

Figure 1. Age vs. Distance for our Observed Sample Stars.

This speckle-noise floor is considerably above the photon noise limit and makes planet detection very difficult. Interestingly, space telescopes such as HST also suffer from a similar limiting speckle-noise floor due to imperfect optics and "breathing" (Schneider *et al.* 2003). Direct detection of extrasolar giant planets requires special new instrumentation to suppress this speckle noise floor and produce photon noise limited images. The VLT, Keck, Subaru, and Gemini are all currently developing dedicated planet-finding cameras which exploit these new instrumentational approaches for speckle suppression. The Simultaneous Differential Imager (SDI), which our team built and installed at the VLT and MMT (see Biller *et al.*, this conference), is one of the first dedicated planet-finding instruments to come online.

2. The SDI survey

3. An Example Dataset

Simultaneous Differential Imaging is an instrumental method which can be used to calibrate and remove the "speckle noise" in AO images, while also isolating the planetary light from the starlight. This method was pioneered by Racine *et al.* (1999), Marois *et al.* (2000), Marois *et al.* (2002), and Marois *et al.* (2005). It exploits the fact that all cool (T_{eff} <1200 K) extra-solar giant planets have strong CH_4 (methane) absorption redwards of 1.62 μm in the H band infrared atmospheric window (Burrows *et al.* 2001, Burrows *et al.* 2003). Our SDI device obtains four images of a star simultaneously through three slightly different narrowband filters (sampling both inside and outside of the CH_4 features). These images are then differenced. This subtracts out the halo and speckles from the bright star to reveal any massive extrasolar planets orbiting that star. Since a massive planetary companion will be brightest in one filter and absorbed in the rest, while the star is bright in all three, a difference can be chosen which subtracts out the star's light and reveals the light from the companion. Thus, SDI also helps eliminate the large contrast difference between the star and substellar companions (Close *et al.* 2005; Lenzen *et al.* 2004; Lenzen *et al.* 2005) The SDI device has already produced a number of important scientific results: the discovery of AB Dor C (Close *et al.* 2005) which is the tightest (0.16") low mass companion known, the first surface maps of Titan (Hartung *et al.* 2004), and the discovery of ϵ Indi Ba-Bb, the nearest binary brown dwarf (McCaughrean *et al.* 2005).

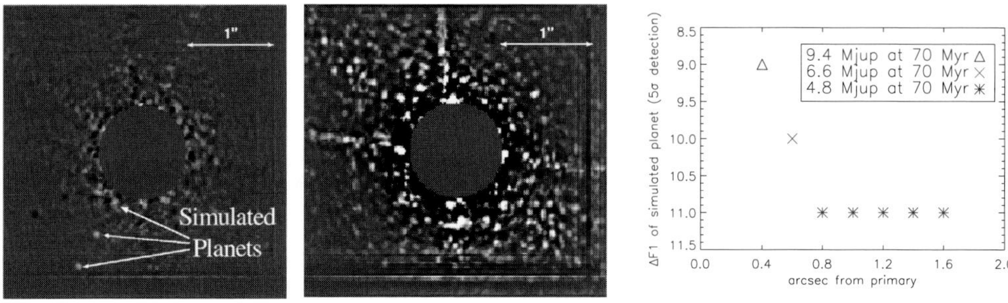

Figure 2. Left: A complete reduced dataset (40 minutes of data at a series of rotator angles – 0°, 33°, 33°, 0°) from the VLT SDI device. Simulated planets have been added at separations of 0.55, 0.85, and 1.35" from the primary, with $\Delta F1(1.575\mu m) = 10$ mag (attenuation in magnitudes in the 1.575 μm F1 filter) fainter than the primary. These planets are scaled from unsaturated images of the example star taken right before the example dataset (and have fluxes and photon noise in each filter appropriate for a T6 object). Past 0.7", the simulated planets are detected with S/N > 10. Center: Standard AO data reduction of the same dataset. Filter images have been coadded (rather than subtracted), flat-fielded, sky-subtracted, and unsharp-masked. Simulated planets have been added with the same properties and at the same separations as before. None of the simulated planets are detected in the standard AO reduction. Additionally, numerous bright super speckles remain in the field. Right: Minimum Detectable Planet Mass (5σ detection vs. Separation for this example dataset (Case B). To determine minimum detectable planet mass as a function of separation, we inserted and then attempted to retrieve simulated planets with a variety of separations and ΔF1 (5σ attenuation in magnitudes in the 1.575 μm F1 filter) contrasts. ΔF1 contrasts were translated into planet masses using the models of Burrows *et al.* 2003. For this particular star, we can detect a 5 M_{Jup} planet 12 AU from the star.

Table 1. Properties of Example SDI Survey Stars and Comparison Stars

Case	Spectral Type	Age	Distance	H	V	ΔF1[1]	ΔH[1]	Exposure Time
A	K2V	30 Myr	45.5 pc	7.1	9.1	10.5	12	40 min
B	K1V	70 Myr	15 pc	4.8	6.9	10.5	12	40 min
C	M3V	30 Myr	24 pc	7.1	12.2	10	11.5	40 min
10 K-M stars[2]	K-M	0-1 Gyr	10-50 pc	6.4-8.7	8-12	8.61	10.31	10-25 min
Gl 86[3]	K1V	10 Gyr	10.9 pc	4.2	6.2	12.8	14.3	80 min

[1] 5σ at 0.5" [2] Masciadri *et al.* IAUC 200 [3] Mugrauer & Neuhäuser 2005

We are currently completing a survey with the SDI device of ~50 young (<300 Myr), nearby (<50 pc) stars. Stars were chosen based on strong lithium absorption features (our best targets have Li equivalent widths of >100 mÅ from the Li 6707 Å line, corresponding to age <100 Myr) and accurate Hipparcos parallax measurements (parallaxes of > 0.02", corresponding to distances <50 pc). Complete datasets have been acquired for 42 stars total –35 stars in the southern sky and 7 stars in the northern sky. Ages, distances, and spectral types of observed objects are presented in Fig. 1. The "average" survey object is a late K star with an age of 120 Myr and at a distance of 26 pc.

Based on realistic scaling laws for the semimajor axis, eccentricity, mass power law, and luminosity of extrasolar planets (statistics from Marcy *et al.* 2003 and Lineweaver and Grether 2003, masses and ages from Burrows *et al.* 2003) and scaling to semimajor axes >5 AU, we can determine a rough detection probability for each program star (read Nielsen *et al.* this conference for more detail). Our program stars have average detection probabilities of ~10–20%. Integraing over the probability distribution of our program stars yields ~4 likely detections in 50 stars.

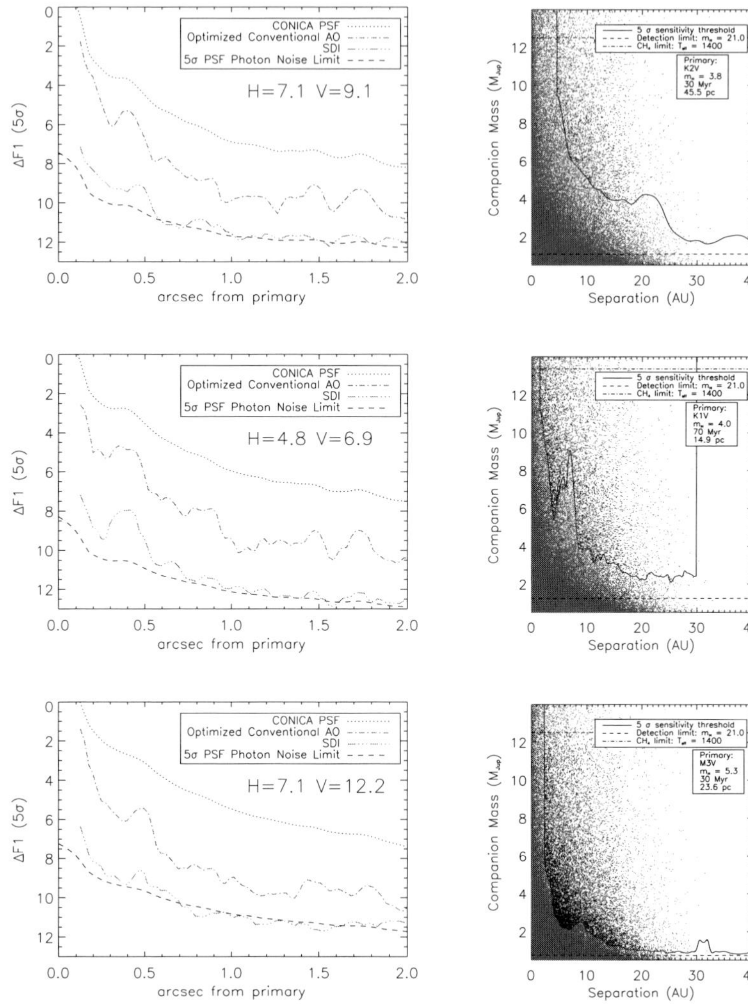

Figure 3. Top: Case A, Middle: Case B, Bottom: Case C. Left: ΔF1 (5σ attenuation in magnitudes in the 1.575 μm F1 filter) vs. Separation for 40 minutes of VLT SDI data for each example star. The top curve is the AO PSF. The next curve is the "classical AO PSF" unsharp masked. The third curve down is 40 minutes of SDI data taken at two different position angles and subtracted (0° data -33° data). The last curve is the theoretical contrast limit due to photon-noise. For each case, at star-companion separations >0.5", we are photon-noise limited and achieve H band star to planet contrasts >25000 (5σ ΔF1(1.575μm)>10 mag, ΔH>11.5 mag for a T6 spectral type) Right: Minimum Detectable Planet Mass vs. Separation – using the models of Burrows *et al.* (2003) and primary star properties from the literature, we can convert our ΔF1 values into a minimum detectable planet mass for each object. Objects above the 1400 K methane cutoff line (horizontal dashed line) are not detected with the SDI device. For details, see Nielsen *et al.* 2005 (this conference).

A number of tentative candidate extrasolar planets (objects which showed CH_4 absorption at appropriate separations from the primary) have been identified so far – followup observations of these candidates are planned to confirm if these candidates share common proper motion with their parent stars. Low resolution follow-up CONICA grism spectroscopy will be performed on any confirmed candidate once the exact planet location is confirmed. With our excellent spatial resolution, we can determine an extremely accurate offset between the planet and thus will be able to precisely align a narrow 0.1" slit on the planet. See Close *et al.* 2005 and Nielsen *et al.* 2005 for details on spectral reduction.

A fully reduced dataset from the VLT SDI device as well as the same dataset reduced in a standard AO manner is presented in Fig. 2. This is 40 minutes of data for AB Dor A, a 70 Myr K1V star at a distance of 14.98 pc (V=6.88). Simulated planets have been added at separations of 0.55, 0.85, and 1.35" from the primary, with ΔF1 = 10 mag (attenuation in magnitudes in the F1 1.575 μm filter) fainter than the primary. These planets are scaled from unsaturated images of the example star taken right before the example dataset (and have fluxes and photon noise in each filter appropriate for a T6 object). In the SDI reduction, the simulated planets are detected with S/N >10 past 0.7". In comparison, none of the simulated planets are detected in the standard AO data reduction and numerous bright super speckles remain in the field.

4. Contrast Limits and Planet Detectability

To determine the range of star-planet contrasts achievable in our SDI young stars survey, we consider three example cases which span the space of our target stars: case A – a high quality dataset (observed with seeing of $\sim$0.5"), case B – AB Dor A, a young solar analogue, and case C, a faint young M star. Properties of each example star (distance, age, spectral type, etc.) are presented in Table 1.

ΔF1 (5σ attenuation in magnitudes in the 1.575 μm F1 filter) vs. separation from the primary is presented for each of the example cases in Fig. 3. For these datasets, we achieved H band star to planet contrasts >25000 (5σ ΔF1(1.575μm)>10 mag, ΔH>11.5 mag for a T6 spectral type) at a separation of 0.5" from the primary star – approaching the photon-noise limit in 40 minutes of data.

Using the models of Burrows *et al.* (2003) and adopting values for the primary star's age (from the Li 6707 Å line), distance, and spectral type from the literature, we can convert our measured attenuations for each object into a minimum detectable mass (see Nielsen *et al.* 2005, this conference). Minimum detectable mass vs. separation for each of the examples is also presented in Fig. 3. Although we achieve similar contrast limits for our example cases (with slightly higher contrasts for brighter targets as one might expect), the mass and separation of objects detectable around each varies strongly with age and distance. Even though case A was our best quality data, we are more likely to detect planets for case B and C, simply because these two objects are closer to the sun, and hence, we can resolve the inner $\sim$20 AU around the star. For case C, we can detect (>5σ) a 3-5 M_J planet at 6 AU from the primary. ΔF1 and ΔH (for a methane object) for each survey case as well as for two other comparison objects are shown in Table 1 – it is clear that the achievable contrast varies according to the magnitude of the object and total exposure time.

To determine what sort of objects we can realistically detect with this level of contrast, we inserted and then attempted to retrieve simulated T6 dwarf planets to the case B dataset with a variety of separations and ΔF1 contrasts. ΔF1 contrasts were translated into planet masses using the models of Burrows et al. 2003. In Fig. 2, we plot minimum detectable planet mass (for a 5σ detection) vs. separation. For this particular star (case B: K1V, 70 Myr, 15 pc), we can detect a 5 M_{Jup} planet 12 AU from the star. In this particular case we were able to detect a non-methane companion (AB Dor C) at 3 different epochs and separations from 0.15" to 0.2" even though ΔH>5 mag (Close *et al.* this conference and Nielsen *et al.* 2005).

5. Conclusions

The novel SDI device at the VLT and MMT has been fully commissioned and is currently achieving attenuations of >25000 (ΔH >12.5 for a T6 spectral type object, ΔF1(1.575μm)>10 at 0.5"). With these contrasts, we can detect a wide range of substellar objects. For instance, for AB Dor A (a 70 Myr K1V star 15 pc away) we can detect (>5 σ) a 5 M_{Jup} planet 12 AU from the star. For a younger closer star (30 Myr age at 10 pc), we can detect a 2-4 M_{Jup} planet at 5 AU.

We have currently observed 42 of the youngest (<300 Myr), nearest (<50 pc) stars as part of a survey of young, nearby stars. We have received time at the VLT for followup observations of 8 tentative candidates found as part of the survey. With a total sample size of ~50 stars, we will be able to place strong constraints on the frequency and semimajor axis distribution of massive extrasolar planets >5 AU from their primaries. From scaling laws derived from the distribution of known radial velocity planets (Marcy *et al.* 2003, Lineweaver and Grether 2003, Burrows *et al.* 2003), we expect to detect ~4 planets for our total sample (see Nielsen *et al.*, this conference). Whether or not we detect planets, our survey will begin to measure the true distribution young massive extrasolar planets >5 AU from their primaries and will provide valuable contraints for theories of planet formation and migration.

Acknowledgements

BAB acknowledges support through the NASA GSRP program. LMC acknowledges support through NSF CAREER and NASA Origins grants.

References

Burrows, A., Hubbard, W.B., Lunine, J., & Liebert, J. 2001, *Reviews of Modern Physics* 73, 719

Burrows, A., Sudarsky, D., & Lunine, J. 2003, *ApJ* 596, 587

Chauvin, G., Lagrange, A.-M., Dumas, C., Zuckerman, B., Mouillet, D., Song, I., Beuzit, J.-L., & Lowrance, P. 2005, *Astronomy and Astrophysics* 438, L25

Close, L.M., Lenzen, R., Biller, B.A., Brandner, W., & Hartung, M. 2005, *Science with AO ESO Workshop*, 136

Close, L.M., Rainer, R., Guirado, J.C., Nielsen, E.L., Mamajek, E.E., Brandner, W., Hartung, M., Lidman, C., & Biller, B.A. 2005, *Nature*, 433, 286

Hartung, M., Herbst, T.M., Close, L.M., Lenzen, R., Brandner, W., Marco, O., & Lidman, C. 2004, *Astronomy and Astrophysics*, 421, L17

Kaisler, D., Zuckerman, B., & Becklin, E. 2003, *ASP Conference Series* 294, 91

Lenzen, R., Close, L., Brandner, W., Hartung, M., & Biller, B. 2004, *Proc. SPIE* 5492

Lenzen, R., Close, L., Brandner, W., Biller, B., & Hartung, M. 2005, *Science with AO ESO Workshop*, 46

Lineweaver, C.H. & Grether, D. 2003, *ApJ* 598, 1350

Marcy, G., Butler, R., Fischer, D., & Vogt, S. 2003, *ASP Conference Series* 294, 1

Marois, C., Doyon, R., Racine, R., & Nadeau, D. 2000, *PASP* 112, 91

Marois, C., Doyon, R., Racine, R., & Nadeau, D. 2002, *Proc. SPIE* 4008, 788

Marois, C., Doyon, R., Nadeau, D., Racine, R., Riopel, M., Vallée, P., & Lafrenière, D 2005, *PASP*, 117, 745

Masciadri, E., Mundt, R., Henning, T., Alvarez, C., & Barrado y Navascues, D. 2005, *ApJ* 625, 1004

Masciadri, E. *et al.* 2006, IAUC 200 proceedings

McCaughrean, M.J., Close, L.M., Scholz, R.-D., Lenzen, R., Biller, B., Brandner, W., Hartung, M., & Lodieu, N. 2004, *Astronomy and Astrophysics*, 413, 1029

Mugrauer, M. & Neuhäuser, R. 2005, *MNRAS*, 361, L15

Neuhäuser, R., Guenther, E.W., Wuchterl, G., Mugrauer, M., Bedalov, A., & Hauschildt, P.H. 2005, *Astronomy and Astrophysics* 435, L13

Nielsen, E.L., Close, L.M., Guirado, J.C., Biller, B.A., Lenzen, R., Brandner, W., Hartung, M., & Lidman, C. 2005, to appear in Astronomische Nachrichten

E. Nielsen *et al.* 2005, this conference.

Racine, R., Walker, G.A.H., Nadeau, D., Doyon, R., & Marois, C. 1999, *Proceedings of the Astronomical Society of the Pacific* 111, 587

Schneider, G., Becklin, E., Close, L., Figer, D., Lloyd, J., Macintosh, B., Hines, D., Max, C., Potter, D., Rieke, M., Scoville, N., Thompson, R., Weinberg, A., & Windhorst, R. 2003, *solicited by STScI in preparation for HST Cycle 12*

Discussion

LAGRANGE: In your companion mass vs. AU plot *(see Figure 3 in this paper)*, which planets are cool enough to have CH_4 and hence be detectable with the SDI device?

BILLER: No planets hotter than 1400 K are considered detected since their CH_4 absorption features are too weak to use the SDI technique (see methane cutoff line at T_{eff}=1400 K in Figure 3.)

All photographs: Laurent Thareau [`l.thareau@free.fr`].

Direct Imaging of Exoplanets: Science & Techniques
Proceedings IAU Colloquium No. 200, 2005
C. Aime and F. Vakili, eds.
© 2006 International Astronomical Union
doi:10.1017/S1743921306009082

Detection of Pegasides: a ground-based approach and preliminary observations

J. Gay[1], J.P. Rivet[1], Y. Rabbia[2] and F. Fressin[1]

[1]Observatoire de la Côte d'Azur, B.P. 4229, F-06304 Nice Cedex 4, France.
email: gay@obs-nice.fr

[2]Observatoire de la Côte d'Azur, Avenue Copernic, F-06130 Grasse, France.

Abstract. Discriminating exo-planetary photons from stellar photons with a ground-based instrument is a physically interesting and challenging issue. We propose an observational technique based upon both spectral and polarimetric coding to discriminate the exo-planetary photons. We also discuss a preliminary observation run performed with the spectrograph "CARELEC" on the 193 *cm* telescope at the Observatoire de Haute provence (OHP).

Keywords. techniques: photometric, spectroscopic.

1. The principle

Several physical phenomena are responsible of the light reflection by the atmosphere of a Pegaside. Among them, the well known Rayleigh scattering may play a non negligible role, especially if the fraction of solid particles in the atmosphere is low enough. Under this hypothesis, the star's light scattered by the planet (called "planetary photons" in this article) should be at least partially polarized, with a maximum of polarization at quadratures (see Figure 1). In this article, we shall assume this hypothesis to be valid.

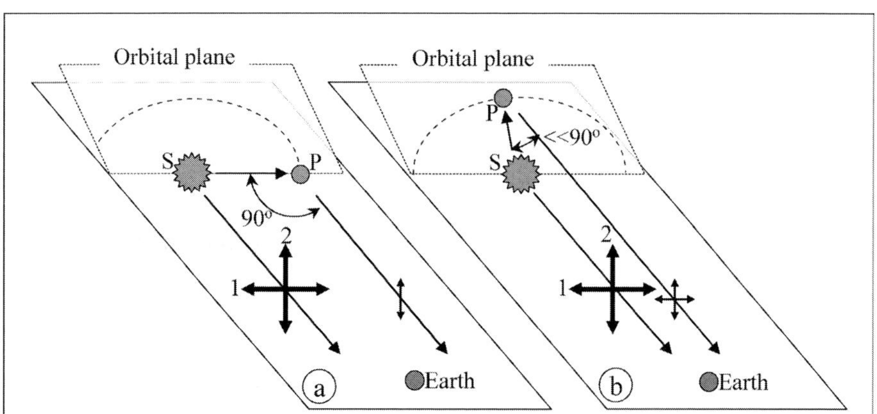

Figure 1. The polarization of the planetary photons, if only Rayleigh scattering occurs. **(a)** At quadrature, the planetary photons are polarized (ideally 100% if only Rayleigh diffusion occurs). The stellar photons are not. Thus, the composite light (star+planet) in polarization "2" contains more planetary photons than the light in polarization "1". **(b)** At or near opposition, the planetary photons are not (or few) polarized, and the composite light (star+planet) in polarization "1" and "2" contains equal amount of planetary photons.

We thus propose to use both the orbital motion induced Doppler shift *and* this possible partial polarization of planetary photons, to distinguish them from the unpolarized stellar photons. If successful, this method could provide for relevant physical and orbital information yet unavailable, such as the on-sky orientation of the nodal line, and the orbital inclination.

The principle of the method is to use a telescope with a spectro-polarimetric device, whose polarization direction can be freely oriented on the sky. This instrument is to provide for pairs of spectra in two orthogonal linear polarization states denoted $A(\theta)$ and $B(\theta)$, where θ is the angle between the polarization direction A and the local north direction.

For bright Pegasides such as υ *And* or τ *Boo*, the orbital motion is well known, and it is in principle possible to determine the epochs for optimal spectro-polarimetric efficiency. This corresponds to a trade-off between the positions of maximum polarization (the quadratures) and the positions of maximum brightness (the oppositions). However, the orientation of the nodal line on the sky is unknown. Thus, several pairs of spectra, with different on-sky orientations have to be taken. A typical sequence of orientations could be: $\theta = 0^o$ and 90°, then $\theta = 30^o$ and 120°, then $\theta = 60^o$ and 150°.

These spectra have to be expressed as functions of the logarithm of the wavelength, for a reason explained in Figure 2. Indeed, the composite spectra in both polarization states involve essentially the spectrum of the stellar light (Figure 2-a), plus a very low amplitude Doppler-shifted replica due to the planetary scattered photons (Figure 2-b). In terms of the wavelength, the Doppler shift is *not* wavelength-independent, whereas it is in logarithmic coordinates (Figure 2-c). From both spectra in $log\lambda$ coordinates, a reference Dirac mask is computed from a selection of the most powerful absorption lines (Figure 2-d).

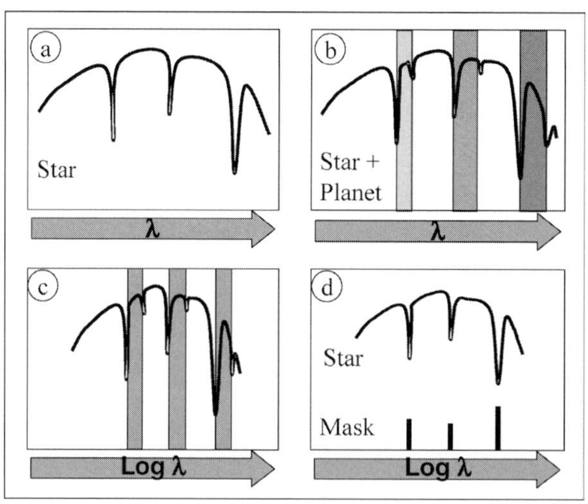

Figure 2. The sketched spectrum of a star and a Pegaside. **(a)** The spectrum of the star alone. **(b)** The spectrum of the star + planet Doppler-shifted. **(c)** The same in $log(\lambda)$ coordinate, so as to make the Doppler shift constant. **(d)** The spectrum of the star and the correlation Dirac mask.

For each on-sky orientation angle θ, The reduction procedure involves the following steps (assuming standard spectral and photometric calibrations have been done):

- Extract the four carefully calibrated spectra $A_\theta(\lambda)$, $B_\theta(\lambda)$, $A_{\theta+90^o}(\lambda)$ and $B_{\theta+90^o}(\lambda)$.
- Compensate for the instrumental photometric difference between the two

polarization channels by computing the two averages $A_\theta(\lambda) + B_{\theta+90^o}(\lambda)$ and $B_\theta(\lambda) + A_{\theta+90^o}(\lambda)$. Indeed, rotating the whole instrument of 90° around its axis is equivalent to permutate the two polarization channels.

- Express the resulting functions in terms of $\log\lambda$ instead of λ.
- Compute the two correlation functions of the two compensated and log-scaled spectra $a(\log\lambda)$ and $b(\log\lambda)$ with the common reference Dirac mask.
- Compute their difference $\Delta(\log\lambda)$. The stellar signature should vanish, and the planetary signature should emerge out of the noise, with the expected Doppler shift (see Figure 3).

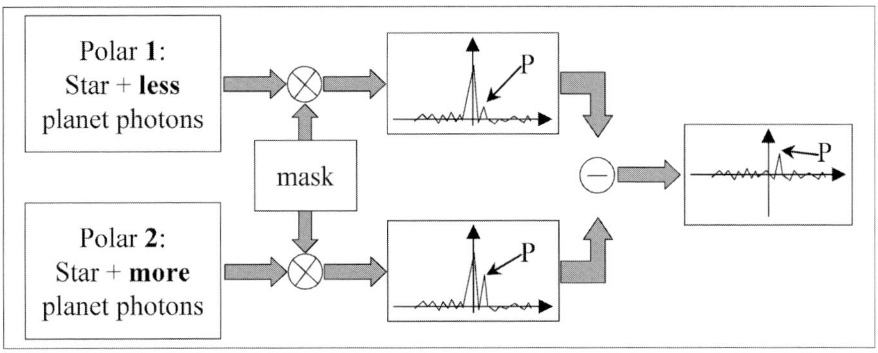

Figure 3. How the planetary signal should appear. The spectra in two complementary polarization states are convoluted with the same Dirac mask. Both correlation functions are dominated by the stellar photons (small Doppler shift). The small planetary Doppler shifted pike is present with different amplitude in both signals. When subtracting one to the other, the stellar features should cancel, and the planetary signature should remain, hopefully emerging out of the residual noise.

2. Preliminary observations at OHP

One of the best candidates for a spectro-polarimetric detection is certainly τ *Boo* (HD120136), a $F6IV$ star at $15.6\,pc$, with $M_v = 4.50$. This star is well known to have a Jupiter-like planet orbiting in 3.3128 days at a distance of $0.0462\,AU$ (see Butler *et al.*(1997), Cameron *et al.*(1999) and Charbonneau *et al.*(1999) for example).

For a first preliminary spectro-polarimetric observation of τ *Boo*, we have chosen the spectrograph CARELEC on the $193\,cm$ telescope at the Observatoire de Haute Provence (OHP). On this instrument, the bonette can be rotated over 360° on the sky, but no polarimetric capabilities are available.

Thus, we had to introduce additional optics in front of the slit of the spectrograph, so as to separate the two complementary linear polarizations. This additional optics is a $1\,cm$ thick calcite ($CaCO_3$) crystal, with a cylindrical diverging lens (see Figure 4). The crystal splits the input beam into two polarized beams. The lens spreads the light of each polarized beam along the slit, so as to produce elliptic rather than circular focal spots. This avoids too rapid a saturation of the CCD pixels, which would result in several short exposures, with a lot of time-consuming readout cycles. With this optics added, each CCD image displays simultaneously two complementary polarized spectra (see Figure 5). The spectral domain under consideration was $[385\,nm\ ,\ 480\,nm]$.

A very careful post-processing pipeline has been designed and implemented to reject all systematic instrumental biases below the photon noise.

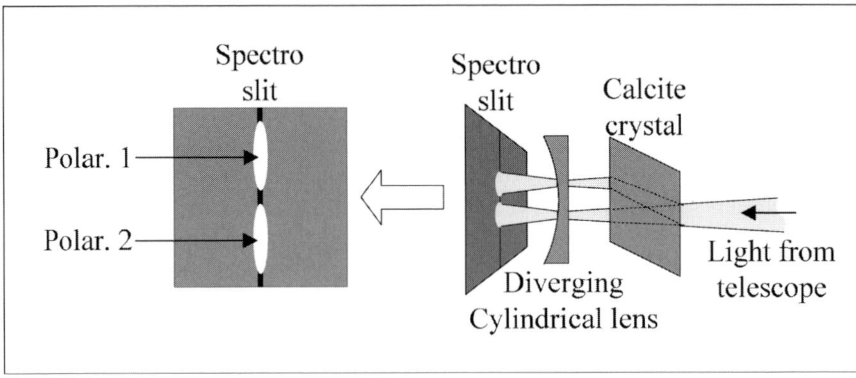

Figure 4. The additional optics to split polarizations and spread the focal spots along the spectrograph slit. A calcite crystal ($CaC0_3$) transforms the convergent beam from the telescope into two convergent beams, one with polarization 1, the other with polarization 2. The cylindrical diverging lens spreads the two spots in the direction of the spectrograph slit.

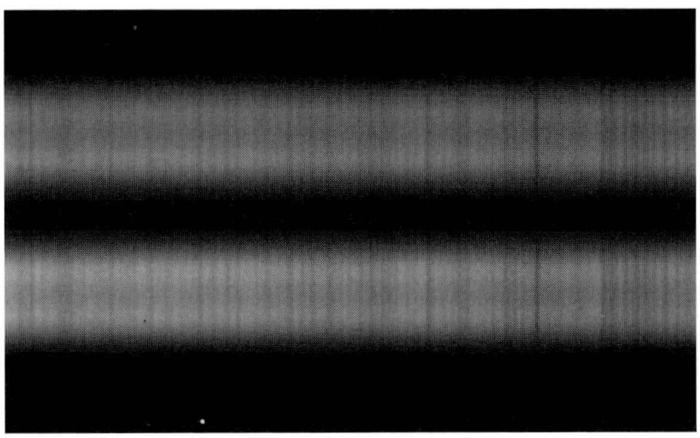

Figure 5. An example of polarized spectra taken with CARELEC+crystal+lens.

3. Conclusion

The preliminary observation run on the modified CARELEC spectrograph at OHP did not yield an effective detection of planetary photons for τ *Boo*. Indeed, the *rms* residual noise on the difference of correlation functions corresponds to a magnitude difference of 8.2, which is definitely not sufficient for any planetary photon discrimination. One of the limiting features is the overall photometric efficiency of the whole instrumental chain (atmosphere, telescope, calcite crystal + lens, spectrograph and camera), which remained below 1 %, that is, far below our most pessimistic assumptions.

Further observations should be performed on more recently designed spectrographs with built-in polarimetric capabilities (no added calcite crystal), such as NARVAL (TBL, Observatoire du Pic du Midi) or ESPADON (CFHT). On such instruments, the planetary signature should emerge within a few hours of exposure time.

References

Butler, R.P., Marcy, G.W., Williams, E., Hauser, H. & Shirts, P. 1997, *ApJ letters* 474, L115

Cameron, A., Horne, K., Penny, A. & James, D. 1999, *Nature* 402, 751

Charbonneau, D., Noyes, R., Korzennik, S., Niesenson, P., Jha, S., Vogt, S. & Kibrick, R. 1999, *ApJ Letters* 522, L145

Direct Imaging of Exoplanets: Science & Techniques
Proceedings IAU Colloquium No. 200, 2005
C. Aime and F. Vakili, eds.
© 2006 International Astronomical Union
doi:10.1017/S1743921306009094

Evolution of the Habitable Zone and Search for Life Around Giant Stars Part II: Feasibility with Darwin/TPF

W. C. Danchi[1], B. Lopez[2], J. Schneider[3], A. Belu[2,4], R. Barry[1,5], J. Rajagopal[1,6], and L. J. Richardson[1]

[1]NASA Goddard Space Flight Center, Exoplanets and Stellar Astrophysics Lab., Code 667, Greenbelt, MD 20771, USA

[2]Observatoire de la Cote dAzur, BP 4229, 06304 Nice Cedex 4, France

[3]Observatoire de Paris, Paris, France

[4]University of Nice - Sophia Antipolis, LUAN, Parc Valrose, 06108 Nice Cedex 2, France

[5]The Johns Hopkins University, Physics and Astronomy Dept., Baltimore, MD 21218, USA

[6]University of Maryland, Astronomy Department, College Park, MD 20742, USA

Abstract. We discuss the evolution of the habitable zone around low mass and intermediate mass stars as they evolve off the main sequence. This work shows that this new class of stars should be included in the search for life because if planets could be found in their habitable zones, and these planets showed evidence for life, it is possible to empirically determine a lower limit to the timescale for the formation of life. This time scale is not well determined from the study of the Earth (or planets around main sequence stars), as life formation initially occurred during a period of heavy bombardment from comets and asteroids during the formation of the solar system. Our initial research was recently published (Lopez, Schneider, & Danchi 2005). We will describe our work in progress, in which we perform calculations and simulations aiming to demonstrate the potential of TPF and Darwin for the search and characterization of planets around evolved stars.

Keywords. instrumentation: high angular resolution, techniques: interferometric, space vehicles: instruments, planets and satellites: general, planetary systems: formation.

---

1. Introduction and Basic Concepts

Low and intermediate mass stars with masses ranging from 0.5 to $\sim$8 $M_\odot$ undergo dramatic changes in temperature and luminosity during their evolution. For example, the luminosity of the Sun, which is located on the Main Sequence of the Hertzprung-Russell diagram, currently is 30 percent greater than its zero-age (initial) luminosity (Gough 1981). Other major changes take place after the Sun evolves off the Main Sequence. Along the Red Giant Branch (RGB) the radius of a 1 $M_\odot$ star progressively increases up to 10 $R_\odot$, and the luminosity increases up to about $10^{1.5}$ $L_\odot$ (Maeder & Meynet 1988). Then, after leaving the RGB a star evolves onto the Asymptotic Giant Branch (AGB). At this stage in its evolution, its radius increases up to a few hundred times RSun and the luminosity up to nearly 10^4 $L_\odot$. The lifetime of a 1 $M_\odot$ star, from the moment it leaves the main sequence until the end of the RGB, is about 2.5×10^9 yr, which can be compared to $\sim 10 \times 10^9$ yr spent on the main sequence. The evolution of a star is accompanied by mass loss processes that play an important role. During the post-main sequence evolution of a 1 $M_\odot$ star nearly 0.4 $M_\odot$ will be lost to the interstellar medium before it ends its life as a white dwarf (Weidemann 1987).

Table 1. Lifetimes (Gyr) of stages of stellar evolution as a function of initial stellar mass

Mass ($M_\odot$)	Main Sequence	Subgiant	First Red Giant	Core He Burning
1.0	7.41	2.63	1.45	0.95
1.5	1.72	0.41	0.18	0.26
2.0	0.67	0.11	0.04	0.10

The habitable zone is conservatively defined as a shell region around a star where liquid water can be found at the surface of a planet. This zone is typically from about 0.95 AU to 1.37 AU for G stars (Kasting *et al.* 1993, Forget 1998). A less conservatively defined habitable zone extends the outer edge of the limit of the habitable zone to as large as 2.4 AU, depending on the radiative properties of CO_2 ice clouds (Forget & Pierrehumbert 1997, Mischna *et al.* 2000). Consequently, as the temperature and luminosity of a star changes during the course its evolution, the radius and width of this zone will also evolve with time.

The habitable zone gradually moves outward after a star leaves the main sequence. Depending on the length of time that is necessary for life to form, we show that it is possible that life may exist or develop on a planet that is located within a range of 2-9 AU from a 1 $M_\odot$ star, during the first stages of its post main-sequence evolution and up to 22 AU from the star during the second stage after the core He flash. It is possible that the effects of natural migration of micro-organisms through meteorites may be a possible trigger for the emergence of life around evolved stars (Arrhenius 1903, O'Keefe *et al.* 1996, Zubrin 2001). Micro-organisms might also transport life from one planet to another planet during the evolution of a star. For example, they could be transported from a planet where life is ending to a more distant planet where favorable conditions for the re-birth of life are encountered because of the sufficiently slow evolution of the habitable zone.

Future space missions, such as TPF (Beichman, Woolf, & Lindensmith 1999) and Darwin (Fridlund & Gondoin 2003), that focus on searches for signatures of life on extrasolar planets primarily around main-sequence stars, should also consider the environments of sub-giant and red giant stars. Pathfinder missions such as EPIC (Clampin *et al.* 2001), Eclipse (Trauger *et al.* 2003), and FKSI (Danchi *et al.* 2003a, 2003b) may also be able to observe planets around red giant stars.

In this short paper we discuss two main issues related to the evolution of the habitable zone. The first is some of the characteristics of the subgiant and giant stars in the solar neighborhood. The second is the feasibility of detection of earth-like planets around these stars by comparing these stars to the current star list for the interferometer version of TPF or Darwin.

2. Characteristics of subgiant and red giant stars in the Solar neighborhood

Figure 1 displays an empirical Hertzprung-Russell diagram for stars in the globular cluster M3 that we have modified from Shu (1982). The focus of our work on the evolution of the habitable zone is on stars that have evolved off the main sequence into the subgiant or red giant branches, and NOT the extremely large and luminous Asymptotic Giant Branch (AGB) stars. Common misconceptions are based on the fact that many astronomers automatically think only of AGB stars in the context of evolved stars. Thus they think that red giant stars are too cool, too luminous, and too large to be worth

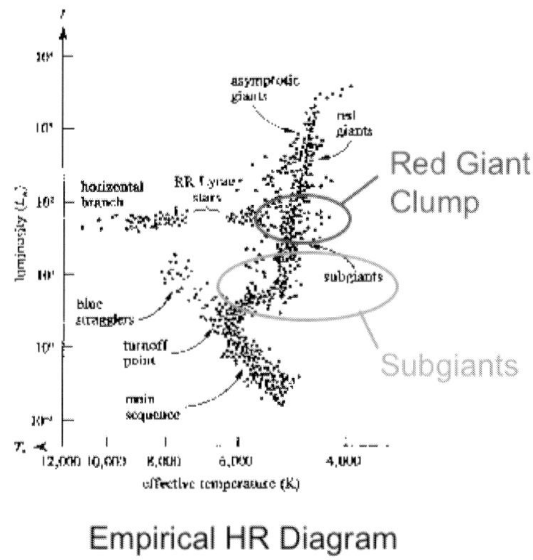

Figure 1. Hertzprung-Russell diagram of the globular cluster M3. Adapted from Shu (1982).

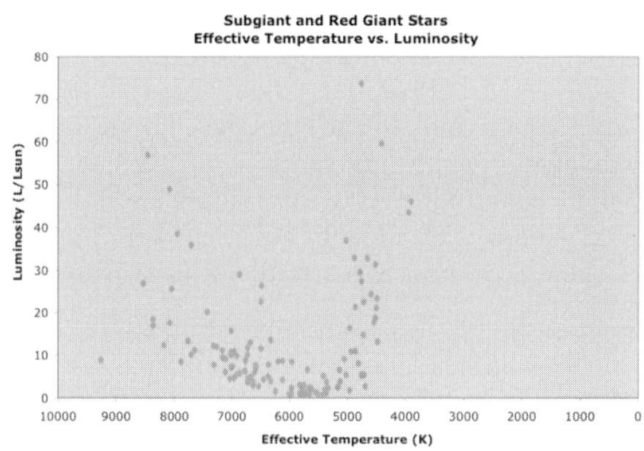

Figure 2. Temperature versus luminosity for the approximately 130 nearby luminosity class III and IV stars within 30 pc of our solar system.

studying in the context of life formation and evolution. As we shall see in the remainder of this section, these misconceptions are false, and a reasonable sample of such stars exists and are observable with TPF or Darwin without modification from what is planned for the main sequence stars.

Figures 2-4 display some of the characteristic features of luminosity class IV and III stars within 30 pc of the Solar system. The sample of stars in these figures was extracted from the catalog of 2350 stars within 30 pc that was created for the Terrestrial Planet Finder mission. This list contains approximately 120 stars and is essentially the same set of stars that were published in Lopez, Schneider, and Danchi (2005) in their Tables II-IV. The master list for TPF, however, compiles much useful physical information, which we have used in this paper to compare with main sequence luminosity class V stars that are the main focus of the TPF mission.

It is easy to see from Figure 2 that the subgiant and red giant stars of interest to this work have only modestly higher luminosities and roughly the same effective temperatures of the main sequence stars. Figure 3 displays a histogram of the luminosities of the

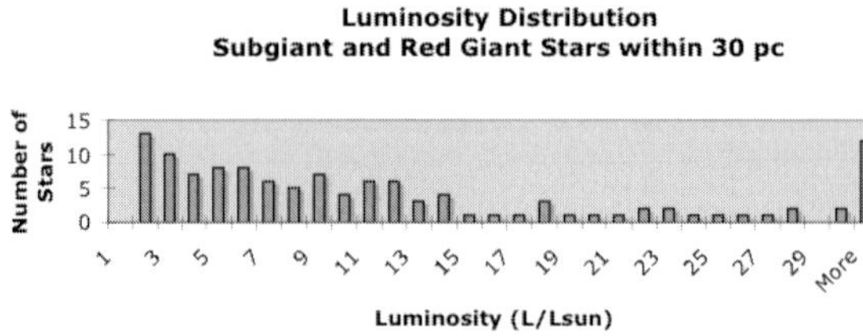

Figure 3. Luminosity distribution for the luminosity class III and IV stars in the solar neighborhood. Note the distribution is peaked at luminosities in the range of 3-5 $L_{\odot}$.

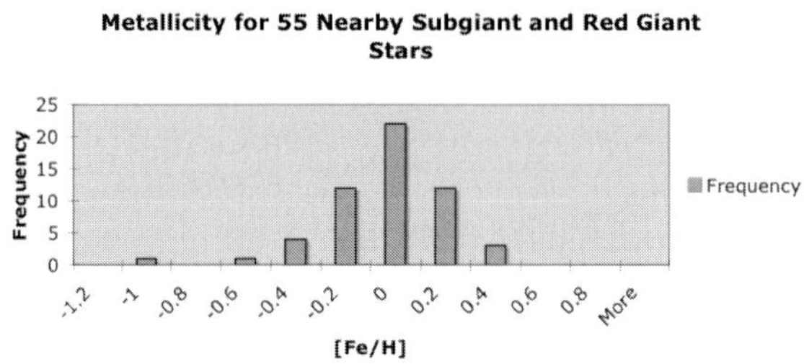

Figure 4. Distribution of metallicities for 55 luminosity class IV and III stars within 30 pc of the solar system extracted from the approximately 130 stars shown in the preceding figures.

selected stars. It is clear that the bulk of the stars have luminosities in the range of 3-10 $L_{\odot}$, and hence their stellar radii are in the range of ~1.7-3.3 $R_{\odot}$.

Figure 4 displays metallicities compared to solar values for 55 stars of the larger sample. These stars were the only ones with metallicities that were available in TPF star list database. These stars show a peak in the metallicity distribution about 0.1 dex larger than solar metallicity.

3. Feasibility of observations with TPF and Darwin

The signal-to-noise ratio for detecting planets around stars using a nulling interferometer depends on a number of factors. The signal term is simply the photon rate for the planet R_{planet} times the integration time, T. Similarly, the noise terms include the stellar leakage signal (R_{SL}), the presence of exo-zodial dust (R_{EZ}), our own local zodiacal foreground (R_{LZ}), and thermal noise (R_{th}) from the telescope and instrument itself, where these quantities are also multiplied by the integration time to obtain the total number of noise photons. The signal-to-noise is thus:

$$SNR = \frac{Planet\ Signal}{Noise} \propto \frac{R_{planet}T}{[(R_{SL} + R_{LZ} + R_{EZ} + R_{th})T]^{1/2}}, \tag{3.1}$$

where we follow the notation of Dubovitsky and Lay (2004). The present architecture studies for the interferometer version of TPF and Darwin indicate that under most circumstances the noise terms for the extra-solar zodiacal noise can be neglected if the total flux is less than about 5 times the contribution from the planet signal (Dubovitsky & Lay 2004). The term for the thermal noise is usually small compared to that from the local zodiacal emission; this is normally done as a requirement on the temperature of the telescope itself, and is why TPF and Darwin need to be cooled to about 35-40 K. These considerations imply that the dominant terms for the noise are the local zodiacal foreground and the stellar leakage. Clearly for the subgiant and red giant stars, the local

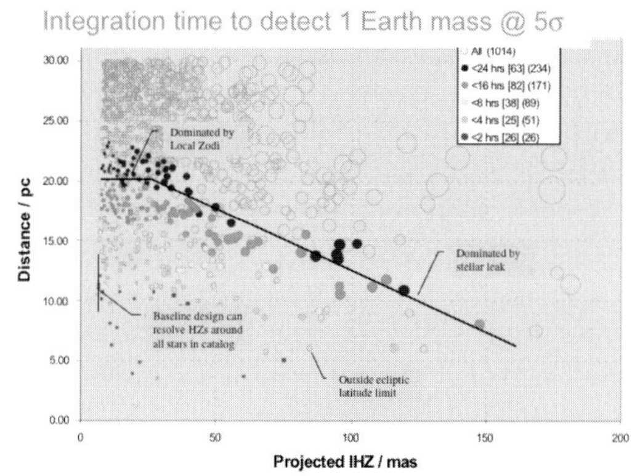

Figure 5. Integration time estimates for the Terrestrial Planet Finder Interferometer (TPF-I) as a function of projected inner habitable zone in milli-arcseconds (mas) and distance in parsecs (pc) from our solar system (Dubovitsky & Lay 2004).

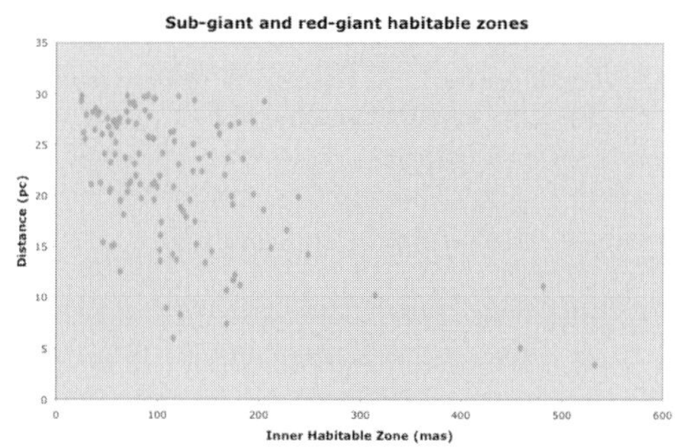

Figure 6. Inner habitable zone radius (mas) versus distance (pc) for the stars in Fig. 3.

zodiacal foreground is not different than for *any* other stars, thus the only term that is different is the stellar leakage.

Studies of the detectability of Earth like planets for TPF and Darwin indicate that the dominant noise term for the bulk of the detectable systems is the local zodiacal foreground emission and not the stellar leakage contribution. This is because the largest number of detectable planets is for stars that are the furthest away from the Sun. As a result, the contribution from stellar leakage is important only for the nearest stars. These conclusions are summarized in Fig. 5, which shows the projected Habitable Zone radius versus distance and displays the breakpoint in dominant noise contributions between stellar leakage and local zodiacal emission (see Dubovitsky & Lay 2004).

Figure 6 displays the inner habitable zone radius versus distance for the subgiant and giant stars in our sample. It is clear from this figure that a similar breakpoint should exist, and therefore Earth like planets in the habitable zone could also be detected for these stars as well. It is also evident from the figure that a larger sample of such stars should be investigated because current designs for TPF have sufficient resolution and sensitivity.

4. Conclusions

We have shown that the evolution of the habitable zone as stars move off the main sequence and onto the subgiant and later the red giant branches of the HR diagram provide a new opportunity for the search for Earth-like planets and life. For these stars life has an opportunity to evolve under conditions with much less bombardment than existed in the early Solar system. We also show there are adequate numbers of such stars in the Solar neigborhood, and that preliminary analysis demonstrates that planets around such stars could be detected without any modifications to present architectural concepts for the interferometer version of TPF and Darwin.

Acknowledgements

We acknowledge support from GSFC Internal Research and Development and Director's Discretionary Funds, a Michelson Fellowship, a National Research Council (NRC) Fellowship, and support from the CNRS.

References

Arrhenius, S.A. 1903, in *The Quest for Extraterrestrial Life*, ed. D. Goldsmith (Mill Valley: University Science Books), 32
Beichman, C., Woolf, N.J., Lindensmith, C., ed. 1999 (JPL Publ. 99-3; Pasadena: Caltech)
Clampin, M., Ford, H.C., Illingworth, G., & Petro, L. 2001 *BAAS*, 33, 1365
Danchi, W.C., Deming, D., Kuchner, M.J., & Seager, S. 2003a *ApJ*, 597, L57
Danchi, W.C., *et al.* 2003b, (ESA Publ. SP-539; Noordwijk: ESA), 83
Forget, F. 1998 *Earth, Moon, Planets* 81, 59
Forget, F. & Pierrehumbert, R.T. 1997 *Science*, 278, 1273
Fridlund, C.V.M. & Gondoin, P. 2003 *Proc. SPIE*, 278, 1273
Gough, D.O. 1981 *Sol. Phys.*, 74, 21
Dubovitsky, S. & Lay, O.P. 2004 *Proc. SPIE*, 5491, 284
Kasting, J., Whitmire, D.P., & Reynolds, R.T. 1993 *Icarus*, 101, 108
Lopez, B., Schneider, J., & Danchi, W.C. 2005, *ApJ*, 627, 974
Maeder, A. & Meynet, G. 1988 *A&AS*, 76, 411
Mischna, M.A., *et al.* 2000 *Icarus*, 145, 546
O'Keefe, J.D. & Ahrens, T.J. 1996 *Science*, 234, 346
Shu, F. 1982 *The Physical Universe*, (Mill Valley: University Science Books), 173
Trauger, J., *et al.* 2003 *Proc. SPIE*, 4854, 116
Weidemann, J. 1987 *A&A*, 188, 74
Zubrin, R. 2001 *J. British Interplanet. Soc.* 54, 262

Discussion

DEN HARTOG: Giants are considerably larger in diameter than MS V stars, and in order to mitigate the extra leakage much shorter baselines are required: don't you run into trouble with collision avoidance requirements? On the other hand, the HZ are located further away from these stars: does the field of view limitation, coming in particular from the fiber coupling not restrict you too much in the assessment of the outer parts of the HZ?

DANCHI: Actually this is a common misconception because many astronomers are thinking of Asymptotic Giant Branch stars, which indeed are very large. However, since the subgiants and red giant stars are only modestly larger in diameter, our studies indicate this is not a problem except for perhaps a few of the closest ones. In answer to your second question, one could use a different fiber with a wider FOV for the closest such stars. For most of the sample it is not an issue.

Direct Imaging of Exoplanets: Science & Techniques
Proceedings IAU Colloquium No. 200, 2005
C. Aime and F. Vakili, eds.
© 2006 International Astronomical Union
doi:10.1017/S1743921306009100

Mid-IR direct imaging of superjupiters around nearby stars

Gabriel Bihain[1,2], Rafael Rebolo[1,2], José Antonio Caballero[1] and Victor Javier Sánchez Béjar[1,3]

[1]Instituto de Astrofísica de Canarias, 38205 La Laguna, Tenerife, Spain
[2]Consejo Superior de Investigaciones Científicas, Spain
[3]Gran Telescopio Canarias Project, Spain
email: gbihain@ll.iac.es

Abstract. Superjupiters are much brigther when they are young and still contracting, and the contrast with the harboring stars is significantly more favorable for their detection. Within the JOVIAN (Jupiter-like Objects in the Visible and the Infrared: Astrophysical Nature) project, a search has been performed for superjupiters around nearby young stars using 4 m class telescopes. Here we summarize the results and present prospects for a future search around young and also relatively older nearby stars with the 10 m Gran Telescopio Canarias and the mid-infrared camera CanariCam.

Keywords. (Galaxy:) solar neighborhood, stars: low-mass, (stars:) planetary systems.

1. Introduction

The sensitivity to detect superjupiters around stars improves when the spatial resolution is higher (i.e. for nearby stars) and the contrast is lower (i.e. for primaries that are low-mass stars and planets that are intrinsically brighter due to youth). Numerous searches for massive planets by direct near-infrared (IR) imaging around nearby young dwarfs have been performed in the past 10 years (Nakajima, Oppenheimer, Kulkarni, *et al.* 1995; Oppenheimer, Golimowski, Kulkarni, *et al.* 2001; McCarthy & Zuckerman 2004; Luhman & Jayawardhana 2002; Neuhäuser, Guenther, Alves, *et al.* 2003; Masciadri, Mundt, Henning, *et al.* 2005; Lowrance, Patrick, Becklin, *et al.* 2005). Only two good planet candidates around dwarfs have been discovered so far (Chauvin, Lagrange, Dumas, *et al.* 2004,2005; Neuhäuser, Guenther, Wuchterl, *et al.* 2005). The very paucity observed of exoplanets at large- to moderate separations (>10 AU) can provide valuable indications on the main formation process of these objects, supposing that their intrinsic evolution (and thus photometric and spectroscopic properties) are well enough predicted by theoretical models (e.g. Baraffe, Chabrier, Allard, *et al.* 2002; Burrows, Sudarsky & Lunine 2003). We summarize the effort made by our group to search for substellar companions using Adaptive Optics (AO) systems and present our plan to exploit the advantages offered in the mid-IR by high planet/star contrast, diffraction-limited and subarcsecond imaging with a 10 m class telescope.

2. Adaptive optics searches

Using Alfa+Omega-Cass at the 3.5-m Calar Alto, AdOpt+NICS at Telescopio Nazionale Galileo and, especially, NAOMI+INGRID at the 4.2 m William Herschel Telescope (WHT), Caballero *et al.* (in preparation) have searched for substellar companions of more than

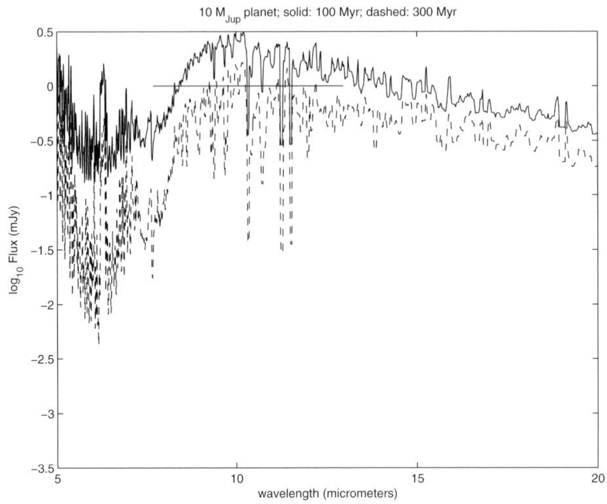

Figure 1. The expected mid-IR spectral energy distributions for a 10 M_{Jup} planet at 10 pc, for ages of 100 and 300 Myr (solid and dashed lines, respectively; Burrows *et al.* 2003). The horizontal line corresponds to the CanariCam 25 σ N-band sensitivity in 30 min of on-source integration.

fifty young nearby stars (d<50 pc) in the near-IR. The stars were selected with spectral types later than solar and features indicative of youth (high lithium abundance, X-ray and/or UV emission, membership to young proper motion associations, etc.). Their ages ranged between 15 and 600 Myr. All of them have been completely analysed, comparing first and second astrometric epochs. Although the data would allow them to discover objects with masses down to 5 M_{Jup} at separations of 4 arcsec around several of the primaries (i.e. at 40 AU for the nearest stars), they have not detected any new substellar companion up to ~1000 AU. This result implies a frequency of substellar companions at intermediate and large separations of $< 4\%$, in agreement with the $1\pm1\%$ obtained by McCarthy & Zuckerman (2004). It should also be compared with the $7.3\pm1.5\%$ of the solar-like stars that have exoplanet candidates discovered at small separations with the radial velocity method.

3. Mid-infrared searches

We plan to perform imaging at 10 μm with the 10 m Gran Telescopio Canarias (GTC)/ CanariCam and with enough sensitivity (0.2 mJy at 5 σ in 30 min on-source integration) to detect planetary companions of masses 10 M_{Jup} around stars of the solar neighborhoud. In Fig. 1 and 2 we illustrate the sensitivity of this instrument. The subarsecond imaging provided by CanariCam will permit us to explore physical separations as small as 10 AU around targets at 10 pc (approximately five times the telescope resolving power 0.2 arcsec).

In addition to the diffraction limited imaging in the mid-IR, we can take advantage of the improved planet/star brightness contrast. For example, a 10 M_{Jup} superjupiter at 10 pc and 100 Myr old has a predicted mid-IR magnitude $N \sim 10.5$ and a near-IR magnitude $J \sim 20$, whereas its K0-type harboring star would have $N \sim J \sim 4$. Therefore the contrast planet/star in magnitude decreases from 16 to 6.5 magnitudes when shifting to the mid-IR, and thus enters in the typical instrumental dynamical range for detection.

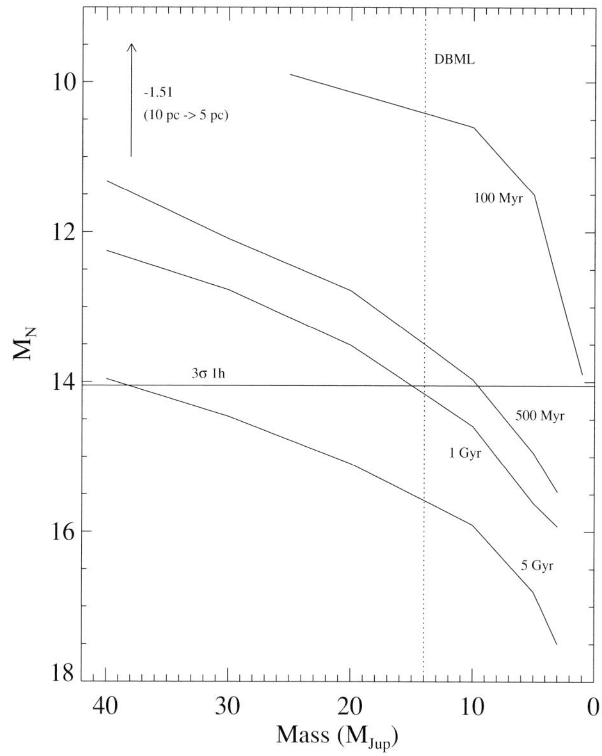

Figure 2. 5-σ N-band sensitivity of CanariCam in 1 hour of on-source integration. Theoretical isochrones from Burrows, Sudarsky, Lunine, *et al.* (1998) at ages 0.1, 0.5, 1 and 5 Gyr for ~1–40 M_{Jup} substellar objects at 10 pc are overploted (solid lines). The arrow indicates the brightness increment when these objects are at 5 pc. Finally the deuterium mass burning limit (DMBL) is traced by the vertical dotted line.

With GTC/Canaricam we will be able to search three types of 10 M_{Jup} superjupiters (see Fig. 2) of different ages and distances: (*i*) 100 Myr at <50 pc; (*ii*) 1 Gyr at <10 pc; (*iii*) 5 Gyr at <5 pc. Objects at 5 pc are 1.5 mag brighter than at 10 pc, and objects at 50 pc are 3.5 mag fainter than at 10 pc. For case (*iii*), there are approximately 60 stellar targets available for the GTC in the North Hemisphere and some would require integration times up to 2.5 hours. Direct mid-IR imaging of stellar systems spanning an age range of 0.1–5 Gyr will permit to gain further insight on the formation and evolution processes of planetary mass objects.

4. Conclusions

The near-IR Adaptive Optics searches that we have performed with 4 m-class-telescopes provided no detection of massive planets at large separations from 15–600 Myr-old nearby stars. In the mid-IR, higher planet/star contrast can be obtained than in the near-IR, and also diffraction-limited images. We plan to search for superjupiters in the mid-IR with the GTC/CanariCam. For 5 Gyr old stars at less than 5 pc, superjupiters of 10 M_{Jup} could be discovered at separations of ~ 5AU.

References

Baraffe, I., Chabrier, G., Allard, F., & Hauschildt, P.H. 2002, *A&A* 382, 563

Burrows, A., Sudarsky, D., Sharp, C., Marley, M.S., Hubbard, W.B., Lunine, J.I., Guillot, T., Saumon, D., & Freedman, R. 1998, *ASPC* 134, 354

Burrows, A., Sudarsky, D. & Lunine, J.L. 2003, *ApJ* 596, 587

Chauvin, G., Lagrange, A.-M., Dumas, C., Zuckerman, B., Mouillet, D., Song, I., Beuzit, J.-L., & Lowrance, P. 2004, *A&A* 425, L29

Chauvin, G., Lagrange, A.-M., Dumas, C., Zuckerman, B., Mouillet, D., Song, I., Beuzit, J.-L., & Lowrance, P. 2005, *A&A* 438, L25

Lowrance, P.J., Becklin, E.E., Schneider, G., Kirkpatrick, J.D., Weinberger, A.J., Zuckerman, B., Dumas, C., Beuzit, J.-L., Plait, P., Malumuth, E., Heap, S., Terrile, R.J., & Hines, D.C. 2005, *AJ* 130, L1845

Luhman, K.L., Jayawardhana, R. 2002, *ApJ* 566, L1132

Masciadri, E., Mundt, R., Henning, Th., Alvarez, C., & Barrado y Navascués, D. 2005, *ApJ* 625, 1004

McCarthy, C. & Zuckerman, B. 2004, *AJ* 127, 2871

Neuhäuser, R., Guenther, E.W., Alves, J., Huélamo, N., Ott, Th., & Eckart, A. 2003, *AN* 324, 535

Neuhäuser, R., Guenther, E.W., Wuchterl, G., Mugrauer, M., Bedalov, A., & Hauschildt, P.H. 2005, *A&A* 435, L13

Nakajima, T., Oppenheimer, B.R., Kulkarni, S.R., Golimowski, D.A., Matthews, K., & Durrance, S.T. 1995, *Nature* 378, 463

Oppenheimer, B.R., Golimowski, D.A., Kulkarni, S.R., Matthews, K., Nakajima, T., Creech-Eakman, M., Durrance, S.T. 2001, *AJ* 121, 2189

Direct Imaging of Exoplanets: Science & Techniques
Proceedings IAU Colloquium No. 200, 2005
C. Aime and F. Vakili, eds.
© 2006 International Astronomical Union
doi:10.1017/S1743921306009112

A simulation code for AO assisted 3D spectroscopic imaging of extrasolar planets

Alessandro Berton[1], Markus Feldt[1], Raffaele G. Gratton[2], Thomas Henning[1], Silvano Desidera[2], Stefan Hippler[1] and Massimo Turatto[2]

[1]Max-Planck-Institut für Astronomie (MPIA), Königstuhl 17, D-69117, Heidelberg, Germany
email: berton@mpia.de, feldt@mpia.de, henning@mpia.de, hippler@mpia.de.

[2]Osservatorio Astronomico di Padova INAF, Vicolo dell'Osservatorio 5, I-35122, Padova, Italy
email: gratton@pd.astro.it, desidera@pd.astro.it, turatto@pd.astro.it.

Abstract. We present here a software for simulations of observations made with an Integral Field Spectrograph attached to an extreme adaptive optics system, with the main goal of simulating direct detection of extrasolar planets and to test the capabilities in detecting planets of an instrument based on IFS on large telescopes. This code, written for IDL, has been conceived within the CHEOPS project, a second generation "Planet Finder" instrument for ESO's VLT; but it has been extended to the case of various ELTs. Here we describe in detail the procedure adopted in order to simulate realistic values of speckle noise, Adaptive Optics corrections, specific instrumental features and the efficiency of a Simultaneous Differential Imaging technique to increase the signal of the planet.

Keywords. Extrasolar Planets, Simulations, Adaptive Optics, Integral Field Spectroscopy.

1. Introduction

Direct detections of Jupiter-like extrasolar planets, with the present technology available, are still a challenging goal to achieve. Difficulties come from the complex structure of the stellar Point Spread Function (PSF), which is covering the light of the planet. This stellar halo is usually brighter than the planet, at every wavelength, and since atmospherical speckles are the most important source of noise that we have to consider in direct detections of planets (Racine *et al.* 1999), an efficient method for speckle noise reduction is required. This noise can actually be reduced in case of an AO system being able to reach very high Strehl ratios S: thus, an extreme AO system is mandatory in order to reduce dramatically the speckle intensity. But even after AO correction, the speckle noise is still orders of magnitude larger than the photon noise; therefore the reduction of the speckle noise to a level comparable to the photon noise can be achieved only using additional techniques.

A special Simultaneous Differential Imaging technique (SDI), applied to Integral Field Spectroscopy (IFS) observations of the star-planet system has been suggested within the CHEOPS project (CHaracterizing Exo-planets by Opto-infrared Polarimetry and Spectroscopy, Feldt *et al.* 2004), a project for a Planet Finder instrument on ESO's VLT. The basic idea is to use the spectral features of the planet in the near-infrared, which are in general broad molecular absorption bands (methane bands for Jupiter-like planets), in order to enhance the planetary signal, and then to subtract at the same time the light from the star and the noises related to it.

In order to test the capability of such a technique in reducing noise and detecting planets, we developed a dedicated software that simulates an observation of an extrasolar

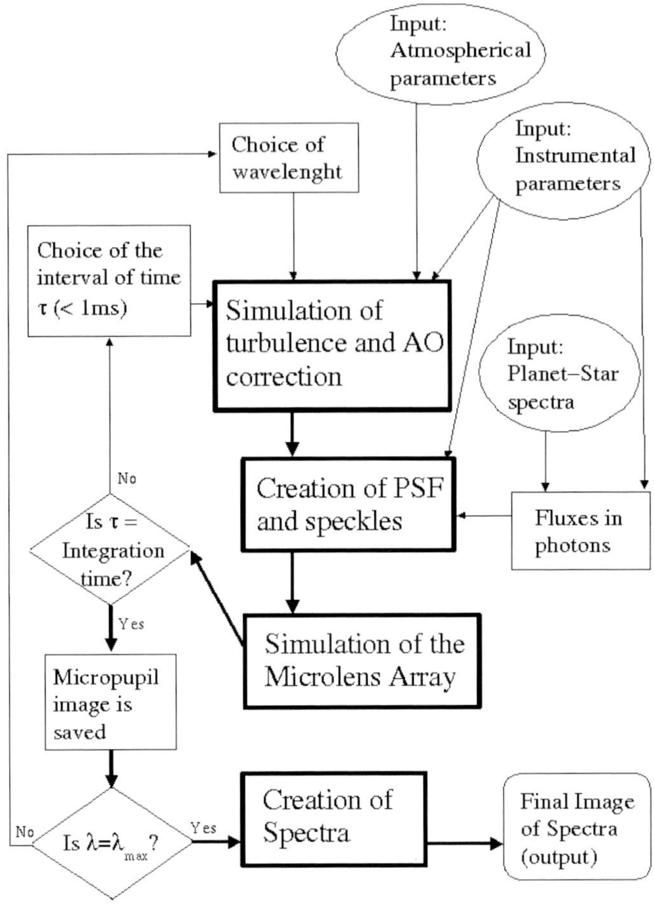

Figure 1. General flow-chart of the simulation code. Details are described in the text.

planet using an IFS. From the analysis of the results of the simulations, it has been possible to understand which kinds of planets are really detectable and characterizable for various telescope diameters, e.g. an 8m-class telescope. The code can be easily adapted to various kinds of telescopes, sites and instrumental features, in particular allowing an interesting analysis of the different results achievable with Extremely Large Telescopes (ELTs) with mirror diameters between 25 and 100 meters. Some results are presented here, and discussed more in detail in Feldt *et al.* (2005).

2. Description of the simulation code

The simulation code we are going to present here was written for IDL in the context of the CHEOPS project, and its general flow-chart is shown in Fig. 1. The IFS consists on a microlens array composed by 254×254 hexagonal lenses covering a field of view of $3''.5\times3''.5$. The disperser is an Amici prism, which can give spectra with constant resolution of 15 in the spectral region between 0.9 and 1.7μm. A detailed description of the CHEOPS IFS can be found in Claudi *et al.* (2004). With our simulation code we simulate accurately speckle patterns in order to estimate correctly the Speckle noise. Then, using these simulated observations we determine which kinds of planets are really detectable with such an instrument, as a function of different parameters: physical (masses,

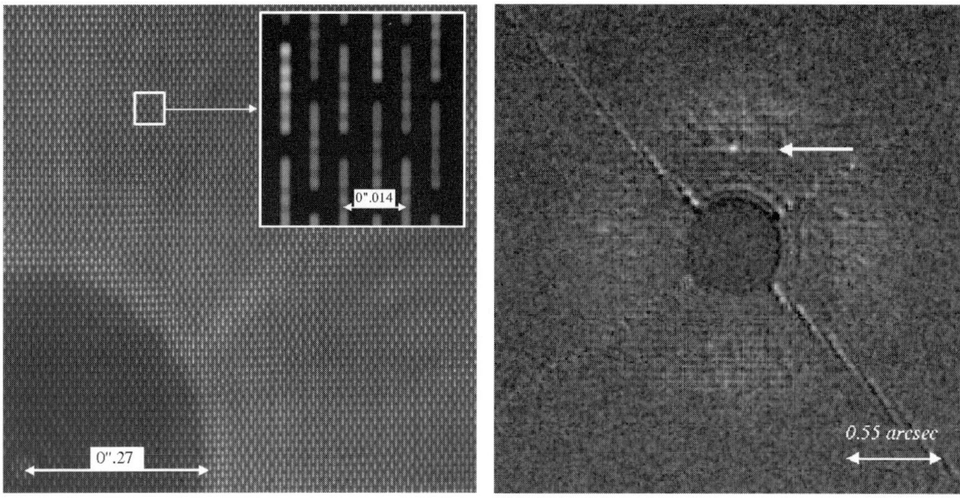

Figure 2. At left: a detail of the image of spectra obtained at the end of the procedure. The star is placed on the left-bottom corner, and it is visible the effect of a simulated coronograph. At right: the detection of a BD of 30 M_J around a 5 Gyr old star at 40 pc after the differential imaging (S/N∼30).

ages, angular separations, distances), instrumental (telescope features, AO, etc.) and the atmospheric seeing conditions.

Here we summarize the main concept and results of the code, a more detailed description would be available in a forthcoming paper (Berton *et al.* 2005). The procedure starts creating an istantaneous wavefront describing the atmospheric turbulence in a specific moment of the observation (in general for a time interval shorter than a millisecond). This wavefront is at first corrected by a simulated AO system, able to reach a Strehl Ratio of 0.7 or higher in J-band; then it is combined with the pupil of the telescope. In this way an instantaneous PSF is generated, including the speckle pattern related to that specific moment. If needed, a coronograph can be introduced in simulations. The simulation of a long exposure (for planet detection several hours can be necessary) is done by combining together many instantaneous PSFs: this allows the representation of the speckle evolution with time. The result of this process is a final monochromatic PSF. The next step is the simulation of the microlens array. This is done by splitting the image in small hexagonal portions which are analyzed and focused on the disperser as micropupils. To simulate the disperser the code must repeat the entire procedure at various wavelengths to recreate the spectral structures of the PSF. This gives as a result a set of micropupil images each one for a specific resolution element, which will be combined on the detector for creating the final spectral image. An example of this final image is shown in Fig. 2 at left.

At this point, in order to apply SDI we extract the information related to specific λ from all the spectra and then we create a set of monochromatic images, each one corresponding to a given resolution element. These images can be combined in different ways, applying different mathematical operations in order to remove the stellar light (and the speckles with it). The images must be scaled in wavelength and flux before the comparison. In Fig. 2 at right the detection of a Brown Dwarf of 30M_J placed at 40pc from the Sun is shown, and it is detectable with a Signal-to-Noise ratio of 30.

3. Preliminary results

We made a complete set of simulations for an 8-meter class telescope, based on VLT, using different models of planetary systems, with different masses (between 1 and 30 M_J), ages (between 0.01 and 5 Gyr), distances (3, 10 and 40 pc), spectral type of the host star (G0V and M0V) and various angular separations, for a fixed exposure time of 4 hours. The models of the planetary spectra were taken form Burrows *et al.* (2004). In general, the speckle noise was found to be reduced of a factor around 10^{-3} by using the IFS, which makes the total noise approach to the photon noise. The results in the different cases have been expressed as signal-to-noise ratios (S/N) and plots of S/N vs. separation. The complete list of results will be presented in Berton *et al.* (2005): at the moment we present here some general conclusion about all the simulations.

First of all, an important result is that a Jupiter-like planet younger than 1Gyr, around a G0V star at 3 pc from the Sun, appears to be visible at a separation of ∼1 arcseconds from the host star after an integration time of 4 hours. This could be close to the case of the planet around ε Eridani, one of the most important targets for the new ground-based "Planet Finder" instruments which are on project now. If we go further, at 10pc, planets with masses $\geqslant 5M_J$ and ages around 1 Gyr can be clearly detected. Passing from a G0V star to an M0V star the S/N increase considerably, because of the lower contrast between star and planet.

Some simulations are ongoing at present also for some ELTs: in particular we applied our code to a 100-meter telescope basing on ESO's OWL project (OverWhelmingly Large telescope, Gilmozzi 2004). We used a model of the entrance pupil (N. Yaitskova, private communication) in order to consider all possible diffraction patterns given by spiders and segmentations of the mirrors. A similar model has been used also for a smaller 50-meter telescope. Some preliminary results are presented and discussed in Feldt *et al.* (2005).

4. Conclusion

Our simulation software is at present sophisticated and complex, and although it may need some higher order improvements, which will be done in the immediate future, it can be considered almost complete. With numerical simulations we evaluated the capabilities of an IFS instrument to reduce speckle noise. One can achieve a reduction factor of 10^{-3}, which can be considered a good result for the SDI technique. It allows the detection, even with an 8-meter telescope, of a vaste range of faint companions around stars in the solar neighbourhood. Beside detection, in case of high S/N with IFS-SDI we can extract the whole near-infrared low resolution spectrum, for a firts characterization of the planet.

References

Berton, A., Gratton, R.G., Feldt, M. *et al.* 2005, *in preparation*

Burrows, A., Sudarsky, D., & Hubeny, I. 2004, *ApJ*, 609, 407

Claudi, R.U., Turatto, M., Gratton, R.G. *et al.* 2004, in: *Ground-based Instrumentation for Astronomy, Proceeding of the SPIE*, Volume 5492, 1351-1361

Feldt, M., Henning, T., Hippler, S. *et al.* 2004, in: *Exploring the Cosmic Frontier: Astrophysical Instruments for the 21st century, ESO Astrophysics Symposia*, 17-21 May 2004, Berlin, Germany

Feldt, M., Berton, A., Henning, T. *et al.* 2005, in: *Direct Imaging of Exoplanets: Science & Techniques*, October 3-7, Nice, France. (This proceeding.)

Gilmozzi, R. 2004, in *Ground-based Telescopes, Proceeding of the SPIE*, Volume 5489, 1-10

Racine, R., Walker, G.A.H., & Nadeau, D. *et al.* 1999, *PASP*, 111, 587

Direct Imaging of Exoplanets: Science & Techniques
Proceedings IAU Colloquium No. 200, 2005
C. Aime and F. Vakili, eds.
© 2006 International Astronomical Union
doi:10.1017/S1743921306009124

Giant Transiting Planets Observations GITPO

C. Afonso[1], Th. Henning[1], D. Weldrake[1], T. Mazeh[2] and S. Dreizler[3]

[1]Max-Planck Institut für Astronomie, Heidelberg, Königstuhl 17, 69117 Germany
email: afonso@mpia.de

[2]Tel Aviv University, Tel Aviv 69978, Israel
email: mazeh@wise.tau.ac.il

[3]Institut für Astrophysik, Universität Göttingen, D-370077 Göttingen, Germany
email: dreizler@astro.physik.uni-goettingen.de

Abstract. The search for extrasolar planets is nowadays one of the most promising science drivers in Astronomy. The radial velocity technique proved to be successful in planet hunting, harvesting more than a hundred planets to date. In these last recent years, the transit method has come to fruition, with the detection of seven Jupiter-mass extrasolar transiting planets in close-in orbits (<0.05 AU). Currently, the radius of planets can only be determined from transiting planets, representing the principal motivation and strength of this technique. The MPIA is presently building the Large Area Imager (LAIWO) for the 1m telescope in the Wise Observatory, Israel. LAIWO will have a field of view of one square degree. An intensive search for extra-solar planets will be performed with the 1m Wise telescope, together with the 1.2m MONET telescope in Texas. We will monitor three fields at a given time during three years and more than 200 nights per year. We expect several dozens of extra-solar planets.

Keywords. instrumentation: detectors, techniques : image processing, stars: planetary systems

1. Introduction

Extrasolar planets searches have gained major importance in the recent years and revealed itself as one of the most promising science drivers in Astronomy. Unexpected results have led to new clues about the formation and evolution of planets. An example is the detection of massive planets, with Jupiter masses, at short periods $P \leqslant 10$ days, indicating that a migration process took place, after acquiring the bulk of their mass. The radial velocity (RV) technique proved to be successful in planet hunting, harvesting more than a hundred planets to date (Mayor&Queloz (1995); Marcy & Butler (1996)). Recently, the Earth-mass range was attained with the detection of several Neptune-mass planets with masses equal to 7.5, 14, 18 and 21 $M_{\odot}$ (Rivera *et al.* (2004), Santos *et al.* (2004), McArthur *et al.* (2004), Butler *et al.* (2004)).

Although this technique is undoubtly successful, the degeneracy between the planetary mass and the inclination angle of the orbit, only allows to obtain a minimal mass for the planet. Another promising method has come to fruition in the recent years, with the detection of seven Jupiter-size transiting extrasolar planets in close-in orbits (<0.05 AU) : HD209458b (Charbonneau *et al.* (2000)), OGLE-TR-56,OGLE-TR-111, OGLE-TR-113, OGLE-TR-132, OGLE-TR-158 (Udalski *et al.* (2002)) and TrES-1 (Alonso *et al.* (2004)). Planetary transits, in addition to the mass and the orbital radius, yield the radius of the planet and the inclination angle, provided the light curve has a high photometric accuracy, a high time sampling and the radius of the parent star is known or a stellar mass-radius relation is assumed. Currently, the radius of planets can only be determined from transiting planets, representing the principal motivation and strength of the transit technique. A radius measurement is an important quantity, since it allows to constrain the evolutionary and migration history of the planet and to infer its composition and atmosphere through evolutionary models. Furthermore, transit surveys do not pre-select the target stars to be monitored, as radial velocity searches do. In the coming years, several hundred transiting planets are expected to be discovered from the ground, with more than 20 experiments on-going or being planned (see reference for Planets Encyclopedia), as well as from space with missions such as COROT and Kepler (see references).

2. The Method - Planetary Transits

The detection method relies on the temporary drop in the brightness of the parent star harboring the planet. If the planetary system is in a favorable orientation relative to the line of sight (nearly perpendicular to the plane of the sky), then once per orbit the planet passes between its star and the observer, causing an occultation or transit that results in a dip in the light curve (see figure 1). Since the fractional change in brightness δF (or transit depth) is proportional to the stellar surface eclipsed by the planetary disk, the planet's size R_p can be inferred from photometric measurements, provided the stellar radius is known ($\delta F = (\frac{R_p}{R_\star})^2$). For Jupiter-sized planets transiting sun-sized stars, the expected transit depth will be about 1%.

If three or more transits can be measured and confirmed to be due to the same planet, the period P can be determined and hence from Kepler's third law $P = \sqrt{4\pi^2 a^3/GM_\star}$, the orbital radius or semi-major axis a can also be inferred. Once the radius of the planet and the period are known, the inclination can be determined from the shape of the transit light curve, as parameterized by the ratio of the duration of the transit's flat bottom to

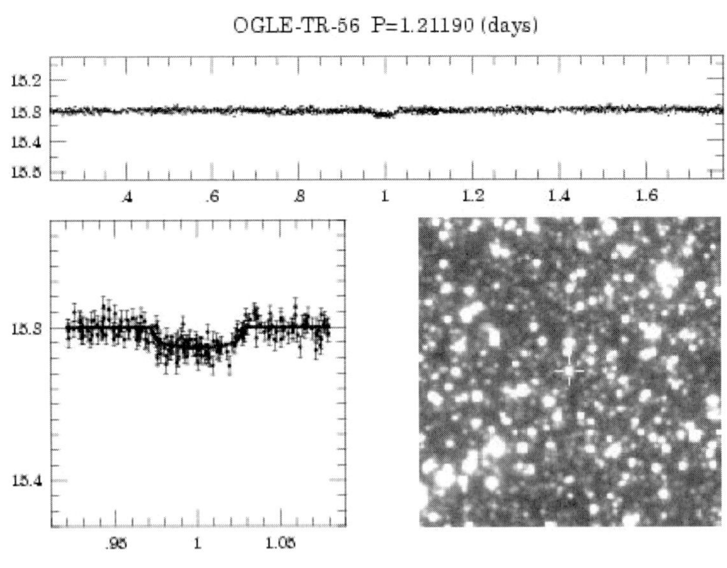

Figure 1. Light curve of OGLE-TR-56, the first planetary system discovered by the transit method. The extra-solar planet has a mass of 0.9 M_{Jup} and a distance of 0.023 AU to its parent star (Udalski *et al.* 2002).

the total transit duration t_T,

$$(\frac{t_{flat}}{t_T})^2 = \frac{[1 - R_p/R_\star]^2 - [(D/R_\star)\cos i]^2}{[1 + R_p/R_\star]^2 - [(D/R_\star)\cos i]^2} \tag{2.1}$$

The total transit duration being equal to

$$t_T = \frac{PR_\star}{\pi a}\sqrt{(1 + \frac{R_p}{R_\star})^2 - (\frac{a}{R_\star}\cos i)^2} \tag{2.2}$$

3. The Project - Giant Transiting Planets Observations - GITPO

The Max-Planck für Astronomie, the University of Tel Aviv and the Sternwarte of Göttingen initiated a transit search program, the Giant Transiting Planets Observations GITPO (see reference), funded by the Max-Planck für Astronomie and the German-Israeli Foundation (GIF). The aim of the project is to detect Jupiter-size extra-solar planets around main-sequence stars with magnitudes down to V=17. Currently, a Large Area Imager for the Wise Observatory (LAIWO) is beeing built at the Max-Planck für Astronomie. LAIWO is composed by an array of four front-side illuminated Lockheed CCDs with 4K×4K pixels, with a pixel size of 15μm (see fig. 2). The camera will be mounted on the 1 m telescope in the Wise Observatory, in the Negev desert, Israel (see fig. 3). The field of view will be one square degree. The observing strategy will consist in the continuous monitoring of three fields at any given time, until 3,000 image are acquired. We anticipate to have more than 200 nights allocated per year during three years, covering a total sky surface of 30 deg^2. Observations will be coordinated with the 1.2m MONET (see reference) telescope located in Texas, USA, operated by the University

Figure 2. The CCD mosaic of the LAIWO camera. The four CCDs as well as the guider CCD at the center are shown.

Figure 3. The Wise Observatory in the Negev desert, 200 km South from Tel Aviv, Israel.

of Göttingen, Germany. The network of these two telescopes will increase the number of measurements, enhancing the expected number of planets. Dozens of transiting extra-solar planets are expected to be found over the three year observation campaign.

References

Alonso, R. *et al.*, *ApJ* 613L, 153
Butler, B. *et al.*, *ApJ* 617, 580B
Charbonneau, D. *et al.*, *ApJ* 529L, 45
COROT, *http://sci.esa.int/home/corot/index.cfm*
Kepler, *http://www.kepler.arc.nasa.gov*
Marcy, G. & Butler, R .P., *ApJ* 464, L147
Mayor, M. & Queloz, D., *Nature* 378, 355
McArthur, B. *et al.*, *ApJ* 614L, 81M
MONET, *http://monet.uni-goettingen.de/cgi-bin/WebObjects/MonetPortal*
Rivera, E. *et al.*, *ApJ* 634, 625
Santos, N. *et al.*, *A&A* 426L, 19S
Planets Encyclopedia, *http://www.obspm.fr/planetes*
Udalski, A. *et al.*, *Acta. Astron.* 52, 115

Direct Imaging of Exoplanets: Science & Techniques
Proceedings IAU Colloquium No. 200, 2005
C. Aime and F. Vakili, eds.
© 2006 International Astronomical Union
doi:10.1017/S1743921306009136

Near-UV to near-IR disk-averaged Earth's reflectance spectra

S. Hamdani[1], L. Arnold[1], C. Foellmi[2], J. Berthier[3], D. Briot[4], P. François[4], P. Riaud[5] and J. Schneider[4]

[1]Observatoire de Haute Provence (CNRS-OAMP), 04870 St Michel l'Observatoire, France
email: hamdani@obs-hp.fr
[2]ESO, Casilla 19001, Santiago 19, Chile
[3]IMCCE, Observatoire de Paris, 77 Avenue Denfert-Rochereau, 75014 Paris, France
[4]Observatoire de Paris-Meudon, 5 place Jules Janssen 92195 Meudon, France
[5]Institut d'Astrophysique et de Géophysique de Liège, Université de Liège, Allée du 6 Août, 4000, Sart-Tilman, Belgium

Abstract. We report 320 to 1020nm disk-averaged Earth reflectance spectra obtained from Moon's Earthshine observations with the EMMI spectrograph on the NTT at ESO La Silla (Chile). The spectral signatures of Earth atmosphere and ground vegetation are observed. A vegetation red-edge of up to 9% is observed on Europe and Africa and ≈2% upon Pacific Ocean. The spectra also show that Earth is a blue planet when Rayleigh scattering dominates, or totally white when the cloud cover is large.

Keywords. Earth, albedo, reflectance, vegetation red-edge, biosignature, biomarker, Earthshine

---

1. Introduction

Since the first measurements of Earth disk-averaged reflectance spectra (Arnold *et al.* 2002, Woolf *et al.* 2002), several attempts have been successful in the same spectral bandwidth (Seager *et al.* 2005, Montañés-Rodriguez *et al.* 2005). Most of these spectra show signatures of Earth atmosphere and ground vegetation. Green vegetation has a much higher reflectivity (factor of 5 typically) in the near-IR than in the visible which produces a sharp feature at $\approx 700nm$, the so-called Vegetation Red-Edge (VRE). Earth disk-averaged reflectance spectrum rises by a few percents at this wavelength. The red part (above 600nm) of the Earth reflectance spectra also shows the presence of O_2 and H_2O, and the bluer part (320nm to 600nm) clearly shows the Huggins and Chapuis ozone (O_3) absorption bands. Spectra show that our planet is blue due to Rayleigh scattering, but also that Earth can be almost perfectly white depending on the importance of the cloud coverage.

2. Data

The observations have been made at the NTT/La Silla telescope (3.5m) on July 24th and September 18th 2004 for the descending Moon and May 31st and June 2nd 2005 for the ascending Moon (table 1). The spectra were obtained with the EMMI spectrograph in the medium dispersion mode in the blue (BLMD mode) and in low dispersion in the red (RILD mode), enabling us to record a complete spectra from 320nm to 1020nm, except for a 20nm gap around 520nm. The spectral resolution is R≅450 in the blue and R≅250 in the red. To record both Earthshine (ES) and sky background spectra

Table 1. Dates of observations

Date	Hour (UT)	Exposure Time (blue)	Exposure Time (red)	MS Phase angle
05/24/2004	23h16(blue)	2×180 sec.	-	117°
09/18/2004	0h15(red)	-	2×100 sec.	137°
05/31/2005	9h30(red) and 10h01(blue)	2×250 sec.	3×120 sec.	101°
06/02/2005	9h25(red) and 9h50(blue)	2×250 sec.	2×120 sec.	126°

simultaneously, EMMI's long slit is oriented East-West on lunar limb (Arnold *et al.* 2002). ES exposures are bracketed by at least two exposures of the bright Moonshine (MS) with the slit oriented North-South. The length of the slit (6-arcmin in blue and 8-arcmin in red modes) allows to sample the Moon spectrum over a large lunar region giving a correct mean of the Moon spectrum. MS spectra are recorded through a neutral density in the blue arm and a diaphragm in the red arm (unfortunately the diaphragm could not be put exactly in a pupil plan, resulting in a strong vignetting - no neutral density was available in the red).

3. Spectra processing method

3.1. *Standard data reduction*

Data reduction is done with dedicated IDL™routines. All spectra are processed for cosmic rays, bias, dark and flat corrections (with dome-flat frames). The sky background spectrum recorded near a ES spectrum is extrapolated and subtracted from the ES spectrum, following the principle of Qiu's method (Qiu *et al.* 2003) used for broad-band photometry. Each ES exposure is bracketed between two MS exposures. To estimate the MS spectrum at the epoch of the ES exposure, the MS is obtained by a mean of the two bracketing MS spectra, weighted with respect to the time elapsed before and after the ES exposure. Each image is binned into an improved S/N ratio spectrum after an accurate correction of the image distortion (caused by the instrument). The color effect of the lunar phase gives an excess of red on the MS spectra. This colour bias is corrected by a 2nd order polynomial least square approximation from Lane's photometry (Lane & Irvine 1973, Kieffer & Stone 2005) normalized at 1020nm. Lane's photometry is available only for phase angle with steps of 10° below 120°, so we used the nearest Lane's data and data for 120° even when Moon phase angle is higher. This correction has a large impact on the overall spectrum slope, i.e. from 22% to 31% between 320nm and 1020nm.

3.2. *Extraction of Earth reflectance (ER)*

When corrected for the Moon's colour phase effect, Earth's reflectance (ER) is given by the ratio (Arnold *et al.* 2002):

$$ER(\lambda) = \frac{ES(\lambda)}{MS(\lambda)}. \tag{3.1}$$

This equation assumes that ES(λ) and MS(λ) are recorded simultaneously. This is the reason why we have to measure MS(λ) before and after ES(λ) to compute a mean MS(λ) to interpolate the difference of Rayleigh scattering due to the changing airmass between the exposures. A quantification of Earth's albedo spectrum can be done in principle by following the procedure used for broad-band albedo measurements (Qiu *et al.* 2003). This requires to measure spectra or brightness only on calibrated areas to properly take into account the Moon's phase function. Our data do not allow to calibrate all fluxes, so

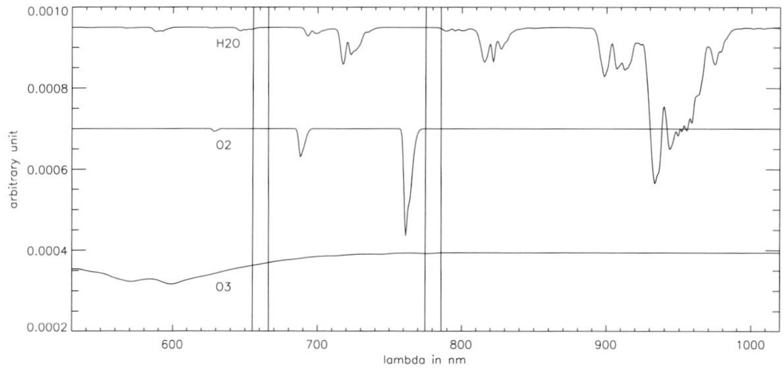

Figure 1. Absorption spectra of main atmospheric components. Vertical lines indicate the spectral bandwidth used to compute the VRE.

our Earth's reflectance spectra are not calibrated. But of course, the shape of the final spectra remains correct and allows for example to identify the Vegetation's Red Edge and measure its correct value.

3.3. *Extraction of surface reflectance (SR) and VRE measurement method*

To detect the vegetation signature, we need to extract Earth's surface reflectance SR(λ) from the ER(λ) which obviously also contains all signatures of the atmosphere. To remove these signatures, ER(λ) is divided by the atmosphere transmittance AT(λ). The AT(λ) spectrum can be obtained from the ratio of two spectra of a calibration star (or MS(λ)) taken at two different airmasses (Arnold *et al.* 2002). It can also be built from spectral databases and adjusted to fit the observed spectrum (Woolf *et al.* 2002). Once SR(λ) is obtained, the Vegetation Red-Edge (VRE) is given by:

$$VRE = \frac{r_I - r_R}{r_R} \tag{3.2}$$

where r_I and r_R are the mean reflectance in the [655:665nm] and [770:780nm] spectral bands. Our ER(λ) spectra are fitted with O_3 spectra from the GOME satellite (Burrows *et al.* 1999), H_2O and O_2 from the MODTRAN model and all convolved to fit our spectral resolution. O_2 and H_2O lines in ER(λ) can be removed quite easily. The removing of the broad Chapuis band of O_3 requires great cares: the band can easily be slightly over-corrected, leading to an apparently smooth Rayleigh scattering but an underestimated VRE (figure 1).

Once the atmospheric lines are suppressed (figure 2), spectra still show the Rayleigh scattering. It is removed by a least squares fit of a $a + \frac{b}{\lambda^4}$ function on two or three continuum zones between 580nm and 650nm (figure 2). The spectrum around 550nm is not used to avoid the bias from a contribution of the vegetation (small bump in reflectance spectrum). The final Rayleigh-corrected Earth's reflectance spectra are obtained after a division by the fitted Rayleigh function. The obtained SR(λ) spectra allow to measure the VRE (figure 3).

4. Results and discussion

Figure 4 shows the ER spectra. During the first observing run, we could not record both blue and red spectra during the same night, but the aim was to record at least a red spectrum of the Earth reflectance showing a minimum of vegetation which was

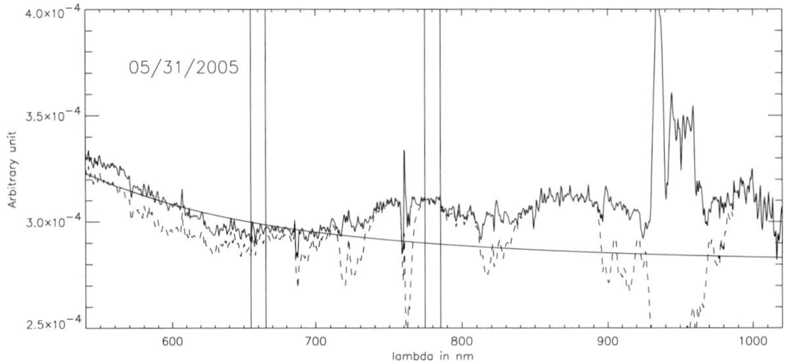

Figure 2. ER(λ) spectrum (doted line) and SR(λ) obtained after removal of atmospheric (also called 'telluric') absorption lines but still showing the Rayleigh scattering. Smooth line is the fitted Rayleigh law. Vertical lines define the spectral domains used for the VRE measurement.

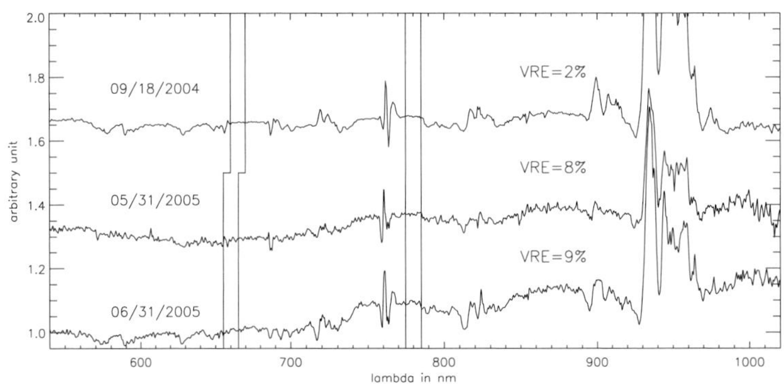

Figure 3. Surface reflectance SR(λ) (i.e. ER(λ) corrected from atmospheric absorption lines of O_2, O_3 and H_2O, and from Rayleigh scattering). Vertical lines define the spectral domains used for the VRE measurement. The plots have been shifted vertically for clarity.

Table 2. Vegetation Red Edge

Date	2004/09/18	2005/05/31	2005/06/02
VRE	2%	8%	9%

the case for the 09/18/2004. The second run allowed to get all spectra from 320 to 1020nm in less than 40 minutes. We have measured the VRE between 2 and 9% (table 2, figure 3).

We observed that the estimation of the VRE is strongly dependent on how both Rayleigh scattering and ozone Chapuis band are estimated in the Earth reflectance (see figure 2). The amount of O_3 is optimized by minimizing the variance between the Surface reflectance and the Rayleigh fitting on identified continuum bands. We noticed that the VRE measured directly from the ER give different values from those measured from SR, pointing out the relevance of telluric lines correction. For example, we have a VRE on ER slightly higher than on SR (10 and 9% respectively) the 06/02/2005, but lower on ER than on SR (5% and 8% respectively) the 05/31/2005.

The suppression of the H_2O absorption lines remains difficult. As shown figure 2, the band between 930 to 960nm is over-corrected while other lines remain under-corrected.

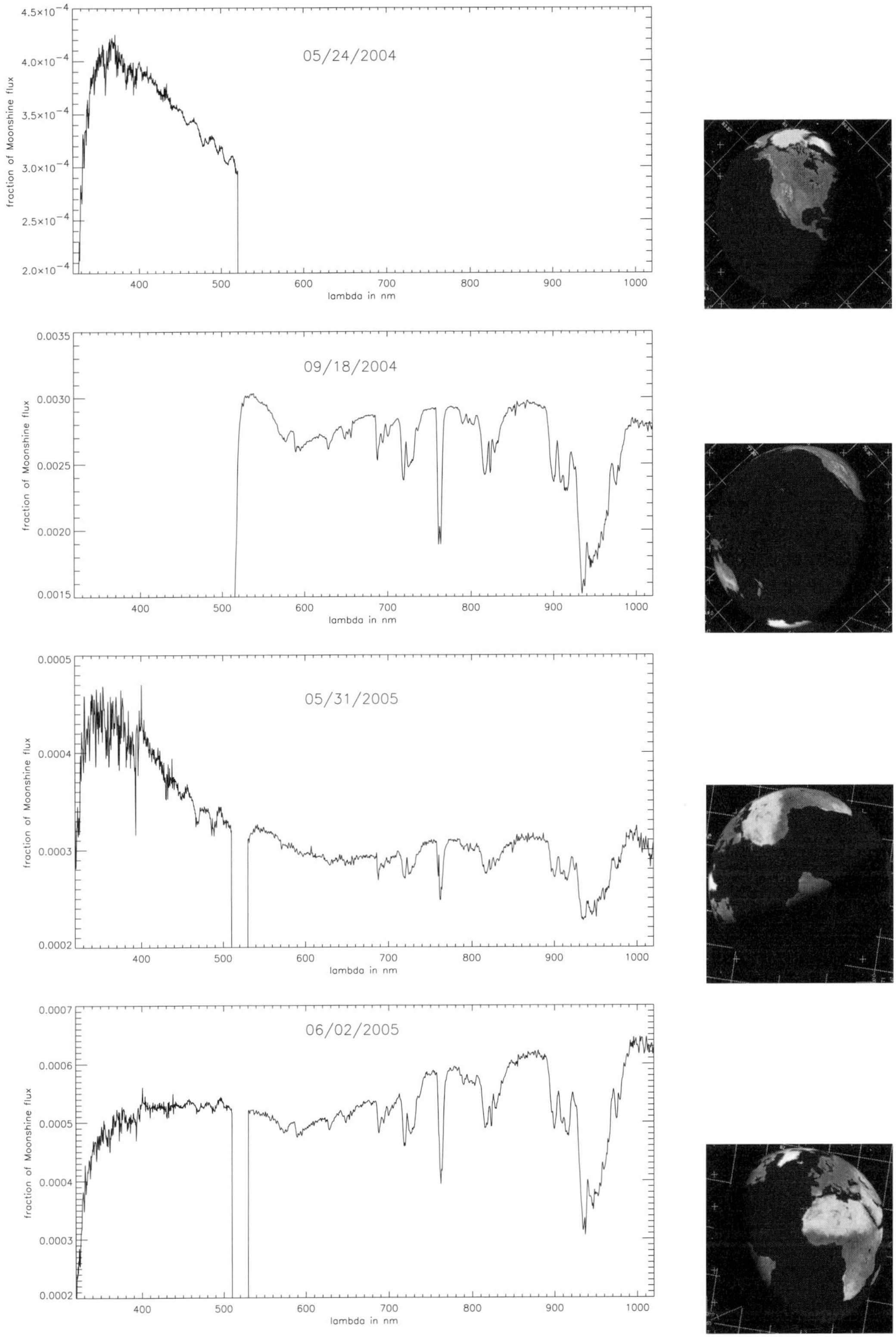

Figure 4. Spectra of the Earth's reflectance in fraction of the Moonshine flux. On the right, the Earth seen from the Moon at the epoch of the observation.

This results in an error in the VRE estimation due to uncorrected (or poorly corrected) H_2O absorption between 555 and 560nm, and also in a reduction of the clean continuum window between 585 and 600nm available to fit the Rayleigh scattering.

To accurately fit the whole spectrum with a Rayleigh scattering law, we can not consider the band between 530 and 560nm as a clean continuum because vegetation reflectance has a small bump in this domain. If this domain is nevertheless taken into account for the Rayleigh fit, the obtained fit is significantly above the blue part of the spectrum. On the other hand if it is not taken into account the fit is slightly below the spectrum between 530 and 560nm, suggesting the presence of vegetation. But measurements are not consistent with the VRE measured at 700nm upon the Pacific Ocean and Africa. Clearly a better model of the atmosphere is needed here. Aerosols scattering following a $\lambda^{-1.5}$ law could help to solve the problem as it increases the blue part of the spectrum, but it is not yet implemented in the code.

The comparison of spectra in figure 4 shows that Rayleigh scattering can be almost absent, pointing out the importance of the cloud coverage. Cloud cover is $\approx$60% over the evergreen forests meaning that the VRE could in principle reach $\approx$20%. But this is unlikely because the cloud cover is always significant above these forest. But the vegetation cover varies in temperate lands with seasons, and simulations or a one-year monitoring should help to determine the VRE seasonal variations and its relevance as a good biomarker.

5. Conclusion

We obtained spectra of the Earth reflectance spanning from 320 to 1020nm for four different nights. Data show significant variations in Rayleigh scattering depending of the cloud cover (Earth 'blue dot' can be white). A spectrum was taken with nearly just the Pacific Ocean while others with a part or the whole Africa and Europe. The Vegetation Red Edge is observed when lands with forests are present, but remains very low otherwise. Measuring the VRE requires great cares in the data reduction and, although measurable, the VRE remains a small feature when compared to O_3, O_2 and water vapor absorption lines. A survey over one year or more with monthly observations would allow to follow VRE seasonal variations and improve our knowledge of the behaviour of this biomarker.

Acknowledgements

S. H. is supported by the Swiss National Science Foundation under grant number PBSK2–107619/1.

References

Arnold, L., Gillet, S., Lardière, O., Riaud, P., & Schneider, J. 2002, *A&A* 352, 231

Burrows, J. P., Dehn, A., Deters, B., Himmelmann, S., Richter, A., Voigt, S., & Orphal, J. 1999, *J. Quantitative Spectr. & Rad. Transfer* 61, 509

Kieffer, H. H., & Stone, T. C. 2005, *AJ* 129, 2887

Lane, A. P. & Irvine, W. M. 1973, *AJ* 78, 267

Montañés-Rodriguez, P., Pallé, E., Goode, P. R., Hickey, J., & Koonin, S. E. 2005, *ApJ* 629, 1175

Qiu, J., Goode, P. R., Pallé, E., Yurchyshyn, V., Hicke, J., Montañés Rodriguez, P., Chu, M.-C., Kolbe, E., Brown, C. T., & Koonin, S. E. 2003, *J. Geophys. Res. (Atm.)* 108, D22, 12

Seager, S., Turner, E. L., Schafer, J., & Ford, E. B. 2005, *Astrobiology* 5, 372

Woolf, N. J., Smith, P. S., Traub, W. A., & Jucks, K. W. 2002, *ApJ* 574, 430

Direct Imaging of Exoplanets: Science & Techniques
Proceedings IAU Colloquium No. 200, 2005
A.C. Editor, B.D. Editor and C.E. Editor, eds.
© 2006 International Astronomical Union
doi:10.1017/S1743921306009148

Target star catalogue for Darwin: Nearby Habitable Star Systems

L.Kaltenegger,[12]†, C.Eiroa[23], A. Stankov[2], and M. Fridlund[2]

[1]Harvard-Smithsonian Center for Astrophysics, 60 Garden St, Cambridge, MA02138, USA
email:lkaltenegger@cfa.harvard.edu

[2]European Space Agency ESTEC, P.O.Box 299, NL-2200AG Noordwijk, The Netherlands
email:Malcolm.Fridlund@esa.int, astankov@rssd.esa.int

[3]Universidad Autonoma de Madrid, Cantoblanco, 28049 Madrid, Spain
email:carlos.eiroa@uam.es

Abstract. In order to evaluate and develop mission concepts for a search for Terrestrial Exoplanets, the European Space Agency has prepared a list of potential target systems. In this paper we present the criteria for selecting potential target stars suitable for the search for Earth like planets. We also establish the habitable zone for the stellar systems. Planets found within these zones would be potentially able to host complex life forms. The Darwin target star catalogue was created from the Hipparcos catalogue by examining the information on distance, spectral classification, multiplicity and stellar variability. Additional information on the targets stars including spectral type, metallicity, X-ray luminosity, rotation and Strömgren photometry from several other catalogues are included to determine the target star sample. Design constraints of the mission are added to derive a final target star catalogue of 262 target stars excluding and 447 target stars including M stars for the Darwin mission. We also discuss the completeness of sample for different classes of stars.

Keywords. extrasolar planets, interferometers, catalogs, planetary systems, Earth, planets

1. Introduction

Among current scientific endeavours, the question if life exists only the Earth or elsewhere in the Universe is arguably one of the most important. Many issues remain to be solved before the "big question" of life in the Universe as contrasted to life in our Solar System can be addressed. One important constraint is: what and where is the sample of stars around which we find planets suitable for life? To model realistic observation scenarios the luminosity, distance and radius of the star and the corresponding habitable zone (HZ) for an Earth-like planet has to be taken into account. The optimization and success of the mission depends critically on those parameters. Analysis of the planetary light requires that the stellar light is suppressed to a high degree. This is done in the IR by a technique called nulling interferometry. Briefly, this implies that achromatic phase shifts are applied to the beams collected by individual telescopes before recombination such that the on-axis light, i.e. the stellar light, is cancelled by destructive interference, while the much weaker planetary light emitted at a certain small, off axis angle interferes constructively. In the visible wavelength range interferometry is currently impractical, NASA′s Terrestrial Planet Finder (TPF) concept in the visible is implemented as a very large coronagraph. Detailed simulations critically depend on correct data on the target stars.

† Present address: CfA, 60 Garden st, MS20, Cambridge, MA02138, USA

2. Target star catalogue

The sample of nearby stars is limited. As a preparatory activity for the future missions, as well as the required ground based programs, both NASA and ESA are preparing input catalogues. It is expected that consortia will form across the whole world, which will start determining the properties of the target stars. In preparation for the Darwin mission a catalogue of stellar systems that are potentially candidates has been prepared (see Kaltenegger *et al.* 2005, Eiroa *et al.* 2003, Kaltenegger 2004). A preliminary list of target stars from the Gliese star catalogue prepared by Ollivier and Leger had been used earlier for calculations for the Darwin feasibility study.

The Darwin target catalogue is based on the Hipparcos catalogue. Using various existing data archives we established a catalogue of well characterized stars that will be used as the Darwin target star list (see Kaltenegger *et al.* 2005). We use distance, spectral classification, multiplicity and stellar variability and adding information including spectral type, metallicity, X-ray luminosity, rotation as well as Strömgren photometry from several other catalogues to establis a preliminary target star sample. For many of the nearby late K and M stars data on multiplicity, rotation, metallicity etc are still missing, thus modifications of the Darwin catalogue will be implemented as such data becomes available. Further developing this sample, and taking several different configurations of the Darwin interferometer into account we have now a final sample, extending out to a distance of 25pc from Earth. As what concerns F, G and early K-type main sequence stars it is considered complete. Later stellar types, which should be numerous, are conspicuous in their absence. This is not expected to impact the Darwin or TPF missions, since ESA′s GAIA mission will fly in 2012 and it is expected to provide parallaxes down to 20 microarcseconds for objects brighter than R-magnitude 22 within $1\frac{1}{2}$ years, thus completing our knowledge of nearby stars out to at least 25 pc and for essentially all of class M.

Four selection criteria were adopted to first generate the "Darwin all sky target catalogue" that does not take the instrument design into account and assumes access to the whole sky. An additional instrument design criterion was added v) based on the instrument design leading to the "Darwin prime target star catalogue": i) FGKM stars within a distance of 25 pc, ii) main sequence stars nature based on the $B - V$ index: $(0.25 < B - V < 1.93)$ and their assigned luminosity class iii) stars with magnitudes smaller than 12 in V, iv) single stars, v) stars within a cone of aperture ± 45 degree of the ecliptic plane.

The Darwin all sky target catalogue consists of 628 including and 387 stars excluding M stars. The current analysis results in a Darwin prime target star list of 447 target stars including and 262 excluding M stars consistent with the selection criteria within 25 pc including instrument limitations see figure 1 and figure 2.

From the color magnitude diagram shown in figure 1, it is obvious that most target stars are still on the main sequence while a few have already evolved off it. The different colors show the different stellar types.

The interferometric systems suggested for Darwin and TPF-I operate in the mid-IR, the coronagraph suggested for TPF-C in the visible. For the former it is thus the thermal emission emanating from the planet that is detected and analysed while for the latter the reflected stellar flux is measured. This means if you want to observe earth-like planets in the HZ around a given star, the flux in the IR will to first order be constant for a given planetary size. The suppression of the primary′s light will, on the other hand, be progressively easier for later spectral types. Surprisingly enough, it may thus be easier to detect a habitable Earth around an M dwarf than around something more akin to our

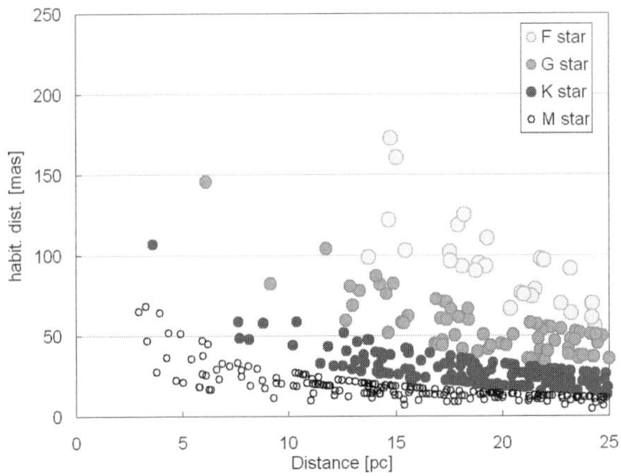

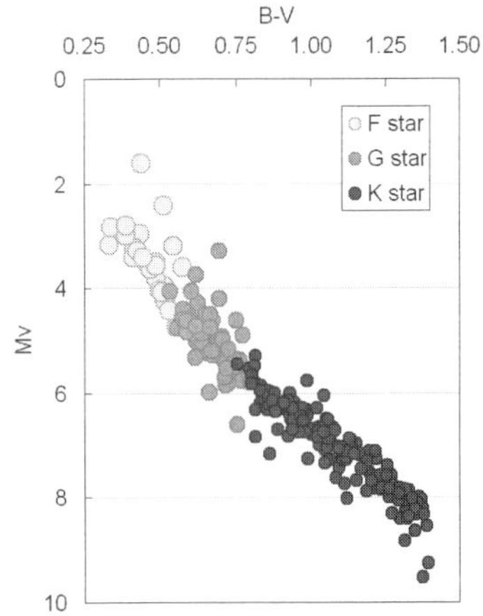

Figure 1. Distance of the HZ in mas (left) Color-magnitude diagram for the Darwin prime target stars (right).

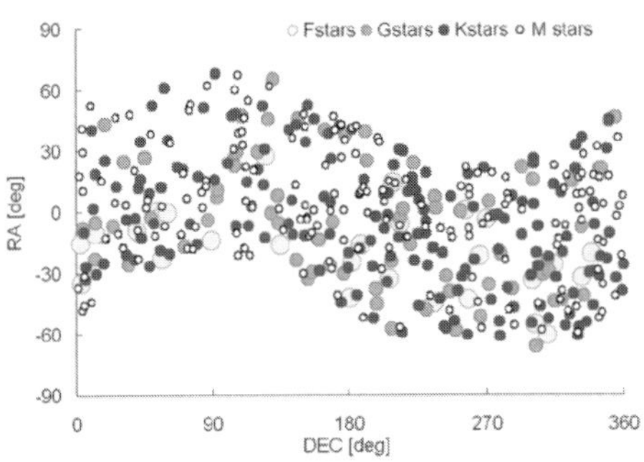

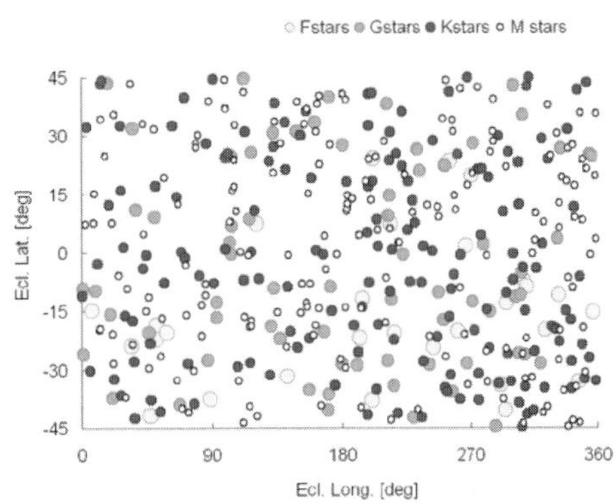

Figure 2. Position for the 45-degree cone Darwin target stars in RA, DEC (left) and ecliptic coordinates(right).

own Sun. The HZ moves closer to the star for later stellar systems thus the interferometric systems has to be adapted to each individual target system for optimum performance, implemented as a free flyer array. The baseline of the interferometers has to increase to resolve M star planetary-systems at larger distances, a constraint that is taken into account for the M target systems at largest distance in the target star catalogue.

The target star list presented is used for the trade off studies of different mission architectures for the proposed Darwin mission (see Kaltenegger & Karlsson (2004), Hartog, Absil, Kaltenegger *et al.* (2004) for the TPF-I architecture see Dubovitsky & Lay (2004)). Limitations imposed by the current instrument design are e.g. the 45-degree observation cone in anti-sun direction, a minimum distance to a close binary companion that cannot be simultaneously nulled while searching for a planet and a minimum observation time of the star in the field of view of the instrument. The 45-degree cone in the selection criteria is given by constraints of the sunshields on each free flyer resulting in a sky-coverage of 70.7%. A new concept is under investigation that would allow the instrument to cover 95% of the whole sky. Considering this we included statistics for an all sky survey as well as for the Darwin 45-degree cone concept. The limitation to a ±45-degree cone in anti-sun direction places no severe constraint on the mission or limits its performance.

2.1. *Completeness of the catalogue*

From the catalogues, some important conclusions can be immediately drawn. Within 25pc, we find about 700 single stars, with well defined Hipparcos distances (see Kaltenegger *et al.* 2005). Of these about half are F, G and K type objects, while the remainder consists exclusively of M dwarfs. The F and G type stars known within a distance of 25pc can be considered to be a complete sample. The same is not true for the later spectral types, especially for the M-dwarfs. As one progresses more than about 10pc from the Earth, the sample of known M-dwarfs dwindle rapidly. The incompleteness is easily seen in that between 10 and 20pc we find 14 F, 35 G, 87 K and 140 M single stars in the Darwin prime target star sample, while for the volume between 20 and 25pc the corresponding numbers are 15, 34, 77 and 45. In the all sky target list, the lack in M stars at further distance can be seen even more severely. Naturally G stars will turn out to be prime targets as they are similar to our sun and the likelihood that under similar conditions, planets with similar life-forms showing similar atmospheric features evolve, seems high. Including other stellar types provides the unique opportunity to characterize planets around different host stars. Extrasolar Giant Planets have been found around different stellar types thus the target sample should reflect those stellar classes. It is essential that we have a certain number of targets stars to derive conclusions for detections or non-detections. Even so the frequency of extra-solar terrestrial planets can only be guessed. We believe that a sample has to consist of a minimum of 100 to 150 stars to be able to derive conclusions.

3. Conclusions

In preparation of the Darwin mission we established a catalogue of stellar systems of nearby stars that are potentially habitable to complex life. The list was created from the Hipparcos catalogue by examining the information on distance, stellar variability, multiplicity, location and spectral classification. The Darwin all sky target catalogue consists of 628 target stars including and 387 target stars excluding M stars, reflecting an alternative mission concept that would allow access to the whole sky. Design constraints of the current architecture of the Darwin mission limits the sky access to ±45-degree in ant-isun direction. Applying these design constraints we derive a prime Darwin target star sample of 262 target stars excluding and 447 including M stars within 25 pc as potential targets for the search for earth-like planets.

References

Dubovitsky, S. & Lay, O.P. 2004, *SPIE* 5491
Eiroa, C., Fridlund, M., & Kaltenegger, L. 2003, *ESA-SP* 539
Hartog, R., Absil, O, Kaltenegger, L., Karlsson, A., & DArcio, L. 2004, *SPIE* 5491
Kaltenegger, L. 2004, PhD thesis, Karl-Franzens University Graz, Austria
Kaltenegger, L & Karlsson A. 2004, *SPIE* 5491
Kaltenegger, L, Eiroa, C., Stankov, A., & Fridlund, M. 2005, *in prep*
Karlsson, A.L., Wallner, O., & Perdigues Armengol, J. M 2004 *SPIE* 5491
Kasting, J.F., Whitmire, D.P., & Reynolds, R.T. 1993 *Icarus* 101,108

Direct Imaging of Exoplanets: Science & Techniques
Proceedings IAU Colloquium No. 200, 2005
C. Aime and F. Vakili, eds.
© 2006 International Astronomical Union
doi:10.1017/S174392130600915X

Evolution of the Habitable Zone and Search for Life Around Red Giant Stars, Part I: Interest of the Study.

B. Lopez W. C. Danchi[2], J.-L. Menut[1], J. Rajagopal[2,3], J. Schneider[4], A. Belu[5], and R. Barry[2,6]

[1]Observatoire de la Cote d'Azur, BP 4229, 06304 Nice Cedex 4, France

[2]NASA Goddard Space Flight Center, Exoplanets and Stellar Astrophysics Lab., Code 667, Greenbelt, MD 20771, USA

[3]University of Maryland, Astronomy Department, College Park, MD 20742, USA

[4]Observatoire de Paris, Paris, France

[5]University of Nice - Sophia Antipolis, LUAN, Parc Valrose, 06108 Nice Cedex 2, France

[6]The Johns Hopkins University, Physics and Astronomy Dept., Baltimore, MD 21218, USA

Abstract. We discuss the evolution of the habitable zone around low mass and intermediate mass stars as they evolve off the main sequence. This work shows that this class of stars should be included in the search for life because if planets could be found in their habitable zones, they will allow to test different hypothesis concerning the conditions of life formation. For instance if these planets show traces for life, it is possible to empirically determine a lower limit to the timescale for the formation of life. This time scale is not well determined from the Earth's study, as life formation initially occurred during a period of heavy bombardment from comets and asteroids during the formation of the solar system. We will summarize (to the best of our knowledge) some of the recent results concerning the early signs of life on Earth. We will present some of our work in progress in which we evaluate the effects of stellar evolution on the habitable zone changes.

Keywords. instrumentation: planets and satellites: general, high angular resolution, techniques: interferometric, space vehicles: instruments.

1. Introduction

Low and intermediate mass stars with masses ranging from 0.5 to ~8 $M_\odot$ undergo changes in temperature and luminosity during their evolution. For example, the luminosity of the Sun, which is located on the Main Sequence of the Hertzprung-Russell diagram, currently is 30 percent greater than its zero-age (initial) luminosity (Gough 1981). Other major changes take place after the Sun evolves off the Main Sequence. Along the Red Giant Branch (RGB) the radius of a 1 $M_\odot$ star progressively increases up to ~10 $R_\odot$, and the luminosity increases up to about $10^{1.5}$ $L_\odot$ (Maeder & Meynet 1988). After leaving the RGB a star evolves onto the Asymptotic Giant Branch (AGB). At this stage in its evolution, its radius increases up to a few hundred times $R_\odot$ and the luminosity up to 10^4 $L_\odot$. The lifetime of a 1 $M_\odot$ star, from the moment it leaves the main sequence until the end of the RGB, is of about 2.5 $\times 10^9$ yr, which can be compared to $\sim 10 \times 10^9$ yr spent on the main sequence. The evolution of a star is accompanied by mass loss processes. During the post-main sequence evolution of a 1 $M_\odot$ star nearly 0.4 $M_\odot$ will be lost to the interstellar medium before it ends its life as a white dwarf (Weidemann 1987).

The habitable zone gradually moves outward after a star leaves the main sequence. Depending on the length of time that is necessary for life to form, we show that it is possible that life may exist or develop on a planet that is located within a range of 2-9 AU from a 1 $M_{\odot}$ star, during the first stages of its post main-sequence evolution and up to 22 AU from the star during the second stage after the core He flash.

Future space missions, such as TPF (Beichman, Woolf, & Lindensmith 1999) and Darwin (Fridlund & Gondoin 2003), that focus on searches of signatures of life on extrasolar planets primarily around main-sequence stars, should also consider the environments of sub-giant and red giant stars. Pathfinder missions such as EPIC (Clampin *et al.* 2001), Eclipse (Trauger *et al.* 2003), and FKSI (Danchi *et al.* 2003a, 2003b) may also be able to observe planets around red giant stars.

Based on a list of selected objects, we show there are adequate numbers of such stars in the Solar neigborhood and that preliminary analysis demonstrates that planets around such stars could be detected without any modifications to present architectural concepts for the interferometer version of TPF and Darwin (see our part II paper, Danchi *et al.* 2006 presented at this conference).

2. The evolution of the habitable zone

From a biological point of view and given the limits of our present knowledge, the existence of life is strongly associated with the presence of liquid water (Brack 1993). Several evaluations of the inner and outer limits of the habitable zone have been proposed, as discussed in the review of Forget (1998). A conservative estimate, minimizing the width of the habitable zone (Forget 1998), assumes that around our own Sun the habitable zone currently ranges from 0.95 to 1.67 AU. The inner limit of this conservative estimate is set by an atmospheric water loss effect occurring in the stratosphere of a planet (Kasting *et al.* 1993, Kasting 1998). The outer limit is set by the lowest temperature at which the liquid/solid phase change of water occurs. The estimate of this outer limit assumes the existence of a greenhouse effect involving CO_2 and H_2O gas (Kasting *et al.* 1993). A less conservatively defined "habitable zone" extends the outer edge of the limit of the habitable zone to as large as 2.4 AU (Mischna *et al.* 2000).

A convenient way of estimating the inner and outer limit of the habitable zone at different stages of stellar evolution is to assume that the planet behaves as a gray-body with albedo, A, and with perfect heat conductivity (the temperature is uniform over the planet's surface). Its radiative equilibrium temperature, T_p, is then determined by:

$$T_p = [(1 - A)L_\star/(16\pi\sigma d^2)]^{0.25} \tag{2.1}$$

where $L_\star$ is the luminosity of the star, σ is the Stefan-Boltzmann constant, and d is the distance from the center of the star to the planet. Typically a value of approximately 0.2 is chosen for the albedo to be representative of an earth-like planet. For these calculations we determine the temperature at the limits of the habitable zone, and from this we calculate the inner and outer distances for a given star.

For the conservative limit, we assume an effective equilibrium temperature of 269 K and 203 K, which for the Sun gives inner and outer limits of 0.95 and 1.37 AU, respectively. For the less conservative limit, the equilibrium temperature is 169 K, so that the outer limit is extended to 2.4 AU. The three equilibrium temperature extrema, defined as the inner and the 2 possible outer limits, can be used to derive the approximate location and width of the habitable zone around a star at different stages of its evolution.

Models describing the stellar evolution are proposed by Maeder and Meynet (1988). For intermediate mass stars having masses from 2 to 5 $M_{\odot}$ the evolution is followed up

Table 1. Lifetimes (Gyr) of stages of stellar evolution as a function of initial stellar mass

Mass ($M_\odot$)	Main Sequence	Subgiant	First Red Giant	Core He Burning
1.0	7.41	2.63	1.45	0.95
1.5	1.72	0.41	0.18	0.26
2.0	0.67	0.11	0.04	0.10

to the end of the early asymptotic giant branch. For stars with masses less than 1.7 $M_\odot$, the evolution is calculated up the He-flash occuring at the end of the RGB part of the HR diagram (Iben 1967).

Figure 1(a) displays the time evolution of the inner and outer limits defining the habitable zone assuming the initial mass of the star is 1 $M_\odot$. The solid curve represents the evolution of the inner limit of the habitable zone, while the dashed curve depicts the conservative outer limit, and the dotted curve represents the less conservative outer limit.

During the course of stellar evolution, the habitable zone is a shell sweeping progressively outward over a wide range of distances from the star as can be seen in Figure 1(c). The dashed curve represents the duration of the habitable zone for the conservative assumptions for the inner and outer limits of the habitable zone. The dotted curve assumes the same inner limit of the habitable zone but the less conservative outer limit is used. Figure 1(b) displays the evolution of the stellar radius (in units of Solar radii). Immediately after the star leaves the main sequence, the habitable zone progressively moves to 2 AU (Figure 1(c)). The duration of the transit at this location is approximately 10^9 years. A bump/plateau is observed in the curve up to 9 AU (for the conservative limits) and up to 13 AU (for the less conservative limits), where the duration of habitable conditions lasts from a few to several 10^8 years. At 10 AU the duration is smaller, around 10^8 years. At 15 AU from the star, the duration of habitable conditions lasts more than 10^7 years, and at the largest distances the duration gradually decreases. Thus the evolution of the habitable zone at distances up to 25 AU has been considered. During this time period, a 1 $M_\odot$ star has not yet reached the AGB.

Although individual masses of red giant stars are not well known in general, and have not been directly determined except in a few cases, Neckel (1975) has shown theoretically that the peak of the red giant mass distribution occurs at progenitor masses of about 1.1 $M_\odot$. An empirical estimate of red giant masses determined by Scalo, Dominy, and Pumphrey (1978) is in agreement with the theoretical estimate, with values 0.8 – 1.2 $M_\odot$. This is fortunate since stars selected in a magnitude-limited or distance-limited survey will therefore mostly likely be low mass stars that have long habitable zone lifetimes similar to those calculated in Fig. 1.

Also, it is important to note that the calculations in this paper are focused on the evolution of the habitable zone from the time period after the star evolves off the main sequence up the first ascent along the Red Giant Branch during which the star is burning hydrogen in a shell around a growing He core, i.e., only to the point of the helium flash. After the helium flash, there is an additional long-lived phase of quiescent core helium burning with nearly constant luminosity.

Scalo and Miller (1979) developed interpolation formulae for the lifetime of a star at its various stages of evolution: main sequence, subgiant, first red giant, and first and second helium burning phases, as a function of initial stellar mass. For completeness we display the lifetimes computed from their formulae for stellar masses of 1.0, 1.5, and 2.0 $M_\odot$, which we display in Table 1.

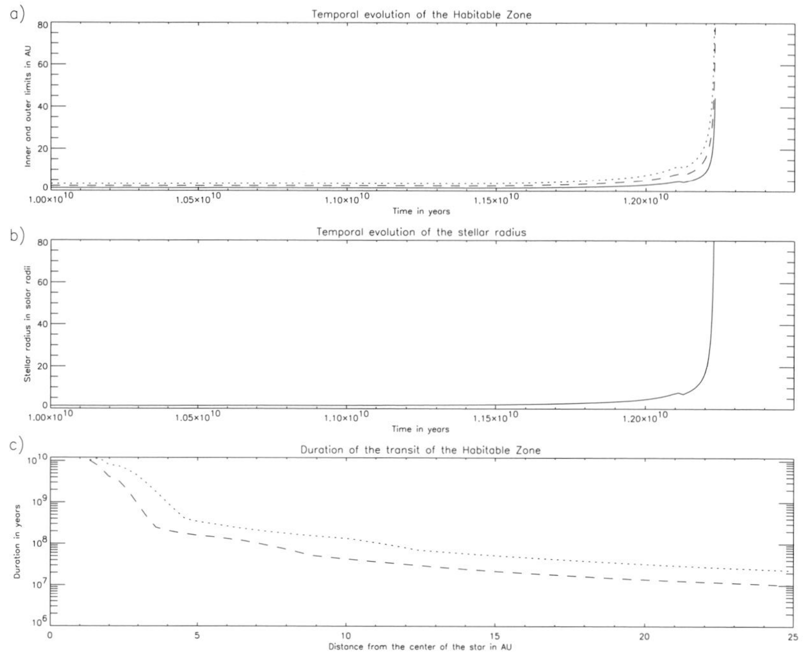

Figure 1. a) The evolution of inner and outer limit of the habitable zone around a post-main sequence star of 1.0 $M_\odot$. The time scale displayed is between 12.5 and 12.8 Gyrs. b) The corresponding evolution of the stellar radius (from Maeder and Meynet 1988). c) The duration of the transit of the habitable zone represented at different distances from the star.

For a 1.0 $M_\odot$ star, core helium burning provides a nearly constant luminosity phase, adding an additional 10^9 years to the duration of the habitable zone for low mass stars, with substantial increases to the lifetime of the habitable zone for higher mass stars as well. The luminosity during the core helium burning phase is approximately 2.3×10^2 $L_\odot$, 9.3×10^2 $L_\odot$, and 2.3×10^2 $L_\odot$, for 1.0, 1.5, and 2.0 $M_\odot$ stars, respectively (Kozlowski & Paczynski, 1975). Thus an additional period of habitability exists for the region between 7 and 22 AU for a 1 $M_\odot$ star during core helium burning.

We summarize our analysis of the evolution of the habitable zone by noting that there are two opportunities for life in the extended habitable zones around evolved stars. Initially life has an opportunity as the star slowly evolves off the main sequence. After the first ascent along the RGB and the He core flash, there is a second chance during the core He burning phase. This latter phase is particularly interesting because of the nearly constant luminosity of the star under these conditions.

3. The possibility of the existence of life around Red Giant Stars

The Earth is known to have formed about 4.55 Gyr ago. Substantial evidence exists that life on Earth occurred at least as far as 3.85 Gyr ago rests on the basis of isotopic ratios of carbon residues within grains of apatite (Mojzsis *et al.* 1997, Holland 1997).

Rocks older than 3.60 Gyr are all metamorphosed at a high grade and most are strongly deformed, thus morphological fossils if originally present are unlikely to be preserved (Rosing 1999). The oldest purported fossils of microorganisms have an age of 3.50 Gyr and may represent photosynthetic cyanobacteria (Schopf 1993, 1994). The findings and interpretation of this work was recently debated (Brasier *et al.* 2002, Dalton 2002).

Based on these considerations, it is possible for life to have existed at least as long ago as 3.85 Gyr, but this is uncertain because of the imprecise knowledge of the ages of the rocks, and ambiguous interpretations of the isotopic $^{12}C/^{13}C$ ratios. Thus, for our discussion, we consider three time scales for life to form, a conservative estimate of 1 Gyr, and a less conservative one of 0.5 Gyr, and an optimistic one of 0.1 Gyr. Recent analyses of isotopic ^{13}C ratios on micron-scale mineralized tubes in Archean pillow lavas by Furnes *et al.* (2004) suggest that microbial life colonized these oceanic volcanic rocks soon after they were formed approximately 3.5 Gyr ago. This provides some further support for the optimistic time scales of a few hundred million years or less for life formation.

Dating of terrestrial fossils reveals that life had evolved on Earth by the end of the period of heavy bombardment. Evidence for the bombardment comes from theoretical studies of planetary formation and directly from lunar cratering. Heavy bombardment may have helped set the stage for the terrestrial origin of life by delivering key biogenic elements (H, C, N) to the Earth's surface (Chyba *et al.* 1994). For its first 800 Myr, the Earth's surface would have been subjected to frequent impacts, possibly delaying the development and spread of life (Maher and Stevenson 1988).

Due to the rapid evolution of a star once it has left the main sequence, the transit time of the habitable zone shortens. For an evolved 1 $M_\odot$ star, the transit of the habitable zone crossing a planet at 2 AU of distance lasts slightly more than 10^9 years, a time scale clearly larger than the time which has been required for the emergence of life on Earth.

At about 5 AU of distance, the duration of habitable conditions lasts from 10^8 years to several 10^8 years. It is about one order of magnitude shorter than the time estimated for occurrence of life on Earth, however this time scale may be long enough for life to develop. It represents an interesting test case for future observations.

Effects of the stellar mass loss phenomena and tidal dissipation between the planets and the red giant star are not negligible in the dynamics of the orbits (Rasio *et al.* 1996). When a Sun like star reaches a radius of about 100 $R_\odot$, only a little more than 10% of the initial mass of the star has been lost, and the effect of this mass loss among the dynamical changes of the planet orbits (for planets more distant than 1 AU not yet subject to tidal effects) is to increase by 10% the semi-major axis. It is interesting to note that the increase of the semi-major axis, if occuring for a planet during a time when it is in the habitable zone (itself expanding), will tend to increase the time duration of habitable conditions for the planet.

4. Conclusions

We have estimated the time transit of the habitable zone around evolved stars of initial mass 1.0, 1.5 and 2.0 $M_\odot$ from the time period after which a star moves off the main sequence until it reaches the point of the He flash (see for more details Lopez *et al.* 2005). The temporal transit of the habitable zone is estimated to be of the order of several 10^9 years for a planet at 2 AU and around 10^8 years for a planet at 9 AU. Furthermore, after the star ascends the Red Giant Branch for the first time and undergoes the He flash, there is an additional long stable period of core He burning $\sim 10^9$ years during which the habitable zone ranges from 7 to 22 AU from the star.

We have shown that the evolution of the habitable zone as stars move off the main sequence and onto the subgiant and later the red giant branches of the HR diagram provide a new opportunity for the search for Earth-like planets and life. For these stars life has an opportunity to evolve under conditions with much less bombardment than existed in the early Solar system. We also show (Danchi *et al.* 2006 this proceeding) there are adequate numbers of such stars in the Solar neigborhood, and that preliminary analysis demonstrates that planets around such stars can detected without any modifications to present architectural concepts for the interferometer version of TPF and Darwin.

Acknowledgements

We acknowledge support from : CNRS; GSFC Internal Research and Development and Director's Discretionary Funds; a Michelson Fellowship and a National Research Council (NRC) Fellowship.

References

Beichman, C., Woolf, N.J. & Lindensmith, C., ed. 1999 (JPL Publ. 99-3; Pasadena: Caltech)
Brack A. 1993, 'Origine of Life', 3, 10
Brasier, M.D. *et al.* 2002, LPI, 33, 1614
Chyba C.F., Owen T.C. & Ip W.-H., 1994, 'Hazards due to Comets and Asteroids', T. Gehrels Editor, The University of Arizona Press
Clampin, M., Ford, H.C., Illingworth, G. & Petro, L. 2001, *BAAS*, 33, 1365
Dalton, J.B. 2002, LPI, 33, 1555
Danchi W.C., Deming D., Kuchner M.J. & Seager S. 2003a, ApJ, 597, L57
Danchi, W.C., *et al.* 2003b, in the Proceedings "Towards Other Earths: Darwin/TPF and the Search for Extrasolar Terrestrial Planets," Heidelberg, Germany, 22-25 April 2003, ESA Publication SP-539, p. 83
Forget, F. 1998, *Earth, Moon, Planets* 81, 59
Fridlund, C.V.M. & Gondoin, P. 2003, *Proc. SPIE*, 278, 1273
Furnes, H., Banarjee, N.R., Muelenbachs, K., Staudigel, H. & de Wit M. 2004, Science, 304, 578
Gough D.O. 1981, Solar Physics 74, 21
Holland H.D. 1997, Science 275, 38
Iben I. 1967, ARAA 5, 571
Kasting, J., Whitmire, D.P., & Reynolds, R.T. 1993, *Icarus*, 101, 108
Kasting J.F. 1998, AAS 193, 5003
Lopez, B., Schneider, J., & Danchi, W.C. 2005, *ApJ*, 627, 974
Maeder, A. & Meynet, G. 1988, *A&AS*, 76, 411
Maher K.A. & Stevenson, D.J., 1988, Nature 331, 612.
Mischna, M.A., *et al.* 2000 *Icarus*, 145, 546
Neckel, H. 1975, A&A, 42, 379.
Rasio, F.A., Tout, C.A., Lubow, S.H. & Livio, M. 1996, ApJ 470, 118
Rosing M.T., 1999, Science 283, 674
Scalo, J.M., Dominy, J.F., & Pumphrey, W.A. 1978, ApJ, 221, 616
Scalo, J.M. & Miller, J.E. 1979, ApJ, 233, 569
Schopf, J.W. 1993, Science, 260, 640
Schopf, J.W. 1994, Proc. Natl. Acad. Sci. USA 91, 6735
Schröder, K.-P. & Sedlmayr, E., 2001, 366, 913
Trauger, J., *et al.* 2003, *Proc. SPIE*, 4854, 116
Weidemann, J. 1987, *A&A*, 188, 74

Direct Imaging of Exoplanets: Science & Techniques
Proceedings IAU Colloquium No. 200, 2005
C. Aime and F. Vakili, eds.
© 2006 International Astronomical Union
doi:10.1017/S1743921306009161

Gravitational instability and the formation of giant planets by core accretion

K. Kornet[1], S. Wolf[1] and M. Różyczka[2]

[1]Max Planck Institute for Astronomy, Königstuhl 17, 69117 Heidelberg, Germany
email: {kornet,swolf}@mpia-hd.mpg.de

[2]Nicolaus Copernicus Astronomical Center, Bartycka 18, Warsaw, 00-716, Poland
email: mnr@camk.edu.pl

Abstract. We investigate the problem of giant planet formation around stars with various masses based on the core accretion/gas capture model. At first, we follow the evolution of gas and solids from the moment when all solids are in the form of small grains to the stage when most of them have reached planetesimal size. We show that the surface density of a planetesimal swarm tends to be higher around less massive stars. Subsequently, we derive the minimum surface density of the planetesimal swarm required for the formation of giant planets, both in a numerical and in an approximate analytical approach. We combine these results by calculating a set of representative disk models, characterized by different masses, sizes, and metallicities. This allows us to quantify the probability of each individual disk model to form giant planets. Furthermore, we take the fact into account, that in some of these models, the outer regions of the disks become gravitationally unstable.

Keywords. protoplanetary disk, planetary systems: formation

1. Introduction

Radial velocity surveys led to discovery of over 150 extrasolar planets around main sequence stars Marcy *et al.* (2005), Mayor *et al.* (2004). Those surveys have been most successful in the case of G dwarf stars. Due to the constant progress in the detection techniques, the observational programs recently started to include also stars with lower masses on a larger scale, namely M dwarfs. Moreover, some of these surveys are now particularly dedicated to lower-mass stars Endl *et al.* (2003), Bonfils *et al.* (2004).

From the theoretical point of view the problem of giant planet formation around M dwarfs was studied recently by Laughlin *et al.* (2004). They addressed it within the core accretion - gas capture model (CAGCM). They conclude that M dwarfs have a limited ability to form Jupiter-mass planets. This is a direct consequence of their assumption that the surface density of the planetesimal swarm out of which planetary cores are formed scales linearly with the mass of the central star. However, the solid component of the protoplanetary disk evolves in a different way than the gaseous component Weidenschilling & Cuzzi (1993). Due to the gas drag a significant redistribution of solids takes place, and in the inner disk their surface density can be substantially enhanced compared to the initial one Weidenschilling (2003), Stepinski & Valageas (1997). An obvious conclusion is that the analysis of the formation of giant planets around stars with various masses should include the global evolution of solids in protoplanetary disks.

2. Methods of calculation

We model the protoplanetary disk as a two component fluid, consisting of gas and solids. The gaseous component is described as an alpha disk by the analytical model of

Stepinski (1998) which describes its properties, as a function of distance a from the star and time t, in terms of a self-similar solution to the viscous diffusion equation.

The main assumptions underlying our approach to the evolution of solids are (1) at each radial distance from the central star the particles have the same size (which in general varies over time), (2) there is only one component of dust, in this case corresponding to high-temperature silicates with the evaporation temperature $T_{\rm evap} = 1350$ K and bulk density 3.3 g cm^{-3}, (3) all collisions between particles lead to coagulation. At each radius, the vertical extent of the solid particle distribution is calculated and is evolved in time, so the effect of sedimentation of solids toward the midplane of the disk is taken into account. All assumptions and approximations are discussed in Stepinski & Valageas (1997), Stepinski (1998) and Kornet *et al.* (2001).

We model the formation of a giant planet *in situ*, so the orbital parameters of the planet do not vary in time. Our procedure is based on the following assumptions: (1) core accretion starts when solids at a given distance a from the star reach radii of 2 km, (2) at each time, the planetesimals are well mixed through the feeding zone of the planet; thus, their surface density $\Sigma_{\rm s}$ is always uniform in space, but usually decreasing with time as planetesimals accrete onto the planet, (3) the planetesimals do not migrate into the feeding zone from outside or inside, and vice versa, but they can be overtaken by the boundary of the feeding zone as it expands due to the growing mass of the planet, (4) to calculate the rate of gas accretion onto the planet, following Ida & Lin (2004a), we use a simplified approach based on fits to the numerical results Bryden *et al.* (2000).

3. Results

In the CAGCM the process of planet formation is naturally split in two main phases. In the first one a planetesimal swarm is formed in the protoplanetary disk. In the second phase planetary cores are assembled which subsequently accrete planetesimals and gas from the disk.

To illustrate how the mass of the central star influences the formation of the planetesimal swarm, we follow the evolution of a protoplanetary disk with an initial mass $M_0 = 0.1\ M_\odot$ and initial outer radius $R_0 = 1000$ AU for three values of $M_\star$ (0.5,1 and 4 $M_\odot$). Initially, the dust is well mixed with the gas, with the ratio $\Sigma_{\rm s}/\Sigma_{\rm g} = 6 \times 10^{-3}$ being constant everywhere in the disk. As the disk evolves the grains grow in size due to mutual collisions and gain inward radial velocities due to the gas drag. If their growth time of is shorter than the timescale of inward migration, they manage to reach sizes of a few km *before* reaching the evaporation radius. Their radial motions are then stopped and the planetesimal swarm attains its final form. Fig. 1 shows the distribution of planetesimals in our model after 10^6 yr from the beginning of its evolution. In all three cases, the outer radius of the planetesimal swarm is much smaller than the initial outer radius of the disk. The difference is larger for models with smaller $M_\star$. This is because small solid bodies evolving in disks around less massive stars gain larger inward velocities and tend to migrate to smaller radii before reaching kilometer sizes. Due to the inward migration of solids and their confinement to much smaller radii, the final surface density of planetesimals is increased locally by factor of a few in comparison with the initial value of $\Sigma_{\rm s}$. This effect is larger in disks around less massive stars.

To quantify the influence of $M_\star$ on the effectiveness of giant planet formation from a planetesimal swarm, we introduce the concept of the minimum surface density $\Sigma_{\rm s,min}$. We define it as the minimum value of the initial surface density of planetesimals $\Sigma_{\rm s,init}$ needed to from a Jupiter-mass ($1{\rm M_J}$) planet in less than the lifetime of the protoplanetary disk $\tau_{\rm f}$. For $\tau_{\rm f}$ we adopt a value of 3×10^6 yr. Fig. 2 shows $\Sigma_{\rm s,init}$ as function of distance

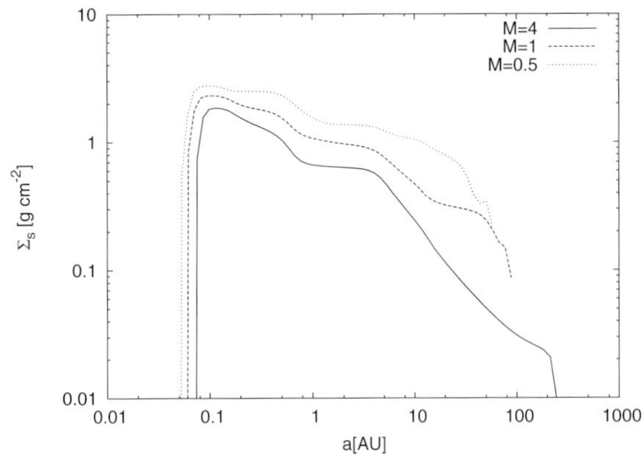

Figure 1. The surface density of solids as a function of the distance from the star in a protoplanetary disk with an initial mass 0.1 $M_\odot$ and outer radius 1000 AU after 10^6yr from the beginning of its evolution. Different curves correspond to different values of the mass of the central star in solar masses as described by the labels.

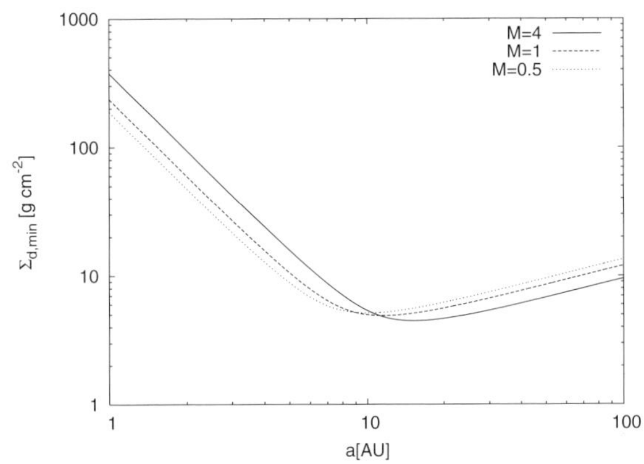

Figure 2. The minimum surface density of a planetesimal swarm needed to from a $1M_J$ planet in less than 3×10^6 yr as a function of distance from the central star. Different curves are obtained for different masses of the central star, as labeled in units of solar masses in the upper right corner.

from the star, for the same values of $M_\star$ as before. Close to the star (a < 10AU), $\Sigma_{\rm s,init} \sim a^{-2}$, so it is a decreasing function of the radius a. For sufficiently large radii (a > 10 AU), $\Sigma_{\rm s,min}(a)$ changes its slope and becomes an increasing function of a. In this regime $\Sigma_{\rm s,init} \sim a^{1/2}$. We also obtained the approximate, analytical formulas for $\Sigma_{\rm s,init}$, which are in a good agreement with the results of numerical calculations.

To investigate the influence of the mass of the central star on the whole process of giant planet formation we calculate a grid of models characterized by different values of the initial disk mass M_0 and outer radius R_0. We choose M_0 in the range between 0.02 and 0.2 $M_\odot$. The range of R_0 is adjusted for every metallicity so that all models in which the formation of giant planets is possible is accounted for.

We follow the evolution of each model until all solids are in the form of planetesimals or are accreted onto the star. Then, we evaluate every model with planetesimals to determine whether the surface density of planetesimals exceeds the minimum surface density $\Sigma_{\rm s,min}$ anywhere in the disk. Models with this property are labeled as *planet bearing*. For every such model we determine the minimum and maximum distance from the star at which

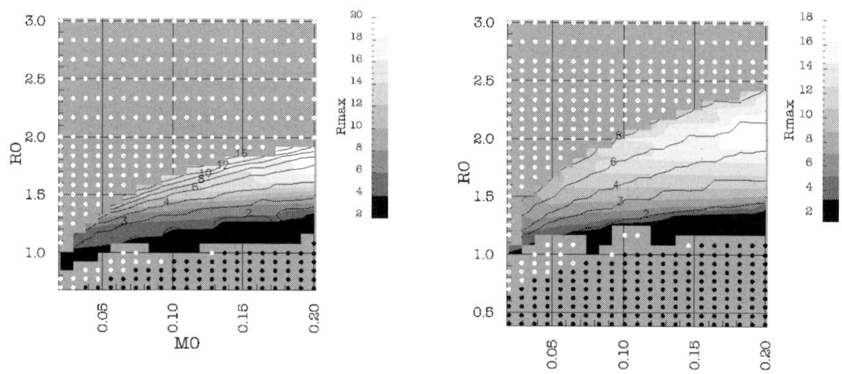

Figure 3. The plane of initial parameters of protoplanetary disk models $[M_0, \log R_0]$. Minimum and maximum distance from a $1M_\odot$ (left panel) and $0.5M_\odot$ (right panel) star at which the formation of a $1M_J$ planet is possible within 3×10^6 yr is indicated by contours and grey scale, respectively. White circles indicate disk models in which the surface density of the planetesimal swarm is everywhere lower than the critical value for planet formation. Black circles indicate disks in which all solids are accreted onto the star.

the surface density of the planetesimal swarm is larger than $\Sigma_{s,min}$. The results obtained for different values of the stellar mass are shown in Fig. 3.

The area of the region occupied on the $[M_0, R_0]$ plane by the planet bearing disks is clearly anticorrelated with the mass of the central star. Solid grains gain higher inward velocities in disks around less massive stars, and the resulting planetesimal swarms have higher surface densities. We see that, reducing the mass of the central star increases the maximum R_0 for which planet formation is possible in disks with the same initial mass M_0. We can also see that in a given disk the inner edge of the planet-bearing region moves inward as we decrease the mass of the star. This effect is mainly caused by the larger surface density of planetesimal swarms produced by disks around less massive stars. Differences in $\Sigma_{s,min}$, while appreciable, are much less important. The outer edge of the planet-bearing region also moves closer to the star. This is because in most cases it coincides with the outer edge of the planetesimal swarm which is more compact around less massive stars. The difference in the minimum surface density $\Sigma_{s,min}$ also tends to decrease the outer radius of the planet bearing region, but it is again a second order factor.

One of the main results of extrasolar planet searches is the discovery that planet bearing stars tend to have higher metallicities than field stars Santos *et al.* (2000), Fischer & Valenti (2003). Ida & Lin (2004b) and Kornet *et al.* (2005) showed that this correlation can be quantitatively reproduced within CAGCM. Herein we extend calculations of Kornet *et al.* (2005) onto disks around stars with various masses. The change in the disk metallicity influences the processes leading to the formation of planets in two ways. Firstly, it changes the structure of the gaseous disk by changing the opacity in the disk. In our models we scaled the opacity by a constant factor Z equal to the metallicity of the disk expressed in solar units. Secondly, the primordial metallicity of the disk determines the initial ratio of dust to gas surface densities. In our models this ratio is initially independent of the distance from the star and is equal to $\Sigma_s/\Sigma_g = 6 \times 10^{-3} Z$. We compute grids of models similar to those described above for eight different values of Z distributed between 0.2 and 3. Additionally, for every value of $[M_0, R_0]$ we check the gravitational stability of the corresponding gaseous disk. In some cases the value of the Toomre parameter Q drops below 1 in the outer region of the disk. We assume that such region

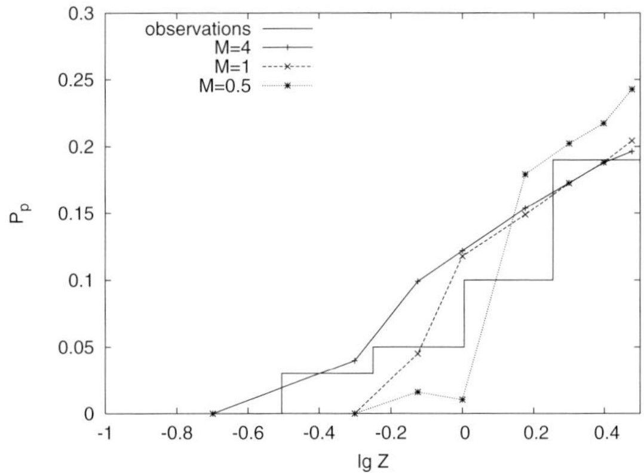

Figure 4. The rate of planet occurrence as a function of the primordial metallicity of protoplanetary disks. Different lines are obtained for models with different masses of the central star, as labeled in solar units in the upper left corner. The histogram shows the observational data compiled by Fischer & Valenti (2003).

fragments and some giant planets are formed there on a very short time scale. In that case, we use modified initial values of M_0 and R_0, which correspond to the mass and outer radius of the stable part of the original disk. Following the procedure described by Kornet *et al.* (2005), for every Z we calculate the area $A_{\rm p,5}$ of the region occupied on the $[M_0, \log R_0]$ plane by disks that form planets at distances smaller than 5 AU from the central star. A measure of the rate of planet occurrence can be defined as $P_{\rm p} = A_{\rm p,5}/C$, where the normalization factor C is chosen in such a way as to reproduce the observed value of $P_{\rm p}$ for $Z = 3$ and $M_\star = 1\ M_\odot$.

The results are presented on Fig. 4, showing the rate of planet occurrence as a function of disk metallicity for three values of $M_\star$. As expected, $P_{\rm p}$ is an increasing function of Z. The minimum value of Z below which no giant planets are formed at orbits smaller than 5 AU decreases with the mass of the central star from $Z_{\rm min} \approx 0.6$ for $M_\star = 0.5M_\odot$, to $Z_{\rm min} \approx 0.2$ for $M_\star = 4M_\odot$.

We see that for Z smaller than $\sim$0.2, $P_{\rm p}$ is an increasing function of $M_\star$. This is because most disks in which formation of planets would be possible around less massive stars have outer parts which are gravitationally unstable, and the amount of solids present in their stable parts is too low to enable subsequent formation of giant planets according to the CAGCM. However, as Z increases, smaller and smaller disks become planet-bearing for every M_0, and the percentage of gravitationally unstable disks in which formation of giant planets is not possible decreases. Consequently, the factors promoting the formation of giant planets around less massive stars as described in previous sections become important, and $P_{\rm p}$ changes into a decreasing function of $M_\star$.

4. Discussion

Obviously, our description of the evolution of solids is very simplified. The basic underlying assumptions like the single-size distribution of solid grains or the neglect of planet migration were already discussed by Kornet *et al.* (2004) and Kornet *et al.* (2005). The main additional assumption introduced in the present paper is the independence of the initial parameters of protoplanetary disks on the mass of the central star. While admittedly *ad hoc*, it seems to be better than that adopted by Laughlin *et al.* (2004), who scaled their initial surface density of planetesimals linearly with the mass of the star. They did not take into account the antecedent evolution of solids leading to the formation of planetesimal swarms, and concluded that the formation of giant planets around low-mass stars is difficult. Our models show that it is a vital factor facilitating the formation of giant planets, whose role should be particularly well seen for low-mass stars.

Acknowledgements

This project was supported by the German Research Foundation (DFG) through the Emmy Noether grant WO 857/2-1 and the European Community's Human Potential Programme trough the contract HPRN-CT-2002-00308, PLANETS. MR acknowledges the support from the grant No. 1 P03D 026 26 from the Polish Ministry of Science.

References

Bonfils, X., Delfosse, X., Udry, S., *et al.* 2004, in Astronomical Society of the Pacific Conference Series, 101
Bryden, G., Lin, D. N. C., & Ida, S. 2000, ApJ, 544, 481
Endl, M., Cochran, W. D., Tull, R. G., & MacQueen, P. J. 2003, AJ, 126, 3099
Fischer, D. & Valenti, J. A. 2003, in Scientific Frontiers in Research on Extrasolar Planets, ASP Conference Series, Vol 294, ed. D. Deming & S. Seager (San Francisco: ASP), 117
Ida, S. & Lin, D. N. C. 2004a, ApJ, 604, 388
Ida, S. & Lin, D. N. C. 2004b, ApJ, 616, 567
Kornet, K., Bodenheimer, P., Różyczka, M., & Stepinski, T. F. 2005, A&A, 430, 1133
Kornet, K., Stepinski, T. F., & Różyczka, M. 2001, A&A, 378, 180
Kornet, K., Stepinski, T. F., & Różyczka, M. 2004, A&A, 417, 151
Laughlin, G., Bodenheimer, P., & Adams, F. C. 2004, ApJ, 612, L73
Marcy, G. W., Butler, R. P., Vogt, S. S., *et al.* 2005, ApJ, 619, 570
Mayor, M., Udry, S., Naef, D., *et al.* 2004, A&A, 415, 391
Santos, N. C., Israelian, G., & Mayor, M. 2000, A&A, 363, 228
Stepinski, T. F. 1998, Icarus, 132, 100
Stepinski, T. F. & Valageas, P. 1997, A&A, 319, 1007
Weidenschilling, S. J. 2003, in Lunar and Planetary Science Conference #34, abstract #1707
Weidenschilling, S. J. & Cuzzi, J. N. 1993, in Protostars and Planets III, ed. E. Levy & J. Lunine (Tuscon: Univ. of Arizona Press), 1031

Direct Imaging of Exoplanets: Science & Techniques
Proceedings IAU Colloquium No. 200, 2005
C. Aime and F. Vakili, eds.
© 2006 International Astronomical Union
doi:10.1017/S1743921306009173

Detectability of planetary rings around an extrasolar planet from reflected-light photometry

L. Arnold[1], and J. Schneider[2]

[1]CNRS - Observatoire de Haute-Provence 04870 Saint-Michel-l'Observatoire, France
email: arnold@obs-hp.fr

[2]Observatoire de Paris-Meudon, 92195, Meudon Cedex, France
email: Jean.Schneider@obspm.fr

Abstract. The next generation of high-contrast imaging instruments will provide the first unresolved image of an extrasolar planet. While the emitted infrared light from the planet in thermal equilibrium should show almost no phase effect, the reflected visible light will vary with the orbital phase angle. We study the photometric variation of the reflected light with orbital phase of a ringed extrasolar planet. We show that a ring around an extrasolar planet, both obviously unresolved, can be detected by its specific photometric signature.

Keywords. Stars: planetary systems – Planets: rings – Extrasolar planet characterization.

1. Introduction

The discovery of extrasolar planets by radial velocity measurements has provided the first dynamical characteristics of planets (orbital elements and mass). The next step will be to investigate physical characteristics of extrasolar (albedo, temperature, radius etc.) and their surroundings. Among the latter are planetary rings. Coming space or ground-based high-contrast imaging instruments will be able to provide the first image of an unresolved extrasolar planet, in the thermal infrared or in the visible light. The emitted thermal infrared light from the planet should show no phase effect assuming the planet is in thermal equilibrium. But the reflected visible light will vary with phase angle, as should be shown by a broad-band photometric follow-up of the planet during its orbital motion.

We thus argue that it is of interest to study how the presence of a ring around a planet would influence its brightness as a function of its orbital position. This paper shows a few examples (more in Arnold & Schneider 2004) of how the reflected light curve of a ringed planet is different from that of a ringless planet, thus revealing the presence of the ring. The specific signature of a ring has been already mentioned and briefly qualitatively discussed (Schneider 2001), but here a more quantitative model is considered, in which the basic optical properties of the planet and the ring are taken into account, together with all geometrical parameters describing the ringed planet.

2. On the relevance of the existence of ringed planets

Although all giant planets of the Solar System have rings, ringlets or arclets, those of Saturn are by far the brightest. But at methane absorption wavelengths, giants planets are fainter than in the visible and their rings appear relatively brighter, increasing their detectability. For Uranus for example, the rings appear with a brightness comparable to the planet (Lellouch *et al.* 2002). Even the Earth might have been surrounded by a ring, but only during 10^5 to several 10^6 years (see Arnold & Schneider 2004 for a detailed bibliography). Although Saturn rings are probably younger than the planet itself, 10^8 years (Cuzzy & Estrada 1998), i.e. one order of magnitude younger than the planet, we consider that it remains relevant to look for rings around extrasolar planets, at least for giant planets.

3. The ringed planet model

To compute the light curve of the ringed planet as a function of the orbital phase angle, we build radiance maps of the object, for a given set of geometrical parameters (Arnold & Schneider 2004). We assume that the planet is an isotropic (lambertian) gray spherical light diffuser. The ring is a planar, homogeneous and anisotropic (non-lambertian) gray scattering layer. The ring brightness is estimated by assuming only single scattering in the ring. Mutual lighting (i.e. planet-shine on the ring, or ring-shine on the planet) is not modeled, but mutual shadowing is computed quite easily. All curves are normalized to the flux of a ringless planet seen in quadrature.

4. Discussion

4.1. *A ringed planet observed at inclination* $i = 0°$

Let's first consider the simple geometrical configuration where an extrasolar system is observed pole-on. This configuration means that the planet always remains in quadrature, thus showing always the same phase. For a ringless planet, it means that its brightness remains constant during the full orbit (assumed circular). Now for a ringed planet with non-zero obliquity, the ring alternately shows its illuminated and dark side, moreover it projects its shadow on the planet. This is illustrated in figure 1 showing strong slope changes at the equinoxes. This dichotomy of the light curves is due to the ring being alternately seen in reflection and in transmission. This is a clear ring signature which is often well visible in other geometrical configurations.

4.2. *Dual-band photometry of a ringed planet*

The planet and the ring may have different chemical compositions, and dual-band photometry (or spectroscopy if the object is bright enough) greatly helps to detect the ring by observing for instance in the (visible) continuum and in a methane absorption band, where a Saturn-like planet becomes much fainter than the ring. The light curves at both wavelengths can be significantly different, as shown in figure 2. In the methane band, stronger slope changes occur at the equinoxes, here again due to the two observation regimes of the ring, either seen reflecting or transmitting the light. Note that when the ring disappears at the equinoxes, the object spectrum may be the spectrum of the planet only, but it can also be composite if a part of the planet is seen through the unilluminated ring.

4.3. *The effect of the longitude of ring obliquity*

The orbital angles for which the brightness extrema occur do not always correspond with those of a ringless planet. It depends on the ring obliquity longitude. It is illustrated in figure 3. We consider this *ϕ-shift* of the light curve a relevant signature of a ring. It is unambiguous for a ringed planet on a circular orbit. A ϕ-shift can be observed if the orbit is elliptical, whether the planet has a ring or not, when brightness variations are induced by distance changes from the planet to its star. But we assume that the astrometry of

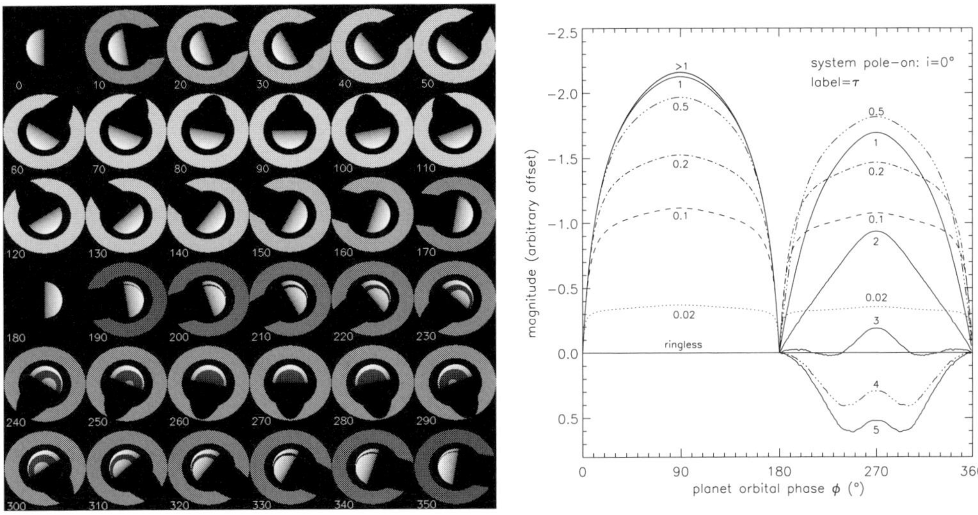

Figure 1. Extrasolar system observed pole-on at different orbital positions (labels, image scale = radiance$^{0.33}$). Inclination $i = 0°$, the planet is always seen in quadrature, while the ring, with a Saturn-like obliquity of $i_r = 26.73°$, shows its illuminated face during the first half-orbit. During the second half-orbit, the back-illuminated ring becomes fainter. Corresponding light curves are given at right for different optical thicknesses τ of the rings (labels).

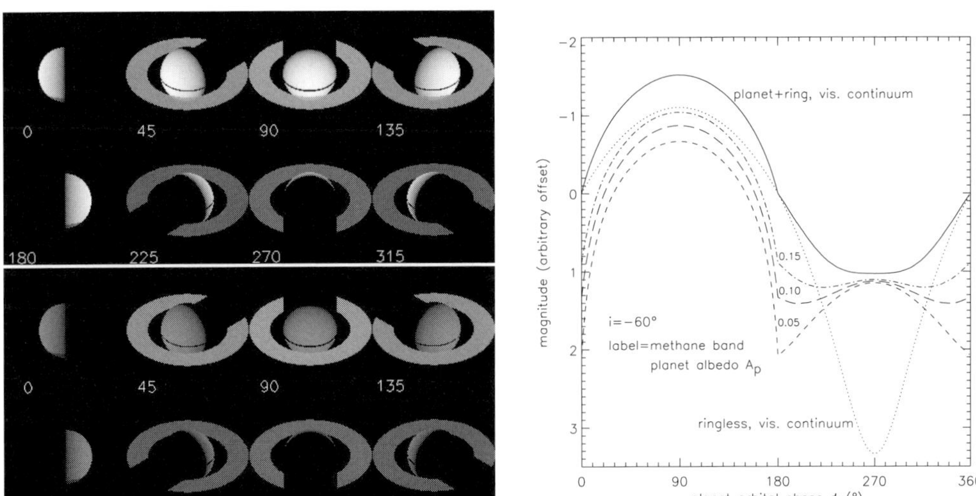

Figure 2. Ringed planet for different orbital positions (labels). Planet obliquity is 5°. The pictures upper left represent the planet in the continuum (albedo: planet 0.34, ring 0.7) and below in a CH_4 absorption band (albedo: planet 0.05, ring 0.7). The corresponding light curves are shown at right, in the continuum and in CH_4 absorption band for three different planet albedo. Light curves have significantly different shapes, especially during the second half-orbit: The ring and planet respective brightnesses vary in opposite directions and by the same order of magnitude, making the light curve almost flat.

the discovered planet will make possible the correction of the measured reflected light photometry for the effect of orbit ellipticity.

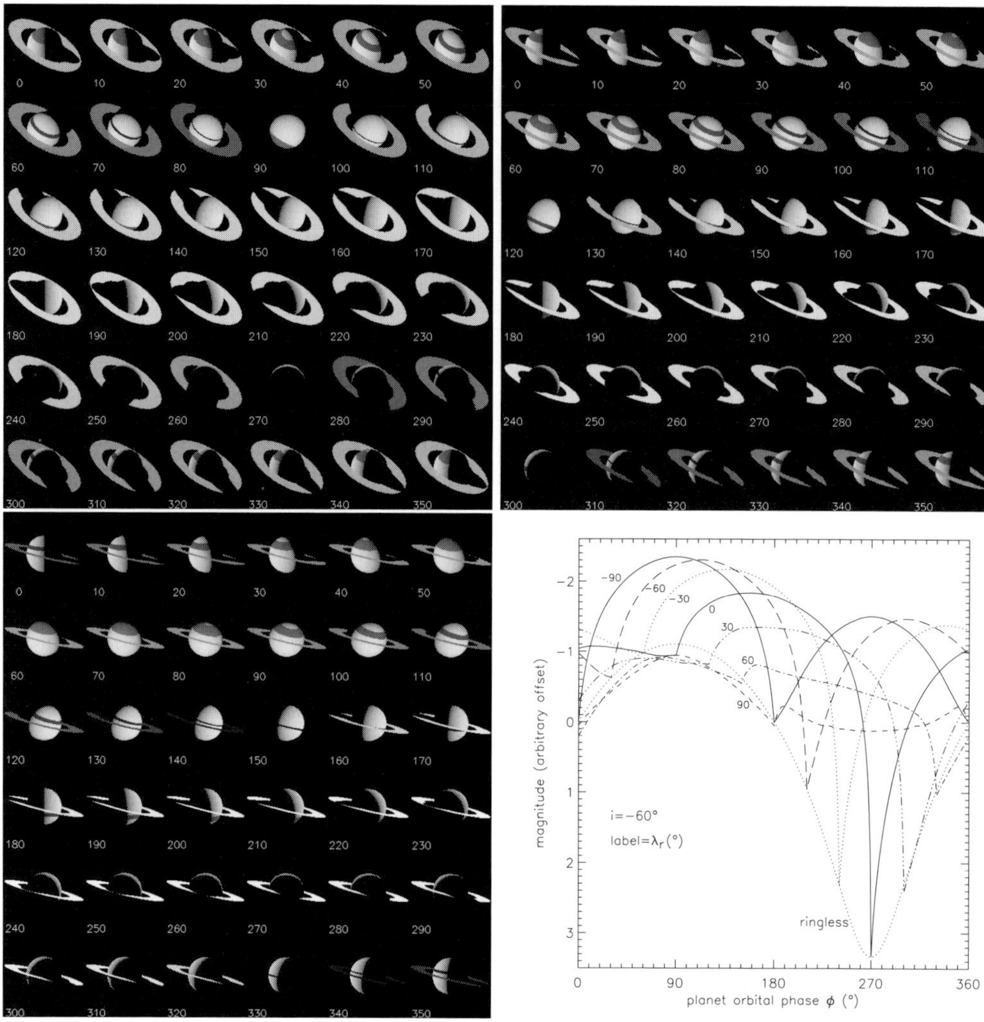

Figure 3. A Saturn-like planet for different orbital positions (labels) and different longitudes λ_r of the ring obliquity: $\lambda_r = 0°$ (above left), $\lambda_r = 30°$ (above right), $\lambda_r = 60°$ (below left). The three sets of images have the same brightness scale. These figures show that respective ring and planet brightness extrema do not always occur simultaneously. For instance, when $\lambda_r = 30°$ (above right), a ring minimum occurs at $\phi = 300°$, while the planet minimum occurs at $\phi = 270°$. This difference induces a shift of the system light curve as illustrated at lower right. The curve for a ringless planet is plotted for comparison. Curves are plotted for $-90° \leqslant \lambda_r \leqslant 90°$. Note that we have the relation: $\text{magnitude}_{\lambda_r}(\phi) = \text{magnitude}_{180°-\lambda_r}(180° - \phi)$, all other parameters being unchanged.

4.4. *A planet with a large ring*

For a given planet obliquity, the smaller the gap between the inner edge of the ring and the planet equator, the lower the latitude of the shadow on the planet. Also, the larger the ring outer diameter, the longer the polar region remains in the shadow of the ring. The ring thus can almost totally hide the planet, as shown by figure 4 around orbital position $\phi = 270°$. Note that when the planet is hidden, the object spectrum is dominated by the spectrum of the ring. We consider that such a strong and quite long extinction occurring at phase angles $\alpha > 90°$ constitutes another specific signature of a ring around

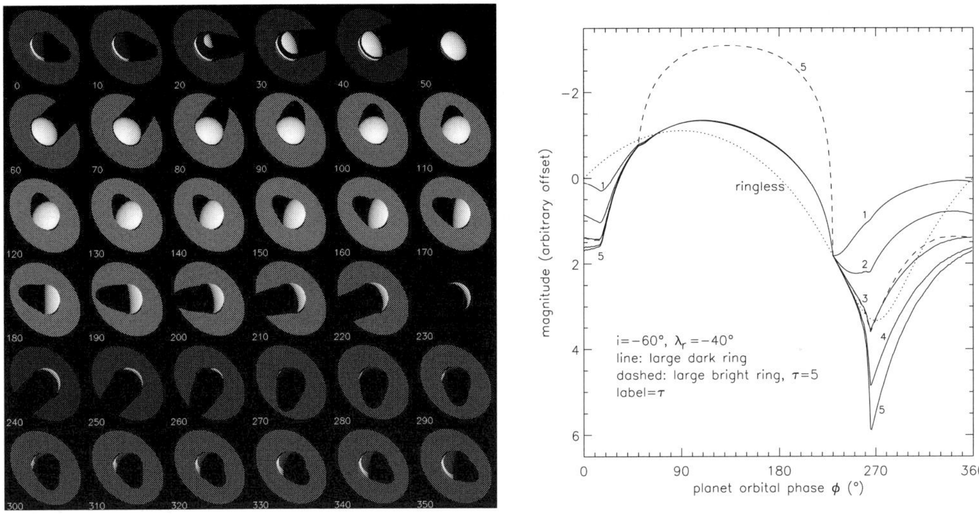

Figure 4. A thick, dark and large ring ($\tau = 3$ and albedo $\varpi_0 = 0.05$) around a planet, seen for different orbital positions ϕ (labels). The ring is so large that it hides the planet around $\phi = 270°$. A curve for a planet with a large, bright and thick ring (albedo $\varpi_0 = 0.7$ and $\tau = 5$) is shown for comparison.

an extrasolar planet. It was qualitatively discussed by Schneider (in Des Marais *et al.* 2002a).

Of course, the larger (and/or brighter) the ring with respect to the planet, the more the ring dominates the light curve. But the ratio of the reflected light from the planet over the stellar flux gives a value of the planetary radius (Schneider 2001, 2002). The ring significant flux contribution can lead to an overestimation of the planet radius by a factor of 10. If the mass of the planet is known (by radial velocity for instance), the overestimated planet radius would lead to an underestimated planet density. Thus a planet with a very low apparent density could be a signature of a ring. Moreover, the planet radius can in principle also be deduced from its infrared thermal flux $F_{p,IR}$, assuming the planet temperature has been inferred from the matching of a Planck function to the observed thermal spectrum (Schneider 2001, 2002). The disagreement between the planet radii at visible and infrared wavelengths could thus be another signature of a ring.

5. Conclusions

The examples shown demonstrate that a ring around an extrasolar planet significantly affects the reflected light curve of an extrasolar planet during its orbital motion. A ring could thus be detected, although both planet and ring would obviously remain unresolved. This may be achieved simply by a photometric monitoring of the planet light. We identified the following signatures, which would require only moderate photometric accuracy to be observed (around $3\sigma \approx 0.5$ mag):

i) *Light curve dichotomy, with strong slope changes at the equinoxes*, due to the ring being alternately seen in reflection and in transmission.

ii) *Light curve dependence on wavelength* in dual-band photometry (methane band and continuum for instance), or *spectral variations* if spectroscopy is possible.

iii) The *ϕ-shift* of the light curve extrema, due to the longitude of the ring obliquity.

iv) *Temporary extinction of the planet* during the orbital motion, due to the rising shadow of the ring on the planet.

v) *High brightness in the reflected light*, leading to an abnormally large planet radius, or/and an abnormally low mass density.

vi) *Disagreement between the planet radii* measured from reflected light and thermal infrared emission.

Although future space missions studies concentrate mainly on infrared instruments for technical and scientific reasons, this work shows the additional interest of shorter wavelengths (visible band) for extrasolar planet characterization.

References

Arnold, L. & Schneider, J. 2004, *A&A* 430, 1153

Cuzzy, J.N. & Estrada, P.R. 1998, *Icarus* 132, 1

Des Marais D.J., Harwit M., Jucks K., Kasting J.F., Lunine J.I., Lin D., Seager S., Schneider J., Traub W., & Woolf, N. 2002a, JPL Publication 01-008, Rev. A, June 2002, p. 37 and 44

Lellouch E., Encrenaz T., Cuby J.G., & Jaunsen A. 2002, ESO Press Release and Photos 31a-b/02, 20 December 2002 (http://www.eso.org/outreach/press-rel/pr-2002/phot-31-02.html)

Schneider, J. 2001, in 'Scientific Highlights 2001', SF2A conference proceedings, Lyon, France, May 28 - June 1, 2001, F. Combes, D. Barret and F. Thévenin, Eds., EDP-Sciences (Editions de Physique), Conference Series, p. 489-492

Schneider, J. 2002, in 'Scientific Highlights 2002', SF2A conference proceedings, Paris, France, June 24-29, 2002, F. Combes & D. Barret, Eds., EdP-Sciences (Editions de Physique), Conference Series, p. 597-602

Direct Imaging of Exoplanets: Science & Techniques
Proceedings IAU Colloquium No. 200, 2005
C. Aime and F. Vakili, eds.
© 2006 International Astronomical Union
doi:10.1017/S1743921306009185

Simulating Extrasolar Planet Populations for Direct Imaging Surveys

Eric L. Nielsen[1]†, L. M. Close[1] and B. A. Biller[1]

[1]Steward Observatory, University of Arizona, Tucson, AZ 85721, USA
email: enielsen@as.arizona.edu

Abstract. As high contrast imaging surveys are being designed and carried out to directly detect extrasolar planets around young, nearby stars it is important to carefully evaluate the criteria for selection of target stars, as well as the predicted success of such a survey based on the sensitivity of the AO system used. We have developed a routine to simulate a large number of planets around each potential target star, and determine what fraction can be reliably (5σ) detected using an AO system's predicted or observed sensitivity curve (the maximum flux ratio between the parent star and a detectable planet as a function of projected radius). Each planet has a randomly assigned semi-major axis, mass, and eccentricity (following extrapolations of detected radial velocity planet power laws), as well as random viewing angles and orbital phase. The orbital parameters give a projected separation for each planet, while the mass is converted into a flux ratio in the appropriate bandpass of the detector using the models of Burrows *et al.* (2003); this allows the simulated planets to be directly evaluated against the system's sensitivity curve. Since this method requires basic parameters (age, distance, spectral type, apparent magnitude) for each target star, a target list can be constructed that maximizes the likelihood of detecting planets, or competing instrument designs can be evaluated with respect to their predicted success for a given survey. We are already employing this method to select targets for our Simultaneous Differential Imaging (SDI) surveys (Biller *et al.* 2004), now underway at telescopes in the northern (MMT) and southern hemispheres (VLT).

Keywords. Stars: Planetary Systems, Techniques: High Angular Resolution.

1. Introduction

With the number of detected extrasolar planets continuing to climb well over one hundred, astronomers are finally in a position to begin to characterize the process of planet formation, as well as give statistical descriptions of planet populations around nearby stars. A significant gap, however, exists in our knowledge for planets with orbits at or beyond that of Jupiter: the widest orbit known for a confirmed extrasolar planet is 6 AU, for 55 Cnc d (Marcy *et al.* 2002). Since this limit is set by the time baseline of radial velocity surveys, this threshold will march out slowly; it will take another 15 years to close an orbit equivalent to Saturn's (9.5 AU). Radial velocity detections of planets in this regime will be further hampered by the fact that the amplitude of the velocity declines with the inverse square root of the semi-major axis.

Direct imaging, then, serves as a completely complimentary detection method to radial velocity surveys, as it is most sensitive to planets with large separations from their parent stars, with no real upper limit to detectable semi-major axes. If the two techniques could be brought together at intermediate separations, we will be able to start building up a complete picture of the distribution of planets. Again, since radial velocity surveys are slow to move outward, direct imaging is well-situated to close the gap.

† Michelson Fellow

SDI, or Simultaneous Differential Imaging, is a technique to achieve very high contrast images at small angular separations. By obtaining images simultaneously through narrow-band filters centered on either side of the 1.6 μm methane bandhead, we can greatly attenuate speckle noise from the star, and achieve very high contrasts for objects with a strong methane signature (most giant planets). We are currently conducting a survey for planets around young, nearby stars using SDI cameras installed at the VLT and MMT. For further information on SDI and the progress of our survey, see Lenzen *et al.* 2004 and Biller *et al.* 2004, as well as the contributions to this proceedings by B. Biller and L. Close.

2. Monte Carlo Simulations

The ability of a survey instrument to detect self-luminous, giant extrasolar planets is typically given in the form of a contrast curve: the faintest magnitude (expressed in relation to the magnitude of the parent star) at which a companion can be significantly detected, given as a function of the angular separation between the two objects. We have assembled a set of four sensitivity curves in order to illustrate the nature of our simulations, as shown in Fig. 1. VLT NACO SDI represents a currently available system, on the 8.2m VLT telescope with a 200 element Shack-Hartmann wavefront sensor, using four-channel SDI optics to reduce speckle noise (Lenzen *et al.* 2004, Biller *et al.* 2004, contrast curve from B. Biller *et al.*, this proceeding). Gemini NICI is expected to come on-line by 2006, utilizing the H85 AO systems, an 85 element curvature wavefront sensor, along with a coronograph and two-channel SDI optics (predicted sensitivity curve from the NICI Request for Proposal (2005)). We also include plausible contrast curves for Extreme Adaptive Optics Coronographs (ExAOC) at an 8m-class telescope, or at the proposed 24m Giant Magellan Telescope (Johns *et al.* (2004)). In either case, our curves consist of a spatially-filtered Shack-Hartmann wavefront sensor (750 element for the 8m telescope, 4400 for the GMT), a coronograph, a focal-plane wavefront sensor for speckle suppression (Codona and Angel (2004)), and SDI optics for added contrasts (Codona *et al.* (2005)).

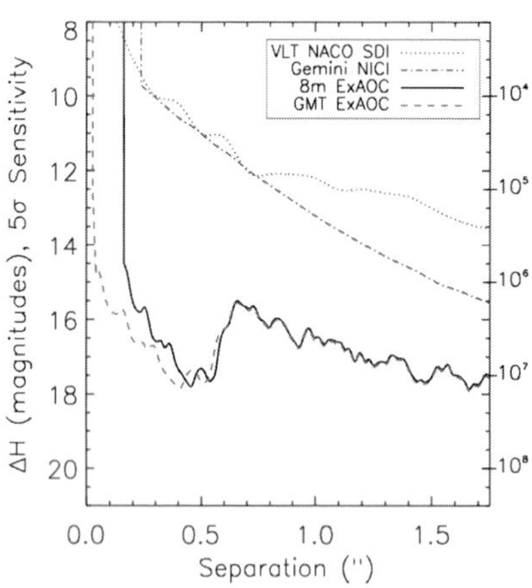

Figure 1. Sensitivity curves for the four planet-finding instruments we consider here. The curves represent the 5σ level of detection for a companion at a given angular distance from the parent star, after an integration of 10,000 seconds.

While the contrast curves of Fig. 1 describe the comparative abilities to image faint companions of the four systems, it is not immediately apparent how these curves translate into the figure of merit about which we're most concerned: the number of planets each system could detect from a survey. In order to evaluate this, we have constructed a Monte-Carlo simulation to create an ensemble of planets, compute their projected separation

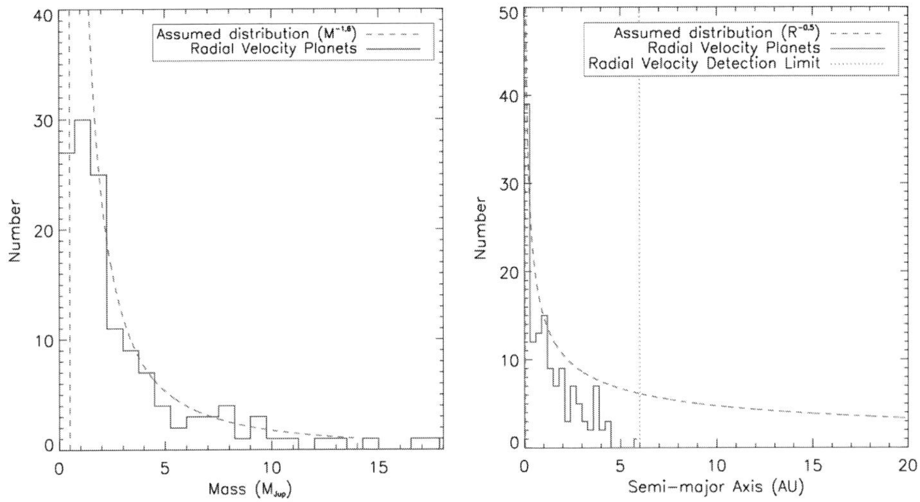

Figure 2. The assumed distributions of mass (power law index -1.6) and semi-major axis (-0.5) of extrasolar giant planets, plotted against histograms of known planets from radial velocity surveys.

and H-magnitude, then compare them to these contrast curves to determine what fraction can be detected with a given system.

For each target star we simulate 100,000 planets, randomly generating six quantities for each planet: mass, semi-major axis, eccentricity, inclination angle, longitude of periastron, and orbital phase. Mass and semi-major axis are assumed to be governed by simple power-law distributions, with indices chosen to fit the population of known radial velocity planets, as shown in Fig. 2 (similar distributions were inferred by Lineweaver and Grether (2003) for mass and Graham *et al.* (2002) for semi-major axis). Distributions of extrasolar planets are taken from California & Carnegie Planet Search Almanac of Planets (2005). High and low mass cut-offs are imposed on the power law distribution at 0.5 and 14 M_{Jup}, and the semi-major axis distribution is truncated at 0.037 AU (orbital radius of the innermost known extrasolar planet, HD 73256) and 20 AU. This outer limit for semi-major axis is uncertain, as the data from extrasolar planet surveys become incomplete beyond ~3 AU, and no data exist for extrasolar planets further out than 6 AU. We take 20 AU (the orbit of Uranus) as a conservative upper limit. The distribution for eccentricity is taken as a simple polynomial fit to the histogram of eccentricities of radial velocity planets.

It should be noted that we have assumed three independent distributions for mass, semi-major axis, and eccentricity of planets, which is almost certainly incorrect. With only 137 known planets, however, the statistics are not sufficient to suggest a more complex distribution. For this same reason, we also assume these planet distributions to be independent of the spectral type of the parent star.

We also assign an inclination angle for the orbit (uniform in cos(i)), longitude of periastron (uniform between 0 and 2π), and orbital phase (uniform random variable is time; between 0 and 1 orbital periods). From these six quantities we can solve for the instantaneous separation between planet and star, as viewed from the observer on earth. The distance to the target star can then be used to solve for angular separation. With the planet models of Burrows *et al.* (2003), the planet's mass (along with the age of the target star) is converted into an H-magnitude, which we use to solve for ΔH, given the

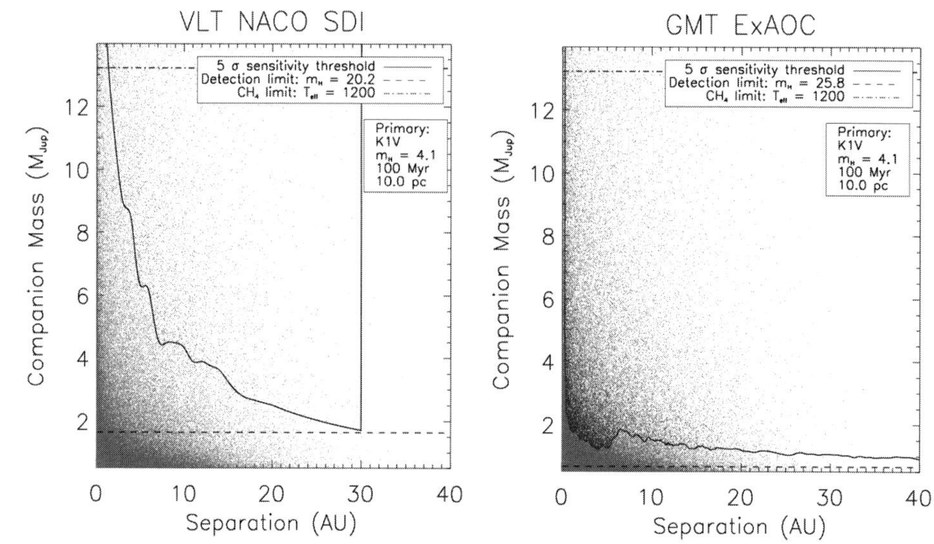

Figure 3. Example simulations for a particular target star with the VLT NACO SDI (left) and GMT ExAOC (right) systems. Each point represents a single simulated planet, out of a total of 100,000 in the simulation run. Detected planets lie above the 5σ sensitivity curve (solid line) and the detection threshold (bottom dashed line), and below the methane limit (top dotted line). For VLT NACO SDI, 6% of these planets are detected; GMT ExAOC can detect 35%.

distance and 2MASS flux of the target star. Then, it is determined whether or not each planet can be detected, given the sensitivity curve of Fig. 1.

We show examples of these simulations for a single target star in Fig. 3. In addition to the sensitivity limit, we consider if the apparent magnitude of the planet is sufficient to be detected by photon noise arguments alone, and whether the planet is too massive (hence too hot) to have a significant methane feature (SDI is most effective when there is a strong drop in the planet's spectrum at the methane bandhead, T_{eff} < 1200K). We can now directly compare two systems by their ability to find planets: VLT NACO SDI should have a 6% chance of finding a planet around the star in Fig. 3, while GMT ExAOC will be at 35%.

3. Target Selection

The ideal target star for a direct imaging survey for self-luminous extrasolar planets is characterized by three traits: the star should be young so the planet's luminosity is larger, the star should be nearby so the angular separation between primary and companion is greater, and the star should be of a late spectral type so that the inherent luminosity of the star does not overwhelm the planet's light. It is not trivial to find a balance between these factors to select an observing list for a survey. In particular, there is a lack of very young stars (<30 Myr) very close (<20 pc) to the sun (especially in the northern sky), so compromises must be made. It is through the simulations described in this paper that we seek to disentangle these target star traits, so that every potential target gets a quantitative ranking. By constructing a survey that maximizes the expected number of planets detected, our results (even null results) become more meaningful.

Another important consideration (especially for higher-order AO systems such as ExAOC) is the limiting magnitude at which the adaptive optics can efficiently operate. Better sampled AO systems (more wavefront measurements, and more actuators to control) require more photons from the guide star, with strehl ratios (and so contrasts) declining faster for fainter guide stars. Since our simulations indicate a preference for younger, further-out stars over older, nearby targets, a planet-finding system with a limiting magnitude that is too bright will exclude many of the best targets. For example, of the 40 best targets from our list of 153 (which we have compiled from the literature), the median V magnitude is 9.75; a next-generation AO system that can't lock onto targets fainter than 9th magnitude will simply not be an efficient planet finder, as half the potential targets are eliminated before the survey begins. This effect can be mitigated by allowing the AO system's performance to be modulated depending on the target star (having the wavefront sensor integrate longer on fainter stars, for example), at the expense of contrasts. In any event, when designing future adaptive optics systems with the primary goal of detecting extrasolar planets, one must be keenly aware of the nature of the available science targets when evaluating potential instrument designs. The number of target stars for direct imaging surveys for self-luminous extrasolar planets is limited, and an instrument that cannot reach most of them would not be competitive.

4. Simulation Results

We now consider the implications of our simulation results for the selection of target stars. In Fig. 4 we examine the fraction of planets that can be detected around target stars of various ages and distances. As expected, younger, nearby targets are more likely to harbor detectable planets. The probability of detecting a planet around an older, or more distant, star drops quickly. We also look at results of the simulation for all 153 of our target stars, plotting the median separation of the detected planets (See Fig. 3) against the percentage of planets that are detectable. The more sophisticated planet-finding systems are more powerful since they minimize the inner working radius, as we would expect most planets to lie at small angular separations from their parent stars.

We also consider the predicted results of a planet-search campaign, directed at our target list, using the four systems. Since the simulations return the probability of detecting a planet around a given star, the sum of these probabilities for all the survey stars should give us the total number of planets we'd expect to detect at the end of a survey. In Fig. 5 we have ordered our targets by detection probability (so the highest-quality targets are observed first), and show the expected yield of planets from surveys of various sizes. Clearly, most of the planets are found initially, as the best targets are observed, and adding lesser-quality target stars to the observing campaign results in a slower gain in planets. The break in these curves, between rapid return of planets and a more gradual increase, depends on the nature of the instrument.

5. Conclusions

With these simple Monte Carlo simulations, extrapolating from populations of known extrasolar planets, we have a method to quantitatively rank stars in a target list for a planet search campaign, or to directly compare competing instrument designs for future planet finding systems. Our basic results suggest that there are a limited number of target stars suitable for direct imaging searches for self-luminous planets, a fact that must be considered when planning a survey or designing a new instrument.

Finally, we consider one of our basic assumptions: that each of our target stars has one planet described by the power laws of Fig. 2. While there is a lack of data for planet populations at large separations, we can make an estimate of total populations based on our initial assumptions and results from radial velocity surveys. Fischer and Valenti

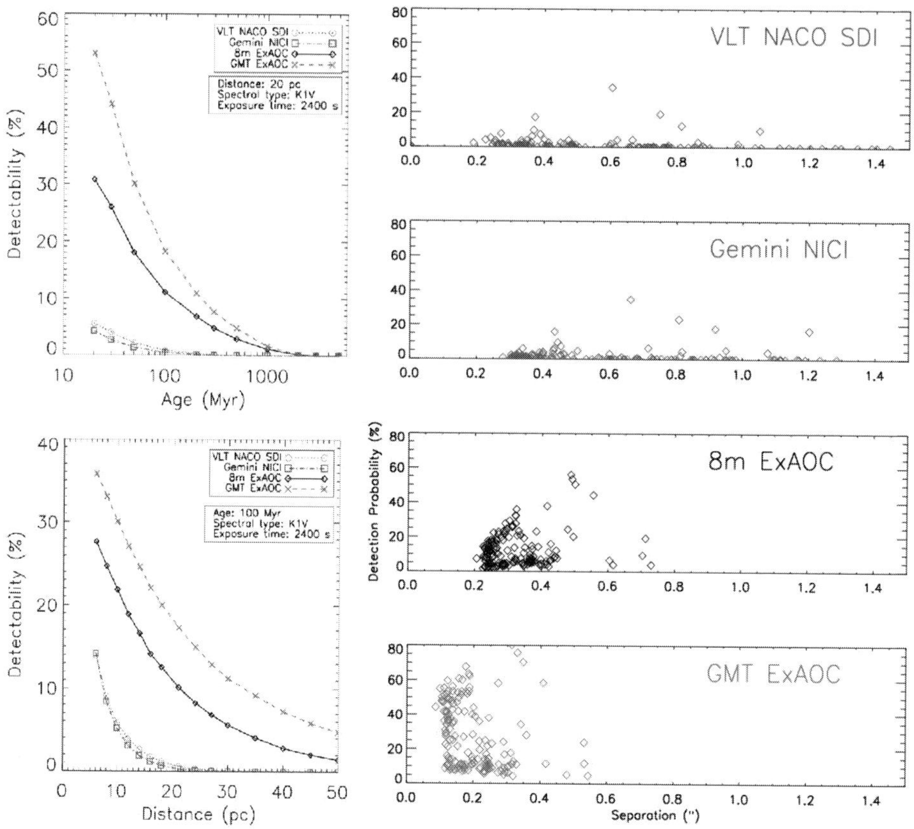

Figure 4. Basic results from the simulation. Top left shows the fraction of simulated planets detected (the detection probability) for target stars of different ages. As expected, younger stars are the preferred targets for finding self-luminous planets, and there is very little value in observing older (>1 Gyr) target stars. Bottom left is a similar plot, only with distance being varied. In the right panel, we consider the properties of the simulated planets that can be detected with each system. For each of the 153 target stars, we plot the fraction of planets that can be detected (y-axis) against the median projected separation (in arcseconds) between primary and companion for those planets that can be detected. The pile-up at small separations underscores the critical importance of the inner working radius of any planet-finding system.

(2005) determine that in a volume-limited sample, about 6% of stars show the radial velocity signature of a massive planet with an orbital period shorter than 4 years (2.5 AU for a solar mass primary). Given this, we can then integrate our assumed semi-major axis distribution ($dN/da = a^{-0.5}$), to give the total number of planets between 0.03 and 20 AU (which we simulate), compared to the fraction found by Fischer and Valenti (2005) between 0.03 and 2.5 AU. From this argument, we'd expect 18% of all stars to have a giant planet, or 40% for stars of solar metallicity or greater. This simplistic argument does not take into account possible effects of changing planet populations (or overall frequency) with spectral type of the parent star, an issue that is still being studied. Nevertheless, as these simulations make real predictions about survey results, they place us in a strong position to interpret any results as they relate to the overall populations of extrasolar planets.

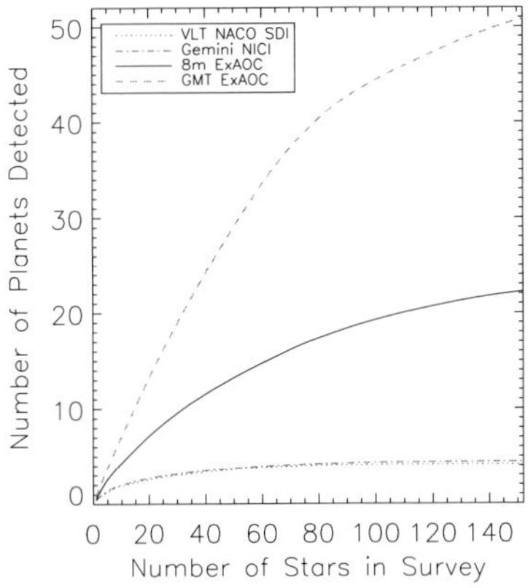

Figure 5. The number of planets expected to be detected, as a function of the number of stars in the survey. In each case, once the best targets are observed, there is a slow gain in planets detected as lower-quality targets are added to the survey. The location of this break, however, depends on the ability of the instrument (in order of increasing power: VLT NACO SDI and Gemini NICI, 8m ExAOC, and GMT ExAOC).

Acknowledgements

This work was performed in part under contract with the Jet Propulsion Laboratory (JPL) funded by NASA through the Michelson Fellowship Program. JPL is managed for NASA by the California Institute of Technology. L. Close thanks the NSF CAREER and NASA Origins program for support.

References

Biller, B.A., Close, L., Lenzen, R., Brandner, W., McCarthy, D.W., Nielsen, E. & Hartung, M. 2004, *SPIE* 5490, 389

Brown, R.A. 2004, *ApJ* 607, 1003

Burrows, A., Sudarsky, D. & Lunine, J.I. 2003, *ApJ* 596, 587

California & Carnegie Planet Search Almanac of Extrasolar Planets 2005

Codona, J.L. & Angel, R. 2004, *ApJ* 604, 117

Codona, J.L., Angel, R., Close, L. & Potter, D. 2005, private communication

Fischer, D.A. & Valenti, J. 2005, *ApJ* 622, 1102

Graham, J.R., Macintosh, B., Ghez, A., Kalas, P., Lloyd, J., Makidon, R., Olivier, S., Patience, J., Perrin, M., Poyneer, L., Severson, S., Sheinis, A., Sivaramakrishnan, A., Troy, M., Wallace, J. & Wilhelmsen, J. 2002, *AAS* 201, 2102

Johns, M., Angel J.R.P., Schectman, S., Bernstein, R., Fabricant, D.G., McCarthy, P. & Phillips, M. 2004, *SPIE* 5489, 441

Lenzen, R., Close, L., Brandner, W., Biller, B. & Hartung, M. 2004, *SPIE* 5492, 970

Lineweaver, C.H. & Grether, D. 2003, *ApJ* 598, 1350

Marcy, G.W., Butler, R.P., Fischer, D.A., Laughlin, G., Vogt, S.S., Henry, G.W. & Poubaix, D. 2002, *ApJ* 581, 1375

NICI Planet Finding Campaign, Request for Proposals 2005

Discussion

QUESTION: Doesn't the conclusion that inner working angle is the most important aspect of the system depend heavily on your choice of distribution for semi-major axis?

NIELSEN: Yes, an alternate distribution with a reservoir of giant planets at large separation would result in less importance being assigned to the inner working angle. We have sought to take a conservative approach, extrapolating only from the known radial velocity planets which suggest, even given the observational biases, a declining frequency of planets with semi-major axis. Though not conclusive, the lack of planets found thus far by current direct-imaging surveys, especially in the large separation regime in which they are most sensitive, suggests that a large semi-major axis reservoir of planets does not exist past 50 AU.

QUESTION: Wouldn't your concern about guide star magnitude go away if a laser guide star system were used?

NIELSEN: For the case of ExAOC, for both the 8m and 24m telescopes, laser guide stars would not be competitive with natural guide stars. The goal of Extreme Adaptive Optics is to achieve strehls >90%, which is not achievable at H-band with a laser system. A sodium laser guide star would be 90 km above the telescope, with the science target at essentially infinity. Both make a cone with the point source at the tip and telescope's primary mirror at the base. If there is a turbulent layer 10 km above the telescope, the cone of the star (basically a cylinder through the atmosphere) would be larger than that of the Na laser beacon's by about 20%: this is turbulence that distorts the wavefront of the science target, which the laser beacon's light never samples. This so-called "cone" effect alone precludes lasers from being a part of an ExAOC system, as typical strehls achieved with lasers are <50% at H-band.

Direct Imaging of Exoplanets: Science & Techniques
Proceedings IAU Colloquium No. 200, 2005
Claude Aime & Farrokh Vakili
© 2006 International Astronomical Union
doi:10.1017/S1743921306009197

Expectations for the Early *TPF-C* Mission

Robert A. Brown

Space Telescope Science Institute, 3700 San Martin Drive, Baltimore, MD 21218, USA;
email: rbrown@stsci.edu

Abstract. We use Monte Carlo techniques to estimate the results and character of the early *TPF-C* mission. Using 10^8 samples to represent the planets of interest, we compute the completeness of the first search observations of prioritized target stars with optimized exposure times that sum to one year total. Assuming simple observing protocols and decision rules for searching, verifying, and characterizing observations, and taking into account ranges of probabilities for confusion sources ($P_{\text{confusion}}$) and the occurrence of planets of interest (η), we compute 10^5 samples of the *TPF-C* schedule and observational outcomes for the first year of exposure time. For example, for Earth-like planets on habitable-zone orbits, assuming no observing overheads or pointing restrictions, and for the values $P_{\text{confusion}} = 0.5$ and $\eta = 0.1$, we find that a median 2.2 planets are found, verified, and characterized in one year of exposure time, with the 68 highest priority stars searched.

Keywords. instrumentation: high angular resolution, space vehicles: instruments, techniques: high angular resolution, (stars:) planetary systems

1. Introduction

The task of finding and characterizing Earth-like planets around other stars is focused and well defined. Because of this sharp definition, the process of mission modeling—estimating results from projected technical performance—is especially useful for planet-finding missions, for evaluating instrument designs, exploring observing strategies, and ensuring robustness against a range of natural uncertainties. This paper discusses expectations for the early *Terrestrial Planet Finder* (*TPF-C*) mission, based on mission modeling.

The concept of completeness is the cornerstone of mission modeling for *TPF-C*. Completeness is the fraction of planets that an instrument can detect in exposures of given depth. Assuming a star has a planet, the number of planets that are found by searching—zero or one—is a Bernoulli random variable with probability equal to the completeness. The expectation value of the number of planets found is also equal to the completeness. For direct-detection techniques, searching a star at different epochs will accumulate completeness. Furthermore, the expectation value of the total result of searching multiple stars is the sum of their accumulated completenesses. Because of this nexus, probability theory and Monte Carlo experiments are emerging as powerful tools for understanding and optimizing the *TPF-C* mission.

The basic mission model for the planet-finding coronagraphic instrument on *TPF-C* comprises five components: (1) the *instrument* (five performance parameters); (2) the *planets* (size, two photometric characteristics, and two orbital characteristics); (3) the *stars* (input catalog and five characteristics related to planetary detectability); (4) the *protocols & priorities* (defining and ranking three types of observations; and (5) the *strategic unknowns* (exozodiacal light, occurrence rate of planets of interest (η), and probability of confusion sources ($P_{\text{confusion}}$)). We combine these components into four structures: (i) *completeness*, based on the star, instrument, and planets; (ii) *exposure time*, based on the star, instrument, planets, and exozodiacal light; (iii) *pure-search results*, based on completeness and exposure time; and (iv) *mission results*, based on completeness, exposure time, protocols & priorities, η, and $P_{\text{confusion}}$.

Instrument. As currently envisioned, *TPF-C* is a visible-light camera with fore-optics that strongly suppress starlight in an annular detection zone around a target star. Currently, the telescope is projected to have a 3.5-by-8 meter elliptical entrance aperture. Planned for launch after 2015, the *TPF-C* spacecraft will be stationed near the second Lagrange point and operate for at least five years. It will search a selected subset of nearby stars for terrestrial planets in the habitable zone, where water might occur in liquid form—an assumed precondition of life. *TPF-C* will obtain the spectra of any found planets, looking for atmospheres compatible with—or even reflecting the presence of—life.

The five most important parameters of the *TPF-C* coronagraph for understanding its early accomplishments are: (1) $A_{\text{eff}} \equiv A\, QE$, the effective area—the product of the area of the entrance aperture and the quantum efficiency of the optical system; A_{eff} governs the information rate of *TPF-C*. (2) *IWA*, the "inner working angle," the angular radius of the central field obscuration, which introduces a selection effect, hiding planets with smaller apparent angular separations. (3) *OE*, the observational efficiency, which is reduced by operational overheads and scheduling constraint. (We assume $OE = 1$ in this paper.) (4) ζ, the residual level of starlight after suppression, which increases the exposure time to achieve the desired signal-to-noise ratio (*SNR*) on a limiting source. (5) The instability of speckles in the residual starlight, which causes them not to be canceled by image subtraction, which introduces systematic error and restricts the achievable sensitivity to the limit $\Delta mag_{0,\max}$, expressed as a limiting delta magnitude with respect to the star. While the values of these performance parameters are not yet fixed, they can nevertheless be reasonably estimated or assumed for purposes of computing *SNR* and completeness.

Planets of interest. The brightness of a planet reflecting starlight depends on the size, derived from the mass (M_p) and average density (ρ), the geometric albedo (p), the phase function (Φ), and the position of the planet. The planets position is specified by the semimajor axis (*SMA*) and eccentricity (*EPS*), the anomaly, and the three random Euler angles orienting the orbit in space.

In this paper, we study two cases of instrumental parameters and the planetary population of interest, which are described in Tables 1 and 2. Case 1 was analyzed in detail by Brown (2005); this paper extends that analysis to include single-visit optimization. Case 2 introduces variable planet sizes, better limiting sensitivity, and a lower *SNRGoal*, the threshold for "detection."

Stars. There are 1408 stars closer than 30 pc with no known close companion that also have main-sequence lifetimes long enough for life to develop under suitable conditions ($B - V < 0.3$). Spread uniformly over the sky, these stars constitute the input catalog for the *TPF-C* mission. Five stellar characteristics are relevant to searching observations with *TPF-C*. (1) V, the visual magnitude, governs the information rate. (2) L, the stellar luminosity in solar units, determines the physical size of the habitable zone. L affects both the relative brightness (Δmag) and apparent physical separation (s, in AU) of planets of interest. (3) d, the stellar distance in pc, determines the resolved fraction of the habitable zone. (4) M_s, the stellar mass, governs the evolutionary time scale of planetary separation and brightness. (This paper treats only initial searches, so M_s is not utilized.) (5) *RA*, *Dec*, the position on the celestial sphere, determines the days of the year that the star can be observed, given the solar-avoidance angle of the telescope. (We ignore solar avoidance in this paper.) Comprehensive information on potential *TPF-C* target stars is available

Table 1. Adopted design parameters of *TPF-C* mission

A	70000π cm^2	area of entrance aperture
QE	0.17	quantum efficiency
$A_{\rm eff}$	11900π cm^2	effective area (A QE)
λ	550 nm	central wavelength of searching passband
$\Delta\lambda$	110 nm	width of searching passband in nm
IWA	0.05672 arcsec	inner working angle
Ω_x	1.18×10^{-15} str	solid angle of critically sampling pixel
n_x	28.6	noise pixels in photometric aperture
ζ	5×10^{-11}	starlight suppression from theoretical Airy peak
ξ	0.001	dark count rate in sec^{-1} pixel^{-1}
C^2	4	read noise equivalent counts in pixel^{-1}
OE	1	observational efficiency
Case 1 (Brown 2005)		
$\Delta mag_{0,\max}$	25	maximum limiting delta magnitude
$SNRGoal$	10	photometric definition of "detection"
Case 2		
$\Delta mag_{0,\max}$	26	maximum limiting delta magnitude
$SNRGoal$	5	photometric definition of "detection"

Table 2. Adopted planetary populations of interest

ρ	planet density	1 Earth density	
Φ	phase function	Lambertian	
Case 1 (Brown 2005)			
p	geometric albedo	0.33	
SMA	semimajor axis	$0.7\sqrt{L} \leqslant SMA \leqslant 1.5\sqrt{L}$ AU	power law = 0 distribution
EPS	eccentricity	$0 \leqslant EPS \leqslant 0.35$	power law = 0 distribution
M_p	mass	$M = 1$ Earth mass	delta-function distribution
Case 2			
p	geometric albedo	0.2	
SMA	semimajor axis	$0.75\sqrt{L} \leqslant SMA \leqslant 1.8\sqrt{L}$ AU	power law = 0 distribution
EPS	eccentricity	$0 \leqslant EPS \leqslant 0.1$	power law = 0 distribution
M_p	mass	$0.33 \leqslant M \leqslant 10$ Earth mass	power law = −1 distribution

in the *TPF* Target List Database, at http://sco.stsci.edu/tpf_tldb/, which was developed by Margaret Turnbull.

Protocols and priorities. Assume the following observing protocols: There are three types of observations: searching, confirming, and characterizing. The searching observation uses one filter; the exposure time at each roll is *tExpOpt*, assumed the same for all filters and equal to the optimal exposure time discussed in the previous section. Three roll angles are searched, due to the elliptical aperture; the total time cost of searching a star is 3×*tExpOpt*. Confirming observations use three filters—to permit disambiguation by both color and motion, for which the total time cost is (1, 2, or 3)×3×*tExpOpt*, because confusion sources can appear at 1, 2, or 3 rolls; a planet can appear only at 1 roll. The three colors will help—along with the relative motion—differentiate speckles and background objects from planets. Characterizing observations cost 70×*tExpOpt*; this assumes the resolving power of the spectrometer is 70 times that of a filter.

Assume the following observing priorities: We prioritize target stars based on their discovery rate. If no confirming or characterizing observation is pending, then we search the highest priority, as yet unsearched star on the prioritized target list. If a searching observation finds a candidate feature—which could be either a planet or confusion source—at any roll angle, then we perform a confirming observation at each such roll. A confirming observation always confirms a planet and eliminates a confusion source. If a confirming observation confirms a planet, then we make a characterizing observation. We continue until one year of exposure time is exhausted.

Strategic unknowns. Three additional—and currently unknown—parameters will influence the course of the *TPF-C* mission. The occurrence probability, η, is assumed the same for all potential target stars. The confusion probability per searching exposure, $P_{\rm confusion}$, can be either astronomical or instrumental in origin. (Confirming observations are needed to disambiguate confusion sources from planets of interest.) The level of zodiacal light around a target star, (μ, in units of the solar-system zodiacal light) increases exposure times. In this paper, we assume $\mu = 3$, as in Brown (2005).

In the following sections, we compute completeness, estimate the yield of planets in one year of exposure time for searching observations, and explore the stochastic evolution of the *TPF-C* mission for ranges of η and $P_{\rm confusion}$.

2. Completeness

As developed by Brown (2004, 2005), there are various types of completeness, distinguished by qualifying adjectives. "Specific" or "ensemble" refer to whether we treat a specific orbital size and shape (*SMA* and *EPS*) or an ensemble of sizes and shapes. "Visit," "program," or "design" indicates whether we treat a single observing visit, a program of observations (between which the planetary positions and brightnesses evolve), or exhaustive observations, assuming unlimited observing time. Finally, "photometric" and/or "obscurational" refers to whether the completeness is a function of planetary flux (i.e., limited by Δmag_0) or a function of apparent separation (i.e., limited by *IWA*)—or both. Brown (2004) treats specific and ensemble obscurational completeness for the visit, program, and design cases (*SVOC, SPOC, SDOC, EVOC, EPOC,* and *EDOC*). Those results apply to any centrally obscured instrument with unlimited photometric sensitivity, in either thermal radiation or reflected starlight. Brown (2005) treats ensemble, visit, photometric and obscurational completeness (*EVP&OC*) for the case of reflected starlight, without optimization. This paper gives some optimized results for *EVP&OC*. Hunyadi, Shaklan, and Brown (2006) treat both optimized *EVP&OC* and program completeness (*EPP&OC*). Brown and Lisman (2006) treat design completeness (*EDP&OC*), which is the best an instrument can do.

To compute photometric and obscurational completeness, we represent the planetary population of each star by a large Monte Carlo sample drawn from the planetary population of interest. Initially, the planets are placed in their orbits at random mean anomalies. Knowing the orbit, we can compute the positions of planets at future epochs, and then we can compute s and Δmag from

$$\Delta mag = -2.5 \log \left(p \; \Phi(\theta) \left(\frac{R}{r} \right)^2 \right) \quad , \tag{2.1}$$

where θ is the planetary phase angle (planetocentric angle between the star and the observer), R is the planetary radius (derived from M_p using ρ), and r is the physical distance between the planet and star. To determine whether a planet is detected or not, we compare s with $d\ IWA$ and Δmag with Δmag_0. The fraction of planets of interest for which $s > d\ IWA$ and $\Delta mag < \Delta mag_0$ is *EVP&OC*, the completeness at the epoch of initial searching. Figures 1 and 2 in Brown (2005) show the computed probability density

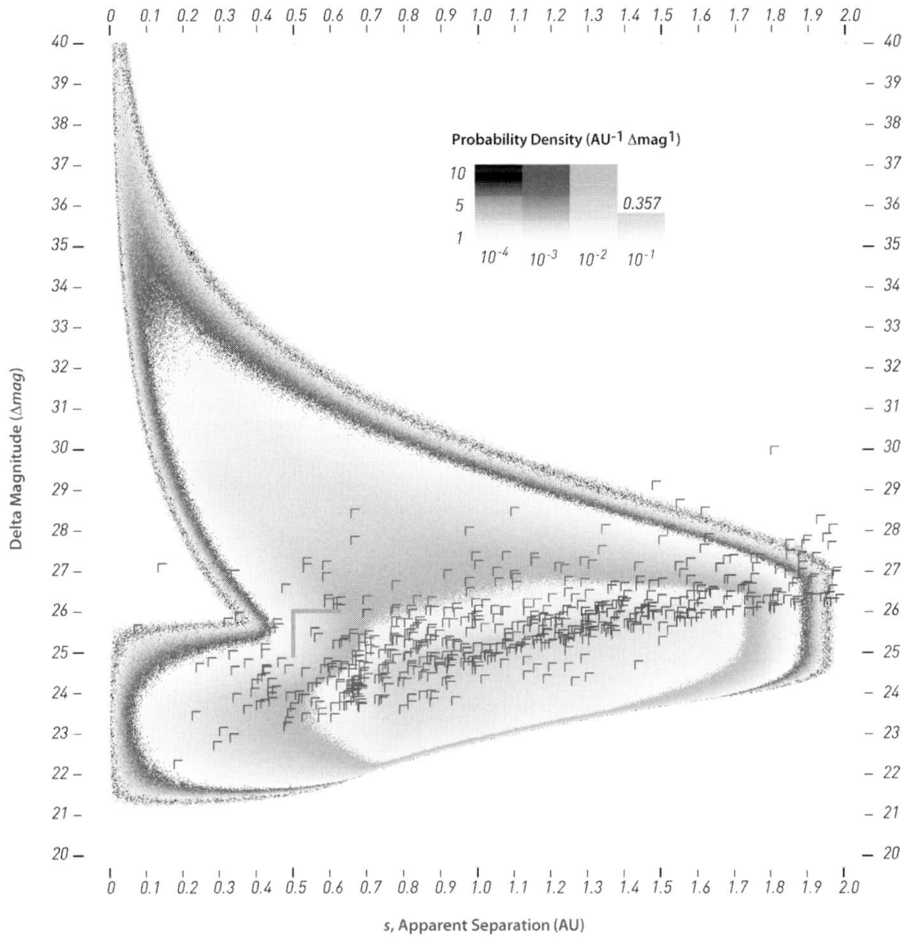

Figure 1. Distribution of probability density vs. s and Δmag for terrestrial planets in the habitable zone in case 2, based on 100 million random planets and shown for $L = 1$. The corners mark the integration zones for computing the completeness on individual valid stars (equivalent positions for $L = 1$). Red: the 198 stars with highest priority in an optimized first search using a total of one year of exposure time. Blue: the 247 lower priority but still valid stars. Orange corner is the fiducial case of Earth at 8.82 pc. The completeness is the probability density integrated below and to the right of a corner. The high priority stars are shown at their optimized values of Δmag_0.

distribution of planets vs. s and Δmag and *EVP&OC* vs. *d IWA* and Δmag_0 for case 1. Figures 1 and 2 in this paper show the corresponding completeness results for case 2.

The number of planets found (0 or 1) is a Bernoulli random variable with probability equal to the accrued completeness times η. This product is also the expectation value of the number of planets found. For multiple observations of multiple stars, the expectation value of the total number of planets found is the sum of the products of accrued completeness times η.

For cases 1(2), only 455(445) stars out of the 1408 in the input catalog are valid targets. "Valid" means that some portion of the habitable zone is unobscured. For each valid star, the search completeness is a unique function of exposure time—unique because the exposure time to achieve *SNRGoal* for any specified Δmag_0 depends on the stellar

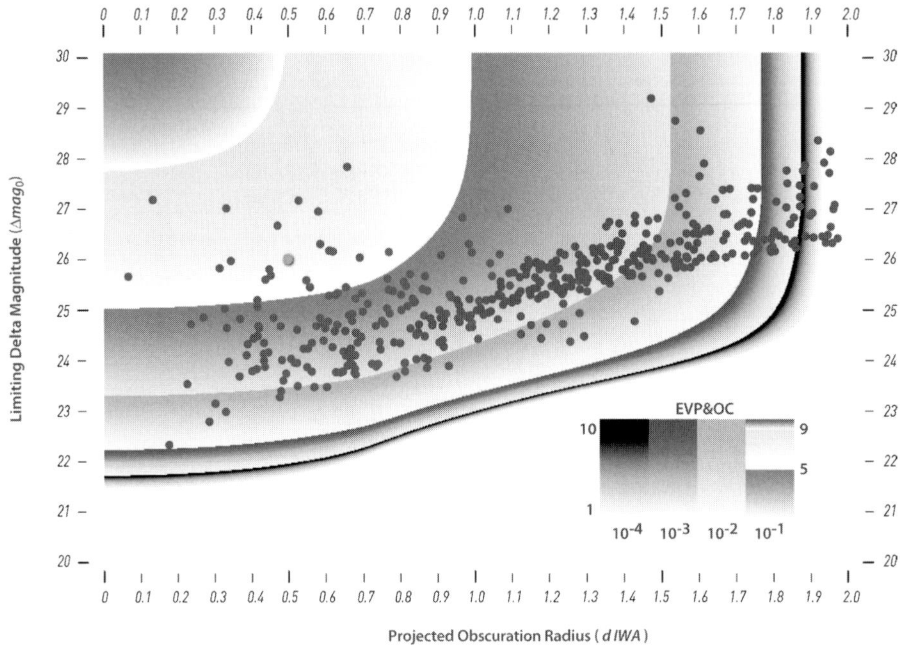

Figure 2. First-visit search completeness (EVP&OC) vs. projected *IWA* and Δmag_0. The points are for valid stars in case 2, based on the corners in Figure 1. The color code is the same as in Figure 1.

quantities V (information rate), d (resolved fraction of the habitable zone, which determines completeness vs. Δmag_0), and μ.

Figure 3 shows completeness vs. exposure time for typical valid stars in case 2. In general, the completeness function has three regimes: an initial time delay before any completeness is accumulated, followed by a monotonic rise, followed by flattening to a terminal value of completeness for exposure times longer than required to achieve *SNRGoal* on $\Delta mag_{0,\max}$, beyond which point the sensitivity is capped by systematics, not limited by photons.

The purely searching observing program described in Brown (2005) is not optimal because it assumed exposing to reach $\Delta mag_{0,\max}$ on all stars. As discussed by Hunyadi, Shaklan, and Brown (2006), this exposure strategy does not make the best use of limited exposure time, because the slope of completeness vs. exposure time on some stars flattens before $\Delta mag_{0,\max}$ is reached. With a limited total budget, it may be advantageous to withdraw some exposure time from such stars and invest it in a new star. An optimized observing program adjusts the exposure times on individual stars and re-prioritizes them dynamically to maximize the sum of completeness for all the stars observed in a given period of time. In the current analyses, the exposure time budget is one year. Figures 1 and 2 show the optimized, single-visit results for individual stars in case 2. The brightest stars are untouched by optimization, because their discovery rates are already high due to short exposure times, and fainter stars show increasingly greater gains in discovery rate from exposure-time optimization.

As shown in Table 3, for one year of exposure time spent purely searching, the non-optimal approach for cases 1(2) produces an expected $31\eta(15\eta)$ discovered planets after searching 117(57) stars. The results for case 2 are lower than case 1, despite increased sensitivity, because the great time costs of the uniformly deeper exposures reduce the

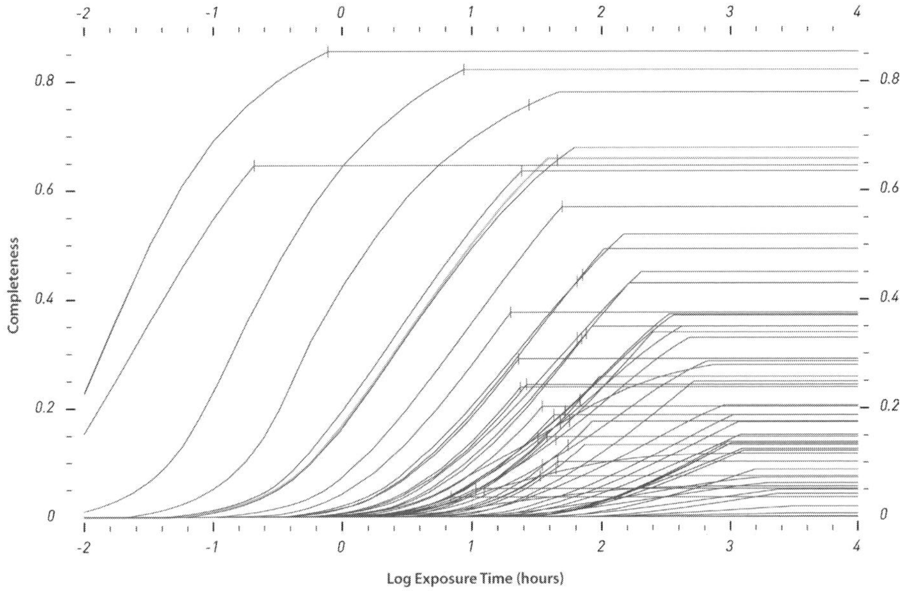

Figure 3. Completeness vs. exposure time for typical stars in case 2. The color code is the same as in Figure 1, except that the fiducial case is now green. Only a random 10% of the 445 valid stars are shown. The marks on high-priority stars show the exposure time after optimization.

Table 3. Pure-search results. Exposure time optimization improves planet yield, moreso for greater Δmag. Planet yield approximately doubles when Δmag is increased from 25 to 26.

Case 1		
Valid stars	455 out of 1408	
Δmag_0	Fixed at 25	Optimized
Planets (stars)	$31\eta(117)$	$33\eta(135)$
Case 2		
Valid stars	445 out of 1408	
Δmag_0	Fixed at 26	Optimized
Planets (stars)	$15\eta(57)$	$56\eta(198)$

number of stars searched. The optimized approach produces an estimated 33(56) planets after searching 135(198) stars. This demonstrates that optimal use of limited exposure time is essential to securing the benefits of greater sensitivity.

3. Mission modeling

We can also use Monte Carlo techniques to model the early *TPF-C* science mission itself, for which we provide the following simple demonstration. In each Monte Carlo trial, we use a Bernoulli random deviate with probability η to assign each target star a number of planets $n_p = 0$ or 1. In each searching observation, the number of planets detected is a Bernoulli random deviate with probability $n_p \times EVP\&OC$. If a planet is detected, we assign it randomly to one of the three rolls. The number of confusion sources detected at each of the three rolls is a Bernoulli random deviate with probability $P_{\text{confusion}}$. Table 4 shows the results of 10^5 Monte Carlo trials for particular values of η and $P_{\text{confusion}}$. Figures 4 and 5 show the histograms of results for the middle entry in the table.

From these results for one year of exposure time, we see that if η approaches unity, the mission becomes spectroscopy dominated, the number of stars searched is ~10% the

Table 4. Results of mission simulations with 10^5 trials for case 1.

η	$P_{\text{confusion}}$ 0.0 planets	0.0 stars	0.5 planets	0.5 stars	1.0 planets	1.0 stars
0	9	135	0	108	0	88
0.05	1.47	115	1.15	73	0.99	58
0.1	2.70	99	2.20	68	1.94	56
0.2	4.69	77	4.02	59	3.65	51
1	11	23	11	23	11	23

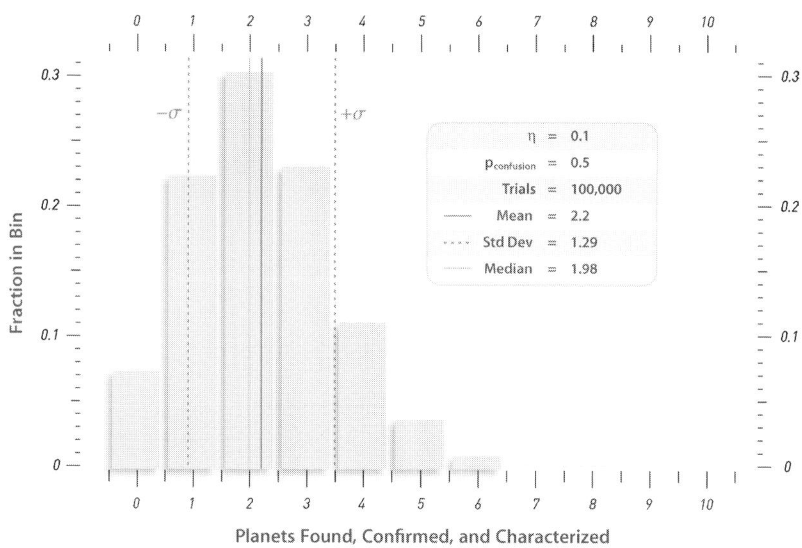

Figure 4. Typical distribution of the number of planets found, confirmed, and characterized by one year of exposure time. Results for the center case in Table 3.

number for small η, and about half the stars searched will have characterized planets. If η approaches zero, the mission becomes search dominated and confusion limited. For intermediate η, the number of planets is proportional to η, but the number of stars searched becomes independent of η as $P_{\text{confusion}}$ approaches unity.

4. Commentary

The completeness formalism provides a useful tool for planning and estimating the search results of the early *TPF-C* mission. We can use discovery rate, which is completeness per exposure time, to optimize exposure times, prioritize target stars, and compute the probability of finding a planet of interest. We can compute completeness for any star from instrumental parameters, the specification of the planetary population of interest, the exposure time, and an assumed value of μ, the level of exozodiacal light.

Because completeness is the expectation value of the number of planets found, it has become the primary performance metric of NASA's *TPF-C* project, used for design trades and verifying that the optical design is compatible with the science requirements.

Mission modeling based on completeness is a useful tool for exploring the character of the *TPF-C* mission for values of the strategic unknowns—η, $P_{\text{confusion}}$, and μ. Including observing overheads and solar avoidance will increase the fidelity of mission modeling.

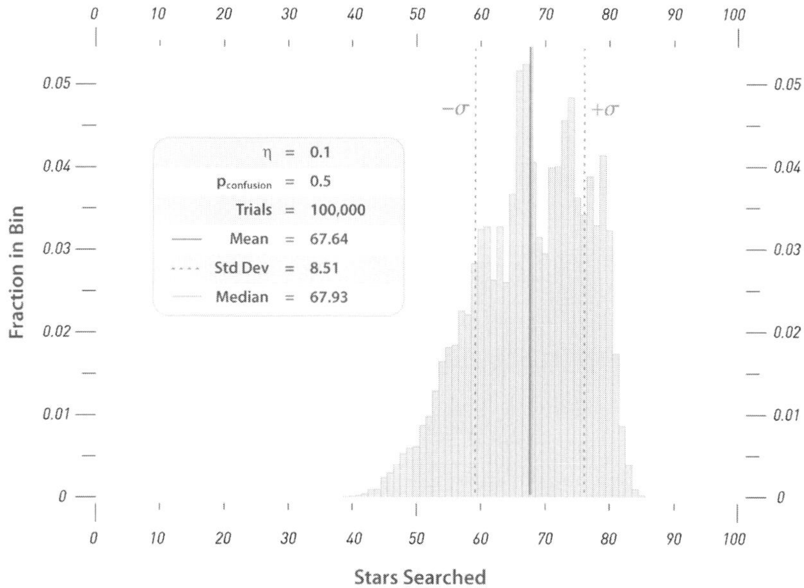

Figure 5. Typical distribution of the number of stars searched for the center case in Table 3.

The *TPF-C* project will use mission modeling to verify that the mission design will meet the science requirements for reasonable ranges of the strategic unknowns.

Acknowledgements

We are grateful for collaborations on completeness and mission modeling with Sarah Hunyadi and Stuart Shaklan of the Jet Propulsion Laboratory. We salute Christian Lallo for his expertise in aiding the computations and his craftsmanship in rendering the results. We thank Sharon Toolan for help in creating the manuscript. JPL contract 1254081 with the Space Telescope Science Institute provided essential support.

References

Brown, R. A. 2004, *ApJ* 607, 1003–1013
Brown, R. A. 2005, *ApJ* 624, 1010–1024
Brown, R. A. & Lisman, D. 2006, *ApJ* (in preparation)
Hunyadi, S., Shaklan, S., & Brown, R. A. 2006, *ApJ* (in preparation)

All photographs: Laurent Thareau [l.thareau@free.fr].

Direct Imaging of Exoplanets: Science & Techniques
Proceedings IAU Colloquium No. 200, 2005
C. Aime and F. Vakili, eds.
© 2006 International Astronomical Union
doi:10.1017/S1743921306009203

Looking for Life in the IR and Neighboring Spectral Regions

N. J. Woolf †

[1]Steward Observatory, University of Arizona, Tucson AZ 85721 and [2]LAPLACE team, NASA Astrobiology Institute
email: nwoolf@as.arizona.edu

Abstract. The paper focuses on the issues for detection of life by remote observations. The problem is divided into the detection of terrestrial planets and the detection of life on them. The problem of detection of planets is compared with the observations of the Hubble Deep Field. The difficulty of detection will likely result in many interesting discoveries before detection techniques are adequate. It is proposed that life is the activity of mutually assisting persistence processes. A persistence process is one that uses the Gibbs Free Energy of the environment for repair and development. It is pointed out that because life produces approximations to equilibrium it is intrinsically difficult to detect by remote observations. Chemical signs of life will be distinctive if the abiotic processes that might produce them are implausible, and preferably absent. The ease of remote detection of terrestrial life is contrasted with problems likely to arise for other planets. It is recommended that initial searches be focused on giant planets. The spectral region of the discovery observations will be set by technical issues which are not yet resolved for any technique. For determining whether life is present, as many spectral regions as possible should be observed.

Keywords. astrobiology.

1. Introduction

The first conference on detecting planets around stars that I attended was at NASA Ames research center in 1976. It was initiated by a plan from SETI to build a giant radio telescope "Cyclops" to search for extraterrestrial communications. I remember Jesse Greenstein organizing the meeting, helped by a young David Black, and Phil Morrison played the role of elder statesman. Now, almost 30 years after that meeting, I am asking some fundamental questions about remote detection of life. What makes life sometimes visible, sometimes not? What does it take to prove that life is present? Are Earth-like planets our best planets for remote detection of life? How rare are Earth-like planets likely to be? Which spectral regions are needed for remote evidence of life, and how convincing can that detection be? How should we best start the planet and life detection process, and how can such a difficult and expensive process proceed? The observer issues involved in the search for extrasolar microbial life are two-fold. There is an astronomical problem of being able to observe earth-like planets around other stars, and there is a bio-geo-chemical problem of trying to interpret evidence about the possible presence of life.

† Present address: Steward Observatory, University of Arizona, Tucson AZ, 85721

2. The Astronomical problem

If we are fortunate, we will find an earth-like planet fairly close, say at a distance of 10 pc. In that case the planet will be comparable in brightness with the faintest stars seen in the Hubble deep field pictures (William *et al.* 1996). The star will be separated from the planet by the width of that faint image, and it will be about 25 magnitudes brighter. Further, the planet image will be confused with other participants in its planetary system. Not only may there be other planets, often with projected separations as low as 0.01 arc seconds, but there is likely to be a strong zodiacal glow, that can only be made less competitive with the Earth signal by increasing the angular resolution.

If we are less fortunate, Earth-like planets may be cosmically abundant (one per galaxy would give 10^{11} planets), but locally rare. The solar system, with its planet orbits having typical eccentricities of 0.06 stands out against the radial velocity planets, the non-roasters with their typical eccentricities of 0.35. Indeed, if the eccentricities of our planet orbits are independent, the probability of the solar system having arisen from the same process as the radial velocity planets is about 10^{-8}. More likely, the eccentricities of planets in a system are interdependent. Or equally there may be a suite of planetary systems like ours, perhaps 30-100 times less common than the currently found radial velocity planets. Sorting out the frequency of systems like ours is a first priority, and perhaps the observations by the Kepler mission will answer this need.

Even though Earths are difficult to observe, and may be modestly rare there are a suite of astronomical problems to be studied on the way to seeing them. Put our goals in perspective. The first stellar parallax observations were made just 1.7 centuries ago. Our conference room at Villefranche-sur-mer is about 4.4 centuries old. *De Revolutionibus* is 4.6 centuries old, and Aristarchus who first looked for stellar parallax lived 23 centuries ago. While astronomy waited for stellar parallax, there were many discoveries along the way, including proper motion, double stars, the velocity of light etc.

Along our way towards other Earths, astronomy can start to ask some fascinating questions. What are the varieties of planetary systems, and the frequency of the various types? How do planetary systems change with time? How can the masses of small planets be estimated, and even better measured? Can giant planets be used to measure the age of systems? How do the abundance and masses of comets change with system characteristics? Can we start to learn about planetary satellites. And there will be all those questions I have not yet thought of because I do not know the varieties of systems that will be found. If we are seeking one specific goal, we are almost guaranteed disappointment. If we are interested in the progress of science, we are almost guaranteed success.

3. Life and terrestrial layered structures

It is not trivial to decide whether some terrestrial structures are produced by life. One kind of layered structure associated with microbial mats is the stromatolite. These heaped up structures arise because microbes grow towards the light by 0.5mm per day, and with some of the sticky substances they produce, they trap sediments. Then variation of growing conditions, day to night, summer to winter etc produce a vertically stratified deposit. In general there is agreement that the younger stromatolites were produced by living organisms, but as they get older, and traces of the organisms are harder to find, there gets to be less and less agreement as to whether they were produced by microbial activity, or are stromatoliths - objects that did not necessarily result from living processes.

One layered structure formerly believed to be fossiliferous was once named the Canadian Dawn Animal *Eozoon canadense* (Hofmann 1971). It was only when a structure like this was found amid igneous material that the idea that it had been some kind of living organism was put to rest.

The Banded Iron Formations (BIFs) that mostly formed from 3-1.8 Gy ago are seen as relics of life processes. Iron, in solution, is green and divalent (ferrous) whereas trivalent ferric iron is red and insoluble. The activity of photosynthetic bacteria either in production of oxygen, or directly by using the iron in solution converted it to insoluble ferric iron or black magnetite (ferri-ferro iron). This settled to the ocean floor. Again, variable production resulted in layered structures, red/black, with milky white chert (SiO_2) layers between, and with the iron layers containing carbon believed to be a residue from the microbes. Such a layered structure could only be produced before the oceans were oxidized and iron ceased to be dissolved. After the Great Oxidation Event (2.2-1.8Gy ago), the iron shows up as redbeds, as in e.g. old red sandstone. There is mostly agreement that BIFs dating back to 3.5 Gy ago were the result of bacterial activity, But 3.7-3.8 Gy layered structures in Greenland are somewhat different, and it is highly disputed as to whether life was involved in their production.

Greenland's Isua BIFs are thinly layered alternations of black magnetite with carbon and a white hydrated magnesium calcium silicate. Here, some suggest that the carbon found with the magnetite is not biological but was produced by the effect of heat and pressure on ferrous carbonate (siderite). This process preferentially leaves ^{12}C rich carbon behind, with the ^{13}C carbon escaping as CO_2.

$$6FeCO_3 = 2Fe_3O_4 + C + 5CO_2$$

On Akilia Island, the layers are different again, with magnetite and carbon found between a highly transparent quartzite. There is a rather small amount of Akilia iron quartzite material, some in smooth uniform layers, but another part has been messed up by a pyroxene dyke passing through, and its study has produced correspondingly odd discussions. But most of the fuss seems to be about whether one can prove that the carbon is an organic residue. The fact that the iron was deposited layered at all in both regions, presents a problem that is more easily resolved by the intervention of living organisms seems to have been lost in the battles. But the point of discussing this here is that if it is so hard to prove that life processes were active on early earth, we are due to have severe problems dealing with the much lower level of evidence we can expect from extrasolar planets.

4. What are we looking for?

Life is characterized by survival (Darwin 1859). The various definitions offered for it do not seem to take this into account, see e.g. (Cleland & Chyba 2002). Indeed, a key need for a definition of life is that it explains the difference between life and death. Biologists seem to want to include evolution in the definition, though it is too slow to be a present observable for most life forms. Chemists want to include the idea that life is a chemical process. It is certainly true that terrestrial life is chemical. But there is no a priori reason why life should be restricted by the absence of chemical interactions, and we will later consider physical life. Likewise, life is not the same as complexity. Indeed if we merely needed complexity to persist, we would never die because our state after a crucial process or organ fails is not appreciably different than before. Besides, complexity is a subjective phenomenon - one person's complexity is another person's mess (high entropy state). Also, life is more than linked cycles (though it is likely to include them), because a number of key life processes, membrane formation, reproduction and evolution for example are not cyclic. Here we will use a hypothesis (to be published) that life is the activity of mutually assisting persistence processes. A persistence process is one that uses the Gibbs Free Energy of the environment for repair and/or development. Key persistence processes include reproduction, metabolism, cellularity and evolution, but the definition of life needs to recognize the similarity that holds these processes together rather than focusing on differences among persistence processes. In this hypothesis death is readily explained as the failure of one process that causes collapse by the domino effect.

Life processes are distinct from those of a stone that merely holds together rather well. And life is different from single persistence process such as fire or crystallization in that linked processes can have benefits of enhancing each other's survival. So reasonably compatible survival processes are likely to stay together after they come together, and to fit their workings together. On the other hand, since environmental resources are limited, there will be competition for them, and associated selection (Malthus 1798, Darwin 1859). Thus cooperation and competition are both inherent in life processes.

The obvious resources of non-equilibrium conditions are in chemical gradients that result from a variation in the oxidation state of material from the inner part of a planet to its surface. Also, the thermal gradient from a hot interior could either drive heat engines in convective fluids, or help maintain an oxidation/reduction gradient. The most distinctive non-equilibrium process on Earth is the thermalization of sunlight. Departure from equilibrium is shown by the presence of the reflection spectrum.

It is also plausible that the influx of complex organic molecules from comets, meteorites, interplanetary dust particles and interstellar dust has played a role in initiation of key prebiotic processes (Deamer *et al.* 2002). These natural products are organized molecules, and so are potentially contributors to the initial organization of living organisms.

The fact that life uses the departures from local thermodynamic equilibrium of the environment implies that life is an equilibrator. And since natural process also tend to produce equilibration, though usually on a slower timescale, life will often be intrinsically difficult to detect.

5. Where and how to look for life

Because life feeds on departure from LTE, the most distinctive features produced by life are likely to be produced in transition regions, where mixing produces departures from LTE. Transitions between liquid solid and gaseous surfaces are likely to acquire the greatest abundance of life activities.

For planets where the atmospheres is not transparent or has a strong haze layer above the surface, leaving it unseen from outside, it will be hard to get evidence about the presence or absence of life, because evidence about physical conditions at the surface will be hard to obtain. On the other hand, for some other planets the absence of an atmospheric shield against strongly ionizing electromagnetic radiation may mean that life needs to hide beneath the surface, and that could make it hard to detect too.

Chemical signs of life will be distinctive if the abiotic processes that might produce them are implausible, and preferably absent. If abiotic processes are potentially present but weakly active, then unless life produces a very strong signal the evidence will be ambiguous. This then shows that the presence of life is unlikely to be determined by the mere presence or absence of some material, but rather by the determination of the quantity of the material coupled to an analysis of potential source and sink rates. Such an analysis will require a suite of good measures that characterize the planet and its environment.

Chemical life like that on Earth has a possibility of revealing itself because it uses reduction/oxidation gradients to manufacture the materials it uses. Earth's interior has materials produced in the reducing environment characteristic of the proto-planetary solar system. As a result, a strongly reducing atmosphere is unlikely of itself to be a useful indicator - it will seem to show that the internal environment of the planet is dominating its exterior. Even if this is untrue, and in the absence of life the exterior would be mildly oxidizing or neutral, that will be extremely hard to prove.

A strongly oxidizing atmosphere on the other hand is less likely. Any volcanic activity will continuously bring up reducing material, which will combine with the oxidizer and remove it. Oxidation is the natural result of photolysis of water and hydrogen escape from the surface of a small planet, so it will be up to the observer to prove that the rate of hydrogen escape is insufficient to cause the observed condition. On Earth at present, water photolysis is about 10^6 times less effective than is oxygenic photosynthesis.

The best chemical evidence for life will be the simultaneous observation of oxidizers and reducers, e.g. oxygen and methane. Then since each destroys the other, there is evidence of simultaneous sources of each, and the prediction of production rates may be possible.

From these comments it can be seen that for very cool planets or large planets which retain hydrogen, proof of the presence of life will require some extraordinary material or process to be present. For most planets and satellites, life may well be present but the evidence of it will rarely be accessible to the remote observer, at least with the minimal telescopes that will make early direct detection of planets. Advanced telescopes either ones that image the planet as a single point, but produce detailed high signal-to-noise ratio spectra or ones that eventually image the surface, or probes that visit the planetary system could have a substantially greater ability to detect life.

Meanwhile, there is a different activity, currently in just a few biochemical laboratories, but likely to become more widespread in the future. Biochemists are exploring the processes of the origin and development of terrestrial life. If the astronomical observations come too slowly, they may be bypassed by our understanding so well how Earth life formed that we can project the likelihood of life developing elsewhere. This will be particularly true if we find that no rare processes occurred in the origin of terrestrial life.

6. Why is life on Earth so readily detectable?

Sagan *et al.* (Sagan *et al.* (1993)) studied Earth from the Galileo spacecraft while it was en route to Jupiter. They noticed three potential signs of life. The first was the suite of strong modulated monochromatic radio signals emitted from our planet, from TV, radar stations etc. The second was the red vegetation edge, especially prominent in Amazonia. The third was the abundant oxygen and weak methane signatures of our Atmosphere.

As we consider the possibility of observing radio waves as with the SETI research, we have to realize how recent is our radio and radar technology. Other technologies such as the steam engine have nearly vanished after only a 200 year reign 10^{-7} of our planet's history. We are at a stage of rapid development in technology, and it would seem that our radio technology using powerful transmissions may well be replaced e.g. by laser transmission over large distances, or by processes we have not yet considered. It may be that there will be signs of advanced technology, but we are liable to confuse the signs of advanced technology with natural noise sources. Shannon has indeed pointed out that an optimum communication will look like noise.

The vegetation red edge is a sign of an internal photoelectric effect, but our experience of such effects have shown some photoelectric effect with a sharp wavelength of cut on, and others where there is a near asymptotic rise below the critical wavelength. The chlorophyll edge that we see could appear at some other wavelength, and the energy requirements for oxygenic photosynthesis operated by coupled processes could permit the photoelectric edge to be at a wavelength of as long as 1 micron - it all depends on the energy loss of electrons as they cascade down a biochemical pathway. Even though for some places on Earth there is a remarkably strong edge, the integrated signal is quite small Arnold *et al.* (2002), Woolf *et al.* (2002), Seager *et al.* (2005). The problem is perhaps that land vegetation requires abundant water, and this is mainly present in regions with frequent cloud cover. The vegetation red edge mainly seems to be a signal to be explored when there is good spatial resolution over a planet such as Sagan *et al.* used.

Finally, we come to the various signatures of oxygen, methane, nitrous oxide etc. To us this is compelling evidence of the presence of life. But it is compelling because we have good evidence of the surface conditions on Earth. It will be hard to know nearly as much about a remote planet. Atmospheric oxygen has a turn over time of about 10,000 years because 99.9% of the oxygen is used up in oxidizing the organic material produced, and so production and destruction rates are close to equilibrium. The high oxygen abundance in Earth's atmosphere occurs because destruction processes are relatively inefficient. In the Carboniferous Era, before white rot fungi had developed as a way of demolishing dead trees, oxygen destruction was even less efficient and the oxygen abundance was as high as 35%. At the extinction event at the Permian Triassic boundary (of uncertain cause) the abundance was for extended periods near 10%. In the Proterozoic Era, when oxygen production was likely within a factor of a few of that today, the oxygen abundance was much lower, perhaps because the small structures of bacteria were more easily oxidized. Even more disturbing is the evidence suggesting that oxygenic photosynthesis could plausibly have been around for 1.5 Gy before the atmospheric oxygen content became appreciable. The transition to a weak oxygen atmosphere is currently an unexplained event, somehow coupled to the great snowball earth events near 2.3 Gy ago, and the late Proterozoic rise to near present levels seems associated with a second set of snowball earth events. Since we plan to use an oxygen atmosphere as an indicator of an inhabited planet, we need to understand why our atmospheric oxygen abundance has responded so strangely to both glaciations and to biological production.

We have Earth as a planet that at present has easily visible effects produced by life. Perhaps this is connected to the presence of oceans. Some of this is because surface life is protected by our atmospheric ozone layer, but our atmosphere is still substantially transparent. Some of the visibility is an effect of recent human technology, part of a rapid transition and therefore of doubtful duration. It is interesting and curious. It is worth looking for such easily observable effects elsewhere (e.g. SETI). The terrestrial signatures phenomena as a group also deserve more study. If a not-Earthlike planet is beyond the solar system, the prospects of being able to identify the activity of biological processes appears to be substantially reduced.

7. Determining the characteristics of a planet from its spectrum

The distinction in the title of this section, which was supposed for the title of this talk is not a very helpful one. The first issue is to detect terrestrial planets rather than to characterize them. En route to that major achievement, we will be learning how to take the planet's spectrum, and the first region where we observe will be set by technical capability rather than a strong scientific preference. Because of the difficulty in proving that we have a planet, either habitable or with life, we need all the information from all spectral regions we can get!

The boundaries of infrared and ultraviolet are set by the sensitivity of the eye and do not match the boundaries of present day or likely future detectors. Likewise the boundary between the reflection spectrum of the Earth and its emission spectrum is not at one end of the infrared but rather near the geometric mean of the infrared spectrum, either might be considered and infrared detection! The main vibrational transitions of molecules are in the mid infrared, but electronic transitions such as in chlorophyll A are at one end of it. Rayleigh scattering is one of our better measurers of atmospheric pressure varies as λ^{-4}. But the $\tau = 1$ point moves longward with increasing pressure, and in the plausible range of terrestrial planet pressures it can move from the ultraviolet through the visible into the infrared.

The optimum wavelength for having some unambiguous measures is the mm. wave region. Millimeter waves would show a continuum which measures ground temperature. The 5mm oxygen line measures both amount of oxygen and ground pressure. The 13.5mm line measures amount of water and also gives ground pressure. A radio interferometer could eliminate star radiation 10^5 times the planet signal, a factor 100 better than for the 9.7 micron ozone line, and there would be a minimal signal from zodiacal dust. Unfortunately the measurement needs several square km of collecting area even with system noise temperature of 10K. So, for the foreseeable future we must consider the IR through UV range.

Each of the 7 octaves of spectrum from 25 to 0.2 μm. has some useful spectral feature to offer! Sometimes strong bands will saturated, and weaker bands will be more informative. Sometimes there will be features overlapped, and additional lines will be needed for confirmation. In general, an identification from a single broad feature is not a good identification, especially if the noise is high. This is why we have the concept of a more sensitive device, Life Finder. Also, there are differences between the emission and reflection spectrum of a planet. All the bands in the emission region of the spectrum are not only telling about the molecules present and their amount, but are also telling about the thermal structure of the atmosphere down to whatever layer develops a high optical depth. Likewise the reflection spectrum is good for telling about the quantities of material in the atmosphere above some reflecting layer, perhaps ground, perhaps cloud, but the temperature of the various layers is not determined, and the size of the planet and hence its mass are ill known. For these reasons examining both spectral regions is desirable.

All bands that we could see in early studies will be in the square root portion of the curve of growth, where the band strength is more determined by the atmospheric pressure than by the amount of the molecule. Though angular resolution is good in the visible region of the spectrum, the near infrared just longward offers a much better view of some key features related to habitability. Each of the 7 octaves of spectrum from 25 to 0.2 μm. has some useful spectral feature to offer! Sometimes strong bands will saturated, and weaker bands will be more informative. Sometimes there will be features overlapped, and additional lines will be needed for confirmation. In general, an identification from a single broad feature is not a good identification, especially if the noise is high. This is why we have the concept of a more sensitive device, Life Finder. Also, there are differences between the emission and reflection spectrum of a planet. All the bands in the emission region of the spectrum are not only telling about the molecules present and their amount, but are also telling about the thermal structure of the atmosphere down to whatever layer develops a high optical depth. Likewise the reflection spectrum is good for telling about the quantities of material in the atmosphere above some reflecting layer, perhaps ground, perhaps cloud, but the temperature of the various layers is not determined, and the size of the planet and hence its mass are ill known. For these reasons examining both spectral regions is desirable.

All bands that we could see in early studies will be in the square root portion of the curve of growth, where the band strength is more determined by the atmospheric pressure than by the amount of the molecule. Though angular resolution is good in the visible region of the spectrum, the near infrared just longward offers a much better view of some key features related to habitability and the search for life (Turnbull *et al.* in press).

8. Physical life

Chemical life may not be the only life in the universe. We already have mobile Mars Rovers with sensors, communication power and a little intelligence at the device end. We could imagine devices with more tactile and actuating ability, more intelligence, sensors to observe their internal states, a capability of making mines and factories to reproduce themselves, and a core program that drives them to "survive". This would be one possible example of physical life. We have not seen self-initiated or non-human initiated physical life elsewhere, but there is always a first observation.

Our chemical complexity tends to produce emotional rather than rational responses. In contrast, electronic life with its logic circuits is inherently rational. It will recognize that diversification of habitat is a better survival technique than developing a large population in one place. Indeed too large a population in one place makes it vulnerable and potentially self-destructive. An astronomically wide diversification of habitat is preferred. For an organism with reliance on electricity, places like Earth with chemically active atmospheres and oceans are to be avoided (do not follow the water!). Low surface gravitational potential is helpful, so moons and asteroids are likely habitats. But since stellar radiant energy will be used for most activity, the inner parts of planetary systems are the most likely places. The organisms would be physical autotrophs, and there would likely be minimal chemical signature.

9. Looking Under the Lamp Post

The illustration above is one of many that could be made to show that our universe may have much life that is not spectroscopically detectable. Indeed as has been shown, Earth is likely an unusual place where we think we could observe it from a distance and be fairly sure that life is there. So if Earth-like planets are sufficiently common to be observable, they are obvious candidates to examine for signs of life, and non-Earth like planets will be harder to interpret for signs of life. If Earth-like planets are not common, our next best option will be to observe planetary systems to understand why Earths are rare, and what causes the rarity. We are looking for life, not where it is, but where we can hope to observe it - if it is present. This is analogous to the man looking for his front door key under the lamp post, because that is the only chance he has for finding it.

The study of Earths will be expensive. Looking for Earths will be the hardest task ever attempted in astronomy, requiring telescopes and auxiliary equipment that is large and very precise. How will we persuade our fellow humans to pursue this task, and if we cannot, is there any alternative?

10. The strategy

We already have a short-term strategy! Every astronomer interested in planets, having access to a large ground telescope, and having a suitable discovery strategy is trying to observe extrasolar planets. Those who have a creative understanding of optics are trying to improve the techniques. The next phase will be to make steadily more difficult observations progressing towards a 25 magnitude differential between the planet and its star for the shorter wavelengths of the reflection spectrum, and differentials of 18 magnitudes at mid-IR wavelengths, all at a separation of a fraction of an arc second. New larger telescopes are being commissioned, built or planned. Some are trying to get large telescopes at unusually appropriate sites such as Antarctica. Space projects are being developed, but the large funds required are keeping the projects mostly about 20 years into the future. And the fate of the Superconducting Supercollider is in all of our consciousnesses. Not only do unlikely processes require excellent verification, but the search for them requires convincing a public that prefers sports and cosmetics, and in at least one country, does not even accept the evidence that evolution is a fact.

Our powers of persuasion are limited, because though the appropriate strategy for long-term survival is to expand a living organism's bases to other worlds, our chemical makeup is ill adapted to novel environments. Indeed our expansion on Earth has been to create our preferred environment on a small scale. But in space, the fall-back strategy has to be to make a doubly protected environment, because the terrestrial fall-back of living in the unprotected environment will not work there, and the protected environment becomes small and cramped. Nobody would want to spend a lifetime in a small cramped environment for the benefit of long-term-future descendants. So we are forced to appeal on the grounds of interest rather than usefulness to humanity.

In the long run we require telescopes so large that current economic processes will not support the work. But fortunately current terrestrial economics do not apply to the distant future. We are in the middle of the second industrial revolution, the automation and artificial intelligence revolution. What physical life forms do on their own, particularly off the surface of the earth is not set by our input. They would create their own energy from the Sun, they could create their own substances by mining and refining. They could create their own labor force by factory reproduction. And they would build their own telescopes. The long term study of life in the universe may well be out of human control.

Acknowledgements

Conversations with Roger Angel have always been stimulating, though he should in no way be blamed for my wildest speculations. Steve Mojzsis has shown me specimens of the Akilia and Isua rocks, and has discussed the problems of their interpretation. Studies have been part of the work of the LAPLACE team of the NASA Astrobiology Institute. Some support for this work has also been provided by JPL under contract nb 1258748.

References

Arnold, S., Gillet, O., Lardire, P., Riaud, J. & Schneider J. 2002, *Astronomy & Astrophysics* 392(1), 231-237

Darwin C. 1859, *The Origin of species by means of natural selection*, John Murray, London. 1859

Deamer, D., Dworkin, J.P., Sandford, S.A., Bernstein, M.P., & Allamandola, L.J. 2002 *Astrobiology* 2, 371-381

Cleland C.E., Chyba CF 2002 *Orig Life Evol Biosph* 32, 387-393

Hofmann H.J. 1971 *Geological Survey of Canada Bulletin* 189

Malthus T. 1798 *An Essay on the Principle of Population* J. Johnson, London 1798

Sagan, C., Thompson, W.R., Carlson, R., Gurnett, D., & Hord, C. 1993 *Nature* 365, 715-721

Seager, S., Turner, E., Schafer, J., Ford, E.B. *Astrobiology* 5, 372-390

Williams, R.E., Blacker, B., Dickinson, M., Dixon, W.V., Ferguson, H.C., Fruchter, A.S.,

Giavalisco, M., Gilliland, R.L., Heyer, I., Katsanis, R., Levay, Z., Lucas, R.A., McElroy, D.B., Petro, L., Postman, M., Adorf, H., & Hook, R.N. 1996 *Astronomical Journal* 112:1335
Woolf, N.J., Smith, P.S., Traub, P.A., & Jucks, K. 2002 *The Astrophysical Journal* 574,430-433

Discussion

VAKILI: You have not explained your choice of spectral region for first observation

WOOLF: The choice of the region for initial observations will be set by technological difficulties not yet fully explored. I have previously stated my reasons for being concerned with issue of vibrations. These issues apply principally to observations at short wavelengths using restricted size space telescopes. Other problems of equal gravity may yet surface.

Direct Imaging of Exoplanets: Science & Techniques
Proceedings IAU Colloquium No. 200, 2005
C. Aime and F. Vakili, eds.
© 2006 International Astronomical Union
doi:10.1017/S1743921306009215

Tracing Planets in Circumstellar Disks

Sebastian Wolf

Max Planck Institute for Astronomy, Königstuhl 17, 69117 Heidelberg, Germany
email: swolf@mpia.de

Abstract. We present simulations which demonstrate to which extent mid-infrared images and millimeter maps can be used to trace the location of giant planets in circumstellar disks. The most promising approach is to look for characteristic signatures in circumstellar disks caused by the interaction of giant planets with the disk. Numerical simulations show that these signatures are usually in size much larger than the planet itself and thus much easier to detect. The particular result of the planet-disk-interaction depends on the evolutionary stage of the disk. Primary signatures of planets embedded in disks are gaps in the case of young disks and characteristic asymmetric density patterns in debris disks. Radiative transfer simulations predict that high spatial resolution observations performed with instruments/telescopes that will become available in the near future will allow to trace the formation and evolution of planets in protoplanetary and debris disks.

Keywords. Protoplanetary disk, Planet formation, Planet-disk-interaction, Radiative transfer simulation, Multi-wavelength observation.

---

1. Introduction

Planets are expected to form in circumstellar disks, which are considered as the natural result of the protostellar evolution, at least in the case of low and intermediate mass stars. While a detailed picture of the evolution of the circumstellar environment, in particular of circumstellar disks, has been developed already, the planet formation process is in major parts still under discussion. There exist two major scenarios for planet formation. The first is characterized by three main phases: a) dust grain growth from submicron-sized particles to centimeter / decimeter-sized bodies via coagulation, followed by b) an agglomeration process that leads to the formation of (sub)kilometer-sized planetesimals, which c) form terrestrial (rocky) and Uranian (icy) planets or the cores of Jovian (gaseous) planets by further accretion of solid material (see, e.g., Pollack *et al.* 1996; Weidenschilling 1997). Alternatively, planetesimals may form via the gravitational instability of solids that have settled to the midplane of a circumstellar disk (Goldreich & Ward 1973; Youdin & Shu 2002). Adequate constraints from observations are required in order to either verify or to rule out existing hypotheses about these planet formation scenarios.

Once (proto-)planets have been formed, they may significantly alter the surface density profile of the disk and thus cause signatures in the disk that are much easier to find than the planets themselves. The appearance and type of these signatures depend on the mass and orbit of the planet, but even more on the evolutionary stage of the circumstellar disk. The most important of these signatures are gaps and spiral density waves in young circumstellar disks (see, e.g., Bryden *et al.* 1999; Kley 1999) and resonance structures in evolved systems, so-called debris disks (see, e.g., Liou & Zook 1999). The importance of investigating these signatures lies in the possibility to use them for the search and the characterization of embedded planets. Therefore, disk features can provide constraints on the processes and timescales of planet formation. Indeed, several debris disks around nearby main-sequence stars show structures and asymmetries that are considered to result from planetary perturbations (Holland *et al.* 1998, 2003; Schneider *et al.* 1999; Koerner, Sargent, & Ostroff 2001).

In this article we discuss selected features resulting from the planet-disk interaction which can be used to trace planets in circumstellar disks. We consider both, high-resolution imaging techniques and spectroscopy as tools to constrain planetary parameters.

2. Young Circumstellar Disks

Hydrodynamic simulations of gaseous, viscous protoplanetary disks with an embedded protoplanet show that planets with masses $\gtrsim 0.1\,\mathrm{M_{Jupiter}}$ produce significant perturbations in the disk's surface density (Bate *et al.* 2003). In particular, they may open and maintain a large gap. This gap, which is located along the orbit of the planet, may extend up to several astronomical units in width, depending on the mass of the planet and the hydrodynamic properties of the disk. Protoplanets also launch spiral waves. However, in the presence of turbulence, these waves appear to be diffused and dissipated (Nelson & Papaloizou 2003). Furthermore, D'Angelo, Henning, & Kley (2002) found smaller-scale spirals in the vicinity of the planet that are detached from the main ones: Along these small spirals the gas orbits the planet, representing a circumplanetary disk (see also Lubow, Seibert, & Artymowicz 1999).

The observability of the outlined features has been investigated by Wolf *et al.* (2002). It was shown, that high-resolution images in the (sub)millimeter wavelength range as to be obtained with the *Atacama Large Millimeter Array* (*ALMA*) will allow to trace the gap, but not the large-scale spiral structure. Wolf & D'Angelo (2005) performed more detailed simulations with the goal to answer the question whether the circumplanetary environment of a giant planet, embedded in a young circumstellar disk, can also be detected. The detection of a gap would already represent a strong indication of the existence of a planet, thus giving information such as planetary mass, viscosity, and pressure scale-height of the disk. However, the detection/non-detection of warm dust close to the planet would provide valuable additional constraints for the temperature and luminosity of the planet and on the density structure of the surrounding medium. It was shown that a detection of the circumplanetary accretion disk will indeed be possible (see Fig. 1). Two major constraints that limit the applicability of this approach to study planet formation were found: First, only for nearby objects at distances of typically less than $\sim 100\,\mathrm{pc}$ the spatial resolution provided by *ALMA* will be sufficiently high to allow to spatially separate the circumstellar disk and circumplanetary material from one another. Second, the emission from the circumplanetary environment is significantly strong only in the case of massive and thus young circumstellar disks. Furthermore, the signal of the radiation from the planet/circumplanetary material will be best distinguishable if a high planet-to-star mass ratio is targeted. We want to remark that in contrast to the most successful planet detection technique so far, based on radial velocity measurements (see, e.g., Mayor & Queloz 1995, Marcy & Butler 2000), the likelihood of detection does not increase with decreasing distance of the planet to the star (within the considered range of orbital radii of the planet of up to 5 AU).

The contribution of the planet to the net flux in the (sub)millimeter wavelength range – by direct or scattered radiation and reemission of the heated dust in its vicinity – is negligible compared to the dust reemission from the entire disk. Furthermore, the planetary radiation significantly affects the dust reemission spectral energy distribution (SED) only in the near to mid-infrared wavelength range. However, since this spectral region is

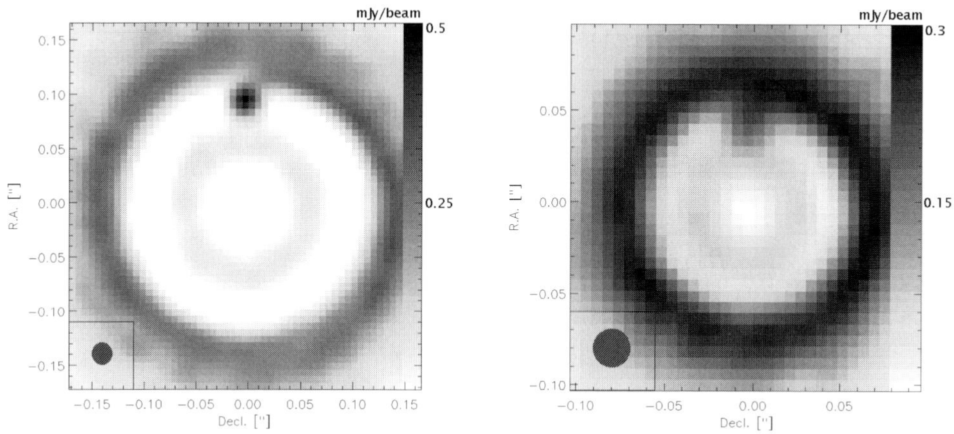

Figure 1. Simulation of *ALMA* observations of a disk with an embedded 1 $M_{jupiter}$ planet around a 0.5 M_{sun} star (orbital radius: 5 AU). The assumed distance is 50 pc (left)/100 pc (right). The disk mass amounts to $M_{disk} = 1.0 \times 10^{-2}\,M_{sun}$ (total disk radius: 300 AU). Only structures above the 2σ-level are shown. The size of the combined beam is symbolized in the lower left edge of each image. Note the reproduced shape of the spiral wave near the planet and the slightly shadowed region behind the planet. *[from Wolf & D'Angelo 2005].*

influenced also by the warm upper layers of the disk and the inner disk structure, the planetary contribution and thus the temperature/luminosity of the planet cannot be derived from the SED alone.

Since the location, depth, and width of a gap depend on the planetary mass and orbital parameters, these quantities can be constrained on the basis of high-resolution images. However, this interpretation will rely on the comparison with numerical simulations and thus on the physical processes considered. For example, the results from magneto-hydrodynamic simulations of circumstellar disks with an embedded planet show, that the gaps are shallower and asymmetrically wider than in the case of purely hydrodynamic simulations, and the rate of gap-formation is slowed (Winter, Balbus, & Hawley 2003).

3. Debris Disks

Debris disks are solar-system-sized dust disks with micron-sized grains produced as by-products of collisions between asteroid-like bodies left over from the planet formation process. In the case of our solar system, the debris of Jupiter-family short-period comets and colliding asteroids represents the dominant source of zodiacal dust located between Mars and Jupiter. A second belt of dust is located beyond the orbit of Neptune (see, e.g., Dermott *et al.* 1992; Liou, Dermott, & Xu 1995). Besides the solar system, optical to mid-infrared images of β Pic (see, e.g., Weinberger, Becklin, & Zuckerman 2003) and submillimeter images of Vega, Fomalhaut, and ϵ Eri (Holland *et al.* 1998; Greaves *et al.* 1998) have revealed spatially resolved debris disks which were first inferred from observations of infrared flux excesses above photospheric values with *IRAS*. Since the mass of small grains in debris disks and therefore the thermal dust reemission from these disks is much smaller than in the case of T Tauri/HAe/Be circumstellar disks, only a very limited sample of observations exists so far.

The much lower optical depth and lower gas-to-dust mass ratio in debris disks (Zuckerman, Forveille, & Kastner 1995; Dent *et al.* 1995; Liseau & Artymowicz 1998; Greaves, Coulson, & Holland 2000) let the stellar radiation – in addition to

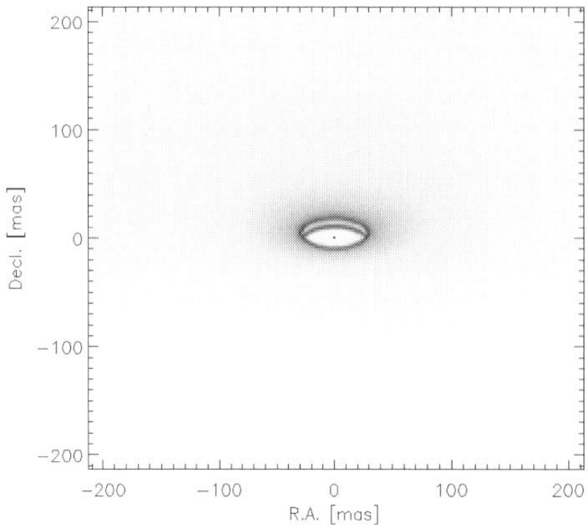

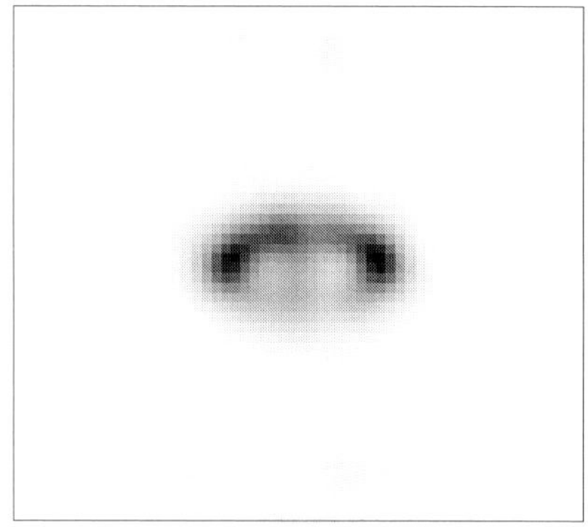

Figure 2. Simulated 10μm image of the inner region of a T Tauri circumstellar disk with a cleared inner region (radius: 4 AU), seen under an inclination of 60° (assumed distance: 140 pc). *Left:* Original image; *Right:* Reconstructed image (not scaled), based on simulated *VLTI* observations combining 4 Telescope (APreS-MIDI/MATISSE concept study; see e.g., Lopez *et al.* 2005). *[from Wolf et al. 2005].*

gravitation – be responsible for the disk structure. The Poynting-Robertson effect, stellar wind drag, and - if existing - gravitational stirring by embedded planets are all important in determining the dust population and disk structure (see, e.g., Grady *et al.* 2000; Moro-Martín, Wolf, & Malhotra 2005). Similar to T Tauri-like disks, however, embedded planets may alter the debris disk structure remarkably (see, e.g., Gor'kavyi *et al.* 1997; Kenyon & Bromley 2001). In evolved, optically thin debris disks, a planet may produce an asymmetric resonant dust belt with one or more clumps and a central cavity void of dust via resonances and gravitational scattering (Ozernoy *et al.* 2000; Liou & Zook 1999). The resulting characteristic density patterns are expected to provide strong indirect hints on the existence of planets in these disks. Kuchner & Holman (2003) pointed out four basic high-contrast resonant structures a planet with eccentricity $\lesssim$0.6 can create in a disk of dust released on low eccentric orbits: (i) a ring with a gap at the location of the planet, (ii) a smooth ring, (iii) a clumpy eccentric ring, and (iv) an offset ring plus a pair of clumps. The appearance/dominance of one of these structures mainly depends on the mass of the planet and the eccentricity of its orbit.

Observed SEDs of selected disks around T Tauri and debris-type disks have been found to show hints of inner cavities (void of small dust grains) such that the inner radius of the disk is apparently much larger than the sublimation radius of possible dust species. Examples among T Tauri disks are GM Aurigae (Koerner *et al.* 1993, Rice *et al.* 2003) and TW Hya (Calvet *et al.* 2002). In Fig. 2 the reconstructed image of a circumstellar disk with an inner hole is shown which could be obtained with the VLTI combining up to 4 Auxiliary Telescopes (Wolf *et al.* 2005). In several prominent debris disks, inner cavities have been found as well: β Pic (inner radius: 20 AU), HR 4796A (30-50 AU), ϵ Eri (50 AU), Vega (80 AU), and Fomalhaut (125 AU) - see, e.g., Dent *et al.* 2000; Wilner *et al.* 2002; Holland *et al.* 2003. Taking into account the physical processes responsible for the spatial dust density distribution in debris disks, scattering of particles by massive planets is assumed to be the major effect in explaining these large inner holes: Dust grains drifting inwards due to the Poynting-Robertson effect are likely to be scattered into larger orbits resulting in a lower dust number density within the planet's orbit. Clearing the inner disk results in a loss of hot dust which is mainly responsible for the near/mid-infrared shape of the SED (see Fig. 3). With increasing gap size the minimum value of

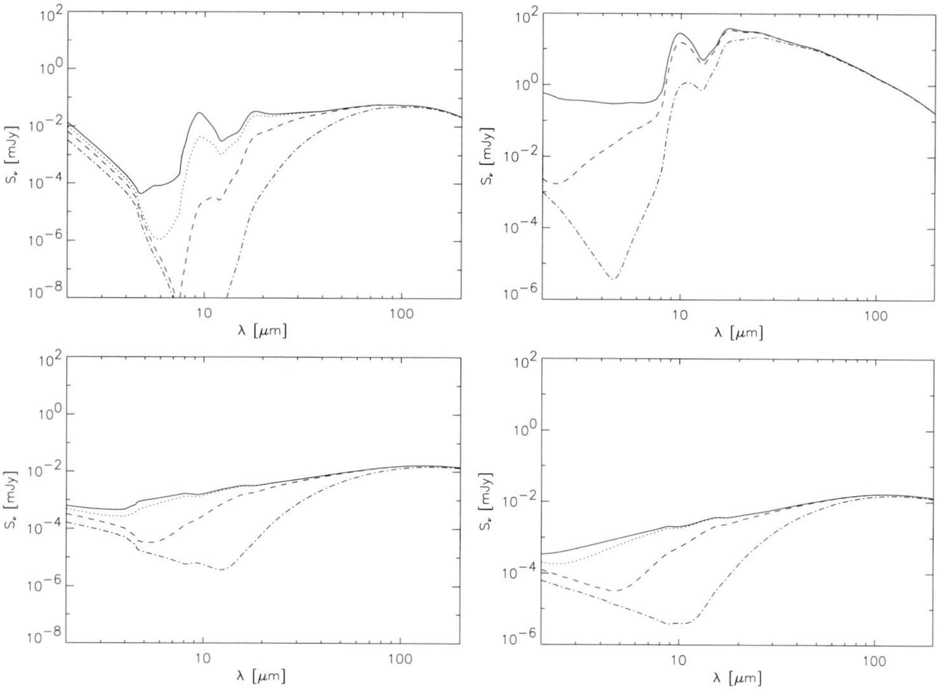

Figure 3. Influence of an inner gap on the SED of a debris disk. Inner disk radius: dust sublimation radius (solid line), 0.1 AU (dotted), 1 AU (dashed), and 10 AU (dash-dotted). Left/Right column: Fe-poor Silicate ($MgSiO_3$)/Fe-rich (Olivine, $MgFeSiO_4$). *Top:* $a = 0.1\mu$m, *Bottom:* $a = 1$mm. Disk mass: $10^{-10}M_{sun}$. *[from Wolf & Hillenbrand 2003]*.

the mid-infrared flux becomes smaller and is shifted towards longer wavelengths. Detailed studies of the influence of planets on the SED of debris disks have been performed by Wolf & Hillenbrand (2003) and Moro-Martín, Wolf, & Malhotra (2005).

4. Summary

Numerical simulations convincingly demonstrate that high-resolution imaging performed with instruments/telescopes that will become available in the near future will allow detection of planets in circumstellar disks. However, scattering of stellar light and thermal emission by small grains (in the case of optically thick disks seen at high inclination also absorption) will make direct planet detection via imaging hardly feasible. But planets cause large-scale perturbations in the disks which will be observable with observatories like the *Stratospheric Observatory For Infrared Astronomy* (*SOFIA*), the *James Webb Space Telescope* (*JWST*), but also ground-based observatories/interferometers, such as *ALMA* or the *VLTI* by spatially resolved mapping of the dust reemission. The particular type of perturbation depends on the evolutionary stage of the disk. Primary signatures of planets embedded in disks are gaps and circumplanetary accretion disks in the case of young disks and characteristic asymmetric density patterns in debris disks. Secondary signatures are, for instance, spiral structures in young disks or warps due to planets on inclined orbits.

Acknowledgements

S. W. was supported by the German Research Foundation through the Emmy Noether grant WO 857/2-1.

References

Bate, M.R., Lubow, S.H., Ogilvie, G.I., & Miller, K.A. 2003, *MNRAS* 341, 213
Bryden, G., Chen, X., Lin, D.N.C., Nelson, R.P., & Papaloizou, J.C.B. 1999, *ApJ* 514, 344
Calvet, N., D'Alessio, P., Hartmann, L., Wilner, D., Walsh, A., & Sitko, M. 2002, *ApJ* 568, 1008
Dent, W.R.F., Greaves, J.S., Mannings, V., Coulson, I.M., & Walther, D.M. 1995, *MNRAS* 277, L25
Dent, W.R.F., Walker, H.J., Holland, W.S., & Greaves, J.S. 2000, *MNRAS* 314, 702
Dermott, S.F., Durda, D.D., Gustafson, B.A., Jayaraman, S., Xu, *et al.* 1992, *in Asteroids, Comets, Meteors* A.W. Harris & E. Bowell (eds.), 153
D'Angelo, G., Kley, W., & Henning, Th. 2003, *ApJ* 586, 540
Goldreich, P., & Ward W.R. 1973, *ApJ* 183, 1051
Gor'kavyi, N.N., Ozernoy, L.M., Mather, J.C., & Taidakova, T. 1997, *ApJ* 488, 268
Grady, C.A., Sitko, M.L., Russell, R.W., Lynch, D.K., Hanner, M.S., *et al.* 2000, *in Protostars and Planets IV* V. Mannings, A.P. Boss, S.S. Russell (eds.), University of Arizona Press, p. 613
Greaves, J.S., Holland, W.S., Moriarty-Schieven, G., *et al.* 1998 *ApJ* 506, L133
Greaves, J.S., Coulson, I.M., & Holland, W.S. 2000, *MNRAS* 312, L1
Holland, W.S., Greaves, J.S., Dent, W.R.F., Wyatt, M.C., Zuckerman, B., *et al.* 2003, *ApJ* 582, 1141
Holland, W.S., Greaves, J.S., Zuckerman, B., *et al.* 1998, *Nature* 392, 788
Kenyon, S.J., & Bromley, B.C. 2001, *ApJ* 121, 538
Kley, W. 1999, *MNRAS* 303, 696
Kley, W. 2000, *MNRAS* 313, L47
Koerner, D.W., Sargent, A.I., & Ostroff, N. 2001, *ApJ* 560, L181
Kuchner, M.J., & Holman, M.J. 2003, *ApJ* 588, 1110
Liou, J.-C., Dermott, S.F., & Xu, Y.-L. 1995, *Planet. Space Sci.* 43, 717
Liou, J.-C., & Zook, H.A. 1999, *AJ* 118, 580
Liseau, R., & Artymowicz, P. 1998, *A&A* 334, 935
Lopez, B., Wolf, S., Dugue, M., Graser, U., Mathias, Ph., *et al.* 2005, *in Proc. of "The Power of Optical/Infrared Interferometry"*, ESO Garching (Germany)
Lubow, S. H., Seibert, M., & Artymowicz, P. 1999, *ApJ* 526, 1001
Marcy, G.W., & Butler, R.P. 2000, *PASP* 112, 137
Mayor, M., & Queloz, D. 1995, *Nature* 378, 355
Moro-Martín, A., Wolf, S., & Malhotra, R. 2005, *ApJ* 129, 1049
Nelson, R.P., & Papaloizou, J.C.B. 2003, *MNRAS* 339, 993
Ozernoy, L.M., Gorkavyi, N.N., Mather, J.C., & Taidakova, T.A. 2000, *ApJ*, 537, L147
Pollack, J.B., Hubickyj, O., Bodenheimer, P., Lissauer, J.J., Podolak, M., & Greenzweig, Y. 1996, *Icarus* 124, 62
Schneider, G., Smith, B.A., & Becklin, E.E. 1999, *ApJ* 513, L127
Weinberger, A.J., Becklin, E.E., & Zuckerman, B. 2003, *ApJ* 584, L33
Wilner, D.J., Holman, M.J., Kuchner, M.J., & Ho, P.T.P. 2002, *ApJ* 569, L115
Winters, W.F., Balbus, S.A., & Hawley, J.F. 2003, *ApJ* 589, 543
Wolf, S., & D'Angelo, G. 2005, *ApJ* 619, 1114
Wolf, S., Gueth, F., Henning, Th., & Kley, W. 2002, *ApJ* 566, L97
Wolf, S., & Hillenbrand, L.A. 2003, *ApJ* 569, 603
Wolf, S., Lopez, B., Augereau, J.-Ch., Berruyer, N., Chesneau, O., *et al.* 2005, "Mid Infrared VLTI Imager – APreS-MIDI/MATISSE Science Case Study", subm. to ESO
Youdin, A.N., & Shu, F.H. 2002, *ApJ* 580, 494
Zuckerman, B., Forveille, T., & Kastner, J.H. 1995, *Nature* 373, 494

Direct Imaging of Exoplanets: Science & Techniques
Proceedings IAU Colloquium No. 200, 2005
C. Aime and F. Vakili, eds.
© 2006 International Astronomical Union
doi:10.1017/S1743921306009227

What can we learn about giant planets from low resolution spectra?

Jonathan J. Fortney[1,2] **and Mark S. Marley**[1]

[1]Space Science and Astrobiology Division, NASA Ames Research Center, Mail Stop 245-3, Moffett Field, CA, 94041
email:jfortney@arc.nasa.gov, mark.s.marley@nasa.gov
[2]National Research Council Postdoctoral Fellow

Abstract. Here we examine the visible spectra of giant planets in anticipation of the science return of missions like the Terrestrial Planet Finder-Coronagraph and proposed Discovery class space coronagraph missions EPIC and ECLIPSE. Our understanding of extrasolar giant planets is already greatly improving because of our studies of old brown dwarfs (which have effective temperatures similar to young giant planets), transiting hot Jupiters, and the planet Jupiter itself. The first data collected on Jupiter-like extrasolar giant planets will likely consist of magnitudes in a few filters or very low resolution spectra. We investigate diagnostics for determining planetary effective temperature, atmospheric chemical abundances, cloud cover, and mass using such limited data. In general, giant planet science is improved significantly if missions in the visible domain extend to wavelengths as long as possible, within engineering constraints.

Keywords. planets and satellites: Jupiter, planetary systems, radiative transfer

1. Introduction

The imaging and spectroscopy of extrasolar giant planets (EGPs) will be a main focus of astronomy in the in the next decade. Optical coronagraphs in space are attractive platforms for detecting stellar flux reflected from EGP atmospheres. It is nearly assured that the first data gathered on old, Jupiter-like EGPs will be in the form of magnitudes in a few well-placed filters, or perhaps very low resolution spectra. It is interesting to address what one could learn about giant planets from this kind of data. Other authors have published fairly hi-resolution spectra of EGPs Sudarsky *et al.* (2003)Sudarsky, Burrows, & Hubeny, Burrows *et al.* (2004)Burrows, Sudarsky, & Hubeny at a variety of orbital separations. Here, we use some of these published models, as well as our own independent spectral calculations, to determine what may be learned about EGPs with only a minimum amount of information. We focus on optical spectra of giant planets older than 1 Gyr and less massive than 10 M_J, and thus already limits our discussion to effective temperatures below ~450 K Burrows *et al.* (2001)Burrows, Hubbard, Lunine, & Liebert.

2. EGP phase space to date

It is usually assumed that Jupiter-like EGPs are a form of *terra incognita*, and while this is for the most part true, there is already available data to help to constrain what these cool, Jupiter-like planets may look like. Our understanding of EGPs is currently progressing on three fronts, as there are three kinds of objects that are currently available for detailed study. These objects help to show us what EGPs are like when they are:

• **Young.** Old brown dwarfs below $T_{\rm eff} = 1000$ K have spectral properties very similar young giant planets. Jupiter's $T_{\rm eff}$ after formation was probably ~700-100 K, and there are already a handful of brown dwarfs that have been found in this temperature range. For example, we now have excellent data and models for Glies 570D, at ∼800 K Geballe *et al.* (2001)Geballe, Saumon, Leggett, Knapp, Marley, & Lodders. In addition, in the very near future even cooler brown dwarfs will be found. Isolated brown dwarfs below ∼500 K are expected to have water clouds in their visible atmospheres. So, in some sense, we understand what the spectra of young giant planets should be.

• **Hot.** Currently, thermal emission had been detected from two Pegasi planets (or "hot Jupiters") with the *Spitzer Space Telescope*. These are HD 209458b, which was seen in the MIPS 24 μm band by Deming *et al.* (2005) Deming, Seager, Richardson, & Harrington and TrES-1, which was seen in the IRAC 4.5 and 8.0 μm bands by Charbonneau *et al.* (2005)Charbonneau, Allen, Megeath, Torres, Alonso, Brown, Gilliland, Latham, Mandushev, O'Donovan, & Sozzetti. For both planets, soon data will be obtained in all 4 IRAC bands and the MIPS band. These planets will be an interesting test-bed for understanding giant planet atmospheres with only a few filters. It is probabilistically likely that cooler transiting planets, reaching perhaps down to 900 K, could be seen with *Spitzer* in the coming few years.

• **Jupiter itself.** While much attention is paid to refining atmosphere models for substellar objects through comparison with brown dwarf data, almost no attention is paid to Jupiter itself. However, it is the substellar-mass object that we understand best. If we aim to to understand EGP atmospheres in the 100-500 T_{eff} range, it makes sense to test our models against Jupiter data that has been in hand for decades.

In summary, at the current time we have data to constrain EGP model atmospheres to temperatures down to $T_{\mathrm{eff}} = 700$ K (from observations of the coolest brown dwarfs) and also at $T_{\mathrm{eff}} = 124.4$ K, from Jupiter. Within several years even cooler brown dwarfs will be found that may reach down to 500 K, where water clouds will likely begin to form in these atmospheres Marley *et al.* (2002)Marley, Seager, Saumon, Lodders, Ackerman, Freedman, & Fan, Burrows *et al.* (2003)Burrows, Sudarsky, & Lunine.

3. Deriving Quantities of Interest from Spectra

Low resolution spectroscopy, while it will not provide the wealth of information that can be derived from high resolution spectra, can be used to determine valuable physical properties for EGPs. As noted earlier, if planetary systems to be studied by TPF-C (or Eclipse or EPIC) are older than 1 Gyr, this already greatly constrains planetary effective temperatures. For planets less massive than 10 M_{J} and greater than 1 AU from their parent star the planet's effective temperature is less than 450 K Burrows *et al.* (2001)Burrows, Hubbard, Lunine, & Liebert. For this range of planets, important physical quantities one could likely understand are detailed below.

• **Effective Temperature.** At visible wavelengths, determining T_{eff} is difficult, but potentially can be done two ways. The best method is to look for thermal emission for young or hot planets at the longest wavelengths one has available. As shown in Fig. 1, at some wavelengths longer than 0.85 μm thermal flux is greater than reflected flux. The ratios of the thermal flux peaks can be used as a temperature indicator down to ∼200 K, where thermal flux is finally swamped by reflected flux. If a mission bandpass can be extended to the near infrared, one could extend this method to even longer wavelengths. One also notices in Fig. 1 that the cooler the planet, the brighter it is in reflected light (at a given distance). This is due to the gradual formation of water and then ammonia clouds, which become optically thicker as a planet cools. To a point, the thicker clouds become more likely to scatter a given photon back into space. Thus, a more subtle indicator may an increasing brightness in the visible as the planet cools.

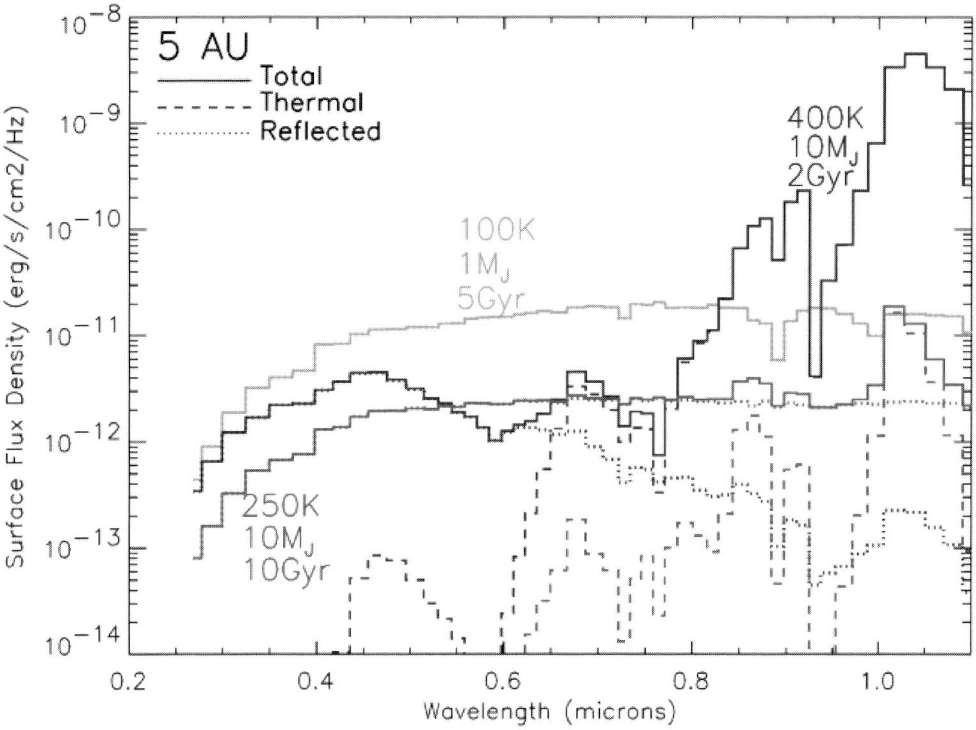

Figure 1. Emergent surface fluxes for 3 planets at 5 AU from the Sun. Thermal (dashed line), reflected (dotted line), and total (solid line) are shown for each planet. Labels indicate the intrinsic effective temperature, mass, and and approximate planet age. A description of our atmosphere code can be found in Marley *et al.* (1996)Marley, Saumon, Guillot, Freedman, Hubbard, Burrows, & Lunine, Burrows *et al.* (1997)Burrows, Marley, Hubbard, Lunine, Guillot, Saumon, Freedman, Sudarsky, & Sharp, and Fortney *et al.* (2005a)Fortney, Marley, Lodders, Saumon, & Freedman.

• **Cloud Properties.** Fig. 1 shows that the slope of the reflected spectrum is a sensitive indicator of cloud optical depths. As ammonia and water clouds are expected to be essentially gray scatterers, a planet with optically thick clouds should show a reflection of the stellar spectrum at wavelengths where the planet's atmosphere has no absorption features. A slope that deviates to the blue from ~0.45 to 0.6 μm indicates holes in clouds or a cloud that is less optically thick. This spectral slope could be determined from a simple wide-band color-color diagram. (See also Metallicity.)

• **Metallicity.** Methane dominates the visible spectra of these objects. However, if one is only able to gauge the depth of one methane band, then it is not possible to determine if an absorption band depth is due to a methane abundance effect to a cloud height effect. However, bands of different strengths are formed at different pressures in a planet's atmosphere. Stellar photons that are absorbed in weak methane bands penetrate deeper into the planet's atmosphere before they are absorbed. In this way, observing the strength of 2 (or hopefully 3 or 4) methane bands, either through medium-bandwidth methane filters or low resolution spectra, can be used to simultaneously determine the methane abundance and cloud top height. The methane abundances in the atmospheres of Jupiter and Saturn are ~3 and 7 times what would be expected for a solar composition atmosphere. This tells of something (that we are still trying to decipher) about the planet formation process in our solar system. Spectral models of Jupiter-like planets with varying methane abundances look nearly identical in wide-band color-color diagrams, so it is quite important to place filters directly on methane features. However, the flux from these bands is necessarily lower, so significantly more integration time may be needed.

• **Mass.** Determining a planet's mass with the help of radial velocity or astrometry is much preferred over any spectral methods. However, a combination of a system's age, a determined effective temperature (as described above) and theoretical cooling models, which are likely accurate for Jovian planets at Gyr ages Fortney & Hubbard(2003), can be used to estimate a planet's mass.

4. Data and Models to Date

4.1. *Jupiter*

Jupiter will long remain our best studied giant planet. The spectrum of Jupiter can be used to test chemical abundance calculations, cloud models, and radiative transfer methods. In Fig. 2(left) we plot flux vs. wavelength for full-disk observations of Jupiter, as seen at opposition, based on data obtained by Karkoschka (1994). This spectrum is computed by multiplying the solar spectrum at 5.2 AU by the Karkoschka (1994) geometric albedo data of Jupiter. We overplot our model atmosphere, with a metallicity of [M/H]=0.5 (3.2 $\times$ solar) which is roughly consistent with Jupiter's atmospheric abundances Atreya *et al.* (2003)Atreya, Mahaffy, Niemann, Wong, & Owen. We use the cloud model of Ackerman & Marley(2001) (with a sedimentation efficiency parameter, $f_{\rm sed}$, of 3) to describe the ammonia cloud particle distribution and optical properties. These spectra are normalized at 0.7 μm. As one can see, at wavelengths longer than 0.5 μm, our model is an excellent match to the overall spectral slope. At wavelengths shorter than 0.5 μm, Jupiter's photochemically derived stratospheric haze layer darkens the planet Rages *et al.* (1999) Rages, Beebe, & Senske. We do not include this haze in our models. One should also notice that our band depths, which are essentially all due to CH_4, are not as deep as observed. This may be because our cloud of condensed NH_3 is too high in the planet's atmosphere. But overall the spectrum is a good match given the lack of any fine tuning.

For an additional comparison, we plot a Burrows *et al.* (2004)Burrows, Sudarsky, & Hubeny spectrum for a Jupiter-like planet a 6 AU from a G2V parent star. Again this spectrum is normalized at 0.7 μm. A main difference between their spectrum and that of Jupiter is the overall visible spectral slope, which in their models is bluer than Jupiter. In addition, their band depths for a solar composition atmosphere are considerably deeper than Jupiter in most bands, but is actually too shallow near 1 μm. These differences could be partially accounted for if in their models, their cloud optical depths at visible wavelengths are not as large as in the real planet. This could be due to a particle size effect. In potential support of this hypothesis, we show our cloud-free "Jupiter" model spectrum, which has a similar slope to the Burrows *et al.* (2004)Burrows, Sudarsky, & Hubeny model from 0.5 to 0.8 μm. A consequence of the bluer Burrows *et al.* (2004)Burrows, Sudarsky, & Hubeny spectral slope in that in wide-band color-color diagrams, while our models plot very close to the actual planet, those of Burrows *et al.* (2004)Burrows, Sudarsky, & Hubeny do not. We not that this comparison has only been performed over the visible spectral range and the agreement between each of the models and Jupiter may be better or worse in the infrared. The main point is that cloud optical properties, rather than gaseous optical properties, dominate the visible spectrum.

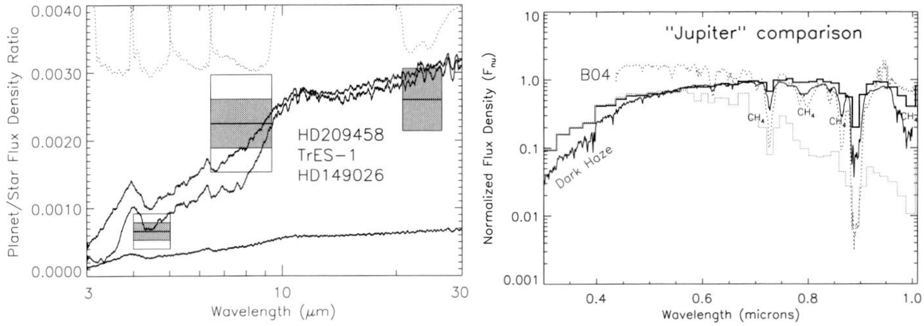

Figure 2. Left: Normalized flux density vs. wavelength for model planets and the observed Jupiter. The actual Jupiter spectrum is the solid high resolution spectrum. A Burrows *et al.* (2004)Burrows, Sudarsky, & Hubeny model of a solar-composition planet at 6 AU from a G2V star is the dotten high resolution line. Our newly computed low resolution models are the solid, histogram-type lines. The dark line is for a 3 times solar model at 5.2 AU including predicted cloud opacity. The gray line is for a solar composition atmosphere, but neglects cloud opacity. **Right**: Planet-to-star flux density ratios for, from top to bottom, HD 209458b, TrES-1, and HD 149026b. Spitzer data are shown—shaded areas are 1 σ error bars, and open boxes are 2 σ error bars. The 4.5 and 8.0 μm data are for TrES-1and the 24 μm datum is for HD 209458b. The four Spitzer IRAC bands and MIPS band are shown as dotted lines at the top. Data will eventually be taken in all of these available bands. There is no data yet published for HD 149026b.

4.2. *Pegasi planets*

The *Spitzer* observations of HD 209458b by Deming *et al.* (2005) Deming, Seager, Richardson, & Harrington and TrES-1 by Charbonneau *et al.* (2005)Charbonneau, Allen, Megeath, Torres, Alonso, Brown, Gilliland, Latham, Mandushev, O'Donovan, & Sozzetti are the best data available for use in understanding the atmospheres of this interesting class of EGPs. Fig. 2 (right) shows how our models of the atmospheres of these Pegasi planets match the available observations. The modeling procedures and a more detailed discussion can be found in Fortney *et al.* (2005a)Fortney, Marley, Lodders, Saumon, & Freedman, Fortney *et al.* (2005b)Fortney, Saumon, Marley, Lodders, & Freedman. The planet-to-star flux density ratios we obtain from our models are in general a good fit. We are within the 1 σ error bar for HD 209458b. For TrES-1, we are at the 1 σ error bar a 4.5 μm, but our models are too dim by 2-3 σ at 8.0 μm. Our model is bluer than a blackbody, but apparently the planet is redder than a blackbody in this spectral region Charbonneau *et al.* (2005)Charbonneau, Allen, Megeath, Torres, Alonso, Brown, Gilliland, Latham, Mandushev, O'Donovan, & Sozzetti. We find the agreement is better for an atmosphere enhanced in metals by 3-5$\times$, or one that has a hot stratosphere, but more data will be needed for a definitive conclusion. Also plotted are the predicted flux density ratios for HD 149026b, a recently discovered transiting planet with a very small radius Sato *et al.* (2005)Sato, Fischer, Henry, Laughlin, Butler, Marcy, Vogt, Bodenheimer, Ida, Toyota, Wolf, Valenti, Boyd, Johnson, Wright, Ammons, Robinson, Strader, McCarthy, Tah, & Minniti. In the coming years, *Spitzer* data on these and other transiting planets will be our best testing ground for trying to understand EGP atmospheres with data in only a few wide filters.

5. Conclusions

Our understanding of EGPs is progressing as we learn more about old brown dwarfs (which have spectra similar to young giant planets), Pegasi planets (which can be observed with *Spitzer*, and Jupiter (which is itself an important data point). Low resolution spectra, or colors from several visible filters, can potentially be used to gauge an object's T_{eff}, metallicity, some aspects of cloudiness, and perhaps to constrain its mass. An instrument spectral range out to at least 1 μm helps these derivations considerably, especially for determining T_{eff}. By the time that old Jupiter-like planets are directly imaged from space in visible light, perhaps with TPF-C or a smaller platform, model atmospheres will be even better constrained by additional brown dwarf and transiting planet data.

References

Ackerman, A. S. & Marley, M. S. 2001, ApJ, 556, 872
Atreya, S. K., Mahaffy, P. R., Niemann, H. B., Wong, M. H., & Owen, T. C. 2003, Planet. Space Sci., 51, 105
Burrows, A., Hubbard, W. B., Lunine, J. I., & Liebert, J. 2001, Rev. Mod. Phys, 73, 719
Burrows, A., Marley, M., Hubbard, W. B., Lunine, J. I., Guillot, T., Saumon, D., Freedman, R., Sudarsky, D., & Sharp, C. 1997, ApJ, 491, 856
Burrows, A., Sudarsky, D., & Hubeny, I. 2004, ApJ, 609, 407
Burrows, A., Sudarsky, D., & Lunine, J. I. 2003, ApJ, 596, 587
Charbonneau, D., Allen, L. E., Megeath, S. T., *et al.* 2005, ApJ, 626, 523
Deming, D., Seager, S., Richardson, L. J., & Harrington, J. 2005, Nature, 434, 740
Fortney, J. J. & Hubbard, W. B. 2003, Icarus, 164, 228
Fortney, J. J., Marley, M. S., Lodders, K., Saumon, D., & Freedman, R. 2005a, ApJ, 627, L69
Fortney, J. J., Saumon, D., Marley, M. S., Lodders, K., & Freedman, R. S. 2005b, ApJ submitted, ArXiv e-prints: astro-ph/0507422
Geballe, T. R., Saumon, D., Leggett, S. K., Knapp, G. R., Marley, M. S., & Lodders, K. 2001, ApJ, 556, 373
Karkoschka, E. 1994, Icarus, 111, 174
Marley, M. S., Saumon, D., Guillot, T., Freedman, R. S., Hubbard, W. B., Burrows, A., & Lunine, J. I. 1996, Science, 272, 1919
Marley, M. S., Seager, S., Saumon, D., Lodders, K., Ackerman, A. S., Freedman, R. S., & Fan, X. 2002, ApJ, 568, 335
Rages, K., Beebe, R., & Senske, D. 1999, Icarus, 139, 211
Sato, B., Fischer, D. A., Henry, G. W., *et al.* 2005, ApJ, 633, 465
Sudarsky, D., Burrows, A., & Hubeny, I. 2003, ApJ, 588, 1121

Discussion

W. Traub: What Jupiter data are you using? The planet's belts and zones vary considerably.

J. Fortney: This is the full-dusk full-phase data from Karkoschka (1994).

M. Liu: Would you care to comment on non-equilibrium chemistry? This has been seen with CH_4 and CO in T dwarf spectra.

J. Fortney: The main reactions will be CH_4/CO, and NH_3/N_2. CO is dredged up from deeper layers, where it is stable, and only slowly is converted to CH_4, relative to the mixing timescale, so the CH_4/CO ratio is smaller than one would expect. Similarly, in the other reaction NH_3 is found in a lower abundance, since N_2 is only slowly converted to NH_3. In reality this issue should not have that great of an effect on cool giant planets.

E. Guidos: Please explain again what kind of cooler transiting planets you expect?

J. FORTNEY: I meant that it is statistically likely that cooler transiting planets will soon be found. However, these cooler planets will be more difficult to detect with *Spitzer*.

W. BENZ: You mentioned our "ground truth" measurements of abundances for Jupiter's atmosphere from the Galileo Entry Probe, and how they mostly agreed with ground-based determinations. Can you comment on the apparent disagreement between Cassini derived abundances and those from the ground?

J. FORTNEY: Well, I would consider Cassini just another remotely sensed observation, and less reliable than an entry probe.

SOMEONE: For the *Spitzer* observations what is actually measured are the brightness temperatures, not the effective temperatures, correct?

J. FORTNEY: Yes, that is true. We compute brightness temperatures for our models and that is what we actually compare to the data. I was quoting effective temperatures for models that match the observed brightness temperatures.

All photographs: Laurent Thareau [l.thareau@free.fr].

Direct Imaging of Exoplanets: Science & Techniques
Proceedings IAU Colloquium No. 200, 2005
C. Aime and F. Vakili, eds.
© 2006 International Astronomical Union
doi:10.1017/S1743921306009239

Terrestrial Exoplanet Light Curves

Eric Gaidos[1], Nicholas Moskovitz[2] and Darren M. Williams[3]

[1]School of Ocean/Earth Sciences & Technology, University of Hawaii, Honolulu, HI 96822 USA
email: gaidos@hawaii.edu

[2]Institute for Astronomy, University of Hawaii, Honolulu, HI 96822 USA
email: nmosko@ifa.hawaii.edu

[3]School of Science, Penn State Erie, The Behrend College, Erie, PA 16563 USA
email: dmw145@psu.edu

Abstract. The phase or orbital light curves of extrasolar terrestrial planets in reflected or emitted light will contain information about their atmospheres and surfaces complementary to data obtained by other techniques such as spectrosopy. We show calculated light curves at optical and thermal infrared wavelengths for a variety of Earth-like and Earth-unlike planets. We also show that large satellites of Earth-sized planets are detectable, but may cause aliasing effects if the lightcurve is insufficiently sampled.

Keywords. stars: planetary systems, techniques: photometric, planets and sateliites: general, Earth, scattering.

---

1. Introduction

The variation in the reflected or emitted light from a planet as it orbits its parent star contains unique information about that body. In 1610, Galileo Galilei used his telescope to discover that the changing brightness of Venus was partly a consequence of its phases (and also overthrowing the Ptolmeic theory of the Cosmos). His anagram to the Tuscan ambassador of Prague (the forerunner to the IAU telegram?) read *Haec immatura a me iam frustra leguntur o.y.* [This immature female has already been read in vain by me] but unscrambled to *Cynthiae figuras aemulatur mater amorum* [The mother of love (Venus) resembles Cynthia (the Moon)] (van Helden 1989).

The first direct detection of an Earth-sized planet around another star would be a scientific triumph of equal significance, yet the ultimate goal is to obtain information about its atmospheres and surfaces, especially those properties related to its ability to support life, or even evidence for life itself. Spectroscopy is a powerful technique to obtain such information (Des Marais *et al.* 2002, Seager *et al.* 2005), however high signal-to-noise is required. Regardless of whether spectroscopy is succesful, direct detection of a planet, confirmation of its companion status, and determination of its orbit require observations at multiple epochs. Photometry at these epochs will produce a partial or complete phase light curve of the planet. Such a light curve is distinct from photometric variability in the planet induced by its rotation and surface features (Ford, Seager & Turner 2001). The latter may be diffiult to observe, but easy to remove by sufficiently long integration times.

A theoretical model of an orbital light curve must include a full description of the star-planet-observer geometry, including 5 orbital parameters plus the obliquity (Figure 1). The light curve will also depend on the scattering and emission properties of the atmosphere and surface modulated by any seasonal effects induced by finite obliquity or eccentricity. The presence of Saturn-like rings around a giant planet may also manifest itself by specific photometric features (Arnold & Schneider 2004). The reflected and emitted light from extrasolar giant planets has been modeled, including effects induced by changes in effective temperature on highly eccentric orbits (Burrows, Hubeny & Sudarsky 2005,Fortney *et al.* 2005, Dyudina *et al.* 2005, Sudarsky *et al.* 2005, Barman, Hauschildt & Allard 2005). Although the complete light curve of a giant planet has not been measured, there are useful upper limits for the reflected light of close-in and transiting planets (Charbonneau *et al.* 1999, Leigh *et al.* 2003, Green *et al.* 2003, Deming *et al.* 2005, Snellen 2005) and detections of secondary eclipses at thermal infrared wavelengths (Charbonneau *et al.* 2005, Deming *et al.* 2005).

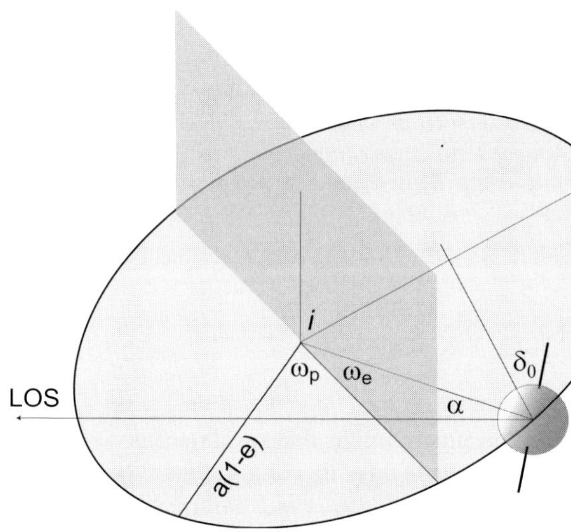

Figure 1. The geometry of the star-planet-observer system for light curve calculations is specified by the inclination of the normal of the planetary orbit plane to the line of sight (i), semi-major axis (a) and orbital eccentricity (e), the longitude of the periastron relative to the ascending node (ω_p), the longitude of the spring equinox of the hemisphere most in view relative to the ascending node (ω_e), and the obliquity (δ_0). α is the phase angle.

2. Reflected Light from Terrestrial Planets

The reflected light from an Earth-like planet will depend on the scattering properties (phase function) of the principle reflectors, i.e., clouds, the atmosphere, ice, and land. Superposed on these effects will be seasonal variation in reflectivity, which for the Earth is as much as 10% (Woolf *et al.* 2002). However, it seems likely that many Earth-sized planets will be non Earth-like, and it is therefore useful to ask whether orbital light curves can distinguish between hypothetical, broad classes of objects characterized by a uniform reflecting surface and a simple scattering law. The simplest case is isotropic (Lambertian) scattering, however Solar System surfaces are better represented by Hapke single-scattering models (Hapke 1993).The phase curves of Mercury and Mars have been successfully reproduced using a representation of the double Henyey-Greenstein function

$$p(\alpha) = \frac{(1+c)(1-b^2)}{2\left(1-2b\cos\alpha+b^2\right)^{3/2}} + \frac{(1-c)(1-b^2)}{2\left(1+2b\cos\alpha+b^2\right)^{3/2}}, \tag{2.1}$$

and parameter values $b = 0.21, c = 0.7$, and $b = 0.18, c = 1.1$, respectively (Warell 2004). The reflected light from a planet with a deep, transparant atmosphere and a dark surface, e.g. a global ocean, will be dominatd by Rayleight scattering (RS). Single RS has the phase function

$$p(\alpha) = \frac{3}{4}\left(1 + \cos^2 \alpha\right). \tag{2.2}$$

Can Hapke- and Rayleigh-scattering planets be distinguished solely on the shape of their reflected light curves (absolute albedos require observations in the infrared)? Figure 2a shows the light curves of hypothetical Lambertian-, single Rayleigh-, and Hapke-scattering planets with uniform surfaces as seen at an orbital inclination angle of 60 deg. The opposition effect is clearly discernable, but the small difference between the Hapke- and Rayleigh-scattering models are unlikely to be distinguishable at observable phase angles. The signal from planets with smooth surfaces, e.g., a cloud-free ocean world Léger *et al.* 2004) with a thin atmosphere, or an ice-covered planet, will include significant specular reflection as well as diffuse scattering. The light curves of such objects can be clearly distinguished (Figure 2b) from those of the previous cases, although increasing cloud cover and Rayleigh scattering by an atmosphere will moderate these effects (Williams *et al.*, in prep).

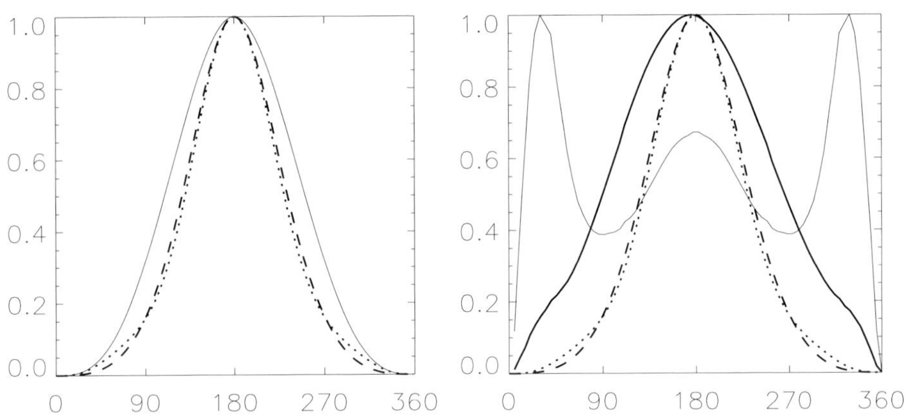

Figure 2. (a) Orbital reflected light curves of three uniform planets observed at $i = 90$ deg. with and different scattering laws; Lambertian (solid line), single RS (dotted line) and a model of Mercury (dashed line) (Warell 2004). (b) Light curves of "smooth" (ocean) planets (solid lines) with 0% (thin) and 38% (thick) cloud cover compared to RS and Mercury models. All curves have been normalized.

3. Emitted Light from Terrestrial Planets

The simplest terrestrial planet case is that of a rotating rocky planet lacking oceans or atmosphere, having a uniform albedo and blackbody emissivity, e.g., Mercury. The thermal inertia j of a solid surface subject to diurnal heating with frequency Ω is $\sqrt{k\rho c}$, where k is the thermal conductivity, ρ the density, and c is the specific heat capacity of the material. Solid basalt ($k = 2$ W m^{-1}K^{-1}, $\rho = 2900$ kg m^{-3}, $c = 1$ kJ kg^{-1} K^{-1}) has $J \approx 2400$ J m^{-2}K^{-1} sec$^{-1/2}$. Dust-covered surfaces on Mars have J values as low as 30 W m^{-2}K^{-1} sec$^{-1/2}$ (Mellon *et al.* 2000). The thermal inertia of the lunar regolith is $\sim$45 W m^{-2}K^{-1} sec$^{-1/2}$, and the large main-belt asteroids have still lower J values

(Müller & Lagerros 1998). The thermal response of a planet can be divided into two regimes, $t_{day} \ll t_{thermal}$ and $t_{day} \gg t_{thermal}$, where the thermal response time is

$$t_{thermal} = \frac{\pi J^2}{4\sigma^2 \bar{T}^6}, \tag{3.1}$$

and σ and $\bar{T}$ are the Stefan-Boltzmann constant and the mean temperature, respectively. For a bare rock surface at 255 K (the emission temperature of the Earth), t_d is 2 months (note the extreme sensitivity to $\bar{T}$). A planet with the thermal properties of the lunar surface has $t_{thermal} = 30$ minutes!

A planets with a fine-grained regoliths and no atmospheres will experience large diurnal temperature variation, and, depending on i, will exhibit significant orbital variation in disk-averaged flux. In contrast, a planets covered with bare rock, e.g. one on which geologic resurfacing continues, will have a relatively small day-night surface temperature difference and orbital variation in its emitted flux would occur only if it had a significant obliquity and seasonality. The modest average obliquity of the Earth ($\delta_0 = 23.45$ deg. may not be typical of terrestrial planets (Laskar, Joutel & Robutel 1993): The obliquities of Mars and Venus are thought to have undergone large excursions over Solar System history and numerical simulations of the final stage of terrestrial planet formation produce planets with an isotropic distribution of primordial obliquities due to the final stochastic accretion of a few large embryos (Agnor, Canup & Levison 1999). Large δ_0 leads to large seasonal and hemispherical differences in incident stellar radiation and a high amplitude light curve (Gaidos & Williams 2004) (Figure 3a).

An ocean and/or an appreciable atmosphere will profoundly affect the emitted flux from a terrestrial planet. The relation between outgoing infrared flux and temperature $I(T)$ for a greenhouse atmosphere deviates significantly from that of a blackbody. Oceans and atmospheres transport heat along thermal gradients and have large thermal inerties. The energy-balance equation governing the surface temperature T of a planet as a function of time t and latitude θ can be modeled using the heat transport equation

$$C\frac{\partial T}{\partial t} = S(1-A) - I(T) + \frac{1}{\cos\theta}\frac{\partial}{\partial\theta}\left(D\cos\theta\frac{\partial T}{\partial\theta}\right), \tag{3.2}$$

where S is the incident stellar flux, A is the top-of-the atmosphere albedo, and C and D parameterize the effective heat capacity of the atmosphere and surface, and meridional heat transport, respectively. The net effect is to obviate diurnal variation and attenuate seasonal variations (Gaidos & Williams 2004) (Figure 3b). Even planets with extremely high obliquities experience habitable temperature ranges over much of their surfaces as a consequence of the modering effect of an ocean and atmosphere (Williams & Kasting 1997, Williams & Pollard 2003). Thus, observation of marked infrared variability alone suggests properties of a planet relating to habitability; lack of an atmosphere or oceans and either a geologically old surface or a high obliquity. On the other hand, *absence* of variability cannot be uniquely interpreted as suggesting habitable conditions. The planet might resemble Earth; it might also be a barren planet with a very low obliquity, or have a thick runaway greenhouse atmosphere like Venus.

4. Terrestrial Planet Satellites

The Earth's Moon is thought to have formed by a low-velocity glancing impact with a Mars-sized body early in its history (Canup 2004). The Moon (or the angular momentum of the Earth-Moon system) stabilizes the Earth's obliquity against chaotic excursions and may be important for habitability (Laskar, Joutel & Robutel 1993). Numerical

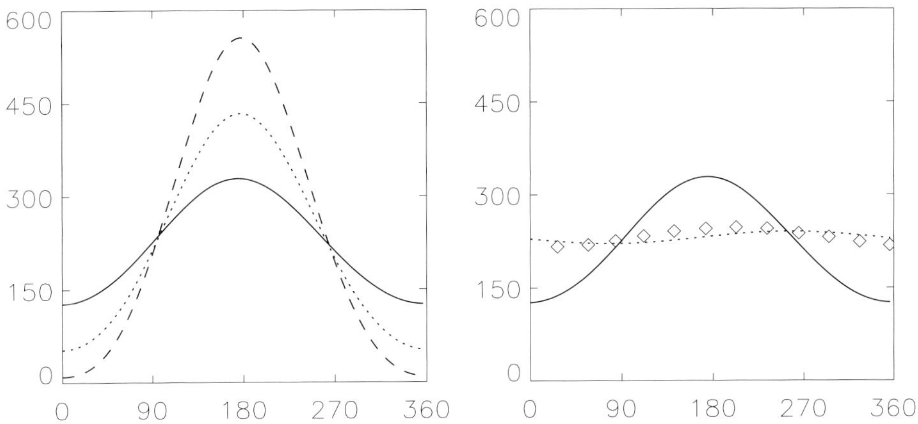

Figure 3. (a) Disk-averaged emitted light curves of planets W m^{-2}) seen at $i = 60$ deg. and with negligible thermal inertia over the orbital period, large thermal inertia over the rotation period, and $\delta_0 = 23.45$ (solid line), 40, and 60 deg. (b) Earth's emitted light curve modeled by a 2-D energy balance model (dotted line) and a 3-D global circulation model (points) (Gaidos & Williams 2004) compared to that of a simlar body with negligible thermal inertia (solid line).

simulations of planet formation show that potentially satellite-forming collisions are common (Ida, Canup & Stewart) and it is dynamically plausible that Mercury and Venus also had satellites but then lost them (Burns 1973, Ward & Reid 1973, Yokoyama 1999). Extrasolar satellites will be unresolved from their parent planet (the Earth-Moon distance subtends 0.25 mas at 10 pc) and only the total signal of the system will be observable.

The theory of impact origin explains the Moon's lack of volatiles by invoking high temperatures in the post-impact circumterrestrial disk; this may be a common property of satellites formed in this manner. There are two consequences of a lack of atmospheres or oceans: First, the satellite will be darker and essentially undectable in reflected light (the Moon-Earth flux ratio is 1.7%). Second, the lower thermal inertia of the satellite (see §3) means that the surface of the satellite experiences larger diurnal temperature variation, and consequently exhibit larger flux variation at thermal infrared wavelengths. Despite the satellite being smaller than the parent planet, the satellite's variation in IR flux may be larger (Figure 4). Satellites of planets may be detectable by sufficiently time-resolved measurements in the infrared. On the other hand, insufficiently time-resolved measurements may suffer from confusion of the two signals and aliasing. Satellites represent another opportunity and a new challenge for the direct detection and characterize terrestrial exoplanets.

Acknowledgements

Support by the NASA Terrestrial Planet Foundation Science program is acknowledged.

References

Agnor, C.B., Canup, R.M. & Levison, H.E. 1999, *Icarus* 142, 219
Arnold, L. & Schneider, J. 2004, *Astron. Astrophys.* 420, 1153
Barman, T.S., Hauschildt, P.H. & Allard, F. 2005, *Astrophys. J.*, 632, 1132
Burnas, J.A. 1973, *Nature* 242, 23
Burrows, A., Hubeny, I. & Sudarsky, D. 2005, *Astrophys. J.* 625, L135
Canup, R.M. 2004, *Annu. Rev. Astron. Astrophys.* 41, 441
Charbonneau, D., Allen, L.E., Megeath, S.T., Torres, G., Alonso, R., Brown, T.M., Gilliland,

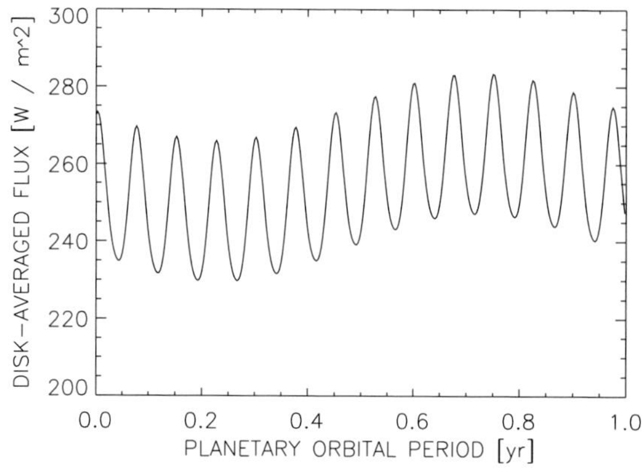

Figure 4. Disk-averaged emitted light curve of the Earth-Moon system (W m^{-2}) seen at $i = 60$ deg. The variation induced by the Moon's phases is larger than that of the Earth.

R.L., Latham, D.W., Mandushev, G., O'Donovan, F.T. & Sozzetti, A. 2005, *Astrophys. J.* 626, 523
Charbonneau, D., Noyes, R.W., Korzennik, S.G., Nisenson, P., Jha, S., Vogt, S.S. & Kibrick, R. I. 1999, *Astrophys. J.* 522, L145
Deming, D., Seager, S., Richardson, L.J. & Harrington, J. 2005, *Nature* 434, 740
Deming, D., Brown, T. M., Charbonneau, D., Richardson, L.J. & Harrington, J. 2005, *Astrophys. J.* 622, 1149
Des Marais, D.J., Harwit, M.O., Jucks, K.W., Kasting, J.F., Lin, D.N.C., Lunine, J.I., Schneider, J., Seager, S., Traub, W.A. & Woolf, N.J. 2002, *Astrobiol.* 2, 153
Dyudina, U.A., Sackett, P.D., Bayliss, D.D.R., Seager, S., Porco, C.C., Throop, H.B. & Dones, L. 2005, *Astrophys. J.* 618, 973
Ford, E.B., Seager, S. & Turner, E.L. 2001, *Nature* 412, 885
Fortney, J.J.,Marley, M.S., Lodders, K., Saumon, D. & Freedman, R. *Astrophys. J.* 627, 69
Gaidos, E. & Williams, D. M. 2004, *New Astron.* 10, 67
Galilei, Galileo *Sidereus Nuncius or the Sidereal Messenger.* 1989, A. van Helden, trans. Univ. Chicago Press, Chicago
Green, D., Matthews, J., Seager, S. & Kuschnig, R. 2003, *Astrohys. J.* 597, 590
Hapke, B. 1993, *Theory of Reflectance and Emittance Spectroscopy*, Cambridge Univ. Press, Cambridge
Ida, S., Canup, R.M. & Stewart, G.R. 1997, *Nature* 389, 353
Laskar, J., Joutel, F. & Robutel, P. 1993, *Nature* 361, 615
Léger, A., Selsis, F., Sotin, C., Guillot, T., Despois, D., Mawet, D., Ollivier, M., Labèque, A., Valette, C., Brachet, F., Chazelas, B. & Lammer, H. 2004, *Icarus* 169, 499
Leigh, C., Collier Cameron, A., Udry, S., Donati, J.-F., Horne, K., James, D. & Penny, A. 2003, *Mon. Not. R. Ast. Soc.* 346, L16
Mellon, M., Jakosky, B.M., Keiffer, H.H. & Christensen, P.R. 2000, *Icarus* 148, 437
Müller, T.G & Lagerros, J.S.V. 1998, *Astron. Astrophys.* 338, 340
Seager, S., Turner, E.L., Schafer, J., & Ford, E.B. 2005, *Astrobiology* 5, 372
Snellen, I.A.G. 2005, *Mon. Not. R. Ast. Soc.* 363, 211
Sudarsky, D., Burrows, A., Hubeny, I. & Li, A. 2005, *Astrophys. J.* 627, 520
Ward, W. & Reid, M. 1973, *Mon. Not. R. Astr. Soc.* 164, 21
Warell, J. 2004, *Icarus* 167, 271
Williams, D.M. & Pollard, D. 2003, *Int. J. Astrobiol.* 2, 1
Williams, D.M. & Kasting, J.F. 1997, *Icarus* 129, 254
Woolf, N.J., Smith, P.S., Traub, W.A. & Jucks, K.W. 2002, *Astrohys. J.* 574, 430
Yokoyama, T. 1999, *Planet. Space Sci.* 47, 619

Direct Imaging of Exoplanets: Science & Techniques
Proceedings IAU Colloquium No. 200, 2005
C. Aime and F. Vakili, eds.
© 2006 International Astronomical Union
doi:10.1017/S1743921306009240

Science case for VLT-Planet Finder

C. Moutou[1], J.-L. Beuzit[2], R. Gratton[3], D. Mouillet[4], F. Allard[5], J.-C. Augereau[2], A. Boccaletti[6], S. Desidera[3], M. Feldt[7], Th. Henning[7], A.-M. Lagrange[2], A. Longmore[8], F. Ménard[2], H.-M. Schmid[9], D.M. Stam[10], S. Udry[11] and R. Waters[10]

[1]LAM, Marseille, France, email: claire.moutou@oamp.fr
[2]LAOG, Grenoble, France
[3]INAF-OPD, Padova, Italy
[4]Obs. Midi-Pyrénées, Tarbes, France
[5]ENS, Lyon, France
[6]Obs. Paris Meudon, France
[7]MPIA, Heidelberg, Germany
[8]ATC, Edimburg, UK
[9]ETH, Zürich, Switzerland
[10]UVA, Amsterdam, The Netherlands
[11]Obs. Genève, Switzerland

Abstract. This paper presents the scientific case for a next generation adaptive optics instrument at the VLT, temporarily named "Planet Finder", that is aimed at detecting and characterizing extrasolar planets through the direct analysis of their emitted photons in the visible and at near-IR wavelengths. We discuss the observational niche of such an instrument to have first light in 2010, in complement to other planet search methods. To improve the efficiency (and consistency) of the search for planets with the PF, the observations will need to be organized in the form of an extensive survey of hundreds of nearby stars, predicted outputs of which are also described here. This summarizes the study phase of the instrument, conducted by two competitive teams and the recent merging of both studies, regarding the scientific impact of Planet Finder.

Keywords. stars: low-mass, brown dwarfs, planetary systems, protoplanetary disks; techniques: high angular resolution; surveys.

1. Introduction

On the road towards Extremely Large Telescopes and space interferometers, one needs to progressively gain orders of magnitude in contrast in the close vicinity of bright stars. This is the motivation for a near-future (2010) near-infrared, extreme adaptive optics instrument on the VLT, equipped with high-performance tools for the removal of coherent and incoherent starlight: interferential coronagraphs and differential imaging. The instrumental design and critical sub-systems are described in Beuzit *et al.*, Schmid *et al.* and Fusco *et al.* (this volume). The current design of the VLT-Planet Finder (PF) instrument is the result of merging two independent instrumental studies. The scientific rationale on which this design was based is also the combined effort of two independent scientific groups. In the following, we discuss the observational niche and potential impact of VLT-PF, the survey strategy and its expected output.

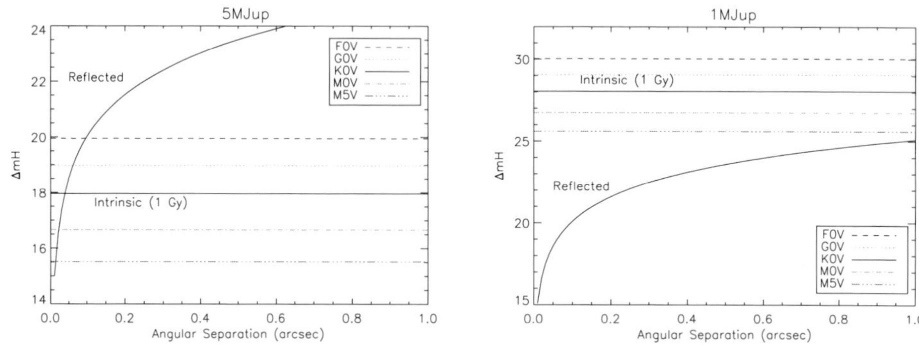

Figure 1. H-band planet-to-star flux ratio in both regimes of planet emission: intrinsic light (horizontal lines) and reflected light, for a system at 10 pc. Left: for a 5 M_J of 1Gy. Right: for a 1 M_J of 1 Gy. The star spectral type induces more than 4 magnitudes variations in contrast, from F0V to M5V, while the reflected light drops by orders of magnitude with increasing orbital distance. The planet radius does not change much in this mass regime; here, 1 R_J is used.

2. Planet finding with VLT-PF

At the time when VLT-PF will be in operation (2010), our knowledge of extrasolar planets will have increased significantly: the mass function and period distribution of planets will have been largely extended by ongoing radial velocity surveys, many planet radii will have been measured by transit observations, direct imaging of some very young planets will have been performed by current high contrast facilities such as NACO (see e.g. Chauvin *et al.* 2005, Close *et al.* 2004). Atmospheric models of giant planets will then have been partly constrained by observations. However, two large sets of planets will probably not be known before 2010: 1) long period planets (unaccessible to indirect methods which require a full orbit coverage), and 2) moderately young to evolved planetary systems (for which the contrast currently achieved in direct imaging is not sufficient). VLT-PF will then enlarge our view on extrasolar planetary systems, in good complementarity to other detection techniques.

When considering direct imaging of extrasolar planets, one must have in mind both types of emission that can be detected:

• Intrinsic emission of the planets, due to residual cooling after their formation. It decreases with decreasing planet mass, increasing system age. The star-planet flux ratio of this emission is independent of the orbital distance (since the planet is self-luminous) but strongly dependent on the star spectral type (later types being more favourable). Due to the planet effective temperature (100–1000K), the intrinsic emission increases towards the mid-infrared. It is also characterized by spectral features, mostly due to CH_4 and H_2O absorption bands, and thus commonly found in spectra over an extended mass range, from Saturn mass up to brown dwarfs (0.3–30M_J). Simulated spectra and calculated luminosities from Allard *et al.* (2001), Chabrier *et al.* (2000) and Baraffe *et al.* (2003) were used to estimate photon counts from the planets in this regime.

• Emission due to the reflection of starlight by the atmosphere of the planet, which strongly depends on the star-planet distance and is modulated by the phase angle. More light is reflected at wavelengths where the star is bright, i.e. in the visible, it depends on the planet radius rather than its mass, and it may be strongly polarized (see Schmid *et al.* this volume, and Stam *et al.*, 2004). Physical processes taking place in the atmosphere of irradiated planets are described in Barman *et al.* (2001) and Sudarsky *et al.* (2000, 2003).

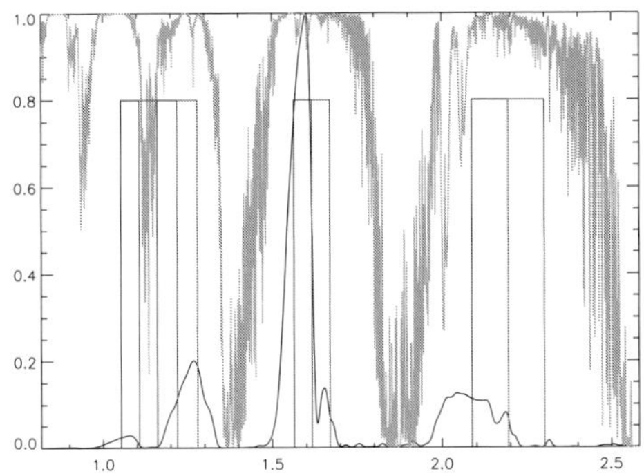

Figure 2. The spectral features of giant planets in the near-infrared (5 $M_J$aged 500 My). In the J-, H- and K-bands, the spectrum is characterized by deep and broad absorption bands. The filters designed for an optimal differential imaging with VLT-PF are superimposed ($\lambda/\delta\lambda \simeq 30$), as well as the terrestrial transmission spectrum.

Figure 1 shows the relative weight of the star-planet contrast due to both types of emission, as a function of the separation. The spectral domain from R to K bands is the best compromise in terms of planet luminosity and sky background for achieving the largest contrast at short separations from the ground.

The challenge of direct imaging with VLT-PF will thus include two types of targets: those where the intrinsic light dominates (young giant planets and evolved planets in the mass regime $\simeq$10 M_J), and those where the reflected light dominates (short-period planets around the most nearby stars). VLT-PF will take advantage of the physical properties of giant planets in its detection techniques: either the spectral features or the polarisation of the planet atmosphere is used to increase the sensitivity through differential imaging methods (Beuzit *et al.*, Schmid *et al.* and Boccaletti *et al.*, this volume). Consequently, the detection of planets with VLT-PF will allow some characterization of their atmospheric components: effective temperature (depth of CH_4 and H_20 features, near-IR low-resolution spectro-photometry), dust grain size and cloud characteristics (degree of polarisation), planet radius and geometric albedo. **VLT-PF will then mainly discover and characterize planets in the mass range 1-10 M_J and period range 1-1000 years, i.e., all gas giants that were expected before the discovery of hot jupiters. The contrast of 10^{-4} to 10^{-8} at angular separations 0.1 to 1.5 arcsec should be achieved**, with a gain of up to 8 magnitudes with respect to current instruments.

3. Survey strategy and output

In order to assess the potential scientific output of VLT-PF, one should answer the following questions: 1) what is the detection performance of the instrument, 2) how many targets are suitable for planet search in direct imaging, 3) how do planetary and stellar characteristics combine in the detection space of VLT-PF? A mutually self-consistent detection performance of each of the VLT-PF focal instruments still has to be estimated from a unique input data set. First, independent, estimates were obtained during the competitive study phase (2002-2004) and are partly exposed in Boccaletti *et al.*, Schmid *et al.* and Beuzit *et al.* (this volume). Table 1 summarizes some of these preliminary results.

Mode	Target	0.2"	0.5"	1.0"	2.0"	Comment
DBI	V= 2	$4.0\ 10^{-6}$	$2.5\ 10^{-6}$	$2.5\ 10^{-6}$	$2.0\ 10^{-6}$	all systematics inc.
IFS	J= 6	–	10^{-7}	$5.0\ 10^{-8}$	$1.2\ 10^{-8}$	not all systematics inc.
Zimpol	I= 6	$6.0\ 10^{-8}$	$3.0\ 10^{-8}$	$2.0\ 10^{-8}$	$1.0\ 10^{-9}$	no systematics inc.

Table 1. Contrast achieved with a SNR of 5 in 4 hours, in each mode of VLT-PF (DBI: Dual Band Imaging; IFS: Integral Field Spectrograph; Zimpol), at separations 0.2–2 arcsec, for various types of targets. Instrumental aberrations were not included in the same way in all simulations (last column). More details and consistent comparisons will be discussed in a further paper.

Three kinds of stars were identified as suitable for planet search with VLT-PF:

• **The most nearby stars (3-20 pc)**: solar-type stars at less than 20 pc are well known and identified in Hipparcos and Gliese catalogs; M dwarfs at less than 12 pc are identified in dedicated surveys (Beuzit *et al.*, 2004 and references therein). These targets would allow discovery of some planets through their reflected light, if their orbital distance is short enough (a few AUs typically) for this emission to significantly contribute. Around the closest M dwarfs, self-luminous planets could be discovered (whatever their orbital distance) since the contrast is favoured by a fainter star. Detailed detection statistics using Monte Carlo calculations and the high performance of the VLT-PF Integral Field Spectroscopy channel have revealed that evolved planets of 1 M_J could be detected around stars within 20 pc, at typical contrasts of 10^6. The output mass distribution of these simulations is shown in Fig. 3. The specific interest of observing the 40 closest M dwarfs accessible from the Southern hemisphere requires the adaptive optics system to operate to full specification up to V magnitudes of 11 if the wavefront sensing is performed in the visible (or H magnitude of 8 with a near-infrared sensing).

• **The youngest stars (20-70 pc, 1-100 My)**: located in star-forming associations, these targets are of great interest since Jupiter-like planets at their earliest stage of formation and migration could be discovered. The mean distance of such associations is about 50 pc, although a few objects are known within 30 pc (see review by Zuckerman and Song, 2004). In total, there is a potential sample of about 130 stars in young associations, in the relevant magnitude range and observable from Paranal. Planets at orbital distances larger than 10 AUs will be searched for around such targets.

• **Intermediate stars (3-50 pc, 0.1-1 Gy)**: relaxing the above constrains on the stellar age and on the distance, one may still find a large quantity of potential targets for VLT-PF. They include stars with planets known by radial velocity measurements, or stars showing a long-term drift indicative of a large distance companion. They also include active stars, that are often discarded in radial velocity surveys but whose activity may indicate a relatively young age (and hence, brighter planetary companions). In the target sample of CORALIE, up to 400 stars could have ages less than 1 Gy. Even when the age-activity relation has been calibrated statistically (Donahue 1993), there are potential additional sources of activity that mean an age estimate for an individual star using this method has considerable uncertainty.

The complementary potential of the three imaging modes of VLT-PF will allow us to probe different ages of the evolution of planetary systems, from few My to 10 Gy, providing clues for planet evolution (dynamical and atmospheric). **Because the frequency of planets in the mass and period ranges accessible to PF is basically unknown, searching for planets with VLT-PF requires an extensive survey on several hundreds of stars, with a volume up to 500 nights over about 4 years.** A dedicated survey will optimize the observing efficiency and ensure database consistency for

1) perform statistical analysis of the results and 2) archive a complete data set as legacy for future missions with ELTs or space interferometers.

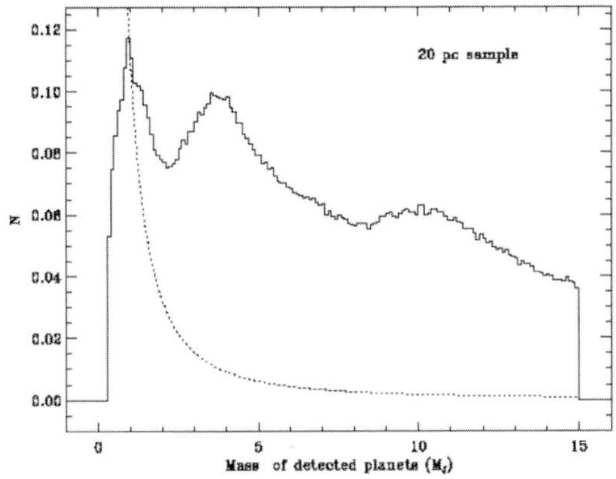

Figure 3. The simulated mass distribution of extrasolar planets that could be discovered by VLT-PF, from Monte-Carlo calculations using the performance theoretically obtained in the IFS channel. The dashed line shows the input distribution from Lineweaver & Grether (2003).

Type	Number	Distance	Age	Mass, Period of 1st interest planets
Very nearby stars	20	< 6 pc	1-10 Gyr	1 R_J, 0.1-10 y
Nearby M stars	40	< 12 pc	1-10 Gyr	1-10 M_J, 1-50 y
Young associations	130	20-100 pc	2-50 Myr	1-5M_J, 10-10^3 y
Active FGK stars	400	< 50 pc	< 1 Gyr	5-20M_J, 1-50 y
Stars with RV planets	> 60	< 50 pc	> 1 Gyr	5-20M_J, 20-1000 y
Total	$\simeq 540$	< 100 pc	all	1-20M_J, 0.1-10^3 y

Table 2. The different types of target stars suitable for planet and brown dwarf searches are listed, with their counts, distance and age ranges. The limiting magnitude is $R \simeq 10$. The typical range of planet mass and orbital period in the parameter space of VLT-PF is shown in the right column. Stars with known radial-velocity planets and active FGK stars both come from the "intermediate class of stars" described in the text. Nearby stars are separated here into spectral types, M dwarfs being specifically interesting nearby targets.

4. Life of stellar and planetary systems

With the most demanding contrast requirement of planet detection, many more science topics may be addressed by the VLT-PF instrument. Apart from marginal work on solar system objects and extragalactic sources (both being faint and extended targets, not optimal for VLT-PF and already efficiently observed with NACO-like instruments), the additional science of VLT-PF concerns the various stages of star life, from birth to death.

- Gas-rich disks: VLT-PF will allow detecting more very young, gas-rich disks around stars in Taurus or Orion clouds, and study the internal part of known disks. The polarimetric mode will be essential for characterizing grain properties in such disks.
- Young stars: physical processes during all stages of star formation will be addressed by observing jets and outflows at high contrast.

• Debris disks: new debris disks are currently been discovered by *Spitzer* (Beichman *et al.*, 2005, intrinsic disk emission); their observation with VLT-PF will provide complementary information on the light scattering properties of the disk grains from polarimetry with ZIMPOL. Disk asymmetries could be detected as possible evidence for planets.

• Evolved stars: the final stages of massive stars, characterized by mass loss, growing envelopes and shocks will be addressed by VLT-PF. Wolf-Rayet stars, novae and supernovae are potential targets.

The science topics mentioned above will benefit from the large contrast at small separations obtained by VLT-PF; some of them, however, less demanding in terms of contrast, will be used as backup programs in median atmospheric conditions.

Note that a workshop entitled "Science with Planet Finder" will be organized in the spring 2006 to help enlarge the future users' community of this instrument and to open discussion on the use of the instrument and survey for additional or alternative proposals.

5. Conclusion

The next generation adaptive optics instrument for the VLT, temporarily called Planet Finder, will be able of conducting a large scale survey on bright and nearby stars, in search for 10^4 to 10^8 times fainter companions at less than 2 arcsec. This will allow a breakthrough in our understanding of the formation and migration scenarii of giant planets and brown dwarfs, and on the physical processes at work in planetary atmospheres, at various evolution stages. VLT-PF could also contribute to understand the brown dwarf desert, if the latter is not present at early ages of the systems or if the non-detection of brown dwarfs is currently limited by sensitivity and period coverage. Gas-rich protoplanetary disks and debris disks will be observed and offer complementary information to *Spitzer* measurements. Polarimetric and spectroscopic facilities foreseen on VLT-PF are great tools to achieve the characterization of detected objects, that will fruitfully complement other future instruments such as JWST/MIRI.

References

Allard, F., Hauschildt, P.H., Alexander D.R. *et al.*, 2001, ApJ 356, 557
Barman, T.S., Hauschildt, P.H. & Allard, F., 2001, ApJ 556, 885
Baraffe, I., Chabrier, G., Barman, T.S. *et al.*, 2003, A&A 402, 701
Beichman, C.A., Bryden, G., Rieke, G. H. *et al.*, 2005, ApJ 622, 1160
Beuzit, J.-L., Ségransan, D., Forveille, T. *et al.*, 2004, A&A 425, 997
Beuzit, J.-L. *et al.*, 2005, this volume
Boccaletti, A. *et al.*, 2005, this volume
Chabrier, G., Baraffe, I. Allard, F. *et al.*, 2000, ApJ 542, 464
Chauvin G., Lagrange, A.-M., Dumas, C. *et al.*, 2005, A&A 438, L25
Close, L., Lenzen, R., Guirardo, J.C. *et al.*, 2004, Nature 433, 286
Donahue, R.A., 1993, PASP 105, 804
Fusco, T. *et al.*, 2005, this volume
Lineweaver, C.H. & Grether, D., 2003, ApJ 598, 1350
Schmid, H.-M. *et al.*, this volume
Stam, D.M., Hovenier, J.W. & Waters L.B.F.M., 2004, A&A 428, 663
Sudarsky, D., Burrows, A. & Pinto, P., 2000, ApJ, 538, 885
Sudarsky, D., Burrows, A. & Hubeny, I., 2003, ApJ, 588, 1121
Zuckerman, B. and Song, I., 2004, ARAA 42, 685

Direct Imaging of Exoplanets: Science & Techniques
Proceedings IAU Colloquium No. 200, 2005
C. Aime and F. Vakili, eds.
© 2006 International Astronomical Union
doi:10.1017/S1743921306009252

Search and investigation of extra-solar planets with polarimetry

H. M. Schmid[1], J.-L. Beuzit[2], M. Feldt[3], D. Gisler[1], R. Gratton[4], Th. Henning[3], F. Joos[1], M. Kasper[5], R. Lenzen[3], D. Mouillet[6], C. Moutou[7], A. Quirrenbach[8], D. M. Stam[9], C. Thalmann[1], J. Tinbergen[10], C. Verinaud[5], R. Waters[9] and R. Wolstencroft[11]

[1]Institute of Astronomy, ETH Zurich, Switzerland
[2]Laboratoire d'Astrophysique, Observatoire de Grenoble, France
[3]Max-Planck-Institut für Astronomie, Heidelberg, Germany
[4]INAF, Osservatorio Astronomico di Padova, Italy
[5]European Southern Observatory, Garching, Germany
[6]Laboratoire d'Astrophysique, Observatoire Midi-Pyrénées, Tarbes, France,
[7]Laboratoire d'Astrophysique de Marseille, France
[8]Sterrewacht, Leiden University, the Netherlands
[9]Astronomical Institute "Anton Pannekoek", University of Amsterdam, the Netherlands
[10]ASTRON, Dwingeloo, the Netherlands
[11]Astronomy Technology Center, Royal Observatory Edinburgh, UK

Abstract. Light reflected from planets is polarized. This basic property of planets provides the possibility for detecting and characterizing extra-solar planets using polarimetry. The expected polarization properties of extra-solar planets are discussed that can be inferred from polarimetry of "our" solar system planets. They show a large variety of characteristics depending on the atmospheric and/or surface properties. Best candidates for a polarimetric detection are extra-solar planets with an optically thick Rayleigh scattering layer.

Even the detection of highly polarized extra-solar planets requires a very sophisticated instrument. We present the results from a phase A (feasibility) study for a polarimetric arm in the ESO VLT planet finder instrument. It is shown that giant planets around nearby stars can be searched and investigated with an imaging polarimeter, combined with a powerful AO system and a coronagraph at an 8 m class telescope.

A similar type of polarimeter is also considered for the direct detection of terrestrial planets using an AO system on one of the future Extremely Large Telescopes.

Keywords. exoplanets, scattering, polarization, instrumentation.

1. Introduction

Direct detection of extra-solar planets from the ground can be achieved with a large telescope, a powerful adaptive optics (AO) system combined with coronagraphy, and a differential detection technique (Angel 2003). A differential detection technique is required because the signal from most planets will be much fainter than the residual AO corrected (and coronagraphic) point spread function (PSF) of the central star. One way to pick out a faint signal of a planet in a strong and variable (speckles!) PSF is to search for spectral features which are present in the planet spectrum but not in the spectrum of the star. Giant planets similar to Jupiter exhibit strong methane bands in the 1.0 μm – 1.7 μm region which are well suited for differential measurements (see Moutou *et al.* 2006). An alternative differential method, which is discussed in this paper, is to search for the polarimetric signal from the light reflected by planets.

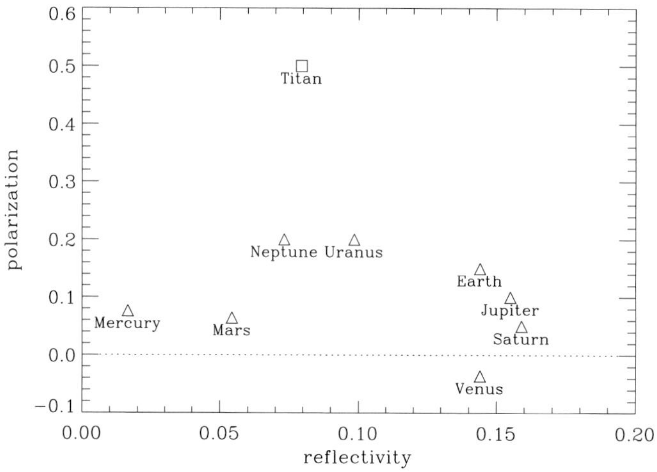

Figure 1. Polarization degree $p(90°)$ versus reflectivity $f(90°)$ for solar system objects at phase angle $\alpha = 90°$ in the R-band.

Reflected light is the dominant emission component for all cold planets (<700 K) in the visual to near IR spectral region. This radiation is generally polarized and the degree of polarization is often particularly high at short wavelengths (<1 μm) due to Rayleigh scattering by molecules and scattering by haze particles in planetary atmospheres. Thus, the visual-red spectral region is well suited for planet polarimetry.

The search for polarized light from extra-solar planets is very demanding, because the signal decreases rapidly $\propto 1/d^2$ with the orbital separation d. Bright, self-luminous, young planets emit predominantly thermal radiation in the near-IR (>1 μm) which is not or hardly polarized. Thus, a successful polarimetric detection of extra-solar planets requires strongly polarized ($\gg$1 %) planets and a very sensitive polarimeter in a sophisticated extra-solar planet instrument at a large telescope.

2. Expected polarization from planets

The expected polarization of extra-solar planets can be inferred from observations of solar system planets. A crucial parameter for the polarization of planets is the phase angle α (angle: central star – planet – observer). For extra-solar planet observations α will be typically in the range $\alpha \approx 60° - 120°$. Rayleigh scattering, and scattering from rocky/sandy surfaces produce for these phase angles a polarization with an orientation perpendicular to the scattering plane. Other α's are not suited equally well for extra-solar planet observations because the amount of polarized light from the planet is often strongly reduced and the separation to the central star is small. Fig. 1 gives the degree of polarization $p(90°)$ and reflectivity $f(90°)$ (reflected flux normalized to the irradiated flux) in the R-band ~0.65 μm for $\alpha = 90°$ for solar system objects. The values $p(90°)$ and $f(90°)$ were collected or derived from the literature.

Figure 1 illustrates that there exists a large diversity in polarimetric properties for solar system planets which is discussed here briefly:

Mercury and Mars: The polarization in the R-band is about $p(90°) \approx 5 - 10$ % and the reflectivity is low (Dollfus & Auriere 1974; Dollfus *et al.* 1983). This is typical for planets (and smaller bodies) with thin or no atmosphere, like Mercury, Mars, or the Moon, where the light is reflected from the rocky/sandy surface.

Venus: The reflectivity is high but the polarization $p(90°)$ is low because the light is scattered by high altitude clouds (Coffeen & Gehrels 1969; Dollfus & Coffeen 1979). Unlike the other planets $p(90°)$ is negative, or parallel to the scattering plane.
Earth: Compared to the other terrestrial planets in the solar system, Earth has a high polarization $p(90°) \approx 15$ % (Dollfus 1957) due to Rayleigh scattering (blue sky). The disk-integrated polarization depends on the cloud coverage together with the exposed areas of oceans, forests, deserts and snow fields.
Jupiter: The polarization is very strong (∼50 %) at the poles where Rayleigh scattering dominates and low (<10 %) at the equator where the light is mainly reflected from clouds (Smith & Tomasko 1984). The estimated disk-integrated polarization for Jupiter is about $p(90°) = 10$ % for equatorial sight lines and somewhat higher for polar sight lines.
Saturn: The polarization pattern is similar to Jupiter with an enhanced polarization at the poles (Tomasko & Doose 1984). However, the overall polarization level is significantly lower than for Jupiter.
Titan: Saturn's moon Titan has a thick Rayleigh scattering haze layer which produces a very high polarization of $p(90°) \approx 50$ % (Tomasko & Smith 1982).
Uranus and Neptune: A high polarization $p(90°) \approx 20$ % is predicted based on ground-based observations of the limb polarization (Schmid *et al.* 2006; Joos & Schmid 2006).

The huge variety of polarimetric properties of solar system bodies suggests that polarimetry may play an important role in the future characterization of extra-solar planets.

3. High precision polarimetric imaging

Very high precision is required with imaging polarimetry to detect a faint, point-like polarization signal of a planet in the residual seeing halo of a bright star. Aperture polarimetry reached already decades ago a very high precision better than 10^{-5} based on fast polarization modulators (e.g. Kemp & Barbour 1981) and a lock-in or photon counting detector. A similar precision can be achieved with array detectors using the ZIMPOL technique. ZIMPOL (Zurich IMaging POLarimeter) uses also a fast polarization modulator and a special CCD camera performing the on-chip demodulation of the modulated signal (e.g. Povel 1995).

The fast modulator, e.g. a ferroelectric liquid crystal working at a modulation frequency of 1 kHz, and a polarizer convert the polarization signal into a fractional modulation of the intensity signal. This intensity modulation is converted back into a polarization signal by a special ZIMPOL CCD camera which measures for each active pixel the intensity difference between the two modulation states. For this every second row of the CCD is masked so that charge packages created in the unmasked row during one half of the modulation cycle are shifted for the second half of the cycle to the next masked row, which is used as temporary buffer storage (the CCD can be equipped with cylindrical micro-lenses which focus the light onto the open CCD rows). After many thousands of modulation periods the CCD is read out within less than one second. The sum of the two images is proportional to the intensity while the normalized difference is the polarization degree of one Stokes component.

Because the measurement is fully differential, systematic error sources are reduced to a very low level. Key advantages of this technique are:

- images for the two opposite polarization modes are recorded practically **simultaneously** (the modulation is faster than seeing variations),
- both images are recorded with the **same pixels**,
- there are **no differential aberrations** between the images for the two opposite polarization due to the atmosphere or the telescope/instrument,

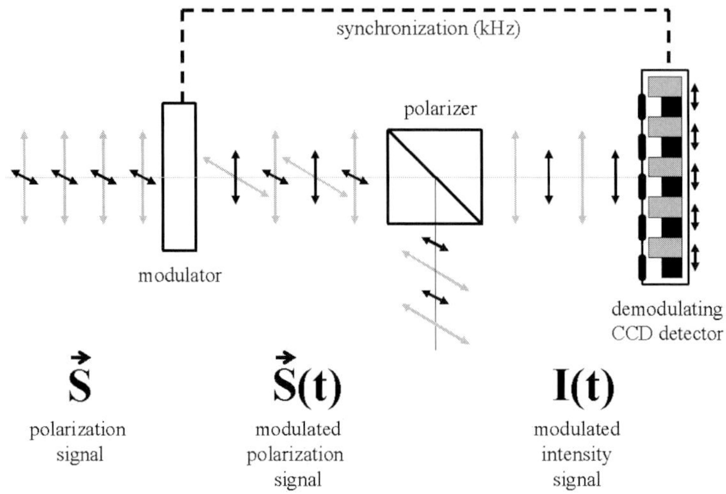

Figure 2. The measuring principle for a fast modulation imaging polarimeter.

• the differential (polarization) signal is **not affected by chromatic effects** due to telescope diffraction or speckle chromatism.

ZIMPOL has proved to be an extremely precise technique for polarimetric imaging. In solar applications it routinely reaches a precision of better than 10^{-5} (e.g. Stenflo & Keller 1996). For the future one might consider other types of demodulating detectors such as CMOS-detectors (Keller 2004) or avalanche diode arrays.

4. Implementation into an exo-planet instrument

ZIMPOL has been considered in combination with an extreme AO system in a feasibility study for the VLT planet finder instrument (Feldt *et al.* 2003; Gisler *et al.* 2004). Key requirements are that the entire instrument must be optimized as far as possible for high precision polarimetric imaging. On the other hand the concept for polarimetry must be adapted in order to avoid any disturbing impact on the performance of the AO system, of the coronagraphic system and of the other instruments.

Figure 3 outlines the basic concept for ZIMPOL within the planet finder which will be further investigated in a design study (Beuzit *et al.* 2006). The imaging polarimeter working in the wavelength range $0.6 - 0.9\ \mu$m is located after the AO system, the beam splitters for near-IR instruments and wave front sensor (WFS), and after the coronagraphic system. A polarization switch (e.g. rotatable half-wave plate) is forseen in front of the AO system to reduce and/or calibrate the instrument polarization. The same concept could also be implemented into an extra-solar planet instrument for an ELT (e.g. EPICS concept for OWL - see Verinaud *et al.* 2006).

5. Expected polarimetric sensitivity for the VLT and the future ELTs

An interesting quantity for the polarimetry of extra-solar planets is the polarization contrast, the ratio between *polarized flux from the planet* to the total flux from the central star:

$$C_{\rm pol} = p(\alpha) \cdot C_{\rm flux} \ = p(\alpha) \cdot f(\alpha)\, \frac{R_p^2}{d_p^2}$$

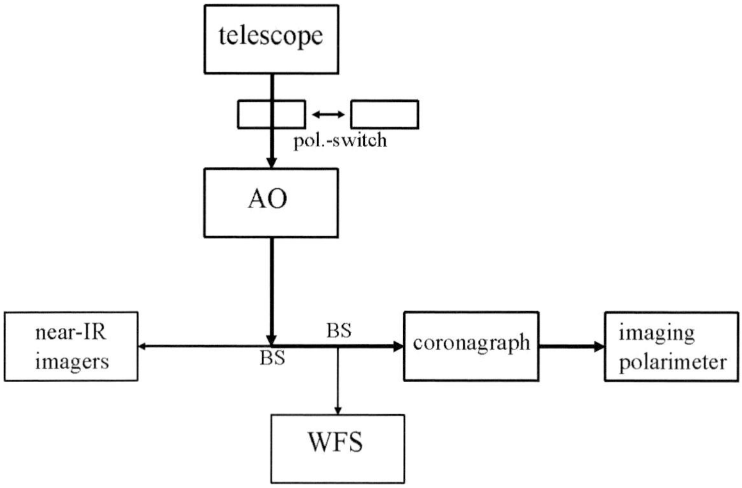

Figure 3. Basic concept of the ZIMPOL polarimeter within the planet finder instrument. The polarization switch is used to reduce and/or calibrate the instrument polarization.

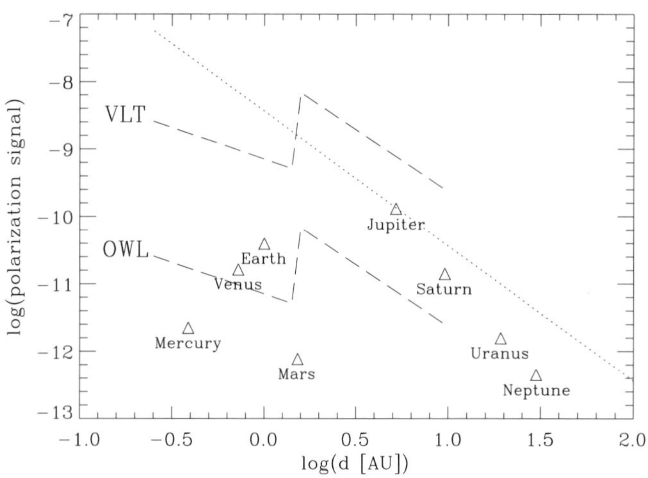

Figure 4. Polarization contrast for solar system planets as function of separation. The dotted line indicates the change in contrast if the separation of Jupiter is changed (but not the other properties). The dashed curves indicate the achievable sensitivity of a planet finder instrument with a band width of 280 nm at the VLT and an Extremely Large Telescope (OWL) for a solar system analog at 5 pc with an integration time of 4 h.

where R_p is the radius of the planet and d_p the separation of the planet from the central star. The polarization contrast for the solar system planets (using the reflectivity $f(90°)$ and polarization $p(90°)$ from Fig. 1) is shown in Fig. 4 as function of the separation. Considering the polarization contrast in the R-band, Earth is the second planet in the solar system after Jupiter. The dotted line through Jupiter indicates $C_{\rm pol}$ if only the separation d_p is changed but reflectivity, polarization and radius remain the same.

Also indicated in Fig. 4 are the estimated detection thresholds for planet finder instruments at the VLT or a 100 m ELT (ESO's overwhelmingly larger telescope OWL) for a solar system analog at 5 pc with an integration time of 4 hours. The adopted AO systems have an actuator spacing corresponding to 20 cm, equivalent to about 40×40 actuators for an 8 m telescope and 500×500 actuators for a 100 m telescope. Achievable Strehl ratios are about $S \approx 0.5 - 0.7$ depending on the adopted atmospheric conditions, AO control loop frequency and other parameters (see e.g. Chelli 2005). For the imaging polarimeter a precision of 10^{-5} is assumed as anticipated for a ZIMPOL system. Due to the ZIMPOL measuring method (see Sect. 3) speckle noise can be neglected.

Figure 4 shows that a Jupiter-like giant planet at 5 pc can be detected with an imaging polarimeter in a planet finder instrument at an 8 m class telescope if the separation is less than about 2 AU. High polarization $p(90°) \approx 30$ % (instead of 10 % as for Jupiter) extra-solar giant planets are also detectable at larger separations with somewhat longer integrations of say 12 hours (instead of 4 hours).

With a planet finder instrument at a 100 m telescope, Jupiter, Saturn, Earth and Venus could be detectable at 5 pc within 4 hours of integration. Uranus and Neptune-type planets will be very difficult targets even for a 100 m telescope. Also the polarization signal from atmosphere-less rocky planets (Mercury type) will be very faint and might only be detectable around the very nearest stars ($d < 3$ pc).

These estimates show that polarimetry is a powerful differential technique for the future investigation of scattered light from extra-solar planets from the gound. The polarization signal provides complementary information on the targets when compared to other techniques. Particularly well suited for polarimetric studies are (old) Jupiter-like and Earth-like planets with substantial Rayleigh scattering.

References

Angel, R. 2003, in: M. Friedlung & T. Henning, *Towards Other Earths: DARWIN/TPF and the Search for Extrasolar Terrestrial Planets*, ESA SP-539, 221
Beuzit J.L., *et al.* 2006, these proceedings
Chelli, A. 2005, *A&A* 441, 1205
Coffeen, D.L. & Gehrels, T. 1969, *AJ* 74, 433
Dollfus, A. 1957, *Suppl. Ann. d'Astrophys.* 4, 1
Dollfus, A. & Auriere, M. 1974, *Icarus* 23, 465
Dollfus, A. & Coffeen, D.L. 1970, *A&A* 8, 251
Dollfus, A., Deschamps, M., & Ksanfomaliti, L.V. 1983, *A&A* 123, 225
Feldt, M., Turatto, M., Schmid, H.M., *et al.* 2003, in: M. Friedlung & T. Henning, *Towards Other Earths: DARWIN/TPF and the Search for Extrasolar Terrestrial Planets*, ESA SP-539, 99
Gisler D., Schmid, H.M., Thalmann, C., *et al.* 2004, in: A.F.M. Moorwood, & M. Iye (eds.), *Ground-based instrumentation for astronomy*, SPIE Conf. Vol. 5492, 1010
Joos, F. & Schmid, H.M. 2006, these proceedings
Keller, C.U. 2004, in: S. Fineschi, & M.A. Gummin, *Telescopes and Instrumentation for Solar Astrophysics*, SPIE Conf. Vol. 5171, 239
Kemp, J.C. & Barbour, M.S. 1981, *PASP* 93, 521
Moutou, C., *et al.* 2006, these proceedings
Povel, H.P. 1995, *Optical Eng.* 34, 1870
Schmid, H.M., Joos, F., & Tschan D. 2006, *A&A* (submitted)
Smith, P.H. & Tomasko, M.G. 1984, *Icarus* 58, 35
Stenflo, J.O. & Keller, C.U. 1996, *Nature* 382, 588
Tomasko, M.G. & Doose, L.R. 1984, *Icarus* 58, 1
Tomasko, M.G. & Smith, P.H. 1982, *Icarus* 51, 65
Verinaud, C., *et al.* 2006, these proceedings

Direct Imaging of Exoplanets: Science & Techniques
Proceedings IAU Colloquium No. 200, 2005
C. Aime and F. Vakili, eds.
© 2006 International Astronomical Union
doi:10.1017/S1743921306009264

Atmospheric signatures by transit of HD209458 with VLT/UVES

Athena Coustenis[1], N. Iro[1,2], C. Moutou[3], M. Mayor[4] and D. Queloz[4]

[1]LESIA – Observatoire de Paris-Meudon, 5, place Jules Jannsen, FRANCE
email: athena.coustenis@obspm.fr

[2]Department of Physics – University of Florida, USA
email: iro@phys.ufl.edu

[3]LAM – Marseille, FRANCE
email: claire.moutou@oamp.fr

[4]Observatoire de Genève, SWITZERLAND
email: Michel.Mayor@obs.unige.ch,Didier.Queloz@obs.unige.ch

Abstract. In our search for clues as to the nature of the exosphere of HD209458 (Moutou *et al.*, 2001 ; Moutou *et al.*, 2003, Iro *et al.*, 2004), we have acquired VLT/UVES data during an ambitious observational campaign performed in June-September 2002 and covering 6 transits of the exoplanet. The resolving power was R=100000 in the 0.475-0.68 micron range. We search for ions and neutral molecules (such as H2O+, CO+, CH+, etc) originating in the planets exosphere and located in the evaporated material around the planet, occulting its primary star. We present in this paper a tentative search in the spectral regions where features of sodium or H_2O^+ can be present.

Keywords. Exoplanets, transits, exosphere, atmosphere, techniques: spectroscopic, (stars:) planetary systems, line: identification.

---

1. Introduction

HD209458b, the first discovered extrasolar planet transiting its star, has been subject to several studies constraining the composition of its atmosphere (see Table 1), in particular Na (Charbonneau *et al.* 2002; Narita *et al.* 2005), HeI (Moutou *et al.* 2003), CO (Deming *et al.* 2005) , H (Bundy & Marcy 2000; Vidal-Madjar *et al.* 2003; Vidal-Madjar *et al.* 2004; Winn *et al.* 2004; Narita *et al.* 2005), O, C and Si (Vidal-Madjar *et al.* 2004), Ca (Bundy & Marcy 2000; Narita *et al.* 2005; Shkolnik *et al.* 2005) and Fe (Bundy & Marcy 2000; Narita *et al.* 2005).

Our group has searched for such signatures since the first detection of 51 Pegasi from the ground and also from space with ISO (Coustenis *et al.* 1997, Rauer *et al.* 2000). We performed observations with UVES in the 0.33-0.67 μm region at high spectral resolution. We searched for ions and neutral molecules originating in the planet's exosphere and located in the evaporated material around the planet, occulting its primary star (Coustenis *et al.* 1997). Iro *et al.* (2004) presented the planet-induced chromospheric activity on HD209458 in these data as well as the He feature at 1.083 μm with data from VLT/ISAAC (Moutou *et al.* 2003). In this paper, we focus on the search for features from Na and H_2O^+.

2. The 2002 UVES campaign

We used VLT/UVES to observe HD209458 in 2002 over 6 of its transits, thus accumulating the time required to obtain a S/B of about 1000, required in order to detect

Constituent		Position	Note	Reference
Na	D	5893 Å	< 0.02 %	Charbonneau *et al.* (2002)
	D1	5896.06 Å	< 0.03 %	Narita *et al.* (2005)
	D2	5890.09 Å		
He	He I	1.083 μm	< 0.5 %	Moutou *et al.* (2001; 2003)
H	Lyα	1215.67 Å	15 % (1 σ)	Vidal-Madjar *et al.* (2003)
	Hα	6563 Å	< 0.1 %	Winn *et al.* (2003)
			< 0.81 %	Narita *et al.* (2005)
	H I	1215 Å	5.3 % (1 σ)	Vidal-Madjar *et al.* (2004)
	Hβ	4861.45 Å	< 3.6 %	Bundy & Marcy (2000)
			< 0.68 %	Narita *et al.* (2005)
	Hγ	4340 Å	< 2.7 %	Bundy & Marcy (2000)
			< 1.57 %	Narita *et al.* (2005)
O	O I	1305 Å	12.8 % (1 σ)	Vidal-Madjar *et al.* (2004)
C	C II	1335 Å	7.5 % (1 σ)	Vidal-Madjar *et al.* (2004)
Si	Si IV	1394 Å	No detection	Vidal-Madjar *et al.* (2004)
Ca	Ca II K	3943 Å	Not quantified	Shkonlik *et al.* (2005)
	Ca II H	3968 Å		
	Ca II & K		< 14 %	Bundy & Marcy (2000)
	Ca I		< 2.14 %	Narita *et al.* (2005)
Fe	Fe I	4384 Å	< 3.3 %	Bundy & Marcy (2000)
			< 1.08 %	Narita *et al.* (2005)
	Fe I	6024.2 Å	< 0.84 %	Narita *et al.* (2005)
CO		2 μm	Very weak (no detection)	Deming *et al.* (2005)

Table 1. Detections of atmospheric constituents in HD209458b.

a possible evaporated exosphere. The observations were spread out between June and September 2002. An equivalent time of observation was dedicated to the observation of the system off-transit, to supply a spectrum of comparison of the primary star equivalent to the on-transit spectrum. Finally, a particular effort was devoted to the calibration of the event, at the same time by inserting frequent observations of a comparison star so as to insure the determination of the telluric contribution, and to provide a high-quality calibration of the instrument.

The data are in the course of analysis since the beginning of 2003. The production of a spectrum of transmission with no telluric or stellar contamination requires a meticulous reduction which we perform ourselves. We note the excellent quality of the majority of the spectra.

3. Data analysis

We averaged the spectra for all the transits in order to obtain ON-transit and OFF-transit spectra with a better S/N ratio. We then calculate the difference between the ON- and the OFF-spectra for the spectral ranges of interests. We note dissymmetric features in the region of H_2O^+ lines at 6544 (see Fig. 1) and 6193 Å (see Fig. 2) (the nominal position of the H_2O^+ line is 6198 Å). Another region that we are investigating is the Na

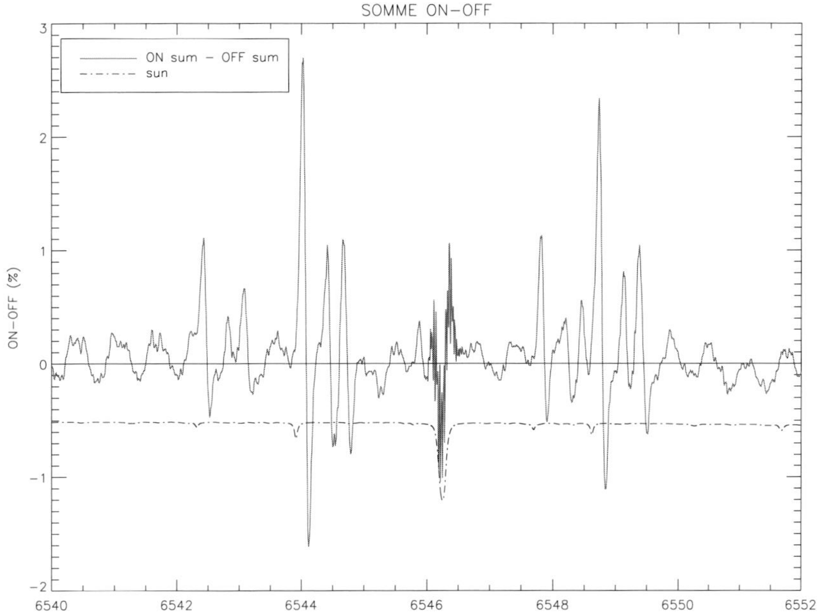

Figure 1. Difference between the sum of all the OFF-transit spectra and the sum of all the ON-transit spectra. The solar spectrum is shown in arbitrary units. We find a strong dissymmetry between ON and OFF at the H_2O^+ line position at 6544 Å.

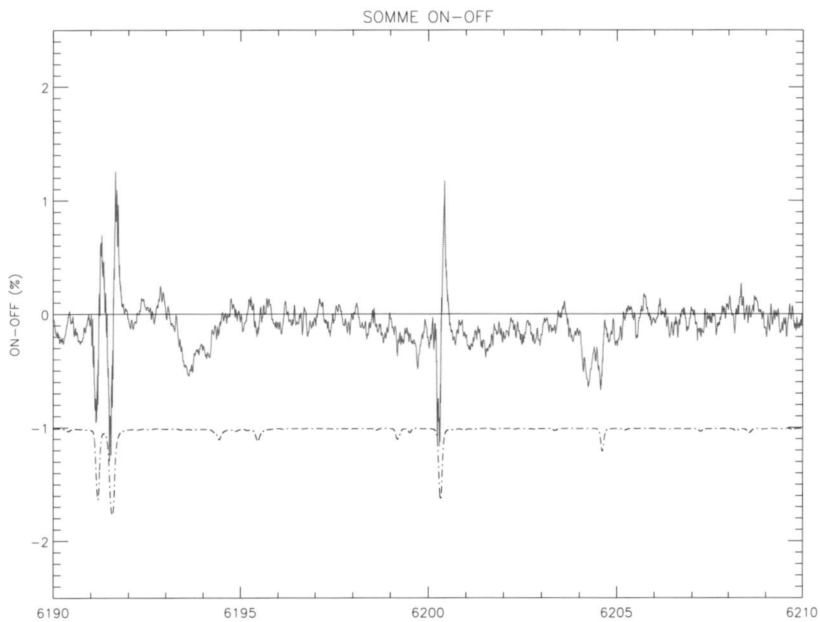

Figure 2. Search for $H2O^+$ (nominally at 6198 Å): zoom. The search in this region is complicated by the presence of a strong Fe line, but there seem to be unidentified features at 6193.5 Å and 6204.5 Å.

line doublet at 5891.9 and 5893.2 Å where we note a dissymetry between the On- and the OFF- transit spectra.

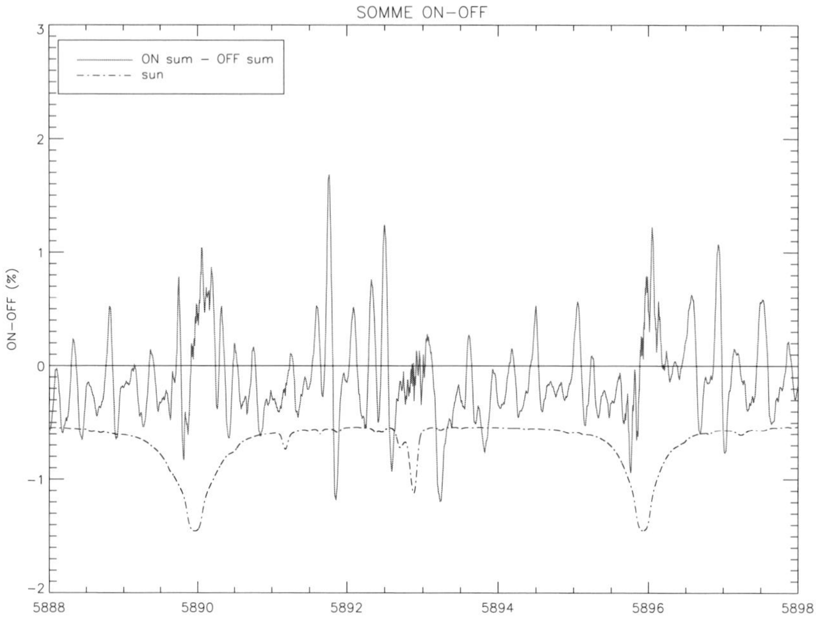

Figure 3. On- and off-transit spectra of HD209458 taken with UVES and the flux difference observed in two different orders within the Na band at 5893 Å. A zoom in the doublet region is shown, with a possible feature at 5893.2 Å.

4. Perspectives

The spectral ranges contained in the data presented in this paper include positions of possible features from Hα (6563 Å) and Hβ (4861 Å), Fe (6024 Å), O I (5577 and 6364 Å), CO^+ (4911 and 5490 Å). The analysis of these spectral regions are in progress.

References

Bundy, K. A. & Marcy, G. W. 2000, *PASP* 112, 1421

Charbonneau, D., Brown, T. M., Noyes, R. W., & al. 2002, *ApJ* 568, 377

Coustenis, A., Scheider, J., Wittemberg, R., *et al.* 1997. ASP 134, 296–303

Deming, D., Brown, T. M., Charbonneau, D., *et al.* 2005, *ApJ* 622, 1149

Iro, N., Coustenis, A., Moutou, C., *et al.* 2004 in *ASP Conf. Ser. Vol. 321, Extrasolar Planets: Today and Tomorrow*, 209

Moutou, C., Coustenis, A., Schneider, J., *et al.* 2003. *A & A* 405, 341.

Narita, N., Suto, Y., Winn, J. N., *et al.* 2005, *PASJ*, 57, 471

Rauer, H., Bockele-Morvan, D., Coustenis, A., *et al.* 2000, *ApJ* 622, 1075

Shkolnik, E., Walker, G. A. H., Bohlender, D. A., *et al.* 2005, *A & A* 355, 573

Vidal-Madjar, A., Lecavelier des Etangs, A., Désert, J.-M., *et al.* 2003, *Nature* 422, 143

Vidal-Madjar, A., Désert, J.-M., Lecavelier des Etangs, A., *et al.* 2004, *ApJ Letters* 604, L69

Winn, J. N., Suto, Y., Turner, E. L., *et al.* 2004, *PASJ*, 56, 655

Direct Imaging of Exoplanets: Science & Techniques
Proceedings IAU Colloquium No. 200, 2005
C. Aime and F. Vakili, eds.
© 2006 International Astronomical Union
doi:10.1017/S1743921306009276

L-band search for substellar companions in the Tucana and β Pictoris moving groups

M. Kasper[1], D. Apai[2], W. Brandner[3], L.M. Close[4] and K. Geißler[3]

[1]European Southern Observatory, Karl-Schwarzschild-Str. 2, 85748 Garching, Germany
email: mkasper@eso.org
[2]NASA Astrobiology Institute
[3]Max-Planck-Institut für Astronomie, Königstuhl 17, 69117 Heidelberg, Germany
[4]Steward Observatory, University of Arizona, 933 N Cherry Ave., Tucson AZ 85721, USA

Abstract. Radial velocity surveys provide evidence that giant extrasolar planets are common, but their detection space is limited to only a few astronomical units from the stars. In order to close this gap, the adaptive optics assisted NIR imager NAOS-CONICA (NACO) at the VLT was used for a deep (15-20 minutes exposure time per target) L-band survey of a sample of closeby young stars. All stars are members of the Tucana and β Pictoris moving groups apart from the the somewhat older star HIP 71395 that has a radial velocity trend suggesting a massive planet in large orbit. The chosen observation wavelength is very well suited for very high contrast imaging of close companions at this age and makes this survey unique. The goal was to detect substellar companions to these stars at distances as close as 5-20 AU and ultimately to detect giant extrasolar planets down to a few Jupiter masses, to measure their frequency, and - by comparison with models - determine their physical properties. This paper presents the results obtained on a subsample of 12 stars that have been observed during ESO P73.

Keywords. extrasolar planets, adaptive optics, L band, NACO, high contrast imaging.

1. Introduction

A major scientific breakthrough of the last years was the discovery of extra-solar planets by radial velocity searches (Mayor & Queloz 1995) which add up to more than 100 such systems known to date (exoplanets.org). Although the sample is still statistically small, it appears that a significant fraction of stars (about 6%) have planetary companions of relatively high mass at close-in orbit. The discovery space of radial velocity searches does not cover planets at a distance larger than a few AU, so few observational results on these objects exist to date. The recent detections of substellar (Liu 2002, McCaughrean *et al.* 2004) or even planetary mass (Chauvin *et al.* 2005, Neuhäuser *et al.* 2005) companions around young stars by direct imaging with AO let this technique appear most promising to close this gap.

Current programs searching for substellar companions are mostly carried out in K-band by broadband imaging (Masciadri *et al.* 2005) or in H-band using simultaneous observations in narrow H-band filters inside and outside the methane absorption feature to efficiently subtract speckle noise (Close *et al.* 2005). However, the models of Baraffe *et al.* (2003) predict colors of H–L≈3 mag and K–L≈4 mag for a 5 Jupiter mass planets at the age of 100 Myr making the L-band an interesting observing window to search for substellar companions. For example, the 5 Jupiter mass companion to the 8 Myr old brown dwarf 2M1207 detected by Chauvin *et al.* (2005) is as red as K–L≈1.75 mag and H–L≈3.81 mag. As additional advantages compared to shorter wavelengths, the L-band image quality delivered by AO is better and temporally more stable reducing speckle

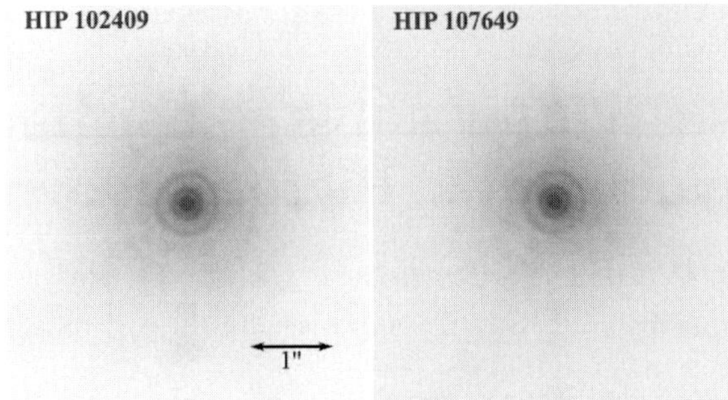

Figure 1. NACO L-band images of HIP 102409 and HIP 107649.

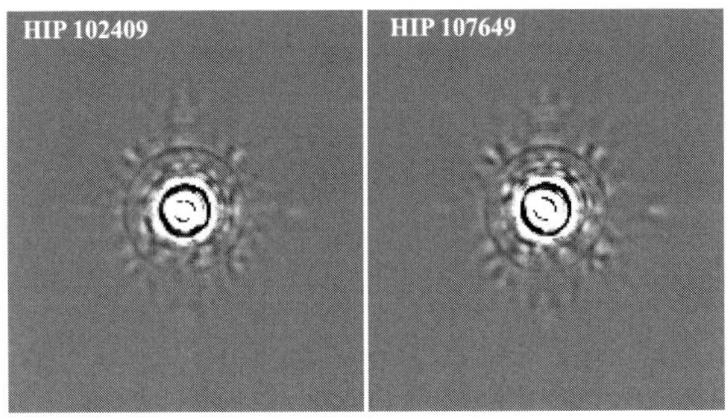

Figure 2. High-pass filtered images of HIP 102409 and HIP 107649.

Table 1. Target stars

Target	Spectral type	Distance (pc)	V (mag)	L (mag)	Age (Myr)	Remarks
HIP 71395	K0	17	7.6	5.6	500-1000	lin. radial velocity trend (Butler *et al.* 2003)
HIP 116748	G6IV	46	8.5	6.9	30	Tucana, 5.3″ binary
HIP 102141	M4.5	10	11.3	5.8	12	βPic, GJ799, 2.8″ binary
HIP 102409	M0	10	8.9	5.2	12	βPic, GJ803, debris disk (Kalas *et al.* 2004)
HIP 107649	G0V	16	5.6	4.2	30	Potential Tucana member
HIP 105441	K2V	31	9	6.8	30	Potential Tucana member
HIP 105441B	?	31	10.5	8	30	Potential Tucana member
HIP 1993	K7V	37	11.5	8.2	30	Confirmed Tucana member
HIP 3556	M1.5	39	11.9	7.9	30	Potential Tucana member
HIP 2729	K5V	46	9.6	6.7	30	Confirmed Tucana member
HIP 4448	K4V	44	9	6.3	30	Potential Tucana member
HIP 1113	G6V	44	8.8	7.2	30	Potential Tucana member
HIP 1481	G0V	41	7.5	6.1	30	Confirmed Tucana member

noise and facilitating PSF subtraction. All these advantages more than compensate for the increased sky background.

2. Targets and observations

The target sample has been compiled of 18 nearby (10-46 pc) young stars which are members of the Tucana association (Zuckerman *et al.* 2001) and the βPic moving group (Zuckerman *et al.* 2001) of which 12 were observed in ESO observing period P73 (2004)

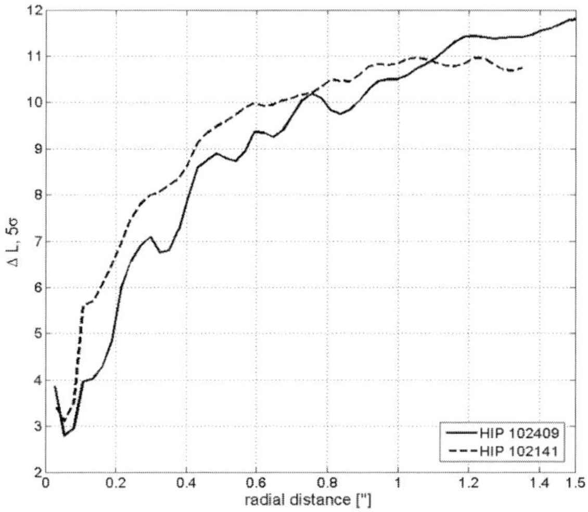

Figure 3. Contrast achieved on HIP 102409 and HIP 102141.

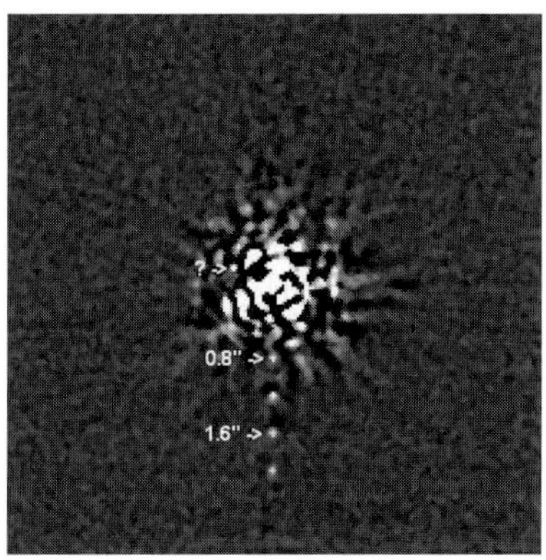

Figure 4. Difference of HIP 102409 and HIP 102141 with artificial test planets (10 mag fainter).

and are listed in table 1. The ages of the member stars were estimated to be between 10 and 30 Myr (e.g., Kastner *et al.* 2003, Stelzer & Neuhäuser 2001) providing an ideal test bed for detecting extra-solar planets and study their formation. The sample was restricted to the later spectral types G, K, and M. Considering the reported V-band luminosities and the colors according to the spectral types, the targets in our sample have estimated L-band luminosities between L = 4.2 mag and L = 8.2 mag which are reported in table 1.

A 100 Myr old, 5 Jupiter mass planet would have luminosities between L = 14.06. mag at 10 pc and L = 17.07 mag at 40 pc (Baraffe *et al.* 2003). The 5σ L-band point source detection limit of NACO in 15-20 minutes exposures is permitting the detection of planetary mass companions for all observed stars.

All targets have been observed in the L′ filter centered at 3.8μm with the L27 camera of CONICA providing a pixel scale of 27 mas. NAOS was counter-chopping, i.e. the field selector of the wavefront sensor followed the motion of the telescope's M2 providing AO

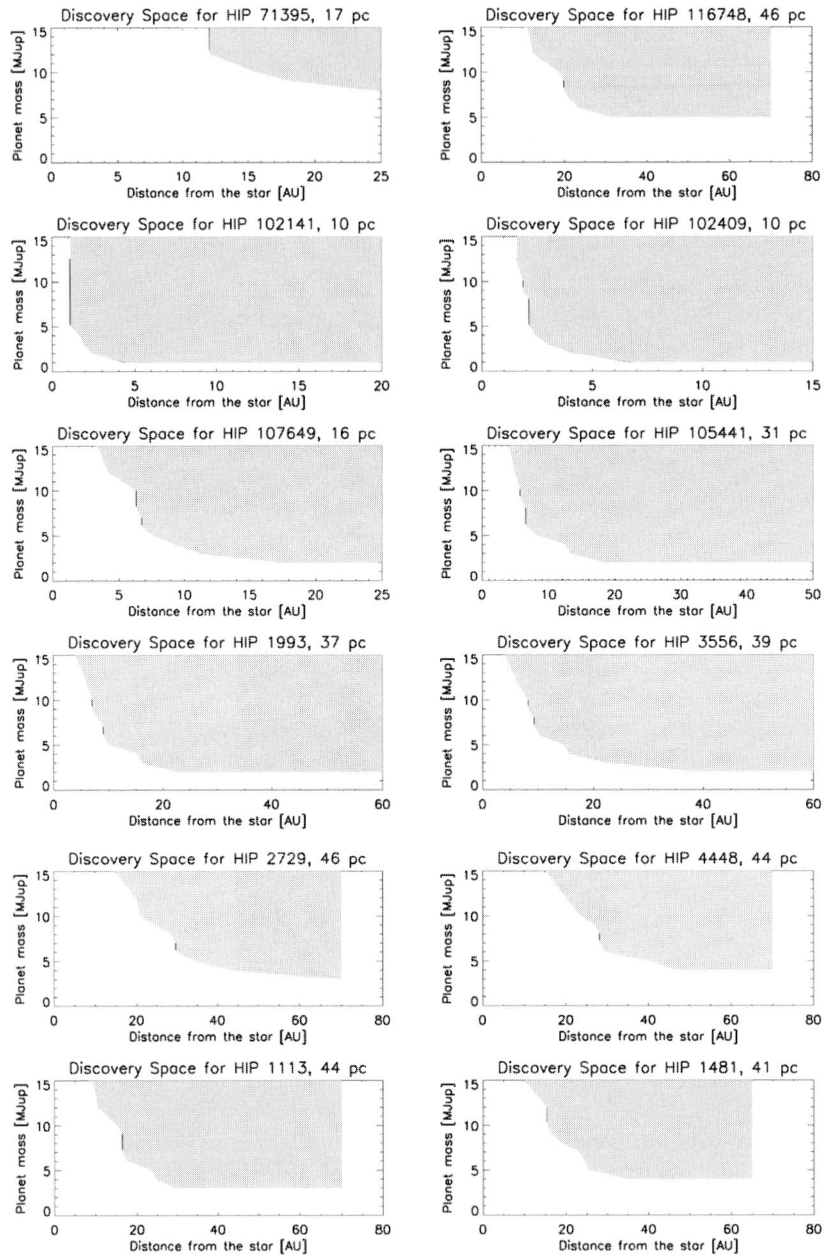

Figure 5. Discovery space for the observed targets.

corrected images at both chopping positions effectively doubling the observing efficiency. Total on-source exposure times for all shown data sets are between 15 and 20 minutes. The targets were observed in pairs that were chosen such that position on sky and brightness of the partners were as close as possible. In this way, the objects could be used to calibrate or subtract each others PSF and no additional observing of calibrator stars was necessary.

Figure 1 shows HIP 102409 and HIP 107649 observed for 15 minutes on source each and separated by about 30 minutes from each other. The image quality is remarkably good (Strehl ratio of about 70%. The PSF core was saturated so the Strehl ratio estimate is based on the intensity in the Airy rings and is pretty coarse) and stable permitting efficient PSF calibration and subtraction.

3. Data analysis

After standard image processing (flat fielding, bad pixel cleaning, sky subtraction), the image pairs were superimposed with subpixel accuracy, normalized and high-pass filtered. The high-pass filtering was done by subtracting from the image the median filtered ($\approx$3 FWHM box-width) image. This procedure efficiently eliminates large scale structure while leaving point sources (such as substellar companions) relatively unaffected. Figure 2 shows a high-pass filtered version of figure 1. Finally the images have been median filtered with a small box width to reduce pixel-to-pixel variations.

Contrast curves were calculated from the difference of a pair of final images that had been observed together by calculating the standard deviation of the pixel intensity in 2 pixel wide annuli of variable radii centered on the stars. Figure 3 plots the contrast at five standard deviations as a function of the angular distance from the stars for two of the observed pairs, i.e., 4 objects. The solid line represents the pair HIP 102409 and HIP 107649 shown in figures 1 and 2. The dashed line shows the contrast of the 2.8$''$ binary HIP 102141. The latter is an interesting case, because the close binary permits very efficient speckle subtraction due to the simultaneous observation. Hereby, angular anisoplanatism is negligible due to the small separation of the stars which is at least 10 times smaller than the anisoplanatic angle of atmospheric turbulence in L-band. The better contrast of about one magnitude at small angular separations indicates the gain achievable by simultaneous observation. The flattening of the curves at larger angular distance is due to the sensitivity hitting the background limit.

A mandatory verification of predicted contrast levels is the insertion of fake planets into the images before performing the data analysis. Figure 4 shows the difference of the two images shown in figure 2 with added fake planets at distances of 0.8$''$, 1.2$''$, 1.6$''$ and 2$''$. Comparing this image with the predicted contrast shown in figure 3, the 5σ criterion is obviously just enough to suspect a detection. Another interesting result is that the added planets are point-like while most speckles show a different mostly elongated morphology. Masciadri & Raga (2004) have developed a method to use this result in order to better discriminate the planets from the speckles. The question mark denotes a non-elongated feature in the image that would be worth further investigations.

4. Results

Knowing the stars age, brightness and distance, the calculated radial contrast for the individual objects can be converted into a discovery space using using planetary evolutionary models by Baraffe *et al.* (2003). The discovery space describes the detectable planet mass as a function of separation from the star in AU and is shown for all target stars in figure 5. Since no clear companion candidates were found, the discovery space provides upper limits on planet masses for a given separation from the host stars. Given the contrast curves are comparable to the ones achieved using other methods such as K-band imaging or H-band differential imaging and the contrast gain of one magnitude and more in L-band due to the smaller brightness difference of potential substellar companions to their host stars, the discussed observations are among the most sensitive observations to date.

Acknowledgements

We thank the ESO VLT astronomers for carrying out our observations in service mode. This material is partly based upon work supported by the National Aeronautics and Space Administration through the NASA Astrobiology Institute under Cooperative Agreement No. CAN-02-OSS-02 issued through the Office of Space Science.

References

Baraffe, I., Chabrier, G., Barman, T.S., Allard, F. & Hauschildt, P.H. 2003, *A&A* 402, 701

Chauvin, G., Lagrange, A.-M., Dumas, C., Zuckermann, B., Mouillet, D., Song, I., Beuzit, J.-L. & Lowrance, P. 2005, *A&A* 438L, 25

Close, L.M., Lenzen, R., Guirado, J.C., Nielsen, E.L., Mamajek, E.E., Brandner, W., Hartung, M., Lidman, C. & Biller, B. 2005, *Nature* 433, 286

Kalas, P., Liu, M.C. & Matthews, B.C. 2004, *Science* 303, 1990

Kastner, J.H., Crigger, L., Rich, M. & Weintraub, D.A. 2003, *ApJ* 585, 878

Liu, M.C., Fischer, D.A., Graham, J.R., Lloyd, J.P., Marcy, G.W., Butler & R.P. 2002, *ApJ* 571, 519

Masciadri, E. & Raga, A. 2004, *ApJ* 611, 137

Masciadri, E., Mundt, R., Henning, Th. & Alvarez, C. 2004, *ApJ* 611, 137

Mayor, M. & Queloz, D. 1995, *Nature* 378, 355

McCaughrean, M.J., Close, L.M., Scholz, R.-D., Lenzen, R., Biller, B., Brandner, W., Hartung, M. & Lodieu, N. 2004, *A&A* 413, 1029

Neuhäuser, R., Guenther, E.W., Wuchterl, G., Mugrauer, M., Bedalov, A., Hauschildt, P.H. 2005, *A&A* 435, 13

Stelzer, B. & Neuhäuser, R. 2001, *A&A* 538, 377

Zuckerman, B., Song, I., Bessell, M.S. & Webb, R.A. 2001, *ApJ* 562, 87

Zuckerman, B., Song, I. & Webb, R.A. 2001, *ApJ* 559, 388

Direct Imaging of Exoplanets: Science & Techniques
Proceedings IAU Colloquium No. 200, 2005
C. Aime and F. Vakili, eds.
© 2006 International Astronomical Union
doi:10.1017/S1743921306009288

Raman Scattering in the non-isothermal giant planet atmospheres

Nadyia Kostogryz

Main Astronomical Observatory of NAS of Ukraine
email: kosn@mao.kiev.ua

Abstract. The influence of non-isothermal atmosphere on the intensity of details of Raman Scattering is computed for spectra of Uranus and Neptune. It follows that the intensity of this scattering depends on the depth of the level formation in an optically homogeneous non-isothermal atmosphere; neglecting of real temperature profiles during the analysis of observation data leads to large errors in the parameters of the atmosphere upper level.

Keywords. Raman Scattering, Planetary nonisothermal atmosphere.

1. Introduction

Raman scattering is the light interaction with matter which is incoherent meaning that the scattered radiation frequency has another value than the frequency of the incident radiation. We can observe the Raman spectra even in molecules which are not observed in infrared spectrum and the best example of Sun molecule is hydrogen. So we can observe the details of Raman scattering in the giant planet Ultra Violet spectrum although the atmosphere of these objects consist of nearly 85% of hydrogen molecules.

In the row of papers (see, for instance, [Dementiev (1992)-Morozhenko(2001)])there were proposed to use the details of Raman Scattering in spectrum of the giant planets where were used isothermal atmosphere for determination of the optical features and vertical structure of the upper layers of the atmospheres, including relations τ_a/τ_R and τ_κ/τ_R (τ_a ,τ_R - scattering part of optical depth of the aerosol and gas, accordingly, τ_κ - absorbing forming optical depth).

2. Calculation Method

Taking into account the effect of Raman scattering was done on the basis of the analysis proposed by J.B. Pollack [J.B. Pollack *et al.* (1986)]. Expressions for single-scattering albedo ω was taking in the form

$$\omega = \frac{\tau_R + \tau_a + \sum \tau_R^i (f_{\lambda_i}/f_{\lambda_0})}{\tau_R + \tau_a + \tau_\kappa + \sum \tau_R^i} = \frac{\sigma_R + \sigma_a + \sum \sigma_R^i (f_{\lambda_i}/f_{\lambda_0})}{\sigma_R + \sigma_a + \sigma_\kappa + \sum \sigma_R^i} \tag{2.1}$$

where $\sigma_R^i(\tau_R^i)$ - averaged of the line of sight the volume coefficient (optical depths) of the Raman scattering by molecules, f_{λ_i} - the intensities of Sun radiation on the wavelengths, from which corresponding to transition in process of the Raman scattering light photon is carried on wavelength λ_i and f_{λ_0} on wavelength λ_0 .

Taking into account that necessary accuracy of the account of the Raman scattering is reached for Stocks rotation S(0), O(2), S(1) and Q1(1) vibration transition [Belton *et al.* (1973), Cochran, Trafton (1978)], and that real atmosphere of giant planets are hydrogen-helium (with relative concentration of these gases about 0.85 and 0.15) then the corresponding sums of (2.1) are:

$$A = 0.85 * (N_0\tau_{S(0)} + N_2\tau_{O(2)} + N_1\tau_{S(1)} + \tau_{Q_1(1)})/\tau_R \quad (2.2)$$

$$D = 1 + 0.85 * [((N_0\tau_{S(0)} + N_2\tau_{O(2)})f_{\lambda_1} + N_1\tau_{S(1)}f_{\lambda_2} + \tau_{Q_1(1)}f_{\lambda_3})/f_{\lambda_0}\tau_R] \quad (2.3)$$

where $f_{\lambda_1}, f_{\lambda_2}, f_{\lambda_3}$ - spectral value of the energy in spectrum Sun on the wavelength, from which rotation $(S(0), O(2), S(1))$ and vibration $(Q_1(1))$ Stocks transition of Raman scattering carries the sun photon on wavelength λ_0 accordingly; $\tau_{S(0)}, \tau_{O(2)}, \tau_{S(1)}$ and $\tau_{Q_1(1)}$ - are optical thicknesses of the Raman scattering of corresponding transition.

In the mentioned above papers where were analysed the data of intensities of the Raman scattering details isothermal atmosphere model were considered while the real atmosphere of the giant planets are characterized by rather complex warm-up profile [Lindal, (1992)]. It is very important that parameter D depends on not only wavelength, but also on temperature. Because when the temperature is changing the relative number of the molecules of the hydrogen in ortho- and para-conditions are changing too. We shall remind the hydrogen molecule has two atoms with alike value of spin (1/2), so the total spin can be equal 1 (symmetrical level or orto-hydrogen) or 0 (asymmetric level or para-hydrogen), for which density populations of the molecules are defined accordingly by expressions

$$N = 3(2j+1)\exp -B * j(j+1)\frac{hc}{kT}; j = 1, 3, 5\ldots \quad (2.4)$$

$$N = (2j+1)\exp -B * j(j+1)\frac{hc}{kT}; j = 0, 2, 4\ldots \quad (2.5)$$

where B - rotation constant, which for hydrogen molecule is equal $60sm^{-1}$, h - Planck constant, c - speed of light, k - Boltzman constant, T- temperature(in Kelvin).

The purposes of our work are using example of the temperature profile in atmospheres of Uranus and Neptune [Lindal, (1992)] to study:

1. The effect of changing Raman scattering depending on efficient depth of the forming of intensities of diffuse reflected radiations.(Figure1)

2. What are the errors when we neglect of real temperature profiles during the analysis of observation data? (table 1)

3. Do we have the differences between manifestations of Raman scattering details in Uranus and Neptune non-isothermic atmospheres? (Figure 3)

4. Can we apply this method to extrasolar planets? (Figure 4)

3. Discussion and Conclusions

We have computed the single scattering albedo for three levels in Uranus atmosphere (with the value of pressures 0.0025, 0.1, 2.309 bar) and in Neptune atmosphere (with the value of pressures 0.0025, 0.141, 6.268 bar)(figure 1).

Figure 1 shows, that the values of single scattering albedo (as parameters A and D) change with depth in the atmosphere. That is why the using of isothermal atmosphere model leads to faults in such determined values $\tau_a/\tau_R, \tau_\kappa/\tau_R$ and τ_κ/τ_S when we analyse the intensity of Raman scattering details. To estimate the probability of errors as observation of data we have calculated using lay expressions ((2.3) and (2.1)) the values of D and ω for level of pressure 0.141 bar and for wavelength where is the Fraunhofer line (F) (λ 397.0 nm) and its "ghost"(D). Then, when we use the isothermal atmosphere model with value of temperature $T = 51.7K$, from the system of equation

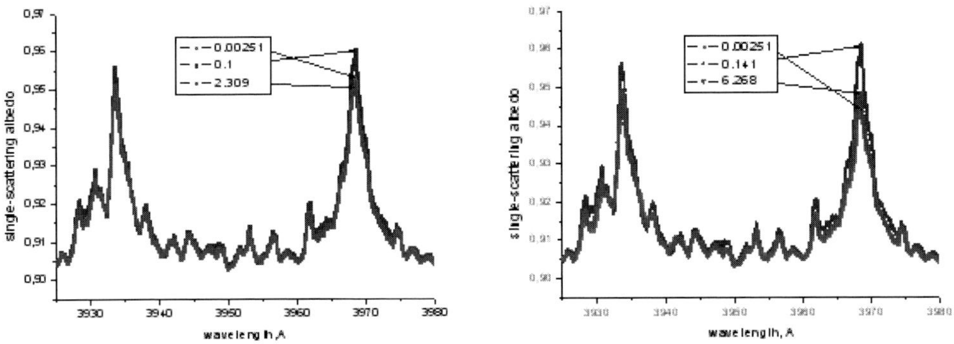

Figure 1. Single scattering albedo versus wavelength for pressure values of 0.0025, 0.1, 2.309 bar in wavelength range from 3925 to 3980 A for Uranus (left) and for pressure values of 0.0025, 0.141, 6.268 bar for Neptune (right).

Table 1. The Comparison of values which were finding from the system of equations with real value.

From system of equations 8				Real values		
P, bar	τ_a/τ_R	τ_κ/τ_R	τ_κ/τ_S	τ_a/τ_R	τ_κ/τ_R	τ_κ/τ_S
0.00251	1.79009	0.28269	0.101	1	0.2	0.1
0.141	1.0459	0.20422	0.0998			
6.268	1.52713	0.257076	0.1017			

$$(D/\omega)^F - 1 = (\tau_\kappa/\tau_R) - [(1/\omega)^F - 1](\tau_a/\tau_R), \tag{3.1}$$

$$(D/\omega)^D - 1 = (\tau_\kappa/\tau_R) - [(1/\omega)^D - 1](\tau_a/\tau_R) \tag{3.2}$$

we determined the values of $\tau_a/\tau_R, \tau_\kappa/\tau_R$ and also

$$\tau_\kappa/\tau_S = \frac{\tau_\kappa/\tau_R}{1 + \tau_a/\tau_R},$$

which are listed in Table 1.

As you can see when we ignore the real temperature profile, the values of single scattering albedo for isothermal atmosphere has the large errors in parameters of atmosphere.

The answer on the question: "Do we have the differences between manifestations of Raman scattering details in Uranus and Neptune non-isothermic atmospheres?" you can receive from figure 3. As you can see the values of single-scattering albedo for two planets are different. In previous papers there were considered single-scattering albedo for the effective temperature which was the same both for Uranus and Neptune. But these values are different for two planets and what is more, as we noted earlier, when we don't take into account real temperature profile, the values of single scattering albedo for isothermal atmosphere have the substantial differences.

Can we use this method to extrasolar planets? This question is very important. If we could use this method to exoplanets, we could define the optical characteristics of there atmospheres. It would be a new method to define very important parameters of atmospheres. As you can see from figure 4 this effect is more visible for low temperature and when the temperature increase, the differences between single scattering albedo

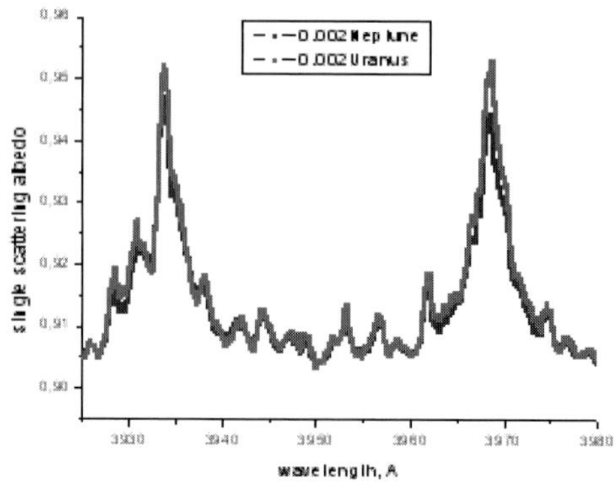

Figure 2. Single-scattering albedo versus wavelength in range of 3925-3980A for Uranus and Neptune atmospheres for the mean of pressure of 0.002bar.

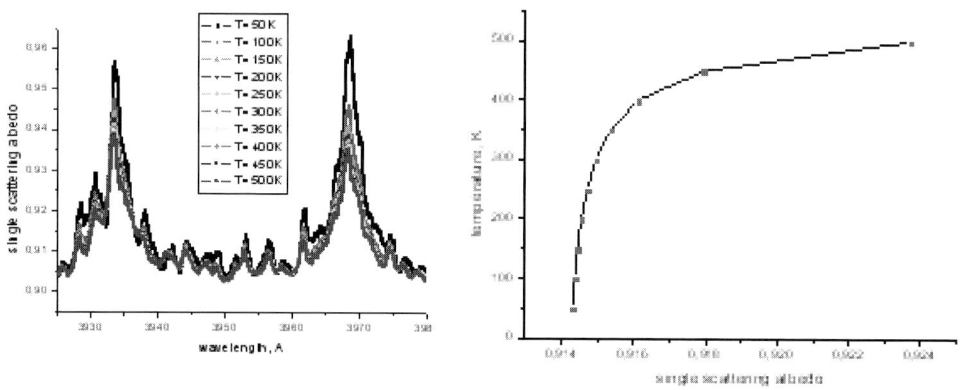

Figure 3. Temperature versus single scattering albedo for 3970A(left) and single scattering albedo versus wavelength for different values of temperature in wavelength range from 3925 to 3980 A for (right).

for this value of temperature decrease. So, we can use this method for not very close exoplanets, not for "hot Jupiter", but we can use it for exoplanets with large orbital period.

References

Dementiev, M.S. 1980, *Kinematics and physics celestial body* 8.- 2.-P.25-35
Morozhenko, A.V. 1997, *Kinematics and physics celestial body* 6.-N.4.-P.22-33
Morozhenko, A.V. 1999, *Kinematics and physics celestial body*, 15. - 2. -P. 110 - 122
Morozhenko, A.V. 1999, *Kinematics and physics celestial body*, 15. - 3. -P. 254 - 264
Morozhenko, A.V. 2001, *Kinematics and physics celestial body* 17. - 3. - P.261 - 278
Morozhenko, A.V. 2003 *Kinematics and physics celestial body* 19. - 6. - P.483 - 500
Belton, M.J.S., Wallace, L., & Price, M.J. 1973, *ApJ* 184. - N3. -P. 143-146
Cochran, W.D. & Trafton, L.M. 1978, *ApJ* 219. - N1. -P.756-762
Lindal, G.F. 1992, *Astron.J.* 103. - N3. - P.967-982
Pollack, J.B., Rages, K., Baines, K.H. *et al.* 1986, *Icarus* 65. - N2/3. - P. 442-466

Direct Imaging of Exoplanets: Science & Techniques
Proceedings IAU Colloquium No. 200, 2005
C. Aime and F. Vakili, eds.
© 2006 International Astronomical Union
doi:10.1017/S174392130600929X

Infrared Light Curves and the Detectability of Close-In Extrasolar Giant Planets

L. Jeremy Richardson[1], Sara Seager[2], Drake Deming[3], Joseph Harrington[4], Richard K. Barry[1], Jayadev Rajagopal[1], and William C. Danchi[1]

[1]Exoplanets and Stellar Astrophysics Laboratory, NASA Goddard Space Flight Center, Mail Code 667, Greenbelt, MD 20771, USA
email: richardsonlj@milkyway.gsfc.nasa.gov

[2]Department of Terrestrial Magnetism, Carnegie Institution of Washingtone
5241 Broad Branch Rd. NW, Washington, DC 20015, USA

[3]Planetary Systems Laboratory, NASA Goddard Space Flight Center, Mail Code 693, Greenbelt, MD 20771, USA

[4]Cornell University, 326 Space Sciences Bldg., Ithaca, NY 14853, USA

Abstract. We compute theoretical infrared light curves for several known extrasolar planets. We have constructed a set of routines to calculate the orbital parameters for a given planet and integrate over the planetary disk to determine the total flux density of the planet as it orbits the parent star. We have further developed a spectral synthesis routine to calculate theoretical spectra of extrasolar giant planets from 3–24 μm. The code requires a temperature-pressure profile as input, calculated by solving the radiative transfer equation; it then calculates continuous opacities and line opacities for water, carbon monoxide, and methane, and finally integrates over the layers of the atmosphere to determine the emergent flux. By integrating the theoretical spectrum over the bandpass of a particular instrument and including realistic instrument noise, we produce a set of multi-wavelength, infrared light curves. Using these light curves, we predict whether a particular known planet can be observed and characterized using the Spitzer Space Telescope, as well as other proposed space-based instruments, such as the Fourier-Kelvin Stellar Interferometer (FKSI) and the James Webb Space Telescope (JWST).

Keywords. methods: analytical; stars: individual (HD 209458).

---

1. Theoretical Light Curves

The following is a brief outline of steps in computing the theoretical light curve (implemented in IDL):

• Read the physical parameters for the star and planet.

• Calculate the orbital elements for the planetary system from spherical trigonometry considerations.

• Treat the planet as a blackbody and assume a non-zero temperature asymmetry between the day and night sides of the planet.

• Integrate over the visible portion of the planetary disk as a function of its orbital position. This gives the emitted radiation of the planet as a function of time.

Example theoretical calculations for HD 209458 b and 55 Cancri b at 24 μm are shown in Figures 1 and 2, respectively. We estimate the photon noise due to the planet and star for the known bandwidth and given integration time (t=10 sec) for Spitzer (85 cm aperture) at the MIPS wavelength of 24 μm Rieke *et al.*, 2004. The *primary eclipse* (when the planet transits the star) and the *secondary eclipse* (when the planet disappears behind the star) are shown for the case of HD 209458 b.

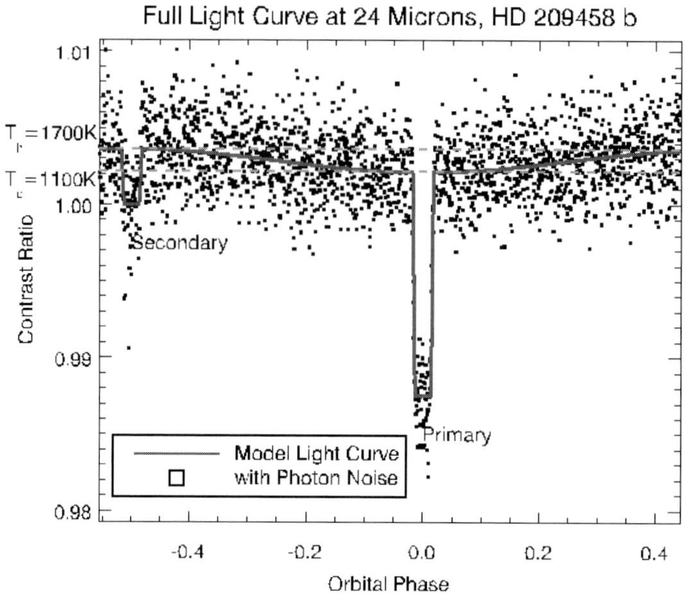

Figure 1. *Theoretical* light curve for HD 209458 b with photon noise included for Spitzer/MIPS. We adopted R_*=1.146 R_{sun}, R_p=1.347 R_J, i=86.68 degrees, M_*=1.107 M_{sun} Brown *et al.*, 2001, and eccentricity $\epsilon = 0.0$ Deming *et al.*, 2005. The dashed lines indicate the assumed temperature asymmetry of 600 K.

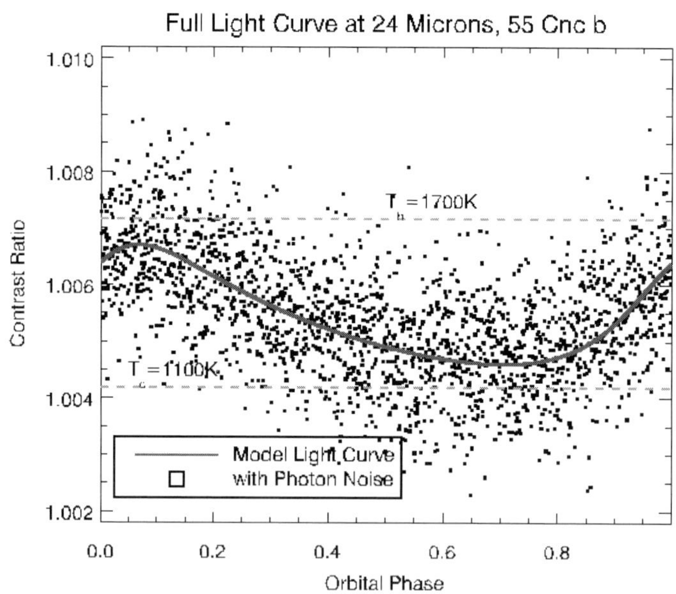

Figure 2. *Theoretical* light curve for 55 Cnc b with photon noise included (also for Spitzer/MIPS). We adopted R_*=0.87 R_{sun} Butler *et al.*, 1997 and assumed M_*=1.107 M_{sun}, i =45 degrees, R_p = 1.347 R_J, and ϵ = 0.03, which was increased to show the effect on the light curve. Again, the dashed lines indicate the assumed temperature asymmetry of 600 K.

We further compare the theoretical light curves to actual data on the eclipse of HD 209458 b. Figure 3 shows that we can match the Spitzer/MIPS data on the primary

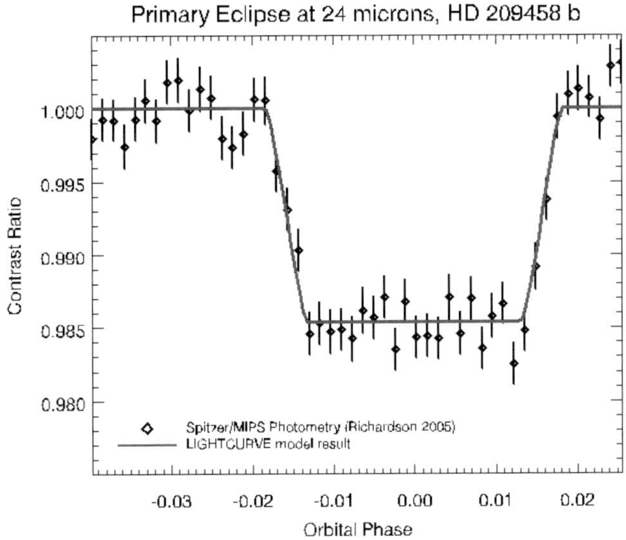

Figure 3. Actual data on the primary eclipse of HD 209458 b at 24 μm from Spitzer, averaged in phase bins of width 0.002. Theoretical light curve (solid line) was calculated for $R_p = 1.29\ R_J$ at 24 μm, and $\chi^2_\nu = 1.035$.

eclipse at 24 μm. Our routine can also reproduce the now-famous HST curve† of the eclipse in the visible Brown *et al.*, 2001, as shown in Figure 4.

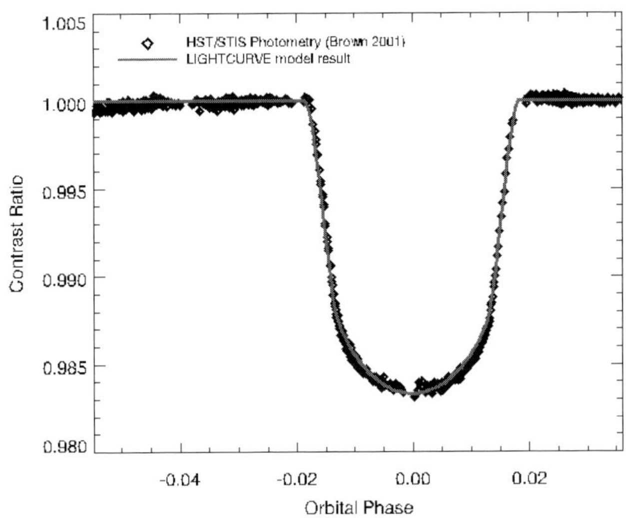

Figure 4. Points are publicly-available data from HST on the primary eclipse of HD 209458 b in the visible. Solid line indicates LIGHTCURVE fit for Brown *et al.*, 2001 parameters; fit for values in eclipse (i.e., between phase of -0.02 and 0.02) has $\chi^2_\nu = 1.969$.

2. Predictions

The utility of this routine for predicting the detectability of extrasolar planets is clear. Given a set of planetary and stellar parameters, we can produce a theoretical light curve,

† Public data can be found at `http://cfa-www.harvard.edu/~dcharbon/papers.html`

and thus determine what features (i.e., temperature asymmetry, transits) may be observable with a given instrument. We integrated this routine with an instrument simulation for FKSI mission concept Danchi *et al.*, 2003, Barry 2006 (these proceedings).

3. Theoretical Planetary Spectra

Here we present a brief summary of IDL routines to calculate theoretical spectra of extrasolar planetary atmospheres.

• Input temperature-pressure profile for the model atmosphere, calculated by solving the radiative transfer equation. We adopt profiles for HD 209458 b from Seager & Sasselov, 1998, and we compare the results to updated models to be described in a future paper (Seager, in prep, 2005).

• Tabulate line opacities for H_2O, CH_4, and CO.

• Calculate continuous opacities due to collision-induced absorption (CIA) Borysow 2002.

• Integrate over the optical depth of the layers of the model atmosphere to obtain the emergent flux density.

4. Future Work

The next step in this program is to combine the spectral synthesis and light curve routines. The light curves were based on the blackbody assumption, which is clearly inadequate to describe the true nature of the planetary atmosphere. The spectral synthesis code must be expanded to include more extensive line opacity data in order to extend the calculation to 24 μm. We can then integrate the spectrum over a given bandpass for a particular instrument, and use this as input to the light curve routine. Finally, we will estimate noise from all sources (e.g., instrument instability, known systematics, and stellar variability), add it to the calculated light curves, and quantify to what extent the light curve is recoverable from the noisy synthetic data, thereby allowing a prediction of what features (e.g., secondary eclipse, temperature asymmetry) are detectable for the known extrasolar planets.

Acknowledgements

LJR is supported by a Research Associateship from the National Research Council at NASA Goddard.

References

Barry, R. K., *et al.* 2006. In C. Aime & F. Vakili (Eds.), IAUC 200, Direct Detection of Exoplanets: Science and Techniques, in press

Borysow, A. 2002, *A&A*, 390, 779

Brown, T. M., Charbonneau, D., Gilliland, R. L., Noyes, R. W., & Burrows, A. 2001, *ApJ*, 552, 699

Butler, R. P., Marcy, G. W., Williams, E., Hauser, H., & Shirts, P. 1997, *ApJ*, 474, L115

Danchi, W. C., Deming, D., Kuchner, M. J., & Seager, S. 2003, *ApJ*, 597, L57

Deming, D., Seager, S., Richardson, L. J., & Harrington, J. 2005, *Nature*, 434, 740

Rieke, G. H., *et al.* 2004, *ApJS*, 154, 25

Seager, S., & Sasselov, D. D. 1998, *ApJ*, 502, L157

Direct Imaging of Exoplanets: Science & Techniques
Proceedings IAU Colloquium No. 200, 2005
C. Aime and F. Vakili, eds.
© 2006 International Astronomical Union
doi:10.1017/S1743921306009306

Polarimetric investigation of Uranus and Neptune

F. Joos and H. M. Schmid

Institute of Astronomy, ETH Zurich, CH-8092 Zurich, Switzerland
email: fjoos@astro.phys.ethz.ch

Abstract. We present ground-based disc resolved polarimetric imaging and spectropolarimetry of Uranus and Neptune in the wavelength range from 530 to 930 nm with the goal to constrain the expected polarization of extra-solar planets with similar atmospheric properties. We find for both planets a high limb polarization as expected for atmospheres with significant Rayleigh scattering layers. The limb polarization is enhanced in the strong methane absorption bands when compared to the adjacent continuum. From the observed polarization features we conclude that extra-solar planets with atmospheric properties similar to Uranus and Neptune are strongly polarized and well suited for polarimetric detection and investigation.

Keywords. Polarimetry, exoplanets, Uranus, Neptune.

1. Introduction

Reflected light from planets is polarized due to scattering. This basic property provides the possibility of detecting and characterizing extra-solar planets with polarimetry. However, it is hard to predict the amount and the wavelength dependency of polarization present in a "typical" extra-solar planet. Currently, observational results on the polarization of planets are only available for solar system targets. Therefore we investigate the scattering polarization of solar system planets in order to extrapolate their polarimetric properties to the case of extra-solar planets.

A typical extra-solar planet is expected to be observed for a phase angle $\alpha \approx 60^\circ$ to 120° (Fig. 1). Observations of our solar system planets for similar α's are desirable, but from Earth all outer planets can only be observed for small phase angles (back-scattering). For Jupiter and Saturn there exist polarimetric observations for $\alpha \approx 90^\circ$ from the Pioneer and Galileo spacecrafts (e.g. Smith & Tomasko 1984; Tomasko & Dose 1984; Braak *et al.* 2002). These data can directly be used as proxy for extra-solar planets. For Jupiter, for example, the polarization for $\alpha \approx 90^\circ$ is up to 50% at the poles and only 5% near the equator (R-band). Polarimetric measurements for small phase angles, $\alpha \leqslant 12^\circ$ taken from the ground show a polarization of about 7% at the poles of Jupiter and only 0.5% at the equator (e.g. Hall & Riley 1976; Joos *et al.* 2005). The polarization for Jupiter is much higher at the poles because they are covered with a Rayleigh scattering layer. At the equator, the radiation is predominantly reflected by clouds and the resulting polarization is small.

For Uranus and Neptune no similar space observations could be found in the literature. For this reason we have taken disc resolved imaging polarimetry and longslit spectropolarimetry of Uranus and Neptune to investigate their polarimetric properties based on the limb-polarization effect.

For disc resolved observations there exists a well known second-order scattering effect in Rayleigh atmospheres, which produces an enhanced polarization at the limbs in a backscattering situation, see van de Hulst 1980. This effect is also found for the first time with our ground based observations of Uranus and Neptune. The measured limb polarization combined with analytical models allows to estimate roughly the amount of polarization of Uranus and Neptune for phase angles around 90°.

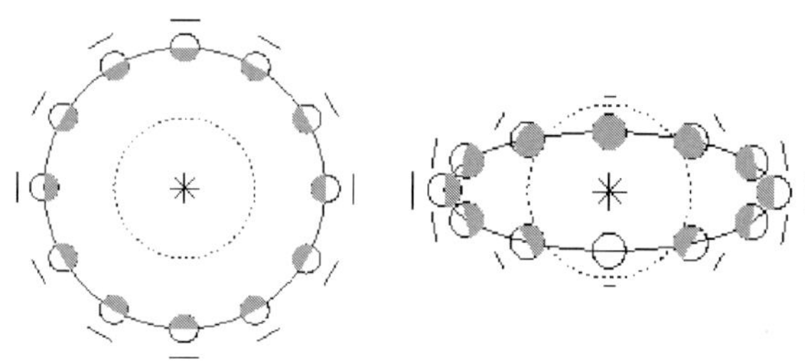

Figure 1. Schematic illustration of the expected scattering polarization (Rayleigh scattering) from an extra-solar planet in a low (left) and high (right) inclination system. The parent star is indicated by an asterisk. For small apparent separations (dotted circle) the planet is too close to the star to be detectable. The bars indicate strength and direction of polarization which is particularly high for scattering angles near 90°.

2. Observations & results

Our polarimetric observations were taken in Nov. 2003 with EFOSC2 attached at the Cassegrain focus of the ESO 3.6m telescope. EFOSC2 is a multi-mode imager and grism spectrograph equipped with a Wollaston prism and a rotatable superachromatic half-wave plate for linear polarimetry and spectropolarimetry. The employed grism provides a coverage from about 5200 to 9350 Å and a resolution of 12.8 Å for a 1" wide slit. The slit was 20" long and 0.5" (Uranus) or 1.5" (Neptune) wide and was placed in North-South direction through the middle of the disc. The pixel scale was 0.157"/pixel and the seeing was better than 1". The diameter of Uranus was 3.5" and for Neptune 2.2". The intensity spectra have been calibrated with the albedo spectrum from Karkoschka 1994. We have binned our spectral data into 30 Å bins in order to reduce the noise. In addition we have taken imaging polarimetry in several filters.

In the spectropolarimetric plots we show data for the *limb* (average of northern and southern limb), the *central region* and the *total slit* of Uranus and Neptune.

For Uranus we detect a peak polarization up to 3% at the limb and about 1% for the entire slit. We also show a cut through the planetary disc in a methane absorption band to point out the spatial features (Fig. 2).

Neptune has a diameter of 2.2", and our spectropolarimetric data contain almost the entire planet except some parts of the eastern and western limbs. We detected for Neptune a peak polarization of about 1% in the methane absorption bands (Fig. 4 left). For both planets we show imaging polarimetry of the Stokes Q flux in the i-band at 0.8 μm (Fig. 3 for Uranus and Fig. 4 right for Neptune).

The limb polarization is for both planets clearly visible in the spectropolarimetric data as well as in the imaging polarimetry, whereas at the disc centers the polarization is close to zero. The imaging polarimetry shows that the polarization is enhanced along the entire limb of Uranus and Neptune and not only at the poles as for Jupiter. Further we find enhanced polarization in the methane absorption bands compared to the adjacent continuum. We also see an enhanced continuum polarization towards the blue.

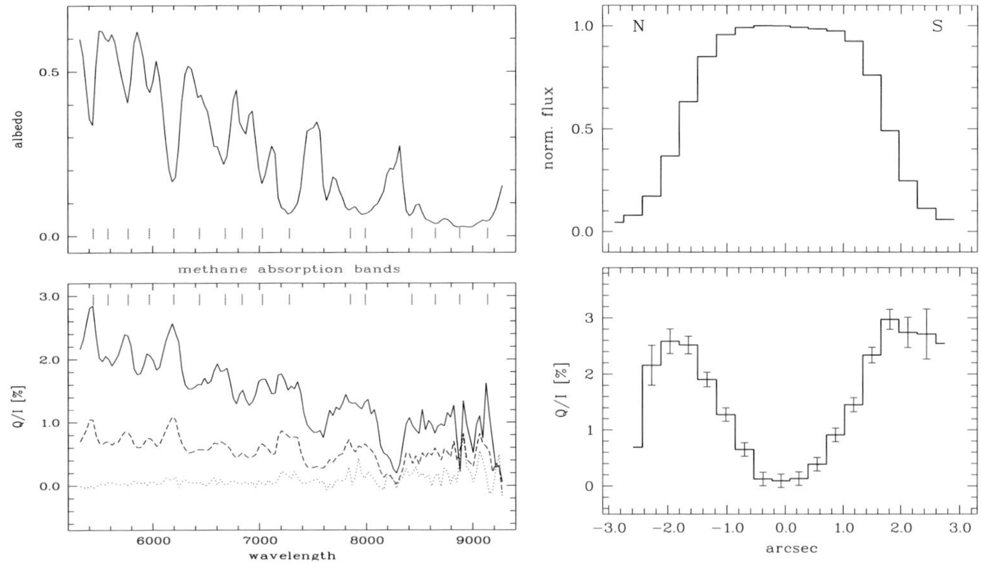

Figure 2. *Spatially resolved spectropolarimetry of Uranus:*
left: The albedo spectrum in the upper and the Stokes Q/I spectra in the lower panel where the solid line shows the *limb* position, the dashed line the *total slit* and the dotted line the *central part* of the planet. The vertical marks indicate the wavelengths of the methane absorptions.
right: Cut through the disk from North (N) to South (S) in the methane band at 6190 Å (±10 Å), normalized flux (upper panel) and Stokes Q/I with error bars (lower panel) versus the radial position on the planet.

Figure 3. *Uranus:* Stokes Q image (I_0-I_{90}) in the i-band, white is positive and indicates a vertical polarization (North-South). North is up and East it left. The greyscale is normalized to the peak intensity and ranges from -0.5% (black) to +0.5% (white). The planetary South pole, the equator and the geometrical limb are marked.

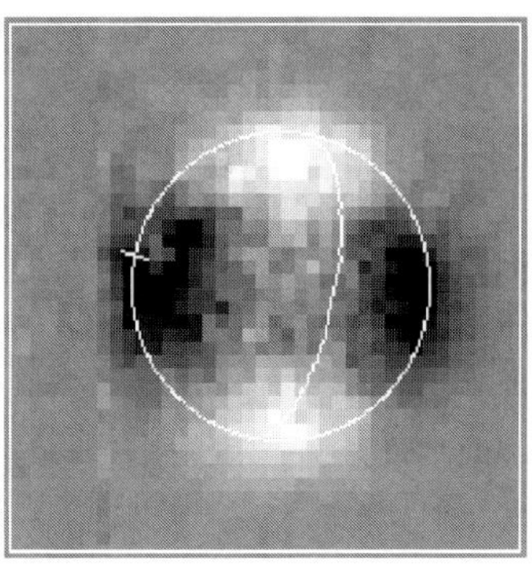

3. Discussion and Conclusions

Our polarimetric measurements of Uranus and Neptune show:

- a high limb polarization is present in both planets along the entire limb. This indicates a high scattering polarization ($\gtrsim$20%) for phase angles near 90° (from comparison with observations of Jupiter's poles and analytic models). Thus extra-solar planets with similar properties as Uranus and Neptune are detectable with polarimetry.
- enhanced polarization in the methane absorption bands compared to the adjacent continuum. A higher polarization in the methane bands seems to be a generic effect for reflected light from planetary atmospheres. This is due to the high absorption in the methane bands which reduces predominantly the multiple scattered light and the

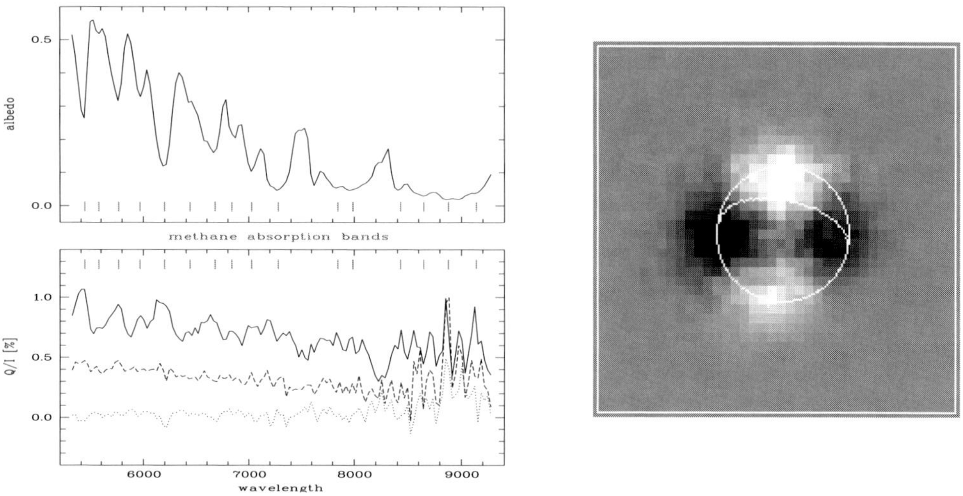

Figure 4. Spectropolarimetry (left) and imaging polarimetry (right) of Neptune. Details are the same as in Fig. (2) and in Fig. (3), respectively but for Neptune.

deeply penetrating photons reflected from cloud layers. Photons reflected in the highest atmospheric layers after one or two scatterings (the latter produce the limb polarization), are much less affected by absorption. This enhancement in the methane bands provides a tool for the verification of a planet detection.

• The enhanced continuum polarization towards the blue is typical for a Rayleigh scattering atmosphere. Since Rayleigh scattering is much stronger for short wavelengths, blue photons will penetrate less deep into the atmosphere and scatter at the high lying polarizing layers, whereas the deeply penetrating red photons have a higher probability to scatter on weakly polarizing clouds.

Based on observations of solar system planets we conclude that Uranus and Neptune-like extra-solar planets will show substantial polarization. The polarization signal is low for planets with high clouds in the atmosphere (e.g. equatorial region of Jupiter). A high polarization, as seen in Uranus and Neptune indicates that there exists an optically thick Rayleigh scattering layer above the clouds. The polarization in molecular bands like methane yield additional information about the composition and structure of the atmospheres. Thus polarimetry provides a powerful tool for the future investigation of extra-solar planets.

References

Braak, C.J., de Haan, J.F., Hovenier, J.W., & Travis, L.D. 2002, *Icarus* 157, 401

Hall, J.S. & Riley, L.A. 1976, *Icarus* 29, 231

Joos, F., Schmid, H.M., Gisler, D., Feldt, M., Brandner, W., Stam, D., Quirrenbach, A., & Stuik, R. 2005, in: Adamson, A., Aspin, C., Davis, C.J., Fujiyoshi, T. (eds.), *ASP Conf. Ser.* (in press)

Karkoschka, E. 1994, *Icarus* 111, 174

Schmid, H.M., Joos, F., & Tschan, D. 2005, *A&A* (submitted)

Smith, P.H. & Tomasko, M.G. 1984, *Icarus* 58, 35

Tomasko, M.G. & Doose, L.R. 1984, *Icarus* 58, 1

van de Hulst, H.C. 1980, *Multiple Light Scattering, 2nd Volume* (Academic Press)

Direct Imaging of Exoplanets: Science & Techniques
Proceedings IAU Colloquium No. 200, 2005
C. Aime and F. Vakili, eds.
© 2006 International Astronomical Union
doi:10.1017/S1743921306009318

Searching for Extrasolar Planets from UNSW

Jessie L. Christiansen[1], M.C.B. Ashley[1], M.G. Hidas[1], J.K. Webb[1], A. Phillips[1], M. Irwin[2] and J. Irwin[2]

[1]Department of Astrophysics, University of New South Wales, Sydney, NSW 2052, Australia
email: jessiec@phys.unsw.edu.au
[2]Institute of Astronomy, University of Cambridge

Abstract. The Automated Patrol Telescope, operated by the University of New South Wales, has been undertaking a search for extrasolar planets using the transit method. We present lightcurves from two recent promising candidates; spectroscopic follow-up using the ANU 2.3m telescope shows that the companions are probably low mass stars rather than planets, although more data will be needed to be certain. Additionally, we outline future improvements to our transit search: a hardware upgrade scheduled for 2006, and the addition of a robust trend-filtering algorithm to the data reduction software.

Keywords. extrasolar planets, transit, spectroscopic follow-up, trend-filtering.

1. Latest Results

The Automated Patrol Telescope is a 50-cm optical telescope sited at Siding Spring Observatory (SSO), NSW, Australia. It is dedicated to an Extrasolar Planet Search, the details of which can be found in Hidas *et al.* (2005). From February to March 2005 we observed a pair of fields centred on 12h RA, $-36°$ dec, obtaining 22 nights of good data (RMS $<$ 4mmag for stars brighter than 10th magnitude). Lightcurves for close to 3000 stars brighter than 13th magnitude were compiled, and two promising candidates designated UNSW-TR-13 and UNSW-TR-14 were discovered.

UNSW-TR-13 displayed 2 complete and 3 partial transits, with a depth of 30mmag and a period of 2.283 days (Figure 1a). UNSW-TR-14 displayed 7 total and 3 partial transits, with a depth of 20mmag and a period of 0.640 days (Figure 1b). For comparison, Figure 1c shows a transit of TrES-1, the first transiting planet discovered with a small, wide field-of-view telescope (Alonso *et al.* 2004). It is clear that the APT is reaching the required level of precision to detect transiting extrasolar planets.

Time and weather constraints resulted in a failure to capture our candidates in transit with the ANU 40-inch telescope at SSO. However, we were able to obtain follow up observations with the DBS instrument on the ANU 2.3m telescope. Spectral classification of the candidate host stars and radial velocity curves are shown in Figure 2. These curves are constrained with the period and phase data taken from the lightcurves. The characteristics derived from the spectra are summarised in Table 1. UNSW-TR-13 is found to probably be orbited by an M dwarf, and UNSW-TR-14 appears to have a brown dwarf companion.

2. Hardware Improvements

With the assistance of a grant from the Australian Research Council, a new CCD camera has been commissioned for the APT, for installation in 2006. Table 2 shows a

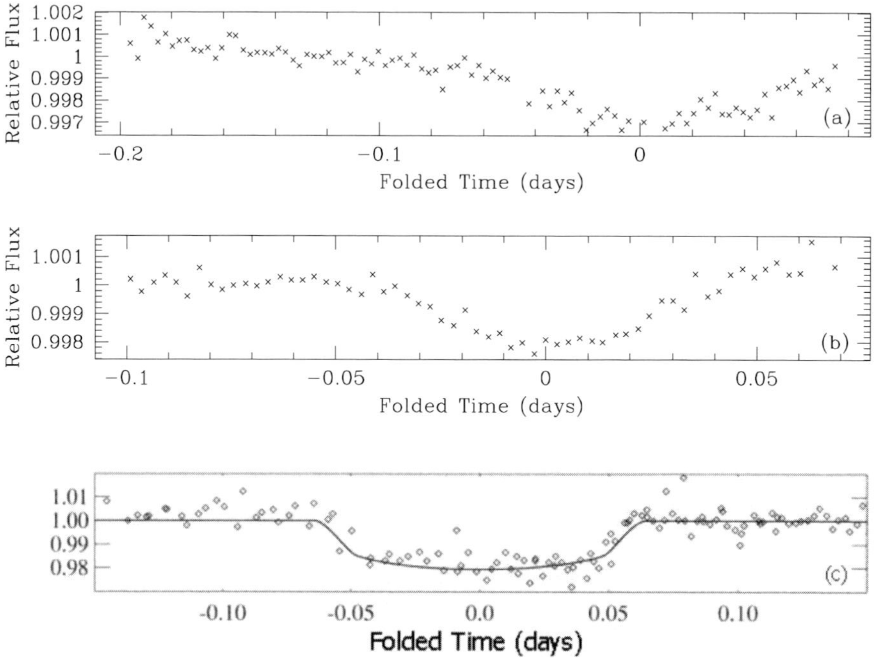

Figure 1. Panel (a) shows a sample transit from the lightcurve of UNSW-TR-13, panel (b) shows a sample transit from UNSW-TR-14 and for comparison, panel (c) shows a transit (taken with the 10cm PSST instrument) of TrES-1, the first planet to be discovered with the transit method (Alonso *et al.* 2004).

Table 1. The derived characteristics of the planetary candidate systems, where K is the relative radial velocity amplitude. We find that UNSW-TR-13 is orbited by a late-type M dwarf; for the quoted mass the orbital distance is 0.034 AU. UNSW-TR-14 is probably orbited by a brown dwarf companion at a distance of 0.015 AU; for the quoted mass. It should be noted that the lower mass limit of the UNSW-TR-14 companion is only $8.5 M_{\rm Jup}$ and we will be investigating this candidate further.

Target	Sp. Type	I Magnitude	$M_1 (M_\odot)$	$M_2 (M_\odot)$	K (km s^{-1})	Period (days)
UNSW-TR-13	G5V	10.7	0.95	0.11 ± 0.05	17 ± 8	2.283
UNSW-TR-14	G2V	9.6	1.00	0.033 ± 0.025	8 ± 6	0.640

comparison of the properties of the current CCD with the new CCD. Due to the increased number of data points, as a result of the larger sky coverage and improved time-sampling (the new CCD quantum efficiency will be three times higher than old, resulting in reduced exposure times), we expect to increase our chances of detecting planets at the current level of precision by a factor of nearly 5. In addition, we would also expect to see higher precision photometry from the reduced undersampling, which would significantly increase our chances of detecting smaller amplitude transits.

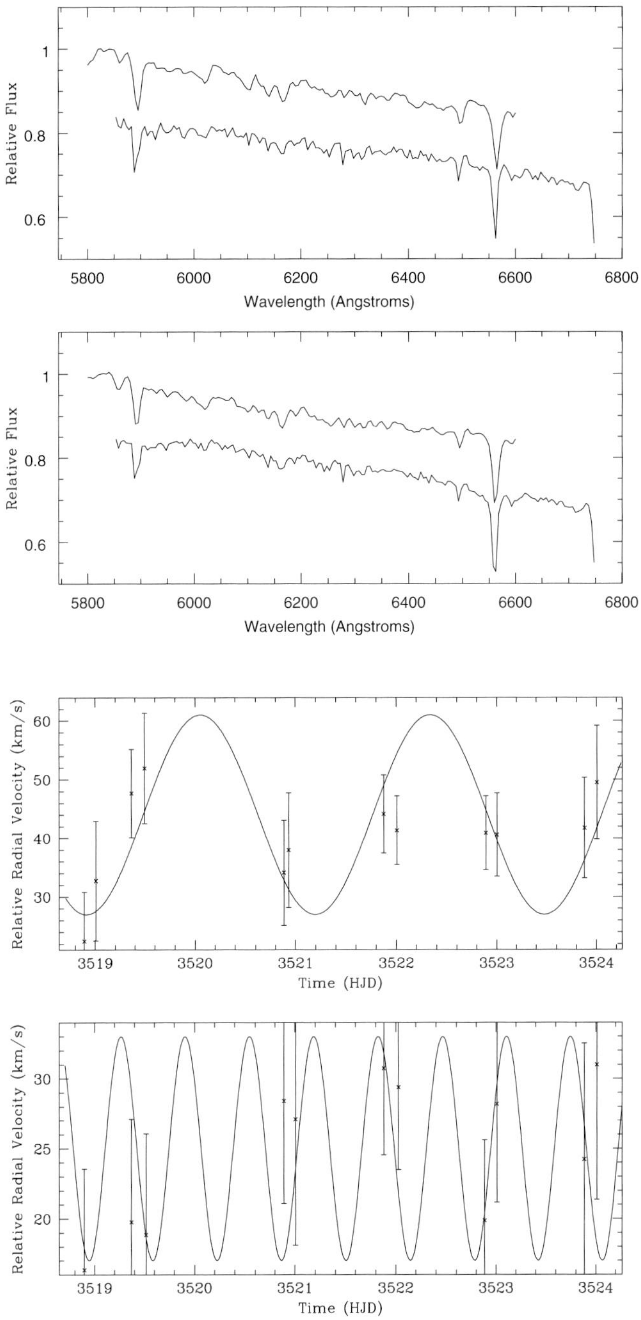

Figure 2. Follow-up results for the two UNSW candidates. The top two panels show data for object UNSW-TR-13. The host star has an I magnitude of 10.7; the stellar spectrum is shown in the left panel, with a model G5V spectrum offset above for comparison. On the right is the derived radial velocity curve with an amplitude of 17 km/s and a period of 2.283 days. The bottom panels show the data for object UNSW-TR-14. The host star has an I magnitude of 9.6. Again, the spectrum is on the left, with a G2V model spectrum offset above, and its radial velocity curve, with an amplitude of 8 km/s and a period of 0.640 days, is on the right. The original stellar spectra have been rebinned to 10Å for comparison with the model spectra, taken from the UVILIB spectral library (Pickles 1998).

Table 2. The attributes of the current CCD being used in the APT are compared to the specified characteristics of the new CCD which is being built for the APT. [a]This is the typical number of stars brighter than 13th magnitude in a field at $b = 15°$. [b]The number of images taken per hour. [c]Although this CCD gives a coverage of 7 by 7 degrees, due to optical limitations the usable field of view of the CCD has a diameter of about 6 degrees.

Attribute	Old CCD	New CCD
CCD Size (pixels)	1152×770	$6k \times 6k$
Pixel Size (μm)	22.5	10
Image Scale	9.43″/pixel	4.19″/pixel
Field of View	$2° \times 3°$	$7° \times 7°$[c]
No. of Stars[a]	~5000	~16000
Cadence[b]	~30	~60

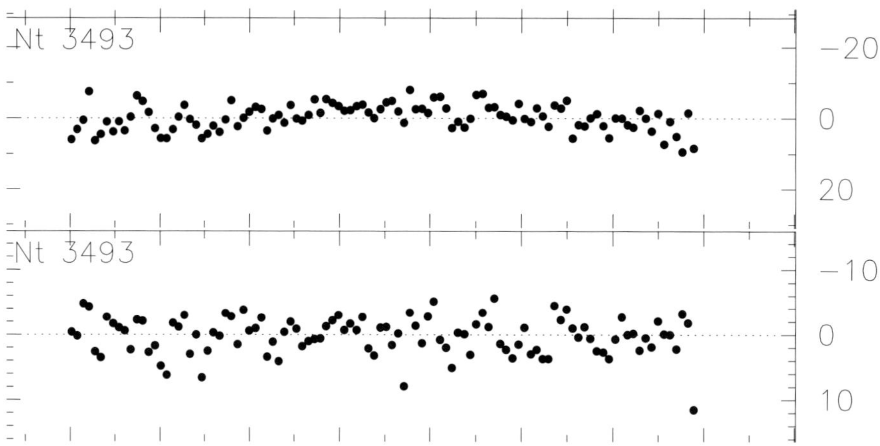

Figure 3. An example of the types of corrections handled by the trend-filtering algorithm. The top panel shows the original lightcurve, and the bottom panel shows the corrected lightcurve.

3. Software Improvements

We are currently developing a trend-filtering algorithm based on one described by Kovács *et al.* (2005) to remove residual systematics from our data. It is generally easy to gauge with the eye which signals are real and which are systematics. However, implementing a period-finding program to detect low amplitude transits when there are significant residual systematics is very inefficient. It generally leads to a high number of spurious detections with periods around n days, where n is a small integer. The basic operation of the algorithm is to use a subset of the stellar lightcurves as templates for construction of the lightcurves of the remaining stars. By subtracting a fitted linear combination of the template lightcurves from the original lightcurves, any systematics common to the stars and the template group should be theoretically removed. Although we are still implementing the final algorithm, preliminary results show a dramatic reduction in systematics, accompanied by a slight increase in the RMS noise, as shown in Figure 3.

4. Conclusions

We are confident that the APT is reaching the precision necessary to detect transiting hot Jupiters, demonstrated by the sample transits presented here. We believe that with the addition of new hardware and software, our chances of detecting these planets will increase by at least a factor of 5, and probably closer to a factor of 10.

Acknowledgements

The authors thank Aliz Derekas for assistance with the ANU 2.3m observations.

References

Alonso, R., Brown, T.M., Torres, G., Latham, D.W., Sozzetti, A., Mandushev, G., Belmonte, J.A., Charbonneau, D., Deeg, H.J., Dunham, E.W., O'Donovan, F.T., & Stefanik, R.P. 2005, *Ap.J.* 613, L153

Hidas, M.G, Ashley, M.C.B., Webb, J.K., Irwin, M., Phillips, A., Toyozumi, H., Derekas, A., Christiansen, J.L., Nutto, C., & Crothers, S. 2005, *MNRAS* 360, 703

Kovács, G., Bakos, G., & Noyes, R.W. 2005, *MNRAS* 356, 557

Pickles, A.J. 1998, *PASP* 110, 863

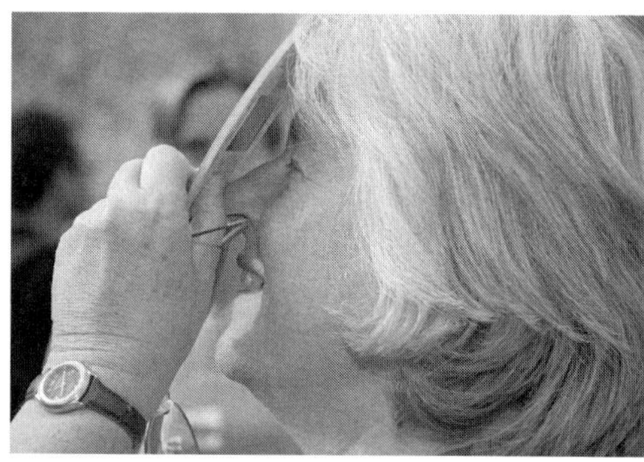

All photographs: Laurent Thareau [`l.thareau@free.fr`].

Direct Imaging of Exoplanets: Science & Techniques
Proceedings IAU Colloquium No. 200, 2005
C. Aime and F. Vakili, eds.
© 2006 International Astronomical Union
doi:10.1017/S174392130600932X

New Worlds Observer: A Novel Mission Concept for Exoplanetary Studies

Jonathan W. Arenberg [1] and Webster Cash[2]

[1]Northrop Grumman Space Technology, One Space Park Drive, Redondo Beach, CA, 90278, USA
email: jon.arenberg@ngc.com

[2]Univeristy of Colorado, Center for Space Astrophysics and Astronomy, Boulder, Colorado 80309-0389 USA

Abstract. This paper describes a novel concept for the direct detection of exoplanets involving two spacecraft. One spacecraft is an occulter, designed to provide adequate starlight suppression within its shadow. The second spacecraft , a conventional telescope of standard quality, is flown into the shadow of the first and used to collect the light from the target planet. The design of the occulter and its expected performance are discussed. It is shown that is possible to simultaneously achieve the necessary contrast ratios and inner working angles necessary to have a scientifically meaningful mission. A brief discussion of the estimated tolerances indicate that such a mission is feasible in the near term.

Keywords. space vehicles, instrumentation: miscellaneous, space vehicles: instruments, telescopes.

1. Introduction

The New Worlds Observer (NWO) mission concept we are presenting is the result of a recently concluded six month Phase I study under the auspices of the NASA Institute for Advanced Concepts (Cash (2005)). NWO separates the planet light from the starlight before it ever enters the telescope, sidestepping the problem of scatter. The concept is based on well established optical and aeronautic practice. The occulter is a large deployable shade that casts controlled shadows on a telescope carried on a separate craft. The telescope on the second craft is required to be of reasonable quality and meter class, but is otherwise unremarkable. The employment of occulters for starlight suppression enable an extendable architecture. With two craft we have New Worlds Observer, capable of mapping planetary systems and following up with spectroscopy and photometry. Additional occulters increases the efficiency of the planet finding mission. Eventually true exoplanet imaging might become possible.

This paper introduces the NWO, a mission designed to detect and characterize exoplanets and exoplanetary systems. The basic operation of the NWO mission is introduced and the viability of the concept shown. We present a design that meets the two main requirements for a successful exoplanetary mission, sufficient contrast ratio to detect terrestrial planets, 10^{-10}, and the ability to have a sufficiently small inner working angle to see in close to the parent star, 40 milliarcseconds (mas). A first look at tolerances for the occulter and station keeping indicates that both are within the state of the art. The paper concludes with remarks about the science that can be expected from NWO.

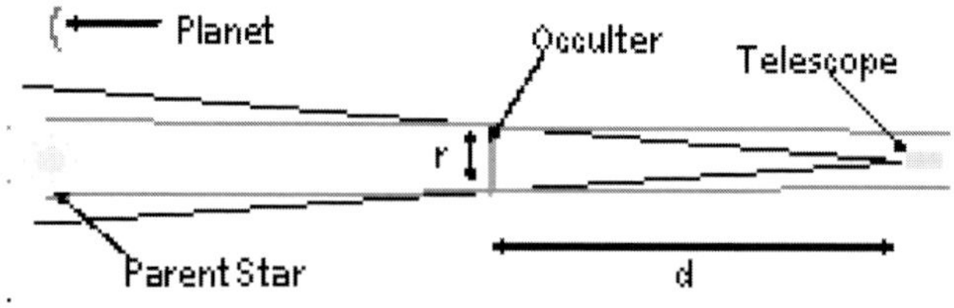

Figure 1. New Worlds Observer Concept.

2. The New Worlds Observer Concept

The New Worlds Observer consists of two spacecraft as shown in Figure 1, the first is an Occulter, and casts a shadow into which the Telescope craft is flown. The idea of an occulter is very simple. When we instinctively raise our hands to blot out the Sun while looking for a fly ball, we are reducing the amount of light that enters our eyes without reducing the signal from the ball. By reducing the amount of light that is potentially scattered, the contrast of the image is greatly enhanced, making it possible to see dim objects, such as planets, near bright ones, the parent stars.

Occulter based systems for the detection of exoplanets have been previously studied (Copi and Starkman (2002)). These previous studies emphasized use of the occulter to improve spatial resolution on various kinds of targets and showed that a simple shape like a circle or a square leads to contrast ratio of about 0.01 diffraction into the dark hole. Other occulter designs that have been studied with Gaussian apodizations result in constrasts of order 10^{-4}. This level of performance is still a full factor of a million short of the requirement. Previous occulter designs achieve their various apodizations by the use of variable transmission. The use of a transmissive element limits the achievable contrast ratio due to wavefront error and scattering.

The breakthrough that makes the NWO a viable mission is a new design that results in acceptable contrast ratios. Furthermore, the NWO occulter design is a pure binary optic, that is the transmission of the occulter is either unity or nil.

The general geometric layout of the occulter is shown in Figure 2. The occulter is central circular obscuration of radius r_1. Around the circumference of the central circle are petals, whose width is an exponential of the form $\exp\left(-\left(\frac{(r-r_1)}{r_2}\right)^n\right)$. The petals cut through the various Fresnel zones, which are indicated by the concentric rings in Figure 2. The effect of the petals is to create a situation where the diffraction from the perimeter is cancelled out, leaving a broad, deep null for on-axis light.

The initial occulter designs were developed via numerical methods. Shortly after the Phase I study was concluded, an analytic result for the contrast ratio, R, was derived (Cash (2006)). The contrast ratio on axis is given by

$$R = \left[n!\left(\frac{d\lambda}{2\pi r_1 r_2}\right)^n\right]^2. \tag{2.1}$$

Many interesting insights can be gleaned about the occulter design and performance from inspection of (2.1). First is that since it we want R to be as small as possible, $r_1=r_2$ which is the design that also resulted from the numerical optimization. Secondly, that R increases with the occulter-telescope distance d. The increase in R with d is easily interpreted, a decrease in the Fresnel number, $\frac{d\lambda}{r_1 r_2}$, results in fewer zones involved in the cancellation, lessening the depth of the null. As we shall see, this latter trend will be important in the demonstration of the viability of the NWO concept.

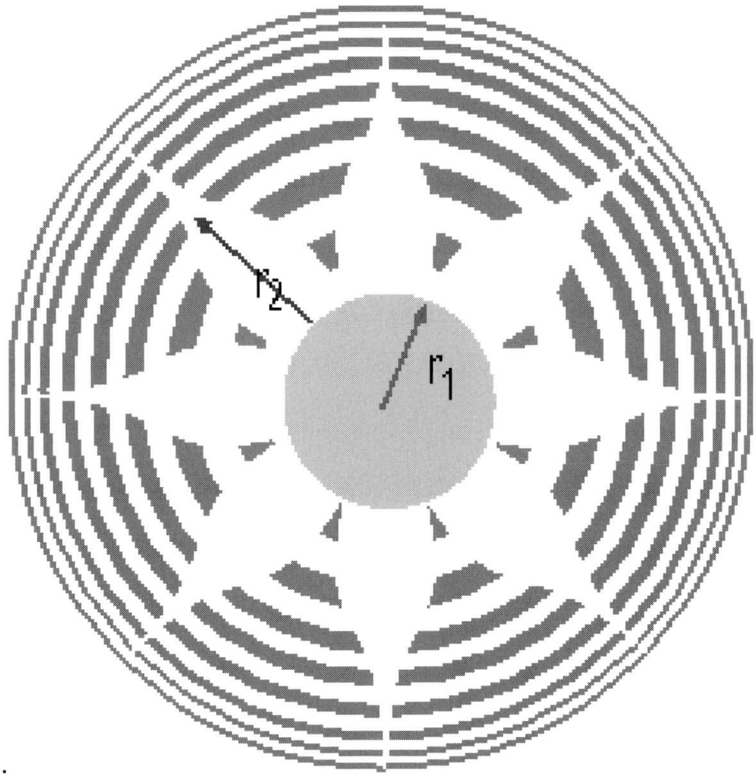

Figure 2. Occulter with Fresnel Zones.

Figure 3 shows numerical results for an occulter operating at $\lambda = 600$ nm with $r_1 = r_2 =$ 12.5 m and at 50,000 km separation. These illustrative results show it is possible to reach $R = 10^{-10}$, and give results orders of magnitude better than previous designs.

Figure 4 shows a simultaneous plot of the inner working angle and contrast ratio as a function of the distance between the occulter and telescope. For the purposes of discussion, this occulter is intended to operate at 1 μm wavelength, and is 40 m in radius, so $r_1 = r_2 = 20$ m. The curve with symbols and growing with d is R. If a requirement of $R < 10^{-10}$ is enforced then d can be no greater than 250,000 km, which is indicated by the solid arrow labeled "contrast". The curve decreasing to the right is the inner working angle, $(r_1 + r_2)/d$, which decreases with increasing d. Taking a typical inner working angle requirement of 40 milliarcesconds, then d must be at least 200,000 km as indicated by the rightward point bold arrow with the label "IWA". As can be seen from Figure 4, that there is a zone, namely between 200,000 and 250,000 km where both contrast and inner working angle requirements are met. This region of viability shows that there is a region in design space where

the seminal requirements can be met, indicating that NWO can see some of the most interesting science targets.

Initial studies of the accuracy of the shape required to achieve the contrast ratio values shown above, imply that the tolerance is of order of the square root of the area (Cash (2006)). Namely, for the occulter discussed above, the areal tolerance is of order 250 cm^2. It is expected that this result will be confirmed on the basis of further (planned) analysis, so that the contour tolerances will be shown to be well within the state of the art.

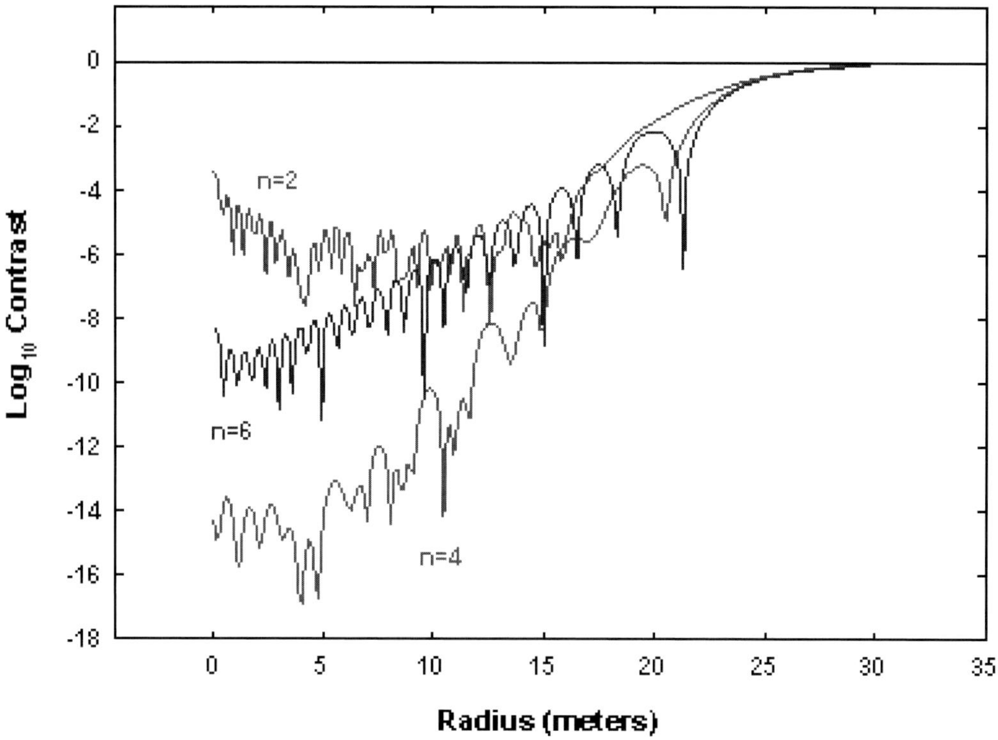

Figure 3. Contrast Ratio.

3. Science Performance

As the telescope moves from outside the hole into the center of the shadow, the view of the planetary system changes dramatically. Before the system is aligned (i.e. before the telescope is fully in the shadow) the view is of an overpoweringly bright star with scattered light all over the field of view. Once inside the hole the star disappears completely, leaving the light from off axis sources, i.e. the planets, visible.

The lateral positioning tolerance for the telescope to the occulter is approximately the radius of the shadow minus the radius of the telescope and is of order meters. Such an alignment is not beyond the current state of the art, and therefore not an extreme challenge or risk to the realization of NWO as a viable mission. As for axial tolerancing, Figure 4, shows that the R changes about a factor of 10 in 50,000 km, so maintaining the occulter telescope distance need only be done to kilometers and is argued to be easy to accomplish.

When the NWO is aligned, the planets can now be studied by the conventional techniques developed by astronomers for the study of faint stars. Long term imaging of the system will show changes in brightness of all the planets simultaneously as their features rotate in and out of our sight. Spectroscopy can be performed on planets by placing a spectrograph slit at the correct position.

The feasibility of the New Worlds missions is entirely dependant on the telescope's ability to achieve sufficient signal to noise from an exoplanet in a short amount of time. Larger telescopes allow for higher photon counting rates, while at the same time increasing resolution. Still, increasing apertures can drive up costs significantly, so we have studied the tradeoffs between the different proposed designs.

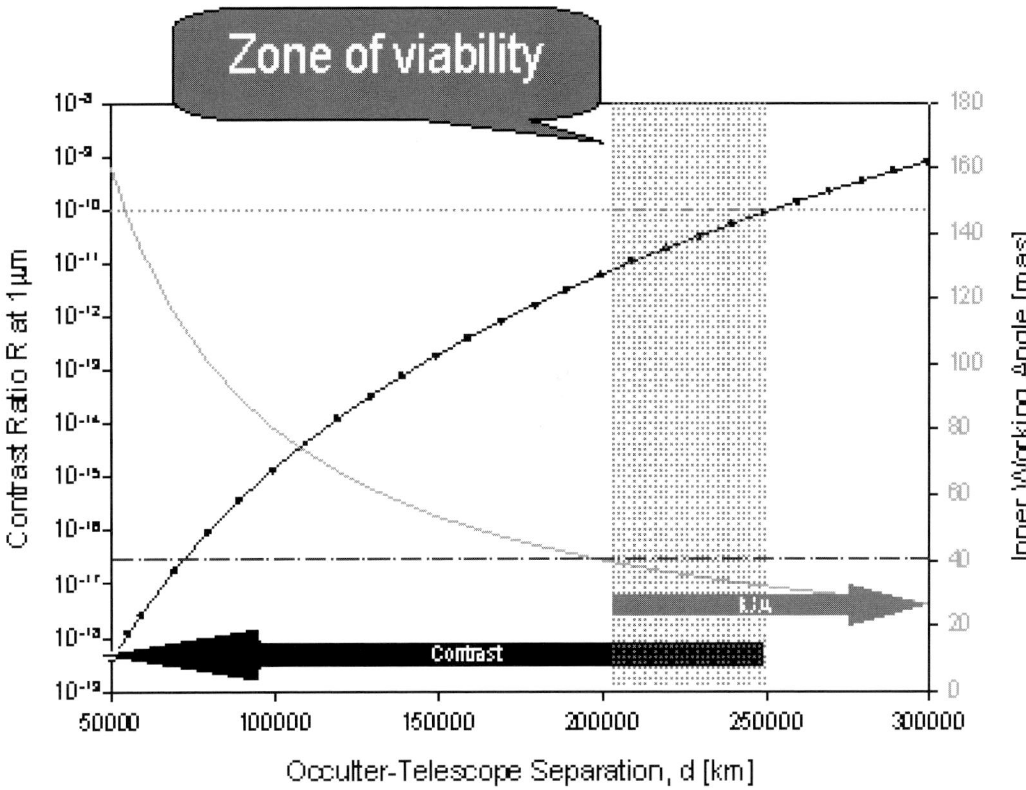

Figure 4. Contrast Ratio and Inner Working Angle versus Occulter to Telescope Separation.

Simulations were run to test NWO's ability to detect extrasolar planets assuming a configuration capable of 10^{10} contrast (Schindhelm *et al.* (2005)). Integration times were studied from 50 kiloseconds (ks) to 400 ks in increments of 50 ks to see signal to noise improvement and which planets are detectable. The occulter based architecture of NWO gives a large advantage in the study of exoplanets, namely the entire planetary system simultaneously since there is no outer working angle limitation.

The simulations include the expected level of detector noise, but some sources of background come from space. A major concern for TPF, operating deep in the IR, is Zodiacal Light. Dust in the planetary system scatters starlight and creates a diffuse glow detectable in the inner planetary system (Hahn *et al.* (2002)). But, in the visible, this is a lesser problem. Our first simulation with zodiacal light added looked no different. With a deep exposure and a ten times increase in zodiacal scattering made the effect visible, but it still did not interfere with the planet studies as shown in Figure 5.

It is also a concern that in many directions there will be background objects such as the deep field galaxies shown so spectacularly by HST (Williams *et al.* (1996)). We simulated frequency and brightness of these faint galaxies and discovered that they will be a nuisance. Figure 5 shows a simulated observation with a typical deep field galaxy included. Over 90% of the time we will not find one inside the habitable zone. During the 10% frequency of unlucky observations, it should usually be apparent by the shape and color that the object is a field galaxy. If the galaxy interferes with our image, then we will have to return in a few months when proper motion of the nearby system has carried it out of the critical line of sight. It should be noted, that the problems of zodiacal,

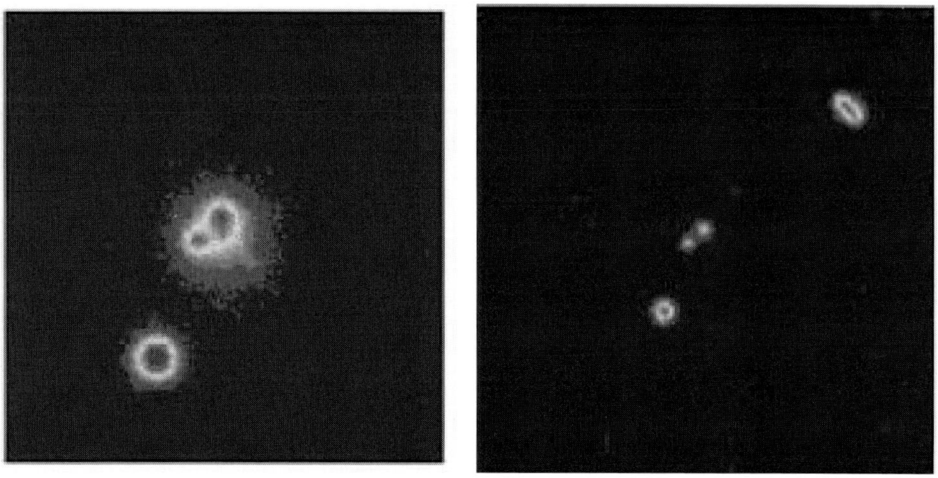

Figure 5. Simulated Imagery from New Worlds Observer. Right panel shows background galaxy.

exo-zodiacal light and background galaxies are common to all direct expolanetary detection methods.

This quick analysis has shown that a telescope of a couple meters in diameter, operating in concert with a high performance New Worlds design occulter, can produce scientifically relevant observations in reasonable amounts of time.

4. Conclusions

The concept for the New Worlds Observer has been introduced as consisting of an occulter and telescope flying in formation. NWO has been shown to be able to achieve sufficient contrast ratios with a binary optic. Also, it was shown that the occulter based system can result in small enough inner working angles for meaningful science. Neither the occulter or the telescope are beyond the start of the art and while requiring some development are relatively low risk technologies

The nature of the NWO also gives some science advantage, since entire systems can be viewed at once, it is expected that observations will give us a profound insight into the nature of our solar system and its place in the galaxy.

New Worlds Observer has been selected for funding as a Phase II NIAC study, which will be concluded in 2007.

References

Cash, W., http://niac.usra.edu/studies/
Copi, C. J. & Starkman, G. D., 2002, *ApJ* 532, 581–592 Schultz *et al.*, Proc. SPIE 4860
Cash, W., in preparation
Schindhelm, E., Cash, W., & Seager, S., 2005, *Proc. SPIE* 5899
Hahn, J. M., Zook, H. A., Cooper, B., & Sunkara, B., 2002, *Icarus* 158, 360-378
Williams *et al.*, 1996, *AJ* 112, 1335

Direct Imaging of Exoplanets: Science & Techniques
Proceedings IAU Colloquium No. 200, 2005
C. Aime and F. Vakili, eds.
© 2006 International Astronomical Union
doi:10.1017/S1743921306009331

Exoplanet imaging: from ELT's to interferometers and hypertelescopes

Antoine Labeyrie

College de France & Observatoire de Calern, 06460 Caussols, France
email: antoine.labeyrie@obs-azur.fr

Abstract. ELT's, interferometers and their multi-aperture direct-imaging form called hypertelescopes are expected to provide images of exo-Earths, and even resolved images in the latter case. Both terrestrial and space versions are potentially usable, although high-performance adaptive optics will be needed on Earth. Current advances in the way of hypertelescopes on Earth and in space are briefly described.

Keywords. exoplanets, interferometer, hypertelescopes.

1. Introduction

Following the discovery of more than 160 exo-planets, with the indirect method of "wobble spectrography", obtaining exo-planet images is becoming a major challenge for astronomy. Three decades after the proposal of a coronagraphic camera for exo-planet imaging with the Hubble Space Telescope (Labeyrie (1974), Bonneau *et al.* (1975), NASA is planning TPF-C a Hubble-class space telescope dedicated for exo-planet coronagraphy, and many groups now explore various ways of achieving enough detection sensitivity for imaging Earth-like planets, also called exo-Earth's.

2. Telescopes, interferometers and hypertelescopes for exo-planet imaging

These three types of instruments, operated on Earth or in space, can potentially serve for exo-planet imaging. Telescopes can have a monolithic mirror, for which the size limit is considered to be of the order of 8 meters, or a mosaic mirror which can be much larger, up to a hundred meters according to the conclusions of ESO's study work for their OWL concept. Interferometers, using several telescopes or separately mounted mirrors, can reach much larger sizes, and therefore attain a higher resolution. The information content of the snapshot exposures can however be very limited if there are few apertures. In principle, however, it can be improved with multiple-exposures, according to the aperture synthesis and super-synthesis techniques. These have been rather successful at radio wavelengths, but not yet much utilized in the optical and near infra-red ranges. Hypertelescopes provide a way of bridging the gap between the telescope and interferometer realms, by using many sub-apertures forming an aperture which can be highly diluted while retaining a capability to produce snapshot images like conventional telescopes. Their imaging capability includes coronagraphic modes and therefore makes them suitable for obtaining multi-resel images of exo-Earths with arrays becoming as large as 100km.

3. Principle and properties of hypertelescopes

These have been described elsewhere (Le Coroller *et al.*, 2004). Here we only recall

4. Types of hypertelescope architectures

The principle of hypertelescopes, described elsewhere, can be materialized in several different practical architectures. These may be classified as using a primary array having either a planar shape, or a paraboloidal shape, or else a spherical shape.

4.1. *Planar hypertelescopes*

These utilize an array of telescopes, each equipped with a Coudé train, possibly fibered, to bring the captured light to a common focus. Coherencing is achieved either by moving the telescopes, or with delay lines such as those installed at the Keck, the VLTI, the CHARA, etc. The high cost of such moving telescopes or delay lines limits the number of apertures which can be used and the baseline span. In space, no such restriction applies since planar telescope arrays can be globally pointed, but the concave mirror arrays discussed below may then provide simplified optical solutions.

4.2. *Spherical hypertelescopes*

This concept, called Carlina, has a primary array of mirrors located on a common spherical surface. On Earth, natural rocky sites such as extinct volcano craters, canyons or sink holes can have suitable shapes to support many fixed mirrors, adjusted co-spherically. A corrector of spherical aberration, typically a two-mirror clam-shell arrangement such as the Mertz corrector which also corrects coma, provides a usable image near the focal surface, located half way to the curvature center.

4.3. *Pointable paraboloidal hypertelescopes*

The corrector of spherical aberration can be avoided if the primary locus of the mirror segments is paraboloidal rather than spherical. But the aberration-free field of view then becomes very narrow around the axial direction. A pointing mount is therefore necessary to attain a useful sky coverage. On Earth, conventional pointing mounts cannot be made much larger than 100m however, and can certainly not reach the kilometric size which would be desirable for high resolution imaging. Figure 2 sketches a putative solution involving mirror segments supported by a net-like structure which is suspended and pointable by pulling cables. This version, called Galanthus, is of interest for the Dome C in Antarctica, where the winds are very low but with a significant ground turbulence justifying a 40m altitude for the mirrors. The suspension can be from balloons, three at least, or from fixed structures such as canyon walls, skyscrapers, pylons, etc...

5. Comparison of hypertelescope architectures

For planar arrays, the cost of telescopes and movable optical elements in the Coudé trains obviously limits the aperture count to a few units. This severely restricts the imaging performance of such instruments, particularly in terms of the allowable object complexity. On Earth, the adoption of a spherical architecture is highly dependant on the availability of suitably shaped sites, also having desirable astronomical characteristics, accessibility, infra-structure, etc. Craters, either extinct volcanoes, meteorite impact craters or karstic depressions, can be exploited like the sink hole where the 330m wide Arecibo radio telescope is nested. Instead of its high pylons, a balloon can serve to suspend the focal optics. Several stars can be observed simultaneously with as many focal correctors, carried independently. The number of mirrors can reach in the hundreds or thousands. Their anchoring to the bed-rock provides a more stable optical adjustment than with other architectures. Three micrometric screws carrying each mirror can be manually adjusted at months intervals if the bed rock is reasonably stable. The size of the focal corrector for spherical aberration increases as the third power of the focal ratio, reaching about 1% of the primary size at F/2. This limits in practice the design possibilities to focal ratios slower than F/1.5. For a 1.5km effective aperture, achievable in principle at sites such as the Taburiente crater (Canary Islands), 12m mirrors are needed for the corrector, and these can be made in the form of a balloons reflective membrane. In space, effective apertures larger than 100km, and probably up to 100,0000km, appear feasible.

Paraboloidal architectures are limited in aperture size, on Earth, to 100m with conventional pointing mounts and perhaps one kilometer with the net suspension structure. The mirror segments have shapes which differ according to their distance from the axis. As experienced with the segments of the Keck telescopes, this makes them more costly than uniform spherical segments. At a flat site such as Antarctica's Dome C, where wind velocities are often exceptionally low, balloons are of particular interest for suspending the net, which can be located above the ground layer, for improved seeing.

6. Directly imaging exo-planets with these instruments

Three decades after the early proposal of an exo-planetary coronagraph attached to the Hubble Space telescope NASA is now planning a dedicated coronagraphic and exoplanetary telescope named TPF-C. Current studies indicate that a 3m aperture telescope, equipped with "extreme" forms of coronagraphic cleaning can provide exo-Earth images at visible wavelengths. But these images will not have enough resolution to show any details of these planets. Low-resolution spectra of the planets integrated light can be expected to provide clues on the presence of gas molecules indicative of life, such as O2,O3, CO2, H2O, but none of these molecules can be considered as a robust criterion: the photodissociation of H2O, possibly catalyzed by certain minerals, can in principle generate O2, O3, and CO2. Resolving planetary details is desirable to reach more robust criteria for the presence of life. Local, at the scale of regions, and seasonal variations of cloud cover and absorbtion bands analog to the Earths chlorophyll features, can provide more useful information. This requires large baselines, beyond 100km at visible wavelengths, which themselves require space instruments.

7. Resolving the details of exo-Earths with hypertelescopes in space: the Exo-Earth Imager

Obtaining resolved images of exo-Earths with enough detail to attempt the detection of life forms requires, in addition to baselines longer than 100 km, a deep rejection of light from the parent star. For fast observations, providing snap-shot images within 30mn, 100 or more apertures are needed and their size should be 3m at least. With hypertelescope imaging, a larger number of smaller apertures can provide a better result at equal collecting area. In terms of the hypertelescope architecture to be used in space, the paraboloidal and spherical types are especially relevant. The former requires global pointing whenever a new star is to be observed, a process which is necessarily long with a formation flight spanning 100 km or more. The latter can have a stationary formation flight of primary mirrors if these are arranged as a full, but diluted sphere. Only the focal optics needs to move in order to explore the focal sphere. However, this

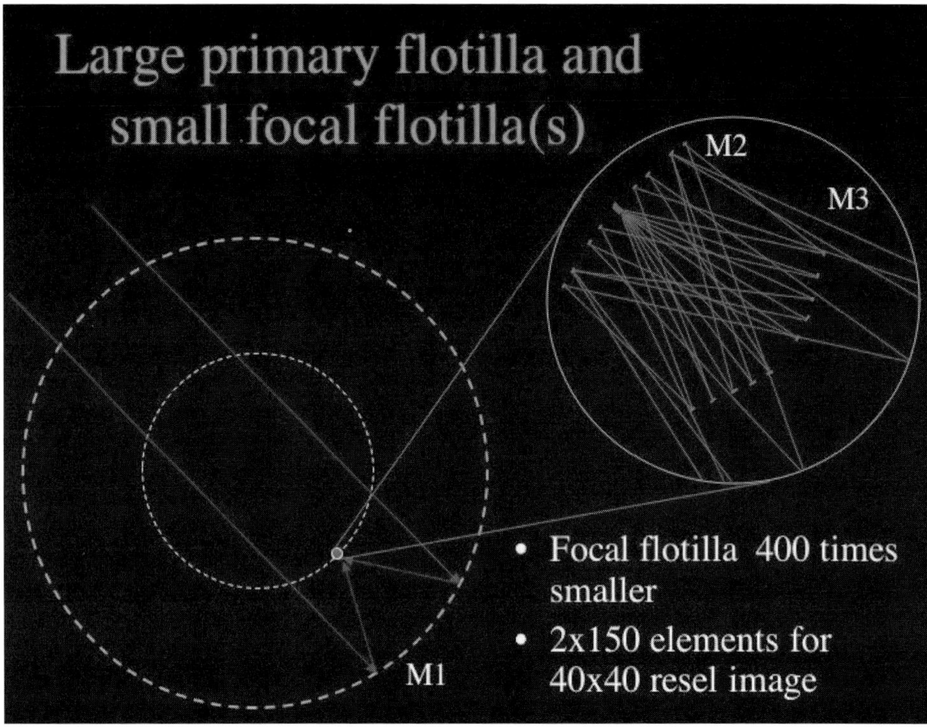

Figure 1. Concept of the Exo-Earth Imager (EEI), a large space version of spherical hypertelescope intended for imaging exo-Earths. It includes 10,000 mirrors of 3m, arranged as a spherical bubble with 400km diameter. The focal sphere, half way to the center, is explored by one or more beam-combiner systems which include an aberration corrector in the form of a smaller formation flight, 1km in size and containing 10cm mirrors. The large diluted spherical mirror is static, and there can be several static focal flotillas as well, dedicated to the permanent observation of favorite exo-planets.

concept requires correcting optics which itself should be made in the form of a smaller formation flight, spanning a kilometer or more, although with mirror elements as small as 100mm. A spherical hypertelescope concept proposed along these lines, called Exo-Earth Imager, is under preliminary study. A precursor instrument, called Luciola, is also studied for developing suitable techniques of formation flight control, while performing high-resolution snapshot imaging of stellar surfaces.

8. Exo-Earth coronagraphy with ELTs, TPF-C and Exo-Earth Imager

Techniques of extreme coronagraphy are needed to obtain images of exo-Earths at visible wavelengths, where their relative luminosity approaches a billionth. Some of the apodizer and coronagraph systems can in principle reach a complete nulling of the residual stellar light, assuming: a- perfect optics devoid of any surface bumpiness; and b- a non-resolved star. But neither condition is fulfilled in practice: a- mirrors, even corrected with high-order adaptive optics, have residual bumpiness diffracting significant levels of residual star light in the form of speckles contaminating the planet's image; b- stars are slightly resolved in telescopes larger than a few meters, at visible wavelengths.

Further cleaning with coherent or incoherent techniques is therefore needed for a deeper nulling of the star's residual speckles. Coherent techniques can use actuators located:

a- before the focal plane: Trauger *et al.* (this conference) have shown that a very accurate adjustment of the actuators can clean the coronagraph output image down to levels suitable for detecting Jupiters. It is unclear however whether the method is affected by the partial resolution of the star. b- after the focal plane, within a coronagraph, using destructive interference to null the residual stellar speckles. This is achievable in different ways using for example a multi-resel version of Bracewells nulling method, with a spatial filter in the form of a fiber bundle, or a Mach-Zehnder interferometer, or a dynamic hologram. These methods relax the needed actuator accuracy. The effect of the star's partial resolution remains to be assessed, and ways of reaching a usable spectral bandwidth have to be developed.

Incoherent cleaning is performed post-detection, typically by subtracting a reference image of a star having no planet. Unlike coherent cleaning, which can in principle achieve a complete null, it leaves in each speckle a residue level, the ultimate value of which corresponds to the photon-noise, amounting to the square root of the number of recorded photons. Incoherent cleaning should therefore be applied once the best possible coherent cleaning has been achieved.

These coronagraphic techniques are applicable to conventional telescopes and ELT's, to interferometers and to hypertelescopes. With the future very large arrays needed to image the details of exo-Earth's, such as the Exo-Earth Imager, coherent cleaning is likely affected by the fact that the planet is highly resolved. But the method can be applied before combining the beams, by using a separate coronagraph for each sub-aperture. On Earth, highly accurate adaptive optics is needed and it is difficult to predict how the planet detection sensitivity of ELTs will compare with that of space instruments.

9. Progress of Carlina hypertelescopes

Following the first fringes obtained on Vega by Le Coroller *et al.* (2004) with a partly completed Carlina prototype, J.Dejonghe has installed a Mertz corrector on the instruments gondola. It has been partially tested on stars. Also, the three ground mirrors have been re- positioned to provide baselines of 10.5m, 9m and 5m.

One of the dolines at Calern Observatory is considered for a larger Carlina II version. A contour map with 10 cm accuracy, generated by aerial photogrammetry, indicates that an effective aperture size of 40 to 60m can be exploited at this site. Canyon sites, in the southern Alps and elsewhere, are also considered for larger effective apertures, reaching a few hundred meters. Some of them can be equipped with a cable stretched across, which can make the balloon unnecessary.

10. Feasibility study of Galanthus, a paraboloidal hypertelescope for Dome C

The Galanthus genus of flowers includes G. nivalis, known as the Snowdrop in English and Perce-neige in French. It does not exist at Dome C, but may lends its name to a flower-like paraboloidal hypertelescope especially suited for this site, having a flat topography and very low wind Figure 2. Unlike the "grounded" Carlina mirrors, the suspended mirrors cannot be expected to keep a sufficiently stable geometry by passive means only. Each mirror is therefore carried by three actuators, under control from a real-time metrology device, such as a laser theodolite installed at the focal gondola. The complete pointable optical structure is expected to be located above the turbulent ground layer, said to be less than 40m thick. The flexural behavior of the net structure, particularly as a function of the pointing elevation, and its oscillations should now be modeled

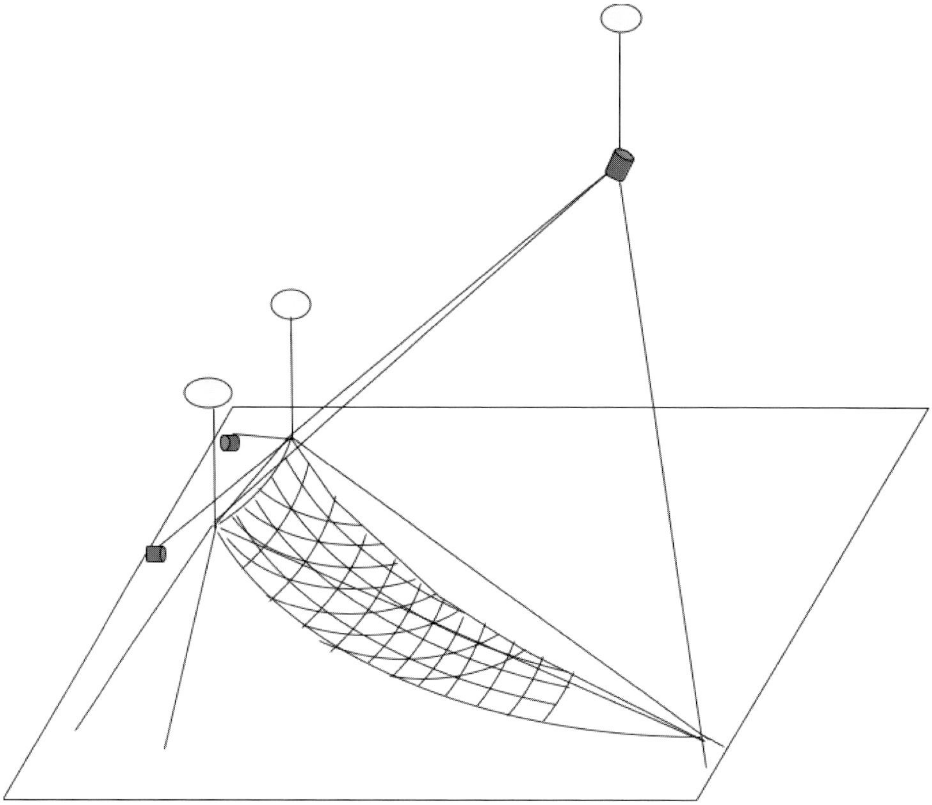

Figure 2. Scheme of Galanthus, a pointable paraboloidal hypertelescope for Dome C, Antarctica. A net-like structure stretched at the bottom of a tetrahedron, both made of high-modulus tensioned cable, is suspended 40m above the snow surface from three balloons. Small lightweight mirrors, figured as paraboloidal segments, are supported at each node of the net by three actuators for active geometry corrections. Tethers, actuated by computerized winches, control the pointing direction. The focal optics, including a pupil densifier, coronagraph, adaptive optics, camera, etc.. is located at the upper tip of the tetrahedron, although part of it can be located at ground level in a Coude laboratory.

with finite- elements code. This is needed to calculate the maximal aperture size, which may be of the order of a kilometer. In comparison with Carlina architectures, Galanthus has the drawback of strongly relying on active corrections, but advantages of not requiring a corrector of spherical aberration nor an exceptional concave site topography. Its focal ratio can also be shorter, F/1 instead of F/1.7 for Carlinas. A delicate issue is to manage the occasional landing of the structure, for storm protection and maintenance. There is some controversy on the statistics of storms at Dome C. Also, the emission of polluting soot particles by the local power station is a major concern for astronomical projects at the site.

11. Design progress for a Luciola hypertelescope in space

Luciola is a "bare bone", and much smaller, precursor version of an Exo-Earth Imager which can be flown in the near future to verify the concept and, if successful, provide direct resolved images of comparatively easy objects such as the nearby stars. The brighter exo-planets, such as the hot Jupiters associated to 51 Peg and Tau Boo are also

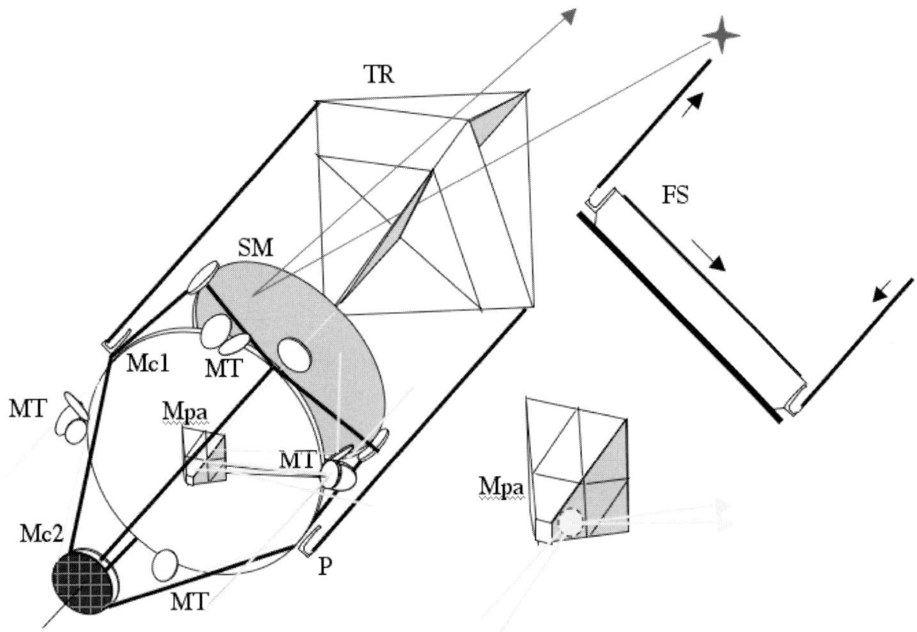

Figure 3. Concept of Luciola spaceships carrying elements of the giant diluted mirror. The low mass structure is translated and rotated by a small rigid solar sail, configured as a Ritchey-Chretien telescope (mirrors Mc1 and Mc2). It is pointed towards the Sun and focuses its image on the faceted pyramid mirror Mpa, mounted at the center of Mc1 and having 16 concave facets which relay pupil images on corresponding facets of the peripheral trihedral mirrors MT. Mc2 is controlled in tip-tilt by two actuators for switching the light output direction, by addressing different facets of Mpa. The tip-tilt actuators can remain nearly inactive during normal observing since the four central facets of Mpa feed those peripheral MT facets which reflect the beam in the anti-solar direction. The resulting torque tends to correct pointing errors passively. The actuators are needed to control all six force and torque components applied to the satellite. In the event of a complete loss of sun pointing, the sun image falls outside of Mpa and it becomes ineffective for attitude control. The tail reflector TR then becomes illuminated and applies a restoring torque, which tends to re-point the Cassegrain optics towards the Sun. Oscillations are damped by a damped flexural swivel FS connecting TR to the main structure. The stellar mirror SM is nearly flat and located in the shadow behind Mc1, also serving as a sunshield. It has one rotation axis, activated by a micro-motor. The second rotation needed to point a star is achieved by rotating the entire satellite about the Sun's direction, using the MT facets which provide a tangential output beam. SM has a central hole with a miniature EM-CCD camera serving for stellar pointing. The satellite has a low mass, typically a few hundred grams for a 20cm SM mirror, allowing sufficient accelerations with solar radiation pressure.

potential targets. On Earth, Luciola is also a genus of nocturnal firefly beetles, seen as dancing swarms with spectacular flashing displays. Here, it alludes to the flashing swarm of free-flyers equipped with pulsed diode lasers. These serve to determine the swarm's geometry,using time of flight measurements.

The segments of the diluted primary mirror are carried by nano-satellites driven by small solar sails for economy, for a long-life, and for avoiding the exhaust plumes of conventional thrusters, which tend to contaminate the cold mirrors. The scale size considered for Luciola is 20cm for the mirrors and a few hundred meters for the array aperture. Figure 3 shows the optical concept, the solar sail part of which is based on a Zemax ray-trace by David Vernet and Alain Blazit.

Each satellite emits encoded light pulses, non directional, and detects pulses from other satellites, using simple non-directional detectors. From the detected signals, the internal geometry of the swarm can be calculated with millimeter accuracy. A laser theodolite installed on the focal satellite, or a white-light interferometer installed at the curvature center can also provide such metrologic information, but can be more costly and difficult to operate. In the latter case, the co-sphericity errors of the primary segments can be measured with nanometric accuracy.

The solar radiation pressure is very weak, slightly less than a microNewton on a 20cm mirror, but the response is highly linear and predictable, with no stick-slip, hysteresis or other problems of earthly actuators. In the absence of wind and the other disturbances affecting Earth based interferometers, it is therefore expected that the sail system can maintain the position of the stellar mirror within Rayleigh's tolerance, and possibly better. If not, a conventional adaptive deformable mirror will be needed in the focal package to achieve proper phasing.

12. Conclusions

Connsiderable progress is under way towards forming images of exo-Earths, and then high-resolution images showing their detail, at the micro-arc-second scale, in multi-resel images. Established spectro-imaging techniques will provide spectra of each resel in such images, and this can help searching life through chlorophyll-like absorbtion bands and their seasonal variation. Current efforts in our group are directed towards building precursor versions of hypertelescopes on Earth and in space. Among these versions, the Carlina II and the Luciola are rather modest in size and cost, but expected to teach us how to build much larger hypertelescopes with effective aperture diameters reaching 1500m on Earth and 100 to 100,000 km in space. These aperture sizes in space are expected to respectively resolve exo-Earths and neutron stars.

References

Labeyrie, A. 1974, *proposal to NASA*

Bonneau, D., Josse, M., & Labeyrie, A. 1975, in: Nieuwenhuijzen & de Jager (eds.), *Image processing techniques in Astronomy*, D. Reidel, (Holland), p. 403

Le Coroller, H., Dejonghe, J., Arpesella, C., Vernet, D., & Labeyrie, A. 2005, *Astron.Astrophys.* vol. 426, p. 721–28

Direct Imaging of Exoplanets: Science & Techniques
Proceedings IAU Colloquium No. 200, 2005
C. Aime and F. Vakili, eds.
© 2006 International Astronomical Union
doi:10.1017/S1743921306009343

Very Deep Nulling Interferometers: limits and advantages

Daniel Rouan[1]
[1]LESIA-CNRS, Observatoire de Paris, 92195 Meudon cedex, France
email: daniel.rouan@obspm.fr

Abstract. Direct detection of a planet around a star by a nulling interferometer, requires to minimize as far as possible the stellar leaks due to the resolved angular size of the star. The original Bracewell configuration features a nulling function in θ^2 which is insufficient in many cases. Several interferometric configurations have been proposed in order to improve the quality of the rejection with a nulling function θ^n with $2 \leqslant n \leqslant 6$. I proposed recently a method to build linear configurations of telescopes that achieve nulling function θ^n for any even value of n, using the Prouhet-Thué-Morse sequence to select those telescopes where a π phase shift is applied. In a first part, I recall the basis of this method and its generalization to 2D configurations and 1D arrays of non-identical telescopes, or even to configurations where the phase shift is not π. In a next step, I evaluate the efficiency of deep nulling interferometers in *real world*, i.e. when nulling is not perfect because of variations of distances or of phase shift between telescopes. I conclude that there is a clear advantage given by the highest order systems that keep a better nulling capability than conventional interferometers, even in severe conditions where parameters driving the nulling performance are highly fluctuating.

Keywords. Interferometry, Exoplanets, Nulling interferometry.

1. Introduction

There is a huge contrast (10^{6-10}) between a star and a planet that orbits it, so that directly detecting the planet means cancelling as much as possible the stellar light. Bracewell, 1978 was the first to propose the concept of a nulling interferometer, where the light collected by two telescopes are coherently combined with however the trick that a π phase shift is applied on one of the optical path. The system of fringes projected onto the sky shows a central dark fringe. If the star image is put on this central dark fringe, it disappears (actually, the stellar photons are all sent in the second output of the interferometer). If the planet is at the same time on a bright fringe, then it can be detected, in principle. The modulation of the fringes with respect to the angular distance is according to a $(1 - \cos\theta)$ function, so that close to the axis the nulling is $\propto \theta^2$. If the interferometer has enough resolution (equivalently a long basis), as it should to separate a planet, then the angular diameter of the stellar disk is no longer small with respect to the fringe period and leaks of stellar light will contaminate the planet flux and severely limit the detection capability.

Several interferometric configurations have been presented in order to improve the quality of the rejection, especially in increasing the exponent of the term θ^n which gives the cancellation efficiency with respect to angular distance to the axis of the central fringe (e.g. Léger *et al.*, 1995). The Angel's Cross (Angel, 1989), Oases configuration or Mariotti / Laurance (Darwin ESA project) configuration and others (Woolf, 1997) do fill this condition of a nulling with an exponent >2. In general, those configurations have been found through some trial method, but, as far as I know, no systematic method has been proposed to reach any given power of θ.

I presented recently (Rouan, 2003-a,Rouan, 2003-b) a method to define linear configurations of telescopes that provide a θ^n nulling function for any even value of n. The principle is based on a remarkable property discovered by Prouhet (Prouhet, 1851) about a peculiar partition in two sets of the $N = 2^L$ first integers, done according to the Prouhet-Thué-Morse sequence. This property is such that, provided that to half of the N telescopes output a π phase shift is applied, it is possible to cancel the (L-1) first terms of the developement in θ of the recombined amplitude, thus leading to the wished θ^{2L} behaviour of the intensity. 1D configurations of identical telescopes is the basic configuration from which one starts to generalize the method for construction of 2D arrays – still of identical telescopes –, then to build 1D and 2D arrays of non-identical telescopes and finally extend the theory to arrays where the phase shift is $2\pi/n$. The various steps of this approach are summarized in the first section.

However, this work would remain a rather formal exercise and more a mathematical trick than a true instrumental concept, if one would not go further. This is the object of section two to assess the actual performances of those Very Deep Nulling Interferometers (VDNI), when non perfect conditions of nulling are considered.

2. Building Very Deep Nulling Interferometers

2.1. *Prouhet-Thué-Morse sequence and Prouhet partition of the integers*

The Prouhet-Thué-Morse sequence is very easily built and can be considered as one of the simplest fractal objects. Let's start from the symbol "+". This is the first term of the series. Now take its inverse, "-", and concatenate it to the first element: this produces the second term of the series, i.e. "+ -". Let's redo the operation: we take the current term "+ -", and concatenate to it its complement, "- +": this leads to the third term, "+ - - +". The following terms are built the same way by concatenating the complement of the previous term:

+ - - + - + + -

+ - - + - + + - - + + - + - - +

+ - - + - + + - - + + - + - - + - + + - + - - + + - - + - + + - etc.

One can note that: *a)* any term of the series never contains more than two identical digits (+ or -) that are juxtaposed (e.g. no +++ nor - - - -); *b)* periodic patterns never appear; *c)* the Lth term of the series contains 2^L digits (considering that the index of the first term is 1). Prouhet, 1851 had the idea to distribute the first 2^L integers into two sets according to the Prouhet-Thué-Morse sequence: the first set corresponds to integers that have the same rank as the (+) in the sequence and the second set to those that have same rank as the (-). This reads: $\mathcal{P} = [1, 4, 6, 7, 10, 11, 13, 16, \ldots, o_k, ...]$ and $\mathcal{M} = [2, 3, 5, 8, 9, 12, 14, 15, \ldots, e_k, ...]$

A property established by Prouhet, is that the sum of the elements of the two series are equal, but much more remarkably, the sum of the squares, of the cubes, etc. are also equal, and this up to the power L-1 when considering the 2^L first integers. In a condensed form this can be written:

$$\sum_{k=0}^{2^L-1}(o_k)^p = \sum_{k=0}^{2^L-1}(e_k)^p, \; \forall \; p < L.$$

2.2. *1D nulling interferometers*

Let's consider an interferometer made with a set of $N = 2^L$ identical telescopes regularly aligned, with a distance d between two adjacent telescopes, as illustrated on Fig. 1. If the telescopes are pointing in a direction θ (measured from the normal to the baseline), then the amplitude a of the wave at the output of the recombiner is:

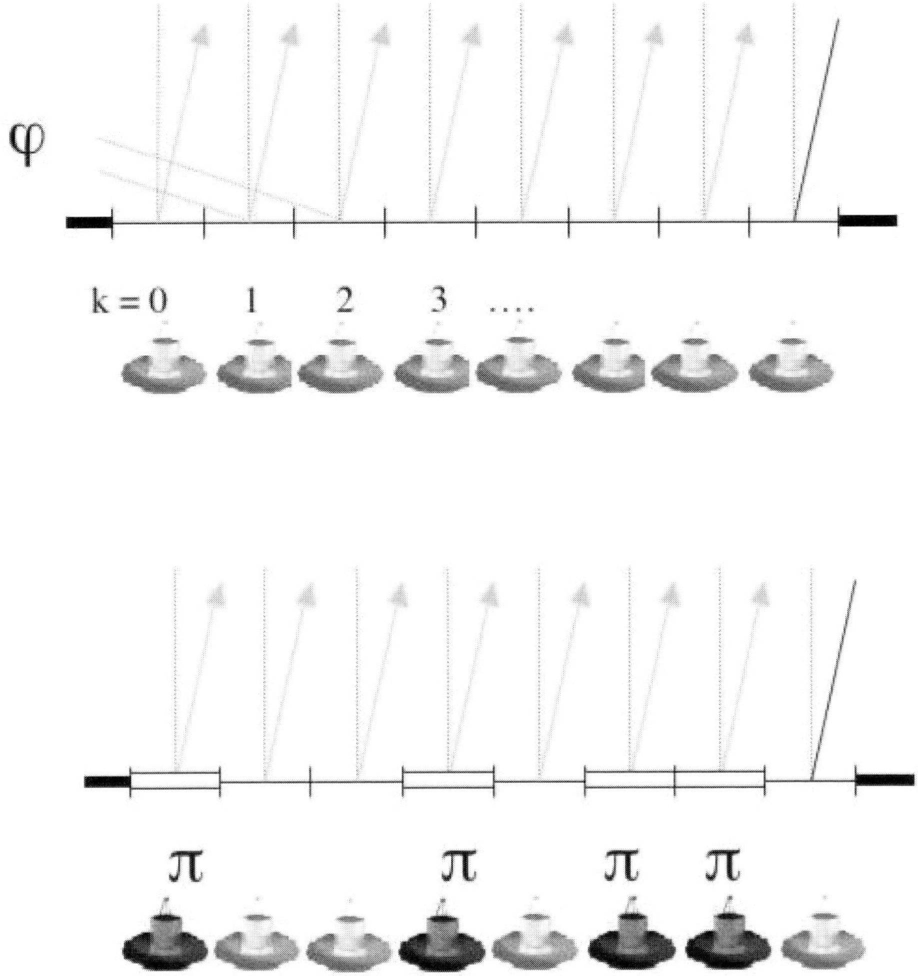

Figure 1. Top: configuration of a linear interferometer with equally spaced telescopes. Bottom: the *fractal* nulling interferometer, where some telescopes (the darker ones) experience a π phase shift.

$a = \sum_{\mathrm{k}=0}^{\mathrm{N}} \exp(j\mathrm{k}\phi)$

Where ϕ is the phase difference between two consecutive telescopes: $\phi = 2\pi\theta d/\lambda$. Let's now add a π phase shift on all telescopes whose rank is a "-" in the Prouhet series, in other word we change the sign of the amplitude. The previous relation becomes:

$a = a_+ - a_- = \sum_{\mathrm{k}\in\mathcal{P}} \exp(j\mathrm{k}\phi) - \sum_{\mathrm{l}\in\mathcal{M}} \exp(j\mathrm{l}\phi)$

where $\mathcal{P}$ and $\mathcal{M}$ stand respectively for the "+" and "-" sets. Considering relatively small values of ϕ, i.e. of θ, one can develop each exponential in series of $(\mathrm{k}\phi)^n$, so that the two terms of the above expression become:

$a_+ = \sum_{\mathrm{k}\in\mathcal{P}} 1 + j\phi \times \sum_{\mathrm{k}\in\mathcal{P}} \mathrm{k} - \phi^2 \times \sum_{\mathrm{k}\in\mathcal{P}} k^2 - j\phi^3 \times \sum_{\mathrm{k}\in\mathcal{P}} k^3 + \ldots$

and

$a_- = \sum_{\mathrm{k}\in\mathcal{M}} 1 + j\phi \times \sum_{\mathrm{k}\in\mathcal{M}} \mathrm{k} - \phi^2 \times \sum_{\mathrm{k}\in\mathcal{M}} k^2 - j\phi^3 \times \sum_{\mathrm{k}\in\mathcal{M}} k^3 + \ldots$

Thanks to the remarkable property found by Prouhet, i.e. $\sum_{\mathrm{k}\in\mathcal{P}} \mathrm{k}^p = \sum_{\mathrm{k}\in\mathcal{M}} \mathrm{k}^p$, the L-1 first terms of each series cancel out mutually, so that the wished property is reached: $a \propto \phi^{\rm L} \propto \theta^{\rm L}$, and the intensity has the behaviour: $I \propto \theta^{2\rm L}$. For instance, the interferometer shown in the lower part of Fig. 1 would produce a nulling function $\propto \theta^6$.

2.3. *Algebra of nulling interferometers*

2D VDNI

Let's consider a square grid of telescopes where on any row or any column the sign of the output of each telescope is set following a Prouhet-Thué-Morse sequence – or its complement –. For instance the grid above would describe a 4×4 VDNI.

+	-	-	+
-	+	+	-
-	+	+	-
+	-	-	+

The intensity varying as θ^{4L}, its nulling capability is in θ^{12}. One can note that: *a)* the case L=1 (2×2 telescopes) corresponds to the Angel's cross (Angel, 1989); *b)* the number of telescopes required to reach a given nulling exponent α (on the intensity) is the same in the 2D array and the 1D array cases: it is equal to $2^{\alpha/2}$.

VDNI with telescopes of different sizes

One shows that by shifting by k steps each line with respect to the previous one, in the 2D array, and by summing algebraically the columns of the resulting matrix, one defines a 1D array of telescopes of different sizes which has the nulling power of the initial 2d-array. For instance, the sheared 2d-array of the previous example

+1	-1	-1	+1						
		-1	+1	+1	-1				
				-1	+1	+1	-1		
						+1	-1	-1	+1
+1	-1	-2	+2	0	0	+2	-2	-1	+1

leads to the 1D nulling configuration at the bottom line (0 means no telescope). One can also build 2D arrays from 1D arrays by simply having a matrix where the first column and the first line are any 1D VDNI and making the algebraic product at the intersection, as illustrated by he following example:

+1	-2	+2	-1
-2	+4	-4	+2
+1	-2	+2	-1

VDNI with phase shift different from π

Finally, one shows that the Prouhet property can be extended to partitions of the first n^L integers in n sets rather than 2, through some permutations at each step to build the terms of the series (Rouan, 2003-b). Provided that the phase shift for each group of telescopes becomes 2 k π/n instead of π, the nulling property in θ^{2L} is as effective as in the bipartition.

3. Comparison between VDNI interferometers in the real world

We compare here the nulling properties of VDNI when a non perfect system is examined, i.e. when distances and phase are fluctuating, so that nulling is no longer constant. Indeed the question of the nulling stability is an issue since the associated noise tends to be the most important contributor in the present studies of space interferometers (Lay, 2004; Chazelas, 2005). For this comparison, I considered only 2D square arrays of the

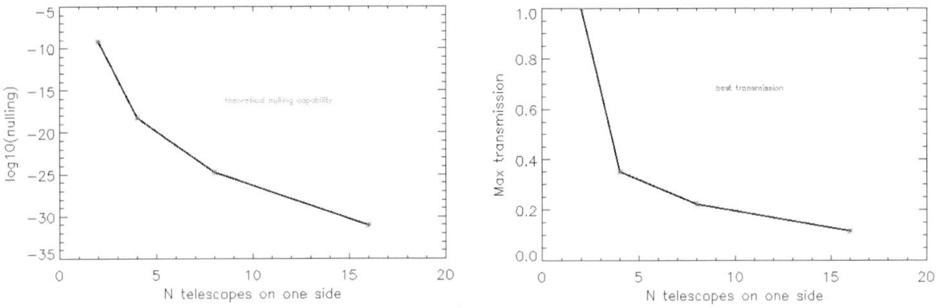

Figure 2. Left: theoretical nulling capability of the four first 2D square VDNI vs the number of telescopes on one side; Right: best transmission of the VDNI.

basic type: 2×2, 4×4, 8×8, 16×16. I assumed that the total surface of telescopes was the same: either few large telescopes or many small telescopes. I have in mind the TBC assumption that in terms of mission cost, this could be equivalent because the mass production of identical small telescopes should be rather cheap (e.g. 64 × 25 cm telescopes vs 4 × 1m telescopes). Simulations were done for an Earth-Sun system: this means that the distance between telescopes is such that the projected first maximum of transmission is at 1 AU and that the star diameter to consider for leaks calculation is .01 AU. Two causes of nulling degradation were considered in this preliminary exercise: *a)* the distances between telescopes are not constant and the rms relative variation is $\delta d/d = 10^{-3}$; *b)* there is an imperfect π phase shift with an rms fluctuation of .01 radians. Fig. 2, 3, and 4 summarize the results of this study.

First, in terms of absolute performances (perfect nulling configuration), the nulling capability of VDNI is indeed excellent and far beyond the required factor of 10^{10} for an Earth detection, since, for instance, it reaches 10^{-32} for the 16×16 array (Fig. 2-a). However the maximum transmission of the planet is decreasing with the order, it is 12% for the 16×16 pattern, while it is 100% for the 2×2 one (Fig.2-b).

The most interesting results is the one giving the nulling performance when fluctuations of distances and of phase are introduced: Fig.3a-b shows that there is a clear improvement when using, for a given level of fluctuation, larger arrays: for instance, even with a very severe phase fluctuation of .5 ^{o} rms, the nulling capability can be maintained at a fairly good level with a 16×16 array, with a gain of 85 over the 2×2 array. If one looks at the detail of the behavior with time of the nulling performance (Fig. 4), it clearly appears that the larger the array, the lesser the amplitudes of fluctuations, and especially the big ones (note the logarithmic scale of Fig. 4): for instance in the case of the 2×2 array, the nulling factor can vary by as much as a factor of 2000, while it is only a factor of 370 for the 16×16 array. One can logically suspect some averaging effect to explain this robustness of the nulling efficiency in the higher order VDNI.

Since the stability of the nulling efficiency is clearly a concern in the design of space nulling interferometer, it seems thus that considering the VDNI concept to improve this performance deserves some attention. More detailed estimates are obviously required.

4. Conclusion

Reaching a good nulling on starlight is mandatory for detecting terrestrial planets. A method to design a nulling interferometer that provides a central null depth of any given even power of θ has been given previously. It is based on a property of the Prouhet-Thué-Morse sequence, and allows to build 1D interferometers where to half of the telescopes

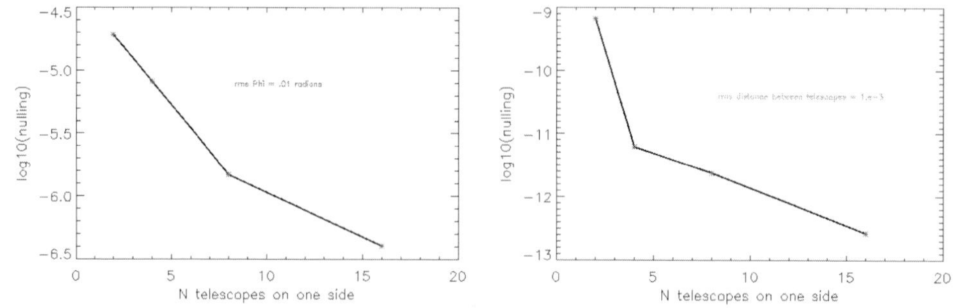

Figure 3. Left: nulling capability of the four first 2D square VDNI when distance between telescopes fluctuates, with rms $\delta d/d = 10^{-3}$; Right: same but when π phase shift fluctuates, with $\sigma_\phi = 1\,\mathrm{deg}$.

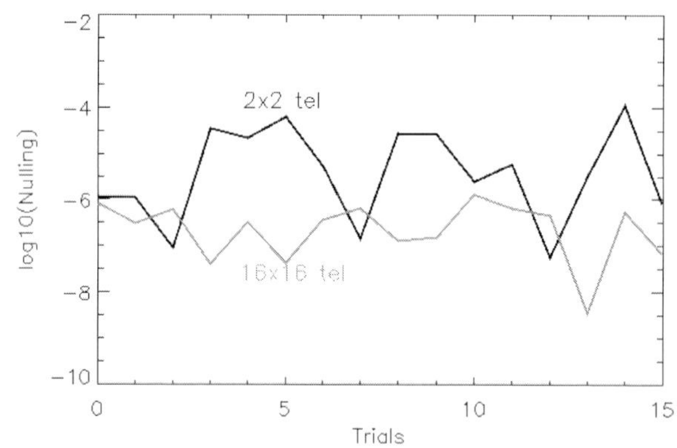

Figure 4. Exemple of nulling stability as a function of time for 2×2 and 16×16 patterns.

is applied a π phase shift. From this basic pattern, one can then build 2D or other 1D patterns, by combining in one dimension or two dimensions those building bricks. In a second step I estimate how those VDNI behave when the basic interferometric parameters (distances and phase) fluctuate, as this happens in the real world. I show that the higher the order of the VDNI, the more robust they are with respect to nulling performances, beating largely conventional nulling interferometers of lower orders. This is especially true when very severe phase variations are introduced. This robustness is likely the result of some averaging effects. Especially the nulling stability noise, which is a very serious concern in the design of space interferometers, is largely reduced with a high order VDNI. This type of interferometer may be a serious alternative to more conventional design, if indeed the nulling stability noise remains the main concern in presently studied missions. For a given collecting surface, the increasing number of telescopes could be compensated, in terms of cost, by the mass production of identical small telescopes.

References

Angel, R., 1989, in The Next Generation Space Telescope, p. 81
Bracewell, R.N., 1978, Nature 274, 780
Chazelas, B., Brachet, F., Ollivier, M., Mennesson, B., *et al.*, 2005, A&A accepted
Lay, O.P., 2004, Applied Optics, 43, 6100–6123
Léger, A., Mariotti, J.-M., Menesson, B., *et al.*, 1996, Icarus 123, 249

Prouhet, E., 1851, Mémoire sur quelques relations entre les puissances des nombres. in: Comptes Rendus des Séances de l'Académie des Sciences

Rouan, D., 2003, Proceedings of the Conference on Towards Other Earths: DARWIN/TPF and the Search for Extrasolar Terrestrial Planets, 22-25 April 2003, Heidelberg, Germany. Edited by M. Fridlund, T. Henning, ESA SP-539, Noordwijk, Netherlands, p. 565 - 568

Rouan, D., 2004, From Fractals to Exoplanets: building ultra-nulling interferometers, EAS Publications Series, 12, 21

Woolf, N.J. & Angel, J.R.P., 1997, "Planet Finder Options I: New Linear Nulling Array Configurations", in "Planets beyond the solar system and the next generation of space missions". ASP Conference Series, ed. D. Soderblom

All photographs: Laurent Thareau [`l.thareau@free.fr`].

Direct Imaging of Exoplanets: Science & Techniques
Proceedings IAU Colloquium No. 200, 2005
C. Aime and F. Vakili, eds.
© 2006 International Astronomical Union
doi:10.1017/S1743921306009355

The Fourier-Kelvin Stellar Interferometer: an achievable, space-borne interferometer for the direct detection and study of extrasolar giant planets

R. K. Barry[1,9], W. C. Danchi[1], L. D. Deming[1], L. J. Richardson[1], M. J. Kuchner[1], V. J. Chambers[1], B. J. Frey[1], A. J. Martino[1], J. Rajagopal[1,7], R. J. Allen[2,9] J. A. Harrington[10], T. T. Hyde[1], V. S. Johnson[10], R. Linfield[4], R. Millan-Gabet[5], J. D. Monnier[6], L. G. Mundy[7], C. Noecker[4], S. Seager[8], and W. A. Traub[3]

[1]NASA, Goddard Space Flight Center, Greenbelt, MD 20771, USA
[2]Space Telescope Science Institute,3700 San Martin Dr., Baltimore, MD 21218, USA
[3]Jet Propulsion Laboratory, 4800 Oak Grove Drive, Pasadena, California 91109
[4]Ball Aerospace, Boulder, 1600 Commerce St., CO 80301, USA
[5]California Institute of Technology, 1201 E. California Blvd., Pasadena, CA 91125, USA
[6]University of Michigan, Astronomy Department, Ann Arbor, MI 48109, USA
[7]University of Maryland, Astronomy Department, College Park, MD 20742, USA
[8]Carnegie Institution of Washington, 5241 Broad Branch Rd.,Washington, DC 20015, USA
[9]The Johns Hopkins University, 3400 N. Charles St., Baltimore, MD 21218, USA
[10]Rutgers University, 607 Taylor Rd., Piscataway, NJ 08854, USA

Abstract. The Fourier-Kelvin Stellar Interferometer (FKSI) is a mission concept for a spacecraft-borne imaging and nulling interferometer for the near to mid-infrared spectral region. FKSI is a scientific and technological pathfinder to the Darwin and Terrestrial Planet Finder (TPF) missions and will be a high angular resolution system complementary to the James Webb Space Telescope (JWST). There are four key scientific issues the FKSI mission is designed to address. These are: 1.) characterization of the atmospheres of the known extra-solar giant planets, 2.) assay of the morphology of debris disks to look for resonant structures characteristic of the presence of extrasolar planets, 3.) study of circumstellar material around a variety of stellar types to better understand their evolutionary state, and in the case of young stellar systems, their planet forming potential, and 4.) measurement of detailed structures inside active galactic nuclei. We report results of simulation studies of the imaging capabilities of the FKSI, current progress on our nulling testbed, results from control system and residual jitter analysis, and selection of hollow waveguide fibers for wavefront cleanup.

Keywords. instrumentation: high angular resolution, techniques: interferometric, space vehicles: instruments, planets and satellites: general, planetary systems: formation, Galaxy: nucleus.

1. Introduction

The high spatial and spectroscopic resolving capability of the FKSI instrument together with its calculated sensitivity will position it as an important facility for the study of a range of astronomical phenomena. Its science objectives are, very broadly, to directly detect extrasolar giant planets (EGP), study debris disks and the evolution of protostellar

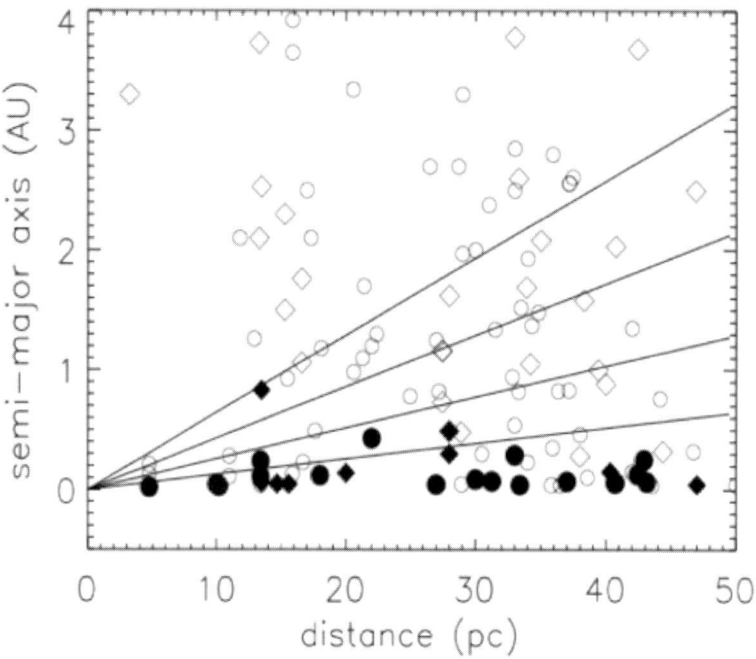

Figure 1. Distance (in parsecs) and orbital radius (semi-major axes, in AU) for 140 of the known extrasolar planets. The planets within at least a 40 degree field of regard, and nominally detectable by FKSI, are plotted with filled symbols. Circles indicate planets within a 20 degree field of regard, and diamonds indicate additional detectable planets if the field of regard is extended to 40 degrees. The lines correspond to the first fringe maximum for nulling interferometers of baselines 8, 12, 20, and 40 m (top to bottom).

and evolved stellar systems, and to facilitate the study of extra-galactic star formation regions and the extended neighborhoods of active galactic nuclei.

The detection and study of EGPs will be greatly facilitated by a dedicated nulling interferometer on a satellite platform operating at infrared wavelengths. To answer key questions about these planets FKSI will be able, depending on configuration, to detect at least 25 known EGPs and make precise determination of their orbital parameters. (See figure 1.) The spatial resolution of the Fourier-Kelvin Stellar Interferometer together with the instrument's $R = 25$ spectroscopic capability will allow the direct detection and analysis of photons from EGP under certain conditions of orbital phase and angular separation. (See Danchi, *et al.* 2003b.) These measurements will enable studies of the environmental conditions to which the planet is subject. They will also provide a means of measuring the composition of the planet's atmosphere.

FKSI exploits the orbital dynamics of close-in planets. Using a spatial dither of the null fringe, FKSI will be able to reconstruct the astrometric orbit of the planet and, in combination with radial velocity studies, derive the mass. In addition to mass, FKSI observes the variation in the planet's infrared spectrum as a function of phase, and facilitates characterization of the planet's thermal state and composition. In favorable cases, data obtained using FKSI could even allow scientists to make inferences about dynamical properties such as winds. This approach of exploiting the infrared brightness of close-in planets builds on the transit and radial velocity studies, which have detected most of the known extrasolar planets.

While direct detection of EGPs is a principal goal of the FKSI mission, study of stellar environments in general will also be greatly facilitated by the instrument's spatial and spectral resolving power. With FKSI, scientists may observe the formative stages of planetary systems, protoplanetary disk structure, the evolution of primordial exozodiacal dust, and debris disks about various stellar types. These observations may then be used to determine the characteristics of circumstellar material that could lead to planet formation or the evolution of the habitable zone about post-main sequence stars. These studies would facilitate refinement of the target lists and science goals of the TPF mission.

2. Design and Mission Concept Approach

Our team, lead by our Principal Investigator, Dr. W. C. Danchi, consists of scientists from a broad array of institutions and, together with engineering support from GSFC, has expended significant effort to develop a range of design options for FKSI. We have studied various beam combination techniques and array architectures in preparation for submission of FKSI as a Discovery-class mission. These design studies were conducted initially at GSFC's Instrument Synthesis and Analysis Laboratory and the Integrated Mission Design Center. These are important functions within GSFC's infrastructure used to facilitate rapid vetting of various mission and instrument design concepts. These studies were then augmented by the work of a larger, focused team of experienced scientists and engineers dedicated to the FKSI mission. The resulting design is a nulling interferometer configuration with an optical system consisting of two 0.5 m telescopes on a 12.5 m boom feeding a symmetric Mach-Zehnder beam combiner. A null tracker and hollow-core fiber for wavefront cleanup further augment the system and allow it to produce the required .01% null of the central starlight. A thorough description of the design study and technology trade process was published recently. (See Danchi, *et al.*, 2003a.)

3. Dynamic Jitter Analysis and the Nulling Budget

An in-depth control system and dynamic jitter analysis has been conducted for the FKSI mission. (See Hyde, *et al.*, 2004) This analysis consisted of several distinct phases. First, a structural finite element analysis was conducted using a NASTRAN model of the spacecraft bus and instrument payload systems. A normal modes analysis was performed and mode shapes and frequencies provided for the integrated model. Important nodes in the model were input degrees of freedom (DOF) at the reaction wheel assemblies (RWA) mounted to spacecraft, output DOFs at each optical element in train, and interface DOFs at the four mounting points of spacecraft to payload. The first six modes of the combined payload/spacecraft are the isolator modes and range from 0.3 to 3 Hz in the nominal design. Higher modes include boom-flapping modes at 5.6 and 7.3 Hz. A mode at 25.5 Hz includes the motion of the combiner relative to the siderostats and strongly influences optical path delay.

The second phase of the analysis consisted of development of an optical model and included tracing of ray bundles from the entrance pupils above the siderostats through the optical train to the angle tracker, fringe tracker, and nuller detectors. This model may be described as a first-order Taylor expansion of the optical path lengths of a grid of rays traced through the system, uniformly spaced at the entrance pupil. The variables expanded upon are the rigid body DOFs for each optical component in the system. All ray tracing for the linear optical model was performed ahead of time to generate linear sensitivities that acted as transfer functions converting rigid body motions of optical components to absolute image motion at the detectors and wavefront error induced due to the misaligned components. Comparing wavefront maps generated from the sensitivities to those generated by ray tracing show less than 1 percent difference for random perturbations of one micron motions or one micro-radian tilts.

The third modeling effort included the normal spacecraft attitude control as well as control of many optical elements in the payload. This model, carried out in MATLAB Simulink, includes the spacecraft attitude control system (ACS) as well as the two types of continuous control loops in the payload, the optical path delay (OPD) and the angle steering mirrors (ASM). The ACS uses star trackers and gyros to estimate the pointing of the spacecraft bus and controls this to about 1 arcsec RMS in roll and pitch. The instrument errors consist of two tilt errors and the OPD. These are sensed by the angle tracker and fringe tracker respectively and are controlled in a feedback loop to the angle steering mirrors and the optical delay line (ODL) fine stage. Sensor noise in the FT accounts for about 5 nm of OPD error. Actuation noise in the ODL fine stage accounts for about 5 nm of OPD error. The low frequency ACS control residual of about 1 arcsec results in about 60 μm of OPD which is reduced to less than 7 nm by pathlength variation control. By including an unloading loop from the ODL command back to the pitch axis of the ACS, the ACS control residuals are further reduced by a factor of 100. This means that the fine stage of the ODL needs only 2 μm of stroke.

A forth element is the passive vibration isolation between the payload and the spacecraft bus. The isolator system reacts to the relative displacement between the payload and spacecraft and generates relative forces that act on the two masses. The transfer function between relative forces and relative displacements between the two masses behaves similarly to a high-pass filter. In the model of this system, the isolator physical parameters are mapped directly from three isolator design parameters that describe the desired isolation behavior. The baseline design has an isolation corner frequency of approximately 0.8 Hz with approximately 40 percent damping.

Finally, the results of the discipline models outlined above were integrated into a single system model. The model integration and subsequent analysis was performed in the Disturbance-Optics-Controls-Structures (DOCS) environment, a suite of tools developed at the Massachusetts Institute of Technology for performing integrated modeling, critical parameter identification, and system design. The influence of each design parameter and anticipated disturbance was explicitly evaluated. Critical parameters were then traded to generate a design that meets requirements optimally with respect to mission cost and risk.

Null depth, a critical final output parameter of the integrated analysis, is affected by alignment, control, and dynamic stability concerns. Science drivers require a 0.01 percent null depth that may then be allocated across the derived null depth loss contributors. Null loss formulae subsequently yield a tolerance on parameter for each subsystem. The high-fidelity analysis allows the budgeted OPD tolerance term to be further broken down into separate component contributions, which are incoherent and are, therefore, root-sum-squared (RSS) rather than directly summed. Finally, these parameters are directly applied to FKSI testbed OPD components as design requirements. The ACS, RWA, and boom noise contributions are RSSed and are equated in the testbed environment to residual seismic and air turbulence perturbations.

4. The Nulling Stellar Interferometry Testbed

The FKSI testbed is now fully funded and is being built in the Horizontal Flow Facility at NASA's Goddard Space Flight Center in Greenbelt, Maryland. This testbed will allow the instrument designers to evaluate technically challenging aspects of the design. In particular, the instrument's symmetric Mach-Zehnder nulling architecture will be examined together with a novel ditherless quadrature phase detector approach. Significant progress has already been made on the testbed, which has now transitioned from visible to monochromatic IR testing.

The nulling architecture of the testbed consists of a symmetric Mach-Zehnder beam combiner. A laser beam is collimated by an off-axis parabola, which is then passed through a mask with two apertures. This mask simulates the beams coming from the siderostats of the proposed two-aperture version of the FKSI design. These beams then enter a pair of mirror-symmetric right angle periscopes, and are rotated 180 degrees resulting in a field reversal, which affects the incident polarization states by reversing the s and p plane reflections. The beams are then relayed through cats-eye retro-reflectors in order to control the interference of the beams in delay space without effecting polarization. After the beams are passed through the delay line, two beamsplitters split both beams, and two more beamsplitters re-combine them. The beamsplitters and the fast steering mirrors are controlled by picomotors for fine adjusting. All four recombined beams are sent to four detector channels, which are used to gather both scientific data and metrology.

After determining the cause and solution to several alignment limitations using a 632 nm source, we have now converted to a 1.15/3.39 μm source and have upgraded several optics. In particular, we have replaced all silver fold mirrors with uncoated gold and replaced commercial off-the-shelf (COTS) visible light beamsplitters with (50/50) IR beamsplitters. Some near-term goals include the integration of hollow, single-mode fibers for wavefront cleanup and the addition of a phase tracking control system. Phase tracking will be performed with a novel ditherless quadrature phase detector and a voice coil or piezo actuator driven in closed loop using a digital signal processing system. The phase tracking control loop will have a high bandwidth to achieve good disturbance rejection capability even at high frequency where the structural resonant modes dominate. Picomotor actuators set on four mirrors, four beamsplitters and the cats-eye phase-delays will be controlled with control system software in order to continuously track and steady the fringe during measurements. Fringe contrast will be further improved by mounting hollow fibers in the detector space to clean the wavefront of aberrations.

The hollow glass waveguide (HGW) fibers chosen for the testbed have been fabricated at Rutgers University and will soon be integrated into the FKSI testbed. These HGWs were fabricated using wet-chemistry methods to first deposit a silver layer on the inside of silica glass tubing and then to form a silver halide dielectric layer over the metallic film. (See Harrington, 2004.) One advantage of these HGW fibers over the more typical solid-core polycrystalline (PC) fibers is that they can be created with a very flat spectral response in the 3 - 20 μm spectral region. In contrast, PC fibers often suffer from mode mixing and higher loss than HGW fibers due to the extrusion pressure necessary to force the preform through the die. This results in the non-circularity of the core and cladding and contributes to transmission of multiple modes. HGW, in contrast, have no Fresnel loss and require no end polishing to attain single mode performance. HGW made using silica tubing such as those chosen for the FKSI testbed are very robust, and, most importantly, have been shown to have more than 20dB of mode rejection between the HE11 and HE12 modes even though their bore size is significantly greater than the wavelength of light. The FKSI testbed team will be testing various lengths of 100, 150, and 180 μm HGW in the near future and will report our findings.

5. The Observatory Simulation Code

The FKSI observatory simulation code, written in IDL, is used to evaluate the detectability of known radial velocity planets. The code models the instantaneous properties of the null fringe with realistic path difference errors, and convolves the disk profile of a limb-darkened star and orbiting planet with the fringe transmission. Repeating this convolution as the interferometer rotates for a specified number of rotations at a specified rate, the code generates synthetic signals, accounting for detector noise, stellar photon noise, and noise from thermal background within the instrument and from zodiacal emission in our solar system. The code can also generate synthetic signals from hypothetical zodiacal dust in the extrasolar system using ZODIPIC, a software tool developed by Marc Kuchner to compute images of exozodiacal clouds. User input to the simulation code can specify the bond albedo of the planet, and a parametric representation of how stellar irradiation is redistributed on the planetary disk. Phase-dependent variations in the planetary spectrum are also included, with orbital parameters from the Doppler data, or specified by the user. This simulation code was used to evaluate the detectability by FKSI of all know extrasolar planets as of July, 2005.

Acknowledgements

This work was supported in part by NASA Goddard Space Flight Center's Internal Research and Development (IRAD) and Directors Discretionary Fund (DDF) programs.

References

Artjushenko, V. G., Lerman, A. A., Litvinenko, E. G., Nabatov, A. O., Kuznetsov, V. I., Plotnichenko, V. G., Pylnov, I. L., Shtein-Margolina, and four other authors. 1992, *Proc. SPIE* 1591L, 83–89

Danchi, W. C., Allen, R. J., Benford, D. J., Deming, D., Gezari, D. Y., Kuchner, M., Leisawitz, D. T., Linfield, R., and eight other authors. 2003a, *Proceedings of the Conference on Towards Other Earths: DARWIN/TPF and the Search for Extrasolar Terrestrial Planets* 89

Danchi, W. C., Deming, L. D., Kuchner, M. J., & Seager, S. 2003b, *ApJ* 597, L57

Harrington, J. A. 2001, *Infrared Fiber Optics*, M. Bass, J. Enoch, E. Van Stryland, and W. Wolfe, eds., (McGraw-Hill, New York)

Hyde, T. T., Liu, K., Blaurock, C., Bolognese, J., Howard, J. M., & Danchi, W. C. 2004, *Proc. SPIE, Modeling and Systems Engineering for Astronomy* 5497, 553–564

Direct Imaging of Exoplanets: Science & Techniques
Proceedings IAU Colloquium No. 200, 2005
C. Aime and F. Vakili, eds.
© 2006 International Astronomical Union
doi:10.1017/S1743921306009367

Long Baseline Nulling Interferometry with the Keck Telescopes: a Progress Report

B. Mennesson[1], R. Akeson[2], E. Appleby[4], J. Bell[4], A. Booth[1], M.M. Colavita [1], S. Crawford[1], M.J. Creech-Eakman[3], W. Dahl[4], J. Fanson[1], C. Felizardo [2], J. Garcia[1], J. Gathright[4], J. Herstein[2], E. Hovland[1], M. Hrynevych[4], E. Johansson[4], C. Koresko[2], D. Le Mignant[4], R. Ligon[1], R. Millan-Gabet[2], J. Moore [1], C. Neyman[4], D. Palmer[1], T. Panteleeva[4], C. Paine[1], S. Ragland[4], L. Reder[1], A. Rudeen[4], T. Saloga [4], E. Serabyn[1], M. Shao[1], R. Smythe[1], K. Summers [4], M. Swain[1], K. Tsubota[4], C. Tyau[4], G. Vasisht[1], P. Wizinowich [4], and J. Woillez[4]

[1]Jet Propulsion Laboratory, California Institute of Technology, 4800 Oak Grove Drive, Pasadena CA 91109-8099

[2]Michelson Science Center, California Institute of Technology, 770 South Wilson Avenue, Pasadena CA 91125

[3]New Mexico Tech, Dept of Physics, 801 Leroy Place, Socorro, NM 87801

[4]W.M. Keck Observatory, California Association for Research in Astronomy, 65-1120 Mamalahoa Highway, Kamuela, HI 96743

Abstract. The Keck Interferometer Nuller (KIN) is one of the major scientific and technical precursors to the Terrestrial Planet Finder Interferometer (TPF-I) mission. KIN's primary objective is to measure the level of exo-zodiacal mid-infrared emission around nearby main sequence stars, which requires deep broad-band nulling of astronomical sources of a few Janskys at 10 microns. A number of new capabilitites are needed in order to reach that goal with the Keck telescopes: mid-infrared coherent recombination, interferometric operation in "split pupil" mode, N-band optical path stabilization using K-band fringe tracking and internal metrology, and eventually, active atmospheric dispersion correction. We report here on the progress made implementing these new functionalities, and discuss the initial levels of extinction achieved on the sky.

Keywords. instrumentation: interferometers, infrared: stars, stars: circumstellar matter.

1. Introduction

The main science objective of KIN is to measure the level of exo-zodiacal dust emission around nearby main sequence stars in the mid-infrared (Serabyn, Colavita & Beichman 2000), a key design parameter for the Darwin (Léger *et al.* 1996) and TPF-I (Beichman *et al.* 1999) space missions. The basic requirement for KIN is to be able to detect exo-zodiacal dust emission down to 30 solar zodi levels around a solar analog located at 10 pc. In this case, the exozodi signal is roughly 1.3 10^{-3} times fainter than the stellar one at 10 microns, and arises mostly from the inner 1 AU region around the star. This clearly indicates that the null depth requirement with KIN is of the order of 10^{-3}, or that the null measurements need to be calibrated down to this level. Unlike other mid-infrared facilities, such as SPITZER for instance, KIN's spatial scales are specifically adapted to the detection of the very inner material around nearby stars. KIN's outer working angle is defined by the smallest of three fields: the unvignetted field of view

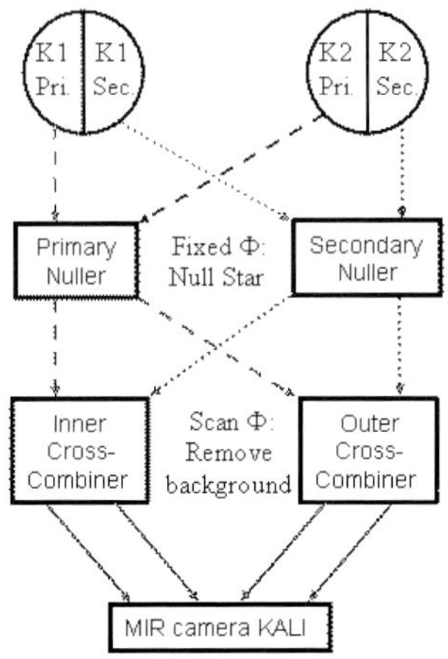

Figure 1. Conceptual KIN measurement architecture (see text).

of the interferometer, the coherent field of view fixed by KIN's spectral resolution, and the final science camera field stop. The latter is the smallest, corresponding to a field of $\simeq$500 mas around the science target. This allows observation of the central 2.5 AU of a dust disk around a star at 10 pc, precisely the region of interest for exo-zodiacal dust characterization. The spatial resolution of KIN is given by the first constructive peak of the sky null transmission pattern, which is $\lambda/2B$=12 mas, when observing at λ=10 microns, with the Keck baseline B=85 m. This corresponds to 0.12 AU at 10 pc, meaning that the inner brighter region of the disk is accessible for all targets of the TPF sample.

In addition to strong stellar light rejection, very precise background estimates are necessary in order to measure accurate null levels from the ground at 10 microns. This calls for a 2-step removal process: starlight suppression and then background subtraction, as illustrated in figure 1. Each of the Keck telescope pupils is divided into 2 sub-apertures, defining primary and secondary beams. The primary beams of each telescope are nulled together, as well as on the secondary side. The residual primary and secondary nulled beams are then coherently mixed together with a modulated phase shift, using the "cross combiner" beam splitters. As a result, the residual light coming from the astronomical source becomes an "AC" signal, easily separated from the incoherent background signal which remains a "DC" signal. The final null measurement is still affected by the background noise, and its fluctuations at the modulation frequency.

It is worth noting that this scheme is fairly similar to the phase chopping technique Woolf and Angel 1997, Mennesson, Léger & Ollivier 2005) considered for the TPF-I and Darwin missions.

2. From the laboratory to the sky

The 4 beam nulling/phase modulation scheme described above was tested and implemented gradually over the past 2 years.

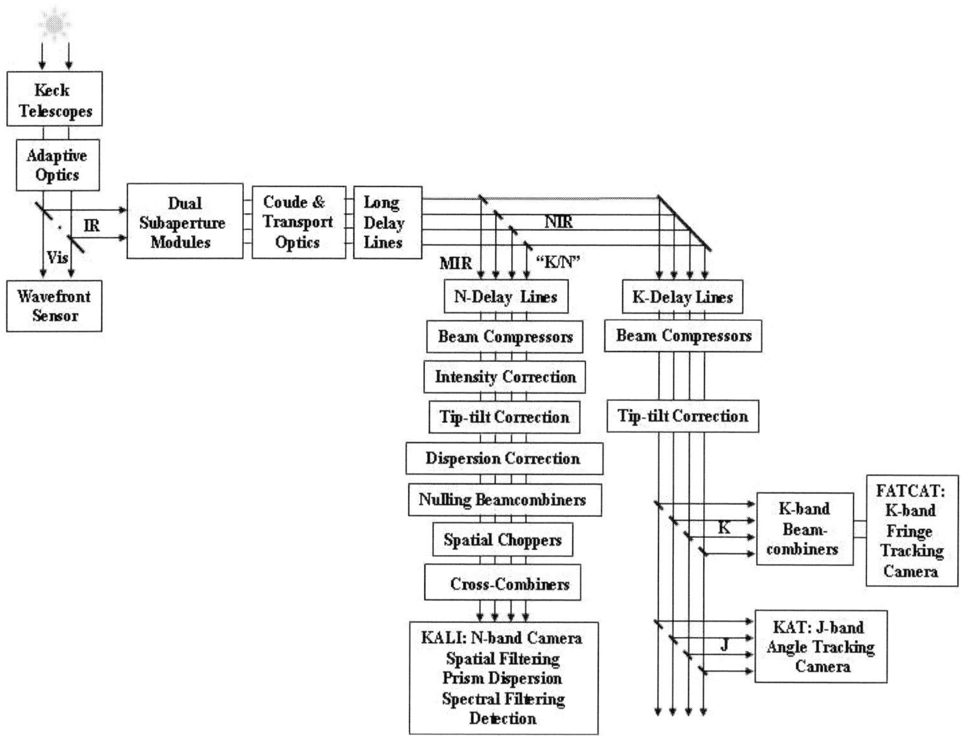

Figure 2. System level block diagram of the KIN, and the starlight path through the system. Not shown explicitely are the MIR nuller (8) and cross-combiner (2) beam splitters, which lie in the boxes labeled "Nulling Beam Combiners" and "Crosscombiners".

A basic physics experiment was set-up at JPL, using a pair of inverted beam splitters for symmetric beam combination (Serabyn & Colavita 2001) and dielectric phase plates for achromatic phase shifting. Deep thermal infrared wide band nulls were obtained in 2003 (Mennesson *et al.* 2003), basically validating the optics approach. A more realistic set-up was then integrated and tested in the laboratory, with long airpaths above various optical tables. It included most of the components needed for nulling at the Keck observatory: 4-N band delay lines, 4 K-band delay lines, a 4 beam nulling bread-board, laser metrology, and a helium cooled science camera (Creech-Eakman *et al.* 2003). In May 2004, wideband (8-10μm) deep nulls ($<10^{-3}$) were obtained and stabilized over 5 minutes, with repeatability maintained over 48 hours (Serabyn *et al.* 2004).

At this point, the nulling components were shipped to Hawaii for integration with the Keck Interferometer optical train, which is summarized in figure 2. The initial phase of the install was completed in August 2004, culminating with the first thermal infrared fringes between the Keck telescopes. These observations were carried out in regular visibility mode. They confirmed the instrument's ability to derive dust shell properties (optical depth profile, spatial extent) in the favorable case of stars with known large infrared excesses, such as the dusty AGB star RS CrB (Mennesson *et al.* 2005).

Subsequent developments targeted nulling more specifically. A split pupil module was installed at the telescope level, in order to define the primary and secondary beams issued from each Keck telescope. Simultaneous fringes between primary beams and secondary beams were obtained in October 2004, as well as cross fringes between beams coming from a given telescope. Further details on the metrology and real time control systems

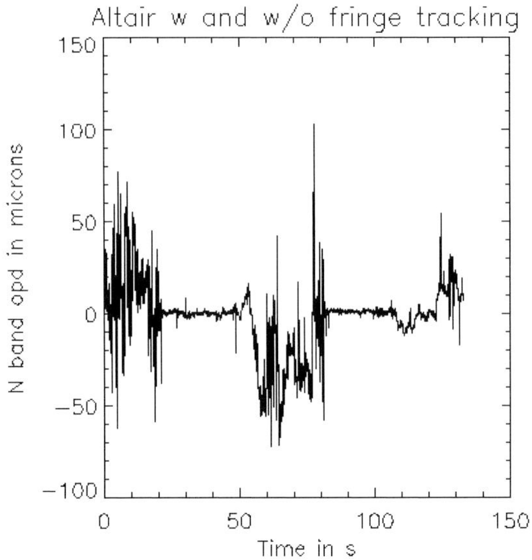

Figure 3. Optical path difference (opd) measured around 10 microns. The "quite" periods with small opd fluctuations correspond to the time intervals when the K-band phase is "fed forward" to the N-band delay line. No LASER metrology used during the sequence.

installed on the KIN can be found in Serabyn *et al.* 2005. A key functionnality for deep nulling is the K-band fringe tracking feed forward capability. KIN's target stars are too faint at 10 microns to allow fast fringe tracking in the N-band. The fast opd stabilization is instead carried out at 2 microns, - where stellar photons are plentiful. The phase information is then "fed forward" to the N-band delay lines, using LASER metrology to monitor the non-common K vs N band airpath fluctuations. The dramatic effect of enabling K-band feed forward is illustrated in figure 3. Only feed forward allows opd stabilization at the level required for deep nulling.

3. Current performance and future improvements

The current level of broad-band extinction achieved with KIN (Serabyn *et al.* 2005) is illustrated in figure 4. It shows the nulls measured on Vega, a bright A0V star (37 Jsky at 10 microns), as a function of wavelength, and for each of the two KALI ports presently used (open and filled squares). The dotted line shows for comparison the geometric null depth expected as a function of wavelength, representing Vega as a 3.28 mas uniform disk using the limb darkened diameter derived by Ciardi *et al.* (2001). It is unclear at this point whether the extra leakage detected with KIN is purely instrumental or of astrophysical origin. Yet, assuming that all of the extra leakage is of intrumental origin, figure 4 shows that a broadband instrumental null depth of 0.01 or better has been achieved between 8 and 10 microns.

Although accessible via accurate calibration, the detection of structures fainter than the geometric stellar leakage - fixed by the finite stellar disk size- is of course difficult. Faint structures will be more readily detectable around stars small enough to provide low geometric nulls in the first place. Achievable null depth and sensitivity are then closely related, as shown in figure 5. Considering naked stars as uniform spherical blackbodies, hotter stars are also less resolved for a given flux, making early type stars more favorable targets. Over the last year, KIN targets have evolved from bright red giants - with strong

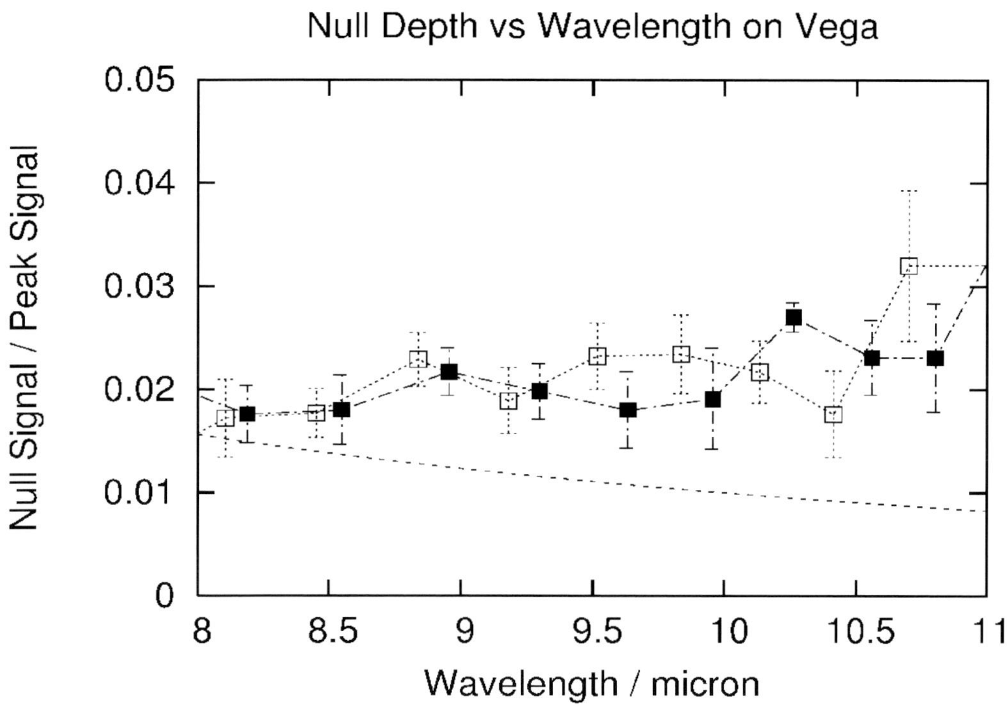

Figure 4. Raw null measurements obtained on Vega.

and easily detectable infrared excesses - to main sequence stars in the 15-30 Jy range. Clearly the main objective of future KIN developments will be to reach the top of the TPF sample region, with stellar fluxes clustered around a few Jy (lower light region of figure 5).

4. Summary

All of the equipment is now in place for deep thermal infrared nulling with the Keck Interferometer. Broadband instrumental nulls of at least 100:1 have already been demonstrated on bright targets of a few tens of Jy.Future efforts will concentrate on achieving similar levels of instrumental nulls on fainter stars. The final goal is to reach a sensitivity of a few Jy, as needed to survey the TPF targets core.

Acknowledgements

The Keck Interferometer is funded by the National Aeronautics and Space Administration (NASA). Part of this work was performed at the Jet Propulsion Laboratory, California Institute of Technology, and at the Michelson Science Center (MSC), under contract with NASA.

References

Beichman, C.A., Woolf, N.J., & Lindensmith, C.A. 1999, The Terrestrial Planet Finder: a NASA Origins Program to Search for Habitable Planets, JPL Publication 99-3

Ciardi, D.R. *et al.* 2001, *ApJ*, 559, 1147

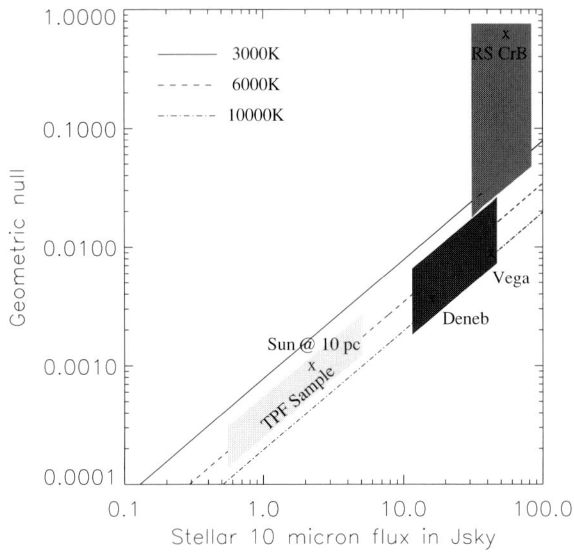

Figure 5. Null as a function of 10 micron flux for a naked star considered as a blackbody at various temperatures.

Creech-Eakman, M.J. *et al.* 2003, SPIE Conf. Series 4841, 330

Léger, A., Mariotti, J.M., Mennesson, B., Ollivier, M. *et al.* 1996, *Icarus*, 123, 249-255

Mennesson, B. *et al.* 2003, in Conf. Towards Other Earths: DARWIN/TPF and the Search for Extrasolar Terrestrial Planets, ed. M. Fridlund & T. Henning, (Noordwijk: ESA-SP-539), 525

Mennesson, B., Léger, A., & Ollivier, M. 2005, *Icarus*, 178, 2, 570

Mennesson, B. *et al.* 2005, *ApJL* in press

Serabyn, E., Colavita M. M., & Beichman C.A. 2000, ASP Conf. Series 196, 357, eds M.L. Sitho, A.L. Sprague & D.K. Lynch

Serabyn, E. & Colavita, M. M. 2001, *Applied Optics*, 40, 1668

Serabyn, E. *et al.* 2004, SPIE Conf. Ser. 5491, 806, ed. W.A. Traub

Serabyn, E. *et al.* 2005, SPIE Conf Series 5905, in press.

Woolf, N.J. & Angel, J.R.P. 1997, ASP Conf. Series 119, 285–292, D. R. Soderblom, ed.

Discussion

Yves Rabbia: What are the next developments expected with KIN?

B. Mennesson: We will concentrate on improving the sensitivity. We plan to do this by lowering our control loop bandwidth on the 10 micron phase. In order to keep the same nulling level, we will need to reduce the K versus N band phase drifts due to atmospheric dispersion and residual differential paths unseen by the present LASER metrology.

Alain Léger: How does your current performance translate in terms of exo-zodi detection capability?

B. Mennesson: In the case of a solar system twin located at 10 pc, only 0.004% of the total 10 micron flux ($\simeq$2 Jy) comes from the exo-zodi. A calibrated null of 100:1 then corresponds to a level of 250 zodis around a solar type star.

Direct Imaging of Exoplanets: Science & Techniques
Proceedings IAU Colloquium No. 200, 2005
C. Aime and F. Vakili, eds.
© 2006 International Astronomical Union
doi:10.1017/S1743921306009379

The prospects of detecting exo-planets with the Ground-based European Nulling Interferometer Experiment (GENIE)

R. den Hartog[1], O. Absil[2], P. Gondoin[1], A. Stankov[1], J.C. Augereau[3], V. Coudé du Forresto[4], D. Mourard[5], M. Fridlund[1], P. Gitton[6], F. Puech[6], R. Wilhelm[6], M. Barrilot[7], and R. Flatscher[8]

[1]Science Payloads and Advanced Concepts Office, ESA, ESTEC, Noordwijk, The Netherlands
email: rdhartog@rssd.esa.int

[2]Institut d'Astrophysique et de Géophysique, Université de Liège, Belgium
[3]Laboratoire d'Astrophysique de l'Observatoire de Grenoble, France
[4]Université de Paris Meudon, France
[5]Observatoire de Côte d'Azur, France
[6]European Southern Observatory, Garching bei München, Germany
[7]Alcatel Alenia Space, Cannes, France
[8]EADS Astrium, Friedrichshafen, Germany

Abstract. The European Space Agency's Darwin and NASA's Terrestrial Planet Finder (TPF) are among the most challenging space science missions ever considered. Their principal objective is to detect Earth-like planets around nearby stars and to characterize their atmospheres. Darwin and TPF-I are currently conceived as nulling interferometers with free-flying telescopes. Within the frame of the Darwin program, the ESA and the European Southern Observatory (ESO), supported by European industries and scientific institutes, have performed two parallel Phase A studies of a ground-based nulling interferometry experiment (GENIE) at the site of ESO's Very Large Telescope Interferometer (VLTI) in Paranal, Chile. GENIE will demonstrate several key technologies required for the Darwin mission. Its science objectives include the detection and characterization of dust disks and low-mass companions around nearby stars. These studies have established detailed instrumental designs, in which GENIE will operate in the L' band around 3.8 microns as a single Bracewell nulling or constructive interferometer, using either two Auxiliary or two Unit Telescopes. The studies were supported by detailed numerical simulations which indicated the possibility of detection and low-resolution spectroscopy in nulling mode of extra-solar giant planets (EGPs) with atmospheric temperatures down to 700 K, provided that a proper calibration of instrumental effects is applied. Detection of circumstellar exo-zodiacal (EZ) dust clouds is possible down to 0.5 mJy, with interesting prospects for the characterization of planet-forming disks.

Keywords. instrumentation: interferometers, techniques: interferometric, planetary systems, protoplanetary disks.

---

1. Introduction

GENIE, the Ground-based European Nulling Interferometry Experiment, is a collaboration between ESO and ESA, and intended for commissioning on ESOs VLTI facility in Paranal (Gondoin *et al.*, 2005). Its objectives are to obtain experience with the design, construction and operation of a nulling interferometer, and as a preparation for the science program of ESA's DARWIN mission (Fridlund and Gondoin, 2003). In this context are foreseen the demonstration of deep nulling with a stellar light suppression more than

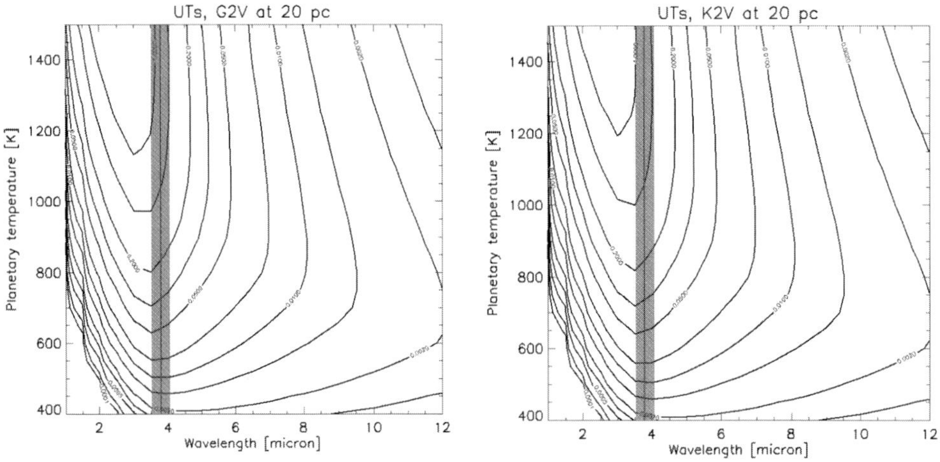

Figure 1. Wavelength trade-off on the basis of (maximum) S/N ratio (obtainable in one second) for the detection of EGPs, for two stellar types. The plots show iso-SNR contours. The SNR peaks in the upper left corner of the diagrams. In particular for lower planetary temperatures, the SNR peaks near the L' band (indicated by the grey band) and then decreases towards lower wavelengths as the star/planet contrast deteriorates, and towards higher wavelengths as the shot noise from the IR background increases.

a factor 1000, and the screening of targets of the Darwin catalogue (Eiroa *et al.*, 2003, Kaltenegger *et al.*, 2005) for the presence of zodiacal emission, and, when possible, the presence of EGPs. EZ clouds with 20 times the density of the solar cloud may seriously hamper the detection of Earth-like planets, and the presence of giant planets in orbits inside the inner habitable zone (HZ) radius may either indicate the absence of Earth-like planets in the HZ, or hamper their detection if present. Paranal provides an excellent vantage point with access to ≈90% of the targets in the Darwin catalogue. An instrument that is capable of detecting faint dust disks or giant planets within a fraction of an arcsec from a star which is 3 to 4 orders of magnitude brighter, is obviously useful to other lines of research as well (Gondoin *et al.*, 2005); GENIE is therefore foreseen to become a general user instrument at Paranal, with a constructive interferometry mode in addition to its nulling capabilities.

During 2004, GENIE was the subject of a parallel Phase A study by two industrial consortia, one led by EADS Astrium Space (Flatscher *et al.*, 2004) and the other by Alcatel Alenia Space (Barillot *et al.*, 2004). Both consortia were asked to select two interferometric concepts, based on VLTI infrastructure and instrumentation specifications, to provide detailed optical and mechanical designs for these concepts, and to assess their performance. Although the instrumental designs forwarded by the two consortia differed in many details, a remarkable agreement was found in their conclusions. The only two feasible concepts were found to be a single Bracewell nulling interferometer, and a split-pupil double Bracewell (i.e. a Keck nuller, see Mennesson *et al.*, 2005, or Serabyn *et al.*, 2004). Both concepts will work with any two UT's or two AT's. The L'-band was selected as the optimal waveband for the detection of both EZ dust and giant exo-planets. An illustration of the wavelength trade-off in terms of signal-to-noise ratios (SNR) for EGP detection is shown in Fig. 1. The performance predictions were also remarkably similar, despite differences in the set-up of the control loops.

During the studies it also became clear that GENIE is depending on a strict realization of the VLTI specifications of its noise environment. A driver for the specifications of the

Table 1. Simulated GENIE nulling performance over 100 sec. observations, for a range of atmospheric conditions.

	worst case	best case
Piston	17 nm @ 20 kHz	6.2 nm @ 13 kHz
Dispersion	17.5 nm @ 200 Hz	4.5 nm @ 300 Hz
Intensity	4% @ 1 kHz	4% @ 1 kHz
Mean total nulling	9.7×10^{-4}	6.2×10^{-4}
Mean instrumental nulling	5.0×10^{-4}	1.5×10^{-4}
RMS instrumental nulling	5.8×10^{-4}	1.6×10^{-4}
Zodi level for SNR=5 (raw)	1800	530
Zodi level for SNR=5 (calibrated)	56	34

Genie control loop for the stabilization of the optical path difference (OPD) between the two beams turned out to be a 750 Hz resonance of the bimorph mirror used in the VLTI adaptive optics system MACAO. Also, the level of differential polarization between two arms, which is currently not known in sufficient detail, is subject to rather stringent requirements. The two consortia have specified detailed tests to resolve this point, which were endorsed by the 123rd meeting of ESA's Astronomical Working Group (AWG), pointing out that the success of the Darwin mission depends critically on a tight control of all polarization effects.

2. Simulated detections

The current analysis of the GENIE science performance is based on computations with the GENIE science simulation software package "GENIEsim" (Absil *et al.*, 2003 and 2005), which has been updated with the results from the two parallel Phase A studies, and the latest information supplied by ESO about VLTI performance.

2.1. *Geniesim*

GENIEsim models the full history of the science photons. The generation in the target system by star or planet is modelled with black bodies, and by the EZ dust cloud through a deprojection of the COBE/DIRBE Interplanetary Cloud Model (Kelsall *et al.*, 1998) with the zodipic package by Kuchner, 2001. The description of the turbulent aspects of the atmosphere are based on the standard Kolmogorov model combined with the Taylor hypothesis and include atmospheric OPD, (water-vapour) dispersion, Strehl, tip/tilt, scintillation and IR background fluctuations. The modelling of the VLTI environment is based on ESO specifications of the MACAO adaptive optics, the PRIMA fringe sensing units, the delay line and, optionally, the IRIS tip/tilt sensing and correction, and includes control loop models as well as additional noise from bimorph-induced OPD and tunnel-induced tip/tilt, plus an IR background contribution from ≈20 mirror surfaces. The GENIE instrument model includes the basic functions of wavefront filtering with a single-mode fiber, nulling with a modified Mach-Zender beam-combiner and detection with a state-of-the-art IR focal-plane array, plus the corrections by closed control-loop systems for OPD, dispersion, intensity mismatch and, optionally, tip/tilt.

2.2. *Control loop performance*

The main output of GENIEsim is a time sequence of detector readings, which can be converted into nulling ratios or SNR's by keeping a detailed administration of the noise

Table 2. Properties of seven short period EGP systems accessible from Paranal.

target	star sp. type	dist. [pc]	M sin$i$ [$M_{\rm jup}$]	planet per. [day]	a [AU]	temp. [K]	$B_{\rm proj}$ [m]	signal to noise ratio in 4 h	in spectral channels 4	8	16	32	64	128
HD 73256	G8 V	36.5	1.87	2.55	0.037	1200	129.5	6.4	5.0	4.1	3.1	2.2	1.4	0.7
HD 83443	K0 V	43.5	0.41	2.99	0.035	1117	128.6	4.4	2.5	1.8	1.3	0.9	0.5	0.3
HD 179949	F8 V	27.0	0.98	3.09	0.045	1383	102.4	4.4	4.4	4.3	4.3	4.1	3.7	2.6
Tau Boo	F7 V	15.0	4.13	3.31	0.047	1388	60.9	3.4	3.4	3.4	3.4	3.4	3.4	3.2
HD 75289	G0 V	28.9	0.42	3.51	0.047	1265	126.7	4.5	4.5	4.4	4.3	4.1	3.5	2.4
HD 162020	K2 V	31.3	13.8	8.42	0.072	784	129.0	3.9	3.1	2.5	1.9	1.4	0.8	0.4
HD 130322	K0 V	30.	1.08	10.7	0.088	724	123.6	3.8	2.7	2.0	1.5	1.0	0.6	0.3

sources. Table 1 shows the expected nulling performance of GENIE during 100 sec. observations, taking into account all turbulence- and VLTI-induced disturbances. The performances of the control loops are summarized together with their repetition frequencies under both pessimistic and optimistic atmospheric conditions (see Absil *et al.*, 2005 for further details). The mean total nulling ratio includes contributions from both the geometric leakage (due to the fact that the stellar disk is resolved by the nulling interferometer) and instrumental leakage (due to disturbance-driven null floor and intensity matching excursions), and is computed as the ratio between the total input and output stellar fluxes (from both telescopes). The next row gives the standard deviation of the instrumental nulling ratio for this 100 sec observation block, a number governing the (im-)possibility of the direct detection of EZ clouds and planets. The last two rows give the density of EZ dust, measured in solar units or zodis, that can be detected to an SNR of 5 around a Sun-like at 20 pc in 1 hour in the full L' band with two UTs. The simulation indicates that with the expected null stability it will be possible to detect EZ light to a useful level only after calibration.

2.3. *Calibration procedure*

The calibration scheme consists of estimation and subtraction of the three main noise sources (IR background, geometrical and instrumental stellar leakage) are estimated and subtracted from the signal, leaving only the planet signal. The IR background can be estimated either with two off-axis fibers that just fit in the field of view adjacent to the on-axis science fiber, or by applying a split-pupil design.

The geometrical leakage is given by $G = (\pi/2)^2(\theta_\star B/\lambda)^2$, and the accuracy of its estimate is limited mainly by the accuracy with which the stellar angular radius $\theta_\star$ can be derived. For the closest stars direct interferometric measurements will provide accuracies $\Delta\theta_\star/\theta_\star < 0.5\%$. For more distant stars such accuracies are possible using empirical relations between (limb-darkening corrected) angular diameters and e.g. effective temperatures (Kervalla *et al.*, 2004). Such methods do require apparent magnitudes to be known better than 0.02 mag, thus calling for an additional photometry programme. In Table 1 a diameter precision of 0.5% is considered best case, and 1% worst case.

The instrumental leakage will be derived from observations of calibrator stars. Calibrator stars should exhibit *a*) negligible emission from dust or planets, *b*) be at least as bright as the target in order to feed the control loops with the same photon flux, *c*) be located in the same part of the sky to ensure similar atmospheric conditions, and *d*) have radii that are either small enough to ignore geometrical leakage, or large enough to allow accurately estimates. Distant late-type giants (typically K III) appear the best

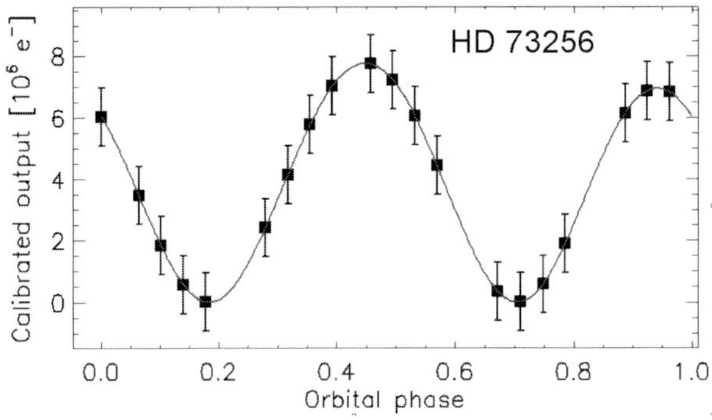

Figure 2. Illustration of the expected results after 20 consecutive nights of observation on HD 73256, assuming a sub-optimal estimate of the stellar diameter ($\Delta\theta_\star/\theta_\star = 1\%$).

candidates. Several catalogues have already been compiled for these type of photometric calibrators (Mérand *et al.*, 2005; Cohen *et al.*, 1999).

3. Detection and spectroscopy of EGP

The detections of seven short-period EGP were simulated and the results are summarized in Table 2. For each target this Table lists the relevant properties of star and planet. The planetary temperatures were derived assuming simple radiation equilibrium, with a planetary Bond albedo of 0.34. All planetary radii were taken to be 1.2 Rjup, the median of 4 transition measurements (cf. Burrows *et al.*, 2004). The simulations assume observations with 2 UT's under standard Paranal atmospheric conditions, a planet crossing a maximum of the transmission map, and a stellar diameter accuracy of 1%. All listed EGP can be detected with a SNR better than 3 in 4 hours of integration. If a series of observations during consecutive nights are combined, as illustrated in Fig. 2, the SNR per observation may be lower and still allow sufficient confidence in the planetary detection. Since the SNR is dominated by the calibration accuracy of the geometrical and instrumental leakage, the L'-band can in most cases be subdivided in several spectral bands without suffering a corresponding decrease in SNR (until the read-out noise or other shot noise terms start to dominate again). Detection and low-resolution spectroscopy ($R \approx 100$) of EGP's with temperatures as low as 700 K ('warm Jupiters') appear feasible.

4. Detection of dust disks and planetary formation

Based on GENIEsim simulations, Fig. 3 shows the expected distribution of levels of EZ emission around Darwin target stars, which are detectable at a SNR of 5 after 1 hour of integration with two UT's, and after application of a calibration scheme as outlined above. The EZ emission is measured along the ordinate in units of Local Zodiacal emission (zodi), and along the abscissa in mJy. The absolute detection limit appears to be roughly 0.5 mJy, and not strongly dependent on stellar type. For cooler stars (which have cooler dust disks, the dust needs to be denser for a detection above the SNR limit. The limits to which EZ dust clouds can be detected in solar units are 30 zodi for F stars, 40 zodi for G stars and 60 zodi for K stars.

Absil *et al.* (2005) have shown that innovative measurements on debris disks are possible using GENIE with the large range of baselines possible with the AT's. Figure 4 shows the nulling ratio (measured as the ratio between nulled and constructive output of the beam combiner) as a function of baseline for various models of the debris disk

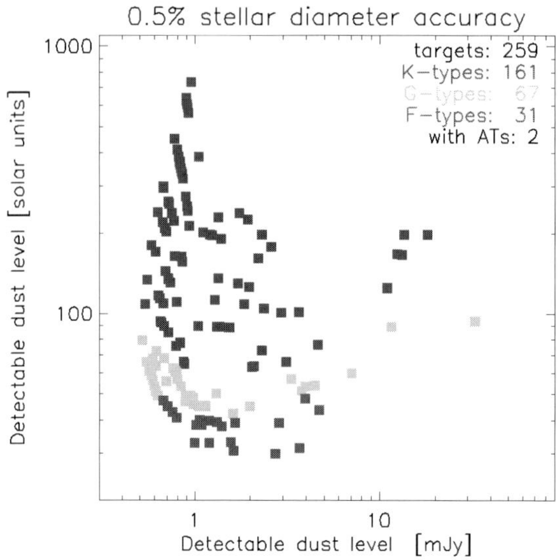

Figure 3. Dust detection limits for Darwin catalogue stars.

around ζ Lep (cf. Augerau *et al.*, 1999). The observations were calibrated using η Lep, a nearby F1V. The nulling ratio is a function of the sublimation radius, the radius inside which dust grains are destroyed. The sublimation temperature of the grains depends on their size and composition. Cold coagulation grains are more porous than silicates, and therefore less efficient in re-emitting thermal radiation. Hence a larger density of porous grains is needed to produce the same mid-IR excess, which results in an enhanced L'-band flux, as is seen in Fig. 4*a*. Smaller grains are more easily destroyed than larger grains, hence for smaller grains the sublimation radius lies further out. The first maximum of the transmission map will coincide with the sublimation radius for smaller baselines, as can be seen in Fig. 4*b*. These figures illustrate that important clues about the structure and composition of planet-forming disks can be obtained with a fairly straightforward set of nulling experiments.

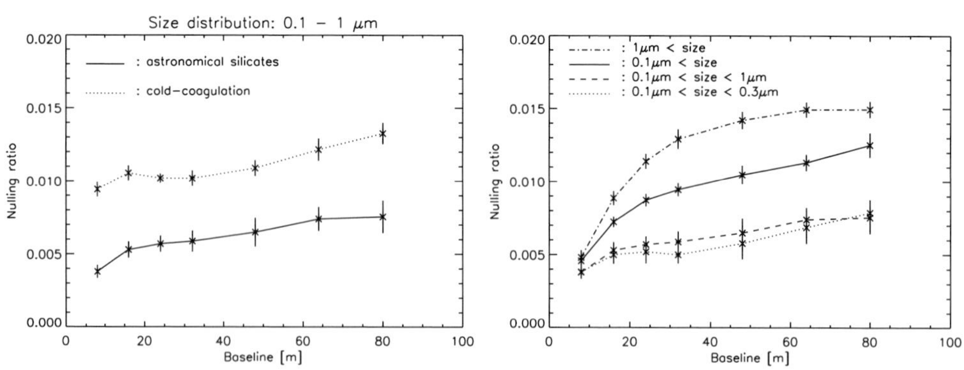

Figure 4. Nulling ratio as a function of baseline for dust disks with, respectively, different grain composition, and different ranges of silicate grain sizes.

References

Absil, O., den Hartog, R., and Gondoin, P., 2003, in: M. Fridlund and T. Henning (eds.) *Proc. of Conf. on "Towards Other Earths: Darwin / TPF and the Search for Extrasolar Terrestrial Planets* (Heidelberg), ESA SP-539, 317

Absil, A., den Hartog, R., *et al.*, 2005, *A&A* in press

Augereau, J.-C., Lagrange, A.-M., *et al.*, 1999, *A&A* 348, 557

Barillot, M., Ruilier, C., *et al.*, 2004, *Alcatel final presentation of GENIE Phase A study to ESA*

Burrows, A., Hubeny, I., Hubbard W. *et al.*, 2004, *ApJ* 610, L53-56

Cohen, M., Walker, R. G., Carter, B., Hammersley, P. L., Kidger, M. R., and Noguchi, K., 1999, *AJ* 117, 1864

Coudé du Foresto, V., Absil, A., Barillot, M., Swain, M., 2005 *These Proceedings*

Eiroa, C., Fridlund, M. and Kaltenegger, L., 2003, in: M. Fridlund and T. Henning (eds.) *Proc. of Conf. on "Towards Other Earths: Darwin / TPF and the Search for Extrasolar Terrestrial Planets* (Heidelberg), ESA SP-539, 403

Flatscher, R., Wallner, O., Ergenzinger, K., *et al.*, 2004, *Astrium final presentation of GENIE Phase A study to ESA*

Fridlund, M., and Gondoin, P., 2003, *Proc. SPIE* 4852, 394

Gondoin, P., den Hartog, R., *et al.* 2005, in: A. Chelli and F. Delplancke (eds.), *Proc. of ESO Workshop on The Power of Optical/IR Interferometry: 2nd Generation VLTI Instrumentation,* (Heidelberg: Springer), in press

Kaltenegger, L., Eiroa, C., Stankov, A. and Fridlund, M., 2005, *These Proceedings*

Kelsall, T., Weiland, J.L., Franz, B.A., *et al.*, 1998, *ApJ* 508, 44

Kervella, P., Thevenin, F., Segransan, D., *et al.*, 2003, *A&A* 404, 1087

Kuchner, M., 2001, *IDL zodipic package,* see http://www.astro.princeton.edu/ mkuchner

Mennesson, B., *et al.*, 2005 *These Proceedings*

Mérand, A., Bordé, P., and Coudé du Foresto, V., 2005, *A&A* 433, 1155

Serabyn, E. *et al.*, 2004, *Proc. SPIE* 5491, 806

All photographs: Laurent Thareau [l.thareau@free.fr].

Direct Imaging of Exoplanets: Science & Techniques
Proceedings IAU Colloquium No. 200, 2005
C. Aime and F. Vakili, eds.
© 2006 International Astronomical Union
doi:10.1017/S1743921306009380

PEGASE: a DARWIN/TPF pathfinder

M. Ollivier[1], J.-M. Le Duigou[2], D. Mourard[3], O. Absil[4], F. Cassaing[5], E. Herwats[4], L. Escarrat[5], B. Chazelas[1], F. Allard[6], R. Clédassou[2], V. Coudé du Foresto[7], M. Delpech[2], P. Duchon[2], P.-Y., Guidotti[2], A. Léger[1], X. Leyre[8], F. Malbet[9], D. Rouan[7] and S. Udry[10]

[1]Institut d'Astrophysique Spatiale (IAS), bâtiment 121, F-91405 Orsay (France), Université Paris-Sud 11 and CNRS (UMR 8617), email: marc.ollivier@ias.u-psud.fr
[2]Centre National d'Etudes Spatiales - TOULOUSE
[3]Observatoire de la Côte d'Azur, Dpt GEMINI
[4]Université de Liège
[5]Office National d'Etudes et de Recherches Aéronautiques
[6]Ecole Normale Supérieure de Lyon
[7]LESIA, Observatoire de Paris
[8]Alcatel Alenia Space, CANNES
[9]Laboratoire d'Astrophysique de l'Observatoire de Grenoble
[10]Observatoire de Genève

Abstract. The space mission PEGASE, proposed to the CNES (Centre National d'Etudes Spatiales = French Space Agency) in the framework of its call for scientific proposals : "formation flying missions", is a 2-aperture interferometer, composed by 3 free flying satellites (2 siderostats and 1 beam combiner), allowing baselines from 50 to 500 m in both nulling and visibility modes. With an angular resolution of a few mas and a spectral resolution of several tens in the spectral range 2.5-5 microns, PEGASE has several goals:
- science : spectroscopy of hot jupiters (Pegasides) and brown dwarves, exploration of the inner part of protoplanetary disks
- technology : validation in real space conditions of formation flying, nulling and visibility interferometry concepts.
PEGASE has been studied at a 0-level. In this paper, we summarize the scientific program and associated technological and mission trade-off coming from this 0-level study. We also discuss how PEGASE can be considered as a TPF/DARWIN pathfinder in an international roadmap towards more complex space interferometry missions such as DARWIN/TPF.

Keywords. Interferometry, Formation flying, Spectroscopy of low mass objects, Protoplanetary disks.

1. Introduction

Both European and American space agencies plan to launch a space interferometer dedicated to the detection and analysis of extrasolar planets and particularly telluric ones, in order to determine the composition of their potential atmosphere and search for evidence of life thanks to gazeous bio-tracers identification (DARWIN and TPF projects, called in the following pages DARWIN/TPF). DARWIN/TPF is based on the nulling interferometry concept (Bracewell 1978), and consists of 3 to 6 free flying telescopes which beams are recombined on a central hub, located at a distance of a few tens of metres from each telescope (Fridlund *et al.* 2000, Beichman *et al.* 1999). The difficulty of such a mission is huge and if we except the SIM mission (astrometric interferometer), no interferometer will ever have flown at DARWIN/TPF launch epoch. In this context, a project of space interferometry precursor has been proposed to the CNES as an answer to its call for ideas on formation flying. This project is called PEGASE and has been studied at a 0-level by the CNES. PEGASE is a space 2 telescope interferometer (two 30 to 40 cm telescopes on the central hub fed by 2 siderostats on separate free flying spacecrafts) allowing baselines from 20 to 500 m. The observation spectral range goes from 2.5 to 5 μm, and the spectroscopic mode should allows spectral resolution of about 60 in the whole spectral range. We present in the next sections the main characteristics of the project.

2. PEGASE: scientific objectives

PEGASE science cases can be addressed in two categories:
- the main program : for which the instrument is particularly designed or suited,
- the secondary program : achievable only if compatible with the choices made during the instrument design phase, or using non classical ways of measurement (using the fringe sensor information for instance).

At the boundary of these two classes, several programs can be considered : the first design of the instrument and the observatory seem to make them achievable, but the demonstration has to be done. This is the case of the exoplanet zodiacal disks which are debris disks responsible for the exo-zodiacal light emission (called later "exo-zodi") study. A back of the envelop calculation of the sensitivity of the instrument lets hope the detection and measurement of exo-zodi until the limit of a few tens solar zodi level (spatially integrated emission of the Solar System zodiacal emission) around solar type stars. However, this point has still to be confirmed and won't be detailed in this paper. In this section we consider only the main program for which specific studies have been done.

The main program is made of two different scientific targets:
- the stellar, substellar and planetary companions,
- the protoplanetary disks.

The observation of stellar, substellar and planetary companions aims at providing a unified theory for the formation of these objects. In the standard model, stars form by gravitational collapse of a gas cloud, while planets form by accretion of a cloud of planetesimals. The question is to know where the limit between these two mechanisms is, and if it corresponds to the the observed "brown dwarf desert" (Marcy *et al.* 2000). Observing stellar companions (and thus gravitationally linked objects) has the advantage of giving an accurate estimation of the mass of the companion (thanks to the observation of the companion trajectory and the three Kepler's laws) and its age (thanks to the spectroscopic study of the main target) compared to free flying object where these parameters have to be estimated using theoretical models. The study of the structure and the atmosphere of all these objects (low mass stars, brown dwarfs and planets) should give a new vision of their formation and evolution. The observation of young systems helps including the time parameter in the description of these objects. Spectroscopy, even at low resolution (typically 60), is sufficient to discriminate the potential role of the clouds and aerosols in the thermalisation of the atmosphere allowing the evolution of present atmospheric models. Compared to the specific method of spectroscopy (or at least spectral band photometry) during the secondary eclipse (when the planets passes behind the star), direct observation by interferometry of the planet at different orbital phases allows observing different parts of the atmosphere and determining its thermal homogeneity (Fig. 1). The scientific programs plans the observation of 20 to 30

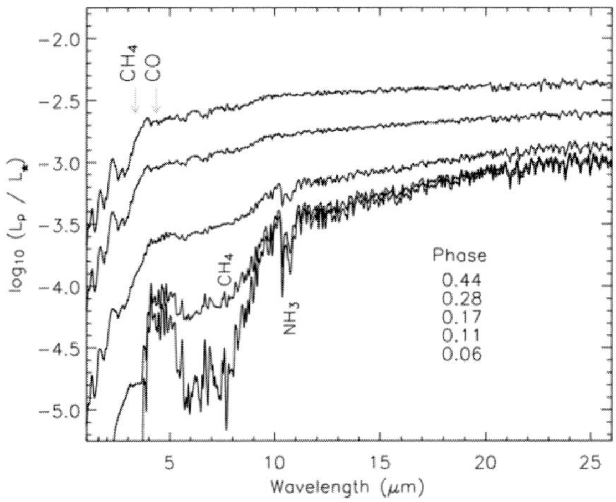

Figure 1. Planet-star flux density ratios for HD209458b at different orbital phase (from Barman *et al.* 2005).

pegasides ("hot jupiters"), and 10 to 20 brown dwarfs and low mass stars, according to their existence and observability.

The observation of protoplanetary disks in visibility mode aim at constraining the models used to describe the objects. Instead of classical visibility measurements as a function of the baseline, we suggest to measure the visibility in spectroscopic mode (as a function of the wavelength). In that observation mode, inhomogeneities in the disks (gaps for instance) appear more distinctly compared to a classical homogeneous disk (Fig. 2). Protoplanetary disks observations require however a strong effort in terms of source modeling.

With a unique observation spectral range from 2.5 to 5 μm, and effective baselines from 50 (maybe less) to 500 m, PEGASE is a unique instrument that has no equivalent on ground, particularly for Pegasides observations and spectroscopy. In addition, PEGASE, if decided, may be the first interferometer in space (SIM launch is not clear at the time these words are written).

3. PEGASE: description of the project

PEGASE is a free flying interferometer made of 3 satellites, 2 simple siderostat satellites with only propulsion, pointing, a metrology capability and a central hub, where the whole formation flying control, the beams combination, the data processing and transmissions are performed (Fig. 3).

The beam combiner allows 2 types of observations on the same instrument: visibility measurements (direct recombination) and nulling interferometry. A concept which allows simultaneously both recombination type is under study at ONERA. The main characteristics of the instrument and the mission, compared to DARWIN/TPF's requirements are summarized on table 1. They mainly come from a trade-off study made by the CNES during the 0-level study.

One of the particularities of the mission, is that some of the requirements are spread over the satellite characteristics and the instrument itself. For instance, the global

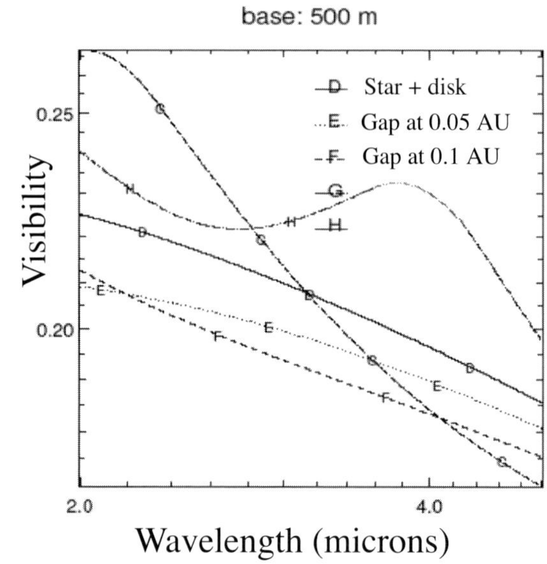

Figure 2. Visibility of a FUOr disk with a gap as the function of the observation wavelength and gap position.

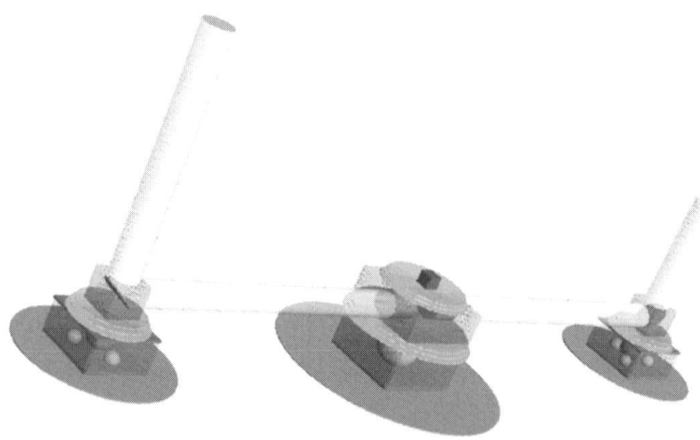

Figure 3. Artist view of the PEGASE observatory (courtesy CNES).

pointing is performed by the satellites themselves while the fine pointing is performed by a moving mirror on the optical path in the instrument. The cophasing of the array is performed by a global positioning of the satellite with an accuracy of 1 cm, and the action of delay lines with a stroke of 1-2 cm. The fact that some of the servo controls of the constellation are performed by the instrument itself requires a strong preliminary work in the laboratory to validate the concept.

4. PEGASE as a DARWIN pathfinder

With different science cases, and as a consequence different optical and technology requirements and reduced cost and complexity, PEGASE cannot be considered as a

Characteritic	PEGASE	DARWIN / TPF
opd control	2.5 nm rms	3 nm rms
nulling ratio	10^4	10^5
APS specification	5 10^{-3} rad	a few 10^{-3} rad
Spectral bandwidth	2.5-5 μm	7-20 μm
Baseline	20-500 m	50 - 500 m
Maximum angular resolution	1 mas (at 2.5 μm)	3 mas (at 7 μm)
Overall transmission (Detector included)	7 %	?
Spacecraft pointing	a few arcsec	?
Fine pointing	20 mas	8 mas
Fine propulsion	Improved cold gas	FEEP
Pupil size	0.3-0.4 m	1.5-3 m
Number of spacecrafts	3	4-8
Optics temperature	100 K $\pm$ 1 K	40 K
Detector temperature	55 K $\pm$ 0.1 K	10 K
Orbit	L2	L2
Launch	1 Soyouz	2 Soyouz

Table 1. Main characteristics of PEGASE's instrument and mission compared to DARWIN/TPF's requirements.

light version of DARWIN, which main goal remains the observation and analysis of telluric extrasolar planets around nearby stars. However, PEGASE can be considered as a useful pathfinder because it will allow validating directly several concepts that cannot be validated from ground (high accuracy delay lines for instance cannot be tested on stellar target because of the atmosphere effects, formation flying...). In the context of a risky, expansive DARWIN concept, a mission comparable to PEGASE or a bit more ambitious project can be a first step towards the spectral characterisation of planetary companions. The keypoints of the trade-off will be :

- the kind of objects that have to be observed (and thus the sensitivity and the design of the mission)
- the spectral range and resolution at which observations are done
- the costs and time schedule of the development and building phases of the project

This point is under study in international space agencies. A collaboration process is also under discussion.

Acknowledgements

The authors would like to acknowledge all the people involved in the project on both scientific side (Co-I of the proposal and all the anonymous small hads) and technical side, mainly at CNES and Onera for the huge work done during the 0-level study in spring 2005.

References

Barman, T.S., Hauschildt, P.H., & Allard, F., 2005, ApJ, 632, 1132-113
Beichman, C.A. *et al.*, 1999, JPL Publication 99-3
Bracewell, R.N., 1978, Nature, 274, 780
Fridlund, M. *et al.*, 2000, ESA-SCI(2000)-12
Marcy, G.W. *et al.*, 2000, Protostars an Planets IV, 1285

All photographs: Laurent Thareau [`l.thareau@free.fr`].

Direct Imaging of Exoplanets: Science and Techniques
Proceedings IAU Colloquium No. 200, 2005
C. Aime and F. Vakili, eds.
© 2006 International Astronomical Union
doi:10.1017/S1743921306009392

Archaeology and direct imaging of exoplanets

John B Campbell[1]

[1]James Cook University of North Queensland, Australia

Abstract. The search for extraterrestrial technology effectively began 45 years ago with Frank Drake's Project Ozma and a radioastronomy start to the search for extraterrestrial intelligence (SETI). Eventually searches began for possible interstellar probes in stable orbits in the Solar System, as well as for infrared excesses from possible Dyson spheres round Sun-like stars. Whilst the Cold War was still underway, some scientists looked for evidence of nuclear waste dumps and nuclear wars elsewhere in the Milky Way. None of this work was carried out by archaeologists, even though by their very nature archaeologists are experts in the detection of ancient technologies. The technologies being searched for would have been partly ancient in age though advanced in techniques and science. The development of ESA's Darwin and NASA's TPF for detection and imaging of Earth-like exoplanets in our galactic neighbourhood represents an opportunity for the testing of techniques for detecting signatures of technological activities. Ideally, both Darwin and TPF might be able to provide spectroscopic data on the chemistry and biochemistry of the atmospheres of Earth-like exoplanets, and thus to detect some of the signs of life. If this can be accomplished successfully, then in theory evidence for pollution and nuclear accidents and wars should be detectable. Some infrared signatures of ETT on or round exoplanets might be detectable. Direct visual imaging of ETT structures will probably not be feasible till we have extremely powerful interstellar telescopes or actually send orbital craft.

Keywords. Search for extraterrestrial technology (SETT), Exoplanets, Archaeology, Pollution, Infrared.

---

1. Archaeology and exoplanets

Archaeology is the scientific study of the physical evidence for human and proto-human behaviour and evolution from the early Palaeolithic of 2.6 My in Africa to modern forensic archaeology and space heritage (Campbell 2004; Gorman 2005). Archaeology could be the study of the evolution, technology, society and impact of any intelligent species or intelligence in the Universe (Campbell 2004). It is limited at any given time by what is feasible.

The scientific search for extraterrestrial intelligence (SETI) began 45 years ago with the work of Frank Drake (see Tarter 2001; Ekers *et al.* 2002). Some SETI research also involved at times a search for other signs of extraterrestrial technology (Shostak 1998; Ekers *et al.* 2002). Having been negatively influenced by the efforts of popular authors like Erich Von Daniken in the 1970s and subsequently to discredit Earth-based ingenuity and to claim that all of Earth's ancient civilizations owe their origins to one form or another of extraterrestrial visitor, most professional archaeologists have avoided widening their associations with the space sciences, other than satellite imaging of Earth and studies in archaeoastronomy, namely studies of ancient or traditional naked-eye observations of the sky as recorded in texts, represented in monuments, represented in rock art or recorded by ethnographers, archaeologists or astronomers working with living Indigenous peoples (see Ruggles and Campbell In Prep.). I prefer the acronym SETT (search for extraterrestrial technology) to SETA (search for extraterrestrial artefacts) as the latter term has been fully adopted by those like Von Daniken who believe that alien artifacts and influences are present nearly everywhere on Earth.

For radioastronomers working in SETI, the intelligence of a technologically capable species elsewhere in the Milky Way is basically indicated by the ability to control radio waves and/or to fire pulsed laser beams. However, defining intelligence in a broader way is also problematic. The social and behavioural science disciplines which look at intelligence in humans have never come to a full agreement on how to define intelligence. Further, as scientists we can not agree on whether certain other Primates (in particular chimpanzees and bonobos), certain species of Cetaceans (especially certain dolphin species), some birds (e.g. New Caledonian crows which invent and construct tools) and some Molluscs (particularly cephalopods, e.g. laboratory-confined octopus which open locks, steal crabs and then hide the evidence of what they have done) are intelligent. The relevance of detecting and understanding intelligence in other Earth-bound species is enormous. If we can not manage it here, how will we be able to recognize and understand a signal from ETI, other than saying we have found a carrier beam. The issues of communication and understanding in interstellar message construction are being addressed by workshops and conferences organized by Doug Vakoch of the SETI Institute (http://www.seti.org).

Now that more than 155 exoplanets (primarily gas giants) have been detected, and the technology is being developed to detect and image smaller Earth-like exoplanets (ESA's Darwin, NASA's TPF etc), it is time for archaeology to reconsider its inactive roles in astronomy and astrobiology (on the latter, see Chyba and Hand 2005).

Thus far the search for astroengineering structures like Dyson spheres and Dyson rings has been unsuccessful (Shostak 1998; Ekers *et al.* 2002). Equally, the search for extraterrestrial probes which might have been positioned at certain Lagrange points or placed in special orbits in the Solar System has been unsuccessful (with the possible exception of the object catalogued as 1991 VG and in heliocentric orbit near Earth, see Steel 1995). Does this then automatically mean that the answer to Enrico Fermi's paradox ("Where is everybody?") is zero, or rather n = 1, the "1" being us, the only intelligent species in the Milky Way, as Ward and Brownlee (2000) and Morris (2003) would have it. Fermi's paradox is more of a puzzle than a paradox, but in any event I do not think we can answer it fully as yet. It is also particularly anthropocentric. For what are supposedly "fifty solutions" to it, see Webb (2002) and for critiques of the reasoning behind Fermi's Paradox, as well as behind Earth-centred astrobiology generally, see Cohen and Stewart (2004). I would argue that our instruments and search strategies have been inadequate, and that they have not allowed for a sufficiently wide range of options or possibilities. Further, many of the earlier searches for extrasolar technology were done during the paranoia of the Cold War and before the current observations and hypotheses on the Universe had been developed (e.g. see Gott 2002). The various sorts of interference, distortion and outright blockage of line of sight detection which are now conceivable are much greater than 20 or 30 years ago. Have there in fact been no signs of extrasolar nuclear accidents, nuclear wars or stellar dumping of nuclear waste in the Milky Way, as it would seem, or have we missed the evidence? I think one can argue for the latter, that is, the evidence could exist but so far we have been unsuccessful in detecting it. It might also still be impossible or highly unlikely for us to be able to detect or recognize a highly advanced civilization (a view espoused early on for example by Freitas 1983; on the so-called "Interdict Hypothesis", see Fogg 1987).

The situation could change with the development of ESA's Darwin, NASA's TPF (Terrestrial Planet Finder) and other space-based telescopes which might be designed, launched and parked at L2 in the Earth-Sun system or elsewhere. If the chemistry and biochemistry of Earth-like exoplanets possibly in orbit round some 492 stars within 25 pc can be detected and analysed (Kaltenegger *et al.* this volume, a), then in theory we could detect some signs of possible technologies (pollution, "greenhouse" effects, infrared excess, variable ozone levels, nuclear accidents or wars etc). On Earth we can distinguish between natural and artificial pollution, although there is an overlap in the characteristics. Our own impact or influence on the creation of a "greenhouse" event is becoming more clearly established, though there are ongoing debates on what is evidence for a natural climate cycle and what is evidence for human interference with nature. The mass rate of extinctions has gone up so much, and is continuing to climb, that future palaeontologists looking back on the present time will possibly interpret it as equal to the K/T boundary event (the Cretaceous/Tertiary boundary extinction of the dinosaurs and many other forms of life at about 65 My). There might eventually be an argument for shifting the boundary of the Holocene/Pleistocene to the present, as the total extinctions will vastly outnumber the extinctions which occurred at the end of the Pleistocene as currently defined (11 Ky).

From archaeological and SETT points of view the development of instruments and software for the direct imaging of exoplanets should allow for detection and analysis of chemical, biochemical and radioactive traces of technological activity (cf. Kaltenegger *et al.* this volume, b), Both Earth-like planets and other exoplanets and portions of targeted star systems should be surveyed. Traces of technological activity (e.g. massive mining) could occur either on Earth-like planets or on other bodies such as asteroids and rocky moons. Massive use of ice and water might also be detected, as might variations of the concept of Dyson spheres and rings. We should also search beyond the galactic habitable zone (cf. Lineweaver *et al.* 2004; Campbell 2004). Eventually, we might investigate some of the neighbouring star systems literally directly with multigeneration spacecraft (Kondo *et al.* 2003).

References

Campbell, J.B. (2004) The potential for archaeology within and beyond the habitable zones of the Milky Way. In Norris, R. & Stootman, F. (eds), Bioastronomy 2002: Life among the Stars. International Astronomical Union Symposium 213. San Francisco: Astronomical Society of the Pacific

Chyba, C.F. & Hand, K.P. (2005) Astrobiology: the study of the living universe. Annual Review of Astronomy and Astrophysics 43: 31-74

Cohen, J. & Stewart, I. (2004) What Does a Martian Look Like? The Science of Extraterrestrial Life. London: Ebury Press/Random House

Ekers, R., Cullers, D.K., Billingham, J. & Scheffer, L.K. (eds) (2002) SETI 2020: a Roadmap for the Search for Extraterrestrial Intelligence. Mountain View: SETI Press

Fogg, M.J. (1987) Temporal aspects of the interaction among the first galactic civilizations: the "Interdict Hypothesis". Icarus 69: 370-384

Freitas, R.A. (1983) Extraterrestrial intelligence in the Solar System: resolving the Fermi Paradox. Journal of the British Interplanetary Society 36: 496-500

Gorman, A. (2005) The cultural landscape of interplanetary space. Journal of Social Archaeology 5: 85-107

Gott, J.R. (2002) Time Travel in Einstein's Universe: the Physical Possibilities of Travel through Time. London: Phoenix/Orion Books

Kaltenegger, L., Eiroa, C., Stankov, A. & Fridlund, M. (this volume, a) The Darwin star catalogue: nearby habitable star systems. In Aime, C., Carbillet, M. & Vakili, F. (eds), Direct Imaging of Exoplanets: Proceedings of IAUC200. Cambridge: Cambridge University Press

Kaltenegger, L., Karlsson, A., D'Arcio, L., denHartog, R. & Fridlund, M. (this volume, b) Characteristics of new proposed 3 and 4 telescope configurations for Darwin and TPF-I. In Aime, C., Carbillet, M. & Vakili, F. (eds), Direct Imaging of Exoplanets: Proceedings of IAUC200. Cambridge: Cambridge University Press

Kondo, Y., Bruhweiler, F.C., Moore, J. & Sheffield, C. (eds) (2003) Interstellar Travel and Multi-generation Space Ships. Detroit: Apogee Books

Lineweaver, C.H., Fenner, Y. & Gibson, B.K. (2004) The galactic habitable zone and the age distribution of complex life in the Milky Way. Science 303: 59-62

Morris, S.C. (2003) Life's Solution: Inevitable Humans in a Lonely Universe. Cambridge: Cambridge University Press

Ruggles, C. & Campbell, J.B. (eds) (In Prep) The Heavens Above: from Archaeoastronomy to Space Heritage. London: UCL Press

Shostak, S. (1998) Sharing the Universe: the Quest for Extraterrestrial Life. Sydney: Lansdowne

Steel, D. (1995) SETA and 1991 VG. The Observatory 115: 78-83

Tarter, J. (2001) The search for extraterrestrial intelligence (SETI). Annual Review of Astronomy and Astrophysics 39: 511-548

Ward, P.D. & Brownlee, D. (2000) Rare Earth: Why Complex Life Is Uncommon in the Universe. New York: Copernicus

Webb, S. (2002) If the Universe Is Teeming with Aliens... Where Is Everybody? Fifty Solutions to the Fermi Paradox and the Problem of Extraterrestrial Life. New York: Praxis/Copernicus

Direct Imaging of Exoplanets: Science & Techniques
Proceedings IAU Colloquium No. 200, 2005
C. Aime and F. Vakili, eds.
© 2006 International Astronomical Union
doi:10.1017/S1743921306009409

Study of the polarization effects in a Nulling interferometer: Design consequences

Bruno Chazelas[1], Claude Valette[1], Yves Rabbia[2], Frank Brachet[1], Jean Gay[2], Alain Labque[1], Alain Lger[1], Yuying Longval[1], and Marc Ollivier[1]

[1]Institut d'Astrophysique Spatiale, Orsay, France
email: bruno.chazelas@ias.u-psud.fr

[2]Observatoire de la cte d'Azur, Dpartement Gmini, France

Abstract. Nulling interferometry is one of the promising techniques for the study of extra terrestrial planets. This technique will be applied in the future space missions Darwin and TPF-I, and from the ground with GENIE. The nulling interferometry techniques require high symmetry of the interfering beams, to obtain the required contrast (typically 10^6 to detect terrestrial exo-planets in the thermal infrared). In this paper we consider the polarization symmetry issue, such as polarization rotation and polarization phase shifts occurring on slightly misaligned optics. We study the consequences of these symmetry requirements on a nulling interferometer design. We find the relation between the misalignment tolerances and the achievable nulling, and we show that this tolerance is highly dependent on the interferometer configuration (the way beams turn right, left, up or down in the interferometer arms). It is typically of the order of the arcminute (not the arcsecond) for a 10^6 contrast. We present a analytical and numerical analyses.

Keywords. Interferometers, Nulling, Darwin.

1. Introduction

Nulling interferometry, is a technique aiming at interferometric coronography. The challenge is to reach the highest and most stable interferometric contrast. The application of such a technique is the search for extra-solar planet: space mission Darwin/TPF-I and Pegase instruments, and ground based instruments such as GENIE. The required contrast to detect extra-solar earths, is 10^6. (baseline requirement for the Darwin mission). This will be the contrast level targeted by this study. The performances of such an interferometer are called nulling ratio N: normalized value of the light leakage of the instrument. Nulling interferometry require a high symmetry in the beam properties of the different interferometer arms. The requirements for polarization symmetry have been described M. Ollivier 1999. The polarization ϵ_θ rotation should be

$$\epsilon_\theta \leqslant 3.10^{-2} rad \tag{1.1}$$

The phase shift difference $\Delta\varphi_{sp}$ between s and p component should be less than:

$$\Delta\varphi_{sp} \leqslant 2.10^{-3} rad \tag{1.2}$$

The mean phase difference between the two beam $\Delta\varphi$ should be:

$$\Delta\varphi \leqslant 2.10^{-3} rad \tag{1.3}$$

These requirement are well established, but it is necessary to study the source of the polarization mismatch in an interferometer. If we ignore the observed object properties

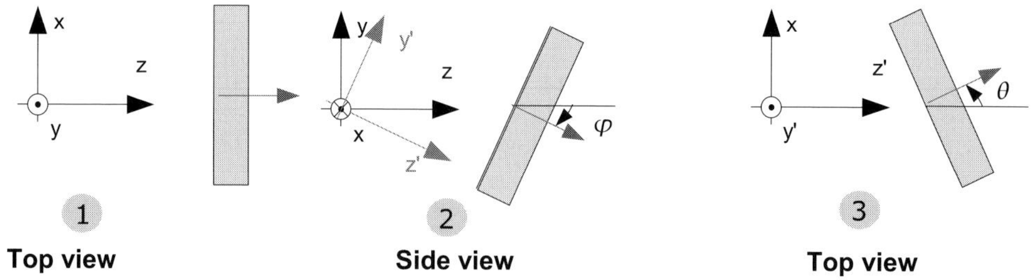

Figure 1. Convention describing the successive rotations to obtain the 3D orientation of a surface.

see]Elias2004, Polarization mismatches come from the reflection and transmission effects in the arms of the interferometer. Traub, 1988 studied some design rules for standard stellar interferometer. Elias *et al.* 2004 also studied the instrumental polarization effects, in particular the misalignment consequences, but the study was independent from the interferometer design (Monte-Carlo simulation on arbitrary optical trains). We will show that polarization mismatches are very dependent from the interferometer design. We have built a numerical model predicting the alignment requirement for a given optical assembly and a nulling requirement.

For the whole study we used Jones formalism see. The electrical field is represented by a 2 component complex vector. It is to be noted that for the whole paper the sign convention for electrical field and the Fresnel equations are the one by Gay, Rabbia 1994.

$$E = \begin{bmatrix} E_x \\ E_y \end{bmatrix} \tag{1.4}$$

In a two beam interferometer the nulling achieved is then:

$$N = \left| \frac{E_1 + E_2}{2E_1} \right|^2 \tag{1.5}$$

2. Polarization effects

The aim of the present work is to determine the alignment requirement of a nulling interferometer, as a function of the nulling requirement and the optical setup design. To describe the misalignment of optics we use the conventions represented in figure 1. We suppose in the whole paper that the *entrance* beam of the interferometer is horizontal, and propagates along the $z > 0$ direction.

2.1. *Effect of the θ rotation*

A rotation around the vertical (y) axis has a quite simple effect (provided the impinging beam is horizontal): It changes the incidence angle. Therefore the reflection or transmission complex coefficients vary. The Fresnel coefficient are function of the incidence angle. Thus the nulling contribution of such a rotation is linked to the Fresnel coefficient variation. It is proportional the square of the alignment error $d\theta^2$.

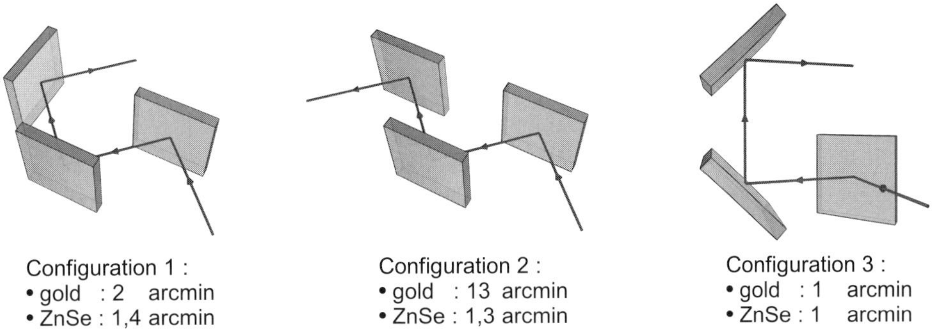

Figure 2. Alignment tolerance for 3 mirrors hypothetic interferometers. We have supposed that both arms of the interferometer are composed of mirrors arranged as shown. The tolerance are calculated for gold coating or bare znse mirrors, at $6\mu m$ Palik, 1985, Tropf, 1995.

2.2. *Effect of the φ rotation*

The φ rotation is more difficult to express. It combines a variation of incidence and a polarization rotation. Around 45 of incidence the incidence variation induced by the φ rotation is of the second order. The major effect is then polarization rotation. For one mirror the polarization rotation is proportional to the φ rotation angle.

3. optical trains modeling

We want to study the effects of misalignment on polarization nulling contribution. Therefore we will suppose that at the output beams of the interferometer are perfectly aligned. That way we separate polarization nulling contribution and tilt nulling contribution.

To perform computations, we build polarization analysis numerical model based on a 3D extension of the Jones Formalism. This program was checked with the commercial simulation software ZEMAX. In simple cases it has been checked with analytical calculations.

4. Results

Most of the numerical result presented here were obtained with Monte-Carlo simulation. Thus the tolerances given are expressing a probability to obtain a certain nulling contribution. The here we give tolerances that ensure to have a 80% of a nulling contribution smaller than 10^{-6}

4.1. *Tolerance variations with configurations*

We have found that the misalignment tolerances are greatly dependent of the mirror configuration in the interferometer. We give the tolerances for 3 interferometer configurations figure 2, with the same total number of mirrors. The tolerance variation can be explained by the geometry. In the configuration 1 the polarization rotation introduced by the 2 first mirrors are compensated by the last mirror. In the two other configurations the last mirror worsen the polarization rotation.

We find that there are two kinds of setup: auto-compensating setups, and setups that add polarization rotations. The rule proposed by Traub, 1988 is good to find out if the setup is auto-compensating or not, at least in plane interferometer setups. In one arm the beam turns should go always in the same direction.

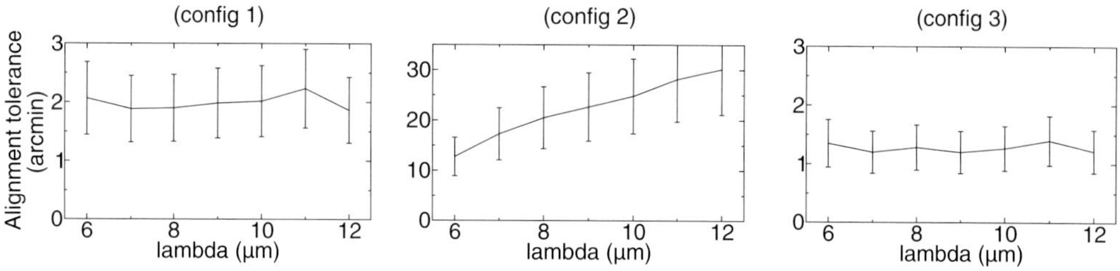

Figure 3. Alignment tolerance versus λ for gold coated mirrors and ZnSe dielectric plates Palik, 1985, Tropf, 1995 for the configuration of figure 2.

4.2. *Materials*

The material on which transmission or reflection occurs have an influence on the alignment tolerances see figure 2. The effect is more sensitive with auto-compensating designs.

4.3. *Wavelength dependence*

The Fresnel coefficients are function of the wavelength. Thus the polarization affect are partly wavelength dependant (see figure 3). This is however untrue for geometrical polarization effects. Only auto-compensating setups have tolerances that vary with wavelength.

5. Conclusion

In this poster we have shown the basic effects of misalignment on the polarization of the beams in a nulling interferometer and the variation of these effects with several parameters. The typical tolerance is around the arcminute for a nulling of 10^{-6}. For the rare autocompensating setups this tolerance can be relaxed of a factor 10.

Acknowledgements

Bruno Chazelas is grateful to *Alcatel Alenia Space* and *CNRS* for funding his grant. Frank Brachet is grateful to *Alcatel Alenia Space* and *CNRS* for funding his grant.

References

R. M. A. Azam and N. M. Bashara 1997, *Ellipsometry and Polarized Light. North Holland Publishing Company*

J. Gay, and Y. Rabbia 1994, *LAMP: a concept for the ESO VLTI fringe sensor*. Second Conference on Ampitude and Intensity Spatial Inteferometry, 195

N. M. Elias, D. W. Draper, and M. C. Noecker 2004, *Polarization and planet-detecting nulling interferometers. In Free-Space Laser Communications IV. Edited by Ricklin, Jennifer C.; Voelz, David G. Proceedings of the SPIE* Volume 5555, pp. 215-230 (2004)., pages 215230, November 2004

M. Ollivier, PhD thesis 1999, *Contribution à la recherche dexoplanètes: Cornographie interférentielle pour la mission Darwin*, University Paris XI

Palik 1985 *Handbook of optical constants of Solids*, Academic Press

W. A. Traub, 1988 *Polarization Effects in Stellar Interferometers. In F. Merkle, ed., NOAO-ESO Conference on High-Resolution Imaging by Interferometry: Ground-Based Interferometry at Visible and Infrared Wavelengths*, volume 29, p. 1029, 1988

W. J. Tropf, 1995 *Optical Engineering*, 34:13691373, May 1995

Direct Imaging of Exoplanets: Science and Techniques
Proceedings IAU Colloquium No. 200, 2005
C. Aime, F. Vakili, eds.
© 2006 International Astronomical Union
doi:10.1017/S1743921306009410

Characteristics of proposed 3 and 4 telescope configurations for Darwin and TPF-I

Kaltenegger L.,[1,2]† and Fridlund M.[2]

[1]Harvard-Smithsonian Center for Astrophysics, 60 Garden St, Cambridge, MA 02138, USA
email: lkaltenegger@cfa.harvard.edu

[2]European Space Agency ESTEC, P.O.Box 299, NL-2200AG Noordwijk, The Netherlands
email:Malcolm.Fridlund@esa.int

Abstract. The Darwin and TPF-I missions are Infrared free flying interferometer missions based on nulling interferometry. Their main objective is to detect and characterize other Earth-like planets, analyze the composition of their atmospheres and their capability to sustain life, as we know it. Darwin and TPF-I are currently in study phase. A number of mission architectures of 3 and 4 free flying telescopes are evaluated on the basis of the interferometer's response, ability to distinguish multiple planet signatures and starlight rejection capabilities. The characteristics of the new configurations are compared also to the former, more complex Bowtie baseline architectures as well as evaluated on base of their science capability.

Keywords. astrobiology, extrasolar planets, Earth, interferometers, nulling.

1. Introduction

The closing years of the 20th century have allowed us, for the first time, to seriously discuss interferometric instruments deployed in space achieving unprecedented spatial resolution. And thus the direct detection of Earth-like exoplanets orbiting nearby stars, and the search for key tracers of life in their atmospheres, are high-priority objectives in the long-term science plan of ESA as well as NASA. Both Agencies are working on the definition of the instruments that will meet this challenge, Darwin and Terrestrial Planet Finder (TPF) respectively with a foreseen launch in 2015+. The baseline mission duration is 5 years, extendable to 10 years in an L2 orbit. Operating in the infrared band requires that all optical components are cooled to roughly 40 K, this is achieved by passive cooling. Only the detector requires active cooling. Darwin is a major element in the Cosmic Vision 2020 program of the European Space Agency. It has the explicit purpose of detecting other Earth-like worlds, analyze their characteristics, determine the composition of their atmospheres and investigate their capability to sustain life as we know it. The Darwin mission is envisioned as four free flying spacecraft including one beam-combining spacecraft. The beam combiner and the telescope spacecraft fly in one plane with each telescope spacecraft at the same distance from the beam combiner. The resolution of the interferometer is adjusted by changing the distance between the telescope spacecrafts. A similar activity has been taking place in the United States within the context of NASA's Origins program. A science collaboration has already been established. The missions are currently in definition phase, here some characteristics of the designs are reviewed. We present the different new configuration architectures under investigation.

† Present address:Harvard-Smithsonian Center for Astrophysics, MS20, 60 Garden Street, Cambridge, MA 02138, USA

2. Nulling Interferometry

The main problem in the direct detection of an exo-planet of a size comparable to our own Earth and located at a similar distance from its own star is contrast and dynamic range. A central star like our Sun (G2V) outshines an Earth-like planet in the visual wavelength range by a factor of at least 10^9. Going to the mid-infrared alleviates this problem, because the planet's thermal emission peaks at 10μm. Even at these wavelengths, the contrast is more than a factor of 10^6. Analysis of the planetary light requires that the stellar light is suppressed to a high degree. In the IR this is done by a technique called nulling interferometry, in essence this means that achromatic phase shifts are applied to the beams collected by individual telescopes before recombination such that the on-axis light, i.e. stellar light, is cancelled by destructive interference, while the much weaker planetary light emitted at a certain off-axis angle interferes constructively. By keeping a star in the center of the image plane, coronography is realized without the presence of a physical mask. The output of the system can be described by an angular transmission map (TM) featuring interference fringes, with a sharp null (destructive interfered area) in the center of the map. The TM gives the interferometer's intensity response as a function of the sky coordinates. The modulation map (MM) is the difference between the recorded outputs on detector A and detector B. The stellar signal is nulled out only on the optical axis. A leakage of photons out of the central null exists because the star has a finite photospheric disk. That leakage is a very important noise source. The actual shape and transmission properties of the pattern are a function of the number of telescopes, configuration, and the distance between the telescopes (see e.g, Absil (2001),Kaltenegger (2004)). Information about the distribution of planets in the target star system can be recovered by modulation of that signal. Figure 1 shows the TM, MM and signal modulation for a simulated Solar System (Earth, Venus, Mars, Jupiter shown) for different configurations. No spatial information is extracted in a single exposure. Rotation modulates the interferometer output intensity as a planet passes in and out of the dark fringes. From the intensity and actual pattern of this modulation one can derive the planet's parameters.

2.1. *Architecture design*

The original Darwin mission concept was optimized for high stellar rejection to focus on observing the closest stars for planetary companions. This led to a baseline concept of a free flying configuration with 6 collector telescopes (Bowtie configuration) of 1.5 m diameter and a central beam combiner (see e.g. Absil (2001) for an overview). Optimizing the mission for nearby stars is not the same as optimizing the mission performance for its overall star sample of a minimum of 165 stars out of the DARWIN target catalogue Kaltenegger, Eiroa, Stankov *et al.* (2005). Recent analyses show that the starlight rejection criteria can be relaxed while still maintaining a target sample of 165 stars. The integration time for nearby stars increases, but the overall mission performance does not degrade significantly. A number of these alternative mission architectures have been evaluated on the basis of interferometer response, achievable modulation efficiency, number of telescopes and starlight rejection capabilities (see Kaltenegger & Karlsson (2004) and Lay & Dubovitsky (2004)). Accordingly, candidate Darwin and TPFI configurations use three (Three telescope Nuller (TTN) see Karlsson, Wallner & Perdigues Armengol (2003)) or four (X-array, see Lay & Dubovitsky (2004) telescopes of 3.5 m diameter, reducing complexity and cost of the mission. Three telescopes is the minimum number of telescopes needed for an interferometer mission that uses rapid signal modulation to detect a planet in the high background noise. An array with 6 telescopes like the Bowtie achieves better performance in starlight rejection than an array with 3 or 4 telescopes, but adds complexity and cost to the mission. Furthermore the maximum modulation efficiency is 100%, 93% while only 70% for the X-array, TTN and Bowtie, repectively. The mean modulation efficiency is 26%, 32% and only 16% for the X-array, TTN and Bowtie, repectively. These numbers show a performance increase for the 3 and 4 telescope architectures in comparison to the Bowtie. Modulation efficiency is a measure of the efficiency with which planet photons are converted into output signal. The maximum modulation efficiency sets the performance for the mission's spectroscopy phase, the mean modulation the efficiency for the search phase. The higher the modulation efficiency, the better the overall performance of the architecture.

The output beam comprises all information from a specific set of sources on the sky, as well as the background. In an ideal scenario with only a single planet around its host star and no other disturbing sources, such as extrasolar zodiacal dust in the target system, the detection of a positive flux would imply that a planet is present, if the star is well and truly 'nulled out'. In real observations, several factors affect the signal to noise in a detrimental way. This has been analyzed by different groups e.g. Lay & Dubovitsky (2004) and Kaltenegger & Karlsson (2004). Figure 1 shows that it is essential to apply spectral channels to distinguish the signal from multiple planets due to their distinct signature as a function of distance from their star, wavelength and rotation of the array (shown here, Venus, Earth, Mars and Jupiter). The simulations of our own Solar system in Figure 1 show that Jupiter's signal is seen at higher frequency over a full rotation of the array than an Earth-like planet. Thus a Jupiter will be easily distinguished. Our simulations show that Venus and Earth are the two planets in our own solar system that are most likely to be confused. Here the different architectures show their capability to disentangle multiple planetary signals. As seen in Figure 1 the rectangular TTN and the X-array perform the best in this respect. Note that these calculations do not take noise into account. By increasing the distance between the telescopes any modulation pattern can be made finer, but that generally increases the stellar leakage. The X-array concept has an independent nulling and imaging baseline (the imaging baseline controls the resolution of the modulation map). It is the distance between the two sub-nulling Interferometers in the X-array concept. Preliminary investigations show that the 4 telescope array (X-array) has a better performance with respect to distinguishing multiple planets due to a finer grid in its modulation pattern Lay & Dubovitsky (2004). The TTN design is a concept that combines the nulling and imaging baseline. Increasing the imaging baseline to gain resolution thus also increases the nulling baseline what decreases the extent of the central null as the whole modulation map scales and thus increases the stellar leakage. Recent simulations by Velusamy & Marsh (2005) and Thiebaut & Mugnier (2005) argue that high spatial resolution is not needed to disentangle multiple planets, avoiding the difference in performance due to leakage increase for the TTN at larger baselines. These results would indicate similar performance of the three and four telescope architectures.

3. Conclusions

A number of mission architectures of 3 and 4 free-flying telescopes are evaluated on the basis of the interferometer's response, modulation efficiency, ability to distinguish multiple planet signatures and starlight rejection capabilities. Figure 1 shows the characteristics of the alternative mission architectures. Even though the starlight rejection properties of the Bowtie configuration is outstanding in the comparison, the higher mean modulation efficiency of the TTN configuration and the X-array counteracts that superiority. An additional factor in mission design and complexity is the number of telescopes

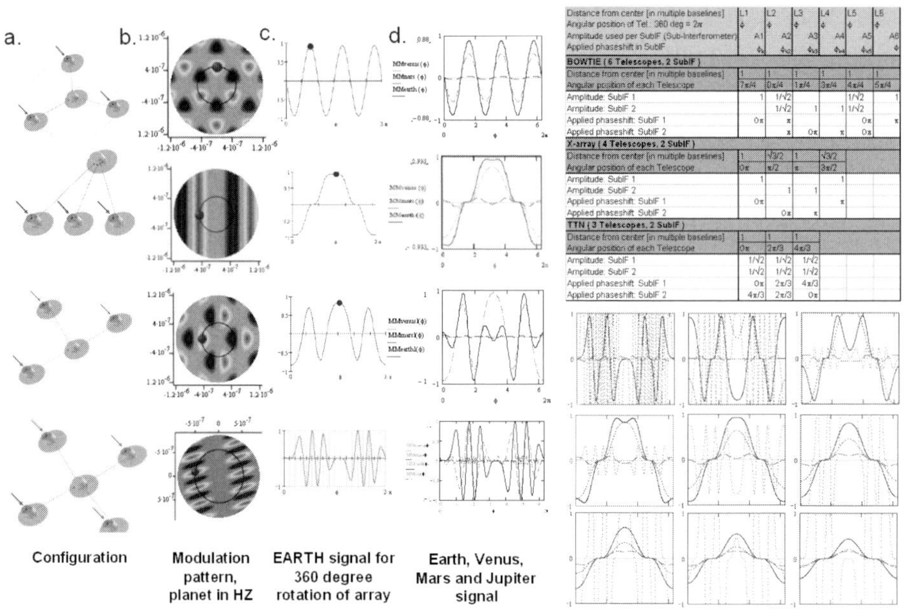

Distance from center [in multiple baselines]	L1	L2	L3	L4	L5	L6
Angular position of Tel.: 360 deg = 2π	ϕ	ϕ	ϕ	ϕ	ϕ	ϕ
Amplitude used per SubIF (Sub-Interferometer)	A1	A2	A3	A4	A5	A6
Applied phaseshift in SubIF	ϕ_{s1}	ϕ_{s2}	ϕ_{s3}	ϕ_{s4}	ϕ_{s5}	ϕ
BOWTIE (6 Telescopes, 2 SubIF)						
Distance from center [in multiple baselines]	1	1	1	1	1	1
Angular position of each Telescope	7π/4	0π/4	1π/4	3π/4	4π/4	5π/4
Amplitude: SubIF 1	1	1/√2			1/√2	1
Amplitude: SubIF 2		1/√2	1	1	1/√2	
Applied phaseshift: SubIF 1	0π	π			0π	π
Applied phaseshift: SubIF 2		π	0π	π	0π	
X-array (4 Telescopes, 2 SubIF)						
Distance from center [in multiple baselines]	1	√3/2	1	√3/2		
Angular position of each Telescope	0π	π/2	π	3π/2		
Amplitude: SubIF 1	1			1		
Amplitude: SubIF 2		1	1			
Applied phaseshift: SubIF 1	0π			π		
Applied phaseshift: SubIF 2		0π	π			
TTN (3 Telescopes, 2 SubIF)						
Distance from center [in multiple baselines]	1	1	1			
Angular position of each Telescope	0π	2π/3	4π/3			
Amplitude: SubIF 1	1/√2	1/√2	1/√2			
Amplitude: SubIF 2	1/√2	1/√2	1/√2			
Applied phaseshift: SubIF 1	0π	2π/3	4π/3			
Applied phaseshift: SubIF 2	4π/3	2π/3	0π			

Figure 1. a) Telescope Array Configuration b) Modulation pattern generated by the array c) modulation of the signal of an Earth-like planet at 1AU over a full 360 degree rotation of the array. d) Modulation of the signal from multiple planets (simulation of our Solar System, planets shown: Venus, Earth, Mars and Jupiter). Different frequency of the signal from different planets allows to disentangle the signal from multiple planets. Configurations shown: upper panel to lower panel: 1) equilateral triangle, 2) linear, and 3) rectangular triangle Three Telescope Nuller (TTN) and 4) X array. Right upper panel: Table with architecture characteristics. Right lower panel: The signal of multiple planets in our solar system (Venus, Earth, Mars and Jupiter) recorded in different spectral channels over a whole rotation of the array for the linear TTN configuration (assuming no noise). The high frequency signal shown is Jupiter's. The panels show the monochromatic signal per channel from 4μm (left) to 20μm (right) in steps of 2μm.

used. Here the Bowtie has a big disadvantage, as it needs 6 telescopes. Figure 1 shows that it is essential to apply spectral channels to distinguish the signal from multiple planets due to their distinct signature as a function of distance from their star, wavelength and rotation of the array (shown here, Venus, Earth, Mars and Jupiter). Here the different architectures show their capability to disentangle multiple planetary signals. As seen in Figure 1 the rectangular TTN and the X-array perform the best in this respect. Recent simulations by Velusamy *et al.* and Thiebaut *et al.* (these proceedings) argue that high spatial resolution is not needed to disentangle multiple planets which would lead to similar performance of the three and four telescope architectures.

References

Absil O. 2001, PhD thesis, University of Liege, Belgium
Dubovitsky, S. & Lay, O.P. 2004, *SPIE* 5491, pp. 284
Lay, O.P. & Dubovitsky, S. 2004, *SPIE* 5491, pp. 874
Kaltenegger, L., Eiroa, C., Stankov, A., & Fridlund, M. 2005, *these proceedings*
Kaltenegger, L. & Karlsson, A. 2004, *SPIE* 5491, pp. 275
Kaltenegger, L. 2004, PhD thesis, Karl-Franzens University Graz, Austria
Karlsson, A., Wallner, O. & Perdigues Armengol, J. M. 2004, *SPIE* 5491, pp. 831
Velusamy, T. & Marsh, K. 2005, *these proceedings*
Thiebaut, E. & Mugnier, L. 2005, *these proceedings*

Direct Imaging of Exoplanets: Science & Techniques
Proceedings IAU Colloquium No. 200, 2005
C. Aime and F. Vakili, eds.
© 2006 International Astronomical Union
doi:10.1017/S1743921306009422

Atmospheric Biomarkers and their Evolution over Geological Timescales

Lisa Kaltenegger[1]†, Jucks K.[1] and Traub W.[1,2]

[1]Harvard-Smithsonian Center for Astrophysics, 60 Garden St, Cambridge, MA 02138, USA
email: lkaltenegger@cfa.harvard.edu; kjucks@cfa.harvard.edu

[2]Jet Propulsion Lab, M/S 301-451, 4800 Oak Grove Dr., Pasadena CA, 91109, USA
email: wtraub@jpl.nasa.gov

Abstract. The search for life on extrasolar planets is based on the assumption that one can screen extrasolar planets for habitability spectroscopically. The first space born instruments able to detect as well as characterize extrasolar planets, Darwin and terrestrial planet finder (TPF-I and TPF-C) are scheduled to launch before the end of the next decade. The composition of the planetary surface, atmosphere, and its temperature-pressure profile influence a detectable spectroscopic signal considerably. For future space-based missions it will be crucial to know this influence to interpret the observed signals and detect signatures of life in remotely observed atmospheres. We give an overview of biomarkers in the visible and IR range, corresponding to the TPF-C and TPF-I/DARWIN concepts, respectively. We also give an overview of the evolution of biomarkers over time and its implication for the search for life on extrasolar Earth-like planets. We show that atmospheric features on Earth can provide clues of biological activities for at least 2 billion years.

Keywords. astrobiology, astrochemistry, extrasolar planets, Earth, atmospheric biomarkers.

1. Introduction

Indirect methods to detect extrasolar terrestrial planets such as radial velocity, astrometry, and transits only allow getting morphological information. The planet is deduced from its effect on the parent star under survey. To understand the properties of a planet we need to collect photons from the planet directly to determine its physical and chemical properties. Spectral analysis of extrasolar planets is a very young science that only recently achieved first results for hot Jupiters. Some atmospheric components like Na were detected in absorption in the upper atmosphere of HD209458b (Charbonneau, Brown, Noyes *et al.* (2002)). The Spitzer telescope recently detected the infrared emission from two transiting hot Jupiters by subtracting the stellar flux when the transiting planet is eclipsed by its star from the combined flux of the star and planet (see Deming, Seager, Richardson *et al.* (2005), Charbonneau, Allen, Megeath *et al.* (2005)).

The first space born instruments able to detect as well as characterize extrasolar planets are DARWIN (see e.g. Kaltenegger, Fridlund & Karlsson (2005)), TPF-I and TPF-C (see e.g. Borde & Traub (2005)). DARWIN/TPF-I are both based on the concept of a free flyer IR interferometer architecture, and TPF-C is envisioned as a coronagraph in the visible. The strategy to search for biological activity on terrestrial planets is based on the assumption that one can use spectroscopy to screen extrasolar planets for habitability. Basing the search for life on the carbon chemistry assumption allows establishing

† Present address:Harvard-Smithsonian Center for Astrophysics, MS20, 60 Garden Street, Cambridge, MA 02138, USA

criteria for habitable planets in terms of their size and distance from their stars (Kasting, Whitmire & Reynolds (1993)). Note that observations over interstellar distances can only distinguish those planets whose habitability and biological activity is apparent from observations of the reflected or emitted radiation. A crucial factor in interpreting planetary spectra is the point in the evolution of the atmosphere and its biomarkers over time. Concentrating on the evolution of our planet we establish a model for its atmosphere and the detectable biomarkers over its evolution history. Figure 2 shows that atmospheric features on Earth can provide clues of possible life forms for at least 2 billion years (see Kaltenegger, Traub & Jucks (2005)).

1.1. *Atmosphere evolution on Earth*

The Earth formed about 4.5 billion years ago. The primitive atmosphere was formed by the release of volatiles from the interior, and/or volatiles delivered during the late bombardment period. This atmosphere was most likely dominated by carbon dioxide, with nitrogen being the second most abundant gas and trace amounts of methane, ammonia, sulphur dioxide, hydrochloric acid and oxygen (see Kasting & Siefert (2002)). Carbon dioxide and/or methane played a crucial role in the development of an early greenhouse effect that counteracted the lower solar output. Two major processes have changed the primitive atmosphere, the reduction of CO_2 and the increase of O_2. A huge amount of CO_2 must have been removed from the atmosphere, most likely by the burial of carbon into carbonate rocks, though the process is still debated. Primitive cyanobacteria are believed to have produced oxygen. After most of the reduced minerals were oxidized, about two billion years ago, atmospheric oxygen could accumulate. Different schemes have been suggested to quantify the rise of oxygen and the evolution of life by anchoring the points in time to fossil finds (see e.g. Owen (1980) Schopf (1993) Ehrenfreund & Charnley (2003)). There are still many open questions. We use a climate model based on work by Kasting & Catling (2003), Pavlov, Hurtgen, Kasting, Arthur (2003), Segura, Krelove, Kasting *et al.* (2003) and Traub & Jucks (2002) to create a schematic atmospheric model of our Earth over geological timescales. We find a surface temperature above the freezing point for all of Earth's history (not considering short term atmospheric changes during glaciation periods). Our radiative transfer model is based on the model by Traub & Jucks (2002) that was developed to successfully analyse data from stratospheric balloon flight campaigns ongoing during the last few years. The same radiative transfer code was used to successfully analyse data from Earthshine measurements in the visible (400 to 900 nm) (see Woolf, Smith, Traub *et al.*) and mid-infrared (700 to 2400 nm) (see Turnbull, Traub, Jucks *et al.* (2005)). It shows excellent agreement with measurements in both the visible and IR wavelength band. We present the spectra of Earth over geological timescale in anticipation of proposed space based terrestrial planet search missions to operate between 500 to 1100 nm in the visible to mid-IR range and 6 to 20μm in the IR. Our model shows a considerable difference in biosignatures that can be detected with those missions. That should be able to constrain an Earth-like planet to its time in evolution.

1.2. *Atmospheres and Biomarkers*

The spectral characteristics of a planet are a very important guide to identifying the best wavelength region to probe for a planet as well as characterize the spectra once it has been detected. The TPF Science Working Group (TPF-SWG) identified the waveband between 8.5μm to 20μm and preferably 7μm to 25μm, respectively, for the search for biomarkers in the mid-IR region and 0.7μm to 1.0μm and preferably 0.5μm to 1.1μm for the visible to near IR region (see Des Marais, Harwit, Jucks, *et al.* (2002)). The

DARWIN project concentrates on the waveband between 6μm to 20μm (see Fridlund, Kaltenegger, *et al.* (2006)). In the thermal part of the spectrum, the shape gives a measure of the temperature of the object examined. Observations from 8μm to 12μm of the H_2O continuum allow estimations of the surface temperature of Earth-like planets. For higher concentrations of water vapor in the atmosphere, the continuum gets lowered due to water absorption.

Biomarkers are features whose presence or abundance requires a biological origin (see Des Marais, Harwit, Jucks, *et al.* (2002), Meadows *et al.* (2005)). They are created either during the acquisition of energy and/or the chemical ingredients necessary for biosynthesis (e.g. atmospheric oxygen and methane) or are products of the biosynthesis (e.g. complex organic molecules and cells). As signs of life in themselves H_2O and CO_2 are secondary in importance because although they are raw materials for life, they are not unambiguous indicators of its presence. Taken together with molecular oxygen, abundant CH_4 can indicate biological processes. Depending on the degree of oxidation of a planet's crust and upper mantel non-biological mechanisms can also produce large amounts of CH_4 under certain circumstances. Oxygen is a chemically reactive gas. Reduced gases and oxygen have to be produced concurrently to be detectable in the atmosphere, as they react rapidly with each other. The 9.6μm O_3 band is highly saturated and is thus a poor quantitative indicator, but an excellent qualitative indicator for the existence of even traces of O_2. Ozone is a very nonlinear indicator of O_2 because the ozone column depth remains nearly constant as O_2 increases from 0.01 present atmosphere level (PAL) to 1 PAL.

N_2O is a very interesting molecule because it is produced in abundance by life but only in trace amounts by natural processes. There are no N_2O features in the visible and two weak N_2O features in the IR. In the IR it can only be detected in regions strongly overlapped by CH_4, CO_2 and H_2O, so it is unlikely to become a prime target for the first generation of space-based missions searching for extrasolar planets that will work with low resolution, but it is an excellent target for follow up missions. Spectral features of N_2O become apparent in atmospheres with less H_2O vapor. In the mid-IR N_2O has a band at 7.9μm, comparable in strength to the adjacent CH_4 band but weak compared with the overlapping H_2O band. The absorption bands of those three species are different. It is not readily separable for low resolution spectroscopy for current Earth, but the methane feature is easily detectable for early type Earth planets according to our models. To detect and study surface properties we can only use wavelengths that penetrate to the planetary surface. On a cloud-free Earth, the diurnal flux variation caused by different surface features rotating in and out of view could be high. When the planet is only partially illuminated, more concentrated signal from surface features could be detected as they rotate in and out of view. Most surface features like ice or sand show very small or very smooth continuous opacity changes with wavelength. Figure 1 shows that the signal detected from a cloud free planet is lower than that from an Earth-like planet. Thus the integration time needed to detect such planets will be longer.

1.3. *The red edge of land planets*

An interesting example for surface biomarkers on Earth is the red edge signature from photosynthetic plants at about 750 nm. Photosynthetic plants have developed strong infrared reflection as a cooling mechanism to prevent overheating and chlorophyll degradation. The primary molecules that absorb the energy and convert it to drive photosynthesis (H_2O and CO_2 into sugars and O_2) are chlorophyll A (0.450μm) and B (0.680μm). Several groups have measured the integrated Earth spectrum via the technique of Earthshine, sunlight reflected from the "dark" side of the moon. Earthshine measurements

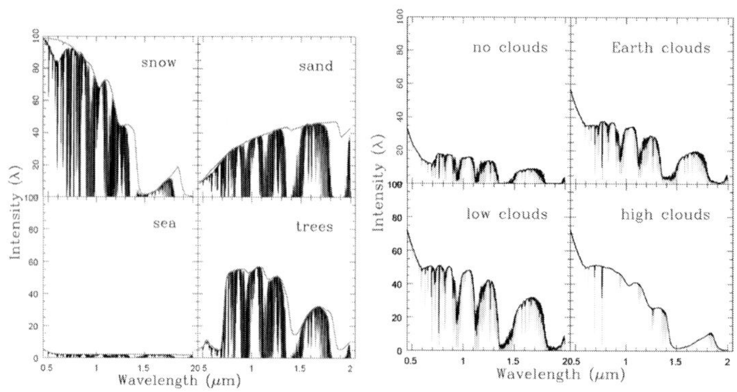

Figure 1. (left) Signal of a current Earth atmosphere considering surface features assuming a clear atmosphere without clouds; Different surface composition (a) snow (b) sea (c) trees (d) sand. (right) Signal of current Earth with realistic surface features and (a) no clouds, (b) 100% cumulus cloud coverage at 4 km and (c) 100% cirrus cloud coverage at 12 km and (d) a mixture of clouds resembling the present Earth.

have shown that detection of Earth's vegetation-red edge (VRE) is feasible but made difficult owing to its broad, essentially featureless spectrum and cloud coverage. Trying to identify such weak, continuum-like features at unknown wavelengths in an extrasolar planet spectrum requires models for different planetary conditions. Our knowledge of different surface reflectivities on Earth, like deserts, ocean and ice, help assigning the VRE of the Earthshine spectrum to terrestrial vegetation (see Figure 2). Earth's hemispherically integrated vegetation red-edge signature is weak, but planets with different rotation rates, obliquities, land-ocean fraction, and continental arrangement may have lower cloud-cover and higher vegetated fraction (see e.g. Seager (2002)).

1.4. *Model results*

Our calculations (see Kaltenegger, Traub & Jucks (2005)) show that the composition of the surface (especially in the visible), the atmospheric composition and temperature-pressure profile can all have a significant influence on the detectability of a signal. The reflectivity of the Earth has not been static throughout the past 4.5Ga (Ga = 10^9 years ago). Oxygen and ozone became abundant about roughly 2.3Ga, affecting the atmospheric absorption component of the reflection spectrum. About 2Ga, a green phytoplankton signal developed in the oceans and about 0.44Ga, an extensive land plant cover developed, generating the red chlorophyll edge in the reflection spectrum. The oxygen and ozone absorption features could have been used to derive the presence of biological activity on Earth anytime during the past 50% of the age of the solar system, while the chlorophyll red-edge reflection feature evolved during the most recent 10% of the age of the solar system. Figure 1 shows the spectral signature of an Earth atmosphere assuming different surface composition and the big impact clouds can have on the spectrum. From the 4 surfaces plotted on the left one sees that an Earth-size ocean planet will be more difficult to detect than a snow covered planet due to its low albedo. The smooth lines in the left panel of Figure 1 show the albedo of the surfaces without overlayed atmosphere. The panel on the right in Figure 1 shows the strong impact clouds can have on the detected signal. Present Earth has about 60% cloud coverage. 100% cumulus (low) and 100% cirrus (high) cloud coverage, respectively, are shown for reference. A planet without clouds or snow has a lower overall albedo and thus requires a longer integration time for detection than a planet with clouds. Figure 2 shows Earth's reflected and thermal emission spectra respectively with its major molecular species (H_2O, O_3, O_2, CH_4, CO_2,

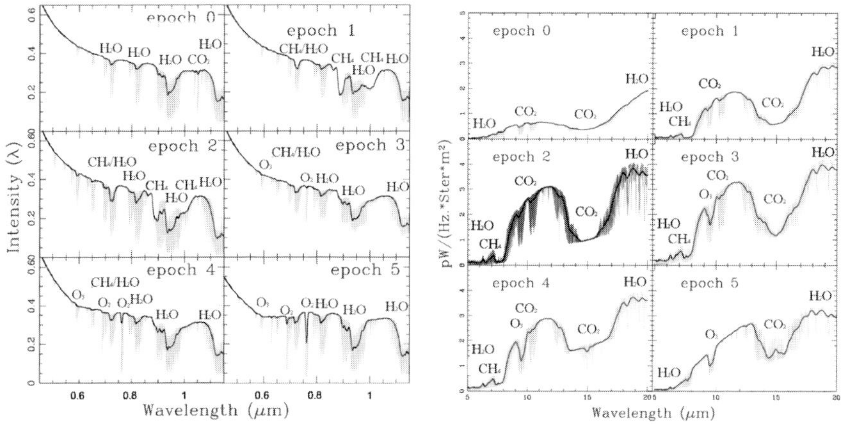

Figure 2. The atmospheric features on an Earth-like planet change considerably over its evolution from a current day atmosphere (epoch 5 = present Earth) over a CO_2/CH_4-rich atmosphere (epoch 3) to a CO_2 rich (3.9Ga ago = epoch 0) in both the visible (left) and the IR range (right). The black lines show spectral resolution of 70 for the visible and 25 for the IR, comparable to the proposed mission concepts.

N_2O) as well as minor contributors (H_2S, SO_2, NH_3, SF_6, CFC-11, CFC-12). The dark lines show a resolution of 70 in the visible and 25 in the IR, as proposed for the TPF-C and Darwin/TPF-I mission, respectively. The atmospheric features on an Earth-like planet change considerably over its evolution from a current day atmosphere (epoch5 = present Earth) over a CO_2/CH_4-rich (epoch 3 around 2Ga) to a CO_2 rich (3.9Ga = epoch 0) atmosphere. O_3 shows a strong feature in the IR for the last 2Ga of Earth's history even at a low resolution of 25. O_2 shows a weaker feature in the visible for the last 2Ga of Earth's history at a resolution of 70 (note that no noise level is assumed for these calculations).

In the mid-IR the classical signatures of biological activity are the 9.6μm O_3 band, the 15μm CO_2 band and the 6.3μm H_2O band or its rotational band that extends from 12μm out into the microwave region (Selsis, Despois, Parisot (2002)). For the visible, a similiar investigation has not been conducted yet. In the same spectral region, the 7.7μm band of CH_4 is a potential biomarker for early-Earth type planets. In the visible to near-IR one can see a strong O_2 absorption feature at 0.76μm, a broadband O_3 absorption at 0.45μm to 0.75μm and a strong H_2O band at 0.94μm. The strongest O_2 feature is the saturated Fraunhofer A-band at 0.76μm that is still relatively strong for significantly smaller mixing ratios than present Earth's. O_3 has two broad features, the extremely strong Huggins band in the UV shortward of 0.33μm and the Chappius band which shows as a broad triangular dip in the middle of the visible spectrum from about 0.45μm to 0.74μm. Methane at present terrestrial abundance (1.65ppm) has no significant visible absorption feature, but at high abundance it has strong visible bands at 0.88μm and 1.0μm, readily detectable in early Earth's history. CO_2 has negligible visible features at present abundance, but in a high CO_2-atmosphere of 10% it would have a significant band at 1.2μm and even stronger ones at longer wavelengths. Especially in the early evolution stage, the weak 1.06μm band can be observed (epoch 0).

2. Conclusions

Concentrating on the evolution of our planet we established a model for its atmosphere and the detectable biomarkers over its evolution history and showed the results for

TPF-I/Darwin and TPF-C. These missions should be able to constrain an Earth-like planet in its evolution. Our calculations show that the atmospheric and the surface composition (especially in the visible) and cloud coverage influence the detected signal considerably (see Figure 1). From the 4 surfaces plotted on the left one sees that e.g. an Earth-size ocean planet will be more difficult to detect than a snow covered planet due to its low albedo (here we consider no clouds). The atmospheric features on an Earth-like planet change considerably over its evolution from a current day atmosphere (epoch 5 = present Earth) over a CO_2/CH_4-rich atmosphere (epoch 3 around 2Ga) to a CO_2 rich (3.9Ga = epoch 0) see Figure 2. Atmospheric features on Earth can provide clues of biological activities for at least 2 billion years. O_3 shows a strong feature in the IR for the last 2Ga of Earth's history even at a low resolution of 25. O_2 shows a weaker feature in the visible for the last 2Ga of Earth's history at a resolution of 70 (note that no noise level is assumed for these calculations). Methane (a potential bioindicator for early Earth) and nitrous oxide (a biomarker with a weak signature in the IR) have features nearly overlapping in the 7μm region, lying in the red wing of the 6μm water band. The absorption bands of those three species are different. It is not readily separable for current Earth at low spectral resolution. N_2O is a very interesting molecule because it is produced in abundance by life but only in trace amounts by natural processes. It is unlikely to become a prime target for the first generation of space-based missions searching for extrasolar planets that will work with low resolution, but it is an excellent target for follow up missions. But if methane is abundant, as it probably was for early Earth, then it is readily detectable, even at low spectral resolution, functioning as an early bio-indicator for young Earth-like planets.

Acknowledgements

Special Thanks to J. Kasting, A. Knoll and A. Segura for constructive discussions.

References

Borde, P. & Traub, W. 2005 *IAUC200 proceedings*
Charbonneau, D., Allen, L.E., Megeath, S.T., *et al.* 2005 *ApJ* 626
Charbonneau, D., Brown, T.M., Noyes, R.W., & Gilliland, R.L. 2002 *ApJ* 568
Deming D., Seager S., Richardson L.J., Harrington J. *et al.* 2005 *Nature* 434
Des Marais, D.J., Harwit, M.O., Jucks, K.W. Kasting, J.F. *et al.* 2002 *Astrobiology*2
Ehrenfreund P. & Charnley S.B. 2003 *ESA SP-539*
Fridlund, M., Kaltenegger, L. *et al.* 2006 *in prep*
Kaltenegger, L., Traub, W. & Jucks, K. 2005 *in prep*
Kaltenegger, L., Fridlund, M.& Karlsson, A. 2005 *IAUC200 proceedings*
Kasting, J.F. & Catling D. 2003 *Annual Rev. Aston. Astrophys.* 41
Kasting, J.F. & Siefert, J.L. 2002 *Science* 296
Kasting, J.F., Whitmire, D.P. & Reynolds, R.T. 1993 *Icarus* 101,108
Meadows, V. *et al.* 2005 *IAUC200 proceedings*
Owen, T. 1980 *Strategies for Search for Life in the Universe* pp. 177
Pavlov, A.A., Hurtgen, M.T., Kasting, J.F., & Arthur, M.A. 2003 *Geology* 31, pp. 87
Schopf, J.W. 1993 *Science* 260, pp. 460
Seager, S. 2002 *astro-ph/0212550*
Segura, A., Krelove, K., Kasting, J. *et al.* *Astrobiology* V.3, 4
Selsis, F., Despois, D. & Parisot, J.-P. 2002 *AA* 388
Traub, W. & Jucks, K. 2002 *AGU* 130, pp369
Turnbull, M., Traub, W., Jucks, K. *et al.* 2005 *ApJ submitted*
Woolf, N.J., Smith, P.S., Traub, W.A., & Jucks, K.W. 2002 *ApJ* 574, pp430

Direct Imaging of Exoplanets: Science & Techniques
Proceedings IAU Colloquium No. 200, 2005
A.C. Editor, B.D. Editor and C.E. Editor, eds.
© 2006 International Astronomical Union
doi:10.1017/S1743921306009434

Nulling interferometry : Lommel's integrals applied to a Fresnel's diffraction effect

Yves Rabbia[1]†, Jean Gay[1], Bruno Chazelas[2], Alain Labèque[2], and Jean-Pierre Rivet[1]

[1]Observatoire de la Côte d'Azur, Av Copernic, 06130 Grasse, France
email: rabbia@obs-azur.fr

[2]Institut d'Astrophysique Spatiale, 91 Orsay, France

Abstract. Nulling Interferometry applied to the search and characterization of earth-like exoplanets requires to eliminate the star's contribution at a rejection level (Rej = collected energy/residual energy)larger than 10^6 over a large bandwidth (6 to 18 μm). Nulling test-benches are in development in several laboratories so as to master such high a rejection. One approach relies on a Mach-Zehnder set-up with Achromatic Phase Shifters (APS). One APS concept is based on the focus-crossing property, providing an intrinsically achromatic phase shift by π. Using a confocal configuration for the focus-crossing approach, a Fresnel's diffraction effect degrades the rejection. Usual optical engineering softwares fail in assessing rejection performance and an analytical approach is needed. We describe the bench optical configuration and the Fresnel's diffraction effect as well as a possible way for correction. Then we describe the analytical method, based on Lommel's integrals, to evaluate the expectable rejection.

Keywords. Nulling interferometry, Exo-Earths, Fresnel's diffraction, Lommel's integrals

1. Introduction

Nulling interferometry, recognized as the relevant tool in the search and spectral analysis for earth-like exoplanets, is the baseline of the ESA-Darwin mission (ESA, 2000). In the framework of preliminary studies, nulling test benches are in development so as to master the rejection process, a key-issue for such a mission. A convenient system is a Mach-Zehnder-like interferometer where Achromatic Phase Shifters are installed so as to provide the required π-phase shift yielding the elimination of the starlight via destructive interferences. Among other approaches (Rabbia et al, 2002) the confocal APS based on focus property (Gouy, 1890) is considered. With this peculiar device a Fresnel's diffraction adverse effect degrade the rejection : Rej = (collected energy)/(residual energy). Eventhough this effect occurs on the bench, not in the flying device, expectable rejection performance on the bench must be calculated. At the very high level of rejection targeted, currently available Optical Engineering Softwares fail to provide reliable results (prohibitively large workspace and sampling rate needed). An analytical approach, using Lommel's formalism (Born & Wolf, 1980) taking advantage of circular apertures and Gaussian beams, yields an alternate approach for calculation of rejection. We describe the generic set-up of the bench, the APS based on the focus-crossing property, the adverse effect and the analytical method for calculation of rejection. Finally expected rejection versus wavelength curves in the effective conditions of operation on the bench are given.

† Present address: Observatoire de la Cote d'Azur, Dpt Gemini, UMR CNRS 6203, Av Copernic 06130, Grasse -France

2. generic design, focus-crossing and associated APS

The set-up basically is a Mach-Zehnder interferometer, fed by a collimated beam from a monomode fiber used as source. The output common beam after recombination by the exit beam splitter is focused on a monomode fiber used as spatial filter.

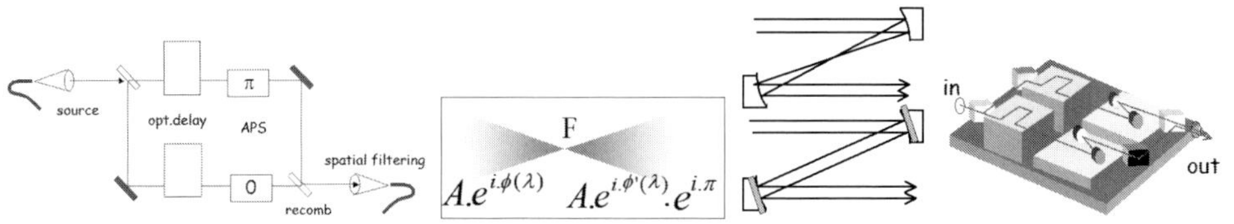

Figure 1. from left to right : principle of mach-zehnder set-up used as nulling test-bench, illustration of the focus-crossing property, principle of the associated APS and global architecture of bench.

In each arm is set an APS component. One component (CF), based on the focus-crossing property, is a confocal assembly of two half-paraboloids. The other component (FF), simply is a couple of flat mirrors set to produce the same optical path and same average incidence as in the other arm.

3. Adverse effect from Fresnel's diffraction and remedies

As seen from the recombination plate at output, there is an axial separation for the images of the input pupilla: the CF unit preserve the location of the input pupilla but not the FF unit. Therefore, because of Fresnel's diffraction effect (caused by unequal propagation length in CF and FF channels) the amplitude distributions set to interfere are not identical what is detrimental to the efficiency of destructive interference, hence to rejection.

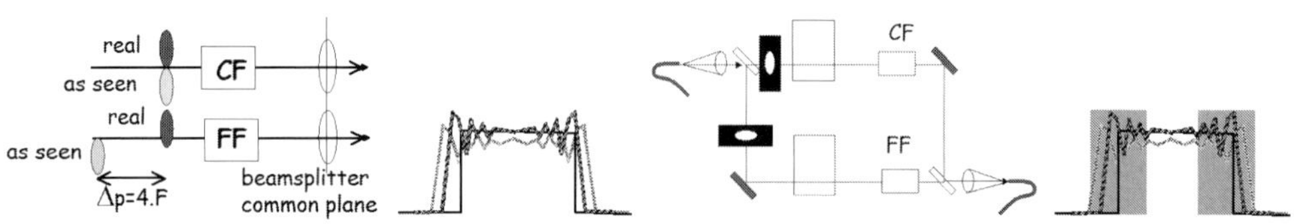

Figure 2. from left : axial separation (F : focal length in CF) for images of input pupilla on each interferometer arm as seen from recombining plate, illustration of resulting wavefront mismatch, insertion of "corrective" stops, "cleaning" effect of stops.

A way to reduce the degradation is to insert aperture stops at different locations in such way that effective pupilla are now seen in the same plane from the recombination plate. Moreover the diameter of stops is chosen so that amplitude distorsions (mainly peripheral) are significantly eliminated (this, at the price of loosing some photons, rather negligible with gaussian beams).

4. Using Lommel's formalism

With a targeted rejection larger than 10^6 wavefronts errors as small as 2.10^{-3} radian induce degradation. Softwares fail providing reliable amplitude maps. Analytical derivation of rejection is needed. Since apertures are circular and input wave is Gaussian, Lommel's integrals provide an analytical approach with sufficient accuracy.

4.1. *propagated complex amplitude*

Notations are in Figure 3 below. Starting from the complex amplitude A_0 (Gaussian amplitude of width "b" at 1/e) coming out of a circular aperture of radius R, the complex amplitude A_p obtained after propagation along the length "p" is given by the usual convolution $A_p(\overrightarrow{r}) = \oint_{r,g'<R,\theta} A_0(\overrightarrow{r}).Z_p(\overrightarrow{r}-\overrightarrow{r'}).r'.dr'.d\theta$ where we use the usual "propagation operator" $Z_p(\overrightarrow{r}) = exp(i.\frac{2.\pi.p}{\lambda}).exp(i.\frac{\pi.r^2}{\lambda.p})$ circularly symmetric.

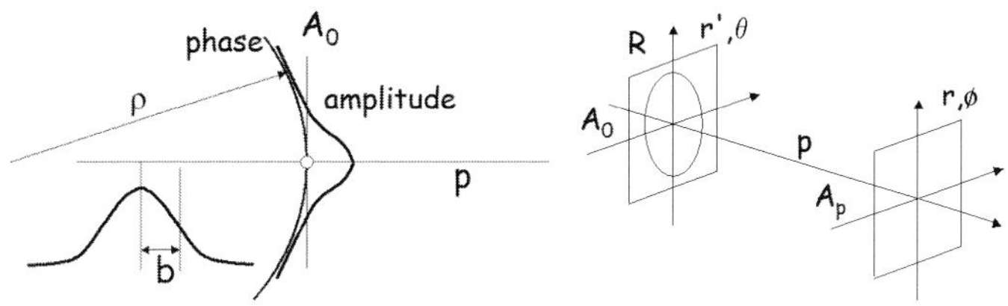

Figure 3. beam and propagation parameters for complex amplitude calculation.

Using a reduced variable $z = r'/R$ and after some algebraic work, we set $v = \frac{2.\pi.r.R}{\lambda.p}$ and $u = 2.\pi.[\frac{R^2}{\lambda.p} + \frac{R^2}{\lambda.\rho} + \frac{i.R^2}{b^2}]$ so that we have the general expression:

$$A_p(\overrightarrow{r}) = \left[2.\pi.R^2.\frac{A_0}{i.\lambda.p}.exp(i.\frac{2.\pi.r^2}{\lambda.p})\right] . I[u,v]$$

with $I[u,v] = \int_0^1 J_0(z.v).\exp(\frac{i}{2}.z^2.u).z.dz$

Depending on the value of the set-up parameters included in "u" and "v", we have:

$I[u,v] = \frac{i}{u}.exp(\frac{i}{2}.u).\left[\sum_{n=1}^{\infty}(\frac{u}{i.v})^n.J_n(v)\right]$ when $\left|\frac{u}{v}\right| < 1$
or
$I[u,v] = \frac{i}{u}.exp(\frac{i}{2}.u).\left[exp(-i.\frac{u^2+v^2}{2.u}) - \sum_{n=0}^{\infty}(\frac{v}{i.u})^n.J_n(v)\right]$ when $\left|\frac{u}{v}\right| > 1$

The complex amplitude is now computed from a sum of Bessel's functions (common software tool), at a precision level (unachievable with optical engineering softwares) set by the number of terms in the sum.

4.2. *calculation of rejection*

The Gaussian complex amplitude at entry (monomode fiber used as source), is given by: $G_{in}(\overrightarrow{r}) = A.\exp(-\pi.\frac{r^2}{b^2}).\exp(\frac{i.\pi.r^2}{\lambda.\rho})$ and the amplitude coming out of the input-pupilla is $A_0(\overrightarrow{r}) = G_{in}(\overrightarrow{r}).P_{in}(\overrightarrow{r})$ with $P_{in}(\overrightarrow{r}) = \Pi(\frac{r}{R_{in}})$ the classical circular pupilla function with radius R_{in}.

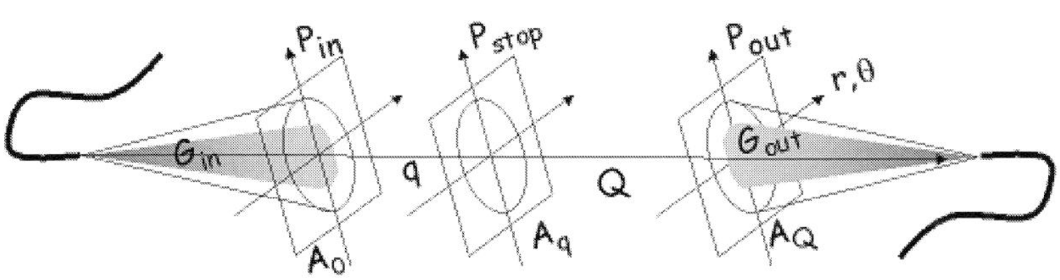

Figure 4. One arm of the bench, with aperture stop on the way from entry to exit.

Calculation of amplitude A_q arriving on the aperture stop at distance "q" from entry is carried as described in the previous section. For writing convenience, now and on, we use the simplified notation $A_q = A_0 * Z_q$. The amplitude coming out of the aperture stop (transmission P_{stop}, radius R_{stop}) simply is $A_q.P_{stop}$. After propagation along the distance "Q", we find the amplitude $[A_q.P_{stop}] * Z_Q$ and we finally have the amplitude $A_Q = ([A_q.P_{stop}] * Z_Q).P_{out}$ transmitted by the output pupilla (radius R_{out}, transmission $\Pi(\frac{r}{R_{out}})$).

To calculate the rejection Rej, we need to derive the algebraic expression of the amplitude driven into the monomode fiber at exit (serving as spatial filter). This expression is given by the "superposition integral" involving the Gaussian mode G_{out} propagated by the fiber:

$$a = \oint_{r<R_{out},\theta} A_Q(\vec{r}).G_{out}(\vec{r}).r.dr.d\theta$$

The point now, is that calculation of A_Q cannot be carried like was done for A_q because the Lommel's approach requires a Gaussian amplitude, what is not the case for $A_q.P_{stop}$. An algebraic derivation pretending to use the expression: $A_Q = ([A_0 * Z_q].P_{stop}) * Z_Q$ is quite tricky and tedious.

Fortunately it is possible to use an alternate approach. The amplitude "a" can be a little more explicitly written as:

$$a = \oint_{r<R_{out},\theta} [([A_0 * Z_q].P_{stop}) * Z_Q].[G_{out}.P_{out}]$$

Using the Parseval-Plancherel relation, we can write:

$$a = \oint_{r<R_{out},\theta} [([\hat{A}_0.\hat{Z}_q] * \hat{P}_{stop}).\hat{Z}_Q].[\hat{G}_{out} * \hat{G}_{out}]$$

Now, playing with brackets and products within the integral and applying again the Parseval-Plancherel relation we find the expression:

$$a = \oint_{r<R_{out},\theta} ([G_{in}.P_{in}] * Z_q).(P_{stop}).(Z_Q * [G_{out}.P_{out}]) = \oint_{r<R_{out},\theta} U_{in}.P_{stop}.V_{out}$$

In the second (compacted) expression are found U_{in} and V_{out} which both can be calculated using the Lommel's formalism (they can be seen as given by direct and reverse propagation respectively). Applying this policy to both arms on the bench (CF: ConFocal unit, FF: Flat-Flat unit) we have the collected and residual amplitudes (a_{FF} and a_{CF}) from which we calculate the rejection:

$$Rej = \frac{collected\ energie}{residual\ energie} = \frac{|a_{FF} + a_{CF}|^2}{|a_{FF} - a_{CF}|^2}$$

5. Exemples of expected rejection curves

Technological constraints, independent of the Focus-Crossing approach have led to split the spectral domain, in two sub-domains. In addition, geometrical constraints prevented to choose the appropriate set of parameters for beams and apertures, which would have provided the required level of rejection over the whole of each spectral interval. Current conditions yields rejection illustrated by the curves in figure 5, next page.

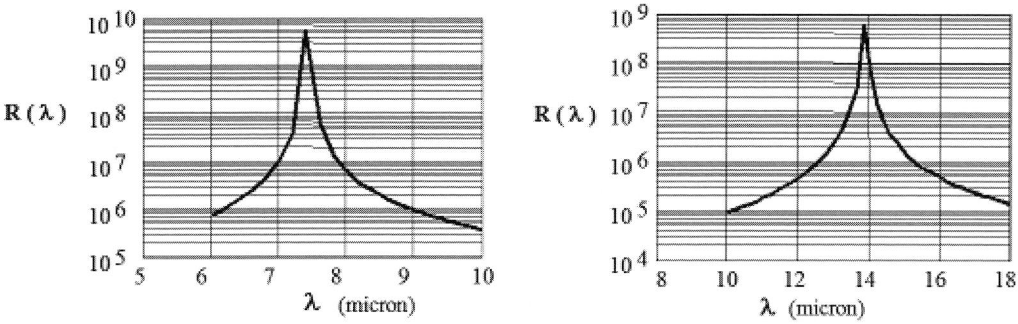

Figure 5. rejection curves obtained with current constraints on bench, for apertures and beam parameters.

6. Conclusion

Calculation of rejection in presence of Fresnel's diffraction adverse effect, and with "corrective" aperture stops in suitable locations, has been carried via Lommel's formalism, more reliable and convenient than usual optical engineering softwares in the present context. This approach made appropriate by the use of circular apertures and Gaussian beams can be applied to other set-ups.

References

Born, M. & Wolf, E. 1980, Principles of Optics, Pergamon Press

ESA 2000 IRSI-Darwin Report, ESA-SCI (2000) 12

Gouy, C., 1890, C.R. Acad. Sci. Paris, vol 110, p. 1251

Rabbia *et al.* 2002, Proc. of Genie Workshop, Leiden, the Netherlands, E. Bakker Editor http://www.strw.leidenuniv.nl/~nevec/GENIE/workshop_2002/abstracts/rabbia.html

All photographs: Laurent Thareau [`l.thareau@free.fr`].

Direct Imaging of Exoplanets: Science & Techniques
Proceedings IAU Colloquium No. 200, 2005
C. Aime and F. Vakili, eds.
© 2006 International Astronomical Union
doi:10.1017/S1743921306009446

Classical versus Bayesian interferometric imaging of exoplanets

Tomasz Sołtysiński[1,2]†

[1]Nicolaus Copernicus Astronomical Centre, Polish Academy of Science
ul. Bartycka 18, 00-716 Warsaw, Poland
email: tomasz_s@camk.edu.pl

[2]Department of Mechatronics, Warsaw University of Technology
ul. Sw. Andrzeja Boboli 8, 02-525, Warsaw, Poland

Abstract. The interferometry is the most promising way to directly observe exoplanets, their spectra and surfaces at optical or infrared domain. The complex imaging process can be described as the extraction of information from the data gathered by the interferometer. This information can be treated to be independend on any *à priori* knowledge or the integration process. In this case it is analyzed in a classical way. In fact, the imaging of exoplanets is not the classical way of the data reconstruction. The best extraction of information requires all accessible *à priori* knowledge. This is the bayesian way. The knowledge gathered during an integration is also contributing as *à priori* information for futher image reconstruction. We disscuss both approaches supporting the analysis with the estimates of information flow through the interferometer and examples of simulations of imaging.

Keywords. instrumentation: interferometers, methods: data analysis, techniques: image processing, astrobiology

1. Introduction

The motivation for this investigation comes from multidisciplinary point of view on exoplanet imaging science. Interferometers seem to be the best tools to image exoplanet (Angel & Woolf (1986), Bracewell (1978)). Serving first as a detectors of exoplanets in IR and visible domain (Léger (1999)), they are expected to provide first resolved images of exoplanets surfaces and their features (Labeyrie (1995)). This will allow the detailed investigation of their properties leading to the determination of life signs (Woolf & Angel (1998), Des Marais *et al.* (2002)). The main reason for this work is to understand the way how the interferometer process the information and hence, how to recover this information, classically or in a Bayesian way. We simulate the interferometric imaging process and investigate the information flow through different steps.

2. Simulations of exoearths interferometric imaging

2.1. *Observing conditions*

We start the simulation with the parameters typical for Earth-like planet placed at 10 pc from the Solar System. The interferometr is pointed towards the planet which is 1 AU far away from its host star. This results in planet angular size of $2.0645 \cdot 10^{-11}$ rad or 8.5 μarcsec. The field of view (FoV) of interferometer was set to 40 μarcsec yielding the ratio FoV/D ≈ 4.7, where D is planet angular diameter. The star flux is completely

† Present address: Dept. of Mechatronics, Warsaw University of Technology, Św. Andrzeja Boboli 8, 02-525 Warszawa

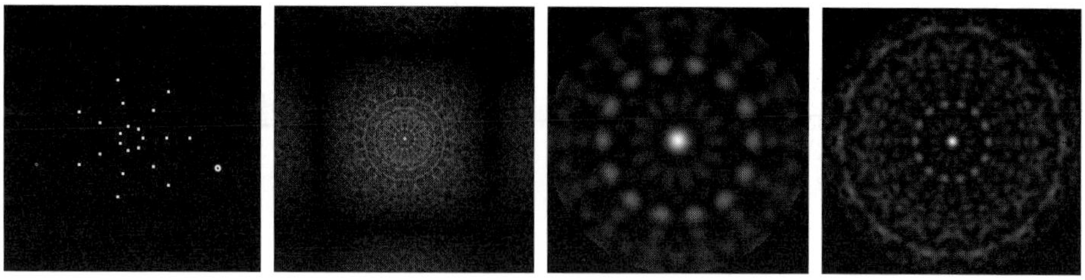

Figure 1. From left to right: the interferometer array design, its Fourier transform, the corresponding optical transmission function or interferometer response function for R and B band.

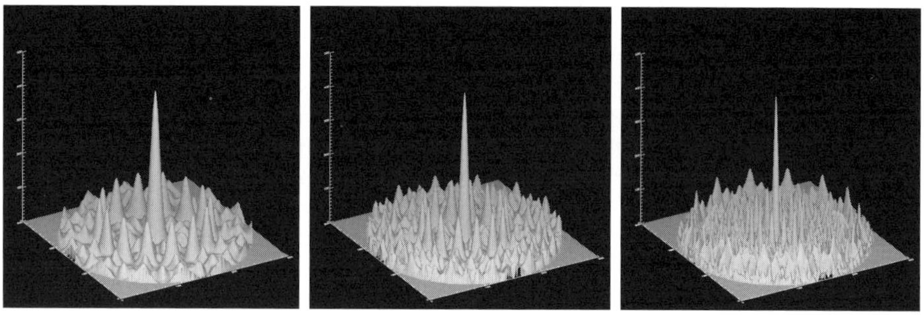

Figure 2. 3D shape of interferometer response function for R,G,B bands.

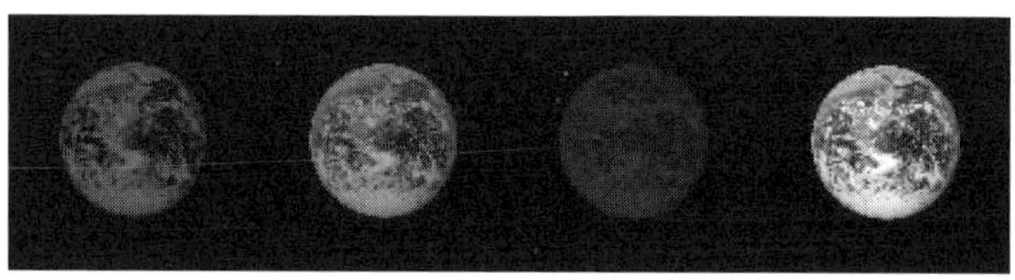

Figure 3. The flux pattern. The multiwavelength image of Earth taken by Apollo crew.

removed and has no influence on the reconstructed image. To simulate the flux coming from the distant planet we adapted the image of Earth as taken from space by Apollo crew (Fig. 3). The natural RGB components are then separated and serving as an image of exoearth for particular bands.

2.2. *Interferometer setup*

We have designed a space interferometer composed of sparse array of free flying telescopes. The array forms a pattern that corresponds to seven arms star with 3 rings (left of Fig. 1). The most distant spaceships are placed at 75 km from the centre, yielding 150 km of synthesized aperture. Such design results in interferometer response function or square modulus of Fourier transform of the initial array as shown on Fig. 1, right. Such optical transmission function yields

the patterns given on Fig. 2, with left picture for R (800 nm) band and next one for B (400 nm). The corresponding interferometer response functions (Absil (2001)) are shown as 3D plots on Fig. 2 for all 3 bands. The interferometer is not allowed to rotate, since we are interested only in its general information processing and not in the detailed (u,v) coverage necessary for accurate imaging. The instrument also stay tuned for the initial baseline length for all bands. No delay lines are assumed nor phase shifts introduced or we assume that all occuring phase shifts are compensated by achromatic phase shifters.

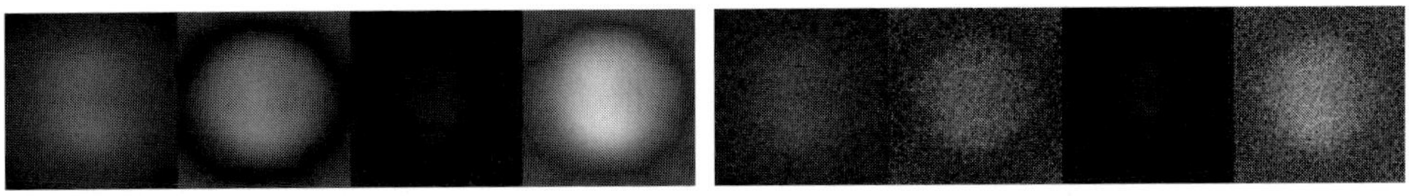

Figure 4. The data provided by the instrument for R,G,B bands and combined multichannel data on the left. The same data affected by instrumental noise with SNR=10.

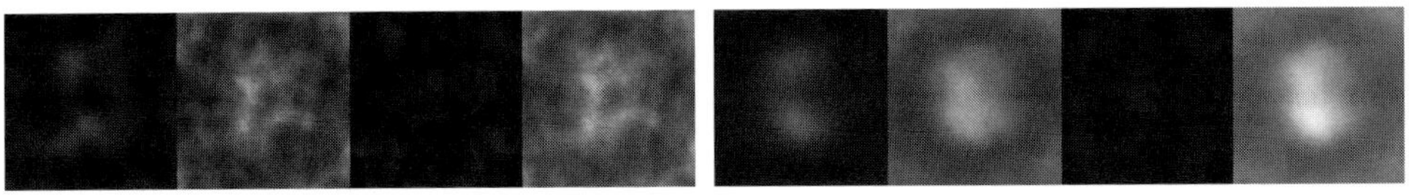

Figure 5. The MEM (left) and ML (right) deconvolved image of noise affected data after 50 iterations.

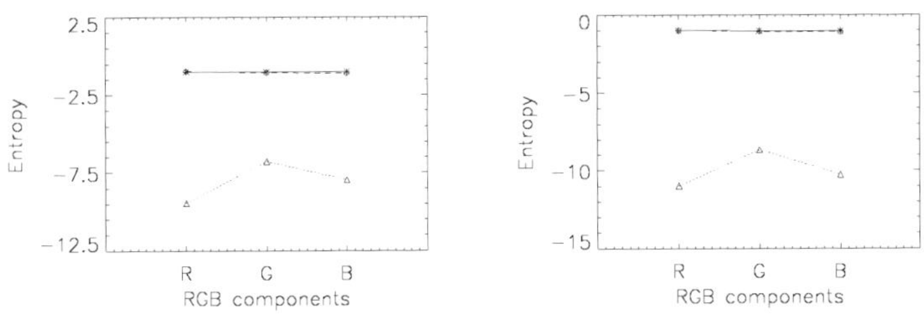

Figure 6. Multiscale entropy $H(X)$ for noise free data reconstruction compared with the multiscale entropy of initial data. Solid line is the original Apollo image, dashed – convolved interferometric data, dotted the deconvolved one, MEM, left and ML, right. All data is normalized in the way that absolute value of Apollo data's maximum yields always 1.

2.3. *Image content processing and extraction of information*

2.3.1. *Standard entropy*

The information content of any signal or image is traditionally measured by its entropy, a term coming from

thermodynamics and communication theory Shanon (1948). It is expressed as

$$H_S(X) = \sum_{k=1}^{N} p_k log(p_k) \tag{2.1}$$

where p_k is the probability of occurence of pixel k among all N pixels of image X and typically taken from histogram.

The information obtained by investigation of entropy may have different interpretation and usually depend on the image content. In astronomy, the flat image has usually very low Shannon entropy and very noisy image has high entropy. We usually look for the level of entropy that combines the high image information content.

or desired information, with the lowest possible level of noise and disturbance coming from point spread function (PSF) of the instrument that served to provide the image. Havin the entropy as common measure of image ordering we use it as a tool to estimate the information flow through the interferometer and reconstruction process. To have another measure of this process we also adapt Burg entropy that has slightly different

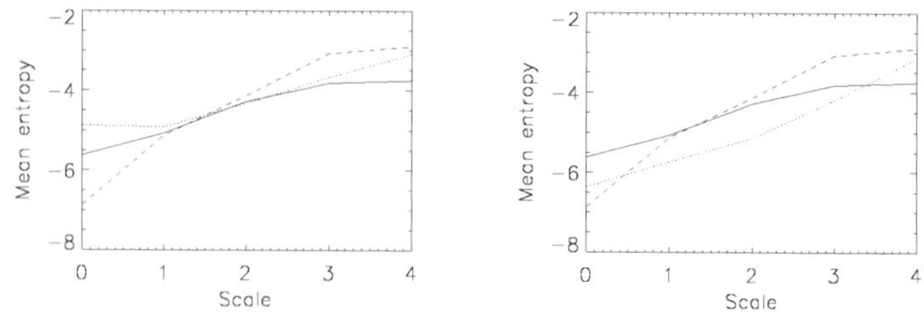

Figure 7. Mean entropy E_j vector for R component for all scales in the case of noise free simulation. Solid line is the original Apollo image, dashed – convolved data, dotted the deconvolved one, MEM, left and ML, right.

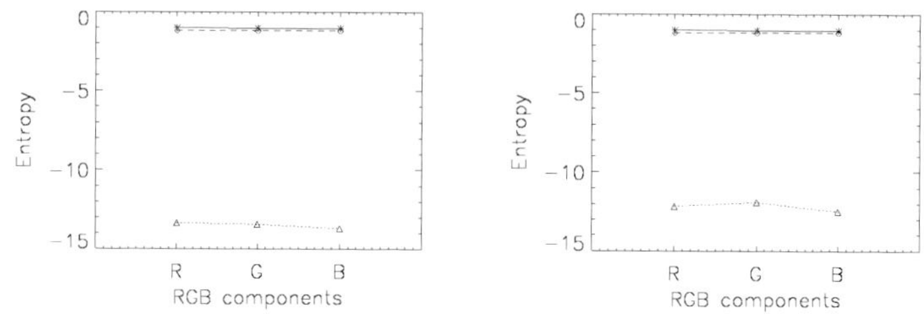

Figure 8. Multiscale entropy $H(X)$ for noise affected data reconstruction compared with the multiscale entropy of initial data. Solid line is the original Apollo image, dashed – convolved data, dotted the deconvolved one, MEM, left and ML, right. All data is normalized in the way that absolute value of Apollo data's maximum yields always 1.

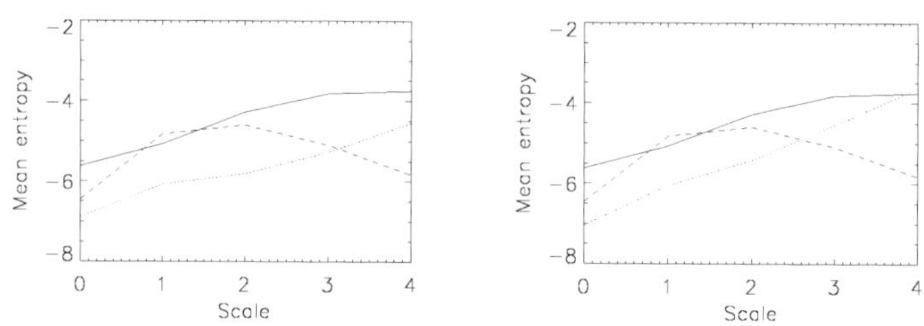

Figure 9. Mean entropy E_j vector for R component for all scales in the case of noise affected simulation. Solid line is the original Apollo image, dashed – convolved data, dotted the deconvolved one, MEM, left and ML, right.

definition

$$H_B(X) = -\sum_k \sum_l ln(X(k,l)). \tag{2.2}$$

Another definitions of entropy are also widely used but we found them less useful for the purpose of our study.

2.3.2. *Multiscale entropy*

Since the introduction of wavelets into contemporary science a powerful tool of signal and image analysis has become accessible. The power of this approach comes from the orthogonality of the wavelet base functions and extremally wide range of their diversity,

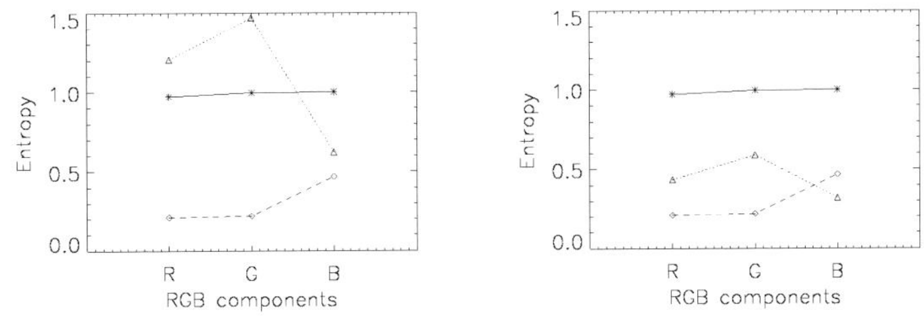

Figure 10. Normalized standard Shannon entropy of noise free case. Solid line is the original Apollo image, dashed – convolved data, dotted the deconvolved one, MEM, left and ML, right. All data is normalized in the way that absolute value of Apollo data's maximum yields always 1.

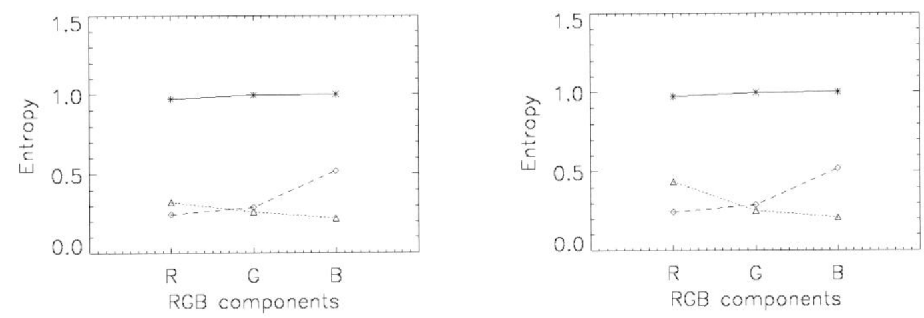

Figure 11. Normalized standard Shannon entropy H_S of noise affected case. Solid line is the original Apollo image, dashed – convolved data, dotted the deconvolved one, MEM, left and ML, right. All data is normalized in the way that absolute value of Apollo data's maximum yields always 1.

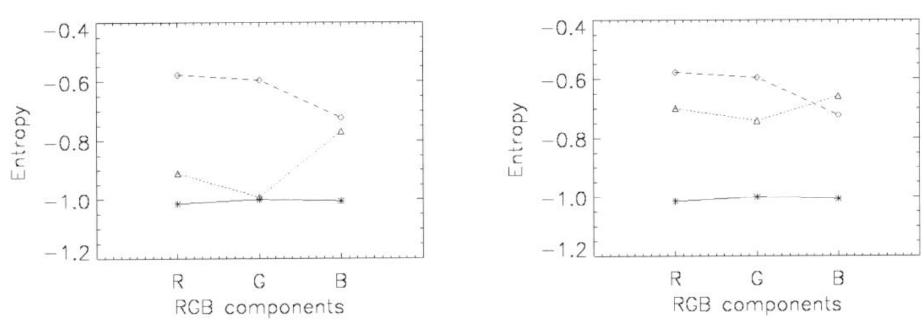

Figure 12. Normalized Burg's entropy H_B of noise affected case. Solid line is the original Apollo image, dashed – convolved data, dotted the deconvolved one, MEM, left and ML, right. All data is normalized in the way that absolute value of Apollo data's maximum yields always 1.

in opposition to Fourier basis. The strength of wavelets comes also from their ability to express the signal or image both in space and frequency domain at the same moment.

A very helpful feature of wavelet based image decomposition is the ability to measure an image multiscale entropy. Such measure provides not only the estimate of image information content but also a very practical estimation of this observable on different scales. As the important information in the image has always multiscale character the mutual dependences between scales are very useful in image restoration, reconstruction and analysis both in classical or Bayesian case. Hence, we also estimate the multiscale flow of information through the ideal interferometer. The formulation of multiscale entropy

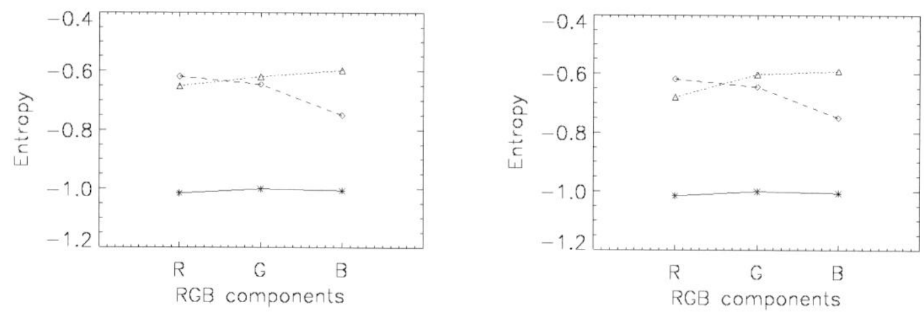

Figure 13. Normalized Burg's entropy H_B of noise affected case. Solid line is the original Apollo image, dashed – convolved data, dotted the deconvolved one, MEM, left and ML, right. All data is normalized in the way that absolute value of Apollo data's maximum yields always 1.

express the formulae:

$$H(X) = \sum_{j=1}^{l} \sum_{k=1}^{N_j} h(w_{j,k}) \tag{2.3}$$

where $h(w_j = ln(p(w_j(k))))$. In order to study the behavior of information at a given scale we use mean entropy vector defined as:

$$E(j) = \frac{1}{N} \sum_{k=1}^{N} h(w_{j,k}). \tag{2.4}$$

It should be stressed that in all cases we are not normalizing the above vector by corresponding noise models. All simulation results were deconvolved with PSF by Maximum Likelihood (ML) and independently by Maximum Entropy (MEM) algorithms.

3. Results

3.1. *Noise free case*

The simulation of a perfect observing run with nonrotating interferometer yields the image shown on Fig. 4, left. There is no noise added on each level of simulation. For once established setup we have performed three observations with changing light wavelenght λ. Due to low resolution coming from poor $((u,v))$ coverage the details of planet surface are hardly resolved. The information flow during noise free imaging is shown on Fig. 6. This picture illustrates multiscale entropy derived after application of ML or MEM algorithm compared to original image. The behavior of mean entropy at different scales is shown on Fig. 7.

3.2. *Realistic noisy imaging*

The noisy case is shown on Fig. 4, right. During simulation we have added noise according to given signal-to-noise ration (SNR) in the same way as Böker & Allen (1999). SNR was set to 10. Then the data was processed by inverse FFT what yielded the simulated observation output, Fig. 4. The image in each particular band was then deconvolved by MEM or ML. The results for both technics are shown on Fig. 5. Then we compared multiscale entropies obtained from MEM and ML as seen on Fig. 8 and their mean vectors Fig. 9.

4. Discussion

4.1. *Classical imaging*

The term "classical imaging" of exoplanet describes well what we really do through this work. It means the observation and "real" data reconstruction from individual "observed" images on particular wavelengths. Even if we apply the Bayesian inference by means of MEM, ML or CLEAN, usually used in interferometric data reconstruction, the entire process should be classified as classical due to no prior knowledge on an observed object. Our results suggest that in general the interferometer lowers H_S (or increase H_B). It can be noticed that MEM (Figs. 10 and 11) is recovering the data to the original data's entropy level introducing some variation between components while ML reconstruction is oscillating around the level of interferometric data entropy. We see on multiscale entropy plots, Figs. 8 and 6 the little difference of H after processing data by an array although it rises up significantly after reconstruction. After reconstruction mean entropy vector E_j reveals significant loss of entropy on higher scales, responsible for details.

However, we have to keep in mind the significant simplifications coming from "naive" design of ideal interferometer and the assumed face-on position of planet that never takes place in real observation where the terminator must be seen.

The information processing in the procedure of interferometric imaging is crucial to understand where the information is being dissipated. Then an effective algorithm of its recovering may be constructed, knowing the precise reasons and the rate of the information leakage, like the importance of noise influence, PSF variations, disturbing signals like zodiacal light or star leakage, unknown error budget, etc. Thus not only the high-level interferometer's model is necessary, but also the model of information processing through the instrument as well as its precise definition. Such model would help in optimization procedure of final instrument design and would lead to support data recovering algorithms. Usually, the initial data is able to be reconstructed from very ill-posed final data taking advantage of some known or predicted structures and knowledge coming from other sources or modalities. In the case of interferometric imaging the other modalities are different wavelength observations. As they all combine the real image being similar in information content on all scales they may serve as a basis for novel Bayesian strategy recovering the data due to the high correlation between different modalities.

4.2. *Bayesian imaging*

The possible Bayesian inference applied to exoplanet imaging can be based on the estimated level of compression of the obtained multichannel data. It is usually the multiscale measure of information content of the image data. Some prior knowledge coming from distant wavelength observation that are going to be done in advance to optical imaging, i.e. those in FIR or NIR, are also going to provide some spectroscopic information that corresponds to some surface features like green or blue spots, clouds, etc. The high level of correlation between multichannel (optical) data and that comig from other modalities may serve as a basement for required methods (Molina, R. & Mateos, J. (1997)). The optimization of all accesible data and the careful analysis of its information content may lead to the Bayesian, most probable model based on the set of the most probable values of parameters. Due to rather poor present understanding of space borned and placed real interferometer, the proper design of its performance as a *noisy channel* that processes, transmitts and looses the information is not known yet but should be important. The first, poorly resolved real observations of exoplanet, will also provide the basis or avaraged data for optimisation of such future Bayesian algorithms.

5. Conclusions

We have discussed the basic concepts of information transfer through the interferometer. There is a leakage of information on higher scales of wavelet decomposed data. More advaced study are necessary to understand the ways of further improvements of information restoration process, both in classical and Bayesian way. The latter seems to be the proper and promising way for future interferometric imaging of exoplanets.

Acknowledgements

The author would like to acknowledge the kind and extremally helpful support from IAU to participate in this IAU Colloquium. The support made attending the conference posssible.

References

Absil, O. 2001, *Diploma thesis*, University of Liege
Angel, J. & Woolf, N. 1986, *Nature* 322, 341
Böker, T. & Allen, R. J. 1999, *ApJ* 125, 123
Bracewell, R. 1978, *Nature* 274, 780
Des Marais *et al.* 2002, *Astrobiology* 2, 153
Labeyrie, A. 1995, *Astron. Astrophys. Suppl. Ser.* 118, 517
Léger, A. 1999, *Astron. Astrophys.* 341, 304
Molina, R. *et al.* 1992, *Astron. Journ.* 104, 1662
Molina, R. & Mateos, J. 1997, *Vistas in Astronomy* 41, 373
Shanon, C. E. 1948, *Bell System Technical Journal* 27, 379
Woolf, N. & Angel, J. 1998, *Annu. Rev. Astron. Astrophysics* 36, 507

Direct Imaging of Exoplanets: Science & Techniques
Proceedings IAU Colloquium No. 200, 2005
C. Aime and F. Vakili, eds.
© 2006 International Astronomical Union
doi:10.1017/S1743921306009458

TPF-I Planet Detection Testbed: Progress in Testing Exo-planet Signal Detection

S. R. Martin, P. Szwaykowski and F. M. Loya

Jet Propulsion Laboratory, 4800 Oak Grove Dr, Pasadena, CA, USA
email: stefan.r.martin@jpl.nasa.gov

Abstract. The Planet Detection Testbed is designed to simulate the architecture and operation of a space-borne four-telescope nulling interferometer. Constructed in the form of a dual chopped Bracewell interferometer together with star and planet sources, it reproduces the principal features of the flight beam combiner designed at JPL for the proposed TPF-I Formation Flying Interferometer. The aims of the testbed are to demonstrate stable four-beam nulling and planet detection at representative star-planet contrast ratios. In the flight design, starlight between 7 and 17 micron wavelength is to be nulled, and 2 to 3 micron starlight is used for fringe tracking and phasing the interferometer. There are also metrology systems and alignment systems which are required for deep and stable nulling. The testbed reproduces these features; 2 to 3 micron light from a thermal source is used for fringe tracking, and nulling and planet detection is performed at 10 microns. The testbed also incorporates laser metrology and other systems enabling continuous control of beam alignment. The ultimate goal is to simulate planet detection at star-planet contrast ratios of order 10^{-7} during full rotations of the telescope array using the phase chopping method. The latest results from the testbed are presented including four-beam nulling experiments at null depths of 10^{-5} and planet signal detections at similar contrast ratios.

Keywords. Nulling, Stellar Interferometry, Exoplanet.

1. Introduction

The detection of extra-solar planets using the radial velocity technique, more than one hundred so far, for example Marcy *et al.* (2005), and more recently the direct optical detection using the NACO instrument (Chauvin *et al.* (2004)) of a giant planet orbiting some distance from its parent star is spurring efforts to detect earth-like planets nearer their stars, within the habitable zone, where temperatures would be conducive to the existence of life. After the challenge of detection amidst the glare from the bright parent star, the characterization of the planet using spectroscopy will require long integrations and stable operation of the telescope. In the thermal infrared between 7 and 20 μm, absorption bands for gases considered to be markers for life processes occur; the thermal spectrum of the Earth viewed from space for example, is modified by the presence of water, methane, ozone and carbon dioxide absorption features in the atmosphere. In the case of the thermal infrared, an earth-like planet orbiting within the habitable zone of a G0 star at a distance of 10 pc has an intensity contrast with the star of a few 10^{-7} and a very small angular separation. Some means of suppressing the bright starlight together with a wide telescope collecting aperture would be helpful in separating the planet light from the star light. To achieve a detection, the concept of a 4-aperture telescope employing 4 m diameter collecting apertures with a baseline length between 60 and almost 500 m, floating in an earth-trailing orbit is being studied. Such a device, the Terrestrial Planet Finder Interferometer (TPF-I), will employ the method of nulling interferometry (Wallace *et al.* (2000)) to suppress the bright starlight and enable the planet photons to be detected and then characterized. To detect the planet will require observation times of 10 hours or more, with greater times needed for spectroscopic measurements. The Planet Detection Testbed is being used to study the combination of beams of infrared light from four telescopes, with an emphasis on achieving deep, stable nulls and performance near to that which will be required for successful operation in space.

2. The Planet Detection Testbed

The Planet Detection Testbed has been described elsewhere (Martin, S.) and consists of a pair of nulling beamcombiners and a cross-combiner configured for nulling and phase-chopping four input beams. The testbed sources simulate a star and a faint planet. The starlight which is to be nulled is produced by a CO_2 laser operating near 10.6 μm. A ceramic filament also provides broadband radiation which is used for fringe tracking, so that the interferometer can be accurately phased on the star. The fringe trackers operate at various wavelengths between about 2.3 and 3.5 μm. The planet source is also a ceramic filament with a dichroic filter to narrow the output to the waveband between 10 and 11 μm. The star and the planet sources are passed through separate choppers so that they can be more readily detected using lock-in amplifiers connected to the output of HgCdTe and InSb infrared detectors. To form the four input beams, the starlight is split once using a broadband 50/50 beamsplitter, and then each beam is split again. The planet light is similarly split once, and then combined with the starlight at the second splitter. This combination method ensures that the planet light has a certain phase relative to the star causing it to be constructively interfered at the nulling beamsplitter when the star is nulled. The recombination of the beams follows the splitting process in reverse; first, pairs of beams are nulled, and then they are combined on the cross-combiner beamsplitter.

3. Null Stabilization

It can be shown that to achieve deep nulls the pairs of input beams must be very similar in terms of intensity and phase. In essence, the aim is to achieve equal electric field amplitudes from each beam at the nulling detector, and carefully control the phase or optical path for each beam. For small deviations from perfect alignment, for example tilt or shear of the beams, the effect is to reduce the coupling to the nulling detector and therefore degrade the null. So, if beam tilt and shear are well controlled, barring other aberrations, only beam optical path remains to be stabilized. A control loop maintains the ideal optical path by means of a lock-in amplifier loop which monitors the nulling signal for an imposed dither signal. The sinusoidal dither signal is placed on the optical path of one of the incoming beams by modulating a piezoelectric (PZT)-driven stage so that the optical path change is approximately one nanometer. When the interference fringe is a small distance away from null the dither frequency appears on the nulling detector output with a phase dependent on whether the optical path difference of the interfering beams is positive or negative. The lock-in control loop software mixes the incoming detector output with the drive signal to the PZT and integrates the result to form an integral error signal. If the result is positive the loop drives the PZT in one direction and if negative it drives the PZT in the opposite direction. With an appropriate choice of loop coefficients the result is a fringe that is stabilized at null or alternatively at the peak. This control technique has been successfully used in previous experiments (Martin *et al.* (2003)) to achieve null depths of one million to one in a smaller infra-red nulling testbed. In the Planet Detection Testbed the optical path is comparatively long, 10 meters, and the beams strike a number of mirrors as they pass from the input to the detector. The optical

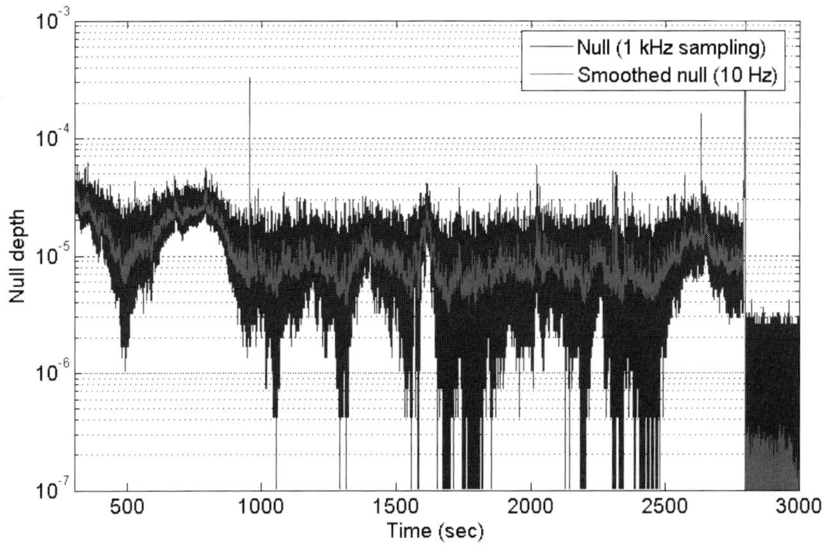

Figure 1. Nulling data.

alignment is subject to drifting because of small mechanical and thermal changes and it was found experimentally that without control of alignment, null depths greater than 10,000 are unlikely to be maintained for more than a few minutes. The drifting of optical alignment also produced a second problem for testbed operations: to emulate TPF-I operation the testbed phase must be maintained by a fringe tracker, operating on two to three micron radiation. This radiation is picked up by detectors positioned on either side of each nulling beamsplitter and on either side of the cross-combiner beamsplitter and the relative intensity of the light on each side of the beamsplitters is used as a measure of the phase. When the beams tilt, the intensity at the detector is affected and an error is introduced into the phase measurement. Because this effect was present in the unstabilized testbed, the fringe tracker could not hold the null fringe precisely although it would remain somewhere near the fringe.

To overcome this problem and to enable deep nulls to be maintained, a two-component dither loop was employed to control the tilt and tip of one of each pair of beams. In the experiments detailed below, two of the incoming beams (beams 1 and 3) were directed at the nulling detector and their directions and phases were not closely controlled. A second beam (beams 2 and 4 respectively) is paired with each of these incoming beams, and dither signals are applied to the second beam to enable control of its phase and angle relative to its partner. To control the beam angle the lock-in method was again used. A circular dither was applied to a tip/tilt stage in beam 2 or 4 so that the angular dither in the horizontal direction would be sinusoidal (for example) and the angular dither in the vertical direction would be cosinusoidal. The effect of the circular dither is to modulate the nulling output, and when the two beams are coaligned the null fringe is at a minimum. Again, the phase of the dither signal can be used to determine the adjustment direction and the tip/tilt control loop can drive the alignment of the pair of beams to an optimum via a simple integral loop.

The next stage in the nulling process is to locate the correct fringe. Since a broadband source is needed for this operation, the fringe tracking light is used to locate the correct optical path setting. The PZT stage is driven over the full extent of its range and the fringe tracker signal analyzed to locate the center of the fringe. Two methods by which this might be located (Pedretti *et al.* (2004)) were tested but it was found that they did not very reproducibly find the desired location on the fringe, possibly because of small alignment changes. Therefore a simple peak detection routine was used instead and this reliably selected a position near the center of the fringe.

Once this location is found the stage is driven to the fringe center and then a dither loop is engaged to drive the interferometer down to the bottom of the null. Once there the fringe tracker loop is engaged to hold the null approximately by maintaining set amplitudes on the pair of fringe tracking detectors placed either side of the nulling beamsplitter. At this stage a shutter can be closed to exclude the residual light from this nulling beamsplitter from the nulling detector and the second nuller can be phased. Once the second nuller is nulling, the shutter can be reopened, the fringe tracking loop discontinued, the dither loop restarted, and the system is now nulling four beams simultaneously. To reach deep nulls the beam alignment must be corrected for any drift and fine-tuned, so the circular dither loop is engaged for beams 2 and 4. When all the loop coefficients have been correctly chosen the nuller is driven below a 100,000 null depth. At the null, four loops are running, 8 Hz and 12 Hz circular dithers to maintain the angular alignments, and 20 Hz and 30 Hz linear piston dithers to maintain the optical paths. These signals all impose small modulations on the nulling detector output.

4. Results

To reach one of our development milestones, a faint planet source at 100000:1 contrast ratio with the star was to be detected while nulling at 100000:1 null depths. The planet could be detected by any reasonable means in the presence of the bright starlight; amplitude chopping of the planet at its source was employed as the detection mode. At the cross-combiner the planet light is incoherently summed; the nullers are not cophased and so the measured input planet light may be easily compared with the output data.

Once nulling deeply on all four beams the trace shown in Figure 1 was obtained. The blue trace shows the raw data (sampling rate 1 kHz) and the red trace the smoothed data (time constant 10 Hz). At about 970 s, the planet signal was admitted into the system; between about 900 s and 2500 s the mean null depth is below 10^{-5}. The apparent noisiness of the trace is mainly due to the control loops which modulate the null at various frequencies. There was also a significant amount of disturbance caused by vehicular traffic outside the building.

The various loops run at these frequencies; 20 and 30 Hz for OPD, and 8 and 12 Hz for tip/tilt. Prominent sum and difference signals are expected at 10 Hz and 50 Hz, and weaker ones at 40 and 60 Hz, with other weaker signals. The planet signal modulation was applied by running the planet chopper at a frequency some 70 Hz below the star chopper. In practice, the frequency difference drifted continuously over several Hz owing to the poor stability of the chopper drivers.

Samples were taken from the trace at 860 to 960 s (a section with no planet), 1100 to 1400 s (a section with a weak planet signal), and 1700 to 2400 s (a section with a stronger planet). These samples were analyzed by taking 1 second time intervals and calculating the magnitude of the Fourier transform for each interval, then summing and obtaining the mean spectrum in the form of a spectral estimate.

Figure 2 shows the results; the red line shows the signal without the planet, the blue line shows the stronger planet and the green the weaker planet signals. The units of the y axis are contrast ratio with the star. The broadening of the planet signal by the

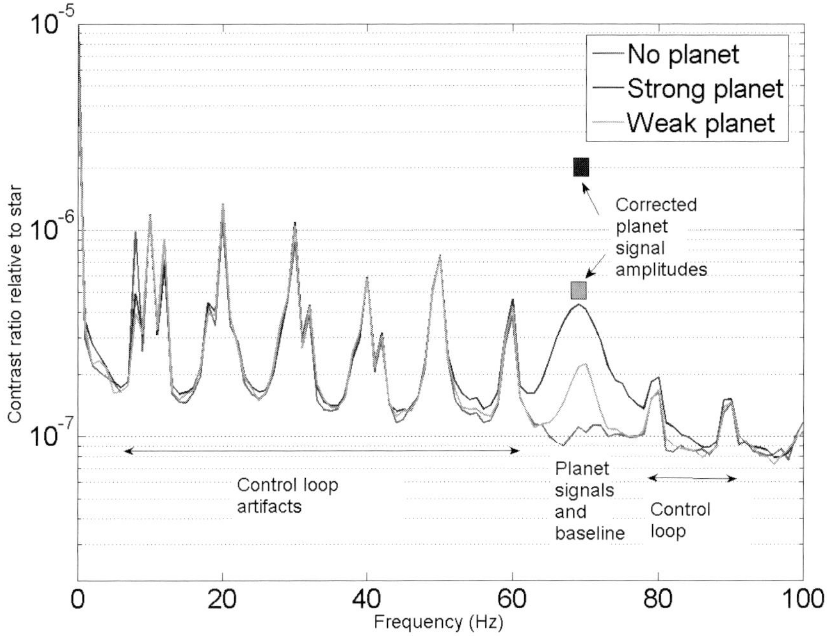

Figure 2. Spectrum of signal from nuller showing the control loop artifacts and the planet signal at 70 Hz.

frequency instability reduces the apparent planet contrast below its true value. To get a better estimate of the signal amplitude the planet signal was integrated over a 7 Hz bandwidth. Then, for the stronger signal, the amplitude is 2.78 10^{-6}, and for the weaker signal it is 1.32 10^{-6}. The noise amplitude in the same passband is 8.2 10^{-7}.

Subtracting the background noise, the ratio of the strong to the weak planet signal is 3.9 which agrees well with direct measurements of the planet signals in the absence of the star source. These measurements showed that the strong planet signal intensity would produce 1.23 10^{-5} V at the detector, and the weak signal 2.76 10^{-6} V, a ratio of 4.5. The planet to star contrast ratio from the direct measurements (strong signal) would be expected to be a little less than 1.97 10^{-6} which agrees well with the spectrally estimated value of 1.96 10^{-6}.

The mean null depth over the period 1700 to 2460 s was 1.37 10^5:1. The star:planet contrast at the strong setting was 5.1 10^5:1, comfortably exceeding the 10^5:1 goal, and at the weak setting it reaches 2.0 10^6:1.

5. Future plans

The next steps in testbed development require deeper nulls of order 10^{-6}, cophasing of the cross combiner, and planet detection at better than 10^{-6} contrast ratio with the star. Phase chopping will be employed at the crosscombiner. To achieve the deeper nulls additional stabilization equipment will be added to the testbed in the form of a laser beam coaligned with the starlight which will be detected by shear and pointing detectors located immediately prior to the nulling beamsplitter. Pointing control to better than 0.01 arc sec and shear control to better than 1 micron across a 25 mm diameter beam have been achieved in testing, which will be adequate for our needs. The second modification for deeper nulling will be the use of a laser OPD metrology system to operate in concert with the fringe tracker, replacing the dither loop now controlling the optical path. The metrology system is needed to compensate for the high-speed vibrations arising from the environment and from the mechanical choppers on the testbed that cannot be followed by the fringe tracker.

Acknowledgements

The authors would like to acknowledge the contributions of the other members of the testbed team: Kurt Liewer, Hong Tang, Randall Bartos, Steve Monacos, Raymond Lam, Boris Lurie and Nasrat Raouf. The work reported here was conducted at Jet Propulsion Laboratory, California Institute of Technology, under contract with NASA. Reference herein to any specific commercial product, process, or service by trade name, trademark, manufacturer, or otherwise, does not constitute or imply its endorsement by the United States Government or the Jet Propulsion Laboratory, California Institute of Technology.

References

Chauvin, G., Lagrange, A.-M., Dumas, C., Zuckerman, B., Mouillet, D., Song, I., Beuzit, J.-L., & Lowrance, P. 2004, *Astronomy and Astrophysics* 425, L29–L32

Marcy, G. W., Butler, R. P., Vogt, S. S., Fischer, D. A., Henry, G. W., Laughlin, G., Wright, J. T., & Johnson, J. A., 2005, *Ap.J*, 619, 570

Martin, S., 2005, *2005 IEEE Aeospace Conf* "TPF Planet Detection Testbed"

Martin, S. R., Gappinger, R. O., Loya, F. M., Mennesson, B. P., Crawford, S. L., & Serabyn, E., 2003, *Proc. SPIE* 5170, 144–154

Pedretti, E., Thureau, N. D., Wilson, E., Traub, W. A., Monnier, J. D., Ragland, S., Carleton, N. P., Millan-Gabet, R., Schloerb, F. P., Brewer, M. K., Berger, J.-P., & Lacasse, M. G., 2004, *Proc. SPIE* 5491, 540–550

Wallace, K., Hardy, G., & Serabyn, E., 2000, *Nature* 406, 700

Direct Imaging of Exoplanets: Science & Techniques
Proceedings IAU Colloquium No. 200, 2005
C. Aime and F. Vakili, eds.
© 2006 International Astronomical Union
doi:10.1017/S174392130600946X

Experimental Progress in Deep and Broadband Infrared Nulling for TPF-I

J. Kent Wallace[1], Randall Bartos, Robert Gappinger Frank Loya, Steve Moser, and John Negron

[1]Jet Propulsion Laboratory, California Institute of Technology, 4800 Oak Grove Dr., Pasadena, CA 91109, USA
email: James.K.Wallace@jpl.nasa.gov

Abstract. Our activities are focused upon the experimental demonstration of deep nulling in the mid-IR over a wide bandpass. Specifically, our near-term goal is demonstrating a contrast of 10^{-6} at 10 μm with a 25% spectral bandwidth. To meet this goal, several areas of technology development are required. These include: single-mode infrared fibers, bright infrared sources, laser path-length and tip/tilt metrology, and improvements to detection sensitivity. Progress in each of these areas of technical development will be reviewed as well as their impact on the overarching technical milestones.

Keywords. techniques: interferometric, instrumentation: detectors, infrared: general

1. Introduction

The challenges for an infrared nulling interferometry mission can be broken down into the ability to control systematic errors and the ability to deeply null over a broad spectral bandpass. This paper will describe the technology development that is a requirement for achieving deep and broadband nulls in the mid infrared independent of both the nulling architecture and the systematic errors.

2. Nulling Requirements

Our specific testbed requirements are driven entirely by a flow down of requirements that in turn are driven by our core science objectives, namely number of target stars. This allocation of factors that contribute to the ultimate achievable null contrast is outside the scope of this paper. This topic has been covered in great detail previously (Noecker, Lay, Ware & Dubovitsky (2003)). Our testbed is tasked with the experimental demonstration of deep nulls that are consistent with the objectives of the mission. The achromatic nulling testbed must demonstrate non-cryogenic nulling to a level of 10^{-6} for a thermal source of bandwidth >25% in the 8 to 12 μm region. This null is to be stable to an order of magnitude greater than the average null. For each of these architectures, deep nulling would not be possible without the development of a bright source, sensitive detector, mid-IR single mode filter and laser metrology for path-length stabilization. That is, deep contrast will only be achieved by a combination of the nulling architectures and the technical developments described herein.

The typical IR thermal source is a resistive element that has a blackbody temperature of about 1500 K. Achieving more photons in the mid IR is a challenge because as the temperature of a blackbody is increased, most of the gain in flux occurs at the shorter wavelengths in the visible portion of the spectrum. The flux at a wavelength of 10 μm grows almost linearly with temperature. Thus, in order to increase the flux by a significant amount, say an order of magnitude, the effective temperature must increase by almost an order of magnitude: 10,000K. We will describe the construction of a source with this effective temperature.

The noise floor of the detector limits the achievable dynamic range at null. Clearly dynamic range is also improved by not only going higher with the source but by going lower with the noise floor of the detection process. The current detector is a single pixel HgCdTe cooled to liquid nitrogen temperatures. One of the most sensitive detectors currently available is a single pixel of a MidIR array. These arrays require a more complicated arrangement: liquid helium operating temperatures, low thermal background environments, and a set of read out electronics to allow slow, full-frame operation as well as rapid readout of a few pixels. Our design for such a system and its current status will be described below.

Two wavefront errors must also be addressed in order to reach deep nulls. The cumulative effect of surface figure errors throughout the beam train serves not only to degrade the imaging quality it also prevents complete interference at the point of recombination. This can be mitigated to some degree by using optics of extremely good wave front quality, or by spatially filtering the wave front before detection. Its not an oversimplification to say that spatial filtering not only reduces the requirements on surface figure errors for the optics, but it makes deep nulling possible. Deep and broadband null would be unreachable were it not for single mode optical filters (fibers). Our solution to this vexing technical problem will be described.

Path length fluctuations due to mechanical vibrations in the experiment, or differential index of refraction fluctuations of air in the two arms of the interferometer will degrade nulls. These path length fluctuations can be monitored and stabilized with a laser metrology system that is independent of the nuller itself and only acts to make rigid the path length differences at the point of recombination. Computing power matured to the point where closed loop operation of a few hundred hertz and data recording at 10kHz is now feasible.

3. Nulling Technologies

3.1. *Arc Source*

Increasing the effective temperature is the only way to get a sufficient quantity of 10 μm photons for a blackbody radiator. We developed a source that was based upon the work of very similar source at NIST (Bridges & Migdal (1996)). Typically, arcs are quite bright in the visible portion of the spectrum, however at 10 μm, the source is not optically thick. A design whereby its possible to look along the length of the arc increases the optical depth and emissivity.

Our design made fundamental improvements to the previous designs (McIntosh, Wofford, Lucchese & Bevan (2001)) namely: improving the flow of chilled water through the housing, active cooling of the electrodes and creating a manifold for the distribution of argon gas. The source also employs a few additional safety features such as a temperature sensor on the body of the source, and a water flow meter for the chilled water supply. Both are interfaced to the electrical interlock such that a failure of either will shut the current source down. The system is also enclosed with a housing that makes accidental shorting of the electrodes virtually impossible. The system readily strikes with a tungsten welding rod. At peak operating conditions the systems consumes 8kW of electrical energy steady-state. The increase in flux from arc source is shown in figure 1.

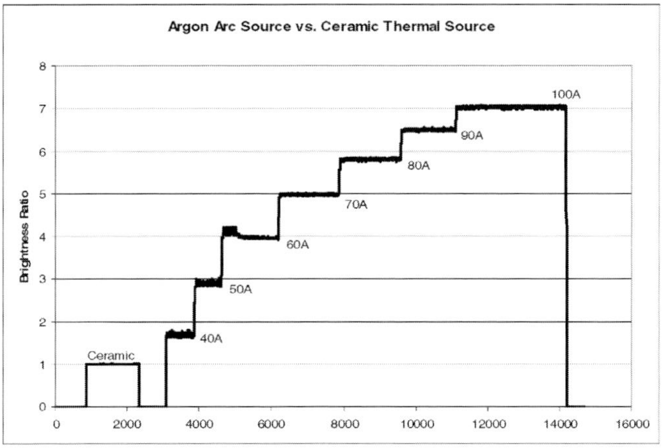

Figure 1. The improvement in IR flux from an argon arc source in comparison to the typical resistive element.

3.2. *MidIR Camera*

The detector of choice for improving the sensitivity of our null measurements is the arsenic-doped silicon, block-impurity band high-flux array from DRS Technologies. This device is 128x128 in size with 75 um pixels. If operates at a temperature of 10K and needs extensive baffling of the thermal background to minimize signal corruption. The dewar has several conveniences. An integral liquid cryogen level sensor for both the LN2 and the LHe will be used in conjunction with the autofill system for both the liquid cryogens. The system is also outfitted with adjustments external to the dewar that permit adjustment of the location of the PSF on the array by pivoting the whole dewar about a point that is also the center of the entrance pinhole. The LHe cold finger also provides a thermal contact to a cold baffle also at LHe temperatures. This provides a very cold solid angle to minimize thermal background. Quick and simple access to the optical mounting surface is accomplished by first removing the bottom dewar housing and then the inner cold shield. Since the electrical feedthroughs are located on the opposite side of this dewar split, there is no additional complication from disconnecting the fanout board. The optical beam train subsequent to the pinhole includes a re-collimating lens, a bandpass filter and low-order direct view prism followed by another lens that forms the image on the array. The array temperature is maintained by temperature sensor and heating element. A total of ten temperature sensors are placed at different locations along the beam train and on the cold plate. Two additional KF-40 flanges are also available in the event that more feedthroughs may be necessary.

3.3. *Single Mode Spatial Filters*

Single mode fibers have long be known for their ability to extract from a distorted wave front its underlying, perfect component (Mennesson, Ollivier & Ruiller (2000)). It is for this ability (to modally filter the output of an optical system) that we view these devices as absolutely essential in allowing us to demonstrate deep nulling. However, it should also be mentioned that a couple of nulling architectures rely upon a single mode input source in order to reach highest possible contrast. Both the through-focus and nulling periscopes require a single mode input source. These architectures perform a geometric flip of the pupil such that at the point of recombination, opposite sides of the pupil are interfering with each other. The only way to insure that this occurs completely is to have an input single mode source whose phase relationship with other parts of the pupil plane

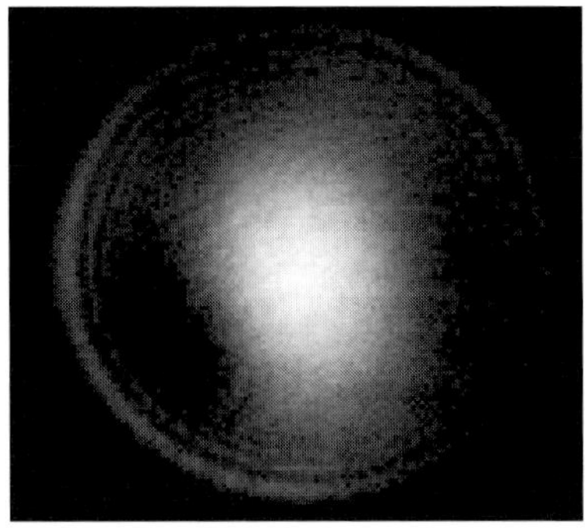

Figure 2. Far field intensity distribution of single-mode chalcogenide Mid IR fiber.

is constant. The spatial filters created for our testbed appear to be single mode fibers. These were procured via a subcontract with Naval Research Laboratories (NRL). Their double-crucible technique for creating near-IR fibers is well documented (Sanghera & Aggarwal (1998)). With a slight change in the glasses that form the core and cladding, they produced fibers which appear to be single mode.

The single mode behavior is predicted based upon a direct measurement of the index of refraction of the core and cladding glasses as well as the geometric radius of the core. The fibers are conveniently connectorized and packaged in a protective jacketing. The intensity distribution in the pupil plane using a single mode fiber (as measured in our experimental setup) is shown in 2. More extensive testing to characterize their single mode behavior is on-going.

Unfortunately, these fibers suffer from large Fresnel losses at each air/glass interface. The glasses employed have a high index of refraction: the result is that after passing through two fibers, the throughput is only 40% of what it would be without them.

3.4. *Laser path-length metrology*

The path length stabilization schemes for the Mid IR require a measurement wavelength that is proxy to the science band to stabilize the optical path internal to the interferometer. Fortunately for us, these laser metrology systems for path length stabilization have a long and fruitful history in interferometry. Our initial attempts at path length stabilization have permitted us to demonstrate sub-nanometer control. This level of control is consistent with our ultimate nulling milestone of one million to one. Unfortunately, piston stability is only part of our technical challenge. When coupling into a single mode spatial filter, point errors couple in as amplitude errors. Therefore tip/tilt metrology is required to maintain a constant angular alignment between the two arms of the interferometer. This angular metrology is accomplished by introducing another pair of linear metrology beams that are collinear with the original pencil beam for piston control. The signals from the three gauges then form a metric for piston, tip and tilt. We have built up a test station where we can test and develop the full tip/tilt/piston metrology system. It is wrapped around a compact nulling interferometer working at 10 μm. This will allows us to fully explore its behavior before integrating it with the rest of the nulling architectures.

4. Conclusions

Several areas of technology development are required in order to experimentally demonstrate deep nulling of mid IR light over a broad optical bandpass. We have described: 1) an argon arc source, 2) a mid infrared camera, 3) single mode spatial filters and 4) laser metrology for path length stabilization. We have made great strides in bringing all of these into reality and are beginning the process of integration. With all of these elements working in concert, we expect to experimentally demonstrate the nulling consistent with requirements for NASAs TPF-I.

Acknowledgements

The work described in this paper was carried out at the Jet Propulsion Laboratory, California Institute of Technology, under contract with the National Aeronautics and Space Administration and supported by the Terrestrial Planet Finder Program TPF.

References

Bridges, J.M. & Migdal, A.L. 1996, *Metrologia* 32, 625

McIntosh, A.L., Wofford, B.A., Lucchese, R.R., & Bevan J.W. 1991, *Infrared Physics & Technology* 42, 509

Mennesson, B., Ollivier, B., & Ruiller, C. 2000, *J. Opt. Soc. Am.* 19, 3, 596

Noecker, M.C., Lay, O.P., Ware, B., & Dubovitsky, S. 2003, *Proc SPIE* 5170, 79

Sanghera, J.S. & Aggarwal, I.D. 1977 in: J.S. Sanghera and I.D. Aggarwal (eds.), *Infrared Fiber Optics* (CRC Press Inc.), p. 325

All photographs: Laurent Thareau [l.thareau@free.fr].

Direct Imaging of Exoplanets: Science & Techniques
Proceedings IAU Colloquium No. 200, 2005
C. Aime and F. Vakili, eds.
© 2006 International Astronomical Union
doi:10.1017/S1743921306009471

Extrasolar Planets with AMBER/VLTI, What can we expect from current performances?

F. Millour[1,2], **M. Vannier**[4], **R. G. Petrov**[1], **B. Lopez**[3] **and F. Rantakyrö**[4]

[1]Laboratoire Universitaire d'Astrophysique de Nice - U.M.R. 6525
Université de Nice-Sophia Antipolis, Parc Valrose, 06108 Nice Cedex 02, France
[2]Laboratoire d'Astrophysique de Grenoble, U.M.R. 5571
Université Joseph Fourier/C.N.R.S., BP 53, F-38041 Grenoble Cedex 9, France
[3]Laboratoire Gemini, U.M.R. 6203
Observatoire de la Côte d'Azur/C.N.R.S., Avenue Copernic, 06130 Grasse, France
[4]European Southern Observatory, Casilla 19001, Santiago 19, Chile

Abstract. We present the current performances of the AMBER / VLTI instrument in terms of differential observables (differential phase and differential visibility) and show that we are already able to reach a sufficient precision for very low mass companions spectroscopy and mass characterization. We perform some extrapolations with the knowledge of the current limitations of the instrument facility.

We show that with the current setup of the AMBER instrument, we can already reach $3\sigma = 10^{-3}$ radians and have the potential to some low mass companions characterization (Brown dwarves or hypothetical very hot Extra Solar Giant Planets). With some upgrades of the VLTI infrastructure, improvements of the instrument calibration and improvements of the observing strategy, we will be able to reach $3\sigma = 10^{-4}$ radians and will have the potential to perform Extra Solar Giant Planets spectroscopy and mass characterization.

Keywords. techniques: interferometric, stars: planetary systems, instrumentation: high angular resolution

---

1. Introduction

In this paper we discuss the current highest performances of Colour-Differential Interferometry (CDI) on the AMBER instrument and compare these performances to signal amplitude we computed from low-mass companion simulations.

CDI is based on *simultaneous* interferometric observations in different spectral channels. As a high-angular resolution and high-dynamic technique, it presents two major advantages. First, the chromatic differences in visibility and phases are much less sensitive to instrumental and atmospheric instabilities, and therefore are easier to calibrate than the absolute complex visibility. Since the beginning of long-baseline optical interferometry with separated apertures, many early astrophysical results have been obtained using this self-calibration feature (Thom *et al.* (1986), Mourard *et al.* (1989)). Second, the colour-differential phase can be measured with an accuracy much better than the angular interferometric resolution λ/B. For objects much smaller than the diffraction limit, it is proportional to the variation of the object photocentre with wavelength.

This paper is placed in the context of Extra Solar Planets characterization with interferometry (Vannier *et al.* (2005)) and is intending to show that this technique has already the potential to get some scientific results on high contrast binaries.

2. Differential Observables computation and error bars

2.1. *The differential phase estimator*

The expression of the coherent flux on tha AMBER instrument is (Millour *et al.* (2004)):

$$C(t,\lambda) = 2N(\lambda)V_i(t,\lambda)V(\lambda)\sqrt{p_1(t,\lambda)p_2(t,\lambda)}\sqrt{\sum_{k=1}^{N_x} a_{1k}(\lambda)a_{2k}(\lambda)} \times e^{\phi_i(t,\lambda)+\phi_p(t,\lambda)+\phi_o(t,\lambda)+\phi_c(t,\lambda)} \quad (2.1)$$

In this equation k is the pixel index (spatial direction), $N(\lambda)$ is the unknown object's flux, $p_1(t,\lambda)$ and $p_2(t,\lambda)$ are transmission coefficients for the two combined beams, and $a_{1k}(\lambda)$ and $a_{2k}(\lambda)$ are related to specific features of each pixel (shape of the beam). $V_i(t,\lambda)$ is the instrumental contrast and $V(\lambda)$ is the amplitude of the complex visibility.

$\phi_p(t,\lambda)$ is the phase induced by the achromatic piston, $\phi_i(t,\lambda)$ is the instrumental-induced phase that varies with time (since the fixed part is already removed by the data reduction algorithm of AMBER as explained in Millour *et al.* (2004)), $\phi_o(t,\lambda)$ is the observed object's phase and $\phi_c(t,\lambda)$ is the chromatic phase induced by other causes (Chromatic atmospheric phase for example).

If we suppose that $\phi_i(t,\lambda) = 0$ (no variable instrumental phase), $\phi_c(t,\lambda) = 0$ (no chromatic effect) and $\phi_o(t,\lambda) = 0$ (unresolved or centro-symmetric object) then we can correct the complex coherent flux from the achromatic piston effect by:

$$C_{\mathrm{nop}}(t,\lambda) = C(t,\lambda) \times e^{\frac{-2i\pi\delta(t)}{\lambda}} \quad (2.2)$$

One can note that we need an estimation of the achromatic piston for each sample of time. This is a part of the Ph.D. thesis of Eric Tatulli (Tatulli (2004)) and it will not be explained in detail in this article. We compute a reference channel for each spectral channel, taking care of non biasing the interspectral term by removing the "work" spectral channel before averaging:

$$C_{\mathrm{ref}}(t,\lambda_k) = \langle C_{\mathrm{nop}}(t,\lambda_i)\rangle_{\lambda_i \neq \lambda_k} \quad (2.3)$$

We compute then the interspectral term between this reference spectral channel and the work spectral channel:

$$W(\lambda_k) = \left\langle \frac{C_{\mathrm{nop}}(t,\lambda_k)C_{\mathrm{ref}}(t,\lambda_k)^*}{|C_{\mathrm{ref}}(t,\lambda_k)|^2} \right\rangle_{(t)} \quad (2.4)$$

Then we compute the differential phase:

$$\phi_{\mathrm{diff}}(\lambda) = \arg(W(\lambda)) \quad (2.5)$$

This expression of the differential phase is quite accurate with a good achromatic piston correction. That is why it can be applied only to high flux sources (for example the star 51 Peg. has a K magnitude of 5, which is sufficient for this type of application). We can express the interspectral term by:

$$W(\lambda) = |W(\lambda)|e^{i\phi_{\mathrm{diff}}(\lambda)} \quad (2.6)$$

Please note that when $\phi_o(t,\lambda)$ cannot be neglected with regards to $\phi_p(t,\lambda)$, then the expression of the differential phase is different and contains a bias related to the interferometric phase, which leads to a bias in $W(\lambda)$. For well resolved objects, this effect has to be taken into account and leads to a specific treatment. Thus it is well beyond the scope of this paper and will be discussed in a futher one.

2.2. *The differential visibility estimator*

Going back to the interspectral term evaluation 2.4, we see that its modulus can be expressed as: $\frac{V(\lambda)}{V_{\mathrm{ref}}}$. If we have an unbiased estimate of this modulus (which we call $\widetilde{V_{\mathrm{diff}}}(\lambda)$), then we can compute the differential visibility. This estimate can be made with the real part of the differential phase correction of the interspectral term:

$$\widetilde{V_{\mathrm{diff}}}(\lambda) = \Re(W(\lambda) \times e^{-i\phi_{\mathrm{diff}}(\lambda)}) \tag{2.7}$$

This estimate of the differential visibility is unbiased as we expect the visibility and phase to be uncorrelated.

2.2.1. *In theory*

Starting from the theoretical estimation of the differential phase, we can express the differential phase and visibility noises from the fundamental photon $\sqrt{N_*}$, thermal $\sqrt{N_{\mathrm{th}}}$ and detector $\sigma_{\mathrm{RON}}\sqrt{n_{\mathrm{f}}\, n_{\mathrm{pix}}}$ noises (Petrov (1989)).

$$\sigma_\Phi = \frac{\sqrt{(N_* + N_{\mathrm{th}} + n_{\mathrm{f}}\, n_{\mathrm{pix}}\, \sigma_{\mathrm{RON}}^2)/2}}{V\,\langle N\rangle} \tag{2.8}$$

$$\sigma_V = \frac{\sqrt{N_* + N_{\mathrm{th}} + n_{\mathrm{f}}\, n_{\mathrm{pix}}\, \sigma_{\mathrm{RON}}^2}}{\langle N\rangle} \tag{2.9}$$

For information, the closure phase noise is given by:

$$\sigma_\psi = \sqrt{3} \times \sigma_\Phi \tag{2.10}$$

2.2.2. *in practice*

For the practical error computation, we perform a statistical dispersion of the observed points and assume a gaussian noise. The expression of the estimated noise for each observable is then the standard deviation of the measurements over an exposure:

$$\sigma_X = \sqrt{\frac{\sum_1^{n_{\mathrm{f}}}(X - \langle X\rangle)^2}{n_{\mathrm{f}}^2}} \tag{2.11}$$

3. Typical expected signal

3.1. *Simulating the companion*

We simulated a standard binary star model as in eq. 3.1 and applied the computation we currently use on the AMBER instrument to extract differential visibilities, differential phases and closure phase, computed the error bars assuming an average instrumental contrast of 50%, 1000 frames of integration, a detector noise of $11e^-$ per pixel and per frame and a total photon count of 4.2×10^7, the same figures as in the following section.

$$C_{jk}(\lambda) = \frac{1 + R(\lambda)e^{-2i\pi\overrightarrow{u_{jk}}\cdot\overrightarrow{\rho}}}{1 + R(\lambda)} \tag{3.1}$$

We used a modeled exoplanet spectrum from Barman *et al.* (2001) convolved with the resolution of the AMBER instrument in LR mode ($R \approx 35$). We scaled this flux ratio to a maximum value of $\approx 10^{-3}$ in order to estimate the signal amplitude of a 10 times brighter companion.

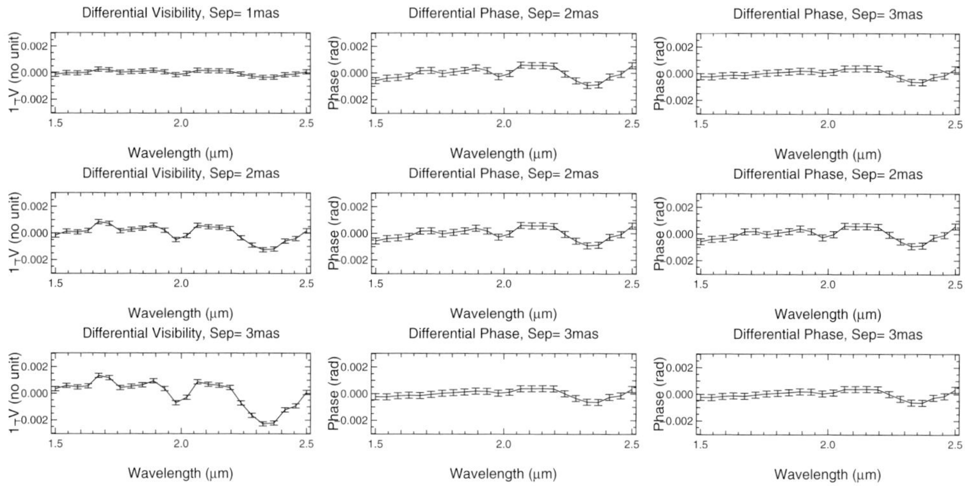

Figure 1. Simulations of an observation at VLTI with UT1-UT3-UT4 showing from left to right the differential visibility, the closure phase and the differential phase and from top to bottom separations ranging 1 to 3 mas. It shows a clear signal for all observables at 3mas but none for the closure phase at 1mas whereas the visibility and phase still have detectable signal. Photon amount is 4.2×10^7, detector noise $11e^-$, number of frames 1000, number of pixels 32 and average visibility 50%. These plots evidence the "super resolution" properties of differental phase and visibility relatively to closure phase.

3.2. *Results*

We tested the obtained signal for several Star/Companion separations (figure 1) and found that the signal detection at 3σ would occur for a separation of only 1mas for differential phase and visibility and 2.5mas for closure phase. This effects is due to the fact that if we perform a 1st order taylor expansion of the phase, we get a linear dependence with $\overrightarrow{u_{jk}} \cdot \overrightarrow{\rho}$ whereas for visibility, we get a squared dependence and for the closure phase we get a cubic dependence. So for small separations, we have $\phi_{\text{diff}} > V_{\text{diff}} > \psi$.

4. Best performances of AMBER

We used the AMBER / VLTI instrument to observe the bright calibration star HD70060 at low spectral resolution during the GTO run of 25 december 2004. We used the technique explained below to compute the differential phases and differential visibilities and computed statistical error bars. These ones were compared to the theoretical ones assuming a detector noise of 11 electrons and a null thermal noise. The figure 2 shows the resulting average differential phases for 5 successive exposures where we selected 50% of the best frames using the finge SNR as selection criterion. Each exposure represents about 20s of observation. They are separated by about 60 s.

The standard deviation σ in each exposure is the statistical dispersion of the differential phase per spectral channel as described in eq. 2.11. We performed weighted averages using the SNR on the fringe signal as weights. n_ϕ is the total number of collected photons per spectral channel and σ_ϕ is the accuracy expected from measured flux, detector noise and a supposed object visibility of 1 (which means the measured visibility is supposed to be only the instrumental one).

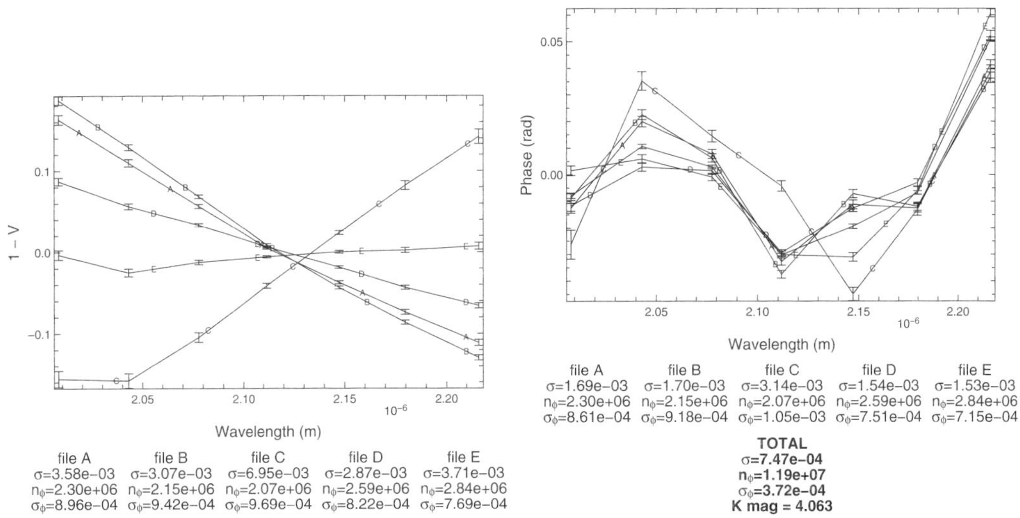

Figure 2. Differential Visibilities and Phases of the calibrator star HD70060 for 5 successive exposures (lines with letters) and the resulting average of the 5 exposures (thick line).

4.1. *Differential Visibility*

The differential visibility values vary only by 4×10^{-3} with time within each exposure but the general slope of the curve changes dramatically between exposure, leading to a variation of about 0.05 radians rms over 5 minutes. This variation is dominated by the changes in the achromatic piston jitter due to seeing and vibrations fluctuations. In the current VLTI situation we have no tool to correct this effect, except including an estimation of the exposure jitter in the model fitting, with an impact on the SNR which cannot be estimated. So the differential visibility is currently not usable for very high accuracy applications. This situation will change dramatically when a fringe tracker is operational. Then we will be affected only by the residual piston jitter after fringe tracking correction.

4.2. *Differential Phase*

The rms variation of the measures within each exposure is of typically 1.8 milliradians (i.e. $\approx 1\mu$arcsecond in colour-differential astrometry). This is about two times the expected rms from fundamental noise. Over the total 5 minutes we get 0.9 milliradians, again very close to twice the photon noise. This shows that the different measures seems statistically independent. An average of 1200 such exposures (15 hours) is needed to reach the 0.5×10^{-4} accuracy needed for the spectroscopy of τ Bootis b. The brighter planet considered in §3 could be observed in 1 hour.

With the improvement of the VLTI (less vibrations, improved overheads) we could expect to use almost all frames instead of only 50% of them with an average instrumental contrast improved by a factor 2. Then the τ Boo observation would be achievable in a couple of hours. We also see a pattern as a function of lambda with a 10^{-2} radians rms over the K band. This pattern is stable over the 5 minutes of observations considered here. That means that it can be eliminated by a fast calibration cycle. We think that it is a mixture of atmospheric dispersion and measurement effects. Measurement effects can be eliminated by beam commutation. Atmospheric dispersion will be eliminated in the closure phase and we plan to try to fit it in the differential phase.

4.3. *Closure Phase*

In this relatively poor quality early data, we have too little frames where the three fringe patterns are good enough. This explains why the closure phase is much more noisy (typically 10^{-2} radians rms) than the differential phases. We are therefore unable to say what part of the differential phase pattern is due to differential chromatic OPD.

5. Conclusion

The preliminary data reduction of bright sources observed in low spectral resolution with AMBER shows that the measured differential phases are accurate and stable enough to achieve the spectroscopy and angular separation of the most favorable Pegasi planets in a few 15 hours observations. This value should be reduced to 2 hours with the foreseen simple improvements of the VLTI. The resulting spectra would be affected by an instrumental term and/or an atmospheric chromatic differential OPD term producing a smooth 10^{-2} radians pattern over the K band.

When the instrumental term will be eliminated by beam commutation, the remaining differential OPD might be possible to fit in the data reduction procedure. However, only a successful use of closure phase guarantees the elimination of the differential OPD. The current quality of the VLTI does not allow accurate closure phase measurements, but this should be improved soon, when the three fringe pattern are better stabilized. We remain very optimistic about the possibility to do spectroscopy of Pegasi planets with AMBER quite sson.

Acknowledgements

The data presented here was taken at the Paranal Observatory in Chile within the AMBER Guaranteed Time.

We thank all the consortium members listed in `http://amber.obs.ujf-grenoble.fr`.

References

Thom, C., Granes, P., & Vakili, F. 1986, *A&A* 165, L13–L15

Mourard, D., Bosc, I., Labeyrie, A., Koechlin, L., & Saha, S. 1989, *Nature* 342, 520–522

Vannier, M., Petrov, R.G., Lopez, B., & Millour, F. 2005, *MNRAS*, Accepted

Millour, F., Tatulli, E., Chelli, A. E., Duvert, G., Zins, G., Acke, B., & Malbet, F. 2004, *New Frontiers in Stellar Interferometry.*, SPIE 5491, 1222

Eric Tatulli 2004, *Ph. D. Thesis*, Université Joseph Fourier

Petrov, R. G. 1989, *Diffraction-Limited Imaging with Very Large Telescopes*, NATO ASIC Proc. 274:, 249

Barman, T. S., Hauschildt, P. H., & Allard, F. 2001, *ApJ* 556, 885–895

Direct Imaging of Exoplanets: Science & Techniques
Proceedings IAU Colloquium No. 200, 2005
C. Aime and F. Vakili, eds.
© 2006 International Astronomical Union
doi:10.1017/S1743921306009483

Exoplanet detection from Dome C, Antarctica: opportunities and challenges

J. S. Lawrence, M. C. B. Ashley, M. G. Burton, and J. W. V. Storey

School of Physics, University of New South Wales, Sydney, NSW, 2052, Australia
email: jl@phys.unsw.edu.au

Abstract. Many methods have been proposed for the detection of exoplanets. The minimum detectable planetary size and mass, and the maximum distance from us at which exoplanets are observable, are determined by both technological and environmental constraints. The unique atmospheric conditions found at Dome C offer significant advantages. Using what is now known about the turbulence profile of the atmosphere above Antarctic plateau sites, we explore the potential these sites offer for detecting exoplanets around nearby stars using various methods.

Keywords. planetary systems, telescopes, instrumentation: adaptive optics, atmospheric effects, techniques: high angular resolution.

1. Introduction

The opening in 2005 of the French/Italian Concordia Station at Dome C opens new opportunities for astronomers. The combination of high altitude, low temperatures and an extremely stable atmosphere leads to major gains in sensitivity and image resolution. Previous studies at the Amundsen-Scott Station at the South Pole have shown that the infrared background on the Antarctic plateau can be as much as two orders of magnitude lower than at temperate sites, while summer-time measurements at Dome C show that the atmospheric transmission in some parts of the 10 micron window (Walden *et al.* 2005) can exceed 99.5%. Above a thin (~36m), turbulent ground layer (Agabi *et al.* 2005; Aristidi *et al.* 2005), the atmosphere is extraordinarily stable: an average seeing of 0.27 arcseconds has been measured over a 6-week period (Lawrence *et al.* 2004). Table 1 lists relevant atmospheric parameters found at Dome C compared with South Pole and typical mid-latitude sites. Although mounting a telescope above the turbulent layer, i.e. on a 20–40 m tower or hill, may at first seem to be a difficult technological challenge, many existing telescopes at other sites are located at similar heights above ground level (e.g. the AAT at 26 m, the CFHT at 28 m, the ESO 3.6 m at 30 m, and the 4 m Mayall at 57 m).

Table 1. Atmospheric parameters relevant for exoplanet detection

Site	Seeing (arcsec)	Coherence Time (ms)	Isoplanatic Angle (arcsec)	M Band Sky Flux (Jy/as^2)	N Band Sky Flux (Jy/as^2)
Dome C	0.27[a]	7.9[a]	5.7[a]	0.5[b]	20[b]
South Pole	1.8	1.6	3.2	0.6	25
Mauna Kea	0.5	2.7	1.9	40	200
Cerro Paranal	0.8	3.3	2.6	>50	>200

[a] Mean measurements 30 m above the surface by Lawrence *et al.* 2004. Measurements by Agabi *et al.* (2005) agree with these values within uncertainties.
[b] Estimates for mid-winter, based on Walden *et al.* (2005) mid-summer measurements.

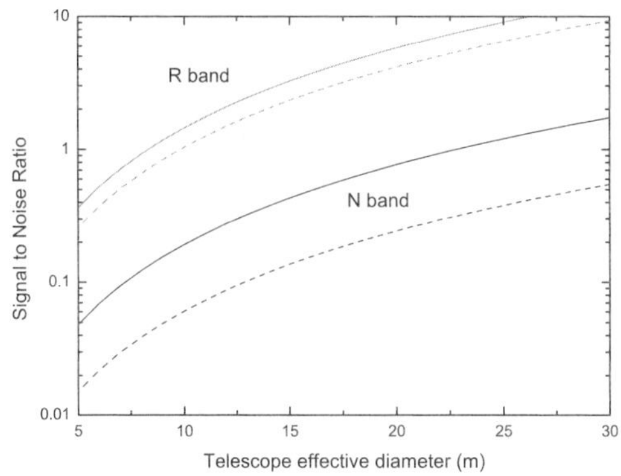

Figure 1. Photon noise limited SNR versus telescope diameter for exo-Earth detection at R (black) and N (grey) bands for Dome C (solid) and Mauna Kea (dashes) in 12 hours integration.

2. Direct Imaging

The planet-to-star flux ratio detectable with a direct imaging high-order adaptive optics coronagraphic system operating at visible or near-infrared wavelengths depends strongly on the atmospheric coherence length and coherence time. Previous models based on predictions of Dome C turbulence by Lardiere *et al.* (2004) and Angel (2003) showed large gains in the achievable SNR compared with mid-latitude sites for a given telescope size. However, recent measurements (Lawrence *et al.* 2004; Agabi *et al.* 2005) show the Dome C atmospheric coherence times may be somewhat lower than expected. Based on these measurements, Guyon (2005) has calculated photon-limited contrast ratios for Dome C and mid-latitude sites.

Provided that wavefront sensing and science imaging are performed at the same wavelength the ratio of photon-limited contrast achievable by a Dome C telescope to that of a Mauna Kea telescope is ~1.9 at visible and near infrared wavelengths. This results in a relative gain in SNR of ~1.4 at R band, as illustrated in Figure 1. While this is a smaller advantage than previously anticipated (Lardiere *et al.* 2004; Angel 2003), it is still significant. A Dome C telescope should be capable of imaging fainter planets around stars at greater distance than a mid-latitude system.

Direct planet imaging in the mid-infrared is limited by atmospheric emission rather than site turbulence characteristics. Although large telescopes or interferometers are required to directly image exoplanets in the mid-infrared, the advantage of Dome C relative to mid-latitude sites is much greater than in the visible or near infrared, as a result of the lower thermal sky emission (Angel 2003). For N band imaging the SNR ratio gain for Dome C relative to Mauna Kea is a factor of 3.2, as shown in Figure 1.

The simple models used for these predictions explore the fundamental limitations to direct imaging of exoplanets arising from photon noise alone. The speckle noise component of the uncorrected stellar halo will be significantly larger than the photon noise component. While techniques have been proposed to completely remove speckle noise, it is currently unknown to what extent they will be successful. Any additional contribution to the SNR from speckle noise, however, will be strongly dependent on the atmospheric characteristics, and thus it is likely that the slower turbulence observed at Dome C will be highly beneficial.

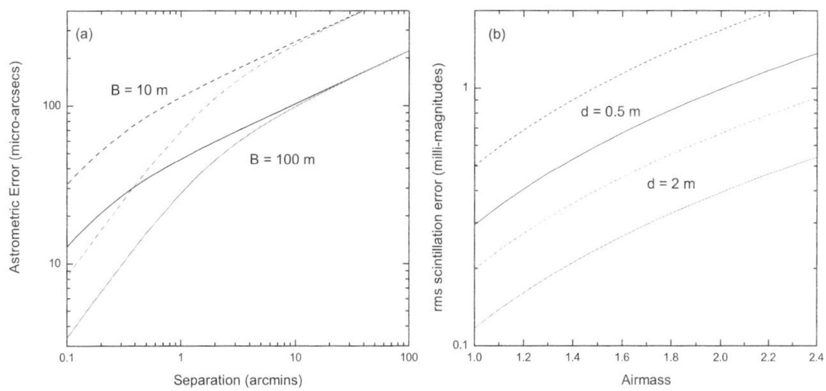

Figure 2. (a) Astrometric error versus angular separation for a 10 m (black) and 100 m (grey) baseline interferometer with a 1 hour integration at Dome C (solid) compared with Mauna Kea (dashes) (b) Scintillation-limited photometric precision as a function of airmass for 0.5 m (black) and 2 m (grey) telescopes at Dome C (solid) compared to Mauna Kea (dashes).

3. Gravitational Microlensing

The detection of exoplanets via gravitational microlensing is not fundamentally limited by any specific atmospheric characteristic or measurement accuracy. Rather, a range of optimum conditions are required to increase the detection probability. The timescale for microlensing light curves generally ranges from a few hours to a few days. Currently, telescopes at multiple sites around the world are required to examine single microlensing events. At Dome C the possibility of continuous uninterrupted observations over long time periods (hundreds of hours) with cloud free skies provides a significant advantage. In order to reduce photometric confusion from observations in crowded fields, microlensing searches require high spatial sampling with high sensitivity over large fields. The natural seeing above the boundary layer at Dome C provides higher spatial resolution than any other ground based site. Additionally, wide field adaptive optics correction at Dome C is much simpler than elsewhere due to the longer coherence time and wider isoplanatic angle.

4. Astrometry

The main limitation to exoplanet detection via ground-based narrow-angle dual beam astrometric interferometry arises from atmospheric tilt anisoplanatism. The benefits of the Antarctic atmosphere in reducing this error were examined by Lloyd *et al.* (2002), who demonstrated that the weak high-altitude turbulence observed at the South Pole should lead to an increase in astrometric precision. The weaker high-altitude turbulence observed at Dome C compared to South Pole should result in a Dome C interferometer achieving an astrometric error ~40% of that achievable from most mid-latitude sites (Shao & Colavita 1992), as demonstrated in Figure 2(a). This reduction in astrometric error is unlikely to allow the detection of Earth mass planets from Dome C, as this would require the precision of 1 micro-arscec possible only from space. It will, however, lead to the capability of detecting much lower mass planets at larger distances than is possible from typical mid-latitude sites.

5. Transit searches

The fundamental limitation to the lowest planet mass detectable with a ground-based planetary transit system is atmospheric scintillation noise, which limits photometric precision to tenths of a milli-magnitude at temperate sites. The precision of ground-based systems is currently within a factor of ten of this limit. The amount of atmospheric scintillation at any site is strongly dependent on high-altitude turbulent layers. At Dome C, the high-altitude turbulence is much weaker than typically found elsewhere; this results in an improvement to the photometric precision, as shown in Figure 2(b). While the reduction in scintillation at Dome C will not be enough to achieve the ($<0.01\%$) precision necessary to detect exo-Earths, planets of significantly lower mass should be detectable than can be seen elsewhere. Additional benefits of Dome C for transit search programs include the exceptionally good clear-sky statistics, the possibility of continuous observations, a more stable point-spread function over wide fields, and lower systematic errors from airmass variations.

6. Conclusions

The Dome C site offers a number of advantages for science programs dedicated to the detection of exoplanets on telescopes of various size. The proposed 2 m aperture PILOT (Pathfinder for an International Large Optical Telescope) will be capable of transit and gravitational microlensing searches (Burton *et al.* 2005). An 8 m aperture telescope would be competitive with the next generation of mid-latitude Extremely Large Telescopes (ELTs) for much of their science, particularly in the area of infrared planet detection. Ultimately, a 25-metre telescope in Antarctica, such as Giant Magellan Telescope Antarctica (Angel *et al.* 2005) would surpass the performance of other ELTs by a wide margin.

References

Agabi, A., Aristidi, E., Azouit, M., Fossat, E., Martin, F., Sadibekova, T., Vernin, J., & Ziad, A., 2005, *PASP*, in press

Angel, J.R.P., 2003, Towards Other Earths, *SP*-539, 221

Angel, J.R.P., Lawrence, J.S., & Storey, J.W.V., 2005, *Proc. SPIE*, 5382, 76

Aristidi, E., Agabi, A., Azouit, M., Fossat, E., Martin, F., Sadibekova, T., Vernin, J., Ziad, A., & Travouillon, T., 2005, *Acta Astr. Sinica Supp.*, in press

Burton, M.G., *et al.*, 2005, *PASA*, 22, 199

Guyon, O., 2005, *ApJ*, 629, 592

Lardiere, O., Salinari, P., Jolissaint, L., Carbillet, M., Riccardi, A., & Esposito, S., 2004, *Proc. SPIE*, 5382, 550

Lawrence, J.S, Ashley, M.C.B., Travouillon, T., & Tokovinin, A., 2004, *Nature* 461, 278

Lloyd, J.P., Oppenheimer, B.R., & Graham, J.R., 2002, *PASA*, 19, 318

Shao, M. & Colavita, M.M., 1992, *A&A*, 262, 353

Walden, V.P, Town, M.S., Halter, B., & Storey, J.W.V., 2005, *PASP*, 117, 300

Direct Imaging of Exoplanets: Science & Techniques
Proceedings IAU Colloquium No. 200, 2005
C. Aime and F. Vakili, eds.
© 2006 International Astronomical Union
doi:10.1017/S1743921306009495

Detection of Exoplanets from the Antarctic Plateau

James P. Lloyd
Cornell University, Ithaca NY, USA

Abstract. The limitation to ground based astronomy is the Earth's atmosphere. The atmosphere above the Antarctic plateau is fundamentally different in many regards compared to the atmosphere at temperate sites. The extreme altitude, cold and low humidity offer a uniquely transparent atmosphere at many wavelengths. Studies at the South Pole have shown additionally that the turbulence properties of the night time polar atmosphere are unlike any mid latitude sites. The lack of high altitude turbulence combined with low wind speeds presents favorable conditions for interferometry. More recent site testing at Dome C has revealed the most superb seeing of any site tested. The unique properties of the polar atmosphere can be exploited for Extra-solar Planet studies with differential astrometry, differential phase and nulling interferometers.

Keywords. Turbulence, Seeing, Adaptive Optics, Interferometry, Antarctica.

1. Introduction

Properties of the atmosphere are the limit for ground based astronomy. Transparency and sky brightness limit the sensitivity of a telescope, and atmospheric turbulence, "seeing", limits resolution. For seeing limited observations, the natural seeing determines the image size, which is of importance not only for resolution, but also for sensitivity as the light from a point source is concentrated into a smaller area, thus suffering less from noise from the contaminating background. For this reason, astronomers have been seeking to deploy telescopes at high altitude sites with stable air for several decades. With the advent of adaptive optics and interferometry, the loss of resolution can be recovered by measuring and tracking the aberrations induced by the atmospheric turbulence. Although it is widely perceived that these techniques can correct for seeing, the seeing enters into the difficulty of the problem in more ways than the image area. The temporal and angular dimensions of coherence of the atmosphere, in addition to the spatial, degrade adaptive optics and interferometry performance. Therefore, these techniques are even more sensitive to degradation by atmospheric turbulence than uncompensated imaging.

It has been recognized for some time that the Antarctic offers the potential for not only exquisitely clear and dark skies, but also for the lowest level of turbulence, with its lowered diurnal variation, high altitude, and remarkably flat topography Gillingham (1993). The combination of superb seeing, low background and high transmission in the thermal infrared allows high levels of performance with an interferometer for the detection and study of extrasolar planets Lloyd *et al.* (2002), Lloyd *et al.* (2003a), Lloyd *et al.* (2003b), Swain *et al.* (2003a), Swain *et al.* (2003c), Swain *et al.* (2003b), Vakili *et al.* (2005), Vakili *et al.* (2004), Swain *et al.* (2004), Fossat *et al.* (2004).

2. The Antarctic Atmosphere

As the optical effect of the turbulence results from the combination of mechanical turbulence with air temperature it is worth examining the processes that generate these. The Palmen-Newton model of the general circulation of the atmosphere (see Figure 1)

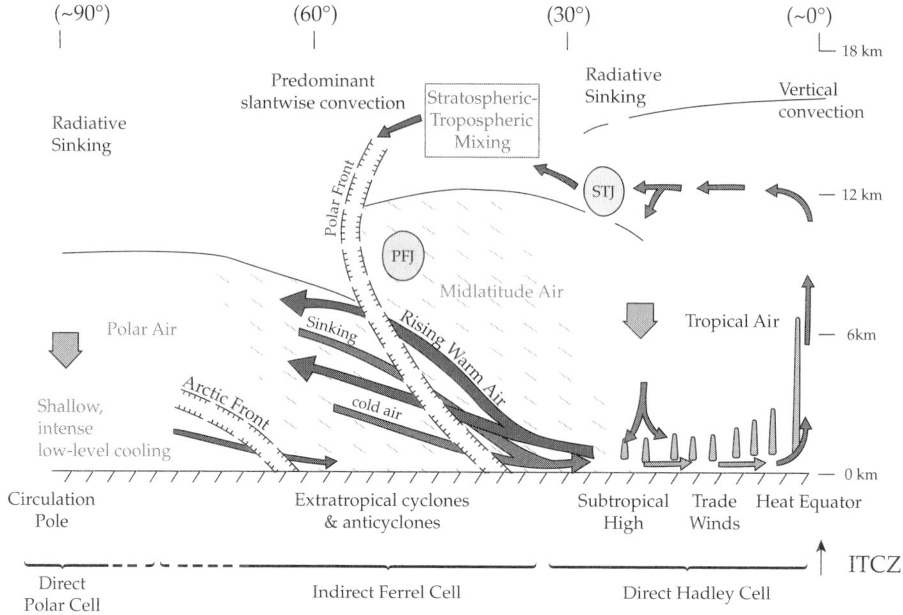

Figure 1. Global circulation and tropopause patterns, adapted from Linacre & Geerts (1997).

illustrates several of the salient basic processes. The heating of the sun drives convective turnover on large scales in the tropics and mid latitudes. This turnover in turn drives strong winds such as the trade winds at low altitudes and the jetstreams at high altitudes. Variations in local topography, such as mountains, oceans and deserts, result in inhomogeneous heating that constantly produces synoptic weather patterns with large scale inhmogeneous air masses. Of particular concern for astronomy are places where the combination of wind shear and temperature inhomogeneities results in strong turbulence, such as the jet stream and fronts between hot and cold air masses. Periodically, the strong turbulence produced at these interfaces degrades the seeing at mid-latitude sites to very poor levels.

By comparison, the atmosphere over Antarctica is remarkably simple. The sole dominant process is the cooling and sinking of air over the plateau, resulting in katabatic winds. As the Antarctic plateau is nearly uniformly flat and white as a result of the ice cap, there is very little potential to drive convection on large scales. In the wintertime, the atmosphere becomes very nearly adiabatic King & Turner (1997). In an adiabatic atmosphere, even the injection of mechanical turbulence will not result in optical turbulence, as the density of the air remains uniform at all altitudes. The consequences of these remarkable conditions for the broad context of interferometry and adaptive optics are discussed in Lloyd (2004).

3. South Pole Seeing

On the basis of these considerations, the seeing at Antarctic plateau sites should be superior to any mid-latitude sites, as has been recognized previously Gillingham (1993), Gillingham (1991). Comprehensive site testing at the South Pole has characterized not only seeing, but background and transparency at a range of wavelengths Travouillon *et al.* (2003b), Travouillon *et al.* (2003a)Travouillon, Ashley, Burton, Storey, Conroy, Hovey, Jarnyk, Sutherland, & Loewenstein, Travouillon *et al.* (2002), Storey *et al.* (2002), Storey *et al.* (2000), Storey (1998), Storey *et al.* (1996), Loewenstein *et al.* (1998). However, the

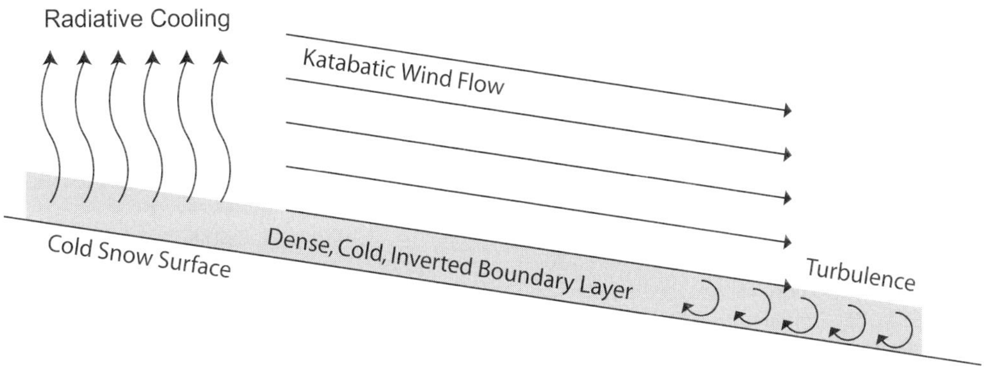

Figure 2. Generation of turbulence in the boundary layer by katabatic wind across the strong wintertime temperature inversion.

seeing at the South Pole was found to be relatively poor, with median seeing of 1.7" at 0.5 μm.

The vertical distribution of the turbulence was extensively studied with balloon measurements Marks *et al.* (1996), Marks *et al.* (1999), Marks (2002) and SODAR Travouillon *et al.* (2003b) and found to be confined to a low altitude layer. The radiative cooling of the ice to space results in a strong thermal inversion in the lower 200 meters of the atmosphere. As the South Pole is significantly downhill from the high point of the plateau, it experiences a constant katabatic wind of several meters per second in the wintertime. The sloshing of the cold boundary layer under the action of this wind produces the optical turbulence that results in the relatively poor seeing. It is worth noting, however, that although this seeing is poor for the purposes of image formation, it is less degrading to adaptive optics and interferometry applications due to larger coherence times and isoplanatic angles than comparable seeing at a mid-latitude site Lloyd *et al.* (2003a). This component of the turbulence, however, is likely to be dramatically lowered in both thickness and strength at the low wind sites on the high plateau. The free atmosphere turbulence measured at the South Pole indicates that under such conditions, exquisite seeing might indeed prevail. Recent observations of the wintertime seeing at Dome C Agabi *et al.* (2005) indicate that Dome C suffers from a dramatically thinner (~30 m), but remarkably turbulent boundary layer. Careful study of the boundary layer is clearly required to verity the seeing conditions.

4. Conclusion

The ambitious goal of direct detection of exoplanets will clearly require large telescopes and intereferometers at the most superb sites both on the ground and in space. The unique conditions that prevail in the Antarctic atmosphere lead to fundamentally different atmospheric turbulence properties, which are particularly favorable for interferometry and adaptive optics.

Acknowledgements

This work has been supported in part by the National Science Foundation under grant AST-0335695.

References

Agabi, A., Aristidi, E., Azouit, M., Fossat, E., Martin, F., Sadibekova, T., Vernin, J., & Ziad, A. 2005, ArXiv Astrophysics e-prints

Fossat, E., Aristidi, E., & Agabi, K. 2004, in EAS Publications Series, 125–133

Gillingham, P. 1993, in Optics in Astronomy: 32nd Herstmonceux Conference, 244+

Gillingham, P. R. 1991, Proceedings of the Astronomical Society of Australia, 9, 55+

King, J. C. & Turner, J. 1997, Antarctic Meterology and Climatology (Cambridge University Press)

Linacre, E. & Geerts, B. 1997, Climates and Weather Explained (Routledge)

Lloyd, J. P. 2004, in New Frontiers in Stellar Interferometry, Proceedings of SPIE Volume 5491. Edited by Wesley A. Traub. Bellingham, WA: The International Society for Optical Engineering, 2004., p.190, 190–+

Lloyd, J. P., Lane, B. F., Swain, M. R., Storey, J. W., Travouillon, T., Traub, W. A., & Walker, C. K. 2003a, in Techniques and Instrumentation for Detection of Exoplanets. Edited by Coulter, Daniel R. Proceedings of the SPIE, Volume 5170, pp. 193–199 (2003)., 193–199

Lloyd, J. P., Oppenheimer, B. R., & Graham, J. R. 2002, Publications of the Astronomical Society of Australia, 19, 318

Lloyd, J. P., Storey, J. W., Swain, M. R., Traub, W. A., & Walker, C. K. 2003b, Astronomy in Antarctica, 25th meeting of the IAU, Special Session 2, 18 July, 2003 in Sydney, Australia, 2

Loewenstein, R. F., Bero, C., Lloyd, J. P., Mrozek, F., Bally, J., & Theil, D. 1998, in ASP Conf. Ser. 141: Astrophysics From Antarctica, 296–+

Marks, R. D. 2002, 385, 328

Marks, R. D., Vernin, J., Azouit, M., Briggs, J. W., Burton, M. G., Ashley, M. C. B., & Manigault, J. F. 1996, 118, 385

Marks, R. D., Vernin, J., Azouit, M., Manigault, J. F., & Clevelin, C. 1999, 134, 161

Storey, J., Ashley, M., & Burton, M. 2002, in ASP Conf. Ser. 266: Astronomical Site Evaluation in the Visible and Radio Range, 524–+

Storey, J. W., Ashley, M. C., & Burton, M. G. 2000, in Proc. SPIE Vol. 4008, p. 1376–1382, Optical and IR Telescope Instrumentation and Detectors, Masanori Iye; Alan F. Moorwood; Eds., 1376–1382

Storey, J. W. V. 1998, in ASP Conf. Ser. 141: Astrophysics From Antarctica, 313–+

Storey, J. W. V., Ashley, M. C. B., & Burton, M. G. 1996, Publications of the Astronomical Society of Australia, 13, 35

Swain, M., Lloyd, J., Traub, W., Walker, C., Stark, A., Lawson, P., Storey, J., Coude du Foresto, V., Fossat, E., Ireland, M., Burrows, A., & Vakili, F. 2003a, American Astronomical Society Meeting, 203,

Swain, M. R., Coude du Foresto, V., Fossat, E., & Vakili, F. 2003b, Societa Astronomica Italiana Memorie Supplement, 2, 207

Swain, M. R., Walker, C. K., Storey, J. W., Traub, W. A., & Lloyd, J. P. 2003c, Astronomy in Antarctica, 25th meeting of the IAU, Special Session 2, 18 July, 2003 in Sydney, Australia, 2

Swain, M. R., Walker, C. K., Traub, W. A., Storey, J. W., Coude Du Foresto, V., Fossat, E., Vakili, F., Stark, A. A., Lloyd, J. P., Lawson, P. R., Burrows, A. S., Ireland, M., Millan-Gabet, R., van Belle, G. T., Lane, B. F., Vasisht, G., & Travouillon, T. 2004, in New Frontiers in Stellar Interferometry, Proceedings of SPIE Volume 5491. Edited by Wesley A. Traub. Bellingham, WA: The International Society for Optical Engineering, 2004., p.176, 176–+

Travouillon, T., Ashley, M. C. B., Burton, M. G., Calisse, P. G., Everett, J. R., Lawrence, J. S., & Storey, J. W. V. 2002, in SF2A-2002: Semaine de l'Astrophysique Francaise, 19–+

Travouillon, T., Ashley, M. C. B., Burton, M. G., Storey, J. W. V., Conroy, P., Hovey, G., Jarnyk, M., Sutherland, R., & Loewenstein, R. F. 2003a, 409, 1169

Travouillon, T., Ashley, M. C. B., Burton, M. G., Storey, J. W. V., & Loewenstein, R. F. 2003b, 400, 1163

Vakili, F., Aristidi, E., Schmider, F. X., Jankov, S., Fossat, E., Abe, L., Domiciano, A., Belu, A., Agabi, A., Daban, J.-B., Robbe-Dubois, S., Petrov, R., Vernin, J., Ziad, A., & Lopez, B. 2005, in EAS Publications Series, 211–217

Vakili, F., Belu, A., Aristidi, E., Fossat, E., Maillard, A., Abe, L., Agabi, K., Vernin, J., Baptiste Daban, J., Hertmanni, W., Schmider, F.-X., Assus, P., Coude Du Foresto, V., & Swain, M. R. 2004, in New Frontiers in Stellar Interferometry, Proceedings of SPIE Volume 5491. Edited by Wesley A. Traub. Bellingham, WA: The International Society for Optical Engineering, 2004., p.1580, 1580–+

Direct Imaging of Exoplanets: Science & Techniques
Proceedings IAU Colloquium No. 200, 2005
C. Aime and F. Vakili, eds.
© 2006 International Astronomical Union
doi:10.1017/S1743921306009501

ALADDIN: an optimized ground-based precursor for DARWIN

Vincent Coudé du Foresto[1], Olivier Absil[2], Marc Barillot[3], and Mark Swain[4]

[1]LESIA, Observatoire de Paris, 92190 Meudon, France
email: vincent.foresto@obspm.fr

[2]Institut d'Astrophysique et de Géophysique, Université de Liège, 4000 Sart-Tilman, Belgium
email: absil@astro.ulg.ac.be

[3]Alcatel Alenia Space, 06150 Cannes, France
email: marc.barillot@alcatelaleniaspace.com

[4]Max-Planck Institut fur Astronomie, 69117 Heidelberg, Germany
email: swain@mpia-hd.mpg.de

Abstract. The ALADDIN concept is an integrated Antarctic-based L-band experiment whose purpose is to demonstrate nulling interferometry and to prepare the DARWIN mission. Because of their privileged location, the relatively modest collectors (1 m) and baseline (up to 40 m) are sufficient to achieve a sensitivity (in terms of detectable zodi levels) which is about twice better than that of a nulling instrument on a large interferometer (such as GENIE at the VLTI), and to reach the 20-zodi threshold value identified to carry out the DARWIN precursor science. These numbers are based on a preliminary design study by Alcatel Alenia Space and were obtained using the same simulation software as the one employed for GENIE. The integrated design enables top-level optimization and full access to the light collectors for the duration of the experiment, while reducing the complexity of the nulling breadboard.

Keywords. Instrumentation: interferometers, antarctic astronomy, techniques: interferometric, planetary systems, protoplanetary disks.

1. Introduction

The DARWIN mission considered by the European Space Agency (detection and characterization of habitable planets around nearby stars using interferometric nulling (Léger *et al.* (1996)) will require a ground-based precursor, whose main objectives are to validate on-sky the technique of nulling interferometry, as well as to carry out scientific observations in preparation of the space mission. Most notably the amount of dust around DARWIN targets needs to be investigated as high levels of exozodiacal light (about 20 times, or more, the amount of the solar level), or an asymmetric distribution of dust, may hamper the detection of Earth-like planets.

The first option for a ground-based precursor is to build an instrument designed for an existing interferometer, such as the GENIE L-band nulling instrument which is being studied for the VLTI (den Hartog *et al.* (2005)). This ties the design and performance of the instrument to those of the suppporting infrastructure. A radically different approach, presented here, is the search for the optimal system at the optimal site in order to achieve the ultimate performance available from the ground.

More specifically, it appears that four factors are key to the success of a ground-based precursor of DARWIN:

• The breadboard should be designed in such a way that its concept focuses on issues relevant to the DARWIN mission (nulling, phase stabilization, data analysis), avoiding as much as possible design and engineering challenges which are circumstancial to the demonstration experiment (such as interfaces to an outside system, control loops not needed in DARWIN etc.);

• Because of the complex nature of an interferometer and the many real time interactions between its separate elements, the ultimate performance can only be achieved if the design is optimized at the system level;

• The quality of the site is critical and drives the final performance of the system. Therefore the experiment has to be located on the best accessible ground-based site. As for any interferometer subject to atmospheric turbulence, the largest possible seeing cell size is required to maximize the number of coherent photons. But most importantly, turbulence needs to be slow as it relaxes the requirements on the control loops and enables lower residuals. Finally, for efficient nulling in the infrared a low and stable background is required;

• The nulling demonstrator is a pathfinder mission and should be conceived as such, integrating development, deployment and operations (for a well defined time frame) into the concept.

The ALADDIN (Antarctic L-band Astrophysics Discovery Demonstrator for Interferometric Nulling) concept results from these considerations.

2. The concept

2.1. *Site selection*

It has been realized in the last few years that the high antarctic plateau, and more specifically the Dome C site (at an elevation 3250 m and −75 deg latitude) provides a place of strong potential for several types of astronomical observations, and most notably for infrared interferometry. This is due to a combination of several unique features, among which:

• Low environmental and sky temperatures (195–235 K) which result in a low and stable emission from the thermal background;

• An extremely dry air (250 μm precipitable water vapor typical) which results in improved infrared transmission, reduced dispersion variations and extended atmospheric windows (for example from 2.8 to 4.2 μm around the L band);

• Low surface winds and low free air turbulence, with a quasi-absence of jet streams as Dome C is located inside the polar vortex.

The dark time atmosphere has been investigated by different techniques (Lawrence *et al.* (2003), Agabi *et al.* (2006)). Most of the turbulence is concentrated in the ground layer (whose typical height is 30 m), above which the seeing has been found to be much lower than at any other measured site (0.27 arcsec median value). Besides, long coherence times ($\geqslant$ 7 ms in normalized conditions) have strong benefits for the performance (lower residuals) of real time control loops (tip-tilt, fringe tracking). The Dome C site is accessible with logistics developed by the French (IPEV) and Italian (PNRA) polar institutes who built the Concordia station there. Most humans reach the site by plane, while heavy equipment is shipped in standard containers by boat to the coast and then by "traverses" (pack trains of Caterpillar trucks) onto the glacier slopes up to Dome C, which thus provides a near space quality environment for infrared interferometry at a small fraction of the access cost.

2.2. *Design*

ALADDIN is conceived to optimize nulling interferometry performance and focus on issues relevant to the DARWIN mission (nulling, phase stabilization, data analysis), avoiding as much as possible engineering challenges which are circumstancial to the demonstration experiment (such as interfaces to an outside system, control loops not needed in DARWIN etc...).

The main features of a preliminary optomechanical design (Figure 1) studied by Alcatel Alenia Space are:

• A pair of 1 m collectors (siderostats) located above the ground layer, which can move on a circular path to maintain the source at the meridian of the baseline. Incidence angles are nearly constant during observations enabling better polarization control;

• An adjustable baseline (4-40 m), optimized for each source to obtain the best compromise between angular resolution and stellar leakage. The baseline can be reverted for calibration purposes;

• A streamlined optical train (off-axis telescopes and only 9 warm surfaces from the sky to the entrance of the bench, vs. 21 for the VLTI) for enhanced throughput and overall emissivity;

• A fixed, cryogenized bench with a single Bracewell nulling beam combiner;

• K-band control and L-band science. Thanks to the dry atmosphere no dispersion control loop is needed. The intensity control loop can also avoided (only a monitoring is needed) due to low d/r_0 ratio. Because of the longer coherence time of the turbulence, tip-tilt and fringe tracking loop rates are considerably reduced (down to a few kHz).

A preliminary analysis shows that, contrary to intuitive expectations, the cost and schedule of ALADDIN could be comparable to those of GENIE, as the effort to build the platform is balanced by the many simplifications and optimizations brought by the integrated design and the Antarctic location. The development plan can be made more robust as it can be parallelized and the reliance on the nominal performance of an external platform disappears. The integrated design enables a prior implementation in a temperate site, with full functional testing, before deployment at Dome C.

3. Performance

The science performance of ALADDIN can be evaluated using the simulation software package "GENIEsim" developed to analyze GENIE (Absil *et al.* (2005)), substituting as input the preliminary design shown here and the Dome C atmospheric properties above the ground layer as measured by Lawrence *et al.* (2003). This protocol enables a direct comparison between the two systems (see Table 1) for a similar integration time (in this case 1800 s).

The quality factor retained for science performance is the $5\,\sigma$ detectivity of exozodiacal light. It is affected by several factors and most notably the instrumental leakage control (thermal background and phase stabilization), the residual shot noise from the (not fully nulled) stellar light, and the calibration of stellar leakage (due to the angular extension of the stellar disk).

Thanks to the quality of the site and its system-level optimization, even with 1 m collectors the sensitivity of ALADDIN is always better than the sensitivity of GENIE at the VLTI. As can be expected, the gain is largest (up to a factor 2.9 for a K0V star at 5pc) for nearby stars which benefit most from the adjustable baseline.

4. Conclusion

It appears that the primary objectives of a DARWIN ground-based precursor can be better served by an integrated experiment located on the Antarctic plateau, which would provide both a more relevant technical demonstration and better preparatory science.

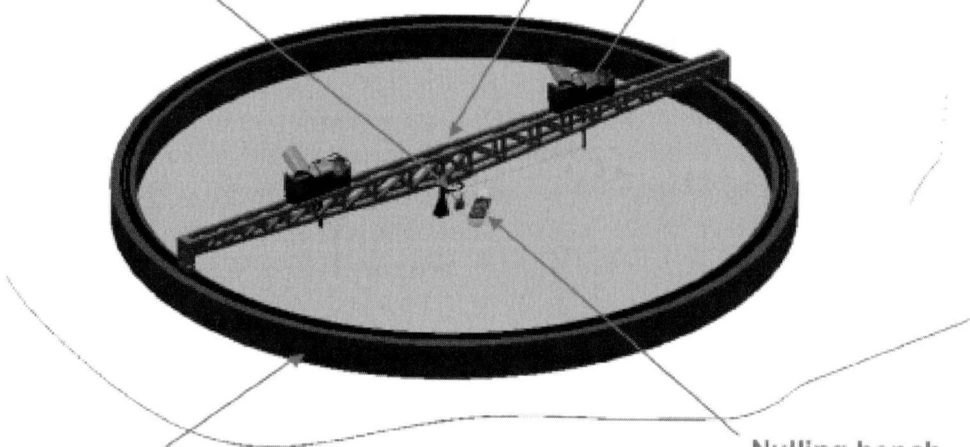

Figure 1. ALADDIN preliminary optomechanical concept.

Table 1. Comparative performance of GENIE and ALADDIN (in units of solar zodiacal light, lower is better) for the detection of exozodiacal light around four fiducial DARWIN targets, assuming a disk diameter known to within 0.5 % for the calibration of stellar leakage. GENIE/VLTI values are derived from ESA's phase A study (den Hartog *et al.* (2005)) and are shown for the pair of collectors (UTs (8 m) or ATs (1.8 m) that provide the best result in each case.

Source	GENIE/VLTI	ALADDIN	Detectivity ratio
K0V at 5 pc	110 (ATs)	38	2.9
G5V at 10 pc	56 (UTs)	26	2.2
G0V at 20 pc	37 (UTs)	20	1.9
G0V at 30 pc	48 (UTs)	23	2.1

While many issues (including critical ones) remain to be investigated, it is worthwhile to further explore this option.

References

Absil, O., den Hartog, R., *et al.* 2005 *A&A*, in press.

Agabi, A., Aristidi, E., Azouit, M., Fossat, E., Martin, F., Sadibekova, T., Vernin, J., & Ziad, A. 2006 *PASP*, in press.

den Hartog, R., *et al.*, these Proceedings.

Lawrence, J. S., Ashley, M. C. B., Tokovinin, A., & Travouillon, T. 2003, *Nature* 431, 278

Leger, A., Mariotti, J. M., Mennesson, B., Ollivier, M., Puget, J. L., Rouan, D., & Schneider, J. 1996, *Icarus* 123, 249

Direct Imaging of Exoplanets: Science & Techniques
Proceedings IAU Colloquium No. 200, 2005
C. Aime and F. Vakili, eds.

© 2006 International Astronomical Union
doi:10.1017/S1743921306009513

Dark fringe interferometry: Multi-beam configuration of the Alcatel nulling interferometer MAII and recent results

Christophe Buisset[1,2]†, Xavier Rejeaunier[2], Yves Rabbia[1], Cyril Ruilier[2], Marc Barillot[2], and Anders Karlsson[3]

[1]O.C.A., Dpt GEMINI, UMR CNRS 6203, Av. Copernic, 06130 Grasse, FRANCE
email: Christophe.Buisset@obs-azur.fr

[2]Alcatel Alenia Space, Cannes, FRANCE
[3]ESA, Estec, Noordwijk, The NETHERLAND

Abstract. Direct detection and spectroscopic characterization of earth-like exoplanets require to eliminate the starlight in the recorded signal at a rejection level (collected flux / residual flux) around 10^6 in the thermal infrared domain. Another need is to reach angular resolution in the range 0.1 to 0.01 arcsec. In this context, nulling interferometry is recognized as the preferred instrumental approach. In the framework of the ESA-Darwin mission, preliminary studies for nulling systems have been initiated, the prime goal to achieve being mastering nulling process. In this purpose, Alcatel Space has developed a nulling test-bench, operating in near infrared (1.55 μm), not only to demonstrate capability of obtaining an appropriate rejection level but also to study and characterize the use of emerging techniques. The Darwin system will ultimately benefit from these results. In this paper, we report the most recent configuration of the Alcatel Space nulling breadboard which uses the multi-beam recombination technique, current baseline for Darwin. The results obtained with this version shows rejection performance at nearly 10^5 (stable level) with a polychromatic source (λ = 1.55 μm, $\Delta\lambda$= 80 nm).

Keywords. Exoplanets, Direct detection, Nulling interferometry, Multi-beam recombination, ESA-Darwin

1. Introduction

In the framework of the DARWIN ESA mission (direct detection and spectral characterization of earth like exo-planets)(ESA(2000)), Alcatel Alenia Space has developed the test bench MAII which first intended to demonstrate the ability of obtaining high and stable rejection ratio $\rho \approx 10^6$. It also serves for testing emerging technologies via ESA contract and acquiring an experience in the field of nulling interferometry so as to develop a cryogenic version of the bench, operating at $\lambda \approx 10\,\mu$m.

The MAII test bench currently operates at $\lambda \approx$ 1,55 μm (telecom spectral domain), fiber and integrated optic technologies being well developed in this wavelength domain. This test bench has already demonstrated the ability of obtaining rejection ratio $\rho \approx 10^5$ (monochromatic) and $\rho \approx 10^4$(polychromatic) using an integrated optic combiner (Weber *et al.* (2004)).

We present a new configuration of the MAII bread board in which the modal filtering and recombination are made using one single-mode fiber. In a first part, we describe the breadboard global architecture. In a second part, we focus on main MAII subsystems

† Present address: Observatoire de la Côte d'Azur, Avenue Copernic, 06130 Grasse, FRANCE

and their functionality and in a third part, we present the nulling results obtained in February 2005.

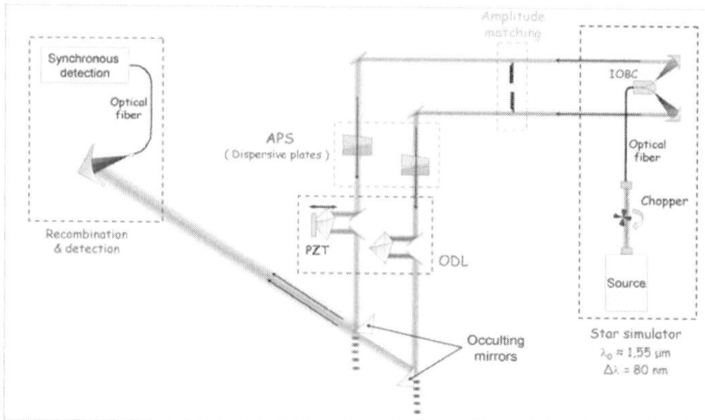

Figure 1. MAII breadboard.

2. The breadboard

The MAII breadboard is represented on Figure 1. It consists of a 2-arm interferometer in which an achromatic π-phase shifting between the fields that will interfere. High contrast interferometry needs for a fully symmetric design Serabyn (2000)), implying that same optical elements must be present along each arm of the interferometer.

The breadboard is composed of three parts : the star simulator, the central part of the interferometer and the recombination and detection part. The central part of the interferometer contains the elements needed for introducing the achromatic π-phase shifting and matching the amplitude and phase of the fields that will interfere. The polarization matching is achieved by an accurate alignment of the optics what leads to a differential rotation angle of polarization of about 0.1 degree. In the following part, we describe the functionality of each subsystem of the breadboard.

2.1. *Star Simulator, amplitude matching and OPD matching*

The light emitted by the source enters the integrated optic component (IOBC) via a single mode optical fiber. The IOBC delivers two beams collimated by off axis parabolas. These beams stand for the output beams of two telescopes. Because of the birefringent behaviour of the IOBC, only simple polarization measurement can be done.

Knive edges are translated inside the beams to reach the amplitude matching. With the use of high precision translation stage, relative flux difference between the beams is less than 10^{-3}.

Two identical Optical Delay Lines (ODL) are installed along each arm of the interferometer. These ODL are equiped with a corner cube instead of the classical rooftop. The OPD matching is reached by translating the corner cube along one arm of the interferometer. On the other arm, the similar but static delay line maintains the symmetry of the interferometer. The OPD is manually and real-time adjusted.

2.2. *π phase shifting and chromatic correction (APS)*

The Achromatic Phase Shifter (APS) is composed of 2 wedged dispersive plates of different materials installed along each arm of the interferometer. A convinient combination of the plates thickness and the use of the ODL yields the achromatic π phase shifting between the arms of the interferometer.

2.3. *recombination and detection*

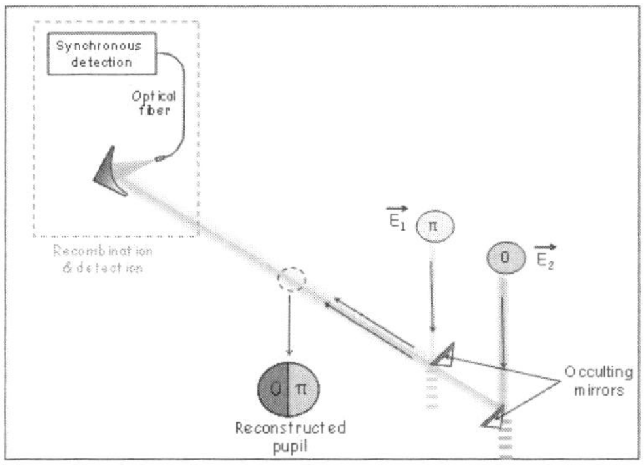

Figure 2. MAII breadboard.

Multibeam mode for recombination consists of:

_ Synthesizing a "reconstructed pupil" from 2 half-pupils. Each half-pupil is taken from each arm of the interferometer by occulting mirrors. Consequently, a π phase shifting remains between each part of the reconstructed pupil.

_ Recombining and filtering the fields with unique single-mode fiber. An off-axis parabola makes the beams focused at the entry of the fiber.

3. Results

In February 2005, we obtained $\rho = 6.7 \times 10^4$ as the best single polarization nulling performance for a polychromatic source with 5% of spectral bandwidth. Values higher than 5.10^4 appear on several time intervals of typically 5 seconds.

4. Conclusion

We have presented the latest version of the MAII breadboard used for demonstrating the feasability of the multibeam-recombination.The results we obtained demonstrate the feasibility of this kind of recombination for obtaining high rejection ratio, compatible with the ESA-DARWIN performance requirement. To go further, replacement of the IOBC by a non birefringent source and automatic control of the OPD should improve improve the null performance and its stability.

References

ESA July 2000, *Darwin, The infrared Interferometer*, ESA-SCI (2000) 12

Weber, *et al.* 2004, *Proc. SPIE Vol. 5491, Astronomical telescopes and instrumentation*, p. 842

Serabyn 2000, *Proc. SPIE Vol. 4006, Interferometry in Optical Astronomy*, p. 328–339

All photographs: Laurent Thareau [`l.thareau@free.fr`].

Direct Imaging of Exoplanets: Science & Techniques
Proceedings IAU Colloquium No. 200, 2005
C. Aime and F. Vakili, eds.
© 2006 International Astronomical Union
doi:10.1017/S1743921306009525

Limiting magnitude for Dome C optical interferometers

T. Elhalkouj[1], **R. G. Petrov**[2], **M. Lazrek**[1] **and Z. Benkhaldoun**[1]

[1]Université Cady-Ayyad, Faculté des sciences, LPHEA, Marrakech, Maroc

[2]Université de Nice Sophia-Antipolis, UMR 6525, Parc Valrose, 06108 Nice, France

Abstract. This paper discusses the potential performances of interferometers at the Dome C site in the Antarctic. The work is focussed on the limiting performances for fringe tracking, considering different fringe trackers architectures. The fringe tracking limiting magnitude is combined with estimates of the isopistonic angle at Dome C to evaluate the possibilities of off axis fringe tracking. The first evaluations presented here indicate that the sky coverage with realistic off axis fringe tracking is quite high at Dome C even with modest individual apertures and this could finally the best and decisive advantage of this site for optical interferometry. This might make a Dome C interferometer an unique deep sky very high angular resolution machine.

Keywords. instrumentation: interferometers, atmospheric effects, site testing.

1. Introduction

The limiting magnitude of a ground based multi aperture interferometer is set mainly by its capacity to cophase the apertures by actively correcting the differences of atmospheric piston. In many cases this cophasing cannot be performed on the science source itself and so far there is no artificial source really working solution. The solution which is being implemented for example on the VLTI is to use an off axis reference star. Given the current limiting magnitude for fringe tracking and the isopistonic angle at Paranal (defined by the angle within which two sources are supposed to be producing the same atmospheric piston) give a very poor sky coverage for the VLTI. We try to investigate what is the limiting magnitude for a large number of apertures and, after combining it with estimates of the isopistonic angle for different sites, we evaluate the conditions for a good sky coverage in the direction of the galactic pole.

2. Limiting magnitude of cophasing

We have considered two cophasing methods (described in figure 1). In the Global one, all fringes peaks in Fourier space are processed and the $N(N-1)/2$ baselines are used to find the $N-1$ piston differences. In the Hierarchic approach we cophase pairs of telescope, then pairs of pairs and so on. The SNR is evaluated from the hight of the fringe peaks above the noise level in Fourier space [Petrov (1989)]. Using the fundamental noises (photon noise from the source photon flux (Γ), thermic noise from the background (Λ), detector readout noise ($\sigma_L = 10e^-$ /pixel)) and the number N of telescopes, we derived the SNR for the Global and Hierarchical cophasing respectively in equations 2.1 and 2.2.

$$(SNR)^{-2} = 2\frac{N-1}{N\Gamma}\left(1+\frac{\Lambda}{\Gamma}+\frac{16N\sigma_L^2}{\Gamma}\right)+\frac{1}{4N^2(N-1)\Gamma^2}\left[1+\frac{\Lambda}{\Gamma}+\frac{16N\sigma_L^2}{\Gamma}\right]^2 \quad (2.1)$$

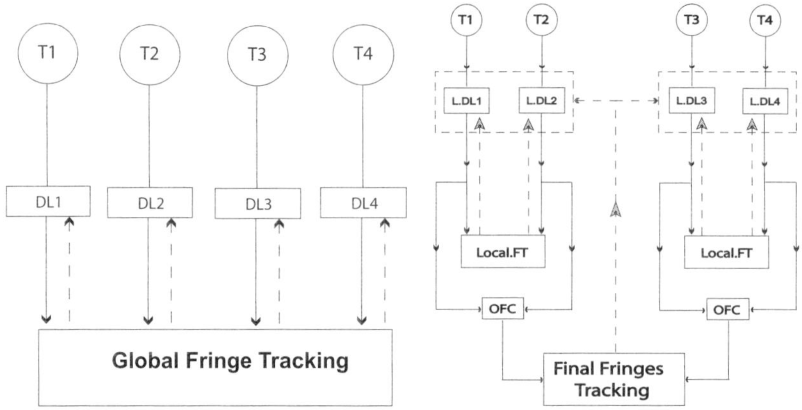

(a) (b)

Figure 1. Global and hierarchical cophasing per pairs. DL: Delay Line; L.DL : Local DL; FT : Fringe Tracking; OFC: Optical Fiber Combinator.

$$(SNR)^{-2} = \frac{Log_2(N)}{\Gamma}\left(1 + \frac{\Lambda}{\Gamma} + \frac{32\sigma_L^2}{\Gamma}\right) + \frac{Log_2(N)}{16\Gamma}\left[1 + \frac{\Lambda}{\Gamma} + \frac{32\sigma_L^2}{\Gamma}\right]^2 \tag{2.2}$$

The background flux Λ has been computed from a black body law with temperature T, the exposure time τ and the full K band. The instrument global efficiency was 2% and the emissivity 95%. The results are given in figure 2 showing the limiting magnitudes at Paranal ($T = 300K$) and Dome C ($T = 200K$). We note that the hierarchical cophasing is better, since the performance decrease much less with the telescope number. In the following, we restrain to that case.

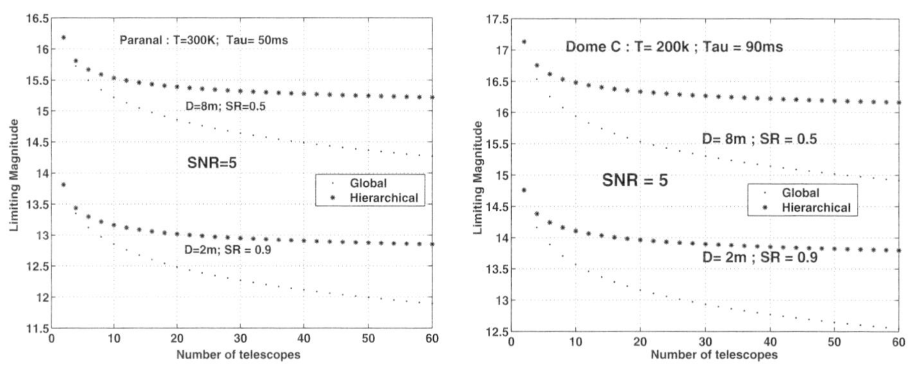

(a) (b)

Figure 2. Limiting magnitude of cophasing for Paranal (a) and Dome C (b) for 8 m (Strehl ratio SR=0.5) and 2 m (SR=0.9) telscopes.

3. Comparing Dome C and Paranal

3.1. *Limiting magnitude and sky coverage*

Figure 3 compares the limiting magnitude at Paranal and Dome C sites for an array of 8 meter telescopes. The gain in thermal background and incoherence time gives a clear advantage to the Dome C. It has been made by computing from equation 2.1 the flux, and therefore the magnitude yielding a $SNR = 5$. Figure 4 shows the radius $R(50\%)$ of the sky patch in which we have a probability of 0.5 to find a star brighter than a given limiting magnitude. If we have an estimate of the isopistonic angle for a site, we can find from figure 4 what is the cophasing limiting magnitude necessary to have a 50% to find a guide star in the isopistonic domain. This curve has been computed using CDS catalogs and counting the number of stars as a function of the diameter of a sky area. The sky direction have been produced randomly for values close to the galactic pole. For a given interferometer on a given site, the limiting magnitude corresponds to an individual aperture diameter.

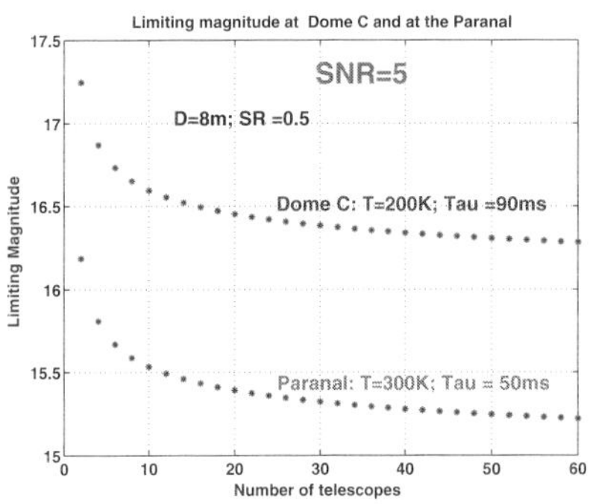

Figure 3. Limiting magnitude at Dome C and Paranal.

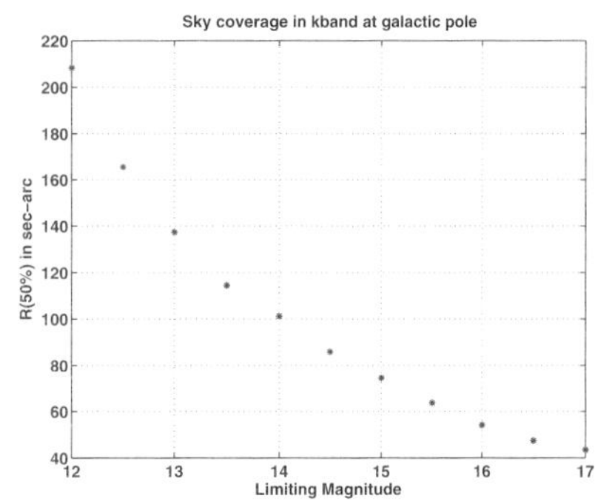

Figure 4. Sky coverage in the galactic pole.

3.2. *Isopistonic angle*

The isopistonic angle is the angle between two sources producing the same piston, with a λ/SNR accuracy. It is deduced from the classical measured isoplanetic angle of the given site according to [Esposito & al (2000)]. In this preliminary work, we have assumed that the ratio between the isopistonic and the isoplanetic angle is the same at Dome C and at Paranal. This basically assumes that the outer scale is the same, which is a conservative estimate. The demonstration showing that this isopistonic/isoplanetic ratio is dominated by the outer scale exceeds the possible length of this paper and is to be published soon. From [Esposito & al (2000)] and the measured Paranal isoplanetic angle (1.9 arcsec in the visible and therefore 9.9 arcsec in K) we found a *Paranal isopistonic angle* = 57 *arcsec* in the K band. The most recent summer time measures made by [Aristidi (2005)] yield a Dome C median isoplanetic angle in the visible of 6.8 arcsec. Extrapolating [Esposito & al (2000)] work therefore yields a *Dome C isopistonic angle* = 121 *arcsec* in the K band. This value should be quite correct for the winter Dome C conditions. It has been shown that the seeing is quite different in winter and in summer, but almost all the difference is produced within the first 30 meters above the ground, which has almost no impact on the isoplanetic nor isopistonic angles. Using figure 4, the isopistonic values at Dome C and Paranal give the cophasing limiting magnitude necessary to have a 50% sky coverage in the direction of the Galactic Pole:

mag(Galactic Pole, K band, Paranal)= 16
mag(Galactic Pole, K band, Dome C)= 13.2

3.3. *individual aperture diameter*

Figure 5 displays the diameter of the individual aperture necessary to achieve the cophasing limiting magnitude needed for a 50% sky coverage near the galactic pole, for a $\lambda/5$

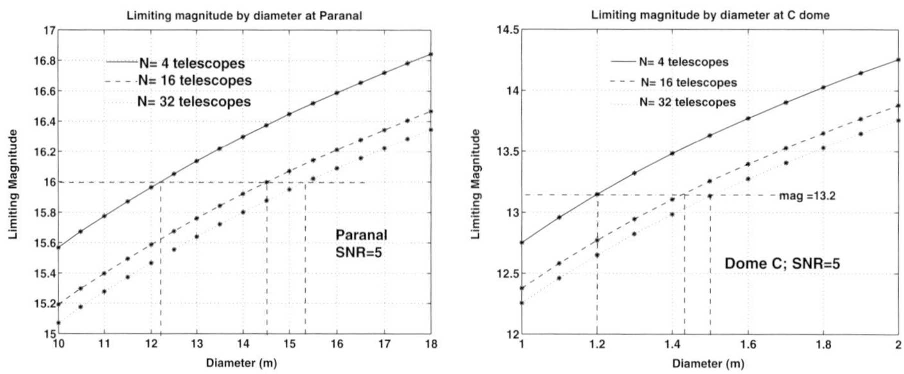

(a) (b)

Figure 5. Limiting magnitude with the diameter at Paranal and at Dome C.

fringe tracking quality. At Paranal, the need to have $K = 16$ yields and individual telescope size ranging from 12 m to 16 m, depending on the number of telescopes of the array. Note that the above computation has been made for the seeing of the best 25% Paranal nights. At Dome C, the needed magnitude is only $K = 13.2$ for a median seeing and the corresponding individual aperture diameter goes from 1.2 m for 4 telescopes to 1.6 m for more than 32 apertures.

4. Conclusion

This preliminary study shows that a Dome C array with typically 1.5 m individual aperture is able to ensure a sky coverage better than 50% near the galactic pole. On a good classical site like Paranal a similar requirement implies using individual apertures in the 16 m class. Therefore, the Dome C might be the single place on earth were a diluted, and therefore very high angular resolution, aperture can achieve very high magnitudes. Indeed, since the fringe tracking is permitted by a straightforward use of off axis stars, the limiting magnitude for the science source will be set only by the total collecting surface of the interferometer. Of course, our study is preliminary. We have to demonstrate more accurately the relation between isopostonic values and measured seeing characteristics and eventually measure directly the in situ isopistonic angle. We have also to consider other parameters, such as the limiting conditions for the adaptive correction of the unit aperture, but at least in the near infrared this should be a minor problem for telescopes smaller than 2 m. If further studies confirm this evaluation, a Dome C interferometer would be an absolutely unique possibility for very high angular resolution deep sky studies.

Acknowledgements

We sincerely thank Dr A. Ziad, for his efficient help concerning the isoplanetic angles, and the Data Center of Strasbourg (CDS,2MASS).

References

Esposito, S. & Riccardi, A. & Femenía 2000, *A.&A. Letter. to.Editor. 353.* L29, L32 (2000)
Aristidi, E. 2005, *Private communication*
Petrov, R.G. 1989, NATO Advanced Science Institutes (ASI) Series C, Volume 274, p. 249
Ziad, A. & al 2000, *Aplied. Optics. Vol.39. No.30. p. 5415*

Direct Imaging of Exoplanets: Science & Techniques
Proceedings IAU Colloquium No. 200, 2005
A.C. Editor, B.D. Editor and C.E. Editor, eds.
© 2006 International Astronomical Union
doi:10.1017/S1743921306009537

A Planet Finder instrument for the VLT

J.-L. Beuzit[1], M. Feldt[2], D. Mouillet[3], C. Moutou[4], K. Dohlen[4], P. Puget[5], T. Fusco[6], P. Baudoz[5], A. Boccaletti[5], S. Udry[7], D. Ségransan[7], R. Gratton[8], M. Turatto[8], H.-M. Schmid[9], R. Waters[10], D. Stam[10], P. Rabou[1], A.-M. Lagrange[1], F. Ménard[1], J.-C. Augereau[1], M. Langlois[4], F. Vakili[11], L. Arnold[12], T. Henning[2], D. Rouan[5], M. Kasper[13] and N. Hubin[13]

[1]Laboratoire d'Astrophysique de Grenoble, BP 53, F-38041 Grenoble Cedex 9, France
email: Jean-Luc.Beuzit@obs.ujf-grenoble.fr

[2]Max-Planck-Institut für Astronomie, Königstuhl 17, D-69117 Heidelberg, Germany

[3]Laboratoire d'Astrophysique de Toulouse et Tarbes, BP 826, F-65008 Tarbes Cedex, France

[4]Laboratoire d'Astrophysique de Marseille, BP 8, F-13376 Marseille Cedex, France

[5]Laboratoire d'Etudes Spatiales et d'Instrumentation Astrophysique, F-92190 Meudon, France

[6]ONERA, BP 72, F-92322 Chatillon Cedex, France

[7]Observatoire de Genève, 51 ch. Des Maillettes, CH-1290 Sauverny, Switzerland

[8]Osservatorio Astronomico di Padova, Vicolo dell'Osservatorio 5, I-35122 Padova, Italy

[9]Institut für Astronomie, ETH Zentrum, CH-8092 Zurich, Switzerland

[10]Universiteit van Amsterdam, Kruislaan 403, NL-1098 Amsterdam, The Netherlands

[11]Laboratoire Universitaire d'Astrophysique de Nice, Parc Valrose, F-06108 Nice, France

[12]Observatoire de Haute-Provence, F-04870 Saint-Michel-l'Observatoire, France

[13]European Southern Observatory, Karl-Schwarzschild-Strasse 2, D-85748 Garching, Germany

Abstract. Direct detection and spectral characterization of extrasolar planets is one of the most exciting but also one of the most challenging area in modern astronomy. For its second generation instrumentation on the VLT, ESO has supported two phase A studies for a so-called "Planet Finder" dedicated instrument. Based on the results of these two studies, a unique instrument is now considered for first light in early 2010, including a powerful extreme adaptive optics system, various coronagraphs, an infrared differential imaging camera, an infrared integral field spectrograph and a visible differential polarimeter. We will briefly summarize the science objectives and requirements, describe the proposed conceptual design and discuss the main limitations and corresponding instrumental issues of such a system. We will also derive the expected performance of the proposed Planet Finder and present the project organization.

Keywords. instrumentation: adaptive optics, techniques: high angular resolution, stars: low-mass, brown dwarfs, (stars:) planetary systems: formation

---

1. Introduction

The prime objective of a "Planet Finder" instrument for the European Southern Observatory (ESO) Very Large Telescope (VLT) will be the discovery and study of new extra-solar giant planets orbiting nearby stars by direct imaging of their circumstellar environment. The challenge consists in the very large contrast between the host star and the planet, larger than 12.5 magnitudes (or 10^5 in flux ratio), at very small angular separations, typically inside the seeing halo. Therefore the whole design of such an instrument shall be optimized towards reaching the highest contrast in a limited field of view and at shortest distances from the central star (Mouillet, *et al.* 2001). Keeping this prime objective in mind, it is obvious that many other research fields will also benefit from the instrument performance, in the domains of brown dwarfs, proto-planetary disks, mass loss of stars and more marginally for Solar System and extragalactic studies.

This VLT Planet Finder instrument will greatly contribute to the next 10 years of studies in the field of extrasolar planets, already very active, particularly by offering direct detections of planets more massive than Jupiter at various stages of their evolution, in the key separation regime 1 to 100 AUs. Migration mechanisms will then be better understood. The complementarities of direct imaging with other detection methods such as radial velocities and photometric transits, in terms of targets, detection biases and measured planetary parameters, and more specifically the combination with results from other projects like HARPS, COROT, VLTI/PRIMA, NGST and Kepler, will offer promising avenues. The present indications that massive distant planets could be numerous will be firmly confirmed or denied by the VLT Planet Finder, if the number of observed targets with relevant detection limits is statistically acceptable, i.e. greater than typically 150-200. This would in particular fully justify a large effort in an extended observational survey of several hundred nights concentrating on the following classes of targets:

- Nearby young sssociations: they will offer the best chance of detecting low mass planets, since they will have brighter sub-stellar companions, and therefore the greatest number of planets per star observed.
- Stars with known planets: especially any that exhibit long term residuals in their radial velocity curves indicating the possible presence of a more distant planet.
- Nearest stars: measuring these targets will probe the smallest orbits and will thus the only opportunities for detecting planets by directly reflected light.
- Stars aged from 100 Myr to 1 Gyr: planets will still be over-luminous as compared to Solar System planets, so mass limit will be lower than for old systems.

The science cases and potential targets of the VLT Planet Finder are discussed in greater details by Moutou, *et al.* (2006) in these proceedings.

2. Science requirements

The key scientific requirements deriving from the science analysis and driving the design of the instrument are summarized below:

- High contrast to reach giant planets 14 to 16 magnitudes fainter than the host star.
- Access to very small angular separations, 0.1" to 3" from the host star.
- Sensitivity and optimal performance for targets up to visible magnitude ~10, for building a potential target list in which the sample volume is consistent with the objectives (more than 500 targets in total, with 200 high priority targets).
- Sensitivity to faint companions down to magnitude H ~24.
- Access to an extended spectral domain at low resolution, for the characterization of the detected objects, at a resolving power ~30.
- Sensitivity to extended sources down to ~17 (respectively 20) magnitudes/$''^2$ at less than 0.2" (respectively 3") from the host star.

3. Instrument concept

To fullfil these science requirements, the proposed instrument will include the following main features:

- A common path optical bench, transmitting the telescope beam via the various correcting elements of the adaptive optics system (pupil tip-tilt and pupil de-rotator, deformable mirror, image tip-tilt mirror, atmospheric dispersion compensator...) and via the coronagraphic unit (apodizer, coronagraphic mask, Lyot stop) to the adaptive optics wavefront sensor and to the science channels. A conceptual design of the Planet Finder instrument is presented in Figure 1.

- An eXtreme AO (XAO) system which will fulfill the following three high level requirements: 1) ensure the measurement and correction of the turbulent phase perturbations, of the telescope and system common optics aberrations and of the non-common path aberrations (main AO loop); 2) ensure an extremely high stability of the optical axis at the level of the coronagraphic mask; 3) ensure the measurement and the correction of any pupil motion. A global trade-off combining error budget analysis, optical design, technological aspects, cost and risk issues has led to the following XAO system main characteristics: a 41×41 actuator deformable mirror, a very fast (>1 kHz) and accurate (0.5 mas) tip-tilt mirror, a spatially filtered Shack-Hartmann wavefront sensor working in the visible at high frequency (>1 kHz), a mixed control law with a Kalman filter law for the tip-tilt control and an optimal modal gain integrator law for the deformable mirror control. In addition an off-line calibration of the non-common path aberrations is foreseen. These aberrations will be compensated in closed loop by a modification of the wavefront sensor references. The expected Strehl ratio will be 90% in H band under reasonably good seeing conditions. This XAO system is described in more details by Fusco, *et al.* (2006) in these proceedings.

- A coronagraphic unit, including a classical Lyot device as well as one or two phase mask devices, allowing to reach very small separations between the host star and the planet candidates. The final choice of these phase mask devices will depend on the results of on-going research and development activities but they are likely to include a 4-Q and an apodized Lyot coronagraph. See for instance Abe, *et al.* (2006), Soummer, Ferrari & Aime (2006) and Mawet & Riaud (2006) in these proceedings for a description of some of these coronagraphic devices.

- An Infra-Red Dual Imaging Spectrograph (IRDIS), the main science module, dedicated to wide-field infrared imaging in one or two simultaneous spectral bands or two orthogonal polarizations as well as low resolution long slit spectroscopy. This main observing mode of IRDIS is based on the principle of differential imaging initially proposed by Racine, *et al.* (1999) and recently demonstrated on the VLT NACO instrument with the SDI device (Biller, *et al.* 2006; Close, *et al.* 2006).

- An Integral Field Spectrograph (IFS), working from 0.95 μm to 1.7 μm and providing low spectral resolution (R$\sim$30) over a limited, 3" $\times$ 3", field of view (Berton, *et al.* 2006).

- A visible dual imaging polarimeter (ZIMPOL), working between 0.65 μm and 0.95 μm (Schmid, *et al.* 2006). Due to its innovative lock-in technique, it can achieve polarimetric precisions better than 10^{-5} on a localized signal measured differentially against a smooth background.

For stability reasons, the instrument will not be directly attached to the telescope Nasmyth rotator but rather installed on the Nasmyth platform. All the above described

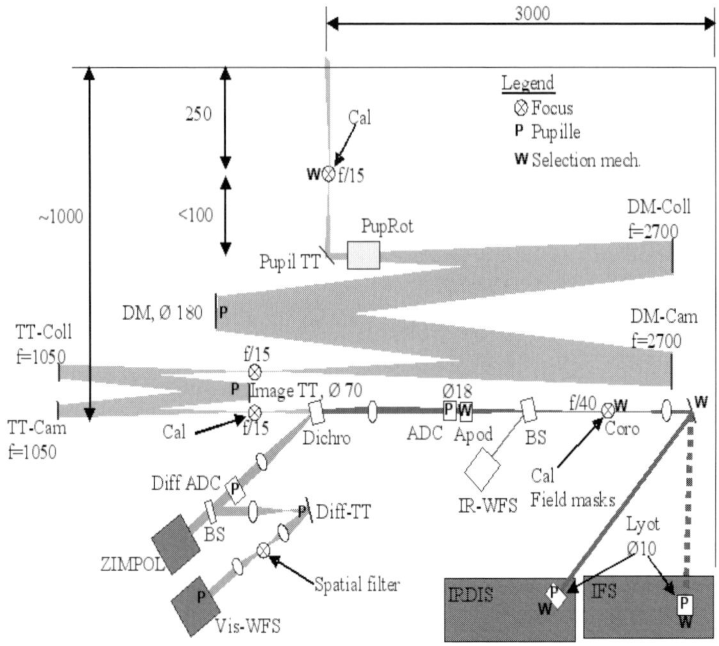

Figure 1. Conceptual design of the VLT Planet Finder instrument.

modules will be located on a fixed bench, covered for protection against dust contamination and controlled for flexures, vibrations and temperature variations.

4. Expected performances

End-to-end simulations of the instrument show that contrasts between 10^5 and 10^6 can be reached depending of the characteristics of the host stars, the foreseen separation and the observing mode. Several papers (Boccaletti, *et al.* 2006; Schmid, *et al.* 2006 and Berton, *et al.* 2006) in these proceedings describe the end-to-end models and discuss the expected performance of the VLT Planet Finder instrument for the various focal modes in more details.

An example of the achieved constrast is given as an illustration in Figure 2, for the dual imaging mode.

5. Project organization

Following two concurrent phase A studies, the ESO Scientific and Technical Committee has recommended in April 2005 to investigate the feasibility and interest of taking benefit from both studies by proposing an optimized instrument including the XAO system, coronagraphic devices and differential imaging camera proposed by one consortium as well as the Integral Field Spectrograph and ZIMPOL visible dual imaging polarimeter proposed by the other consortium. The on-going post-phase A study shall validate the interest of such an approach in terms of scientific return and its feasibility in terms of system analysis and project management.

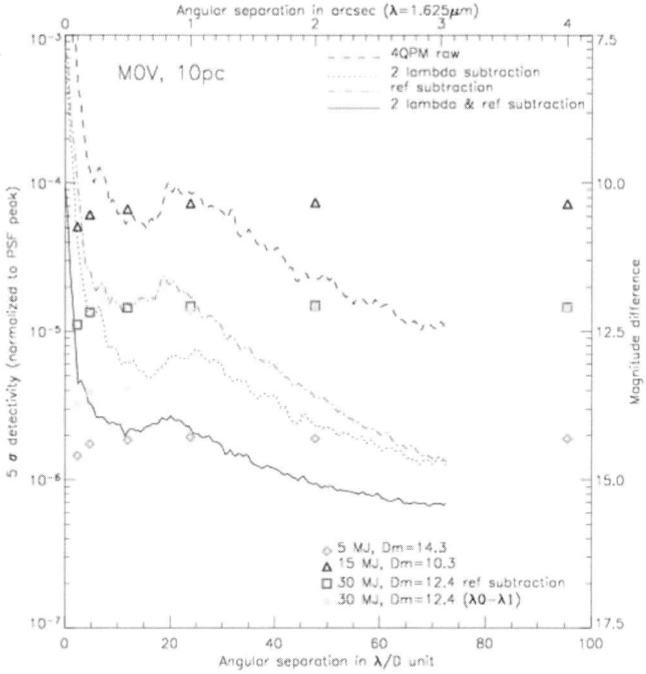

Figure 2. Typical contrast achieved by the proposed VLT Planet Finder instrument in the case of a 1 Gyr M0 star at 10 pc, observed in median seeing conditions using the dual imaging mode.

The current development plan, based on a kick-off date beginning of 2006, foresees a first light for the VLT Planet Finder by the spring of 2010. The consortium includes several European institutes, namely: Laboratoire d'Astrophysique in Grenoble (P.I. institute), Max-Planck Institüt für Astronomie in Heidelberg (co-P.I. institute), Laboratoire d'Astrophysique in Marseille, Laboratoire d'Etudes Spatiales et d'Instrumentation en Astrophysique de l'Observatoire de Paris, ONERA, Observatoire de Genève, Osservatorio Astronomico di Padova, Institute of Astronomy of the Zurich College of Technology, Astronomical Institute of the University of Amsterdam, ASTRON and ESO, with contributions from Laboratoire Universitaire d'Astrophysique in Nice, Observatoire de Haute-Provence and Osservatorio Astronomico di Capodimonte in Napoli. The total cost of the instrument is estimated around 9 MEuros.

6. Conclusion

The considered Planet Finder instrument, optimized for very high contrast imaging around an extensive sample of stars, and its operation model including in particular a large survey strongly supported by the building consortium will allow the direct detection of a sample of giant planets in a variety of conditions. Such a return in the proposed schedule (first light in 2010) will provide a timely and critical contribution to the highly competitive research field of extra solar planets: formation, evolution and characterization. This observational approach will provide specific information, complementary to other observational techniques (radial velocities, photometric transits, thermal IR imaging, etc.) and absolutely necessary to prepare for the foreseen next generation challenges, from the ground with Extremely Large Telescopes of from space with coronagraphic imaging telescopes (TPF-C) and interferometric instruments (DARWIN).

References

Abe, L., Tamura, M., Nakagawa, T., Enya, K., Tanaka, S., Fujita, K., Nishikawa, J., & Murakami, N. 2006, these proceedings

Berton, A., Feldt, M., Gratton, R., Henning, T., Hippler, S., & Turatto, M. 2006, these proceedings

Biller, B., Close, L., Lenzen, R., Brandner, W., McCarthy, D., Masciadri, E., Nielsen, E., & Hartnung, M. 2006, these proceedings

Boccaletti, A., Mouillet, D., Fusco, T., Baudoz, P., Cavarroc, C., Beuzit, J.-L., Moutou, C., & Dohlen, K. 2006, these proceedings

Close, L., Guirado, J., Nielsen, E., Biller, B., Brandner, W., Lenzen, R., Hartnung, M., Lidman, C., & Mamajek, E. 2006, these proceedings

Fusco, T., Rousset, G., Beuzit, J.-L., Mouillet, D., & Dohlen, K. 2006, these proceedings

Mawet, D. & Riaud, P. 2006, these proceedings

Mouillet, D., Beuzit, J.-L., Chauvin, G., & Lagrange, A-M. 2001, in J. Bergeron & G. Monnet (eds), *Scientific Drivers for ESO Future VLT/VLTI Instrumentation*, ESO Astrophysics Symposia, p. 258

Moutou, C., Beuzit, J.-L., Gratton, R., Mouillet, D., Allard, F., Augereau, J.-C., Boccaletti, A., Desidera, S., Feldt, M., Henning, T., Lagrange, A.-M., Longmore, A., Ménard, F., Schmid, H.M., Stam, D., Udry, S., & Waters, R. 2006, these proceedings

Racine, R., Walker, G., Nadeau, D., Doyon, R., & Marois, C. 1999, PASP 111, 587

Schmid, H.M., Beuzit, J.-L., Feldt, M., Gissler D., Gratton, R., Henning, T., Joos, F., Kasper, M., Lenzen, R., Mouillet, D., Moutou, C., Quirrenbach, A., Stam, D., Thalmann, C., Tinbergen, J., Vérinaud, C., Waters, R., & Wolstencroft, R. 2006, these proceedings

Soummer, R., Ferrari, A., & Aime, C. 2006, these proceedings

Direct Imaging of Exoplanets: Science & Techniques
Proceedings IAU Colloquium No. 200, 2005
C. Aime and F. Vakili, eds.
© 2006 International Astronomical Union
doi:10.1017/S1743921306009549

Direct Explorations of Exoplanets with the Subaru Telescope and Beyond

Motohide Tamura[1], Lyu Abe[1]
and Subaru HiCIAO/SPICA coronagraph/JTPF teams

[1]Exoplanet Project Office, National Astronomical Observatory,
Osawa 2-21-1, Mitaka, Tokyo 181-8588, Japan
email: hide@subaru.naoj.org

Abstract. We present a Japanese "roadmap" on direct extrasolar planet studies spanning from the current ground-based telescope to future IR/Opt space-based telescopes. Several searches for young planets and disks have been conducted with the Subaru 8.2m telescope with adaptive optics (AO) infrared coronagraph, CIAO. The instrument will be soon upgraded to a new AO and a coronagraph with simultaneous spectral and polarimetric differential imaging modes (HiCIAO), which will significantly improve the contrast performance and hence the capability of young planet detection. A sensitive unbiased survey for extrasolar zodiacal emission around nearby stars will be conducted with the ASTRO-F space mission (0.7m telescope, ~2–200 μm) to be launched around the beginning of 2006. A successor space mission, SPICA (3.5m, 5–200 μm), is also planned; its high sensitivity will enable the detection and characterization of outer-most planets around nearby stars, if any. For the studies of extrasolar terrestrial planets, a high contrast space telescope (HCST; 3.5m, ~0.3–2 μm). We are also seeking for collaborations with or are considering to join to foreign missions. We describe an outline, status, and role of each project on the extrasolar planet studies.

Keywords. instrumentation: high angular resolution, planetary systems: formation, planetary systems: protoplanetary disks, stars: low-mass, brown dwarfs.

---

1. A roadmap

Figure 1 is a roadmap showing our approach to the explorations of exoplanets in Japan.

Our ground-based direct explorations are concentrated on the fully-operated and successful Subaru 8.2m telescope at Mauna Kea, Hawaii. Making use of the good observing site and the excellent image quality, the infrared coronagraph CIAO (Coronagraphic Imager with Adaptive Optics) has been used for various kinds of surveys, which is described in detail in the next section. Its description and scientific results are introduced in §2. Several indirect efforts are also sought on the Subaru telescope: an international collaboration on the N2K survey on the Doppler searches has been most productive (Sato *et al.* 2005). HiCIAO (Next generation High Contrast Imager with Adaptive Optics for the Subaru telescope) is the Subaru next generation coronagraph which is used with the new AO and its near-future upgrade to an extreme-AO system. This will be used for the more extensive surveys of young exoplanets, as described in §3. Regarding space-based explorations, the MIR and FIR space telescope, *ASTRO-F*, is developed by JAXA/ISAS and will be launched soon (Murakami 2000). This telescope has a similar size to the *Spitzer* telescope, but is optimized to all-sky survey observations. Although its low spatial resolution is not suitable for direct explorations, it will conduct an unbiased survey for faint disk emission. This enables us to make census of exo-zodi and disk evolution with a large sample of stars. As a next project in the OPT/IR community in Japan after Subaru/ASTRO-F, SPICA is currently heavily discussed, a cooled large single mirror

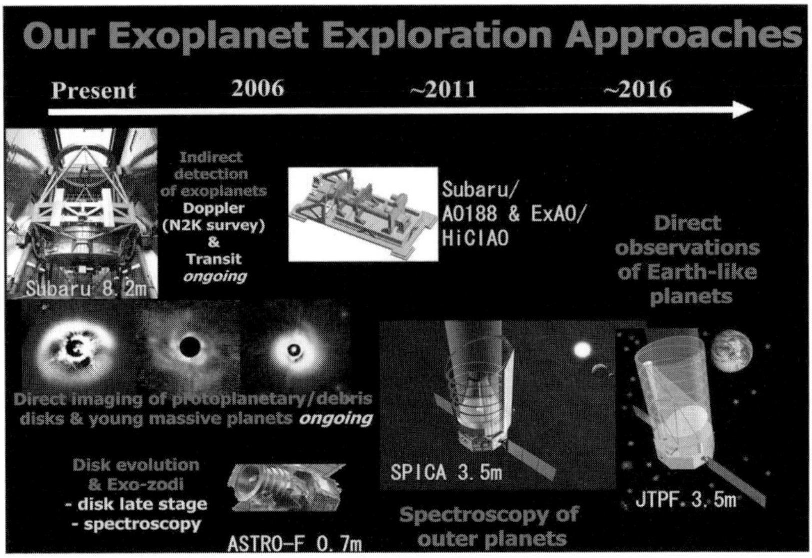

Figure 1. A roadmap of exoplanet explorations in Japan.

space-telescope. Because of its high sensitivity and acceptable resolution, the mission will serve for direct exoplanet explorations. Its coronagraph and exoplanet science are discussed in §4. A space-telescope dedicated for direct explorations of Earth-like planets is also under discussion in the OPT/IR community in Japan, as described in §5.

2. Subaru CIAO – First dedicated infrared coronagraph on 8–10 m class telescopes

Observationally, the direct imaging of both exoplanets and protoplanetary disks is extremely difficult: it requires a high sensitivity, a high resolution, and a high dynamic range or contrast at the same time. Such high sensitivity and high resolution are now simultaneously realized with the 8–10 meter class telescopes with adaptive optics (AO). The AO corrects for atmospheric perturbations in real-time, by the means of wavefront sensors and deformable mirrors. In the current Subaru AO system, 36-elements sensors and 36-elements mirrors are used. With the Subaru adaptive optics and infrared camera we regularly achieve a resolution of 0.07–0.10 arcsec and a sensitivity of 22 mag (5 σ in 1 hour) around 1.5–2 μm (H and K bands). However, a high contrast, that is, a capability to observe faint objects near bright objects, is another thing.

CIAO is the first among those equipped with a full cold coronagraph mode (with various cold occulting masks and Lyot stop optics) on the world 8-m class telescopes. Compared to previous coronagraphs, it has unique features of near-infrared operation and small occulting masks (down to 0.2 arcsec in diameter). See Tamura *et al.* (1998, 2000) for the instrument details.

CIAO has a comparable or higher contrast than other coronagraphs on the 8-m class telescopes (Table 1). See Nakajima *et al.* (2005) for a detailed performance discussion of CIAO.

Table 1. Contrast comparison (from Nakajima *et al.* 2005)

Telescope/Instrument	Contrast	Separation
Subaru/CIAO	ΔK=13	r>2.5″
Subaru/CIAO	ΔK=11	r=1.2″
VLT/AO+Mask	ΔK=13	r>3.0″
VLT/AO+Mask	ΔK=10	r=1.0″
VLT/AO+SDI	ΔH=9.5	r=0.5″

2.1. *CIAO disk searches*

CIAO has been successful to *directly* reveal the morphological diversity of protoplanetary disks, such as rings, spirals, and even others, with a high spatial resolution of ~0.1″. As shown in an inset of Figure 1, besides a well known ring-like disk around GG Tau, a clear spiral disk with several arms has been revealed around AB Aur, and a new-type of disk morphology, banana-split disk, has been recently discovered around HD 142527 (Fukagawa *et al.* 2004 and 2005). In addition, first 2-μm imaging polarimetry of β Pic has demonstrated that the disk is composed of several planetesimal belts with a gap at r ~ 100 AU (Tamura *et al.* 2005).

2.2. *CIAO companion searches*

Companion searches are greatly merited if the targets are younger; the younger companion is brighter and the contrast between the primary and the companion becomes lower (e.g., Burrows *et al.* 2001). This is one of the reasons why we explore YSO companions in nearby star forming regions.

DH Tau is a classical T Tauri star with an age of 0.1-4 Myr and a mass of 0.25-0.5 solar mass. CIAO coronagraphic images have revealed a companion of 15 mag at *H* at 2.3″ (330 AU) from the central star. This star is originally selected because it has a relatively large infrared polarization (Tamura & Sato 1989), suggestive of the circumstellar structures. However, interestingly, no significant circumstellar disks or envelopes are detected except for the companion.

The companion shows the same proper motion of the Taurus cloud members, distinct from the background stars. Its near-infrared spectra show deep absorption band due to water and other metal features, all indicating low effective temperatures, but not as low as those of planetary mass objects (T_{eff} ~ 2700 K). With recent theoretical models, the companion is concluded as a companion young brown dwarf of about 40 M_{JUP} (Itoh *et al.* 2005). This source is more or less similar to the recenly detected companion around GQ Lup and other handful companions around YSOs (Neuhauser *et al.* 2005; Liu 2005 in this proceedings).

Similar surveys are also nearby M stars using CIAO (Nakajima *et al.* 2005).

3. HiCIAO – the Subaru next high contrast instrument

HiCIAO is a new high-contrast instrument for the Subaru telescope, which is currently developed by NAOJ and University of Hawaii. NAOJ PI is M. Tamura and UH PI is K. Hodapp. HiCIAO will be used in conjunction with the new adaptive optics system (188 actuators and/or its laser guide star - AO188/LGSAO188) at the Infrared Nasmyth platform. It is designed as a flexible camera comprising several modules that can be configured into different modes of operation. The main modules are the AO module with its future extreme AO capability, the warm coronagraph module, the high contrast optics module, and the cold infrared camera module. HiCIAO will be the first instrument on the 8-m class telescopes which can combine coronagraphic techniques with simultaneous polarization and spectral differential imaging modes which minimizes the common path errors. The basic concept of such differential imaging is to split up the image into two or more images, and then use either different planes of polarization or different spectral filter bandpasses to produce a signal that distinguishes faint objects near a bright central object from scattered halo or residual speckles. This enables us to achieve a contrast improvement of at least 10 times better than before. The instrument specification is summarized in Table 2.

Table 2. HiCIAO overall specifications.

Platform	Subaru Nasmyth focus (IR)
Combined adaptive optics	188-elements curvature-sensing AO
Wavelength coverage	$zJHKs$ (optimized at H)
Detector	2048 × 2048 HgCdTe (Rockwell HAWAII-2RG)
Detector controller	ASIC (Rockwell "Sidecar")
Pixel scale	$0.01''$/pixel (proper PSF sampling at J-band)
Spatial resolution	$0.03''$ (J), $0.04''$ (H), $0.055''$ (Ks) FWHM
Strehl ratio	0.6 (J), 0.7 (H), 0.8(Ks) w/ $0.5''$ optical seeing
Sensitivity	$H > 23$ mag (5σ, 1 hr)
Contrast goal	10^4 at $0.1''$, 10^6 at $1''$ (w/ SDI mode + coronagraph)
DI mode	Direct imaging
	$20'' \times 20''$ FOV, general purpose imaging
	High throughput, w/wo coronagraph
PDI mode	Polarimetric (o-ray and e-ray) dual imaging (Wollaston prism)
	$20'' \times 10''$ FOV, polarized target imaging
SDI mode	Spectral simultaneous differential imaging (Double Wollaston prism)
	$6'' \times 6''$ FOV, unpolarized target imaging
Coronagraph*	
Occulting masks	3 sizes with real-time remote choice
Lyot stops	Stop with spider/center-hole blocks & rotation compensation

* Starting with the classical Lyot coronagraph. More sophisticated coronagraph will be implemented later.

The main sciences with Subaru/HiCIAO are summarized as follows.

• For directly detecting young (*age* ∼ 100 Myr) extrasolar planets and brown dwarfs, HiCIAO can be used to conduct a survey of ∼120 nearby relatively young dwarfs. About 60 stars younger than 700 Myr with spectral types between A and K are accessible from Subaru. These 60 stars and the 350 Myr M dwarf sample (∼60 stars) will serve as the prime targets for the brown dwarf/planet. The inner radius of search will be reduced from 25 AU to 10 AU for a target at 10 pc by the transition from CIAO to HiCIAO.

• For directly detecting very young (*age* ∼ 1–10 Myr) extrasolar planets and brown dwarfs, HiCIAO can be used to conduct a survey of ∼200 T Tauri stars in nearest star forming regions accessible from the northern hemisphere and to detect young Jupiter-mass companions down to an inner radius of 30 AU. With the availability of the laser guide star, the target can be extended to the optically faint sources including Class I YSOs.

• HiCIAO will be efficiently used to detect or set a limit on scattered light from disks around the young stars in nearby star forming regions. The spatial resolution is $0.03''$ (80 AU at the distance of the Taurus dark cloud) in the J-band, which is better than the

HST resolution. The large sample of disk data will be used to trace the morphological evolution and diversity of YSO disks at the age of age <1 Myr to 10 Myr.

• HiCIAO will be extremely useful to detect or set a limit on scattered light from the known and new, large samples of Vega-like disks from the *Spitzer* and the *ASTRO-F* missions, which requires both high sensitivity and high contrast. The achievable spatial resolution is 0.6 AU at the distance of β Pic. The polarization differential imaging is crucial to study the small-scale and inner disk structures and dust grain properties. A synergy observation with the *ASTRO-F* will be considered.

• The targets for TPF are well pre-surveyed with HiCIAO for characterizing their circumstellar matter and detecting relatively "low"-contrast objects including background stars and galaxies.

An extensive survey dedicated to these sciences on the Subaru telescope will be a significant contribution to astronomy before the high contrast space telescope era. The application to other targets might include not only other galactic objects such as late-type stars but also extragalactic objects such as AGNs and quasars, especially with the laser guide stars, where the AO application has been relatively limited. A high throughput direct imaging will be useful for such objects.

4. SPICA coronagraph

The SPICA mission plans to launch a cooled (4.5K) large (3.5m) single-mirror telescope into a halo orbit around the Sun-Earth L2 orbit around the beginning of 2010s (Nakagawa 2000). Its unprecedented sensitivity at MIR (and FIR) and its simple telescope pupil make SPICA to be one of the best platforms to implement coronagraph instrument for exoplanet studies. The preliminary specification of the SPICA coronagraph is summarized in Table 3.

Table 3. SPICA coronagraph prelliminary specification.

Wavelength coverage	5–27 μm
Spectral resolution	<200
Inner working distance	$3\lambda/D$
Contrast goal	10^6
Detector	1Kx1K Si:AS IBC
Adaptive Optics	Cold Tip-Tilt mirror (to correct for the jitters)
	Cold Deformable Mirror (to correct for the mirror wavefront errors)

SPICA will target direct observations of self-luminous planets at r> a few to ~20 AU of nearby (<10pc) stars. The detectable planets depend on their mass, age, and separation. If we assume the inner working distance of $3\lambda/D$, then at λ=5 μm we can detect 1 Gyr, 2 M_{JUP} planets at r>9 AU around 10 pc stars (~30 G-M type stars). At λ=20 μm we can detect 5 Gyr, 2 M_{JUP} planets at r>18 AU around 10 pc stars (~150 G-M type stars). Very young planets and sub-brown dwarfs in nearby star forming regions and cold brown dwarfs are also good targets.

5. JTPF

JTPF (Japanese Terrestrial Planet Finder) is a future space mission whose main science driver is the extrasolar terrestrial planet studies. Th project is officially approved both by NAOJ and ISAS. In NAOJ, its activity is regarded as one of the main themes of the Phase A project which has started in April 2005. In JAXA/ISAS, the JTPF Working Group has been approved by the ISAS Science Steering Committee in 2002. Discussion has been made among both astronomy and planetary communities in Japan, and regarded as one of the important programs for optical and infrared astronomy. It could be either an independent mission or a joining collaboration to TPF/Darwin missions.

At present, an optical high contrast space telescope (HCST) which realizes clean and stable images is under main discussions for the JTPF architecture among JTPF WG (see Figure 1). JTPF/HCST will make full use of the SPICA bus system with an optimization to shorter wavelengths. The current plan of HCST is shown in Table 4.

Table 4. JTPF/HCST preliminary specification.

Telescope	3.5m off-axis single aperture
Wavelength coverage	Optical and near-IR w/ limited UV capability
Orbit	Sun-Earth L2
Instruments	Opt-NIR coronagraph, Wide-field NIR/Opt instruments

Various JTPF-ralated acitivies in Japan include studies of:

• combinations of nulling interferometer and modified pupil (Nishikawa et al 2005),

• a chromatic nulling of on-axial light to detect off-axial light by means of 3D common-path interferometer (Tavrov *et al.* 2002), and

• polarization differential objective spectroscopy with nulling coronagraph (Murakami *et al.* 2005).

References

Burrows, A., Hubbard, W. B., Lunine, J. I., & Liebert, J. 2001, *Rev. of Modern Physics* 73, 719
Fukagawa, M., Hayashi, M., Tamura, M. *et al.* 2004, *ApJ* 605, L53
Fukagawa, M., Tamura, M., Hayashi, M. *et al.* 2005, *ApJ* submitted
Itoh, Y., Hayashi, M., Tamura, M. *et al.* 2005, *ApJ* 620, 984
Murakami, H. 2000, *ISAS Report SP* 14, 267
Murakami, N., Baba, N., Tate, Y. *et al.* 2005, *PASP* submitted
Nakagawa, T. 2000, *ISAS Report SP* 14, 189
Nakajima, T., Morino, J., Tsuji, T. *et al.* 2005, *Astronomical Notes* in press
Neuhauser, R, Guenther, E. W., Wuchterl, G., Mugrauer, M., Bedalov, A., Hauschildt, P. H. 2005, *A&A* 435, L13
Nishikawa, J., Kotani, T., Murakami, N., Baba, N., Itoh, Y., & Tamura, M. 2005, *A&A* 435, 379
Sato, B. *et al.* 2005, *ApJ* in press
Tamura, M., Suto, H., Takami, H. *et al.* 1998, *SPIE* 3354, 845
Tamura, M., Suto, H., Itoh, Y. *et al.* 2000, *SPIE* 4008, 1153
Tamura, M., Fukagawa, M., Kimura, H. *et al.* 2005, *ApJ* submitted
Tamura, M. & Sato, S. 1989, *AJ* 98, 1368
Tavrov, A. *et al.* 2002, *Opt. Lett.* 27, 2070

Direct Imaging of Exoplanets: Science & Techniques
Proceedings IAU Colloquium No. 200, 2005
C. Aime and F. Vakili, eds.
© 2006 International Astronomical Union
doi:10.1017/S1743921306009550

Current status of the coronagraphic mode for the 3.5m SPICA space telscope

L. Abe[1], M. Tamura[1], T. Nakagawa[2], K. Enya[2], S. Tanaka[2], K. Fujita[3], J. Nishikawa[4], N. Murakami[4], H. Kataza[2] and The SPICA Working Group

[1]Infrared and Optical Astronomy Division & Extrasolar Planet Project Office, National Astronomical Observatory of Japan, Ōsawa 2-21-1, Mitaka-city, 181-8588 Tōkyō, Japan
contact email: abe@optik.mtk.nao.ac.jp

[2]Department of Infrared Astrophysics, Institute of Space and Astronautical Science, Japan Aerospace Exploration Agency Yoshinodai 3-1-1, Sagamihara, Kanagawa, 229-8510, Japan,
contact email: enya@ir.isas.jaxa.jp

[3]Graduate School of Science & Technology, Kobe University, Nada Kobe 657-8501 Japan

[4]MIRA Project, National Astronomical Observatory of Japan, Ōsawa 2-21-1, Mitaka-city, 181-8588 Tōkyō, Japan

Abstract. As of early ~2010's, the Japanese SPace Infrared telescope for Cosmology and Astrophysics (SPICA) space observatory will be launched. This actively cooled, cryogenic (4.5K), 3.5m telescope will operate in the mid and far infrared spectral regions. With its very high sensitivity, one of SPICA's aims will be the direct detection and characterization of extra-solar outer planets of nearby stars. The goal contrast ranges from 10^5 to 10^6 up to an angular separation of ~5 arcsec. The relatively low angular resolution at MIR (5 to 20 μm) requires an efficient and robust coronagraphic mode working at cryogenic temperatures. In this presentation we describe several envisaged preliminary designs and assess their performance against the science goals and host telescope specifications. These are compared against numerical simulations and instrumental environment considerations, such as the need for an actively corrected wavefront.

Keywords. Instrumentation: adaptive optics, Instrumentation: high angular resolution, Technique: high angular resolution, Telescopes.

1. Introduction

Japan have put efforts in this direction and established a long term roadmap toward earth-like planet direct observation and characterization with its JTPF (Japanese Terrestrial Planet Finder) project. But the efforts have already started in that direction some years ago with the ground based coronagraphic facility CIAO installed at the focus of the 8.3m Subrau telescope in Hawaii. The roadmap gathers both ground based and space based observatories. In a very near future (from year 2007), the CIAO coronagraph will be replaced by the next generation HiCIAO instrument (see Tamura *et al.*, these proceedings) which will benefit both from a new adaptive optics system (188 actuators, curvature sensor WFS), and from the excellent seeing atop Mauna Kea. Spectral and polarization simultaneous differential imaging is expected to provide contrasts up to 10^6 between below 1.0" in the nIR bands (JHKs). Further ahead, the 3.5m SPICA space telescope will hopefully host a dedicated coronagraph instrument which studies have just started in late 2004. This mIR to sub-millimeter (5μm to 200μm) cryogenic telescope will have an equivalent sensitivity to JWST below 25μm and will benefit from a monolithic pupil.

2. The SPICA telescope

The SPICA telescope will be launched in the early 2010's onboard an HIIA rocket (Nakagawa, 2000, Tamura, 2000). Many efforts were dedicated to the study of the cryogenic system of the telescope to cool it down to 4.5K. These cryogenic temperature implies a number of constraints on the optics and especially for the coronagraphic part which needs a high quality wavefront prior to enter the coronagraph itself. The primary optics (primary and secondary mirrors), as well as supporting structures will be made of either SiC or C/SiC (Carbon/Silicon carbide). Cryogenic tests of deformation at 4.5K are undergoing at ISAS. Moreover, the cryo-cooler will generate some vibrations producing a large pointing error of about 3" which is much too large for the coronagraphic requirements. To solve this problem, a cryogenic tip-tilt mirror will be used which is expected to correct the tip-tilt jitter down to 0.03". Further details can be found in Enya *et al.*, these proceedings.

3. The SPICA mission

SPICA will be used to study many astrophysical domains, ranging from proto-planetary systems to extragalactic sources. Amongst these goals, the coronagraphic instrumentation will be dedicated to the direct detection and characterization of outer extrasolar planets. For stars within 10pc, the exploration field will be set by the spectral band, and by the coronagraph potential in terms of Inner and Outer Working Angles (IWA and OWA, see below for coronagraphic devices). With a 10pc distance, at the lowest wavelength (5μm), the diffraction limit corresponds to a planet distance of 3AU. Considering that the coronagraph IWA is at most $1.5\lambda/D$, the smallest possible IWA will be around 5AU. Under the current specifications for the telescope optics, the OWA will probably limited by the DM specification and/or by the coronagraph type to be used (see below). At the present, the baseline OWA is set to around $10\lambda/D$, which corresponds to 30AU for a 10pc at 5μm and more than 100AU at 20μm (but the IWA will also increase to 12AU in the best cases). Provided a contrast of 10^6 can be reached within this exploration field, the possible targets are self luminous planets of ages between $1 \sim 2$Gy and around a few Jupiter masses.

4. Coronagraph requirements

Due to the limited angular resolution of the telescope from 5 to 20μm, the coronagraph needs to be sensitive at low *inner working angles* (IWA). But the high residual tip-tilt jitter does not allow certain coronagraphic concepts such as AIC (Gay & Rabbia, 1996) or Roddier&Roddier (Roddier & Roddier, 1997) coronagraphs. The tip-tilt jitter will be mainly due to the vibrations caused by the cryo-cooler devices. The estimated pointing stability is only 1 arcmin, whilst the envisaged tip-tilt mirror correction will provide a 60mas residual ($\lambda/10$ @ 10μm). The current telescope design includes a 20% central obscuration associated with a cross-shaped spider support. Under these conditions, only a few solutions can be envisaged. Most very efficient coronagraphs (e.g. AIC, Phase Mask) are too sensitive to tip-tilt jitter (30 mas) to provide a stable high contrast coronagraphic image. The four quadrant phase mask is also too handicapped by the large central obscuration and maybe too sensitive to the residual tip-tilt jitter. Our choice focused on coronagraphs which are poorly or moderately sensitive to tip-tilt errors, and with the lowest possible inner working angle. The most challenging concept which is currently planned for tests is the Multi-Stage Apodized Lyot Coronagraph (Aime &

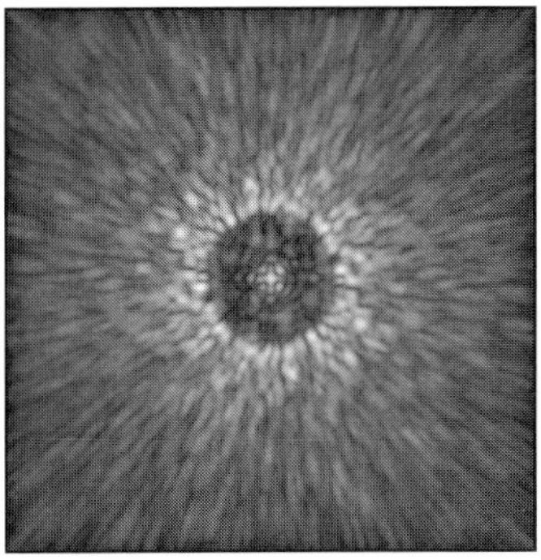

(a)

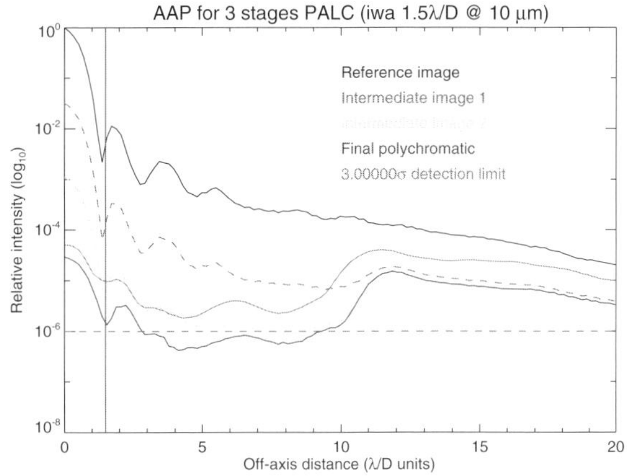

(b)

Figure 1. Numerical simulation of the SPICA with a 3S-PALC coronagraph between 9μm and 10μm: (a) final polychromatic image (non linear intensity scale) and (b) azimuth averaged profiles including intermediate stages profiles. The 3-σ limit includes the residual speckle intensity fluctuations.

Soummer, 2004). The optical complexity and marginally the throughput are the main drawbacks of this concept (see next section for further details). Backup solutions are the shaped pupil masks (Vanderbei (2003)) which are of the most simplest implementation. The huge drawback of this solution is the poor inner working angle and possibly the associated outer working angle which may be a serious problem regarding its behavior against phase aberrations. Other solutions such as the focal plane band limited masks (Kuchner & Traub, 2002) may also be worth considering for the SPICA coronagraph (although the pupil shape may severely reduce the throughput).

5. Performance estimation

Some preliminary simulations were performed including the actual telescope specifications: telescope pupil shape, primary mirror surface roughness/bumpiness, tip-tilt jitter. The mirror surface accuracy is currently set to 0.35μm RMS, which is roughly the Rayleigh criterion for diffraction limited images at the lowest wavelength (5μm). Under these conditions, an additional deformable mirror (DM) is mandatory to correct for these imperfections with the additional difficulty of an operation temperature of 4.5K. ISAS is planing some cryogenic tests with MEMS DM from Boston Micromachines Corporation (36 element DM matrix). The electrostatic actuation is more suitable for cryogenic environment, contrarily to piezo actuators which stroke severely decreases at these temperatures. The simulation shown on Figure 1 is a preliminary result obtained with a 3 stages PALC and a 1.6 λ/D mask radius. The wavefront has a RMS bumpiness of 0.35μm which spatial power spectrum has an inverse power law ($f^{-2.5}$). It is corrected by a DM with 400 actuators (20 across the pupil) and a RMS accuracy of 8nm. The number of actuators of the DM allows spatial frequencies to be corrected up to 10λ/D. This determines the OWA as long as the residual phase errors beyond these spatial frequencies are higher than the corrected part of the spectrum which is currently the case (the corrected part of the phase spectrum is flat except for the residual phase bumpiness). The high specified bumpiness of the primary mirror (0.35μm) actually limits the coronagraphic performance in an absolute way. Reducing the mirror polishing specification is also one of the current discussion point for the coronagraphic mission.

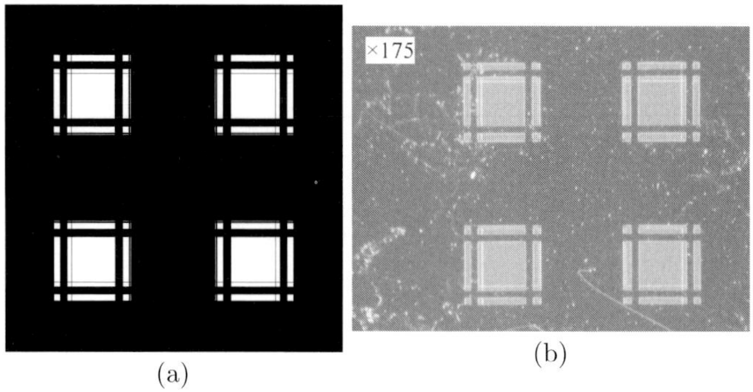

Figure 2. Theoretical (a) and manufactured (b) binary pupil mask. The defects of (b) appear larger that what they actually are due to the slight defocus and light saturation at these location.

6. Laboratory experiments

6.1. *Binary shaped pupils masks*

Binary shaped pupil (Jacquinot & Roizen-Dossier, 1964) masks were invented and largely experimented in the 60's as ways to strongly attenuate the diffraction halo in optical systems by diffracting the light along preferential directions. More recently, binary shaped pupils of various kinds appeared, some of them recalling binary versions of linear apodizers which are commonly used in laser experiments in order to convert gaussian beams into flat beams for example (so called *beam shaping* techniques). Binary shaped pupil masks offer the advantage of being achromatic, meaning that the effect is in principle similar regardless of the wavelength. The other advantage of pupil masks is that they are very poorly sensitive (marginally sensitive) to tip-tilt errors/pointing errors. Indeed, in this case, and contrarily to other "focal plane" coronagraphs for example, the coronagraphic image is here invariant by translation, meaning that the convolution product between the object intensity distribution and the coronagraphic impulse response is preserved. The binary masks are usually designed as the solution of a system of non linear equations under a minimization/maximization criteria: maximize the throughput of the system, and minimize the intensity level in a given spatial area in the image plane. Some mathematical tools have been developed to numerically solve this kind of problems. For example, Vanderbei (2003) uses the AMPL mathematical programming language in association with the LOQO solver (Vanderbei, 1999). A variety of "unusual" pupil shapes can be obtained through this process as shown for example in Kasdin *et al.* (2003).

One of these masks have been fabricated by the *National Institute of Advanced Industrial Science and Technology* using ion etching process. The chosen test design was chosen to be of the type "checkerboard" or "bar-code" (Vanderbei *et al.*, 2004). This mask has a very large inner working angle (8 λ/D) because it was designed to match the current 30% central obscuration of the SPICA telescope. Future mask designs will consider 20% central obscuration telescope which enables a smaller IWA. The prototype mask (first attempt) is shown on Figure 2. Although many small defects are present, the PSF exhibits peak-to-valley contrast ratio of 10^{-5} along the diagonal direction relative to the mask orientation (horizontal direction of Figure 4). Note that these results were obtained after some optical setup tunings, and after the IAU meeting was held, thus improved by a factor of 10 from the previous results, as shown in Enya *et al.* (these proceedings). Improvements of the optical setup as well as the wavefront compensation by a DM is envisaged as future upgrades. A more detailed study of optimal checkerboard mask design for the SPICA coronagraph is described in Tanaka et al. (these proceedings).

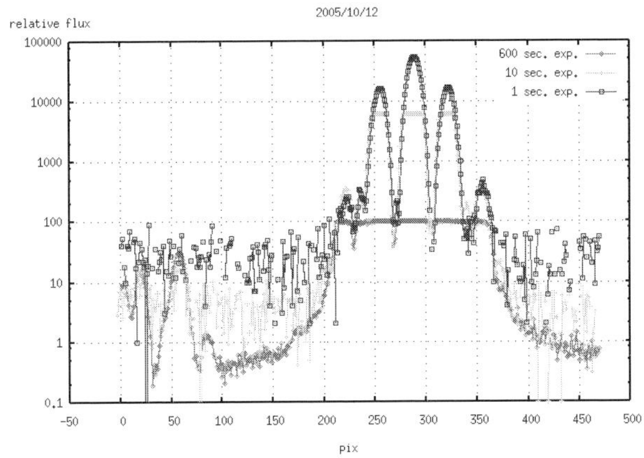

Figure 3. AIST mask PSF profiles along the diagonal of the mask orientation.

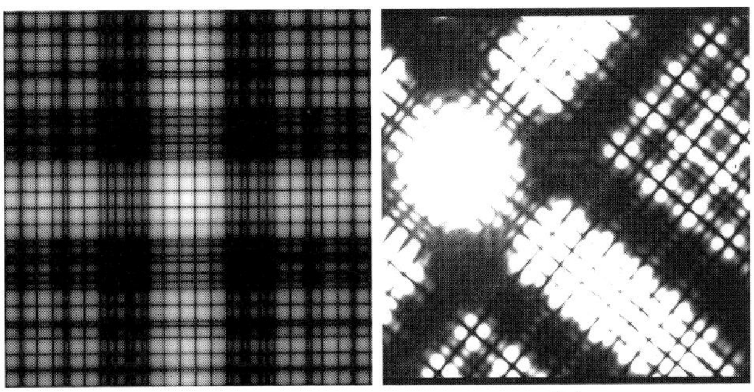

(a) (b)

Figure 4. (a) Theoretical PSF for the mask of Figure 2-a (logarithmic gray level stretch). (b) Actual laboratory recording of the binary pupil mask PSF obtained with the AIST mask (Figure 2-b) displayed as a linear gray level scale. The corresponding pupil mask is shown on Fig. 2. The theoretical PSF achieves a 10^{-7} contrast ratio in the 4 "corner areas" whilst $\sim 10^{-5}$ was obtained in the experiment.

6.2. *Multi-stage Pupil Apodized Lyot Coronagraph*

The prolate apodized Lyot coronagraph has been proposed by Aime *et al.* (2002). This elegant method is based on the mathematical properties of prolate spheroidal functions (Slepian & Pollak, 1961) which are invariant by Fourier transform, even if the original function is truncated (i.e. has a finite definition domain as it is the case for a given optical system pupil). The formalism of prolate apodized Lyot coronagraphy is extensively described in Aime *et al.* (2002), Soummer *et al.* (2003) and won't be detailed here. These authors also evidenced that the residual amplitude distribution in the Lyot plane (coronagraphic pupil) for a matching (unique) combination of apodizer prolate function and Lyot focal plane occulting mask diameter, is also a prolate function, and is still valid when a central obstruction is present (Soummer, 2005). Moreover this amplitude residual is the same prolate function as the entrance pupil one, but with a reduced amplitude, corresponding to the efficiency of the coronagraph. This very nice property directly leads to the multi-stage setup, where only one entrance pupil apodizer is used, but with a cascading number of identical Lyot occutling masks/Lyot stops combinations. Each stage provides the same rejection efficiency $\mathbf{Q}$, leading to a final rejection $\mathbf{Q}^{\mathrm{N}}$ if N stages are used. Consequently, if several stages can be used, then some very small occulting masks can be used, as small as 2 λ/D in diameter, as we plan to test in laboratory.

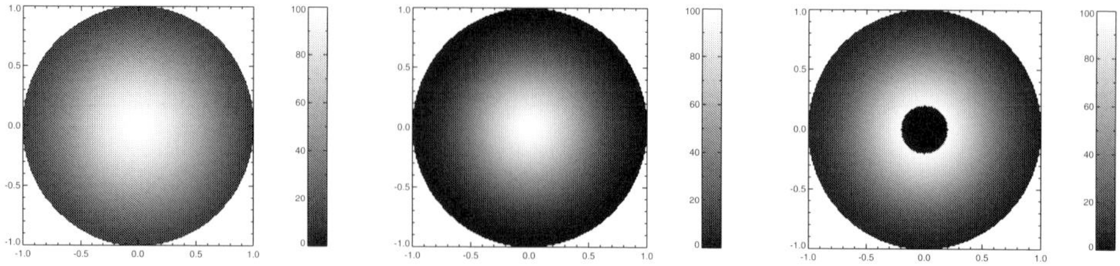

(a) (b) (c)

Figure 5. Three configurations of MS-PALC apodizers that are planned for laboratory tests. (a), (b) and (c) are respectively associated to $2\lambda/D$, $3\lambda/D$ and $3\lambda/D$ occulting mask diameters.

Finally, the MS-PALC offers several advantages but at the expense of some drawbacks so that a tradeoff must be found. The NAOJ and ISAS are setting up an experiment to validate this concept in laboratory. Several masks have been designed to test several configurations. The focal mask sizes vary from 2 to 3 λ/D in diameter, providing effective inner working angles of 1 to 1.5 λ/D respectively. Furthermore, the Lyot stops do not need to be undersized (in theory) so the overall throughput is defined by the entrance pupil apodizer only. Our apodizers throughput range from 20 to 35%.

The masks are currently being manufactured by Canyon Materials Inc. with the HEBS-glass technology. The HEBS substrate contains a superficial layer which can be opacified when bombarded with an electron beam. Details about this fabrication process can be found on the company website (Canyon Materials Inc.). Some encouraging coronagraphic results were already obtained with HEBS-glass focal plane masks (see Balasubramanian *et al.* in these proceedings). Although we might experience some problems with the phase delay introduced by the HEBS-glass pupil apodizer (Halverson *et al.*, 2005), this should be fixed by the future use of a DM (e.g. the already mentioned BMC mirror, or with a higher number of actuators).

References

Aime, C., Soummer, R., & Ferrari, A. 2004, *A&A* 389, 334
Aime, C. & Soummer, R. 2004, *SPIE Proc.* 5490, 456
Canyon Materials Inc., http://www.canyonmaterials.com/
Gay, J. & Rabbia, Y. 1996, *Comptes Rendus de l'académie des Sciences* t332, SérieII, 265
Guyon, O. 2004, *A&A* 404, 379
Halverson, P., Ftaclas, M., Balasubramanian, K., Hoppe, D., & Wilson, D. 2005, *SPIE Proc.* 5905
Jacquinot, P. & Roizen-Dossier, B. 1964, *Prog. Opt.*, 3, 29
Kasdin, J. 2003, *ApJ* 582, 1147
Kuchner, M. & Traub, W. 2002, *ApJ* 570, 900
Lyot, B. 1931, *Comptes Rendus de l'Académie des Sciences* 193, 1169
Nakagawa, T. 2000, *ISAS Report SP* 14, 189
Roddier, F. & Roddier, C. 1997, *PASP* 109, 815
Slepian, D. & Pollak, H. O. 1961, *Bell Sys. Tech. J.*, 40, 43
Soummer, R. 2005, *ApJL* 618, 161
Soummer, R., Aime, C., & Falloon, P.E. 2003, *A&A* 397, 1161
Tamura, M. 2000, *ISAS Report SP* 14, 3
Vanderbei, R. J. 2003, *ApJ* 590, 593
Vanderbei, R. J., Kasdin, J., & Spergel, D. 2004, *ApJ* 615, 555
Vanderbei, R. J. 1999, LOQO Users Manual, Version 3.10, *Optimization Methods and Software* 12, 485

Direct Imaging of Exoplanets: Science and Techniques
Proceedings IAU Colloquium No. 200, 2005
C. Aime and F. Vakili, eds.
© 2006 International Astronomical Union
doi:10.1017/S1743921306009562

Terrestrial Planet Finder Coronagraph Observatory Summary

Virginia Ford[1], Marie Levine-West[1], Andy Kissil[1], Eug Kwack[1], Tim Ho[1], Phil Dumont[1], Doug Lisman[1], Peter Feher[1], and Terry Cafferty[2]

[1]Jet Propulsion Laboratory, California Institute of Technology, Pasadena, CA 91109, USA
email: Virginia.G.Ford@jpl.nasa.gov, Marie.B.Levine@jpl.nasa.gov, Andrew.Kissil@jpl.nasa.gov, Eug-Yun.Kwack@jpl.nasa.gov, Timothy.Y.Ho@jpl.nasa.gov, Philip.J.Dumont@jpl.nasa.gov, P.D.Lisman@jpl.nasa.gov,
[2]TC Technology

Abstract. Creating an optical space telescope observatory capable of detecting and characterizing light from extra-solar terrestrial planets poses technical challenges related to extreme wavefront stability. The Terrestrial Planet Finder Coronagraph design team has been developing an observatory based on trade studies, modeling and analysis that has guided us towards design choices to enable this challenging mission. This paper will describe the current flight baseline design of the observatory and the trade studies that have been performed. The modeling and analysis of this design will be described including predicted performance and the tasks yet to be done.

1. Introduction

NASA is funding a mission called Terrestrial Planet Finder (TPF) that intends to find and characterize terrestrial (or rocky) planets that might harbor life. This mission will look for Earth-like planets around nearby stars in the zone where liquid water could be present on the planet surface. In order to meet this criterion, the planets must be orbiting in the spherical region around a star called the habitable zone where water will be liquid.

When a planet is found in this zone, by studying the light from that planet, the presence of life can be detected from the spectrum of gases that are present in its atmosphere. The presence of life-indicating gases such as water, oxygen, ozone, carbon dioxide, nitrous oxide, methane, and chlorophyll in land plants can be detected in the spectrum of light reflecting off of a planet's atmosphere. The challenge of this mission is to suppress the parent starlight adequately to detect and characterize the fainter light from the planet.

In order to perform this mission, two instruments are being proposed: a visible coronagraph and an infrared interferometer—each with different methods of suppressing starlight. The coronagraph, called Terrestrial Planet Finder Coronagraph (TPF-C), shown in Figure 1, will be the focus of this report. Both instruments provide complementary data in different optical wavebands to establish the presence of life on any planets that may be found and studied.

2. Technology Research and Development

The most challenging technology areas needing development for this project are related to optical wavefront accuracy. Extreme accuracy is needed in order to block out the parent starlight adequately so that the orbiting planet can be detected. In the visible wavelength range of TPF-C, the light from the planet is reflected from its parent star and

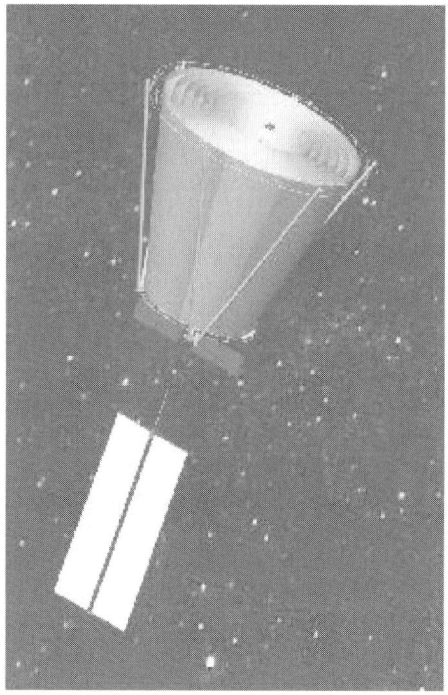

Figure 1. Terrestrial Planet Finder Coronagrapher-Flight Baseline 1 Configuration (Ho).

is 1010 times fainter than the parent star. This intensity reduction factor of 10 billion times requires a deeper understanding of light propagation physics including the effect of optical components on polarization, diffraction, phase, amplitude and wavelength and the crosstalk between these parameters. Test beds, component fabrication, measurements and modeling developments form the core of the technology development efforts that are underway.

Currently, the most mature test bed is called the High Contrast Imaging Testbed (HCIT). It is described in detail in the previous paper in this conference (Trauger 2005); but, in summary, consists of a large, isolated vacuum chamber containing an optical bench with a Lyot-style coronagraph. In addition, there are arrangements to test alternative coronagraph components and systems. The star-simulating source consists of a laser or white-light fed fiber-optic. Also in the system are a deformable mirror, and a camera. To date, this test bed has created a half-dark hole formed using only one DM to correct both phase and amplitude. An average contrast of 8×10^{-10} has been attained that is repeatable and stable in the half area from $4\lambda/D$ to $10\lambda/D$; but in the half area from $4\lambda/D$ to $5\lambda/D$, the testbed has only achieved a contrast of 2×10^{-9}. The team is working to drive down the contrast in that critical smaller area.

Several areas of study support the HCIT and the TPF-C system design by theorizing, modeling, fabricating, and measuring characteristics of components in the HCIT. The goal is to measure component properties and model accurately the optical propagation effects of these components. The model will be validated using HCIT performance; then will be used to explore the best methods of starlight suppression for TPF-C. Some results are presented in a poster in Session 1 (Balasubramanian 2005) of this conference. There is a coordinated effort involving laboratories, NASA centers and universities that has led to increased understanding of electro-magnetic wave propagation through coronagraph components, and has given us a more accurate understanding of methods of suppressing diffracted and scattered starlight. Some of this work is represented in an upcoming paper in this conference (Kuchner 2005) and in a poster (Belikov 2005).

In addition to the technology studies related to attaining starlight suppression, a technology demonstration mirror (TDM) is being fabricated to demonstrate the surface quality, spatial frequency requirements, and light-weight construction required for the TPF-C primary mirror. Figure 2 shows some mirror core segments during fabrication. The TDM has a diameter, 1.8 meters, selected to represent the lowest spatial frequency of interest. Fabrication of the TDM will demonstrate that state-of-the-art technology can meet the spatial frequency requirements. Understanding mirror measurement and interpreting the measured data to ensure compliance with requirements also will be learned, along with materials and coating issues. Trade studies have resulted in the selection of light-weighted, fuse-bonded ULE as the substrate material. ITT and Corning are fabricating this mirror.

Alternative concepts for starlight suppression for this mission are also being studied. A pupil remapping study (Guyon 2005), a visible nulling concept (Shao 2005), and a comparison of coronagraph forms (Shaklan 2005) are presented orally and in posters in this conference.

3. Observatory Design Status

The current design is called the Flight Baseline 1 design concept. The mission and spacecraft choices for the FB1 design are shown in Table 1 and provide the frame work that defines the FB1 observatory requirements. These choices are not considered to be technology drivers—in fact, an attempt was made to stay within the limits of existing technology where possible so that resources at this phase of the project could focus more on areas that require technology development or non-heritage engineering.

The observatory is designed to examine stars in nearly the entire anti-sun semi hemisphere—a 5° margin is included. The field of regard includes all stars located in

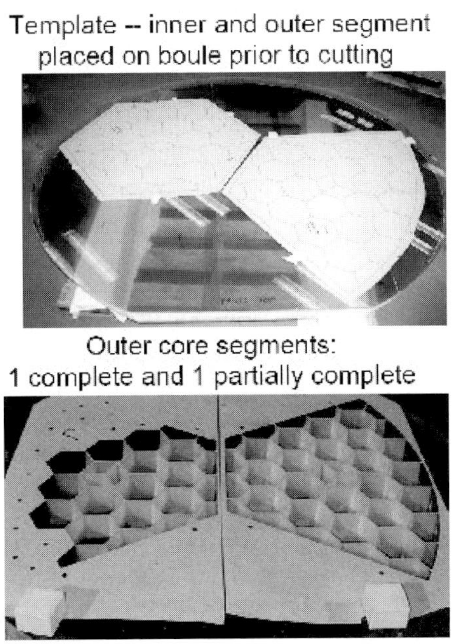

Figure 2. TDM core—made from 6 outer and one inner segment (Cohen, ITT).

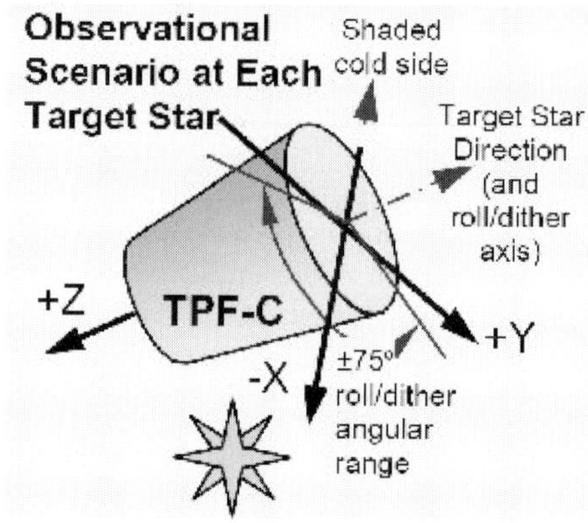

Figure 3. Observational scenario.

the cone defined as greater than or equal to 95° away from the sun. As the observatory travels around the sun, this field-of-regard will sweep the entire sphere of the universe, allowing observation of all star targets of interest during nearly 5 months of the year. This field of regard provides the opportunity to view target stars multiple times during one year so that planets can be seen at a favorable position in their orbit—not behind or in front of the star.

During each star observation, the observatory will point at the chosen target star. Once the dynamics are stabilized, the observatory will collect light. Using adaptive optics, the wavefront errors will be reduced until the starlight is suppressed adequately and an image will be taken. Next the observatory will "dither" about its pointing axis by 30 degrees. Once the dynamics are stabilized, the observatory will take an image in this new position. This image will be subtracted from the previous image to eliminate residual light scattered from the observatory (which rolls with the dither). Any planets present would not roll and would be detectable. Because the primary mirror is oblong, it is most sensitive along its long axis. In order to completely study the habitable zone around a star, the long axis has to be rotated to positions that are ±60° away from the starting point. This is accomplished by a "roll" along the pointing axis. At each new roll position, the adaptive optics reset the wavefront and then the image gathering is repeated, including the dither. With 30° dithers around three 60° roll positions, the total angular rotation around the target direction axis is ±75° as shown in Figure 3 and 4. This observational scenario defines the thermal environmental perturbations that need to be investigated. The observatory must be stable enough between dither maneuvers to retain precise optical wavefront stability. After each roll, the wavefront correction is reset, so stability between roll positions is not as critical. Figure 5 shows a schematic of the design concept and Figure 6 shows the observatory design. The payload is surrounded by a large deployable conic-shaped v-groove sun shade which insulates the payload from the changing sun angles. The sunshade is structurally attached to the spacecraft through deployable arms and booms, so dynamic snaps or warping of the sunshade structures will be filtered through the spacecraft before reaching the sensitive payload.

The optical path enters the telescope baffle and reflects off an 8m×3.5m elliptical off-axis parabolic primary mirror. The light next reflects off the secondary mirror, towards a tertiary fold mirror that directs the beam into the coronagraph starlight suppression assembly. A pick-off mirror sends the outer portion of the beam to a general astrophysics instrument.

Table 1. FB1 Mission & Spacecraft Choices (Lisman & Feher).

	Parameter	Value	Comments
Mission	Duration required/goal	5/10 years	Resources for 10 years
	Orbit	L2	Direct trajectory
	Field of Regard	Sun angles > 95°	Potential earth/moon/planet constraints
	Required ΔV	60 m/s	
	Launch Energy (C_3)	-0.69 km^2/s^2	
	Launch Vehicle	EELV	
	Launch Fairing	5 m diameter	limits primary mirror short axis to ~3.5 m
	Launch Mass	9200 kg	
	Time to reach operating orbit	109 days	
	Ground Station	34m DSN Ka-Band	
	Downlink Data Rate	64Mbps	
Spacecraft	EOL Power	3kW	provided by solar arrays
	Reaction Wheels	6 Goodrich wheels	
	Propellant	242 kg Hydrazine	
	Thrusters	12 20N	
	Hi Rate Downlink Frequency	Ka-Band	avg duration 2.5 hours per day
	Engineering Downlink Frequency	X-Band	
	Uplink Frequency	X Band	
	Transmitter Power	50W	
	Hi Gain Antenna	43dB	0.5m patch array

The primary mirror assembly is mounted semi-kinematically with a hexapod to an aft metering structure (AMS). Behind the primary mirror, the AMS supports thermal control heaters and the secondary mirror assembly deployable tower and the tertiary mirror assembly. Laser sources for laser metrology between the primary mirror and the second mirror are also mounted on the AMS. The laser metrology system will provide continuous feedback on the secondary mirror position. This will be used to adjust thermally-induced relative motions between the primary mirror and the secondary mirror. The aft metering structure mounts to a payload support structure (PSS) through three bi-pods. Further description of the telescope assembly is given in a poster (Ohl 2005) in this conference. Figure 8 shows the Science Payload Assembly excluding the secondary mirror tower.

The coronagraph starlight suppression assembly, all instruments, most of the payload electronics, and thermal control hardware mount to the PSS. Heaters maintain room-temperature conditions. Heat pipes conduct heat from the electronics to a "warm" radiator and from the instrument detectors to a "cold" radiator. The entire lower portion

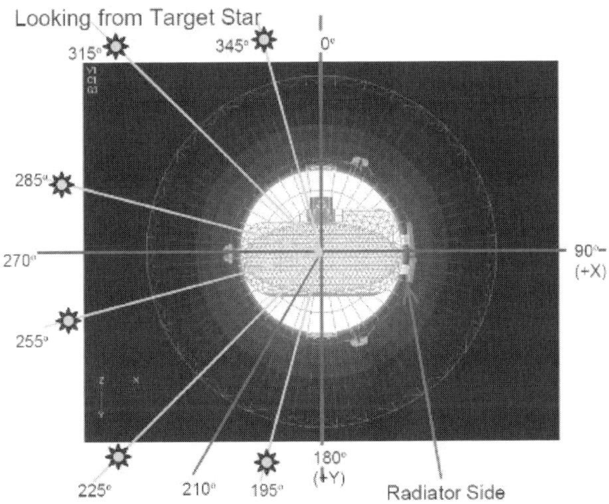

Figure 4. Observation angle nomenclature.

of the payload is surrounded by a thermal enclosure mounted to the PSS. The PSS attaches to the spacecraft through three bi-pods. On the spacecraft side of the interface, at the end of each bi-pod, dynamic isolation is provided—either passive or active—both options were analyzed. The spacecraft carries thruster clusters, orbit maintenance fuel tanks, communications antennas, and reaction wheels. The sunshade mounts off of deployable tension truss, arms and booms. The solar panels and solar sail are also mounted on the spacecraft with deployable structures.

The final important part of the design is how it stows in launch configuration. The volume constraints of the launch shroud pose significant design limitations. Figure 9 shows the FB1 observatory stowed in its launch shroud. The Sun shade folds up and is stowed behind the payload around the spacecraft. The secondary tower folds along 4 joints to fit vertically alongside the primary mirror. The Solar Panels and Solar Sail fold up under the spacecraft, and the antenna arms fold alongside the spacecraft. Launch latches and launch support structure all release and fold away from the observatory upon deployment.

4. Performance Analysis

Several observational completeness models have been created that optimize the observing scenario to maximize planet discovery completeness. A paper (Brown 2005) earlier in this conference describes the approach.

Optical performance modeling is focused on developing propagation models that include all effects of the optical elements on the wavefront. Effects that are included are: diffraction, optical path differences, beam walk on non-perfect optical surfaces, spot

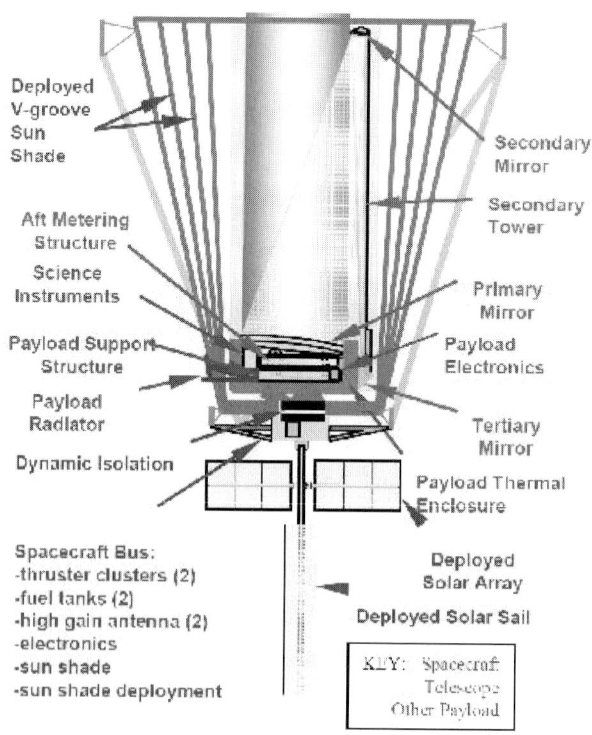

Figure 5. Flight Baseline 1 Schematic.

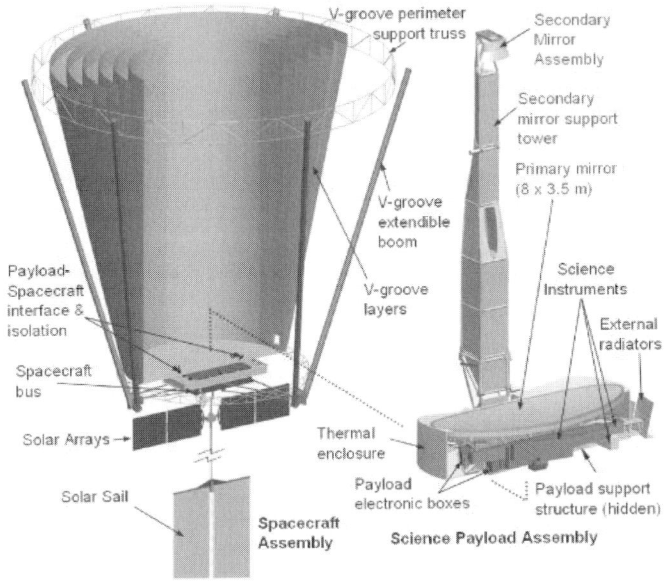

Figure 6. Observatory Design.

diagram centroid errors at occulting mask, jitter effects, cross-talk between polarization, phase, amplitude, and wavelength effects to develop a sensitivity matrix used to calculate contrast resulting from wavefront deformations. The modeling is on-going, but preliminary results are comparable to the error budget analysis that has been developed.

Thermal analysis has included modeling and comparing results from several thermal tools to determine tool accuracy, tool problems and develop solutions to permit achieving the accuracy needed for this observatory. Figure 10 shows the full thermal model using TMG software, and results of the temperature changes for the worst case dither scenario. Even in worst case conditions, the thermal changes meet the error budget allocation.

Structural models form the basis of the thermally induced wavefront aberration calculations and the pointing and control and jitter impact. Structural analysis predicts launch loads. Figure 11 shows the structural models, launch load results, and worst case thermally induced wavefront error. Figure 12 shows a high fidelity structural models of the primary mirror and the acoustic and static launch loads in the mirror. The stresses in the mirror are too high requiring a modification of the launch mounting scheme in the next design cycle.

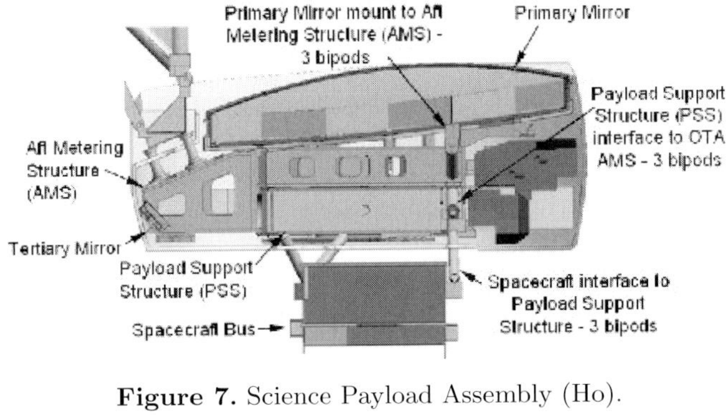

Figure 7. Science Payload Assembly (Ho).

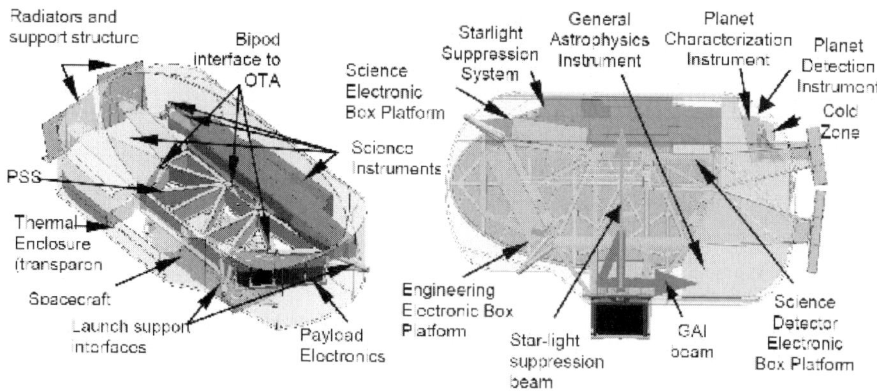

Figure 8. Science Payload excluding Telescope (T. Ho).

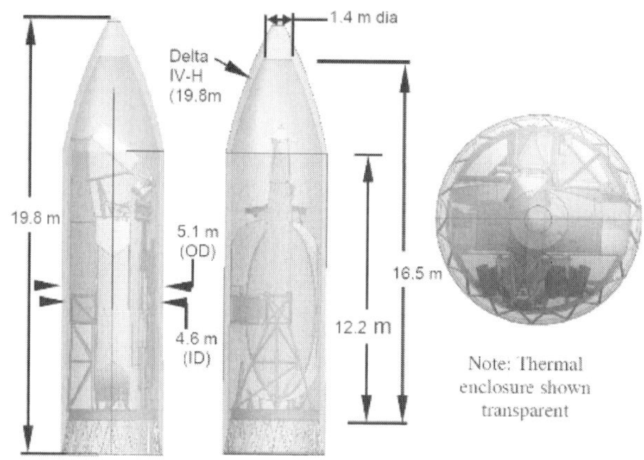

Figure 9. Flight Baseline 1 Launch Configuration (Ho).

The pointing control system concept is based on flight heritage sensors, reaction wheels, actuators, and control system. Two dynamic isolation systems were developed and analyzed for the suppression of reaction wheel disturbance: passive and active. The passive system currently does not meet requirements, but could be tuned with high likelihood of success. The active system meets requirements with large margin. Both designs are important for creating design space, and possibly developing redundancy. Figure 13 shows the worst case motion of the system optics as the reaction wheel speed changes. The top plots show translation and the bottom plots show rotation in three axes. The symbols at each optical element show the wheel speed at which the worst case response occurs. The passive system plot is on the left and the active system plot is on the right. The passive system optical responses are significantly over requirements for the SM, but within requirements for the downstream optics The passive system shows large angular excursions of the Fine Steering Mirror, produced by the control system as it keeps the target star positioned on the mask. The active system response is well below requirements in translation and rotation.

In conclusion, the design concept for Terrestrial Planet Finder Coronagraph continues to evolve with promising results. It appears that the environmental stability required for this project will be achievable. The modeling activities need to continue to develop to be sure light propagation effects are captured fully and understood so that the correct starlight suppression system can be maintained. There are some areas of the design that

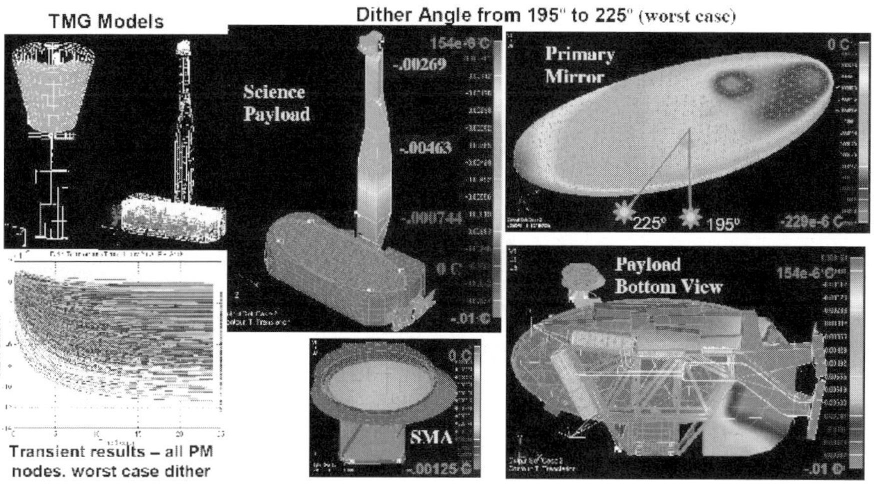

Figure 10. Thermal models and worst case dither temperature change results.

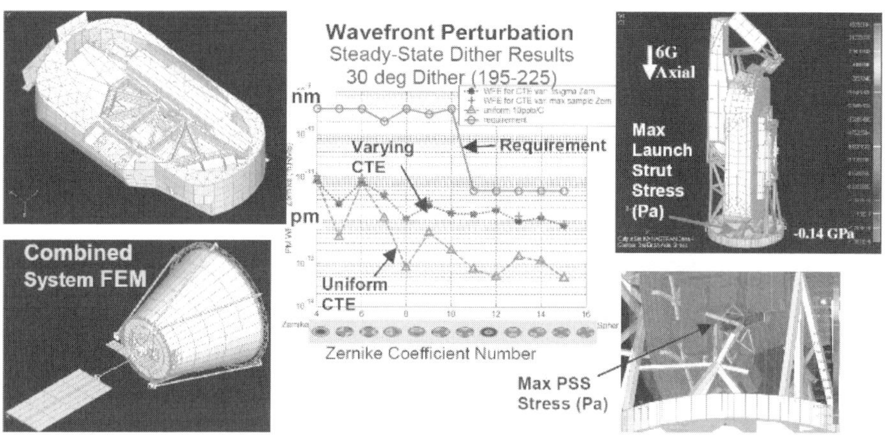

Figure 11. System structural model, Thermally-induced wavefront perturbation, and System Launch Stresses.

need to be addressed and improved, but they appear to be in the realm of engineering that can be accomplished.

Acknowledgements

I would like to thank the following people for their contributions to this paper: Marie Levine-West, Andy Kissil, Tim Ho, Eug Kwack, Phil Dumont, Dave Palacios, Jim Alexander, Sandra Irish, Chuck Engler, Lou Fantano, Alice Liu and Carl Blaurock.

This work was performed at the Jet Propulsion Laboratory, California Institute of Technology, under a contract with the National Aeronautics and Space Administration.

Reference herein to any specific commercial product, process or service by trade name, trademark, manufacturer, or otherwise, does not constitute or imply its endorsement by the United States Government or the Jet Propulsion Laboratory, California Institute of Technology.

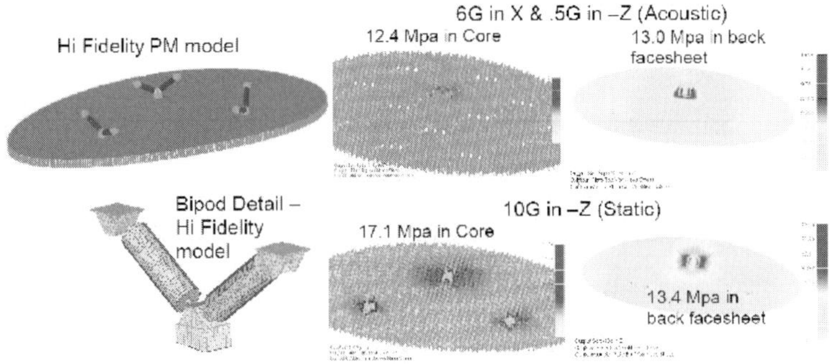

Figure 12. Hi Fidelity Primary Mirror model, detail of bipod model, and launch load results from both acoustic loading and static equivalent loading.

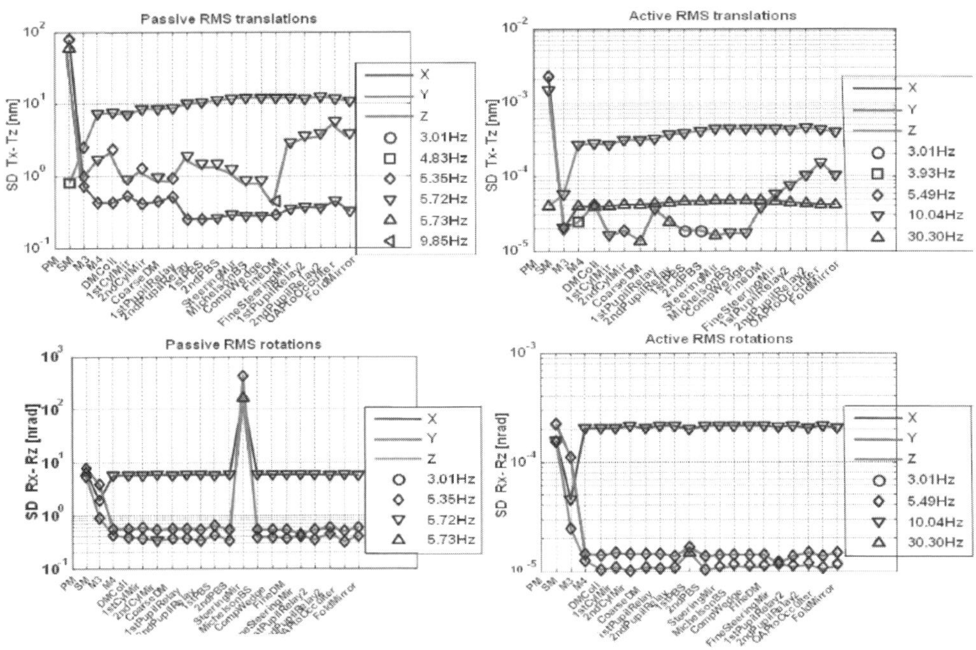

Figure 13. Dynamic response.

References

Trauger, J., "A case study in high contrast coronagraphy for planet discovery: the Eclipse concept and supporting laboratory experience," IAU 200, 6 Oct. 2005

Balasubramanian, K., "Occulting Focal Plane Masks for Terrestrial Planet Finder Coronagraph—Design, Fabrication, Simulations and Test Results," IAU 200, 4 Oct. 2005

Kuchner, M., "Eighth-order image masks for terrestrial planet finding," IAU 200, 6 Oct. 2005

Belikov, R., "Towards 10^{10} Contrast for NASA's Terrestrial Planet Finder Mission: Demonstration of High Contrast in a Shaped-Pupil Coronagraph," IAU 200, 6 Oct. 2005

Guyon, O., "The pupil remapping coronagraph," IAU 200, 6 Oct. 2005

Shao, M., "The Visible Nulling Coronagraph—Architecture Definition and Technology," IAU 200, 6 Oct. 2005

Shaklan, S., "Terrestrial planet finder coronagraph Instrument Design," IAU 200, 4 Oct. 2005

Ohl, R., "The Optical Telescope Assembly for the Terrestrial Planet Finder Coronagraph," IAU 200, 6 Oct. 2005

Brown, R., "Expectations in the Visible Domains," IAU 200, 4 Oct. 2005

Direct Imaging of Exoplanets: Science and Techniques
Proceedings IAU Colloquium No. 200, 2005
C. Aime and F. Vakili, eds.
© 2006 International Astronomical Union
doi:10.1017/S1743921306009574

Visible Nulling Coronagraphy for Exo-Planetary Detection and Characterization

Richard G. Lyon, Mark Clampin[1], Robert Woodruff[2], Gopal Vasudevan[3], Mike Shao, Martin Levine[4], Gary Melnick, Volker Tolls[5], Peter Petrone, Peter Dogoda[6], Julia Duval[7], and Jian Ge[8]

[1]NASA/Goddard Space Flight Center, Greenbelt, MD
email:Richard.G.Lyon@nasa.gov, 301-286-4302

[2]Lockheed-Martin, Civil Space, Denver, CO

[3]Lockheed-Martin, Advanced Technology Center, Palo Alto, CA

[4]NASA/Jet Propulsion Laboratory, Pasadena, CA

[5]Harvard/SAO Center for Astrophysics, Cambridge, MA

[6]Sigma Space, Lanham, MD

[7]CNES, France

[8]University of Florida

Abstract. Visible Nulling Coronagraphy (VNC) is the proposed method of detecting and characterizing exo-solar Jovian planets (null depth 10^{-9}) for the proposed NASA's Extrasolar Planetary Imaging Coronagraph (EPIC) Clampin & Lyon 2004 and is an approach under evaluation for NASA's Terrestrial Planet Finder (TPF) mission. The VNC approach uses a single unobscured filled-aperture telescope and splits, via a 50 : 50 beamsplitter, its re-imaged pupil into two paths within a Mach-Zender interferometer. An achromatic PI phase shift is imposed onto one beam path and the two paths are laterally sheared with respect to each other. The two beams are recombined at a second 50 : 50 beamsplitter. The net effect is that the on axis (stellar) light is transmitted out of the bright interferometer arm while the off-axis (planetary) light is transmitted out of the nulled interferometer arm. The bright output is used for fine pointing control and coarse wavefront control. The nulled output is relayed to the science camera for science imagery and fine wavefront control. The actual transmission pattern, projected on the sky, follows a θ^2 pattern for a single shear, θ^4 for a double shear, with the spacing of the successive maxima proportional to the inverse of the relative lateral shear. Combinations of shears and spacecraft rolls build up the spatial frequency content of the sky transmission pattern in the same manner as imaging interferometer builds up the spatial frequency content of the image.

Keywords. Nulling coronography – exo-solar planets.

1. Introduction

The VNC approach has some significant advantages over more conventional mask coronagraphic approaches in that (i) its inner working angle is $2\lambda/D$ (assuming a 25% shear) allowing a smaller aperture than conventional coronagraphic methods that typically work at $4\lambda/D$, (ii) it has significantly less sensitivity to low order amplitude and wavefront aberrations – the shear is approximately equivalent to a lateral wavefront derivative which gives an f^2 multiplier on the wavefront PSD, thereby allowing for significantly lower quality optics in the telescope, (iii) coupling of the VNC nulled beams to a fiber bundle with a lenslet array on the front and back sides and a faceted micro-mirror array (MMA) within the nuller, with each MMA facet matched to each fiber, gives direct simultaneous

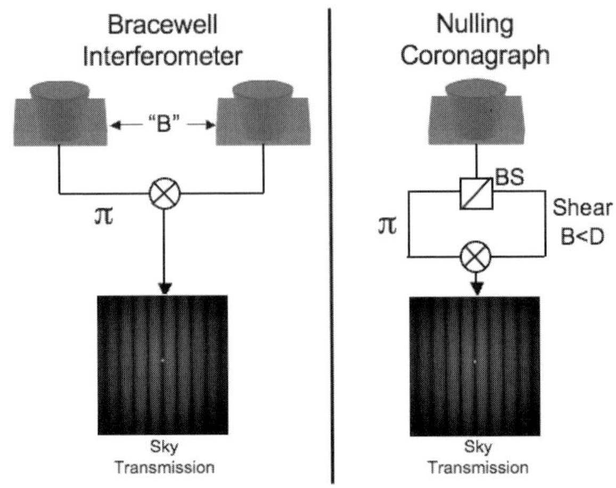

Figure 1. Bracewell and Visible Nulling Interferometer.

correction of both amplitude and phase with a single MMA per nuller, and, each fiber is a natural spatial filter on higher order wavefront errors. This also has a significant advantage in that focal plane mask errors in a conventional coronagraph are dispersive in the following pupil plane, making it difficult for any pupil plane deformable mirror to correct mask errors.

2. Principle of Visible Nulling Coronograph

Figure 1 shows simplified schematics of a Bracewell Nulling Interferometer (BNI) and a Visible Nulling Coronagraph (VNC). In a canonical BNI configuration 2 telescopes, separated by a baseline of B, are coherently interfered with a π phase shift giving a sky transmission, projected on the sky of $T^{\{2\}} = sin^2\left(\pi\frac{B}{\lambda}\theta\right) \sim \theta^2$ where θ is the sky angle in the direction of the baseline and λ is the wavelength. In a VNC the same pattern also arises except that B is now the relative pupil shear.

A VNC effectively emulates a BNI with only a single aperture. The condition for central nulling is given by

$$\sum_{j=1}^{N} e^{i\Phi_j} = 0$$

where Φ_j represents the relative phase shift due to the jth aperture in the BNI and represents the jth shear in VNC for a given set of N apertures or shears.

Figure 2 shows a conceptual optical configuration for a double sheared VNC, which is first sheared in the X direction followed by a shear in the Y direction. Light from the optical telescope assembly (OTA) is incident on a $50:50$ beamsplitter which splits it into two paths. Light from the upper path (shown in red on top of Figure 2) reflects off a $90°$ corner reflector of which one mirror is a multiple mirror array (MMA) (Figure 3) and through two glass plates and to a second beam splitter. The MMA consists of MEMS based array of flat micro-mirrors, each movable in piston, tip and tilt. Light traverses the lower path (blue on top figure in Figure 2) after the first beamsplitter and reflects off a $90°$ corner reflector mounted on a two degree of freedom translation stage. This stage

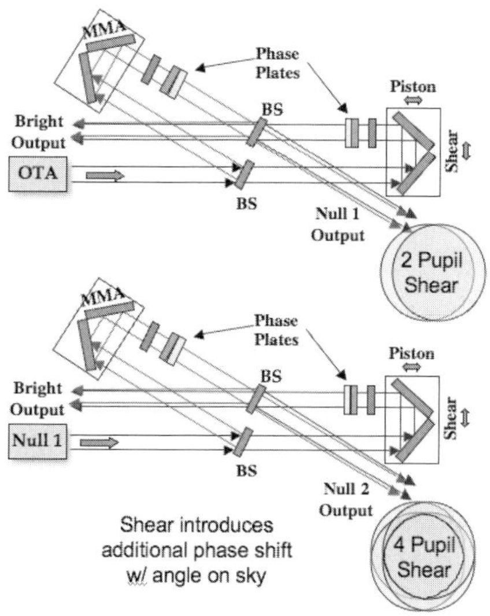

Figure 2. X/Y VNC Configuration.

can be moved in piston and to shear the lower paths beam with respect to the upper paths beam. The two beams are recombined at the second beamsplitter.

Setting the piston to yield a π phase shift allows the on-axis light to pass out through the bright output and the off-axis light to pass through to the nulled output. At the nulled output the two beams are laterally sheared which also introduces an additional phase shift resulting in nulling of the on-axis light (star) and transmission of off-axis light (planet). The output of the "X" nuller is fed into a second nuller which shears the beams orthogonal to the first nuller's shear direction, i.e. the "Y" direction and results in 4 sheared beams as shown in the lower right of Figure 2.

The output of the second nuller is passed into a spatial filter array (SFA) consisting of an array of lenslets, followed by an array of fibers and again by an array of lenslets, one lenslet per fiber (Figure 4). The net effect of the SFA is to allow correction of both amplitude and phase errors. From the two nullers are 4 sheared beams, overlaid, at the entrance aperture of the spatial filter array. The lenslet array divides the pupil into ~1200 hexagonal sub-apertures and thus 4 beamlets, each with relative phase shifts and amplitude shifts are injected into each fiber. At the output of each fiber it is desired that no light emerges. If light is observed at the output then the 2 MMA facets which are mapped to a given lenslet/fiber combination are moved in piston, tip and tilt. Moving

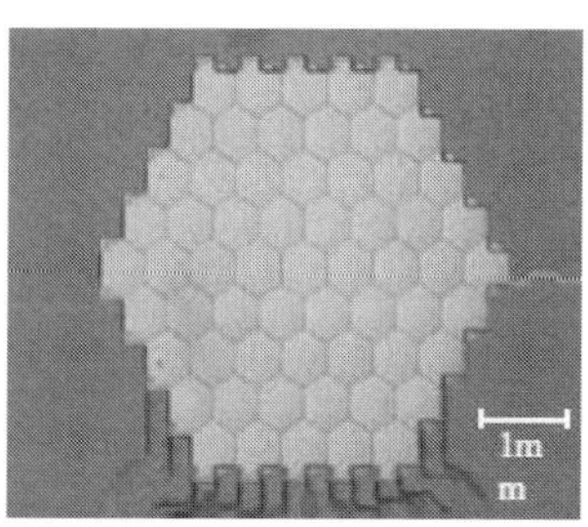

Figure 3. MMA.

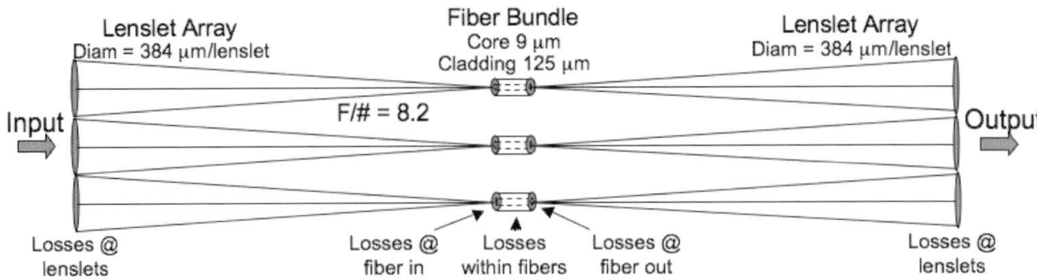

Figure 4. Schematic of SFA.

each MMA facet in piston compensates for non-common path length errors introduced in the two nullers to effectively main the condition for central nulling. Motion of the MMA facets in tip/tilt steer the sub-aperture point spread function (PSF) on the ends of the fibers. Since there is a MMA in only one arm of each of the two nullers, one can in principle always match the amplitudes. Thus the combination of the SFA and MMA correct for amplitude and phase errors.

Figure 5 shows a simulation of the intensity entering the input lenslet array of the SFA (left side) and the intensity entering the fiber bundle. Note that there is one MMA micro-mirror per lenslet/fiber. The speckled pattern is due to residual wavefront errors from the OTA and non-common path errors introduced in the X and Y nullers.

Light injected into a fiber will ultimately evolve into an eigenmode of the fiber. At the fiber output, independent of the functional form of the input field, the output can be characterized by only amplitude and phase. The planetary light, entering the OTA slightly off-axis, but within the numerical aperture of the fiber, will emerge with a stairstep phase as shown on the rightside of Figure 6. The net effect is that the fiber bundle coupled to the lenslets will cause grating like sidelobes in the planet image, rightside of Figure 6. The sidelobes are periodic and will occur at angular sky spacings of $\sim N \times \lambda/D$, where N is the number of fibers across a diameter of the bundle, λ is the wavelength of light and $D$is the OTA entrance aperture diameter. These sidelobes effectively limit the field of view of the VNC and result in $\sim$2% loss in the planetary flux.

The effective planetary sky transmission pattern is built up from several spacecraft rolls and several shears of the VNC. Figure 7 shows an example of a possible transmission patter with 3 rolls of 0, 15 and 30 degrees and 4 shear combinations for a passband of 90 nm centered on 600 nm. It can be see that with a relatively small number of shear/roll combinations high contrast transmission is possible through the planetary search space. Figure 8 a simulation of a Jupiter and Saturn planetary detection.

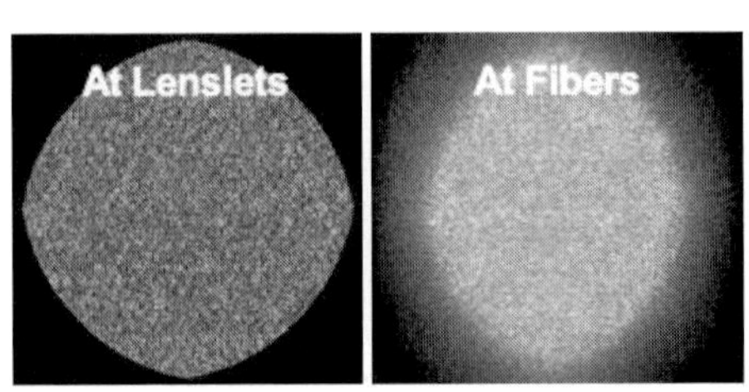

Figure 5. Simulation of Intensity entering SFA lenslet array and at fiber bundle entrance.

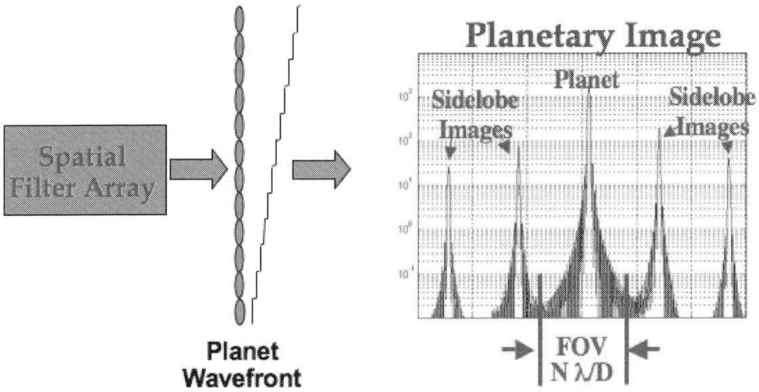

Figure 6. Planetary Image.

3. VNC Sensitivities and Initial Error Budgeting

For the X and Y shear configuration for an resolved stellar on-axis source the scalar electric field at the output of the 2 nullers can be represented as:

$$E_T(x,y) =\approx t\frac{E_0}{4}\sum_{j=1}^{4}(-1)^{j-1}\left(1+\delta A_j(x,y)\right)\left(1+i\Phi_j(x,y)\right) \tag{3.1}$$

where we have expanded to first order the phase and amplitude variations and "t" represents the throughput. Assuming ensemble averages over uncorrelated Gaussian statistics yields for the intensity:

$$I_T(x,y) \approx \frac{1}{16}PSF_0(\theta)\left[\sum_{j=1}^{4}\left\langle\mid \delta A_j(\theta)\mid^2\right\rangle_{\delta A} + \sum_{j=1}^{4}\left\langle\mid \Phi_j(\theta)\mid^2\right\rangle_{\Phi}\right] \tag{3.2}$$

where θ represents the angle on the sky and the angled brackets represent ensemble averages. PSF_0 is the point spread function without errors. If we assume an asymptotic form for the error free point spread function and evaluate the ensemble averaged terms in equation 3.2 for a linearly filtered random process then 3.2 can be evaluated in terms

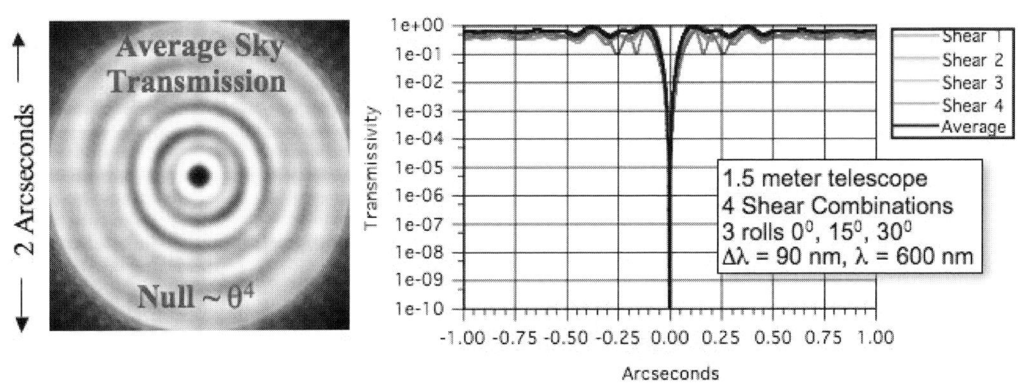

Figure 7. Sky Transmission Pattern.

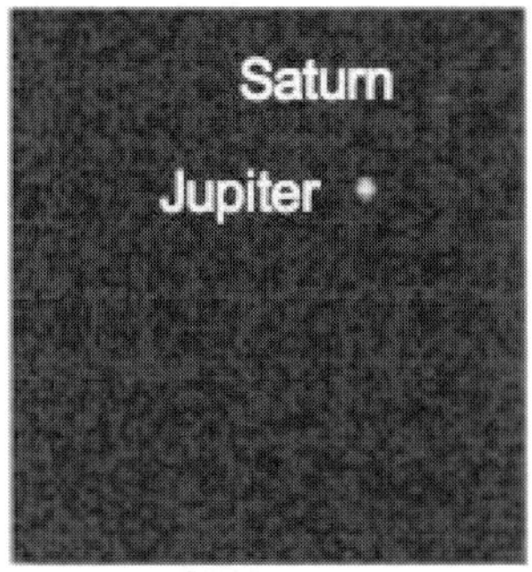

Figure 8. Simulated Jupiter/Saturn Detection.

of power spectral densities to yield:

$$I_{Star}(\theta) \approx \frac{1}{16}\frac{L_{Star}}{1+\frac{\pi^4}{8}\theta^3}\left[\left(\sigma^2_{N1,\delta A}+\sigma^2_{N2,\delta A}\right)PSD_{\delta A}(\theta)+\left(\sigma^2_{N1,\Phi}+\sigma^2_{N2,\Phi}\right)PSD_{\Phi}(\theta)\right] \tag{3.3}$$

where L_{Star} represents the stellar luminosity, θ the sky angle in units of λ/D, and $\sigma_{N1,\delta A}$, $\sigma_{N2,\delta A}$ represent the fractional rms amplitude errors in nullers 1 and 2 respectively and $\sigma_{N1,\Phi}$, $\sigma_{N2,\Phi}$ the rms phase errors (in radians) in nullers 1 and 2 respectively. In the above formalism the asymptotic PSF, $\left[1+\frac{\pi^4}{8}\theta^3\right]^{-1}$ is normalized to unity at the origin as are the amplitude and phase error power spectral densities. Note that in equation 3.3 the error PSDs are multipliers on the stellar leakage, implying that for no errors no starlight gets through at any point in the focal plane.

The off-axis planetary light is given by:

$$I_{Planet}(\theta) = \frac{L_{Planet}}{1+\frac{\pi^4}{8}(\theta-\theta_p)^3} sin^2\left(2\pi S_x\theta_{xp}\right) sin^2\left(2\pi S_y\theta_{yp}\right) \tag{3.4}$$

where $\theta_p = (\theta_{xp}, \theta_{yp})$ is the planet location in units of l/D and (S_x, S_y) is the fractional pupil shear in the X and Y directions respectively. Note that the transmission pattern is a multiplier on the planets asymptotic PSF, but the coordinates of the transmission pattern depend on the location of the planet.

To calculate the contrast of the planet to the stellar leakage, at the location of the planet we only need to ration equation 3.4 to equation 3.3 and evaluate the result at the location of the planet to yield the contrast of:

$$Q = 16\frac{L_{Planet}}{L_{Star}}\frac{1+\frac{\pi^4}{8}\theta_p^3}{\left[\left(\sigma^2_{N1,\delta A}+\sigma^2_{N2,\delta A}\right)PSD_{\delta A}(\theta_p)+\left(\sigma^2_{N1,\Phi}+\sigma^2_{N2,\Phi}\right)PSD_{\Phi}(\theta_p)\right]} \tag{3.5}$$

Equation 3.5 assumes that the nuller shears are set such that the sky transmission function is unity (shear matched condition). From 3.5 it is evident that Q is inversely proportional to the variances of the amplitude and phase errors and increases as approximately the 3rd power of the planet to star angular separation.

Assuming that the error PSDs all have the form of $PSD(\theta) = \frac{1}{1+(\theta/\theta_{knee})}$ and assuming, e.g., that all the error occurs as phase (wavefront) error and solving equation 3.5 for the error yields:

$$\sigma_T = 4\sqrt{\frac{1}{Q}\frac{L_{Planet}}{L_{Star}}\left(1+\frac{\pi^4}{8}\theta_p^3\right)\left(1+\left(\frac{\theta_p}{\theta_{knee}}\right)^3\right)} \tag{3.6}$$

Equation 3.6 is plotted in Figure 9, showing the total rms wavefront error, in units of waves, versus the desired inner working angle for three values of "Q". The planet to stellar luminosity ratio is assumed to be 10^{-10}. It is seen that a desired inner working angle of $4\lambda/D$ requires residual wavefront errors $\sim\lambda/500$ rms and an inner working angle of $2\lambda/D$ requires residual wavefront errors of $\sim\lambda/5000$ rms to detect terrestrial planets.

4. Lab Visible Nulling Coronograph

In order to realize the EPIC science mission the NASA technology readiness levels (TRL) of the VNC must be advanced enough to yield a stable achromatic (15% passband) null (10^{-9}) for 1000 seconds. In order to achieve this goal we have been developing a laboratory testbed (Figure 10). The testbed consists of a source module consisting of a fiber injected source reflecting off a thin 50 : 50 pellicle to an off-axis parabola (OAP), to a reflecting flat, back to the off-axis parabola and transmitting through the pellicle, to a small collimating off-axis parabola (left side of Figure 10) and into the lab VNC (rightside of Figure 10). The OAP and flat of the source module are specially designed and manufactured to comprise a matched pair such that where ever there is a bump/pit on the parabola there is a pit/bump on the flat Kestner & Witebsky 2004. The resultant wavefront quality from the source module is 1.2 nm rms wavefront error. The surface maps and combined maps of these optics are shown in Figure 11. Lateral translation of the flat, away from the ideal aligned position, allows control of the wavefront error from $\sim\lambda/500$ to $\sim\lambda/20$ to demonstrate holding of the null while varying wavefront errors.

5. Summary and Conclusions

The VNC approach has some significant advantages over more conventional mask coronagraphic approaches in that (i) its inner working angle is $\sim 2\lambda/D$ (assuming a 25%

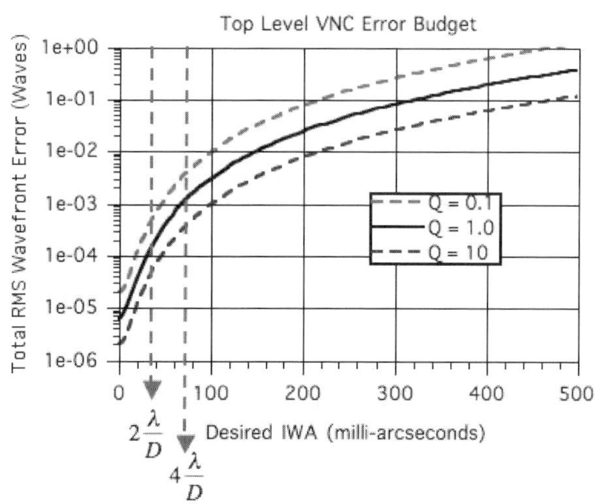

Figure 9. RMS WFE vs IWA.

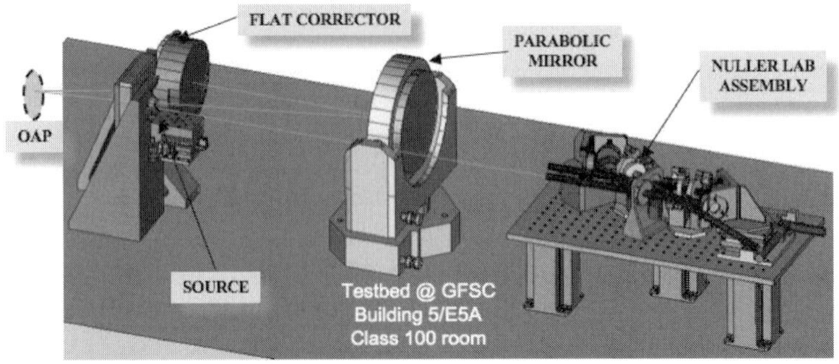

Figure 10. Visible Nulling Coronagraph.

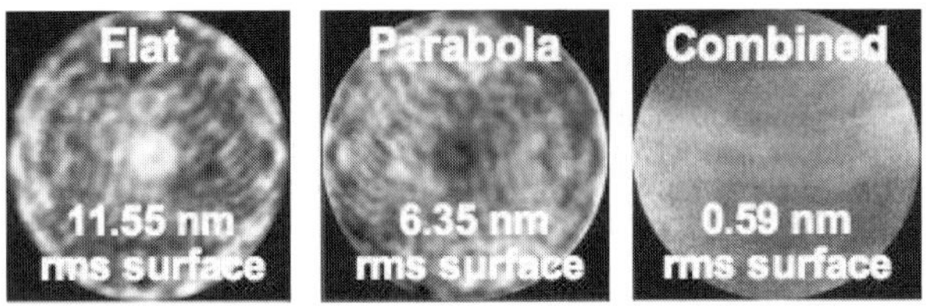

Figure 11. Source Module Surface Maps.

shear) allowing a smaller aperture than conventional coronagraphic methods that typically work at $\sim 4\lambda/D$, (ii) it has significantly less sensitivity to low order amplitude and wavefront aberrations – the shear is approximately equivalent to a lateral wavefront derivative which gives an f2 multiplier on the wavefront PSD, thereby allowing for significantly lower quality optics in the telescope, (iii) coupling of the VNC nulled beams to a fiber bundle with a lenslet array on the front and back sides and a faceted micro-mirror array (MMA) within the nuller, with each MMA facet matched to each fiber, gives direct simultaneous correction of both amplitude and phase with a single MMA per nuller, and, each fiber is a natural spatial filter on higher order wavefront errors. This also has a significant advantage in that focal plane mask errors in a conventional coronagraph are dispersive in the following pupil plane, making it difficult for any pupil plane deformable mirror to correct mask errors.

References

M. Clampin, R. G. Lyon, G. Melnick, P. Nisenson, R. A. Woodruff, M. Harwit, D.Y. Gezari, L. Petro, H. Ford, R. Mauk, H. Smith, *The Extra-solar Planetary Imaging Coronagraph*, Proceedings of SPIE 5487, Glasgow Scotland, June 21–25, 2004

Kestner, R., Witebsky, C., Clampin, M. Lyon, R., Woodruff, R., Irregulary Figured Plates Correct Wavefront Errors, Laser Focus World, Volume 40, Issue 12, December 2004

Direct Imaging of Exoplanets: Science & Techniques
Proceedings IAU Colloquium No. 200, 2005
C. Aime & F. Vakili, eds.
© 2006 International Astronomical Union
doi:10.1017/S1743921306009586

Recent progress of visible light nulling interferometry and first 1 million null result

Edouard Schmidtlin[1], J. Kent Wallace[1], Rocco Samuele[2] B. Martin Levine[1], and Michael Shao[1]

1 Jet Propulsion Laboratory, California Institute of Technology, Pasadena, CA 91007
2 Northop Grumman Space Technology, Redondo Beach, CA 90278
email: edouard.schmidtlin@jpl.nasa.gov

Abstract. In order to detect directly faint planets around other stars it is important to develop techniques to cancel the central starlight. We report progress in terms of null depth on the visible nulling experiments at JPL, essential for NASAs future space missions as well as for an upcoming sounding rocket based experiment. A level of 1 Million to 1 nulling ratio (1e-6 rejection ratio) has been obtained for the first time for visible laser light, and a level of 100000 to 1 nulling ratio has been obtained for a relatively large (5%) bandpass at 650nm. Using a combination of a fiber array and deformable mirror a null of 1e-6 would correspond to a contrast of 1e-9 at a distance of 2-3λ in the airy disk plane, only 10x away from TPFs goal. The configuration used was a fiber fed Mach Zender type interferometer, using 2 or 3 mirrors in each two arm, placed in an enclosed air container. For broadband light we used tilted dispersion glass plates in the nuller arms and an avalanche photon diode module to cover the large dynamic range. We describe a variety of conditions that have to be met and optimized to reach very deep nulls and sub-nanometer optical path difference stability, such as optical alignment, symmetry in the two arms, mechanical stability and vibration isolation.

Keywords. extra solar planet detection, interferometry, nulling, optical path stabilization

1. Introduction

Since 1995 [1] about 150 exo-planets have been detected around other stars, using various methods, but a common fact is that these detections were *indirect* [2,3]. A few recent direct claims have been made but there is ambiguity, such as lack of proof of gravitational relationship, moreover the planets tend to be much larger than Earth types. It still remains a formidable challenge to observe an exoplanet directly, whether the light comes from the planet itself or is a reflection of light coming from the star. In order to detect and observe a faint planet near a bright star it is necessary to cancel the central starlight to high rejection ratios which leads to both a contrast and an angular resolution challenge. It is currently believed that such a challenge is out of the question from the ground because of atmospheric seeing and scatter, even with state of the art adaptive optics, so space observation is mandatory. In terms of contrast the planet to star ratio is 1e-9 for a Jupiter and is 1e-10 (Δmag=25) for an Earth like planet in the visible. In the infrared the contrast is about 1000 times easier, but the search is more difficult. Assuming a theta at the 4th power dependency, a null of 1e-7 would correspond to a contrast of 1e-10 at a close distance of 1 airy spot ($1\lambda/D$).

A nulling interferometer (aka coronagraph) can consist of either a single large aperture telescope (with sheared pupil interferometer or equipped with a mask with two apertures), or an interferometric array of multiple telescopes [4]. In terms of resolution very large telescopes are not necessary: for an Earth at 10parcsec (0.1arcsec) D can be only 1.5m, whereas for a Jupiter 0.3m will suffice. Whereas a classic Michelson interferometer with one of its mirror displaced by $\lambda/2$ can achieve a null of no better than .0005 and .002 for a bandpass of respectively 5% and 10%, a dedicated Achromatic Nulling Interferometer is needed for deeper nulls over larger bandwidths of astronomical interest. Efforts have been conducted in visible, near infrared, and mid infrared for many years [5] addressing issues like control of optical path, high symmetry and matching of angles, pupil orientations, polarizations, spectrum, dispersion, etc. In the mid infrared [6] components availability is more difficult, however the foremost parameter of optical path control is easier to achieve due to 15x longer wavelength. For JPLs future space missions such as TPF (D$\sim$4m) [7] and for the more recent space project PICTURE (a sounding rocket experiment scheduled for 2007) and other applications, we developped the visible nulling testbed at JPL. The experiment has gone through continuous technological and experimental progress over the years [8,9,10]. More progress and deeper nulls have occurred within the last year, which is the subject of the present paper.

2. Nuller experimental description

2.1. *The RoofTop setup and the new W setup*

The visible nulling testbed at JPL is primarily a modified Mach Zender type interferometer, with single pupil input, described in many papers over the years, particularly in the form of the RSI (Rotational Shear) and field flip approach [8,9,10] using rooftop (2 mirrors) until 2004. A benefit of rooftop setups is that OPD and shear can be adjusted independently. In early 2005 we built a W configuration (3 mirror nuller, with an additional reflection), see figure 1, with lower incidence angles to accommodate well a future deformable mirror for future projects and for tank operation. The old rooftop configuration and the current W configuration will soon be merged into yet another configuration (the *Boomerang*) for the rocket experiment, to merge both benefits.

The nuller is essentially an interferometer with two beam splitters and a set of mirrors on each side, defining two arms of high symmetry, connected to an input single mode fiber to a choice of light sources, and connected to an output single mode fiber to a choice of detectors, all of this with remote control and computer acquisition. The heart of the nuller is a pair of quality beam splitters with dielectric coatings and with very symmetric specifications. It is important to note that wavefront quality is not critical since the output fiber (core) acts as a single mode spatial filter. Geometrically the beam diameter was 20mm, angles of incidence $\sim$15deg everywhere, and each arm $\sim$25 inches of path. In terms of polarization (not critical for white light) we had the capability to insert (zero deviation) polarizers at the input or at the output OAP levels or both. The nuller was also equipped with two shutters. To equalize OPDs, the adjustment was coarse (picomotor of range 25mm) to locate the white light fringe, then fine (piezzo PI stage P752 fed by a 100Volt generator) to explore and null the central fringe. Figure 1 shows elements for both setups.

2.2. *Sources*

We used laser and filtered white light (broadband) sources. For laser sources we used our old 650nm laser diode, then in 2004 a stronger laser diode module from Melles Griot emitting at 638nm with a pigtailed power capable of 11mW but tuned to 4mW (3.75V, 80mA) for better stability. Lasers showed in fact linewidths of $\sim$1nm ie 0.15% bandpass. The white light source used was of incandescent type equipped with filters centered at

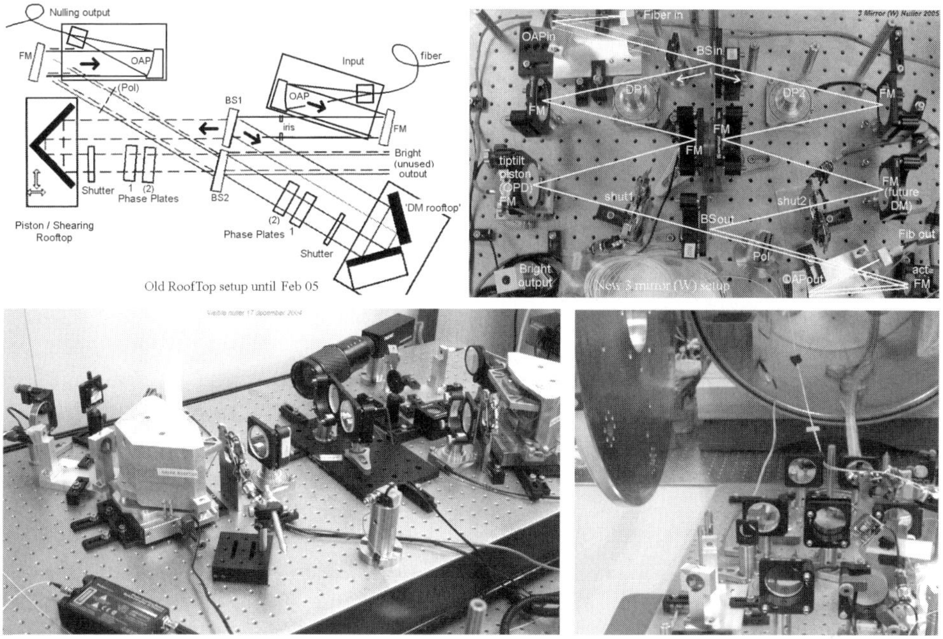

Figure 1. layout and photo of: the old RoofTop nuller (left), and new W nuller (right).

650nm and with relatively large bandpass of 5% and 12%, and sometimes 20%, verified by spectrometer (after the nuller).

2.3. *Detectors*

In order to see very low broadband flux typical of future starlight we had to use an Avalanche Photo Diode hooked up to a high speed counter. The APD detector module used was type SPCM AQR from Perkin Elmer with an advertised QE of ~65% and noise of ~30 cps (counts per second). Due to high dynamic ranges encountered it was important to apply corrections, also to take precautions not over illuminate and burn the APD. The raw count was first noise corrected (dark current typically ~50cps depending on situations), then for higher counts (more than ≈100Kcps) we applied a correction due to the non linear behavior. For laser nulling we could use more traditional photodetectors such as model 2151 from New Focus for high (KHz) speed, and the duochannel multimeter 2532C from Newport for convenient auto-ranging (largest dynamic range from miliWatts to sub nanoWatts). Usually no corrections were applied for these detectors as confirmed with neutral density filter ramp tests.

2.4. *Data processing*

The overall experiment was controlled with a PC under LabView giving the capability to read and control almost every instrument remotely. Typically a dataset would last several minutes and would contain the many features: intensity balance (i1-i2-i1-i2 switching), max intensity (top), manual surfing of the null (taking advantage of nanometric drifts we could do many passes to 'sense' the null), dark, midfringe. For broadband nulling we also recorded the +1 and -1 fringes. Nulling data was acquired over sampling rates of 25Hz or 10Hz with a N.I. DAQ under 16 bits.

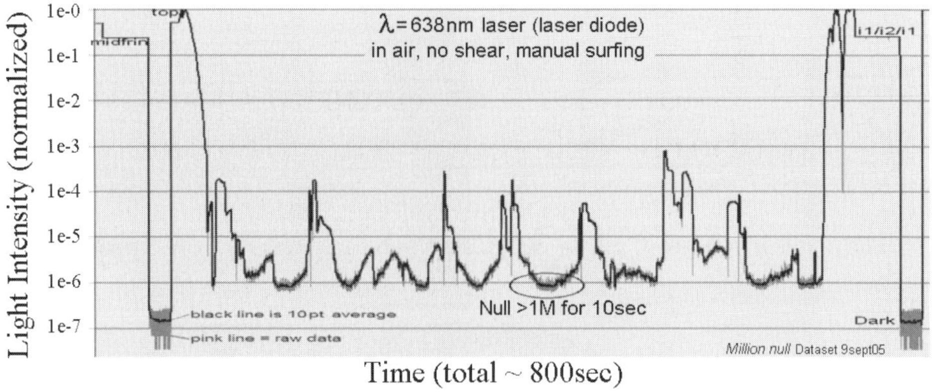

Figure 2. first 1 Million null under visible (638nm) laser light using a Mach Zender type nuller. The plot shows light intensity (log scale) versus time while OPD manual readjustements are done. The steady null is slightly better than 1M for 10seconds. This results was obtained in air.

3. Laser results and diagnosis

Continuous progress and improvements over the years led to laser nulls in the 100K range and broadband nulls in the 10K ranges [8,9,10]. In spring 2005 we have reached 300K laser null. Only recently, on June 10th 2005, we officially reached the milestone of 1 Million average null (ie 1e-6 rejection ratio, in this paper we use both notations equivalently) under monochromatic (laser) light, in air, tank closed, with a polarizer at the output. To this date we have reached slightly better (1.2 Million average raw null over many seconds, 1.4 Million corrected), see figure 2. To our knowledge this is the best null ever in the visible.

A note about the terminology of *transient nulls* (single datapoints): transient nulls have been quoted many times in past papers and can be very deep, but they are merely a consequence of low flux statistical noise. In this paper and for future nulling papers we declare that transient nulls are not scientifically meaningful of the steady null, and should be phased out, we recommend doing an average over at sufficient number of consecutive points (here 10 points ie 1sec) to claim a null result.

A variety of conditions had to be met and meticulously optimized to reach very deep nulls. It takes only *one* to degrade an otherwise deep null. The top two factors that were easy to access were *OPD noise* and *power imbalance*. OPD noise was easily deduced from midfringe (PSD/Fourier) analysis, recording light signal at mid interference using a fast rate (KHz) photodetector. The (quadratic) rule of thumb was that 0.1nm RMS OPD limits the null to ~2.5e-7 solely from this contribution (a 0.1nm side shift causes 1e-6). For the power imbalance (mismatch) the light trace i1/i2/i1/i2 was recorded while the shutters were clicked alternatively. Here the rule of thumb was that 0.15% intensity mismatch limits null to 2.8e-7 solely from this contribution. To fine tune intensities to the 0.1% or better level we used a fine tip tilt control (picomotor) within or outside the nuller (before the output OAP). The latter operation mitigated the Airy spot positions on the core to intercept equal amounts of light from each arm. A technique of a thin wire occultator is now tested to give possibly better stability and resolution, reminding us of the past when we used a venitian blind technique in our Rotational Shear Interferometer [9]. Analyzing our 1M null we found 0.1nm RMS for the OPD noise and 0.15% for the power mismatch, see figure 3. Therefore the sum would be ~5e-7 versus 7e-7 measured, so there was more contributor(s) in the null depth. If these were identified and reduced the current experiment could possibly get us to 1e-7 nulls.

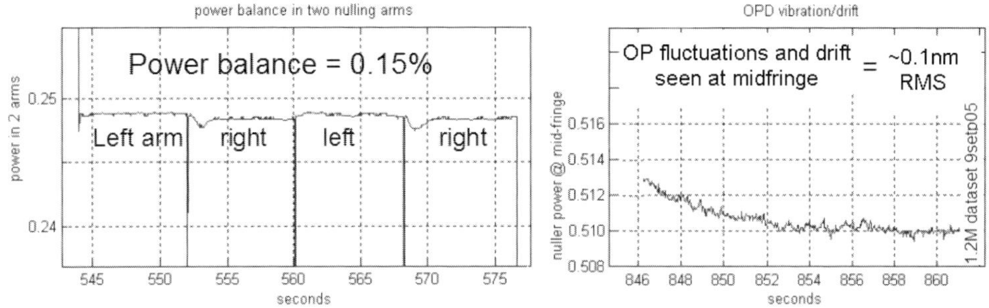

Figure 3. diagnosis of the 1 million null in terms of intensity mismatch and OPD noise.

A third factor under recent investigation is angular/polarization mismatch This parameter is not easy to sense but we have ideas such as high precision polarimetry or precise alignment. From a geometric point of view we need to align the angle of incidence on optics (~15deg) to the arcminute level, as well as nuller planarity and wedge orientations, also make sure about mirror coatings and pairwise matching of coatings for the beam splitters. We are currently rebuilding the nuller under an autocollimator to the arcminute level. If we find a null better than a million then it is the nuller that was not well aligned.

4. Environment

Environmental conditions are often under estimated and have not been described enough in previous papers, but play a major role in very deep (more than 100K) nulling experiments. Many conditions must be perfect or optimized simultaneously, with the foremost condition of zero optical path difference between the two arms. A rule is that any ambient condition or device generating noise and vibrations will have to be stopped, removed or located far away from the nuller bench or enclosure. This was easier said than done: typically air conditioning, always present in American buildings, would cause significant noise at the subnanometer level. Computer fans would cause noise acoustically and through cables coupling, with the constraint that is was not possible to place the computer very far away to the nuller and enclosure. In the nuller itself a major disturbance was air turbulence: devices thought to be quiet such a CCD camera or a shutter were found to generate a tiny amount of heat sufficient to cause slow micro turbulence and fluctuations around the paths at the 1M level (see dips in fig. 3 left). We plan to install a zero heat shutter. A non obvious and in fact paradoxical disturbance was that the piston piezzo itself, normally used to zero the OPD, would show a lot of creep at the subnm level. Definitely the internal pzt servo, normally sought to combat piezzo creep and hysteresis, would add servo noise, we therefore never used this servo when deep nulling, obligeing to manually track and surf the null, hence the irregular aspect of the null bottom on the plots shown in this paper. For tip tilt and OPD control we used the popular *picomotors*, aka the piezzo that turns a screw, from New Focus (for some years now). The fact that they have a small step size (10-20 nm), large range, and can hold a position without power (heat) or vibrations, is much an advantage compared to some drawbacks such as step non repeatability or the need of high voltage pulse generators. Following tests we have never evidenced any picomotor noise or drifts between positions (between tics), an essential ingredient to deep nulling. The nulling bench was resting on a couple of elastic sorbothane pucks which were confirmed to dramatically cutoff frequencies in the Hz-KHz range, also the main optical table was on traditional floatation. At

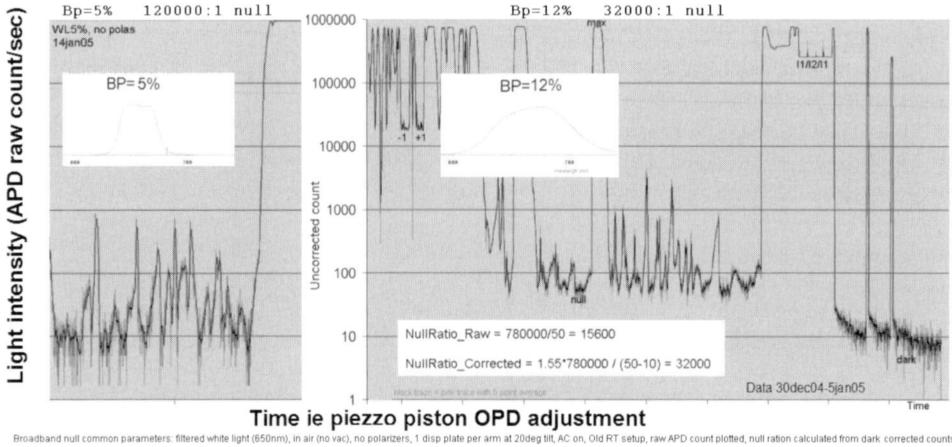

Figure 4. broadband nulls at 650nm of about 120000 and 32000 with BP=5% and 12%.

least 30 minutes of settling time was needed to start any deep null. For the old rooftop setup [8,9,10] the nuller was enclosed in *the aquarium*, a 2x1x1m box of plexiglass with inner foam panels. The aquarium has not been cutting enough acoustics to reach 1M nulls, especially air conditioning noise and vibrations. For the new setup (W) the nuller bench was inserted in a small vacuum chamber in metal. By placing the *fringe-o-phone* (an apparatus with a headphone connected to a photodiode with amplifier built to diagnose opto mechanical instability) to the output fiber, during a laser null sitting near mid fringe, ambient noises such as conversations were heard into the headphones showing that acoustic noise was directly coupling into the OPD. We deduced that any enclosure (plexi or metal) drastically reduced micro air currents in the arms but had almost no effect on acoustic attenuation. The ultimate remedy for this would be vacuum. When we started vaccum tests we verified, doing midfringe analysis and using the fringe-o-phone, that the nuller was very quiet, nevertheless we saw drifts (plate relaxation, thermals, etc). So far we have always seen better results operating in enclosed air, with door closed, in evenings and quiet hours, rather than operating under vacuum. We have obtained the 1M null (fig. 2) in air, with tank door closed.

5. Broadband nulling

Whereas nulling interferometry is easiest for monochromatic light (laser), scientific and astronomical interest lies only in broadband (star light) capability. Much of an improvement occured in fall 2004 when we installed two small Off Axis Parabola (OAP) mirrors (see fig. 1), replacing the doublet lens collimators, to correct longitudinal chromatic aberration. Each OAP was polished to a figure of $\lambda/20$ PV and mounted in a near monolithic aluminum mount design for stability and compactness. The nuller was equipped with tilted glass plates (dispertion plates) as described in the past [8,9,10] to create varying glass effective thickness between the right and the left arm. For broadband visible performance the effective thickness difference was calculated to be 20micron, with ~29micron of extra air in the other arm to compensate. Empirically we adjusted the angles of the plates, using rotary picomotors, to symmetrize the +1 and -1 fringe on each side of the null (symmetry of the fringe packet). When exploring the *rotation fringe* space, we found deeper nulls and higher order nulls.

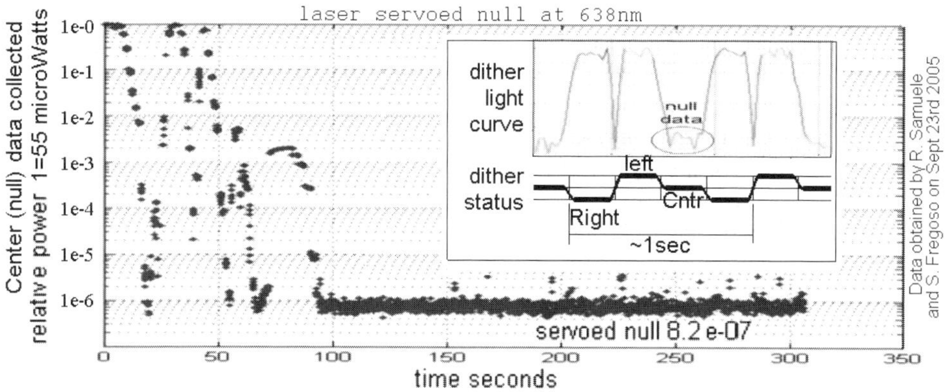

Figure 5. servo ie tracked laser null of 1.2M ie 8.2e-7 over 200s (average, corrected).

Using properly tuned plates in the old (rooftop) setup, in air, without polarizers, with filtered red light at 650nm, we obtained a null of 120000 for 5% bandpass, and 32000 for 12% bandpass, see figure 4. Unfortunately we have not attempted 18-20% so a direct comparison with our past result [10] is not possible. Note that in the left of the 12% plot fringes +1 and -1 were well equalized, also note a large (2x) difference between raw and corrected due to low flux. We have experimented only one glass type per arm, but have plans for 2006 to install two glasses per arm to augment bandwidth. With two glasses the procedure will be much more complex and the difference of effective thickness would theoretically be ~400 microns to give 10M over 20%.

6. Servoed null

A recent (fall 2005) result is that we were able to track a 1 Million laser null, see figure 5, which is a significant improvement over our last tracked result of a little bit better than 10000 uncorrected obtained in 2000 [10]. We implemented a servo technique based simply on the signal itself, using the weak flux around null, as opposed to other schemes typically using heterodyne laser metrology to monitor paths (in fact we implemented this a few years ago), or a quadrature output [11]. We applied a small OPD dither cycle around the null with 3 plateau positions called left, center (null) and right and monitored the light level at these 3 points. From any difference between the right and left levels we could easily tell any drift away from the null. Under LabView, with the proper conversion factor, we fed back the positive or negative error signal into the piston piezzo to bring the null back. In terms of servo parameters we used a relatively slow cycle rate of 1Hz given the low flux (55 microWatts on top of the fringe, 55picoWatts at null), we had a duty cycle of ~25% (8 null points out of 33), and a dither excursion of ~10nm. The closed loop operation added a minimal amount of noise and did not declassify the original 1Million null. We finally tested correct function of the closed loop by applying small OPD disturbances and verifying that the null would (slowly) come back despite of the very low flux captured at the left and right points.

7. Conclusion

Using a modified Mach Zender interferometer as our nuller tesbed we have developed technologies and experimental progress to reach very deep nulls needed for future NASA planet detecting missions such as TPF, or PICTURE (an upcoming sounding rocket experiment for launch in 2007), and more generally for the American and European planet finding communities. Under visible laser light (638nm) we have reached the 1 Million null ratio milestone (ie 1e-6 rejection ratio) on June 10th 2005, in fact to this date we have obtained a slightly better null of 1.2 Million (average raw null over many seconds). To our knowledge this is the first time that a visible Mach Zender nuller has surpassed the Million. To reach this type of null we have worked on many conditions such as optical path difference stability, optical alignment, symmetry in the two arms, mechanical stability and vibration isolation. In terms of broadband nulling, of more astronomical interest, under unpolarized filtered white light at 650nm, and using one glass plate in each arm, we obtained a null of 120000 with a bandpass of 5%, and 32000 with a larger bandpass of 12%. Moreover we report a latest result of closed loop nulling: we have been able to track the 1 Million laser null, with a slow bandwith and with minor noise degradation. We are glad to report that all four results were obtained in air, without pumping vacuum yet. With a 1e-6 rejection ratio we are only 10 times away from TPFs goal of 1e-10, indeed using a combination of a fiber array and deformable mirror a null of 1e-7 would correspond to a contrast of 1e-10 at a distance of 1 Airy spot ($1\lambda/D$). Future plans are to add shear, progress on deep laser nulling (goal of 2M soon, then eventually 10M) as well as an effort on broadband nulling up to 20% bandpass (goal of 1M at 5% soon), which likely needs 2 glasses per arm. In parallel to these efforts we are starting to introduce certain devices for upcoming projects such as a deformable mirror with an associated array of lenslets and single mode fibers [7], a downstream addition of a calibration system, and more remote control for a more end to end system.

8. Acknowledgments

We thank TPF and PICTURE for the support of this work. We thank S.F. Fregoso for help with data acquisition and LabView programmation. This work was performed at the Jet Propulsion Laboratory of the California Institute of Technology, under contract with the National Aeronautics and Space Administration.

9. References used

[1] M. Mayor, D. Queloz, Jupiter-mass companion to a solar-type star, Nature 378, 355 (1995)

[2] A. Burrows, A theoretical look at the direct detection of giant planets outside the solar system, Nature, vol. 433, 261, (2005)

[3] Jean Schneiders exo planet list, daily updated, www.obspm.fr

[4] R. Bracewell, Detecting nonsolar planets by spinning infrared interferometer. Nature 274, 780–781 (1978)

[5] E. Serabyn, Nulling Interferomtry Progress, Proc SPIE, 4838 (2003)

[6] S.R. Martin, R.O. Gappinger, F.M. Loya, B.M. Menesson, *et al.*, A mid infrared nuller for TPF: design, progress, results, proc SPIE, vol 5170, (2003)

[7] B.M. Levine, M. Shao et al, Visible nulling coronograph architecture definition and tech development, this proc

[8] Serabyn, E., Wallace, J. K., Hardy, G. J., Schmidtlin, E. G. H., Nguyen, H. Deep nulling of visible laser light, Appl. Opt. 38, 7128–7132 (1999)

[9] Serabyn, E., Wallace, J. K., Nguyen, H. T., Schmiditlin, E. G. H., Hardy, G. J. in Working on the Fringe: Optical and IR Interferometry from Ground and Space (eds Unwin, S. and Stachnik, R.) 437–442 (1990)

[10] K. Wallace, G. Hardy, E. Serabyn, Deep and stable interferometric nulling of broadband light, Nature, August 2000, 406, 700–702 (2000)

[11] E. Serabyn, Nanometer-level path-length control scheme for nulling interferometry. Appl. Opt. 38, 4213–4216 (1999)

Direct Imaging of Exoplanets: Science & Techniques
Proceedings IAU Colloquium No. 200, 2005
C. Aime and F. Vakili, eds.
© 2006 International Astronomical Union
doi:10.1017/S1743921306009598

Subwavelength gratings for phase mask coronagraphy: the 4QZOG and AGPM coronagraphs

Dimitri Mawet[1] **and Pierre Riaud**[1]

[1]Institut d'Astrophysique et de Géophysique de Liège, University of Liège, Allée du 6 Août, 4000, Sart Tilman, Belgium
email: mawet@astro.ulg.ac.be

Abstract. We present two new phase mask coronagraphs implemented with subwavelength diffractive optical elements. The first one is an evolution of the four quadrant phase mask coronagraph (FQPM), which resolves the π phase shift chromaticity issue: the four quadrant zeroth order grating (4QZOG). The second one is a totally new design consisting of an optical vortex induced by a space-variant grating: the annular groove phase mask (AGPM) coronagraph is fully symmetric and free from any "shaded zones". The potential performances of the 4QZOG and AGPM coronagraph are very good, ensuring, for instance, a theoretical contrast of 1.4×10^{-7} at $3\lambda/D$ over the whole K band. These coronagraphs could be used alone on single-pupil telescopes either in space or on the ground (with an adaptive optics system) to detect exoplanets.

Keywords. High angular resolution, Phase mask coronagraphy, Achromatic phase shifters, Subwavelength gratings, Space-variant gratings, Optical vortices

1. Introduction

Direct detection of exoplanets around nearby stars is very difficult due to the large flux ratio between them. For example, an Earth-like exoplanet is typically 6×10^9 times fainter than its host star in the visible and 7×10^6 times fainter in the thermal infrared. Since 1996, various coronagraphs have been proposed: amplitude coronagraphs deriving from the original Lyot's design and phase-mask coronagraphs, first considered by Roddier & Roddier 1997. Amplitude coronagraphs possess a major inherent weakness: the physical extension of the opaque zone occults a significant region centered on the optical axis and thus all the sources behind it. For example, let us mention the case of the recent Notch-Filter Mask (Kuchner & Spergel 2003): this design presents a large central opaque zone extending in the focal plane up to $3\lambda/D$ (D is the telescope diameter whereas λ is the wavelength) where the potential companion is still attenuated by at least 50%. Of course this weakness reveals to be an advantage as soon as sensitivity to low order aberrations is more critical than the inner working distance. Rouan *et al.* (2000) proposed the so-called four-quadrant phase mask coronagraph (FQPM). The FQPM principle has been validated on the bench in monochromatic (Riaud *et al.* (2003)) and polychromatic light (D. Mawet *et al.*, submitted to A&A) and on the sky with a monochromatic component installed on the NAOS-CONICA instrument at the VLT (Boccaletti *et al.* (2004)). As for every phase mask, its Achilles'heels are the chromaticity issues and the presence of "dead zones", in this case on the quadrant transitions (the central obscuration is a common issue of almost every coronagraphs). Some solutions for the phase masks chromaticity issue have been presented like, for example, the dual zone phase mask coronagraph (DZPMC) which presents only a first order achromatization and must be used with a pupil apodization scheme (Soummer *et al.* (2003)). In the phase coronagraph family, let us also mention the special cases of the achromatic interfero coronagraph (AIC) that provides achromatic π phase shift in a Michelson Interferometer (Gay & Rabbia (1996)). Finally, we must also cite pupil apodization schemes like the phase induced amplitude apodization (PIAA) where the starlight is not attenuated but redistributed (Guyon (2003)).

In this paper, we propose to remedy the two main limitations of the FQPM. First, we present the 4QZOG where the subwavelength grating technology provides a powerful and flexible solution to the chromatism issue. Secondly, we propose a new design of a phase-mask coronagraph derived from the 4QZOG that totally suppresses the annoying dead zones: the annular groove phase mask coronagraph (AGPM).

2. Subwavelength gratings

When the period Λ of the grating is smaller than the wavelength of the incident light, it does not diffract as a classical spectroscopic grating. All the incident energy is enforced to propagate only in the zeroth order, leaving incident wavefronts free from any further aberrations. The subwavelength gratings are therefore often called zeroth order gratings (ZOGs). Whether a diffraction order propagates or not is determined by the well-known grating equation from which a "ZOG condition" on the grating period to wavelength ratio can be derived:

$$\frac{\Lambda}{\lambda} \leqslant \frac{1}{n_I \sin\chi + \max(n_I, n_{III})} \tag{2.1}$$

where χ is the angle of incidence, n_I and n_{III} are the refractive indices of the incident (superstrate) and transmitting (substrate) media, respectively. This type of gratings behaves like homogeneous media with unique characteristics, which can be used to synthesize artificial birefringent achromatic waveplates (Kikuta *et al.* (1997); Nordin & Deguzman (1999)) or monolithic anti-reflective structures (for example, Karlsson & Nikolajeff (2003)). Quarterwave or halfwave plates are extensively used in astrophysics for polarimetric studies. Subwavelength gratings constitute an elegant and flexible solution to produce them. The key point is that by carefully controlling the geometry of the grating structure (via the grating parameters: the period Λ, the depth h and the width of the grating ridges $F\Lambda$ with F the filling factor) we are able to tune the so-called form birefringence $\Delta n_{TE-TM}(\lambda) = n_{TE}(\lambda) - n_{TM}(\lambda)$ where n_{TE} and n_{TM} are the two effective indices associated with the subwavelength structure, one for each polarization state: TE (tranverse electric) and TM (tranverse magnetic). Intuitively, one can understand this artificial anisotropy and the existence of two distinct effective indices: the incident light sees two different media as its vectorial components vibrate parallel or orthogonal to the grating lines. The idea is to make the form birefringence proportional to the wavelength in order to achromatize the subsequent differential π phase shift between the two polarization states TE and TM

$$\Delta\phi_{TE-TM}(\lambda) = \frac{2\pi}{\lambda} h\, \Delta n_{TE-TM}(\lambda) \approx \pi \tag{2.2}$$

with h, the optical path through the birefringent medium. This is a good example of refractive index engineering. To simulate grating responses and calculate the form birefringence $\Delta n_{form} = \Delta n_{TE-TM}$ in the subwavelength and resonant domain ($\Lambda \approx \lambda$), scalar theories of diffraction dramatically fail. The vectorial nature of light must be taken into account implying a resolution of the Maxwell equations by the so-called rigorous coupled wave analysis (RCWA; Moharam & Gaylord (1981)). The RCWA gives the full vectorial diffractive characteristics of the created medium.

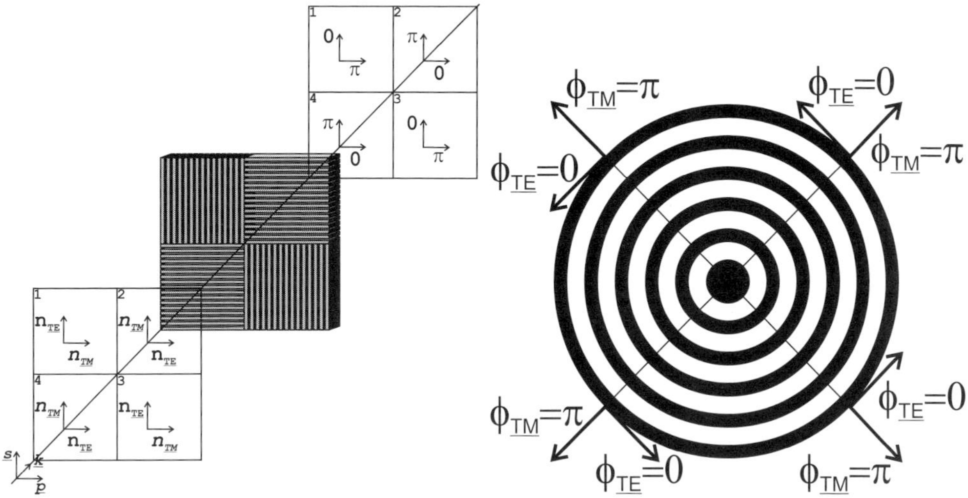

Figure 1. Left: 4QZOG implementation, where the four gratings engraved on a unique substrate are strictly identical and implemented in the following way: two of them in two quadrants along one diagonal are rotated by 90 degrees around their normals with respect to the two others. Right: AGPM implementation. The AGPM consists of a concentric circular surface-relief grating with rectangular grooves with depth h and a periodicity Λ.

3. ZOG implementation in phase-mask coronagraphy

In this section, we will first focus on the ZOG ability to induce a precise achromatic phase shift and a controlled intensity balance in order to create a destructive interference as dark as possible. Their implementation into the FQPM and after on the AGPM coronagraphs will then be discussed.

3.1. *Preliminary considerations*

The design issue consists both in the optimization of the π phase shift quality within a given spectral range for a well-chosen material and the equalization of the interfering fluxes for each polarization. The merit function to be minimized is the null depth $N(\lambda)$. The latter is directly related to the phase shift error with respect to π, $\epsilon(\lambda) = \Delta\phi_{TE-TM}(\lambda) - \pi$ and the flux ratio $q(\lambda) = \eta_{TE}(\lambda)/\eta_{TM}(\lambda)$:

$$N(\lambda) = \frac{\left(1 - \sqrt{q(\lambda)}\right)^2 + \epsilon(\lambda)^2\sqrt{q(\lambda)}}{\left(1 + \sqrt{q(\lambda)}\right)^2} \tag{3.1}$$

The ZOG technology allows to tune the form birefringence in order to minimize the phase-shift error $\epsilon(\lambda)$ but also the difference between the effective indices and the external medium indices that induces unequal TE/TM-Fresnel reflection coefficients at the interfaces. We have investigated several ways to ensure this proper flux balance, i.e. $q \approx 1$. The principle is always the same and consists in trying to minimize the index discontinuities (impedance matching) at the interfaces. The most efficient solution is to cover the ZOG with a single $\lambda/4$ layer. The latter settles at the bottom and on the top of the grating ridges. This leads to an original configuration, a 3-layer grating whose complex behavior is diffraction-governed. We demonstrate here below that this particular structure can be optimized to give very interesting results.

3.2. *4QZOG and AGPM*

In the case of the FQPM implementation, the four gratings engraved on a unique substrate are strictly identical and implemented in the following way: two of them in two quadrants along one diagonal are rotated by 90 degrees around their normals with respect to the two others. This anti-symmetrical configuration achieves the FQPM particular focal plane π-phase distribution (Fig. 1). This new component will be referred as to the Four Quadrant-ZOG, or 4QZOG (Mawet *et al.* (2005b)). The AGPM space variant ZOG (Fig. 1) synthesizes a spiral phase plate which creates an "optical vortex". Indeed, at the center of the components, the phase possesses a screw dislocation inducing a phase singularity, i.e. an optical vortex. The central singularity forces the intensity to vanish by a total destructive interference, creating a dark core. This dark core propagates and is conserved along the optical axis. It can be demonstrated that a focal plane vortex leads to a total rejection in the ideal case (Mawet *et al.* (2005a)). Furthermore, the ZOG unique properties permit a broadband use as in the 4QZOG. Indeed, the ZOG parameters (period, depth, filling factor) are the same in the 4QZOG and AGPM configurations. As seen before, the only difference concerns the geometry of implementation which suppress the dead zones.

4. Simulation results

Presently, subwavelength gratings are more easily manufactured for the infrared domain (near and mid) because there are still technical difficulties to imprint sub-micron patterns. Fortunately, the near-term potential applications (VLT-Planet Finder, for instance) of this technology concern this wavelength domain. We will focus on the astrophysical bandfilters H ($1.475 - 1.825\,\mu m$), K ($2 - 2.4\,\mu m$) and N ($9.3 - 11.6\,\mu m$). We have now to chose among the restricted list of infrared materials keeping in mind that the ZOG technology is sufficiently flexible to accommodate the majority of them. We have selected diamond, zinc selenide ($ZnSe$) and silicon (Si). This choice is justified in two ways: these materials are common in infrared applications and their etching processes are well-known. Results in terms of null depth are excellent either for the H, K or N bands (Fig. 2). Indeed, the mean null depth performances are close to the 10^{-5} level over the whole spectral range of the three considered band filters. These optimal results have been calculated by implementing the simplex search method with our RCWA algorithm.

5. Expected coronagraphic performance

To assess the performance of the 4QZOG and AGPM coronagraphs, we performed numerical simulations that imply three different stages: 1- a RCWA stage where the form birefringence of the local grating is calculated; 2- an analytical treatment giving the spatial distribution of the output polarization field; 3- a scalar far-field propagation Fourier coronagraphic code with assumed wavefront qualities of $\lambda/70$ rms at 632.8 nm (current state of the art). For example, the coronagraphic profile in the K band shows a peak-to-peak null depth better than about $\approx 10^{-5}$. The speckle level of $\approx 10^{-7}$ is quickly reached at a few λ/D (see Fig. 3 for the AGPM). The AGPM coronagraphic behavior is very similar to the achromatic 4QZOG coronagraph, but with a total symmetry. As mentioned before, the FQPM/4QZOG quadrant transitions induce a non-negligible attenuation of the circumstellar features lying on them. These dead zones represent quite a significant portion of the focal plane (about 10% at $6\lambda/D$). Thanks to the perfect

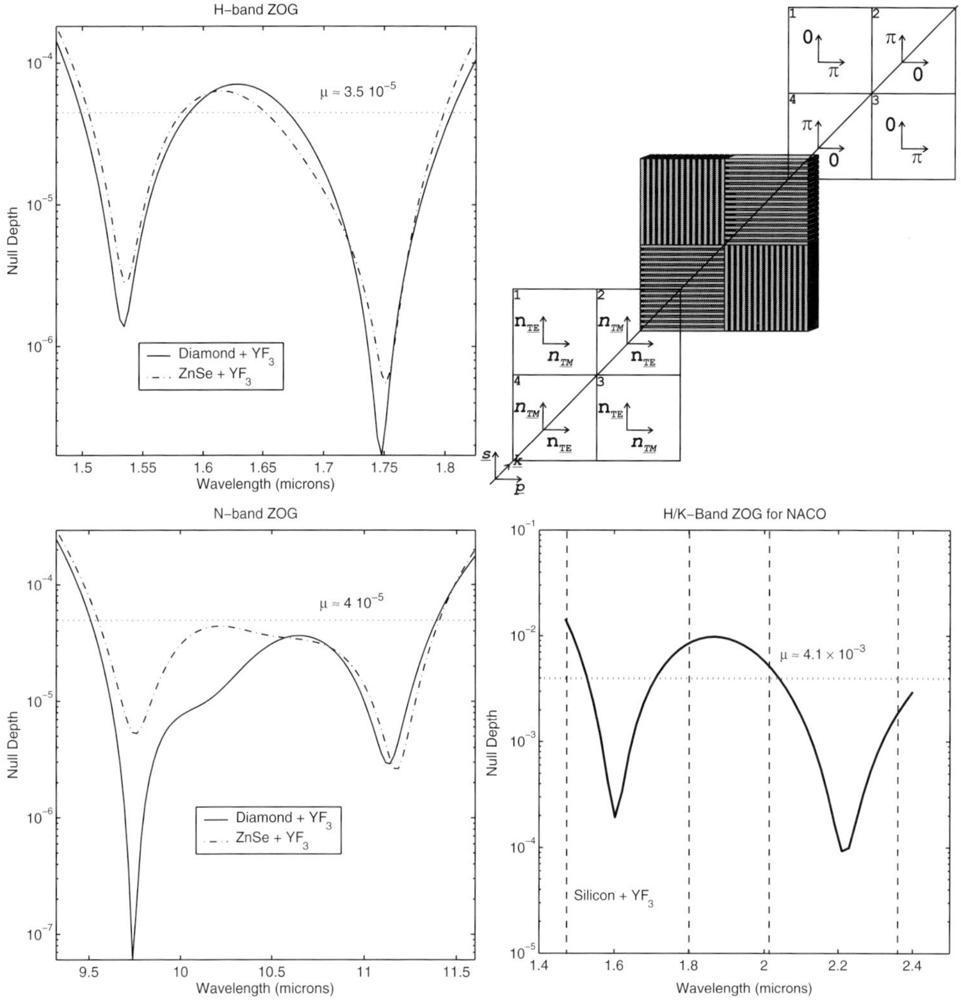

Figure 2. Top left: H-band ZOG null depth versus wavelength. The continuous curve is for the diamond YF_3 AR coated 4QZOG/AGPM. The dashed curve, for the $ZnSe$ YF_3 AR coated one. $\mu \approx 3.5 \times 10^{-5}$ is the mean null depth over the whole H band. Top right: K-band ZOG null depth versus wavelength. $\mu \approx 1.7 \times 10^{-5}$ is the mean null depth over the whole K band. Bottom left: N-band ZOG null depth versus wavelength. $\mu \approx 4 \times 10^{-5}$ is the mean null depth over the whole N band. Bottom right: H/K-Band silicon YF_3 AR coated ZOG null depth versus wavelength. $\mu \approx 4.1 \times 10^{-3}$ is the mean null depth over both the H and K bands.

AGPM circular symmetry, this problem does not exist any more. These performances are comfortably one order of magnitude above the H-band/K-Band VLT-PF specification for example. Indeed, the main limitation for ground-based observations comes from the atmospheric turbulence residuals after adaptive optics corrections of the incoming fluctuating wavefronts. Let us emphasize that the Inner Working Angles (IWA) of the 4QZOG and AGPM are very good, peering well under λ/D. As far as stellar leakage is concerned, numerical simulations show that the AGPM null depth degrades as $\theta^2_{\lambda/D}$ close to the optical axis (Mawet *et al.* (2005a)), just as the FQPM/4QZOG (with $\theta_{\lambda/D}$, the angle from the optical axis).

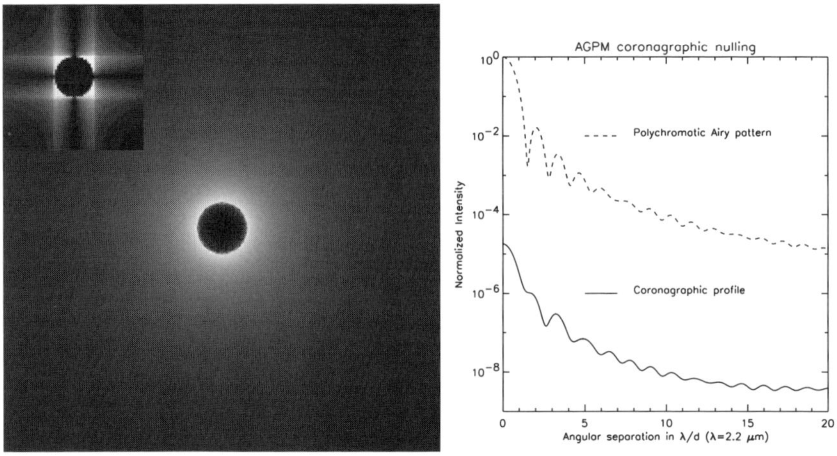

Figure 3. Left: AGPM pupil image showing the perfect annular symmetry. To be compared with the FQPM. Right: coronagraphic simulation in a realistic case with $\lambda_{vis}/70$ rms wavefronts taking into account residual chromatism and amplitude errors.

6. Conclusions

In this paper, we have presented two new phase mask coronagraphs that are inherently quasi-achromatic thanks to the subwavelength grating technology. Results in terms of global null depth are very good ($\approx 10^{-5}$). The potential performance of the 4QZOG and AGPM coronagraphs ensures, for instance, a theoretical contrast of 1.4×10^{-7} at $3\lambda/D$ over the whole K band and with an inherent perfect symmetry in the AGPM case. Thanks to the ZOG design flexibility, the 4QZOG and AGPM coronagraphs can accommodate a large variety of materials and wavelength bands thus making them an attractive solution for future instruments dedicated to high resolution/high contrast space- or ground- based imaging in the context of exoplanet detection and characterization.

Acknowledgements

D. Mawet and P. Riaud acknowledges the financial support of the Belgian "Fonds pour la formation à la Recherche dans l'Industrie et dans l'Agriculture" and the "Pôle d'Attraction Inter-Universitaire", respectively.

References

Boccaletti, A., Riaud, P., Baudoz, P., Baudrand, J., Rouan, D., Gratadour, D., Lacombe, F., & Lagrange, A.-M. 2004, *PASP* 116, 1061
Gay, J. & Rabbia, Y. 1996, *CR. Acad. Sci. Paris* 332, 265
Guyon, O. 2003, *A&A* 404, 379
Karlsson, M. & Nikolajeff, Fr. 2003, *O*pt. Express 11 (5), 502
Kikuta, H., Ohira, Y., & Iwata, K. 1997, *Applied Optics* 36, 1566
Kuchner, M.J. & Spergel, D.M. 2003, *ApJ* 594, 617
Mawet, D., Riaud, P., Absil, O., & Surdej, J. 2005a, *A*pJ in press
Mawet, D., Riaud, P., Baudrand, J., & Surdej, J. 2005b, *A*pplied Optics in press
Moharam, M.G. & Gaylord, T.K 1981, *JOSA* 71, 811
Nordin, G.P. & Deguzman, P.C. 1999, *O*pt. Express 5 (8), 163
Riaud, P., Baudrand, J., Boccaletti, A., & Rouan, D. 2003, *PASP* 115, 712
Roddier, F. & Roddier, C. 1997, *PASP* 109, 815
Rouan, D., Riaud, P., Boccaletti, A., Clénet, Y., & Labeyrie, A. 2000, *PASP* 112, 1479
Soummer, R., Dohlen, K., & Aime, C. 2003, *A&A* 403, 369

Direct Imaging of Exoplanets: Science & Techniques
Proceedings IAU Colloquium No. 200, 2005
C. Aime and F. Vakili, eds.
© 2006 International Astronomical Union
doi:10.1017/S1743921306009604

Apodized Pupil Lyot Coronagraphs: Concepts and application to the Gemini Planet Imager

Rémi Soummer[1]†, Claude Aime[2], André Ferrari[2], Anand Sivaramakrishnan[1], Ben R. Oppenheimer[1], Russell Makidon [3], and Bruce Macintosh [4]

[1]American Museum of Natural History, 79th St at Central Park West, New York, USA
email: rsoummer@amnh.org, anand@amnh.org, bro@amnh.org

[2]Laboratoire Universitaire d'Astrophysique de Nice, Parc Valrose, Nice, France
email: Claude.Aime@unice.fr, Andre.Ferrari@unice.fr

[3]Space Telescope Science Institute, 3700 San Martin Drive, Baltimore, USA
email: makidon@stsci.edu

[4]Lawrence Livermore National Laboratory, 7000 East Avenue, Livermore, USA
email:bmac@igpp.ucllnl.org

Abstract. This paper summarizes the general theory and properties for Apodized Pupil Lyot Coronagraphs which consist of a classical hard-edged Lyot coronagraph with an upstream pupil apodization. The ideal apodization function can be determined from an integral eigenvalue problem which solutions are prolate spheroidal functions. Solutions exist for any geometrty, including rectangular, circular, or elliptical. Formal solutions can be extended to the case of arbitrary apertures, using generalized prolate spheroidal functions for centrally obstructed apertures, spiders, or segmented telescopes. The properties of these coronagraphs enable the possibility of multiple stage coronagraphy and achromatization. The new instrument Gemini Planet Imager (GPI) will include such a coronagraph.

Keywords. instrumentation: high angular resolution, instrumentation: adaptive optics.

1. Introduction

Direct imaging and spectroscopy of extrasolar planets is an exciting and ambitious goal, which requires advanced starlight cancellation techniques, including coronagraphy. The Gemini Planet Imager (GPI), Macintosh *et al.* (2004), will aim at imaging and spectroscopy of Extrasolar Giant Planets (EGP). Classical methods like the Lyot Coronagraph are not efficient enough to reach the necessary contrast for this science goal, Burrows *et al.* (2004), and more advanced techniques have to be considered. The Apodized Pupil Lyot Coronagraph has been selected for this new instrument. We summarize in this communication the theory and properties of these coronagraphs detailed by Aime *et al.* (2001); Aime *et al.* (2002); Soummer *et al.* (2003); Soummer (2005), we also include the possibility of multiple stage APLC, Aime & Soummer (2004), achromatization, Aime (2005) and discuss their fabrication.

† Michelson Fellow

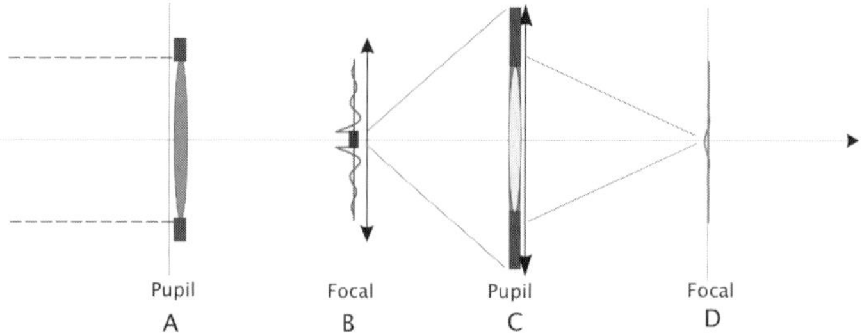

Figure 1. The basic layout of an APLC is similar to a classical Lyot coronagraph, but adding an upstream apodized pupil in Plane A. A hard hedged focal maks is set in the focal plane B, and a Lyot stop *identical* to the entrance pupil shape in plane C. A remarquable difference between APLC and classical Lyot is that APLCs don't require to undersize the Lyot stop.

2. Apodized Pupil Lyot Coronagraphy with arbitrary aperture shapes

Following the notation of Aime *et al.* (2002) and Soummer *et al.* (2003), we briefly recall the general formalism of coronagraphy with apodized pupils. The general layout of an APLC is recalled in Fig. 1, and is similar to a classical Lyot coronagraph, but using an upstream pupil apodization. The telescope aperture function with the position vector $\mathbf{r}$ is denoted by $P(\mathbf{r})$, and $\Phi(\mathbf{r})$ is the apodizer transmission (1 without apodization). At the entrance aperture, the wavefront amplitude is:

$$\Psi_A(\mathbf{r}) = P(\mathbf{r})\,\Phi(\mathbf{r}). \tag{2.1}$$

A mask of transmission $1 - \epsilon M(\mathbf{r})$ is placed in the focal plane. M is a function that describes the mask shape, equal to 1 inside the coronagraphic mask and 0 outside ($\epsilon = 1$ for Lyot and $\epsilon = 2$ for Roddier).

At the Lyot stop plane, the wavefront is then:

$$\Psi_C(\mathbf{r}) = P(\mathbf{r})\,\Phi(\mathbf{r}) - \frac{\epsilon}{\lambda^2 f^2}\int_P \Phi(\mathbf{u})\,\widehat{M}\left(\frac{\mathbf{r}-\mathbf{u}}{\lambda f}\right)\mathrm{d}\mathbf{u}, \tag{2.2}$$

where $\widehat{\ }$ is the Fourier Transform. This general relation is valid for any arbitrary aperture shape (rectangular, circular or elliptical, with or without central obstruction and secondary mirror supports), and any mask shape, Soummer (2005). In this equation we assume that the focal lengths of the successive optical systems are identical (if not, an appropriate change of variables leads to the same result), we re-orientate the axis in the opposite direction and omit the -1 proportionality factor for better readability. The formal problem is to find the mask $M(\mathbf{r})$ and the apodizer $\Phi(\mathbf{r})$ that provide the best cancellation possible inside the Lyot stop, identical to the entrance pupil in this case.

A perfect solution (total extinction) is obtained if the wave diffracted by the mask (the convolution integral) is equal to the initial pupil amplitude. In more mathematical terms, this means that $\Phi(\mathbf{r})$ must be the eigenfunction of the convolution integral in Eq.2.2. This problem is equivalent to the generalization of the uncertainty principle, Papoulis (1968), with the search of band-limited functions having the maximum encircled energy within a given spatial domain. In coronagraphic terms, the solution is the pupil apodization which produces the most concentrated star light behind a given focal plane mask (and thus blocked). This formal problem has been solved analytically for rectangular and circular unobstructed apertures, Aime *et al.* (2001), Aime *et al.* (2002), Soummer *et al.* (2003), involving linear and circular prolate functions, Slepian & Pollak (1961), Frieden (1971), Papoulis (1981). In the general case, formal solutions exist, Slepian (1964), Slepian

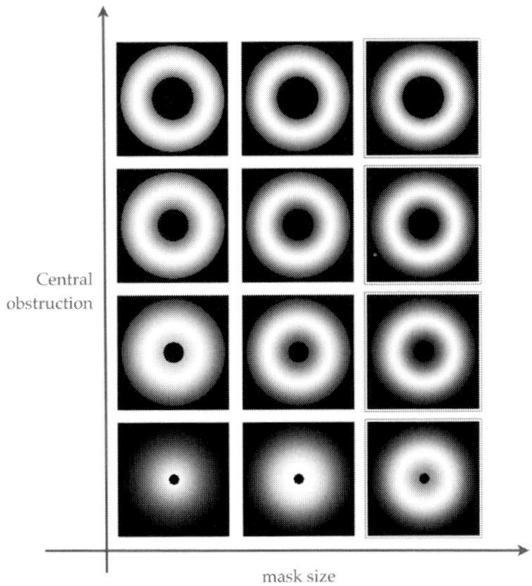

Figure 2. A given telescope geometry defines an integral eigenvalue problem in which for any mask size (eigenvalue), an apodization can be computed (eigenfunction). This figure illustrates the apodization that can be obtained for a few telescope geometries varying the size of the central obstruction (y-axis), and as a function of the maks size (x-axis).

(1965), Papoulis (1968), Soummer (2005), and are the eigenfunction of the 2-D integral equation:

$$\int_P \Phi(\mathbf{t})\, \widehat{M}(\mathbf{r}-\mathbf{t})\, \mathrm{d}\mathbf{t} = \Lambda\, \Phi(\mathbf{r}). \tag{2.3}$$

The solution $\Phi(\mathbf{r})$ corresponds to the largest eigenvalue Λ_0, and to the maximum encircled energy behind the focal plane mask $e = \Lambda_0$. The mask size is directly related to the eigenvalue, and therefore, for a given mask size, a unique apodization function exists. In Fig. 2, we show the apodizer as a function of the mask size and for a few telescope geometries, varying the obstruction size.

The expression of the residual wavefront inside the Lyot stop for circular and rectangular geometries is the same for the general case:

$$\Psi_C^+(\mathbf{r}) = \Psi_A(\mathbf{r})(1 - \epsilon\, \Lambda_0), \tag{2.4}$$

(the $^+$ signifies this is just after the Lyot stop). As detailed in previous work, and illustrated in Fig. 3, the wavefront is therefore simply proportional to the initial apodization function, with a constant of proportionality $(1 - \epsilon\, \Lambda_0)$.

3. Mutiple Stage APLCs and achromatization

3.1. *Multiple Stage*

As described above, the Lyot stop wave amplitude is proportional to the entrance apodized pupil amplitude. This enables the possibility for multiple stage coronagraphs described by Aime *et al.* (2002) and Aime & Soummer (2004) (Fig. 4), which could help ambitious projects such as TPF to reach their required extinction. A multiple stage APLC only requires a single apodizer in the entrance pupil. The Lyot stop plane is naturally apodized, and can be used as the entrance pupil of a second coronagraphic stage without further loss of throughput due to apodization (Fig. 4). Eq.2.4 can be straightforwardly

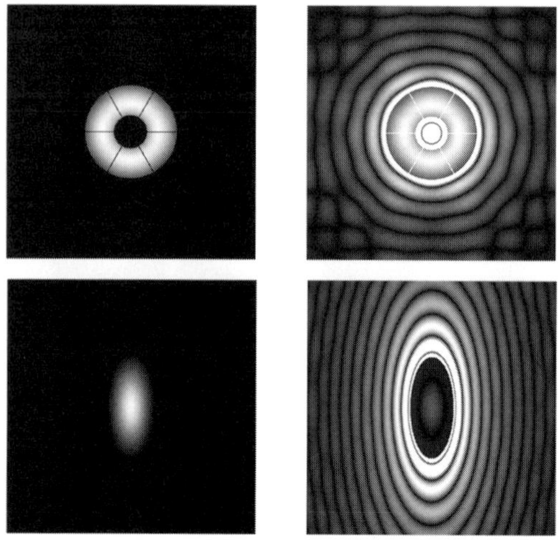

Figure 3. Examples of APLCs for two geometries: the Overwhelmingly Large Telescope (OWL) (top) and the Terrestrial Planet Finder (TPF) (bottom). The left images correspond to the entrance pupil apodization, and the right images to the Lyot stop plane (plane C) wave amplitude. The field amplitude inside the aperture domain is proportional to the initial entrance pupil apodization, since the apodizer is the eigenfunction of the problem. This property enables the possibility of multiple stage APLCs for any telescope geometry.

extrapolated to the multiple stage case, and the residual amplitude in the second Lyot stop plane (plane E of Fig. 4) is:

$$\Psi_E^+(\mathbf{r}) = \Psi_A(\mathbf{r})(1 - \epsilon\,\Lambda_0)^2. \tag{3.1}$$

In principle, several identical coronagraphs can be used in a cascading mode. Each coronagraph produces the same reduction factor of the star light. Denoting Q the efficiency of one coronagraph, the efficiency of N cascaded coronagraph should be Q^N.

3.2. *Achromatization*

The main limitation of APLCs comes from the chromaticity due to the PSF size, which is proportional to the wavelength. In the simple case of a focal plane mask made by means of a hole in a mirror, its size is identical at all wavelengths. Therefore, the mask is too large at short wavelengths, and too small at long wavelengths, which translates into a leak through the coronagraphs. Specific designs can accommodate these limitations, such as the Dual-Zone coronagraph, Soummer *et al.* (2003). In the case of APLCs, a mitigation of these chromatic effects can be obtained by optimizing the mask size for an effective wavelength with optimum efficiency over the actual bandpass, Soummer (2005).

For a given physical mask, each wavelength corresponds to a different mask size in resolution element units (λ/D). Therefore, it is possible to compute the required apodization as a function of the wavelength since an apodization corresponds to a given mask size, Aime (2005).

A perfect achromatic solution can be obtained with an apodizer producing this variable apodization with wavelength. In this case the result is close to that of an achromatic APLC at the shortest wavelength of the bandpass. Aime (2005) has proposed an interferometric method to produce such a wavelength-dependent apodizer, using dispersive optics in the arms of an interferometer. Interferometric methods to produce apodizers have been studied by Aime *et al.* (2001). The results of a laboratory experiment validating this concept is given in this volume by El Azhari *et al.* (2005).

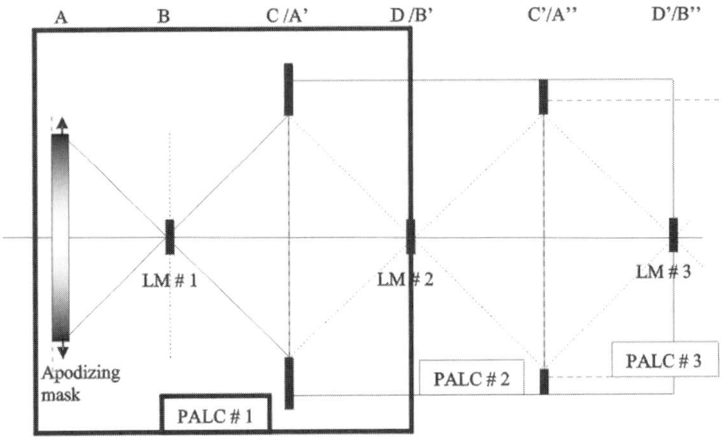

Figure 4. Schematic representation of a multiple-Lyot coronagraph. The light goes from left to right. The first square box, entitled $PALC\sharp 1$, for Prolate Apodized Lyot Coronagraph number 1 is the classical Lyot coronagraph with entrance aperture apodization. The successive planes are denoted A, B, C and D. A is the entrance aperture (or image of the aperture) with the prolate apodization, B denotes the focal plane with the first Lyot mask ($LM\sharp 1$), C is the image of the aperture with the Lyot stop, and D is the image of the focal plane after the coronagraph (D is the final focal plane in the case of a classical apodized coronagraph. The second stage of the coronagraph consists of planes A', B', C' and D'. Planes A' and B' merge respectively with plane C and D. A Lyot mask, identical to the one in plane B, is set in plane B' (or D). A Lyot stop in plane C', identical to the one in plane C, completes the second stage. The image is observed in plane D'. The coronagraph $PALC\sharp 2$ produces the same starlight rejection as the $PALC\sharp 1$, because the amplitude in C/A' is proportional to the amplitude in A without the need of another apodizer. For a perfect system, this rejection is obtained without loss of useful light for the planet. The first part of a third coronagraph ($PALC\sharp 3$) is drawn in the figure.

A more simple method, though which might not be able to provide perfect achromatization, is to manufacture an apodizer using an absorbing material which approximates the required wavelength variation. First tests using High Energy Electron Beam Sensitive glass (HEBS glass) have shown that this material improves the performance of the APLC in the infrared by a factor of 5. This preliminary result can be further improved by adjusting the chromaticity of the material itself (Soummer etal. in preparation).

4. Application to the Gemini ExAOC

Thanks to the recent results on the understanding of the theory of APLCs in the presence of spiders and central obstruction, Soummer (2005), Sivaramakrishnan & Lloyd (2005), and recent lab results on the fabrication of apodizers for the infrared using HEBS glass, the APLC has been selected as the baseline concept for the Gemini Planet Imager. As compared to a classical Lyot for similar mask size ($4.5\lambda/D$ to $5..5\lambda/D$) and because APLCs do not require Lyot Stop size reduction, both the coronagraphic rejection and the throughput can be improved compared to a classical Lyot (Fig. 5). This instrument will include an Extreme Adaptive Optics system, an APLC, an Integral Field Spectrograph, and an active calibration system. The predicted dynamic range limits at the 5-σ level (Fig. 6) are compatible with the main goal of detecting Extrasolar Giant Planets around nearby stars.

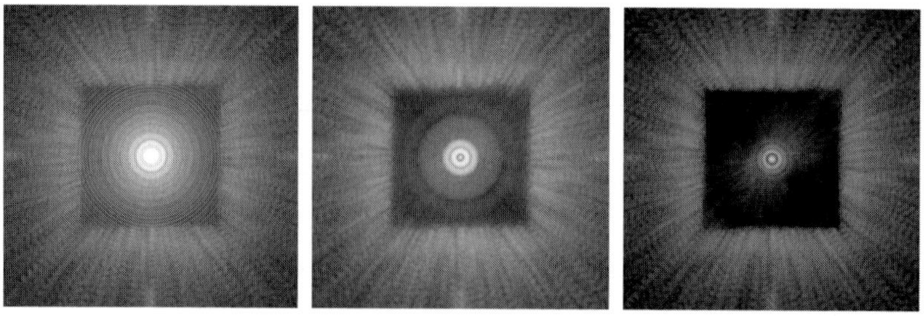

Figure 5. From Left to right: AO-corrected PSF, Lyot Coronagraphic PSF, APLC coronagraphic PSF. The square dark region corresponds to the area of the field corrected by the spatially filtered AO system. The halo clearing is improved over the classical Lyot, using the same mask size ($4.7\lambda/D$). Thanks to the properties of the APLC, its overall transmission (55%) is higher than that of a comparable classical Lyot (50%), and for a better rejection.

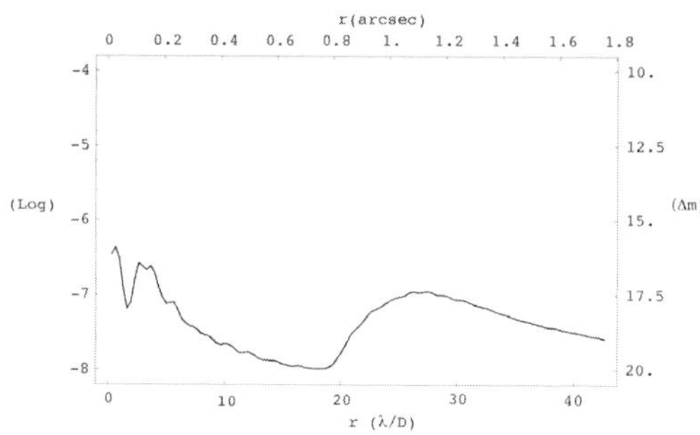

Figure 6. Dynamic range simulation including speckle and photon noise for a 7th magnitude star, in 1000s in Hband for the Gemini ExAOC. This type of dynamic range plots can be readily used in Monte-Carlo simulations to study the completeness of the observations.

Acknowledgements

Rémi Soummer is supported by a Michelson Fellowship, under contract with the Jet Propulsion Laboratory funded by NASA. JPL is managed for NASA by the California Institute of Technology.

References

Aime, C. 2005, *PASP*, 117, 1112
Aime, C. & Soummer, R. 2004, *proc. SPIE*, 5490, 456
Aime, C., Soummer, R., & Ferrari, A. 2001, *A&A*, 379, 697
Aime, C., Soummer, R., & Ferrari, A. 2002, *A&A*, 389, 334
Burrows, A., Sudarsky, D., & Hubeny, I. 2004, *ApJ*, 609, 407
El Azhari, Y., Azagrouze, O., Martin, F, Soummer, R., & Aime, C. 2005, *Proc. IAU Colloquium No. 200*, 609, 407
Frieden, B. R. 1971, *Progress in Optics*, ed. E. Wolf, Vol. 9
Macintosh, B. A., *et al.* 2004, *Proc. SPIE*, 5490, 359
Papoulis, A. 1968, *Systems and transforms with applications in optics* (McGraw-Hill Series in System Science (reprint 1981))
Papoulis, A. 1981, *Signal Analysis*, ed. M. G. Hill
Sivaramakrishnan, A. & Lloyd, J. P. 2005, *ApJ*, 633, 528
Slepian, D. 1964, *Bell Syst. Tech. J.*, 43, 3009
Slepian, D. 1965, *J. Opt. Soc. Am.*, 55; no.9
Slepian, D. & Pollak, H. O. 1961, *Bell Syst. Tech. J.*, 40, 43
Soummer, R., Aime, C., & Falloon, P.E. 2003, *A&A*, 397, 1161
Soummer, R., Dohlen, K., & Aime, C. 2003, *A&A*, 403, 369
Soummer 2005, *ApJ*, 618, 161

Direct Imaging of Exoplanets: Science & Techniques
Proceedings IAU Colloquium No. 200, 2005
C. Aime and F. Vakili, eds.

© 2006 International Astronomical Union
doi:10.1017/S1743921306009616

The optical differentiation coronagraph

José. E. Oti[1], Manuel P. Cagigal[1] and Vidal F. Canales[1]

[1]Departamento de Física Aplicada, Universidad de Cantabria, Avda. Los Castros s/n, E-39005 Santander, Cantabria, Spain
email: otije@unican.es

Abstract. Direct detection of exoplanets is a topic of increasing interest since the first exoplanet discovery by indirect methods. It represents a formidable task because of the small angular separation and large contrast ratio between planet and parent star. We present a novel family of stellar coronagraphs based on the standard coronagraph design but transformed to perform optical differentiation. The proposed coronagraphic masks are used to perform the first or the second derivative of the incoming field. This concept offers a new method to detect exoplanets providing both, deep starlight extinction and high angular resolution. To perform optical differentiation the coronagraph's occulting disk is replaced by an especially designed mask. A further improvement on the coronagraph performance is made by adding a gaussian profile to the differentiation mask in order to reduce the amount of diffracted light. The theoretical rejection rate of our coronagraph is infinite. Computer simulations carried out for the ideal case show that it achieves deep starlight reduction corresponding to a gain of at least 37 mag (10^{-15} light intensity reduction). To take full advantage of the capabilities of our coronagraph atmospheric distortions must be reduced by the use of extreme adaptive optics systems or by its use on space telescopes.

Keywords. techniques: high angular resolution, astrobiology, instrumentation: adaptive optics.

---

1. Introduction

The detection of faint sources around bright objects is a topic of increasing interest since the first indirect discovery of an exoplanet orbiting a Sun-like star (Mayor & Queloz (1995)). Since then, direct detection has risen as a probably viable method to uncover exoplanets in the visible. The goal of direct exoplanet detection is to obtain images of Earth-like planets around solar-type stars and to perform spectroscopy of its atmosphere to look for habitability conditions or even to search for signs of life (Woolf & Angel (1998)). However, the detection of such faint objects presents a major difficulty because of the small angular separation and the large contrast ratio between a Sun-like star and the Earth-like planet; 10^{-6} in the mid-IR and $10^{-9} - 10^{-10}$ at visible wavelengths (Beichman *et al.* (1999)).

In the past years, new coronagraphic designs based on the Lyot coronagraph were described. These proposals include the use of specially designed focal-plane phase masks to achieve better starlight background reduction. Roddier & Roddier (1997) proposed a coronagraphic phase mask covering the inner part of the star's image providing better background starlight reduction than the standard coronagraph. Further starlight extinction can be achieved by the Achromatic Interfero-Coronagraph (AIC) (Gay & Rabbia; Baudoz *et al.* (2000)) or by the Four Quadrant Phase Mask coronagraph (FQ-PM) (Rouan *et al.* (2000)), which uses an alternative phase-step mask arranged in a quadrant-scheme. Kuchner & Traub (2002) (see also Kuchner & Spergel (2003)) described other technique that uses band-limited masks, presenting promising results. Theoretically

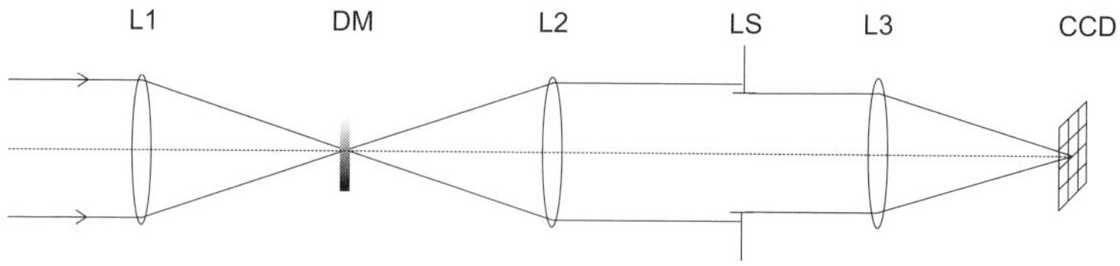

Figure 1. Set-up of the optical differentiation coronagraph. It is composed of three achromatic lenses (L1, L2, and L3), the Lyot stop (LS), the differentiation mask (DM) and a CCD camera.

these techniques, as the optical differentiation coronagraph described here, achieve perfect cancelation of the starlight.

Here we describe a whole new family of coronagraphic masks based on the optical differentiation technique that are implemented on a standard Lyot coronagraph. Theoretically, these differentiation masks allow perfect starlight extinction. Moreover, they provide high angular resolution so they can be used in the detection and discovery of Earth-like exoplanets. As in other coronagraphic designs, the presence of wavefront phase distortions or pointing errors degrades the performance of our coronagraphic technique. Therefore, it is mandatory to reduce atmospheric distortions and jitter in order to take full advantage of the capabilities of our coronagraph.

In § 2 the principle of operation of our coronagraphic design is presented, the mask used to perform optical differentiation is described briefly and the detection procedure is also introduced. Simulations of our coronagraph, in the perfect case with no errors, are presented in § 3. As a simulation example, our coronagraph is used to successfully detect an Earth-like exoplanet. Next, in section § 4, the new designs of coronagraphic masks are introduced and some of their properties are described. Finally, in § 5 some conclusions are given.

2. The optical differentiation coronagraph

The main idea on this new coronagraphic concept is the use of the optical differentiation technique to reduce the light of on-axis stars. In fact, under ideal conditions a centered unresolved star presents a plane wavefront so its derivative is zero. On the other hand, a planet orbiting this star has a tilted wavefront and consequently its derivative is not zero. This coronagraphic concept has been presented in a previous work (Oti *et al.* (2005)). Next, we briefly describe the principle of the optical differentiation technique. From the definition of the Fourier transform (FT), the derivative of a one-dimensional function $f(x)$ can be expressed as:

$$\frac{df(x)}{dx} = \int_{-\infty}^{\infty} 2\pi u_x i \cdot F(u_x) \cdot e^{i2\pi u_x x} du_x \tag{2.1}$$

$F(u_x)$ represents the FT of $f(x)$, u_x is the spatial frequency coordinate in the x direction and i is the imaginary unit. The basic set-up required to carry out the optical differentiation operation consists of a telescopic system made of two achromatic lenses. From (2.1) follows that if a mask of the form $2\pi u_x i$ is placed on the common focal plane of this telescopic system it performs the optical differentiation of the incoming field along the direction of the mask slope (Iizuka (1987); Oti, Cagigal & Canales (2003); Oti, Cagigal & Canales (2005a)). Hence, the idea is to transform a standard Lyot coronagraph into a system able to perform optical differentiation to improve its capacity for exoplanet detection. This is accomplished by substituting the occulting disk by a differentiation mask as shown in figure 1. For a incoming electric field $E(x, y) = E_0 e^{i\phi(x,y)}$ at the entrance pupil of the coronagraph, where $\phi(x, y)$ is the wavefront phase and E_0 is the constant amplitude, the detected intensity at coronagraphic pupil, $I(x, y)$, can be expressed as:

$$I(x, y) = |FT\left[2\pi u_x i \cdot FT\left[E(x, y)\right]\right]|^2 = \left|\frac{\partial E(x, y)}{\partial x}\right|^2 = |E_0|^2 \left|\frac{\partial \phi(x, y)}{\partial x}\right|^2 \tag{2.2}$$

where FT stands for the operation of Fourier transformation. The coronagraph performs a perfect cancelation if the phase $\phi(x, y)$ is constant so its derivative is also equal to zero and no intensity is detected at the output. However, a tilted source presents a phase different than zero that provides a derivative estimate different that zero. Therefore, the light coming from the centered star is removed allowing the search for a potential faint source in the circumstellar environment.

2.1. *Mask description*

The mask used to obtain the first derivative of the incoming field has been described in detail in Oti *et al.* (2005b). It is made of two different complements. First, an amplitude transmittance mask proportional to $|2\pi x|$, and a phase mask consisting of a phase step of $-\pi/2$ for the negative valued coordinates and of $+\pi/2$ for the positive ones. The phase step is required to obtain the complex unit, i, and the correct sign of the differentiation mask. Although this first derivative mask produces a high level of starlight extinction, the diffraction generated by the sharp edges increases the starlight background making hard the detection of faint sources as they are also affected by absorption in the mask. To overcome this limitation we propose to remove the sharp edges of the mask by the use of a transmittance gaussian profile so the differentiation masks ends smoothly. This gaussian profile varies only along the differentiation direction. Furthermore, the gaussian profile concentrates the diffracted light on a narrow area near the pupil border and as a consequence, it can be removed more efficiently by the Lyot stop. Extinction rates may vary largely depending on the width of the gaussian profile and the size of the mask. In general, the larger the differentiation-mask size, the deeper the extinction rate achieved.

2.2. *Off-axis sources detection procedure*

To make a complete survey of all possible off-axis sources locations it is necessary to rotate the coronagraph between 0 and $\pi/2$ and take images during this rotation process. As stated in (2.1), the differentiation mask provides the derivative along the mask slope direction. Consequently, if the exoplanet is located along the perpendicular of the mask-slope direction the measured exoplanet intensity will be zero. In addition, as the differentiation mask rotates, or equivalently the coronagraph rotates, the exoplanet intensity will increase until it reach a maximum corresponding to a position parallel to the mask-slope. This is the optimum position to perform spectroscopy of the exoplanet. Like the rest of the coronagraphic designs our coronagraph is sensitive to pointing errors. A procedure to minimize pointing inaccuracy is to monitor the on-axis starlight and search for a minimum of the detected light intensity. Once this minimum is reached, a rotation of the mask (or coronagraph) should not produce changes in the detected on-axis light intensity.

3. Computer simulations

To check the performance of our coronagraph we carried out computer simulations of an ideal system using the fast Fourier transform routine (Press *et al.* (1995)). To achieve a good sampling and hence, to avoid aliasing effects we use arrays of 1024×1024 data

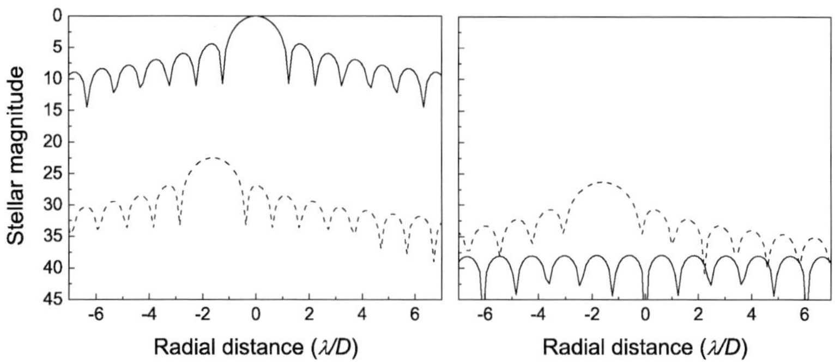

Figure 2. Simulated cross sections of the optical differentiation coronagraph for a successful exoplanet discovery. a) The parent star (solid curve) completely overwhelms the exoplanet (dashed curve) if no coronagraphic system is present. b) With our coronagraph the exoplanet clearly appears in the simulation.

samples. The coronagraph pupil is simulated with a data sampling of 128×128. Figure 2 show the results of a simulation of our system for the perfect wavefront case applied to an Earth-like exoplanet detection example. The mask used is the standard differentiation mask multiplied by the gaussian profile. It has a radius in the x direction of about $50\lambda/D$ and the gaussian profile has a variance of $\sigma^2 = 100(\lambda/D)^2$. The Lyot stop size has been properly adjusted so the extinction rate achieved is the best without a large decrease in exoplanet light throughput. We have found that in this case the optimum Lyot size is 85% of the entrance pupil of the coronagraph. The intensity ratio between the parent star and the exoplanet is 10^9 and the angular separation is $1.5(\lambda/D)$. The starlight intensity reduction achieved is 15 orders of magnitude, corresponding to about 37 mag, hence, the exoplanet clearly appears in the final simulated image.

4. Second derivative differentiation mask

From (2.1) it is easy to show that using a mask of the form $(2\pi u_x)^2$, instead of the linearly increasing transmittance mask, an optical differentiation set-up gives the second derivative of the incoming field. Working out this second derivative for an incoming field $E(x,y) = E_0 e^{i\phi(x,y)}$ follows:

$$I(x,y) = |E_0|^2 \left[\left| \frac{\partial \phi(x,y)}{\partial x} \right|^4 + \left| \frac{\partial^2 \phi(x,y)}{\partial x^2} \right|^2 \right] \tag{4.1}$$

As in the first derivative case presented in § 2, this equation implies that the on-axis light is removed since it presents a plane wavefront. This second derivative mask is pure amplitude mask whose amplitude transmittance increases along the derivative direction quadratically. Consequently it does not require the use of a phase step avoiding problems in the manufacture of an achromatic phase mask. The exoplanet detection procedure is similar to one described in § 2.2. Then, again it is necessary to rotate the mask to make a complete survey of all possible planet locations. To avoid this procedure we propose a further development in the second derivative differentiation mask by making it a symmetric mask of the form $(2\pi)^2(u_x^2 + u_y^2)$, being u_x and u_y the spatial frequency coordinates in the x and y direction correspondingly. With this differentiation mask

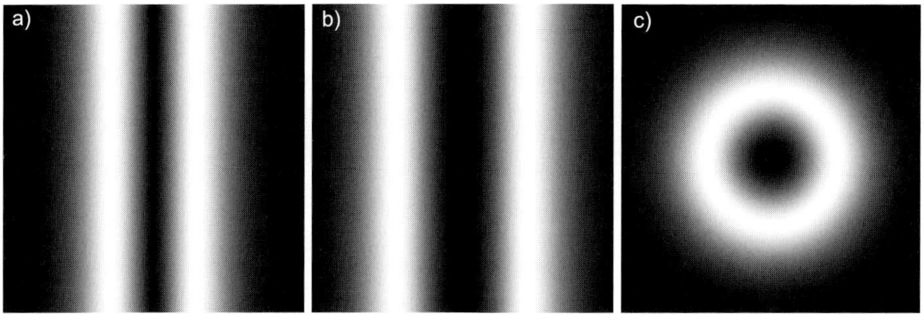

Figure 3. Plot of the proposed differentiation masks. a) Modulus of the first derivative mask of the form $2\pi u_x i$. b) The second derivative mask $(2\pi u_x)^2$. c) And the symmetric second-derivative mask $(2\pi)^2({u_x}^2 + {u_y}^2)$.

placed on the coronagraph the detected intensity can be derived as:

$$
\begin{aligned}
I(x, y) &= \left| \frac{\partial^2 E(x,y)}{\partial x^2} + \frac{\partial^2 E(x,y)}{\partial y^2} \right|^2 \\
&= |E_0|^2 \left[\frac{\partial^2 \phi(x,y)}{\partial x^2} + \frac{\partial^2 \phi(x,y)}{\partial y^2} \right]^2 + |E_0|^2 \left[\frac{\partial \phi(x,y)}{\partial x} + \frac{\partial \phi(x,y)}{\partial y} \right]^2 \qquad (4.2)
\end{aligned}
$$

Once more, the on-axis light is removed without removing the off-axis exoplanet light. The first term in this expression will vanish in all cases because both the on-axis star and off-axis exoplanet second derivative are zero. A plot of all the differentiation masks described here is shown in figure 3.

These second derivative masks are less affected by pointing errors since the central region of the masks presents a larger darker area than in the first derivative case, as seen in figure 3. To check the last statement we perform computer simulations of the coronagraph for a star with various displacements from axis. These displacements are performed along the mask slope direction of the first derivative differentiation mask. The results of the simulations are displayed in figure 4. For the first derivative mask the pointing error seriously degrades the performance of the coronagraph (solid curve) since small pointing errors produces a large starlight leakage. On the other hand, the second derivative mask presents a slower degradation of its performance (dashed curve).

5. Conclusions

We described a novel family of coronagraphic masks based on the optical differentiation technique. The masks presented here perform the first and second derivative of the incoming field and are implemented on a standard coronagraph transforming it into a system able to carry out optical differentiation. Therefore, the light from an on-axis source is successfully removed since its wavefront is plane. On the other hand, the light from an off-axis faint source is not greatly attenuated as it presents a tilted wavefront. In order to improve the starlight extinction achieved further, we multiply the differentiation mask by a gaussian profile. Then, the starlight diffracted by mask is reduced as the masks ends smoothly. To check the capabilities of our coronagraph we have performed computer simulations for the ideal case with no errors. The achieved starlight reduction is as high as 15 orders of magnitude (37 mag). The phase step is not required for the second-derivative mask making its manufacture easier. Furthermore, a symmetric second derivative mask has been developed to modify the exoplanet search procedure so it is no

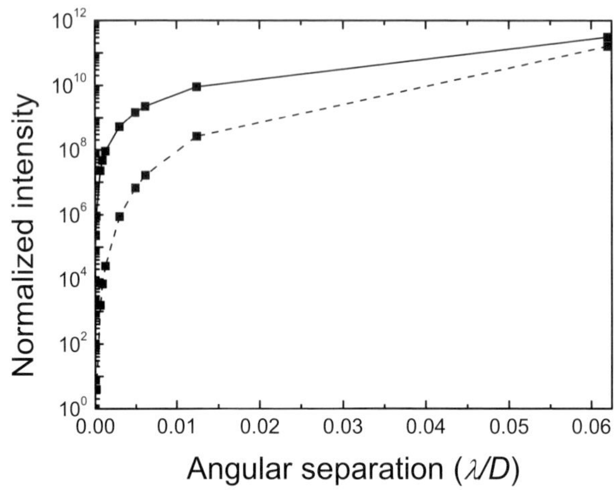

Figure 4. Normalized intensity obtained with our coronagraph for the first (solid curve) and second (dashed curve) derivative differentiation mask as a function of the star displacement from the optical axis. The intensity is normalized by the value of the detected intensity of the on-axis star.

longer necessary to rotate the mask. The coronagraphic designs introduced here can be very useful in the search of Earth-like exoplanet due to its high starlight extinction rate. Nevertheless, more checks of our coronagraphic design should be made under more realist conditions (with phase distortions and jitter errors) in order to completely characterize its performance.

Acknowledgements

This work was supported by Ministerio de Ciencia y Tecnología AYA2004-07773-C02-01. José E. Oti was supported by a FPU grant of Ministerio de Educación y Ciencia.

References

Baudoz, P., Rabbia, Y., & Gay, J. 2000, *Astron. Astrophys. Suple. Ser.* 141, 319

Beichman, A., Wolf, N. J., & Lindensmith, C. A. (eds) 1999, *Terrestrial Planet Finder (TPF)* JPL Publications 99-003, (Pasadena: Jet Propulsion Laboratory)

Gay, J. & Rabbia, Y. C. R. 2000, *C. R. Acad. Sci. Paris* 322 Srie II b

Iizuka, K. 1987, *Engineering optics* (Berlin: Springer-Verlag)

Kuchner, M. J. & Traub, W. A. 2002, *Astrophys. J.* 570, 900

Kuchner, M. J. & Spergel, D. N. 2003, *Astrophys. J.* 594, 617

Mayor, M. & Queloz, D. 1995, *Nature* 378, 355

Oti, J. E, Cagigal, M. P., & Canales, V. F. 2003, *Opt. Express* 11, 2783

Oti, J. E, Cagigal, M. P., & Canales, V. F. 2005a, *Mon. Not. R. Astron. Soc.* 360, 1448

Oti, J. E., Cagigal, M. P., & Canales, V. F. 2005b, *Astrophys. J.* 630, 631

Press, W. H., Teuklosky, S. A., Vetterling, W. T., & Flannery B. P. 1995, *Numerical recipes in C* (2nd ed.; Cambridge: Cambridge Univ. Press)

Roddier, F. & Roddier, C. 1997, *Pub. Astron. Soc. Pacif.* 109, 815

Rouan, D., Riaud, D., Boccaletti, A., Clnet, Y. & Labeyrie, A. 2000, *Pub. Astron. Soc. Pacif.* 112, 1479

Woolf, N. & Angel, J. R. 1998, *Annu. Rev. Astron. Astrophys.* 36, 507

Direct Imaging of Exoplanets: Science & Techniques
Proceedings IAU Colloquium No. 200, 2005
C. Aime and F. Vakili, eds.
© 2006 International Astronomical Union
doi:10.1017/S1743921306009628

Accessing Small Inner Working Angles with a Rotating Sub-aperture Nuller

E. Serabyn and B. Mennesson

Jet Propulsion Laboratory, California Institute of Technology, Pasadena, CA 91109, USA
email: gene.serabyn@jpl.nasa.gov

Abstract. A new approach to high contrast observations near bright stars with a single-aperture telescope is discussed, which is based on the idea of a rotating separated-aperture nulling interferometer. The approach can be described as a rotating sub-aperture nuller, because it nulls two or more sub-apertures within a single telescope's pupil, and uses baseline rotation to modulate the signals from off-axis sources in a manner similar to that of potential space-based nulling interferometers. The sub-aperture beams can be combined in a number of ways, including a fiber nuller and a rotational shearing interferometer. Such a rotating nulling coronagraph has two great advantages. First, it can be used on a ground-based telescope to test signal reconstruction approaches pertinent to potential space-based nulling interferometers. Moreover, it also has the potential to enable ground-based coronagraphic observations of faint off-axis companions very close to bright stars.

Keywords. coronagraphy, nulling interferometry.

1. Introduction

To enable the direct detection of faint companions very close to bright stars, the development of techniques such as coronagraphy on single-aperture telescopes, and nulling interferometry between multiple telescope apertures is required. However, nulling interferometry differs from normal long-baseline interferometry in one important respect - the need to maintain a null at fixed phase implies that the complex visibility of the source brightness distribution is in general not directly accessible. Instead, what is easily measurable is the flux transmitted by the instantaneous nulling fringe pattern, so that a direct Fourier inversion to the source plane is not possible. More indirect inversion techniques such as cross-correlation (Angel and Woolf 1997) are thus likely necessary.

While image reconstruction with a nulling interferometer may be more difficult, the nature of the measurement is clear. As the nulling interferometer rotates, the nulling fringe pattern centered on the star rotates with it (Bracewell 1978). As the fringe pattern rotates, the fringes sweep across the position of the off-axis companion (Fig. 1), modulating its transmitted flux. For larger off-axis source offsets, in general more fringes cross the planet's position, and so higher harmonics are generated in the detected signal. The harmonic content of the signal thus determines the companion's angular offset from the nulled star, while the azimuth of the flux maximum determines its azimuth.

Given the novelty of the rotating long-baseline nuller approach under consideration for space-based nulling interferometers, and the ambiguity inherent in image reconstruction without full complex visibility data, it would clearly be very valuable to have a way to demonstrate both the viability of the approach and the robustness of signal reconstruction algorithms by means of a ground-based experiment prior to launch. These considerations led us to the possibility of implementing a rotating nuller on a single ground-based telescope, by establishing a baseline between separated subapertures. In the course of

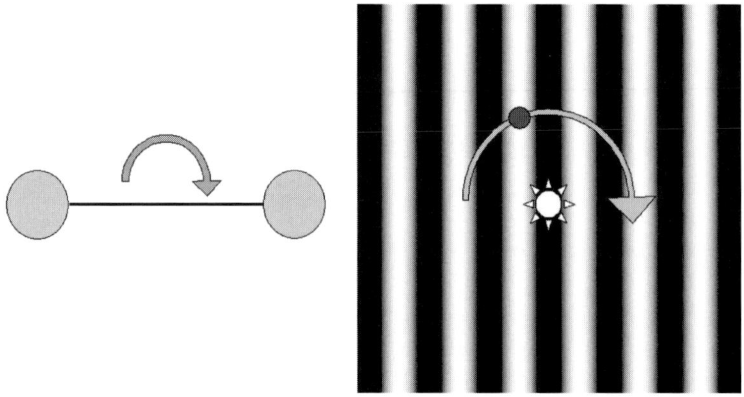

Figure 1. Schematic of a rotating long-baseline nuller and the accompanying fringe pattern. Destructive interference is maintained at the stellar position as the baseline and fringe pattern rotate. The companion's flux is modulated as the transmission fringes sweep past.

these considerations, it became clear that such an approach would also provide unique coronagraphic capabilities. In the following, we describe the possible implementations and potential advantages of this idea.

2. The Idea

Whatever optical implementation is chosen to define the sub-apertures within the telescope pupil, and to combine the individual beams into a nulling baseline or baselines, it will be desirable to maximize the subaperture areas and the spectral bandwidth, in order to maximize the signal to noise (SNR) ratio. It will also be desirable to maximize the baseline length, in order to maximize the angular resolution. However, one could in principle trade off subaperture diameter against subaperture number, in order to maintain a high SNR, and to increase the field-of-view (FOV), as discussed below. Of course, at the outset it will be very important to select an optical implementation which is both simple and amenable to pathlength control, and so here we concentrate mostly on the single baseline case.

Although in principle a number of different beam-combination schemes can be used to combine and null the light from two subapertures lying within a given telescope pupil, including linear and rotational shearing interferometers, perhaps the simplest approach is the fiber nuller. This beam combiner (Fig. 2) is based on the injection of a number of beams into a common single-mode fiber in the focal plane (Wallner *et al.* 2004). Deep nulling of monochromatic light to close to the 10^{-6} level has now been demonstrated with a fiber nuller at JPL (Haguenauer and Serabyn 2005). For more broadband nulling, an achromatic phase shifter to introduce π radian (or other) phase shifts between the beams is needed. In principle, solutions for glass dispersion correctors do exist. However, a fiber nuller beam combiner has not yet been used for on-sky observations. In addition, with only two beams, the combiner injection efficiency is low (Fig. 3).

In the simplest possible optical configuration with a fiber nuller beam combiner, a rotating dual-aperture mask in the pupil plane would select the two beams to be combined at the fiber tip (Fig. 4a). In this approach, the achromatic phase shifter would either need to rotate with the subaperture mask, or it would be necessary to accept a large dead zone in rotation angle, as the subapertures transition across the phase step. Neither of these solutions is ideal. These disadvantages can be obviated if a pupil rotator is inserted

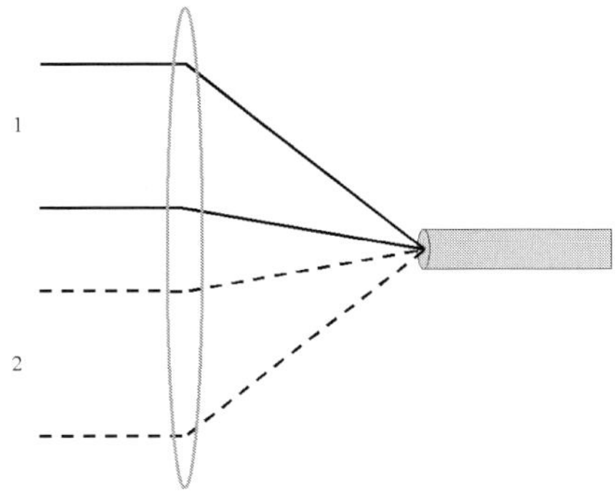

Figure 2. Schematic of a fiber nuller. Two spatially-separated beams are focused by a common optic onto a common single mode fiber. A relative phase shift of π radians is applied upstream.

into the beam train, in which case a fixed dual-subaperture mask can be used to define the subapertures (Fig. 4b). On the other hand, such an approach brings with it the need for a very planar pupil rotation.

As mentioned, a dual-beam fiber nuller has low injection efficiency (Fig. 3). This efficiency loss can be recovered by instead using a more classical beamsplitter-based beam-combiner. In this regard a rotational shearing interferometer (RSI) also provides several additional advantages (Fig 4c). First, since RSIs combine two pupil images rotated by 180 degrees, all baselines centered on the pupil are generated simultaneously. Second, any given baseline can then be selected by means of a single off-center sub-aperture. Third, the light in this single sub-aperture can be focused with high efficiency into a fiber. Fourth, each baseline is duplicated on the opposite side of the center in the recombined pupil. This duplicate baseline can be made use of for e.g., a second waveband, or phase measurement and control. Finally, a number of baselines can be accessed simultaneously by means of different off-axis subapertures feeding different fibers, thus achieving pupil efficiencies approaching unity. Of course, all of these baselines can be rotated simultaneously by means of a common pupil rotator ahead of the RSI. Thus, inclusion of an RSI brings a number of significant advantages, and provides perhaps the best overall approach, although it is also the most complex of the approaches listed in Fig. 4

3. Performance

How well might such a rotating subaperture nuller perform? Other than sensitivity, the two main areas of interest are the achievable contrast, and the inner working distance (IWD), i.e., the smallest angular offset at which a companion can be detected. The IWD is given by IWD $= \frac{\alpha\lambda}{2b}$, where λ is the wavelength, b the baseline length, $\frac{\lambda}{2b}$ is the angular offset to the first constructive fringe peak, and α is a constant of order unity, determined by the amount of sensitivity loss and position information loss that one is willing to accept. Different potential choices for α range from e.g., the fringe half-power point ($\alpha = \frac{1}{2}$), to just under unity ($\alpha \approx 0.85$) if the centralmost region where positional information is lost is to be avoided.

Since the maximum baseline length available is $D-s$, where D is the telescope diameter and s the subaperture diameter, we have IWD $= \frac{\alpha\lambda}{2(D-s)}$. For e.g., $s = D/3$, we then have IWD $= \frac{3\alpha}{4}\frac{\lambda}{D}$. Thus, independent of the exact value of α, this approach provides an IWD significantly smaller ($\sim\frac{1}{2}\frac{\lambda}{D}$) than that of a classical coronagraph (several λ/D).

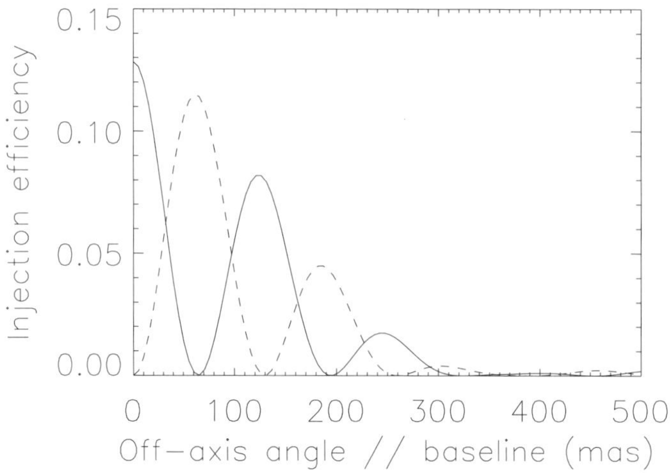

Figure 3. Injection efficiency of a pair of off-center beams into a single mode fiber for the case of 1.5 m subapertures and a 3.5 m baseline. For this plot, the off-axis companion is in the direction parallel to the baseline. Smaller subapertures would access additional fringes further off-axis. The low injection efficiency is due to the underfilling of the fiber mode.

The FOV is set by the width of the subaperture beam coupling to the single mode fiber. In the RSI combiner case, with diffraction-limited coupling, we have FOV $\approx\lambda/s$, so that smaller subapertures imply an increased FOV. In general, FOV/IWD $= \frac{2(D-s)}{s\alpha}$. In the case of $s = D/3$ discussed earlier, FOV $\approx 3\lambda/D$, and FOV/IWD $= 4/\alpha$. The accessible angles are thus somewhat restricted, but they cover a range otherwise inaccessible.

The instantaneous stellar null depth, N (the ratio of fluxes in the destructive and constructive states), will be limited mainly by the phase offset, ϕ, between the two subaperture beams, and is given by $N = (\phi/2)^2$. The expected null depth can therefore be estimated (in a worst case sense) from the root-mean-square phase across the full aperture, $\phi_{\rm rms}$, obtainable from the Strehl ratio, S, by means of the relation $S = 1-\phi^2_{\rm rms}$, which applies for small $\phi_{\rm rms}$. Thus, $N \approx (1 - S)/4$. For a typical current-generation adaptive optics (AO) system, with $S \approx 0.6$ at K-band, and $\phi_{\rm rms} \approx 200$ nm, the average null is then expected to be $\sim$0.1. However, due to path length fluctuations, the null depth will upon occasion reach much deeper levels, making the lower envelope of the null depth fluctuations the quantity of interest.

The detailed performance expected with a basic subaperture nuller behind a current-generation AO system is shown in Figure 5. These simulations of instantaneous null depth reproduce the expected 0.1 average null level (the upper envelope of the traces), and also show that even without further phase correction, off-axis sources can be detected by means of their effect on the lower envelope of the null depth fluctuations. At each rotation angle of the baseline, one simply measures the null over a long enough time interval for the lower envelope of the null depth fluctuations to show the expected variations.

However, with the next generation of AO systems, with Strehl ratios $\approx$0.95 at K-band and $\phi_{\rm rms} \approx 75$ nm, the average null (the upper envelope of the traces in Fig. 5) can be expected to drop to 0.01, thus improving performance by an order of magnitude. Indeed, such an AO performance level is actually already available today with the Palomar off-axis subaperture reimager (Serabyn *et al.*, these proc., Haguenauer *et al.* 2005), and can be made use of if a rotating nuller were to follow such a subaperture reimager.

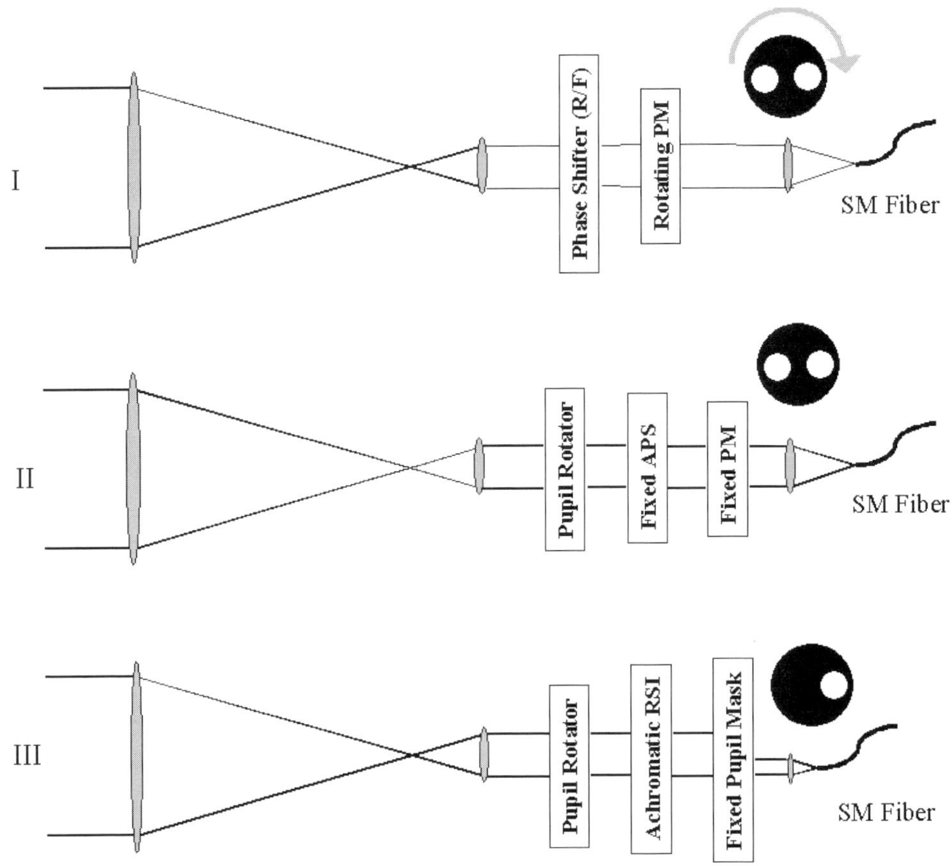

Figure 4. Three potential layouts for a rotating nulling coronagraph. R/F means "rotating or fixed", PM is short for "pupil mask", APS is "achromatic phase shifter", and RSI is "rotational shearing interferometer." In version III, only a single subaperture in the PM is necessary, although more subapertures can be used to increase efficiency or to lay out different nulling configurations, if each subaperture/baseline is coupled to its own fiber through its own lens.

However, a more active improvement is also possible. Since the null depth is limited by the phase fluctuations between the two large subapertures, an accurate measurement and correction for this phase difference can directly improve the null stability. Because of the large size of the subapertures, the SNR of such a measurement can be high. Thus, with a second, post-AO, layer of phase correction, significant improvement in the null depth stability can be achieved.

4. Conclusions

A rotating sub-aperture nuller coupled to a single-aperture ground-based telescope can be used both to test the rotating nulling interferometer approach being considered for space-based interferometers, and also to generate a coronagraph with a small IWD. This type of nulling coronagraph can be implemented in a number of ways, from a simple rotating pupil mask with a fiber combiner, to a more complex but more capable system which includes an RSI as well as a pupil rotator. Likewise, a number of possible

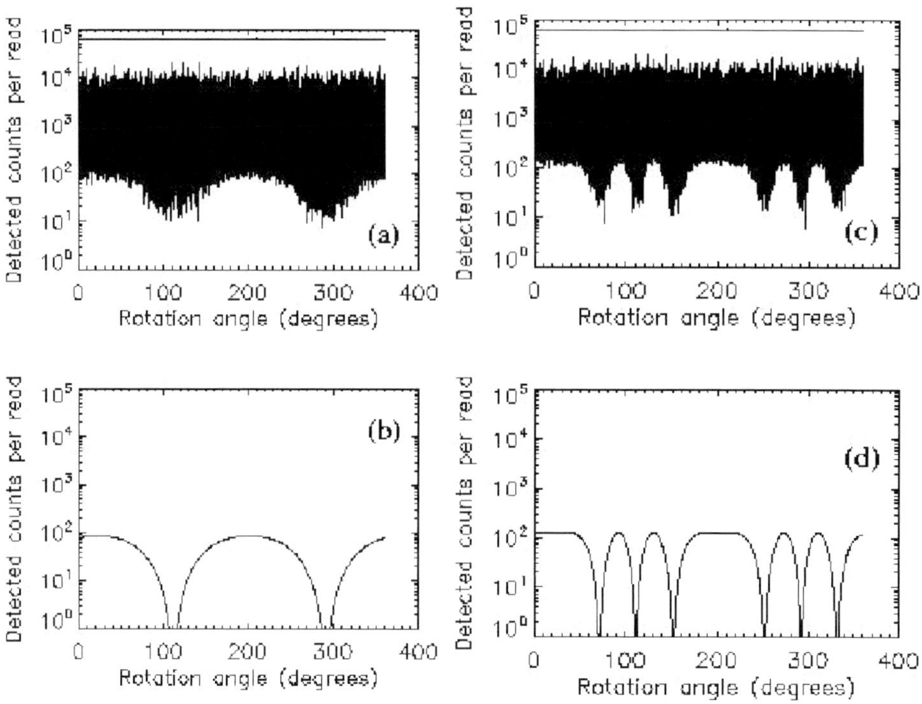

Figure 5. Expected performance of a rotating sub-aperture nuller on the Palomar 200 inch telescope for an $m_K = 3$ star, without any phase stabilization beyond that provided by the Palomar AO system. Panels b and d show the expected null depths vs. rotation angle for a pair of binaries of contrast 500, for separations of 40 mas and 200 mas, respectively. Panels a and c show simulations of the null depths measured, assuming efficiencies representative of a realistic fiber nuller, and a Strehl ratio of 0.7. The data consist of 100 rotations of 10 s, with 1000 reads per rotation. The constructive state is shown by the horizontal lines in the panels. The bottom of the null depth fluctuation envelopes follow the theoretical curves. Phase stabilization between the subapertures will improve the performance.

subaperture configurations are possible, allowing for either an increase in efficiency, for phase correction, or for the testing of different potential pupil configurations.

Acknowledgements

This work was carried out at the Jet Propulsion Laboratory, California Institute of Technology, under contract with the National Aeronautics and Space Administration

References

Angel, J.R.P. & Woolf, N.J. 1997, *ApJ* 475, 373
Bracewell, R.N. 1978, *Nature* 274, 780
Haguenauer, P. & Serabyn, E. 2005, *Applied Optics* in press
Haguenauer, P. *et al.* 2005, *Proc. SPIE* 5905, 59050S-1
Wallner, O., Perdigues Armengol, J.M., & Karlsson, A.L. 2004, *Proc. SPIE* 5491, 798

Direct Imaging of Exoplanets: Science & Techniques
Proceedings IAU Colloquium No. 200, 2005
C. Aime and F. Vakili, eds.
© 2006 International Astronomical Union
doi:10.1017/S174392130600963X

The Phase-Induced Amplitude Coronagraph (PIAA)

Olivier Guyon[1], Eugene A. Pluzhnik[1], Stephen T. Ridgway[2], Robert A. Woodruff[3], Celia Blain[1], Frantz Martinache[4], and Raphael Galicher[5]

[1]Subaru Telescope, National Astronomical Observatory of Japan, 659 N. A'ohoku Pl., Hilo, HI 96720, USA
email: guyon@naoj.org
[2]National Optical Astronomical Observatories,USA
[3]Lockheed Martin Space Corporation, USA
[4]Cornell University, USA
[5]Ecole Normale Superieure, Paris, France

Abstract. The Phase-Induced Amplitude Coronagraph (PIAAC) uses a lossless beam apodization, performed by aspheric mirrors, to produce a high contrast PSF. Thanks to the lossless apodization, this concept offers a unique combination of high theoretical throughput (≈100%), high angular resolution (λ/d), small inner working angle (≈$1.5\lambda/d$), low chromaticity and low sensitivity to pointing errors or angular star diameter. Together, these characteristics make the PRC an ideal choice for direct imaging of extrasolar terrestrial planets (ETPs) from space. We show that a visible telescope smaller than 4m would then achieve the goals of the TPF mission, while other coronagraphs considered for TPF require telescope diameters typically 2 to 3 times larger. On a large size (8m) space telescope, ETPs can be searched for around a significantly larger sample of stars, thus enabling a much higher scientific return.

Keywords. techniques: high angular resolution, planetary systems.

---

1. What is the PIAAC?

1.1. *Lossless apodization by pupil remapping*

The most convenient way of obtaining a high contrast PSF is to amplitude-apodize the telescope pupil. This is usually done by **selective absorption**: light is removed mostly from the outer edges of the pupil (see for example Kasdin *et al.* 2003). Unfortunately, to deliver a 10^{10} PSF contrast with this technique, at least 80% of the light needs to be removed. Since this light is removed from the edges of the pupil, the angular resolution of the telescope is also seriously affected: the Inner Working Angle (IWA) is 3 to 4 λ/d.

An alternative to apodization by selective absorption is apodization by **pupil remapping** (also referred to as Phase-Induced Amplitude Apodization, PIAA): the geometry of the pupil is "distorded" to obtain the desired amplitude distribution (Guyon 2003, Traub & Vanderbei 2003, Galicher *et al.* 2005). To obtain a gaussian-like pupil amplitude profile, the central part of the pupil is to be geometrically shrunk (this increases the surface brightness in the pupil) while the edges of the pupil are streched (decreases the surface brightness).

This operation can be performed by 2 aspheric mirrors which can be placed in a collimated relay beam, as illustrated in figure 1. The combination of the 2 mirrors is required to maintain a flat wavefront for an on-axis source.

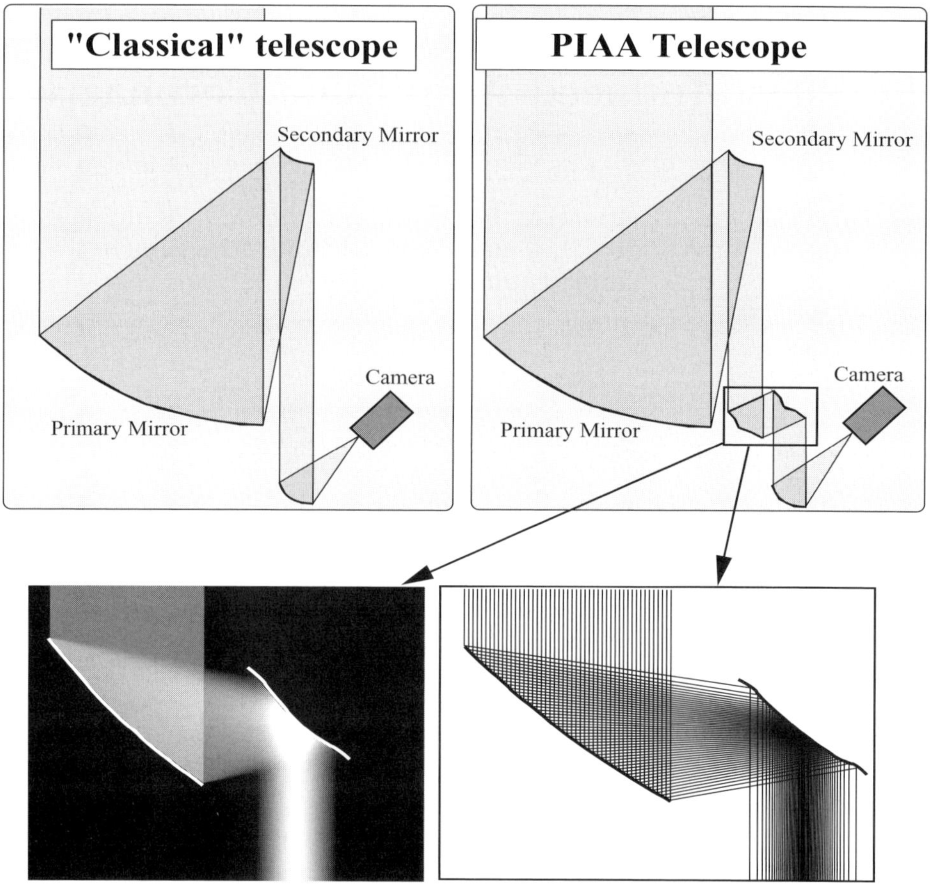

Figure 1. Illustration of the PIAA technique (bottom): a collimated unapodized beam can be amplitude-apodized without light loss thanks to 2 aspheric mirrors. The system preserves the flatness of the wavefront of an on-axis source.

1.2. *The PIAAC design*

The pupil remapping technique described previously is at the heart of the PIAAC design. It is used to produce a gaussian-like (more exactly a prolate spheroidal) output amplitude distribution in the pupil plane. Since the PIAA unit maintains the wavefront flatness of a on-axis point source, the corresponding PSF offers high contrast: most of the light is concentrated in a narrow diffraction core, and the PSF wings can be made fainter than 10^{-10} everywhere else.

Unfortunately, the pupil geometric distortion introduced by the PIAA unit is the source of large off-axis PSF aberrations. While the central PSF is sharp and high-contrasted, off-axis sources have PSFs with large coma-like wings. The solution to this problem is to restore the original pupil geometry with a "inverse" PIAA unit after having removed the on-axis light with a focal plane stop. This concept is described in more detail by Guyon (2003), Vanderbei & Traub (2005) and Guyon *et al.* (2005).

The light of an on-axis source will be concetrated onto the focal stop by the first PIAA unit. The light of an off-axis source will miss the focal stop, and the second inverse PIAA unit will cancel the effect of the first PIAA unit, therefore maintaining a

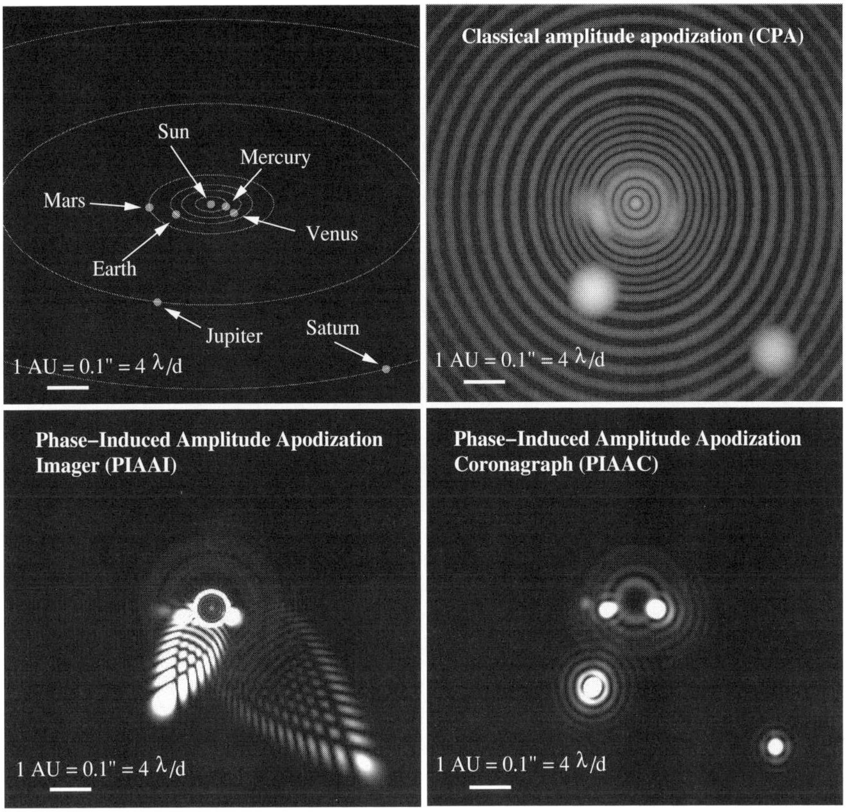

Figure 2. Simulated 0.5 μm images of the solar system at a distance of 10 pc with a 4m diameter telescope. The same apodization profile was produced by classical apodization (upper right) and PIAA (lower left). In the PIAAC (lower right) the original pupil geometry was restored before imaging. For the CPA and PIAA, an focal plane occulting mask was used to block the starlight, and the images shown are acquired in the reimaged focal plane. In all 3 images, the angular size of the star was taken into account and the faint concentric rings centered on the optical axis are residual starlight.

sharp unaberrated PSF across the field of view. Since this design incorporates a focal plane coronagraphic stop, we refer to it as the PIAAC. Figure 2 illustrates the image quality improvement brought by the PIAAC design, and also compares it to a classical apodization technique.

1.3. *The PIAAC hybrid design*

There is however a serious issue with the concept presented above: on the first mirror, the edge exhibits a very sharp "bend", necessary to spread the light which is in the very outer portion of the incoming beam into a much larger area in the exit pupil. This sharp feature brings several problems:

- **Optics manufacturing**. The "bend" is critical to the apodization but extremely difficult to polish
- **Alignement sensitivity**. Since the mirror shape changes over small distances at this "bend", optics alignement tolerances become very strict.
- **Chromaticity**. This problem, unlike the others, goes beyond engineering considerations, and can fundamentally limit the PIAAC's performance. The "bend" is sharp enough to introduce diffractive (as opposed to geometrical) effects (Vanderbei 2005).

Since these effects are chromatic, they would restrict the spectral bandwidth over which the required contrast level can be maintained.

The presence of this "bend" is due to the PIAA unit attempting to produce, in the output pupil, extremely dark areas (at the edge of the apodized pupil). A practical solution is therefore to share the apodization between a PIAA unit and classical apodizer mask. In this scheme, refered to as the "hybrid PIAAC", the PIAA unit does most of the apodization, but leaves some light in the outer parts of the beam (typically 1% of the surface brightness at the center of the beam). This makes the bend much smoother and also spreads it over a wider area. A classical apodizer can then "touch up" the outer parts of the beam to obtain the desired amplitude profile.

Since the classical apodizer mainly works in parts of the PIAA-apodized beam that are already faint, only a few percent (typically 5%) of the light is lost. The performance loss is therefore minimal, while the practical gains are significant.

2. Performance

The PIAAC combines many desirable advantages:

- **Small IWA**. Efficient detection is possible as close as 1.5 λ/d for a 10^{10} contrast.
- **Full throughput**. As explained above, the only losses are due to the "touch up" classical apodizer (typically 5%) and unavoidable reflections on mirrors.
- **Achromaticity**. The apodization relies on geometrics optics, and is therefore achromatic. We note that this is not exactly true since diffraction propagations effects do happen. However, we have verified that, with a good hybrid PIAAC design, chromatic effects are well below the 10^{-10} PSF contrast level over large (20% or more) spectral bandwidths. Of course, other sources of chromaticity typically exist in a full system (telescope optics), and might restrict the useful bandwidth to less than 10%.
- **360 degree search area**. All of the field is used at once.
- **Low sensitivity to low order aberrations**. This is especially important since stars are partially resolved. If uncorrected telescope pointing error are especially large, the PIAAC's focal plane occulter diameter can be increased.

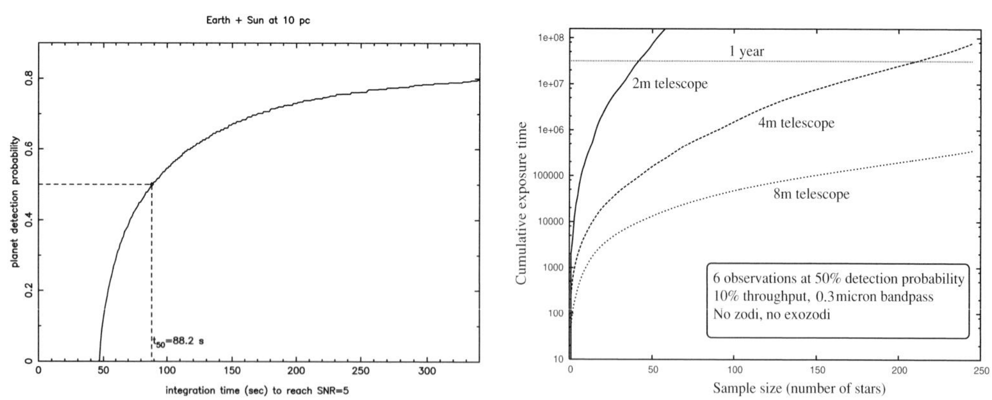

Figure 3. Left: Evolution of the planet detection probability (with a signal to noise ratio $SNR = 5$) as a function of exposure time for an Earth-Sun system at 10pc ($m_V = 5$, $1\,mas$ angular diameter, $100\,mas$ maximum angular separation) observed by a 4m telescope. The bandwidth is $\Delta\lambda = 0.3\,\mu m$ and the central wavelength is $\lambda = 0.5\,\mu m$. Right: Cumulative exposure time required for the survey of a sample of up to 250 stars for ETPs.

2.1. *Direct imaging survey of ETPs with a PIAAC*

We summarize here results described in a recent paper by our team: Martinache *et al.* (2005).

We first consider the observation of a Sun-Earth system at 10pc. We assume that the planet position along its orbit is unknown: the exposure time necessary for detection (defined by SNR=5 on the image) cannot be computed. Instead, we can compute the **planet detection probability** as a function of integration time. Results for a PIAAC system on a "perfect" 4m space telescope are shown in figure 3 (left). The combination of small IWA, high throughput, 360 degree search area results in a 50% detection probability in just over one minute of integration time.

Likewise, the integration time necessary to reach a 50% detection probability can be computed for each star of a target list. As shown in figure 3 (right), using somewhat realistic assumption (6 individual observations, each at the 50% detection probability, required per target; 10% system throughput; 1 year total integration time available), we find that Earth-like planets could be imaged around about 40 stars with a 2m telescope, and around more than 200 stars with a 4m telescope.

2.2. *Simulated images of a nearby system*

In this section, simulated images for the observation of HIP4569, a Sun-like star at 8.7pc, are shown. An Earth-like planet was artificially placed 1.04 AU from the star (distance where the stellar flux recieved by the planet equals the solar flux received by the Earth). The simulated images include the following assumptions:

- Stellar diameter was taken into account (slightly increases coronagraphic leaks).
- An exo-zodiacal cloud similar to our zodiacal cloud was placed around the star.
- Photon noise was included
- Telescope+camera throughput = 0.25
- 0.2 μm bandpass centered at 0.55 μm.
- 2hr exposure time per image.

The images shown in figure 4 illustrate the ability of a small size telescope to directly detect Earth-like planets with a PIAAC. Visual inspection of the images shows that the detection of HIP4569's hypothetical companion requires either a PIAAC on a 2m telescope or a classical apodizer on a 6m telescope. This factor of 3 in telescope size is due to the lower throughput of the classical apodizer (10% vs. 90%: a PIAAC on a 2m telescope collects as many photons as a classical apodizer of a 6m telescope) and larger inner working angle (4 λ/d vs. 1.5 λ/d).

As the telescope size increases, the central star becomes partially resolved: on an 8m diameter telescope, the stellar radius is 4% of λ/d. The coronagraphic leaks then increase for the PIAAC, and concentric diffraction rings appear in the image. To mitigate this problem, the diameter of the focal plane occulter in the PIAAC may be slightly increased (which slightly increases the IWA). These feature are however fainter than the exoziacal background: even in the 8m telescope case without oversizing of the focal plane occulter, the integrated residual starlight is 6 times fainter than the integrated residual exozodiacal light. It is therefore unlikely that stellar angular size would limit the detection efficiency.

Figure 4 also shows that the PIAAC's improved angular resolution allows a better separation between the planet's light and the exozodiacal light. This benefit is especially important for systems with high zodi content, or inclined orientations. Another advantage of any high throughput coronagraph is to allow efficient sensing of the unwanted speckles in the focal plane. In an instrument aimed at achieving 10^{10} PSF contrast, it is critical to quickly sense speckles at this level, so that a command can be issued to the system's

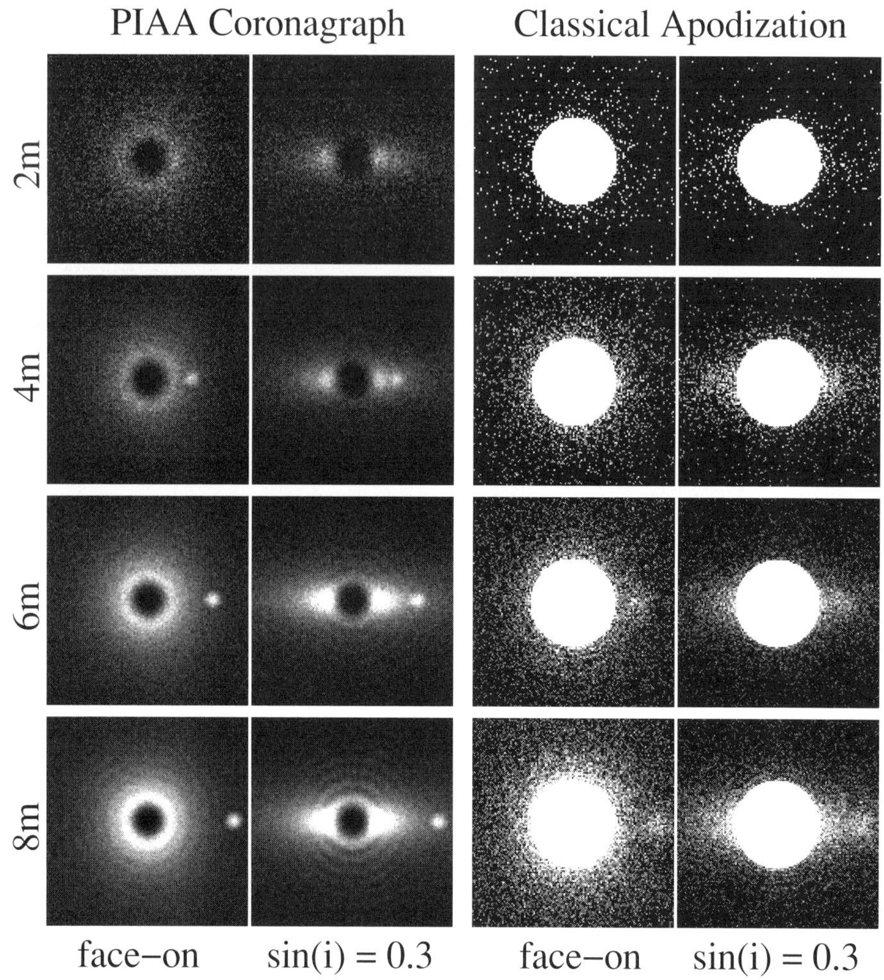

Figure 4. Simulated 0.55 μm images of HIP4569 and an hypothetical Earth-type planet at maximum elongation for telescope sizes ranging from 2m to 8m. For each coronagraph/telescope size combination, images were produced in the optimistic face-on case (left) and the inclined sin(i)=0.3 case (right). See text for details.

deformable mirror to cancel this speckle. With the PIAAC's nearly-100% throughput, the loop can be closed faster: the wavefront correction is more efficient.

3. Conclusion

We have shown in this paper that the PIAAC is theoretically highly efficient for direct imaging of ETPs from space (TPF-C mission). Its performance are close to what may be theoretically the "limit" of what a coronagraph can achieve: the off-axis PSF is diffraction-limited (λ/d width) at nearly-100% throughput very close to the optical axis (from $\approx 1.5\lambda/d$).

Although the design shown in this paper is targeted at observation at $\approx 10^{10}$ contrast, the PIAAC is also a very good candidate for ground-based observation with efficient adaptive optics systems, where the PSF contrast goal is more modest ($\approx 10^6$ at most). In this case, the main advantages if the PIAAC is to allow efficient detection of faint compations at smaller distances than what would be possible with other coronagraphs. Its low sensitivity to tip-tilt errors is especially attractive since the performance of other small-IWA coronagraphs is often limited by uncorrected tip-tilt residuals.

Acknowledgements

This work was carried out under JPL contract numbers 1254445 and 1257767 for development of technologies for the Terrestrial Planet Finder Mission, with the support and hospitality of the National Astronomical Observatory of Japan.

References

Galicher, R., Guyon, O., Otsubo, M., Suto, H., & Ridgway, S. 2005, *PASP*, 117, 411

Guyon, O., Pluzhnik, E.A., Galicher, R., Martinache, F., Ridgway, S.T., & Woodruff, R.A. 2005, *ApJ* 622, 744

Guyon, O. 2003, *A&A* 404, 379

Kasdin, N.J., Vanderbei, R.J., Spergel, D.N., & Littman, M.G. 2003, *ApJ* 582, 1147

Martinache, F., Guyon, O., Pluzhnik, E.A., Galicher, R., & Ridgway, S.T. 2005, *Accepted for publication in ApJ*

Traub, W.A. & Vanderbei, R.J. 2003, *ApJ* 599, 695

Vanderbei, R.J. & Traub, W.A. 2005, *ApJ* 626, 1079

Vanderbei, R.J. 2005, *astro-ph/0506550, accepted for publication in ApJ*

Discussion

: Why use an image image mask ?

GUYON: The image mask is needed to remove the starlight in the "intermediate" focal plane of the PIAAC. Without the focal plane mask, the inverse PIAA and direct PIAA would cancel each other, and an Airy disk (with its rings) would be left in the final focal plane.

: You seem more optimistic about the broadband performance than Roger Angel showed us in his talk.

GUYON: This is still work that is being done. I just showed you our latest designs, which are better than the ones we got a few weeks ago: we can now maintain a "null" better than 10^{-10} over a 20% bandwidth.

RUSLAN BELIKOV: Is a Fresnel analysis accurate enough to produce all the diffraction problems you might encounter ?

GUYON: The diffraction propagation we performed did not make the Fresnel approximation, so it should be quite accurate. Even if we've missed something, it should be small (much less $\lambda/1000$), and well within the range of a wavefront control system.

CLAUDE AIME: In your coronagraph performance comparative slide, you showed the Roddier&Roddier phase mask has high throughput and small IWA, but it doesn't work for a large bandwidth

GUYON: Yes, I did not mention about chromaticity in coronagraphs. The Roddier&Roddier actually has another issue which makes it unsuitable for direct imaging of ETPs: it is very sensitive to tip-tilt. The angular diameter of a star is sufficient to prevent high contrast in the Roddier&Roddier.

All photographs: Laurent Thareau [`l.thareau@free.fr`].

Direct Imaging of Exoplanets: Science and Techniques
Proceedings IAU Colloquium No. 200, 2005
C. Aime and F. Vakili
© 2006 International Astronomical Union
doi:10.1017/S1743921306009641

Pupil Replication: Implementation Aspects

Frank H. P. Spaan[1], Erwan Prot, Vincent Mourai, and Alan H. Greenaway[1]

[1] School of Engineering and Physical Sciences, Heriot-Watt University, Edinburgh, United Kingdom.
email: F.Spaan@hw.ac.uk, A.H.Greenaway@hw.ac.uk

Abstract. In this paper we first introduce Pupil Replication. We then show simulation results for a pupil replicated classical coronagraph. Furthermore, we show experimental results as a proof of concept as well as a preliminary design for the replication optics.

Keywords. (stars:) planetary systems, instrumentation: high angular resolution, techniques: interferometric, techniques: miscellaneous, telescopes.

1. Introduction

In a previous paper Greenaway *et al.* (2005) we introduced the principle of Pupil Replication. This new optical technique decreases the diameter of the image of a star on, or very near to, the optical axis. Used as a supplementary system on a telescope with a coronagraph, an increase in the suppression of the star light is possible. The principle is shown in figure 1. A de-magnified image of the telescope entrance pupil is formed and several replications of the incoming wave at the entrance pupil are created optically and joined together side by side prior to being focused by an imaging system. For an on axis source this set up will act like a telescope with an increased diameter (see figure 1; dark grey lines), reducing the size of the original on-axis star image. For an off-axis source (light grey lines), like a planet to be detected, the width of the point spread function (PSF) is about the same as the width of the PSF without replication.

A currently important type of high dynamic range imaging is interferometric nulling. Applying pupil replication to such a nulling system does not lead to a straight forward improvement of the system. For another promising technique, pupil apodisation, including pupil replication in the system does lead to an increase in performance. Figure 2 shows an example where a 10^{10} times weaker planet becomes visible in most of the field beyond $1.6\lambda/D$. The extra peaks visible are due to chromatic smearing which stays within the diffraction limit of the telescope as shown in Greenaway *et al.* (2005).

Both pupil apodisation and interferometric nulling lead to promising simulation results, but the requirements for building such systems are high. In this paper we look at a system that has lower requirements: a classical coronagraph with a hard-edged star-stop and Lyot-stop combined with pupil replication.

2. Pupil Replicated Classical Coronagraph

We look here at the possible implementation of Pupil Replication to a classical (Lyot) coronagraph. Since the (on-axis) star image size is reduced by pupil replication, an improvement in the star light suppression is expected. We simulated in two dimensions a

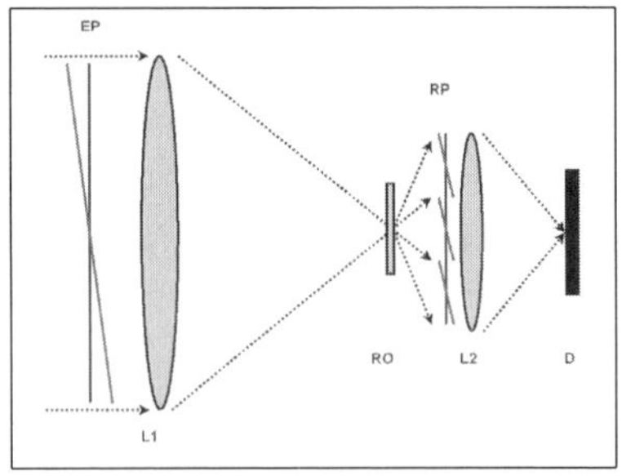

Figure 1. A schematic overview of the principle of Pupil Replication, shown here with 3 replications. The wave front, after passing through the entrance pupil (EP), is focused by the telescope (L1) and then divided equally in amplitude 3 ways using the replication optics (RO). Thus 3 replicated images of the entrance pupil are formed side by side in the replication plane (RP), each 1/3 of the intensity of the original pupil image. If the source is an axial point (dark grey lines) the replicated wave fronts form a continuous plane wave front. If the source is off axis (light grey lines), the replicated wave fronts form a disjointed wave front. The resulting wave front is passed through an imaging system (L2) and an image of the point object is formed through this lens on the detector (D).

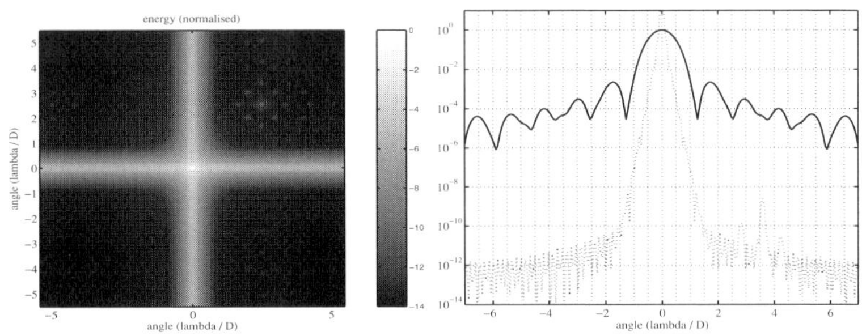

Figure 2. Simulation of a pupil apodisation system incorporating $8 * 8$ pupil replication. In the 2-D image (left) a planet (10^{-10}) is visible in the upper right corner. The right figure shows the diagonal crosssection. The system was simulated in the V-band and had an efficiency of 64%.

coronagraph with a hard edged image plane stop and Lyot stop, including Pupil Replication for a 3.5 meter telescope for two cases: a square input pupil in the V-band (which was simulated by 8 weighed wavelengths) and a hexagonal input pupil at $600nm$. None of this was optimised; the size of the Lyot stop was chosen to yield an efficiency of 50%.

We first chose a square pupil and a replicated pupil existing of $8 * 8$ replicated input pupils; such a set up would be possible with a cascade of 6 of the replication optics described in section 3. The simulation results were obtained using a square star stop with a width of $2\lambda/D$ yielding an inner working angle of $1\lambda/D$. The PSF after the coronagraph for an on-axis source is shown in figure 3 (left). Cross sections through the PSF both for a system without (left) and with (right) pupil replication are shown in figure 4. For a square pupil the results in the different parts of the field are not the same: the diagonal cross section shows an attenuation that is overall much higher than the horizontal cross section. One can see on the left in figure 3 that most of the search area (over 90% in the image shown) has the characteristic of the diagonal cross section and can be used for search of high dynamic range objects.

For a hexagonal pupil we simulated a 7-fold replication system at one wavelength ($600nm$). As can be seen on the right in figure 3 the field is more homogeneous due to the more centro-symmetrical input pupil. The system we simulated had a hard edged

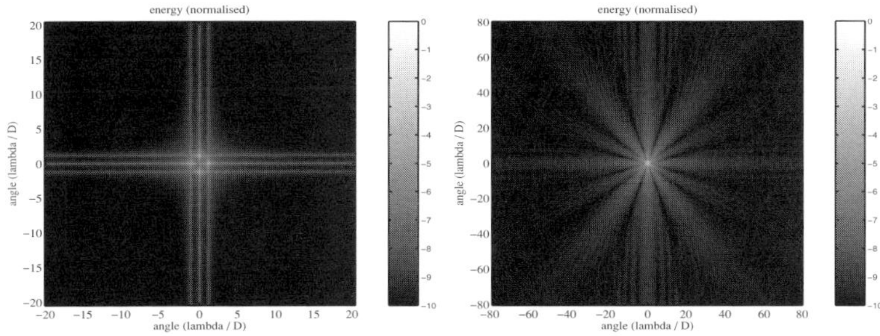

Figure 3. The PSF of the classical coronagraph with an on-axis source; using an $8*8$ replicated square pupil (left, simulation in the V-band) and using a 7-fold replicated hexagonal pupil (right, simulation at 600 nm).

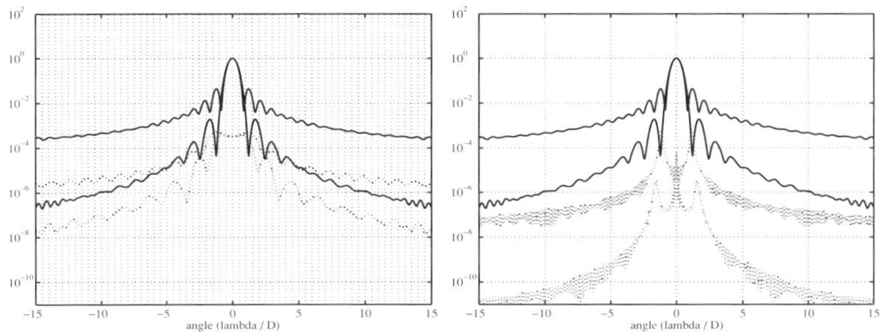

Figure 4. $8*8$ square pupil replication in the V-band. Cross sections through the PSF: left without and right with pupil replication. Solid and dotted lines: without and with star- and Lyot stop respectively. Both horizontal and diagonal cross sections are shown.

round star stop with a width of $2\lambda/D$, yielding a inner working angle of $1\lambda/D$ and a hard edged round Lyot stop. The horizontal and diagonal cross sections shown in figure 5 are now closer together. The absolute value of the attenuation is overall lower than in the example shown before because the number of replications is lower (7 instead of 64) and therefore the star image larger and the suppression of the star light less efficient.

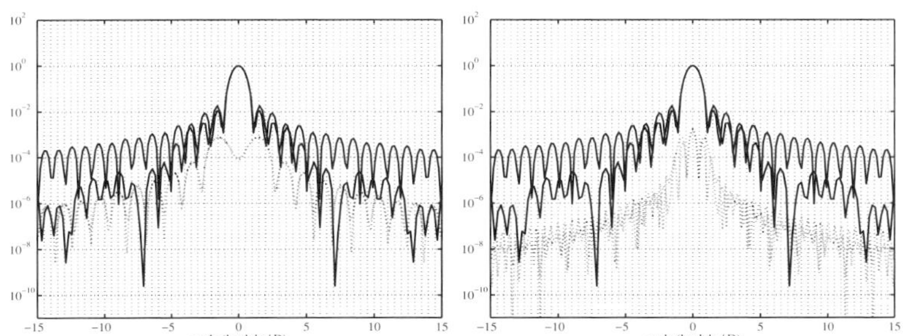

Figure 5. 7-fold hexagonal pupil replication at 600 nm. Cross sections through the PSF: left without and right with pupil replication. Solid and dotted lines: without and with star- and Lyot stop respectively. Both horizontal and diagonal cross sections are shown.

To make a first assessment of the influence of errors, we shifted one of the replicated hexagonal pupils (width 161 pixels) horizontally by 1 pixel (which is 0.6%) to the outside, leaving a gap along three hexagon sides. The result is shown in figure 6. We see the degradation in attenuation performance: the horizontal cross section does not get below 10^{-6} within $15\lambda/D$. In Riaud *et al.* (2005) simulation results are given for a pupil apodised system including pupil replication which show a similar value: about 10^{-6} at $15\lambda/D$ and 0.5% rms shift error. Even though the performance is lower, the system seems to be not exceptionally sensitive to such an error taking into account that we have introduced a relatively high error that would be for a $16mm$ replication size a gap of 100μ width and $2.8mm$ length.

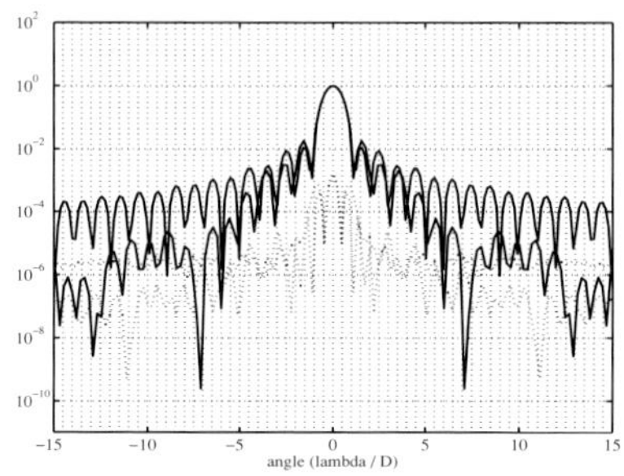

Figure 6. Results from a simulation of a classical coronagraph with pupil replication using a hexagonal pupil with 7 replications but with an error in the replication (see text).

3. Replication Optics

One of the parts of Pupil Replication that have not been assessed so far is the design for the optics that produce the replications of the pupil. A possible design for 2-fold replication of a rectangular input pupil is given here, see figure 7. The light entering from the top is split using a beam splitter. The two beams are then guided further using mirrors to be recombined. The last two mirrors are fixed to a translating element to change the beam separation. If we estimate the general efficiency of a mirror as 97%, the overall efficiency is about 88%. For a cascade of 6 replication optics to produce the 64 replications used in section 2 this would be 48%. Such optics can be put into solid glass (except for the last two mirrors). Furthermore, using the last two mirrors as shown, it is possible to eliminate a gap or overlap error. Finally, this system can be cascaded to obtain 2^n replicas.

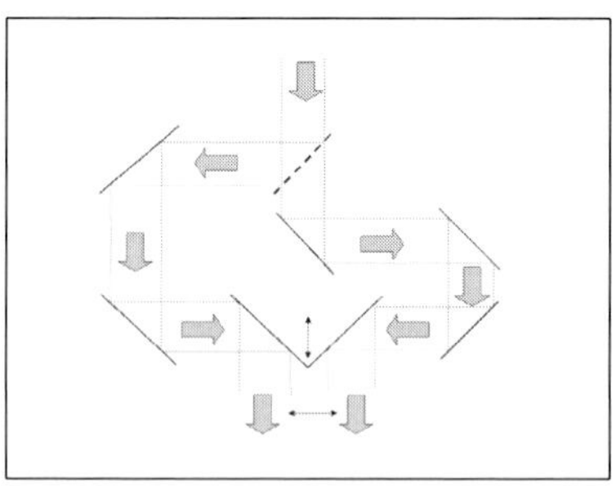

Figure 7. A design for the replication optics.

4. Experiment

An experiment was set up to test our theoretical model of pupil replication, see figure 8. Although pupil replication is meant to replicate the pupil in a telescope, we used for simplicity a laser beam instead of a telescope pupil to create the incoming wave front.

The basic set up is that of a Mach-Zehnder interferometer, but instead of combining the two interferometer beams at the exit pupil so that they overlap, the beams were sheared with respect to each other. This way two disk shaped monochrome wave fronts are created at the exit pupil. These represent the replicated pupils in a telescope with pupil replication. In the experiment the two beams are then both imaged onto a detector through a single lens, as would be the case when applied to a telescope.

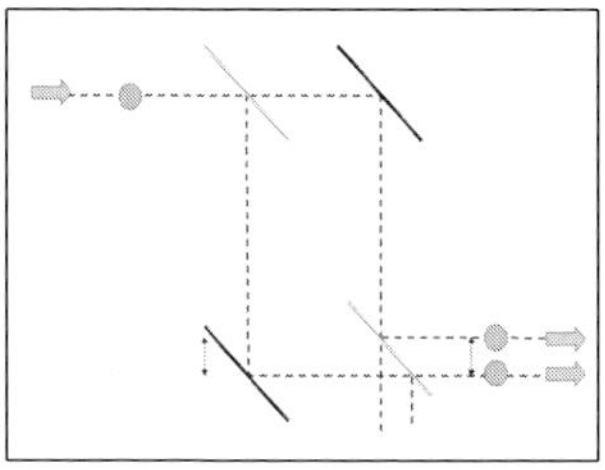

Figure 8. The experimental set up.

Figure 9 shows a typical set of images that were obtained. The top left picture shows part of the beam at the exit pupil, the poor quality of the beam is visible; the optics used were of moderate quality. The interference in the right part of this picture is due to the other pupil replica. On the top right the cross section through this pupil image is shown. The second and third rows show the unreplicated PSF of each of the beams. The fourth row shows the decreased size of the PSF in the replicated system (shown later in more detail). The last image is saturated to show the side lobes of the replicated PSF.

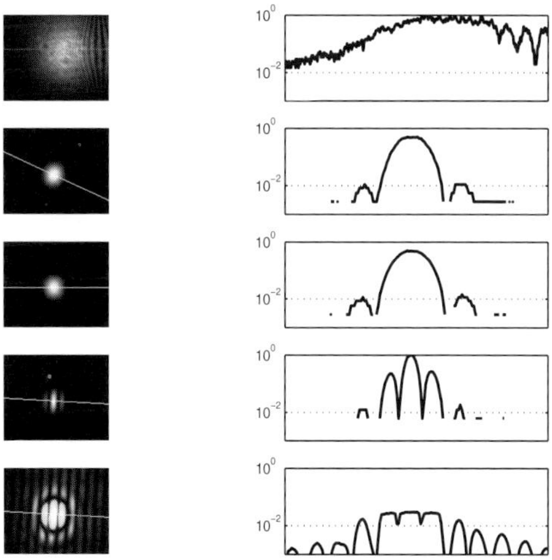

Figure 9. Experimental pictures (see text).

We also simulated the experimental set up and then compared the simulation results with the experimental results. We show here two cases at different values for the separation between the two pupil replicas (shear of the beams in the interferometer). In the figures only the cross sections through the two dimensional PSFs are shown.

In the left figure of figure 10 the separation is zero, that is the two discs are just touching each other. For the central and first order peaks the agreement between simulation and experiment is good; the third peaks come close to the lowest level of the CCD detector and the match here is less good; they are also subject to greater error.

On the right in figure 10 the separation is half the width of the beam, that is the two discs half overlap. Again, for the central and first order peaks the agreement between simulation and experiment is good, and for the third peaks less good. We also see that the central peak is now broader and the side lobes lower; this is because the separation is less than in the previous case.

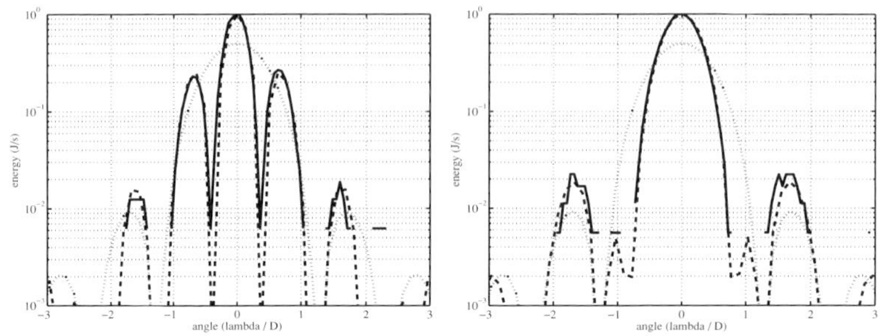

Figure 10. Left: Pupil replicas in contact (see figure 9): comparison with simulation results. Cross sections are shown of the simulated unreplicated PSF (dotted), of the simulated replicated PSF (dashed) and of the experimental replicated PSF (solid). Right: the same but now with half overlapping pupil replicas.

Finally some remarks about the experiment. The beam shape had a soft edge so that the experimental determination of the beam separation was difficult. In practise we adjusted the separation in the simulation to give a best fit of the PSFs. In our simulation we used a beam with a hard edge; a soft edged beam did not give better results. The saturated images we obtained gave more information on the higher order peaks, but a good fit with the simulation could not be obtained. The quality of the optics and other equipment used limited the results in this way. To get a better fit than shown, the optical defects would have to be taken into account in the simulation. Although imperfect, these experimental results give us confidence in our simulation.

Acknowledgements

We would like to acknowledge the funding by PPARC.

References

Greenaway, A. H., Spaan, F. H. P., & Mourai, V. 2005, *Astrophysical Journal Letters*, 618-2, L165.

Riaud, P., Mawet, D., & Absil, O. 2005, *Astrophysical Journal Letters*, 628-2, L81.

Discussion

We have shown simulation results for a pupil replicated classical coronagraph, which indicated the increase in performance when using a pupil replicated system. We also showed an experimental proof of principle and a design for the replication optics. These results overall bring actual implementation of pupil replication in a coronagraphic system a step nearer.

Direct Imaging of Exoplanets: Science & Techniques
Proceedings IAU Colloquium No. 200, 2005
C. Aime and F. Vakili, eds.
© 2006 International Astronomical Union
doi:10.1017/S1743921306009653

Transit, Astrometric, Coronagraphic and Interferometric Exo-planet Studies - Synergy and Complementarity

W. A. Traub[1], S. T. Ridgway[2], C. A. Beichman[3], K. J. Johnston[4], J. Kasting[5], and M. Shao[6]

[1]Jet Propulsion Lab, M/S 301-451, 4800 Oak Grove Dr., Pasadena, CA, 91109, USA
email: wtraub@jpl.nasa.gov

[2]NASA Headquarters, 300 E St SW, Washington, DC 20546
email: sridgway@hq.nasa.gov

[3]Michelson Science Center, California Inst. of Technology, Mail Code 100-22, 770 South Wilson Avenue, Pasadena, CA 91125, USA
email: chas@ipac.caltech.edu

[4]U.S. Naval Observatory, 3450 Massachusetts Ave, NW, Washington DC 20392, USA
email: kjj@astro.usno.navy.mil

[5]Department of Geosciences, 443 Deike, Penn State University, University Park, PA 16802, USA
email: kasting@essc.psu.edu

[6]Jet Propulsion Lab, M/S 301-486, 4800 Oak Grove Dr., Pasadena, CA, 91109, USA
email: mshao@huey.jpl.nasa.gov

Abstract. The goals of the Navigator Program at NASA are to find Earth-like planets around nearby stars, to determine if they are habitable, and to search for signs of life. Three strategic missions are planned to carry out this program: the Space Interferometer Mission Planetquest (SIM), the Terrestrial Planet Finder Coronagraph (TPF-C), and the Terrestrial Planet Finder Interferometer (TPF-I). These missions, along with the PI-class Kepler project, will each discover unique knowledge about extrasolar planets, synergistically building on the other missions.

Keywords. techniques: high angular resolution, interferometric; stars: planetary systems.

1. Introduction

Over 169 extrasolar planets have been discovered since 1996. These planets were detected using Doppler spectroscopy instruments that are sensitive to large, massive objects like our Solar System's giant planets. Other types of instruments will be needed to detect "terrestrial planets" like Venus, Earth, and Mars. NASA† and the US astronomy community currently plan four distinct missions to search for terrestrial planets. This set of missions is designed to return maximum scientific benefit, and in particular to tell us, beyond reasonable doubt, if nearby planets can be found that are habitable and show signs of life. These missions have a synergistic relationship – while each can stand alone, the scientific impact will be reduced substantially if fewer missions are launched.

† For simplicity, the following analysis refers to the current NASA suite of planned major exo-planet missions. This does not exclude other possible NASA missions, or changes in scope of those discussed. Of course, a variety of other missions (COROT, GAIA, DARWIN) capture some or even considerable elements of this science – NASA and ESA consider that DARWIN and TPF-I could be a single joint mission.

2. Missions to Detect and Characterize Extrasolar Planets

At the highest level, the logical relation among the methods of exo-planet detection and the study of Earth-like planets can be understood as follows. Radial velocity programs show, by detection of many planets, that there is indeed a science of exo-planets. Transit detections of large numbers of low-mass planets (Kepler) will show that there is indeed a science of exo-Earths (though the detected objects will be too distant for detailed study). SIM will survey nearby systems, and find exo-Earths that can be studied in detail. TPF-C and TPF-I will show that exo-Earth habitability can be studied.

Kepler will detect planets indirectly by measuring the dimming of a star when a planet transits in front of the stellar disk. Kepler will tell us the size of the planet, how close it is to the star, and the length of its year, using the measured depth, duration, and repeatability of dimming. It will do this with a single small telescope, staring at a distant field of stars (not nearby stars) for several years. Kepler will detect up to hundreds of terrestrial planets in the habitable zone, as well as determining the albedo, size, mass, and density (the latter two when augmented with planned RV data) of short period giant planets. Kepler will tell us for the first time whether terrestrial planets are common or rare – something about which we can only guess today.

SIM uses interferometers to measure stellar positions with respect to neighboring stars, and repeats these measurements over the full sky, many times over several years, to detect planets indirectly by measuring the astrometric wobble of a star. SIM will tell us the mass of each planet, its orbital distance, its orbital shape and orientation, and the length of its year. SIM will be most sensitive to planets that have periods of up to several years. SIM will tell us which nearby stars have such planets, as well as the relative numbers of heavy- and light-weight planets, down to Earth-mass range around some stars. If Kepler finds that Earths are plentiful, SIM might search a smaller number of systems to a deeper limit, while if Earths are rare, SIM might conduct a larger, shallower survey. SIM could search for gas giants in systems for which Kepler finds terrestrial planets.

TPF-C uses a single, large-aperture telescope and a coronagraph to suppress starlight. It will detect and characterize planets directly with a visible-wavelength snapshot image of a star-planet system. TPF-C will be able to tell us the visible brightness of each planet, its orbital distance, its orbital shape and orientation, the length of its year, and the amounts of life-related atmospheric gases (oxygen, water vapor, ozone, carbon dioxide, methane). In favorable cases, TPF-C potentially will be able to tell us the length of the planet's day, the amount of plant life on the surface, an estimate of ocean/land ratio, and a measure of cloud variability. TPF-C will be most sensitive to Earth-like planets in the habitable zone that are in orbital positions such that their day-lit portions are visible from the Earth. TPF-C will be able to observe planets that are, on average, smaller than the planets that Kepler, SIM, and radial-velocity techniques can reveal. TPF-C will be designed to search for habitability and signs of life on extrasolar planets.

TPF-I is the infrared analog to TPF-C and uses an array of four formation-flying large, cooled telescopes with interferometric nulling beam combination to suppress the star's emission. TPF-I will detect and characterize planets directly, using the planet's own infrared emission, independent of the orientation of the day-lit side. TPF-I will be able to tell us the infrared brightness of each planet, its orbital distance, its orbital shape and orientation, the length of its year, the amounts of life-related atmospheric gases (water vapor, ozone, carbon dioxide, methane, and nitrous oxide). If the planet's atmosphere is transparent enough, TPF-I may be able to tell us the length of the planet's day, an estimate of ocean/land ratio, and a measure of cloud variability. Together, TPF-C and TPFI are an ideal pair to search for habitability and signs of life on extrasolar planets.

3. Mission Connections and Synergy

The four missions, Kepler, SIM, TPF-C, and TPF-I, approach the search for terrestrial planets from different perspectives, and all perspectives are needed to establish mission requirements, to determine habitability and to search for signs of life. The rest of this discussion focusses on the missions which conduct detailed study of nearby exo-planets – SIM, TPF-C, and TPF-I.

3.1. *Stable orbit in habitable zone*

Each mission can measure an orbit and determine if it lies within the habitable zone (where the temperature may permit liquid water on the surface of the planet).

(1) A SIM detection of a terrestrial-mass planet could provide each TPF with targets to be characterized and the optimum times for observing them, thus increasing the TPF early-mission characterization yield.

(2) Where SIM finds a planet, of any mass, in almost any orbit, TPF-C and TPF-I will want to search as well, because we expect that planetary multiplicity may well be the rule (as in our Solar System).

(3) For cases where SIM detects a planet marginally, a TPF can verify or reject the detection. Such verifications can mutually enhance the significance of results from SIM and TPFs, no matter in what order the missions are carried out.

3.2. *Planet temperature and temperature variability due to distance changes*

A planet's effective temperature can be roughly estimated by noting its distance from its star, and by assuming a value for the albedo. TPF-I can observe directly the thermal infrared emission continuum at several wavelengths (i.e., infrared color), and use Planck's law to calculate the effective temperature. For a planet with a thick or cloudy atmosphere (like Venus), the surface temperature is different from the effective temperature, but might be inferred from a model of the atmosphere. With all three missions combined, the orbit, albedo, and greenhouse effect can be estimated, and the surface temperature as well as temperature fall-off with altitude can be determined cooperatively and more accurately than with any one mission alone.

Each mission alone can observe the degree to which the orbit is circular or elliptical, and thereby determine if the temperature is constant or varying. The measurement of a terrestrial planet's orbital eccentricity using combined missions (SIM plus TPF) can be much more accurate than from any one mission alone, because complementary sensitivity ranges in planet mass and distance from star combine favorably.

3.3. *Planet radius and albedo*

Planet radius and mass, or equivalently density, are very important for determining the type of planet (rocks, gas, ice, or combination), its habitability (solid surface or not; plate tectonics likely or not), and its history (formed inside or outside of ice line). SIM measures planet mass, from which we can estimate radius to within a factor of 2 (assuming a value of density, which in the Solar System spans a factor of 8). TPF-C measures visible brightness, which along with an estimate of albedo, can give a similarly rough estimate of radius. TPF-I measures infrared brightness and color temperature, which with Planck's law gives a more accurate planet radius. With SIM's mass, and one or both TPF brightness measurements, we can dramatically improve the estimate of planet radius.

The albedo is important because it controls the planet's effective temperature, which is related to its habitability. SIM and TPF-C combined can estimate possible pairs of values of radius and albedo, but cannot pick which pair is best. We can make a reasonable estimate of albedo by using TPF-C to measure the planet's color, then appealing to the planets in our Solar System to convert a color to an absolute albedo. By adding TPF-I measurements we can determine radius (above), then with brightness from TPF-C we can compute an accurate albedo, and therefore effective temperature and potential habitability.

3.4. *Planet mass and surface gravity*

SIM measures planet mass directly. TPF-C and TPF-I depend on SIM for planet mass. If TPF-C and TPF-I do not have a SIM value for planet mass, then they will use theory and Solar System examples to estimate masses. SIM plus TPF-C and TPF-I are needed to distinguish among rock-, ice-, and gas-dominated planet models, and to determine with confidence whether the planet could be habitable. SIM will typically be able to find nearby heavy-Earth planets, pointing the way for TPF to find Earth-mass and Mars-mass planets. We can understand these mass values qualitatively, as follows. SIM can most easily detect a stellar wobble when the planet is heavy, and the instrument's astrometric threshold is sufficiently small. TPF-C can most easily detect a planet when it has a large reflecting surface area per unit mass (i.e., smaller diameters), and the instrument contrast threshold is sufficiently small. TPF-I can have a smaller inner working angle than TPF-C, and so could see planets closer to the parent star. Several synergies result.

(1) TPF might initially avoid a case for which SIM finds a massive planet in an orbit that would make it impossible for a terrestrial planet to exist in the habitable zone.

(2) Since SIM can survey many stars relatively quickly, it can produce a list of stars where habitable-zone planets are strongly or even marginally detected, and hand off this list to TPF for confirmation, and characterization, thereby increasing the early-mission efficiency of the TPF missions.

(3) For SIM detections, TPF-C and TPF-I can confirm the detection and measure the planet's colors, to build up a catalog of extrasolar planet properties.

(4) For SIM detections that have low signal-to-noise ratios, TPF-C and TPF-I can improve the orbit parameters.

Surface gravity is important to habitability because massive, dense planets are more likely to have plate tectonics (a crucial factor in Earth's evolution), and retain an atmosphere (also crucial for Earth). The planet's surface gravity is calculated directly using mass from SIM and radius from TPF-C and TPF-I. Cooperative measurements are the only way to obtain this data.

3.5. *Atmosphere and surface composition*

The TPF missions are designed to measure a planet's color and spectra, from which we can determine the composition of the atmosphere and surface. For the atmosphere, TPF-C can measure water, molecular oxygen, ozone, the presence of clouds for a planet like the present Earth, and in addition it can measure carbon dioxide and methane for a planet like the early Earth or a giant planet. For the surface TPF-C can measure vegetation using the "red edge" effect. TPF-I will add to this suite of observations by measuring carbon dioxide, ozone, water, methane, and nitrous oxide using different spectroscopic features, and in general probing a different altitude range in the atmosphere. SIM is important to this interpretation because it provides planet mass, crucial to interpreting atmospheric measurements.

Both TPF-C and TPF-I are needed in order to determine whether a planet is habitable, because they make complementary observations, as follows (assuming an Earth-like planet). Ozone has a very strong infrared (TPF-I) feature, and a weak visible (TPF-C) one, so if ozone is abundant, both can be used to extract the abundance as well as the thermal structure of the atmosphere – if ozone is weak, only the TPF-I feature will be useable. Water as seen by TPF-C will be near the surface of the planet, but as seen by TPF-I it will be in the upper atmosphere; together both give a more complete picture of the atmosphere. Methane, carbon dioxide and nitrous oxide, in small amounts (as for the present-day Earth) can be detected by TPF-I. Methane and carbon dioxide, in large amounts (as for the early Earth), can be detected by TPF-C. For large amounts of methane or carbon dioxide, TPF-I will see mainly the amount in the upper atmosphere, but TPF-C will see mainly the amount in the lower atmosphere – both are needed for a complete picture. In addition to these overlap topics, only TPF-C can measure oxygen, vegetation, and the total column of air (Rayleigh scattering); likewise only TPF-I can measure the effective temperature.

Both TPF-C and TPF-I have water absorption features in their spectra, so if water vapor is present in the atmosphere, we will be able to measure it. However habitability requires liquid water on the surface, which in turn requires a solid surface as well as a temperature that permits the liquid state; only with the help of a value of mass from SIM will we be able to know the radius, and when TPF-I is launched, the temperature. To know whether liquid water is present on the surface of a planet, we need mass data from SIM, and spectroscopic water data from TPF-C or TPF-I.

Both TPF-C and TPF-I potentially can measure changes in color and the strengths of spectral features as the planet rotates. These changes can tell us the length of day on the planet, and can indicate the presence of large oceans or land masses (with different reflectivities or emissivities). Superposed on this time series of data could be random changes from weather patterns, possibly allowing the degree of variability of weather to be measured. The TPF missions can potentially measure variability of composition over time, which we know from our Earth to be an indicator of habitability.

3.6. *Other planets, comets, asteroids, and zodiacal dust*

All three missions can detect several planets around a star, within their ranges of sensitivity. Thus there may be a planet close to the star that SIM can detect, but is hidden from TPF-C. Likewise there may be a distant planet that TPF-C or TPF-I can detect, but has a period that is too long for SIM. For the more subtle issue of whether the planets have orbits in or out of the same plane, SIM will do the best job. In general, each of the three missions will detect some but not necessarily all of the planets that might be present in a system, so the combination will deliver a complete picture of what planets are present, their masses, their orbits, and how they are likely to influence each other over the age of the system, including co-planarity.

Comets and asteroids have very low masses compared to planets, so will not be detected by SIM, but because they have a large surface area (for their mass) they might be seen collectively, if their numbers are high, by TPF-C in reflected light and TPF-I in thermal emission. The combination of TPF-C (visible) and TPF-I (infrared) measurements can tell us about the average albedo and numbers of solid objects (comets and asteroids), as well as ground-up material (exo-zodiacal dust). The detection of sub-planet material will produce a picture of a planetary system's history and present interactions.

3.7. *Atmospheric and surface biomarkers*

The simultaneous presence of an oxidized species (like oxygen or ozone) and a reduced species (like methane) is considered to be a sign of non-equilibrium that can indicate indirectly the presence of life on a planet. The presence of a large amount of molecular oxygen, as on the present Earth, may also be an indirect sign of life. Since water is a prerequisite for life, as we consider it here, the presence of liquid water (indicated by water vapor and an appropriate temperature) is needed. Together these spectroscopically-detectable species are our best current set of indicators of life on a planet. These markers will be measured exclusively by TPF-C and TPF-I, but to know that we are observing an Earth-like planet will require SIM data on mass.

The red edge is a property of land plants and trees whereby they are very good reflectors of red light (just beyond the long-wavelength limit of our eyes). This is a useful feature for measuring plant cover on Earth. If extrasolar planets have developed plant life like that on Earth, and if the planet is bright, has few clouds, and a lot of vegetated land area, then we may use this feature to detect living vegetation.

3.8. *Beyond exo-planet studies – general astrophysics and technology*

Each mission, SIM, TPF-C, and TPF-I, offers a uniquely new and powerful capability in angular resolution, in deep and high contrast imaging, or in both. Vigorous programs of general astrophysics are expected to develop their own synergies between these missions and other observational platforms. Examples include stellar evolution (from star forming regions to late stars), cosmology (improved measurements of H_0), study of dark matter in the local galaxy group and beyond, and massive compact objects (by enabling microlensing studies from widely separated stations).

Each new mission builds on the technological heritage of earlier related missions, and in some cases that inheritance is essential. SIM has developed exquisitely precise methods of measuring the relative locations of optical elements in a train of optics and of accounting for wavefront uncertainties in optical systems, a technological heritage that will be beneficial to the TPFs. SIM will be the first large interferometer flown in space, and will be the first space instrument to demonstrate aperture synthesis imaging, paving the way for the next generation of dilute-aperture missions, including TPF-I.

4. Conclusions

The exploration of extrasolar planetary systems is a rich and diverse field. It calls for measurements with many kinds of instruments, as well as theoretical studies and numerical modeling. To discover and characterize extrasolar planets that are habitable and may show signs of life, and to be sure beyond a reasonable doubt that we can detect life, we need to measure the statistical distribution of planet diameters, the masses of nearby planets, and the spectra at visible and infrared wavelengths. Kepler, SIM, TPF-C, and TPF-I can carry out these measurements. In short, Kepler is needed for statistical information on the existence of terrestrial planets. SIM is needed for identification of nearby target systems and planets and measurement of planet masses. TFP-C and TPF-I are needed to evaluate the atmospheres for habitability. All are needed to characterize exo-planet systems. Not only does each mission provide its own compelling science, but together these missions form a coherent approach that will advance our understanding better than any single one by itself.

Acknowledgements

We are grateful to the members of the SIM Science Team, the TPF-C Science and Technology Definition Team, the TPF-I Science Working Group, and to Kepler PI William Borucki, for their contributions to this study.

Direct Imaging of Exoplanets: Science & Techniques
Proceedings IAU Colloquium No. 200, 2005
C. Aime and F. Vakili, eds.

© 2006 International Astronomical Union
doi:10.1017/S1743921306009665

Occulting Focal Plane Masks for Terrestrial Planet Finder Coronagraph: Design, Fabrication, Simulations and Test Results

Kunjithapatham Balasubramanian[1], Daniel J. Hoppe[1], Peter G. Halverson[1], Daniel W. Wilson[1], Pierre M. Echternach[1], Fang Shi[1], Andrew E. Lowman[1], Albert F. Niessner[1], John T. Trauger[1], and Stuart B. Shaklan[1]

[1] Jet Propulsion Laboratory, California Institute of Technology 4800 Oak Grove Drive, Mail Stop 301-451, Pasadena, CA 91109-8099, USA
email: kbala@jpl.nasa.gov

Abstract. Occulting focal plane masks for the Terrestrial Planet Finder Coronagraph (TPF-C) could be designed with continuous gray scale profile of the occulting pattern such as $1 - \text{sinc}^2$ on a suitable material or with micron-scale binary transparent and opaque structures of metallic pattern on glass. We have designed, fabricated and tested both kinds of masks. The fundamental characteristics of such masks and initial test results from the High Contrast Imaging Test bed (HCIT) at JPL are presented.

Keywords. instrumentation: high angular resolution, instrumentation: adaptive optics, instrumentation: miscellaneous.

1. Introduction

Occulting focal plane masks operate in conjunction with a Lyot stop to suppress the starlight in a coronagraph (Kuchner & Spergel 2003). In the limit of Fourier Optics, these masks can easily be designed to reject the starlight in the regions of interest to below 10^{-10} (-100 dB), the nominal Terrestrial Planet Finder (TPF) requirement (Kuchner & Traub 2002). Band-limited gray scale and binary masks have been designed, fabricated and tested to meet such a requirement. In this paper, we describe the design, fabrication and test results.

2. HEBS Mask Design and Fabrication

Analog gray scale masks based on a special high energy electron beam sensitive (HEBS) glass can be designed to work as focal plane occulting masks to achieve the required suppression of star light through iterative development of the material, design, and fabrication. Wilson *et al.* (2003) and Halverson *et al.* (2005) have reported on the fabrication and characteristics of HEBS masks for TPF applications. Early experimental results with HEBS masks were also reported by Trauger *et al.* (2004). Band limited focal plane masks with $1 - \text{sinc}^2$ transmission profiles discussed by Kuchner & Traub (2002) could be designed and fabricated on HEBS glass with electron beam exposure techniques. Such masks designed to operate in the high contrast imaging test bed at JPL were fabricated at the Micro Devices Lab (MDL) in JPL; transmission image and cross section profile of one such mask fabricated and tested are shown in Figure 2. Several masks have been

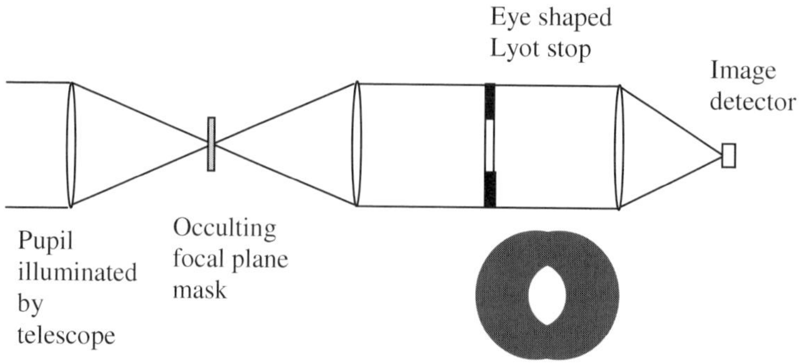

Figure 1. Conceptual drawing of a coronagraph with focal plane mask and Lyot stop.

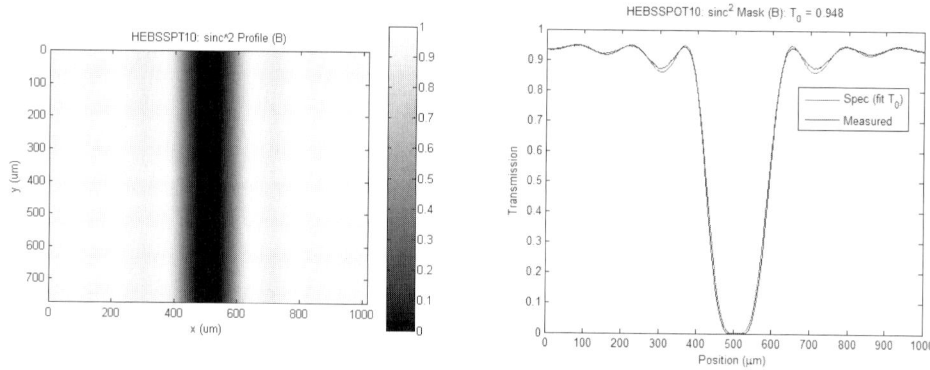

Figure 2. Typical transmission image (left panel) and profile (right panel) of $1 - \mathrm{sinc}^2$ HEBS mask profile for 785 nm.

fabricated with profiles given by:

$$T(x) = T_0 \left[1 - \left(\frac{\sin(\pi x/w)}{\pi x/w} \right)^2 \right]^2 . \tag{2.1}$$

The width parameter w for one of the masks is 141.9 μm, $T(x) = 0.5T_0$ at $x = 81.3$ μm. $T(x)$ was set to 1.0 for all $|x| > 1.20$ mm; $T_0 = 0.90$. A similar profile with $w = 157.7$ μm and $T(x) = 0.5T_0$ at $x = 90.3$ μm has also been fabricated. These profiles were designed for $F = f/D = 28.55$ and $\lambda = 785$ nm.

3. High Energy Beam sensitive Glass (HEBS) Mask Fabrication

The HEBS glass as produced by Canyon Materials Inc.† consists of a 10 μm thick high energy electron beam sensitive layer on both sides of a 90 mil (2.29 mm) thick glass plate. The sensitive layer gets darkened to various levels of optical density when exposed to high energy electron beam irradiation at chosen doses. Specific transmission profiles can be obtained on such a glass when the electron beam is scanned on the glass to dosage levels preprogrammed to produce the darkness of the required mask transmission profile (Wilson *et al.* 2003; Halverson *et al.* 2005). Our masks were fabricated with a high precision electron beam lithography system (JEOL 9300FSZ Electron Beam Lithography

† Canyon Materials, Inc. San Diego, CA http://canyonmaterials.com

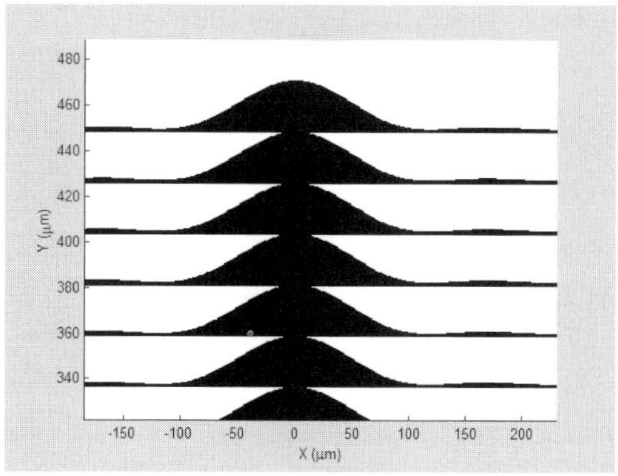

Figure 3. Detailed view of binary $1 - \mathrm{sinc}^2$ mask pattern.

at Micro Devices Lab) with 100 kV beam at 1 nA current. Calibration runs guide the choice of e-beam exposure parameters such as current, exposure time, scan step etc to achieve a desired optical density at a specific location on a mask. Effects of convolution of the e-beam spot size, scan steps and proximity effects are taken in account in calibrating the exposure and scan parameters.

4. Binary Mask Design and Fabrication

In the case of binary masks, an analog band-limited function such as $\sin^2$ or $1 - \mathrm{sinc}^2$ is sampled and a series of apertures is created that mimics the throughput of the analog mask. Sampled and continuous $\sin^2$ and sinc^2 type masks have been designed and fabricated. The mask is referred to as binary since it consists only of perfectly blocking metal and totally transmitting empty apertures (0/1). The binary masks are chosen to be periodic in the non-search direction (y), with a period of $F\lambda$, where $F = f/D$ is the focal length/diameter ratio of the telescope and λ is the operating wavelength. To match the performance of the above HEBS mask designs, binary masks were designed for the same F and λ. Figure 3 shows the design of a binary mask made to mimic the $1 - \mathrm{sinc}^2$ transmission profile of the HEBS mask. The binary mask patterns are generated by e-beam lithography and lift off process with nominally 150 nm thick aluminum film coated by evaporation on fused silica glass.

5. High Contrast Imaging Testbed (HCIT)

The HEBS and binary masks described above were tested for performance in the HCIT at JPL. The test bed set up has been described by Trauger *et al.* (2004) and in the proceedings of this conference (IAUC200). A recent addition to the test bed optics is a calcite polarizer before the final image detector. This polarizer separates the two orthogonally polarized images into slightly offset locations at the final image plane. The deformable mirror (DM) can be optimized to minimize the speckles in either polarization independently, thus estimating the best contrast achievable in one polarization while allowing the other to float. While the HEBS gray scale mask does not exhibit polarization sensitivity, the binary mask is inherently sensitive to polarization.

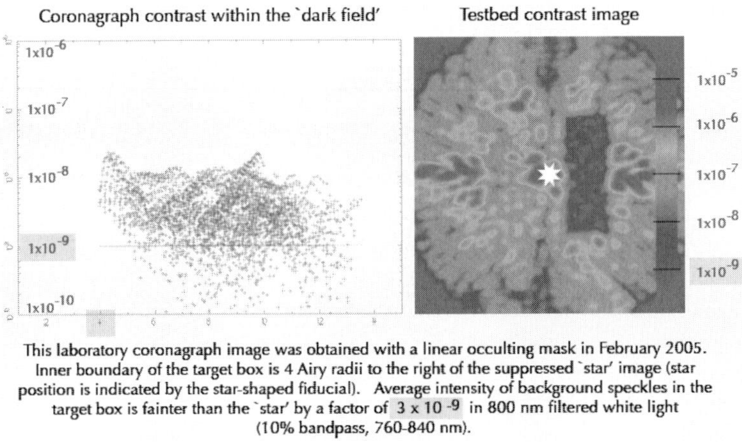

Figure 4. HEBS mask HCIT contrast vs. Airy radius in the target box of 4 to 10 λ/D in x and -10 to $+10\lambda/D$ in y.

6. Experimental Results

Contrast images were obtained with a HEBS $1 - \mathrm{sinc}^2$ linear occulting mask with 785 nm laser source in the HCIT. An average contrast of 9×10^{-10} was achieved with a stability of about 5×10^{-11} per hour. Similar experiments with the same mask with filtered white light source with 10% band pass centered at 800 nm achieved average contrast of 3×10^{-9} in the target box. Figure 4 shows the contrast image and plot for this result. Latest experiments have yielded results in the 10^{-10} scale as presented by Trauger *et al.* at this conference. The binary $1 - \mathrm{sinc}^2$ mask was tested in the HCIT under similar conditions as with the HEBS mask. With incident 785 nm unpolarized laser light and split into orthogonal polarizations after the mask and before the final image, a contrast of 4×10^{-9} was obtained in the target box for a chosen polarization that was nulled with the DM while the other polarization was left floating (Figure 5). The experiment in the HCIT collected two images (Figure 5, left) corresponding to the two orthogonal polarization states after the calcite crystal. Theoretical simulations with Fresnel diffraction models compare closely with the experimental results; observed differences between theory and experiment particularly with respect to the polarization behavior of binary masks may be explained as due to model assumptions and experimental conditions which are being further analyzed.

Acknowledgements

This work was performed at the Jet Propulsion Laboratory, California Institute of Technology, under a contract with the National Aeronautics and Space Administration. Reference herein to any specific commercial product, process or service by trade name, trademark, manufacturer, or otherwise, does not constitute or imply its endorsement by the United States Government or the Jet Propulsion Laboratory, California Institute of Technology. The authors thank Brian Kern of the Caltech Jet Propulsion Laboratory for his help with reviewing and editing this paper in LaTeX format to suit publication.

References

Kuchner, M. J. & Spergel, D. N. 2003, "Notch Filter Masks: Practical Image Masks for Planet-Finding Coronagraphs," *ApJ* 594, 617

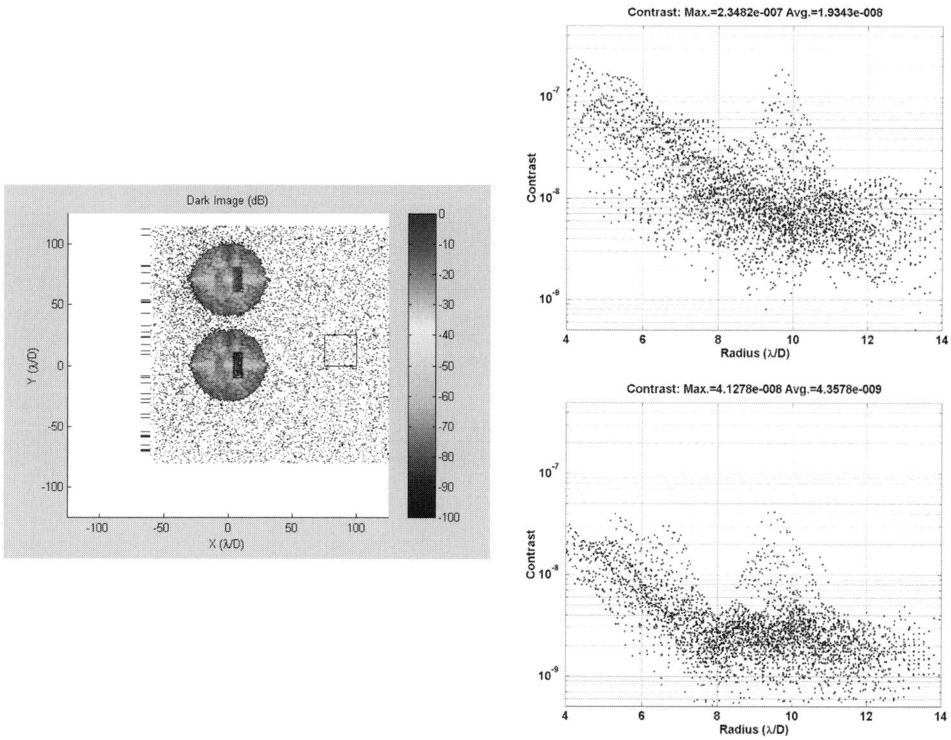

Figure 5. LEFT: Final experimental images showing two polarizations of binary mask. The experimentally measured average contrast in the polarization being nulled (bottom of image) is 4.3×10^{-9}, whereas the average contrast in the orthogonal field not nulled (top of image) by DM is 1.9×10^{-8}. TOP RIGHT: Contrast plot of polarization *not being nulled* by DM settings. BOTTOM RIGHT: Contrast plot of polarization *being nulled* by DM.

Kuchner, M. J. & Traub, W. 2002, "A coronagraph with a band-limited mask for finding terrestrial planets," *ApJ* 570, 900

Wilson, D. W. *et al.* 2003, "Eclipse apodization: realization of occulting spots and Lyot masks," *Proc. SPIE* 4860, 361

Halverson, P. G. *et al.* 2005, "Measurement of phase and optical density of TPF coronagraph occulting mask materials," *Proc. SPIE* 5905

Trauger, J. T. *et al.* 2004, "Coronagraph contrast demonstrations with the High Contrast Imaging Testbed," *Proc. SPIE* 5487, 133. Also latest results were presented in this IAU colloquium.

Wilson, D. W. *et al.* 2005, "Electron-beam lithography for micro- and nano-optical applications," *Proc. SPIE* 5720, 68

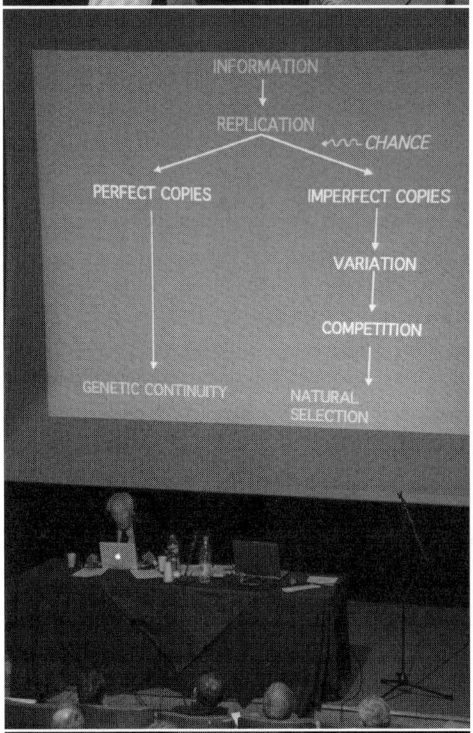

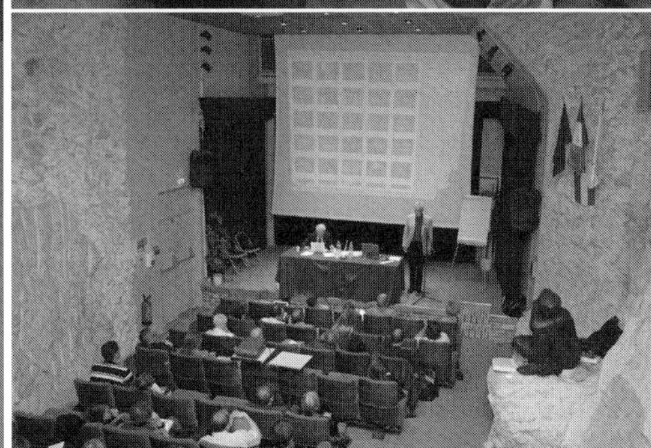

All photographs: Laurent Thareau [l.thareau@free.fr].

Direct Imaging of Exoplanets: Science & Techniques
Proceedings IAU Colloquium No. 200, 2005
C. Aime and F. Vakili, eds.
© 2006 International Astronomical Union
doi:10.1017/S1743921306009677

Towards an Optimal Observation Strategy for Direct Detection Exoplanet Searches

Jonathan W. Arenberg and Todd E. Schuman

Northrop Grumman Space Technology, One Space Park Drive, Redondo Beach, CA, 90278, USA
email: jon.arenberg@ngc.com

Abstract. This paper presents a look at ongoing investigations into the strategies needed to characterize the orbits of observed exoplanets in an optimal manner. Also introduced is an efficient manner of the selection of possible orbits consistent with the observation of an exoplanet. Specifically, this work looks at the ability to conclude that a planet's orbit is resident in the habitable zone (HZ) on the basis of two sequential observations. The role of the range of possible semi-major axes is introduced.

Keywords. methods: statistical, methods: numerical, methods: data analysis,planetary systems: methods.

1. Introduction

This paper is a status report on our investigations regarding the characterization of exoplanetary orbits.

This method is based on a Monte Carlo technique to determine the universe of possible planets consistent with two sequential observations. Eventually, this method will lead to the optimization of the interval between observations. So that the minimum number of observations are required to determine if an orbit is or is not permanently resident in a given zone, typically the so-called habitable zone, where liquid water can exist on the planet's surface.

Our method of investigating the evolution of knowledge of the orbit of a planet is based on sequentially decimating a set that initially represents the universe of possible planets. The universe of possible planets of interest is called Ω. Ω is represented as finite set of orbital elements used to calculate the planet's apparent positions on the plane of the sky. For an observation with a given outcome (planet seen or not seen), only those members of Ω that have the same outcome as the observation are kept. The edited or decimated set is denoted Ω_1. The subscript $_i$ indicating that all members of Ω_1 are consistent with the first observation. The positions of the planets that are members of Ω_1 are projected forward in time, at which time a second observation is made and the set is processed by the results of the second observation to yield Ω_2 and so on.

To produce the individual members of Ω, denoted ω_i, a Monte Carlo method is used to select the various parameters of the orbit, with the orbital inclination, θ, rotation of the orbital plane ϕ and true anomaly α are selected uniformly from the interval $[0,2\pi)$. The eccentricity ε and semi-major axis a are selected to be uniform for their apastrons and periastrons, which result in a and ε being distributed in the characteristic camur or bent delta shape.

2. Random Orbits Consistent with an Observation, Determination of the ω_i

Simple random generation of numbers to get elements that are consistent with an observation of Ω_1 will result in very few close matches to a given observation of a single apparent radial location, resulting in unacceptable computational inefficiency.

The system of equations for x, y, and z for an ellipse of semi-major axis a and eccentricity ε whose orbital plane is inclined to the x-axis by θ and rotated about the z-axis by ϕ is

$$x(\alpha) = -\sin\theta\cos\phi\left(\frac{a(1-\varepsilon^2)}{1+\varepsilon\cos\alpha}\sin\alpha\right) - \sin\phi\left(\frac{a(1-\varepsilon^2)}{1+\varepsilon\cos\alpha}\cos\alpha\right), \tag{2.1}$$

$$y(\alpha) = -\sin\theta\sin\phi\left(\frac{a(1-\varepsilon^2)}{1+\varepsilon\cos\alpha}\sin\alpha\right) + \cos\phi\left(\frac{a(1-\varepsilon^2)}{1+\varepsilon\cos\alpha}\cos\alpha\right), \tag{2.2}$$

$$z(\alpha) = \cos\theta\left(\frac{a(1-\varepsilon^2)}{1+\varepsilon\cos\alpha}\sin\alpha\right). \tag{2.3}$$

The process used to generate the subset Ω_1 for case 2 involves the use of the observed apparent position of the planet, the residency requirement, and Equations (2.1), (2.2), and (2.3). The planet's position is described by five parameters: θ, ϕ, α, a, and ε. A single observation does not provide enough information to solve for all five parameters, resulting in an undetermined system. To solve the system, some parameters are chosen randomly within the constraints established by the observation and the residency requirement. The locus of possible values for the remaining variables, which cannot be determined, is a volume in parameter space. It is this space that is sampled via Monte Carlo methods to yield each member of Ω_1.

The apparent position of the planet gives a measurement for y and z. The apparent radial separation r_{app} is given by

$$r_{app} = \sqrt{y^2 + z^2}. \tag{2.4}$$

The true separation ρ is

$$\rho = \sqrt{x^2 + y^2 + z^2}. \tag{2.5}$$

For the planet to be of interest, ρ must satisfy the residency requirement and is restricted to the interval [IHZ, OHZ]. Since the magnitude of the third coordinate is still unknown, the true separation must be at least as large as the apparent position, so

$$\rho \geqslant r_{app}. \tag{2.6}$$

Combining the condition (2.6) with the residency requirement yields a set of constraints on the random values for ρ

$$\rho \geqslant \max(IHZ, r_{app}) \tag{2.7}$$

and

$$\rho \leqslant OHZ. \tag{2.8}$$

Once ρ is chosen randomly and satisfying (2.7) and (2.8), x can be calculated using Equations (2.4) and (2.5) to solve for x, resulting in

$$x = \pm\sqrt{\rho^2 - r_{app}^2}. \tag{2.9}$$

The positive or negative root is chosen randomly to ensure complete sampling of the parameter space.

Once values have been established for {x, y, z}, it is possible to generate orbital parameters for orbits that pass through this point. Two additional random variables are selected, r_1 and r_2—representing, respectively, the periastron and apastron of the possible planet's orbit. The periastron r_1 is selected from the interval [IHZ, ρ] and the apastron r_2 from [ρ, OHZ].

In order to evaluate Equations (2.1), (2.2) and (2.3), values for a and ε are needed. Expressed in terms of r_1 and r_2

$$a = \frac{1}{2}(r_1 + r_2) \tag{2.10}$$

and

$$\varepsilon = \frac{r_1 - r_2}{r_1 + r_2}. \tag{2.11}$$

Substituting Equations (2.10) and (2.11) into Equations (2.1), (2.2), and (2.3) yields a system of three equations and three unknowns (the remaining orbital parameters θ, ϕ, and α). This system is solved numerically for θ, ϕ, and α. The process is then repeated as many times as desired to create a suitably large Monte Carlo set for Ω to be used in the analysis of case 2.

The r_1 and r_2 variables are the randomized representations of the d and D. That is, we want all orbits that lie between d and D, so the nearest point r_1 is selected to be greater than d and less than r_2 and r_2 is less than D.

3. Evolution of Knowledge of Orbit with Two Observations

A case where two observations of a planet is considered as an example of this method. Both observations have an apparent position within the HZ, since observation outside would remove a planet from consideration. The question at hand is how likely is the true orbit to lie permanently within the HZ? Is this probability strongly dependent on the universe of possible planets considered. This universe (parent set) can be characterized by the magnitude of the apastron, since all closed obits are considered.

For this current example case, a clone of the sun is the parent star, and for this work the HZ is on the interval of [0.62 AU,1.52 AU]. It should be noted that the general method works for any interval, as is useful for restricting the description of the orbit of any planet to a specific interval, not just terrestrial like planets.

The initial observation of the planet occurs at an apparent separation of 1 AU. From this apparent position the set Ω_1 is generated. The right hand column shows traces of all the orbits consistent with this observation. At time t later, the system is again observed and Ω_1 is filtered to give the set of possible orbits consistent with the first and second observations. This set can then be plotted to identify the fraction of possible orbits that are within the habitable zone, those inside the camur delta. The sets Ω_1 and $\Omega_2$are calculated for various values of the maximum apastron in the parent set Ω. In this work, the apastrons considered are 2 and 5 AU.

The plot shown in Figure 1 represents the evolution of knowledge of planets orbit based observation. When no observations have been made, the location of the orbit in the (a,ε) plane is any closed orbit with an apastron of less than 5 AU, this set is represented by the small solid (black) dots. After the first observation, Ω_1, is indicated by the small open circles. Ω_1 is nearly as widely distributed as Ω with the exception that some very small a orbits have been ruled out. On the second observation, Ω_2 the remaining orbits, show a degree of organization, as indicated by the larger solid (red) dots. All of the orbits with (a,ε) consistent with being in the HZ are the ones inside the region defined by the

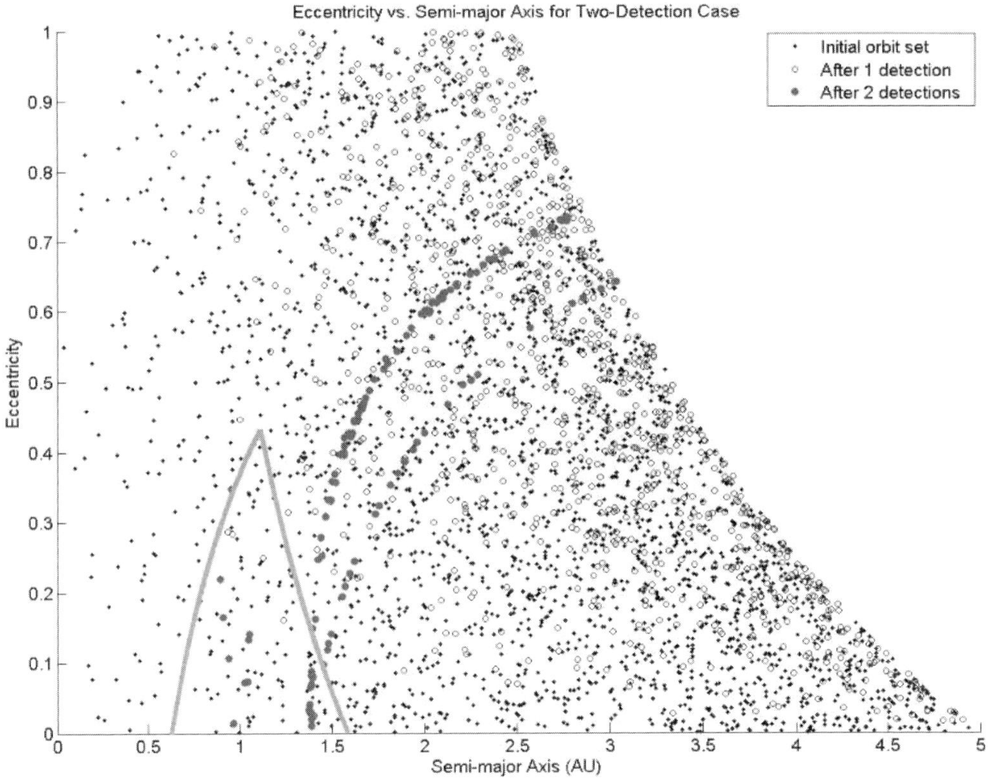

Figure 1. Plot of Ω, Ω_1 and Ω_2 in the (a,ϵ) plane.

triangle in the lower left quadrant of the plot. Thus, while two observations can tell some data of the planets orbit, it is insufficient to claim residency in the HZ.

Future work is directed at developing algorithms to time the observations to make the determination of in or out of the HZ as efficient as possible.

Acknowledgements

This work was performed by Northrop Grumman on internal funds. The authors are grateful for the support.

Direct Imaging of Exoplanets: Science & Techniques
Proceedings IAU Colloquium No. 200, 2005
C. Aime and F. Vakili, eds.
© 2006 International Astronomical Union
doi:10.1017/S1743921306009689

Towards 10^{10} Contrast for NASA's Terrestrial Planet Finder Mission: Demonstration of High Contrast in a Shaped-Pupil Coronagraph at Princeton

Ruslan Belikov[1]†, James Beall[2], Michael Carr[1], Amir Give'on[1], Jason Kay[1], Taofik Kolade[1], Michael Littman[1], Frank Mycroft[1], Laurent Pueyo[1], Robert J. Vanderbei[1], and N. Jeremy Kasdin[1]

[1]Princeton University, Princeton, NJ
[2]National Institute of Standards and Technology, Boulder, CO

Abstract. Our group at Princeton University is developing the Shaped-Pupil Coronagraph (SPC) as a solution to the high contrast imaging requirements for NASA's Terrestrial Planet Finder mission. At the heart of the SPC is a specially designed shaped mask at the pupil plane, and a star occulter at the image plane. We report a measurement of 10^5 contrast at 4 λ/D and 10^6 at 7 λ/D, with no adaptive optics corrections. This contrast is maintained at laser wavelengths of 532, 594, and 632nm, and for broadband light from at least 550nm to 750nm. The contrast is almost certainly limited by wavefront aberrations in the mirrors. Indeed, the level and general structure of the speckles in the high contrast region is consistent with statistical simulations of our optics.

Keywords. space vehicles: instruments, methods: laboratory, planets and satellites: general, stars: planetary systems, instrumentation: high angular resolution, instrumentation: adaptive optics, astrobiology, extraterrestrial intelligence.

1. Introduction

NASA's Terrestrial Planet Finder Coronagraph (TPF-C) is expected to be the first instrument to directly image and characterize an Earth-like extrasolar planet, sometime in the next decade. It will consist of a space telescope with a high-contrast imaging system required to have 10^{10} contrast at inner working angles of 4 λ/D in visible wavelengths for the TPF-C 8x3.5m primary. Many groups are prototyping extreme coronagraph concepts with the goal of demonstrating high contrast in the lab (see, for example, Trauger 2005, Chakraborty 2004, Crepp 2005, Wilhelmsen 2004). Most notably, the group at the Jet Propulsion Laboratory (JPL, Trauger 2005) have demonstrated a contrast level of 10^9 in monochromatic light with a variant of the conventional Lyot coronagraph and an adaptive optics (AO) wavefront control system in a vacuum chamber. Over the past 5 years, our group at Princeton University has been pioneering the Shaped-Pupil Coronagraph (SPC) as an alternative to the more traditional methods. We favor the SPC because it is very simple to manufacture, is inherently broadband, and has minimal sensitivity to aberrations and off-axis effects (Green 2004), while maintaining competitive performance with respect to other approaches. More details can be found in our recent paper (Kasdin

† Present address: Princeton University, MAE Dept., Olden St, Princeton, NJ08544 email: rbelikov@princeton.edu

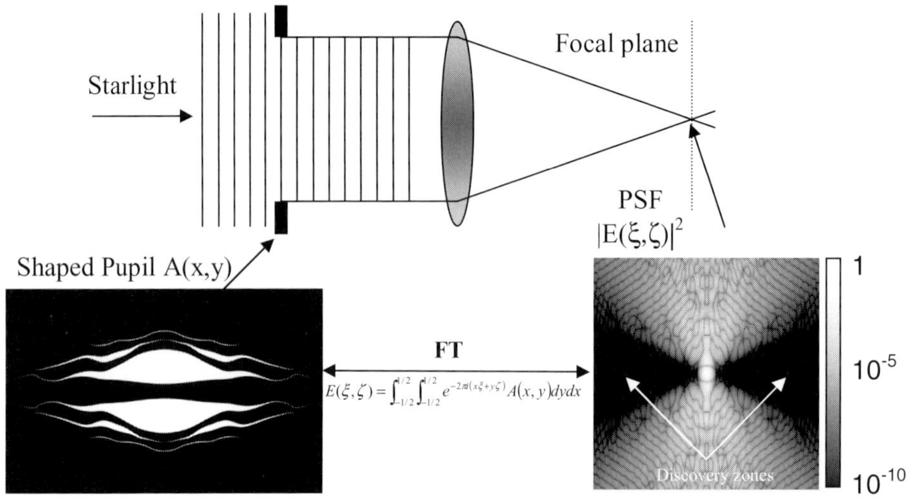

Figure 1. The Shaped Pupil Coronagraph concept. Shown are our chosen shaped pupil design (nicknamed "ripple") and the corresponding PSF.

2005), which contains a general overview of the design, manufacturing, and experimental results.

In this paper, we focus on our laboratory efforts and results. In section 2, we give a brief overview of the shaped pupil concept and describe our lab and the optical setup. In section 3, we describe our results to date, most notably the demonstration of 10^5 contrast at 4 λ/D, which gradually drops to 10^6 level at 7 λ/D, and continues dropping as working angle is increased. It should be noted that these results were achieved in the absence of any wavefront control system.

2. Laboratory Setup

At the heart of the SPC is an appropriately shaped mask at the pupil plane, as shown in Figure 1. With a shaped pupil in place, the point spread function (PSF) of the idealized system is no longer an airy pattern, but the Fourier Transform of the shaped pupil. (More precisely, the PSF is the magnitude squared of the Fourier transform of the indicator function of the shaped pupil). One can design this shaped pupil to yield a concentrated PSF which has high contrast regions, or discovery zones, where a dim planet can be found (Kasdin 2003). After some design effort (Vanderbei 2003), we converged on our currently favorite design shown in Figure 1, nicknamed "ripple", based on what we think is a good tradeoff between throughput and extent of the discovery zone, as well as the simplicity of manufacture. The PSF corresponding to the ripple design has the required 10^{10} contrast ratio within 2 discovery zones, each having a 45 degree opening and a working angle range of 4 to 100 λ/D, where D is the long axis of the ripple mask. The Airy throughput is 30%. (Throughout this paper, by "contrast" we mean the ratio of the peak PSF intensity to the discovery zone intensity, and use the term "suppression" to mean the inverse of contrast. "Airy throughput" is the energy that falls into the main lobe of the PSF measured relative to the total energy conveyed through a fully open aperture.) We have manufactured a mask based on the ripple design at the National Institute of Standards and Technology in Boulder, CO, out of a 76mm diameter double-side polished silicon wafer (Kasdin 2005).

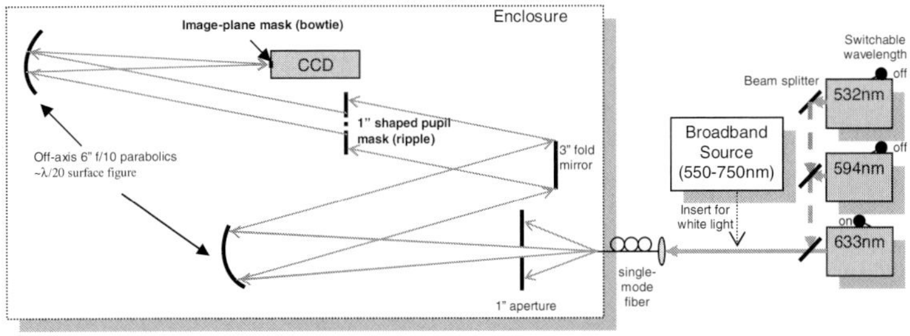

Figure 2. Laboratory setup.

Our lab consists of a 1.2x5m vibration-isolated optical bench with an enclosure to eliminate thermal convection, particulate contamination, and stray light. We also use HEPA filters to further reduce particulate contamination and an AC system to stabilize the temperature and humidity. The optical bench setup is shown in Figure 2. It is meant to recreate the simple schematic of Figure 1 with the goal of imaging the PSF of the shaped pupil and measuring the contrast in the discovery zones. Several sources were set up as inputs to our system: a 5mW doubled YAG laser pointer head at 532nm; 2 He-Ne lasers at 594nm (2mW), and 632nm (5mW) respectively; and a Tungsten lamp white light source with color temperature of 3200K. These sources are coupled into a single mode fiber which enters the enclosure. In the enclosure the fiber serves as a point light source to simulate a star. The diverging Gaussian beam from the fiber first goes through a 1" aperture, gets collimated by a 6" off-axis parabola, redirected by a 3" flat fold mirror, and finally illuminates the shaped pupil. The collimated beam is now 3" in diameter (which is determined by the 1" aperture and is matched to the 3" fold mirror). The beam illuminating the shaped pupil has approximately a flat wavefront and a uniform amplitude profile, especially over the 1" area of the shaped pupil. The beam goes through the shaped pupil, and is focused onto a cooled CCD by a second off-axis parabola. The CCD then records the PSF, as shown in Figure 3a, with exposures typically lasting 1 second. Due to the high 10^{10} dynamic range of the PSF, the bright core of the PSF will overwhelm the CCD circuitry and scatter inside the camera, creating undesirable artifacts in the discovery zones. To avoid these artifacts, we manufactured an image plane mask to block all of the PSF except for the discovery zones. This mask, which we call "bowtie", sits 1-2mm away from the sensitive area of the CCD chip, which is still essentially at focus. A simulation of the portion of the PSF blocked by the bowtie is shown in Figure 3b.

Since we cannot measure the intensity of the core with the bowtie in place, we have to estimate the PSF core intensity some other way in order to properly calibrate the contrast measurements. There are two methods for doing this. In the first, direct method, we take an unsaturated exposure of the PSF core (no bowtie, low power and short integration time), and use that measurement to extrapolate the intensity of the PSF core in full power exposures, where the core is blocked by the bowtie. An alternative, easier method uses the portions of the PSF beyond the outer working angle (OWA) of 100 λ/D instead of the main core. This "OWA structure", shown in 3a and b, corresponds to the portion of the ideal PSF beyond the discovery zone, and thus has a known shape and contrast (which peaks at about 10^6). Note the good correspondence of the theoretical and experimental OWA structures in Figure 3b. Measurements of the dark zone contrast can thus be directly obtained by using the OWA structure as a reference, as long as it is not saturated.

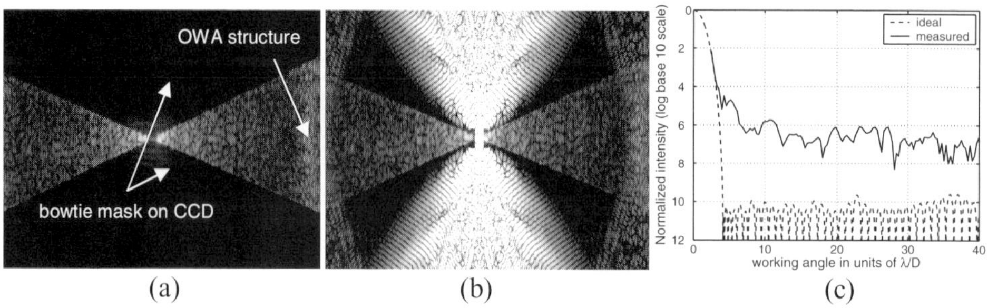

Figure 3. (a) Logarithmically stretched CCD image of the discovery zones, at 632nm, showing a region from roughly -100 λ/D to 120 λ/D (b) Same as (a), but with a superimposed simulation of what the bowtie is blocking. (c) A horizontal slice through the image in (a) from 0 to 40 λ/D.

3. Results

As can be seen in Figure 3, the measured discovery zones contain a lot of speckle. A horizontal slice through the center of the image is shown in Figure 3c. The suppression (inverse of contrast) is about 3x10^{-5} at 4 λ/D, gradually dropping to slightly higher than 10^{-6} after 7 λ/D. (By 100 λ/D, it is down to almost 10^{-8}). These values can change if the mask is laterally shifted, because in that case slightly different areas of the mirrors with different aberrations are effectively being used. There are some positions of the shaped pupil where the suppression drops to 10^{-5} at $4\lambda/D$, for example, but at the expense of worsening the contrast at 7 λ/D.

To determine whether aberrations in the optics are indeed the dominant cause of the speckle in our images, we conducted a series of tests. First, we noted that the speckle pattern is quite stable. Exposures taken seconds apart agree to at least 1 part in 100, and exposures taken an hour apart agree to about 1 part in 10. (And this discrepancy may simply be due to a technical issue such as an instability in laser power). Furthermore, the speckle pattern remains qualitatively the same over days or even weeks until the setup is perturbed somehow. Thus, aberrations due to air or mechanical instabilities are not dominant. We then looked at the effects of transversely shifting the shaped pupil. When it is shifted by as little as 1mm, the entire speckle pattern changes. As mentioned above, what we are really varying by transversely shifting the shaped pupil is the region of the mirrors being used. Since all the speckles change, they must be due to the mirror aberrations. The only part of the image that does not change is the OWA structure, because it in fact is a function of the shaped pupil itself and not the mirror aberrations.

Finally, we conducted a simple statistical simulation of the speckle for our optics, based on their peak-to-valley surface figure rating of $\lambda/20$. The simulation consisted of the following steps: (1) create a random phase variation across the plane of the shaped pupil with power spectral density of $1/f^{\alpha}$, where α is a free parameter to be fitted; (2) scale this phase error to correspond to the combined surface figure of the 3 mirrors; (3) multiply by the shaped pupil; (4) Fourier Transform. The results are shown in Figure 4. Qualitatively, the speckle pattern in the simulation looks about the same as in the measurements, and quantitatively, on the graph, the general level of the simulated speckle seems to match the measured speckle, at least to the extent that two random processes can match. We conclude that our contrast is dominated by wavefront error of the optics, which requires adaptive optics for correction.

So far, the data we have shown has been at 632nm. In Figure 5, we show the images and contrast measurements at 3 different wavelengths. It is apparent from Figure 5 that the PSF dilates with wavelength, as it must. The graph on the right of Figure 5

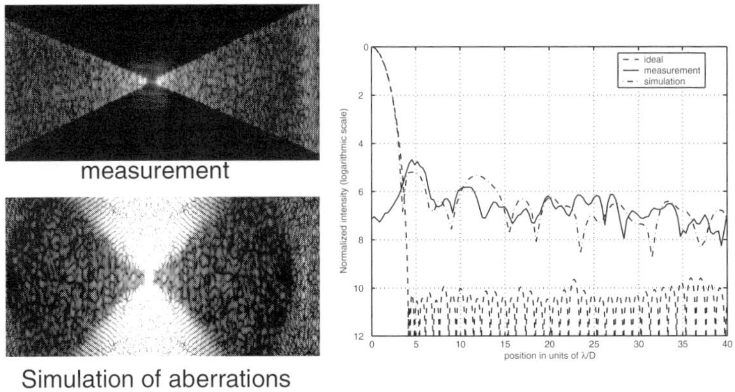

Figure 4. Simulation of random phase aberrations (phase only) based on the quality of our mirrors.

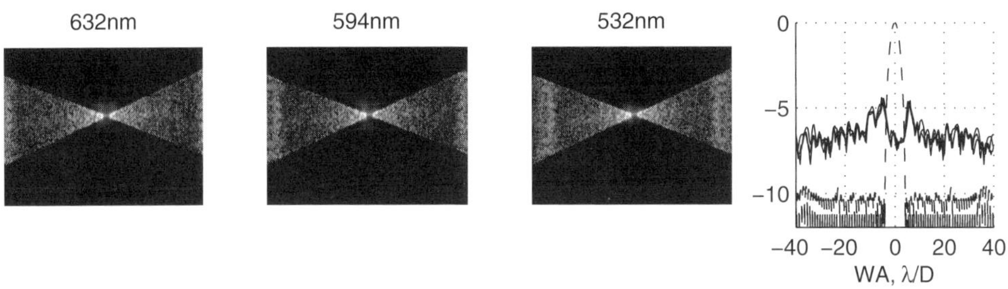

Figure 5. Measurements at 3 different wavelengths. Right: superimposed slices for the three measurements (dashed line is the ideal PSF).

shows superimposed horizontal slices through the centers of the 3 images, going out to 40 λ/D. These traces have been stretched along the abscissa to reverse the dilation due to wavelength, so that all are plotted on the same λ/D scale. Two things on this graph are noteworthy. First, the contrast is the same for different wavelengths, showing the broadband capability of shaped pupils. Second, the speckle structure itself looks very similar for different wavelengths, especially for the lower working angles. However, they are not completely identical. We expect the speckle to be slightly higher at shorter wavelengths because the phase aberrations are effectively greater, but such a difference is almost imperceptible across a scale of 12 orders of magnitude and cannot account for the variations apparent in the plot. There are many reasons why the traces can exhibit these differences, such as: wavelength-dependent amplitude aberrations, higher order speckle effects such as frequency folding, different behavior in the fiber, etc. Tracking down the exact reasons will be important for understanding how to correct them.

In Figure 6, we present the results of our measurements in broadband light. The spectrum of our light source is shown in Figure 6b, and goes from approximately 550nm to 750nm. The speckle image appears rather noisy because the white light power is very low, due to the inherent inefficiencies in coupling a spatially incoherent light source into a single mode fiber. Because of this, electronic noise limits the measurements of speckle level to about $4\text{x}10^{-7}$. Nonetheless, it is still clear from Figure 6c that the contrast in broadband light remains the same as in monochromatic light and that even the speckle structure is very similar, at least for low inner working angles. This once again shows the broadband capabilities of shaped pupils.

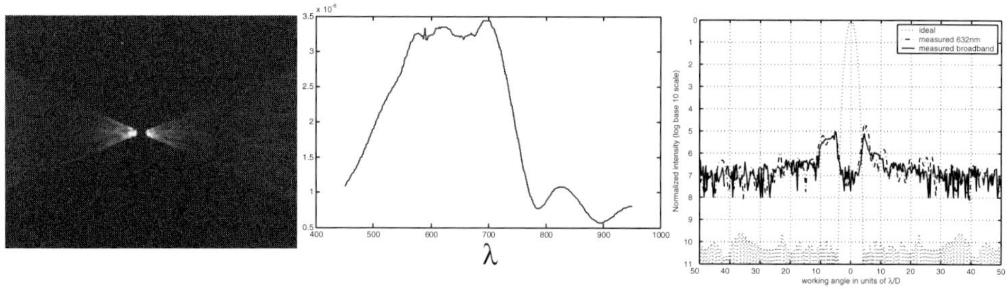

(a) (b) (c)

Figure 6. (a) CCD image in white light (1 hour exposure); (b) Measured spectrum before the fiber; c Slice through the CCD image.

4. Final Remarks

We have demonstrated high contrast in a prototype Shaped-Pupil Coronagraph, which is currently limited by aberrations in the optics. This contrast is maintained over different wavelengths and over broadband light. However, it is still far from the designed 10^{10} value. In order for the full 10^{10} contrast to be reached, aberrations need to be corrected by adaptive optics. It is believed that correction to $\lambda/10^5$ rms level would be required. Furthermore, amplitude aberrations and wavelength dependency of aberrations cannot be neglected. Work in several groups, including ours, is energetically proceeding on this problem (see, for example, Pueyo 2004, Borde 2005, Give'on 2005, and Redding 2002).

Acknowledgements

We gratefully acknowledge the support of the National Aeronautics and Space Administration through the Jet Propulsion Laboratory, and the Michelson Science Foundation, California Institute of Technology. We also acknowledge the National Institute of Standards and Technology for the use of their facilities.

References

Borde P. & Traub W., *astro-ph/0511326*, 2005

Chakraborty, A. & Thompson, L. A., *Optics Express* Vol. 13, No. 7, pp. 2394-2402, 2004

Crepp, J., Ge, J., Debes, J., Kuchner, M., & McDavitt, D., *Proc. of SPIE* 5487 (1383C), 2005

Give'on, A., Kasdin, N. J., & Vanderbei, R. J., *Proceedings of IAUC 200*, 2005

Green, J. J., Shaklan, S. B., Vanderbei R. J., & Kasdin, N. J., *Proc. of SPIE* 5487(184), 2004

Kasdin, N. J., Vanderbei, R. J., Spergel, D. N., & Littman, M. G., *The Astrophysical Journal* 582, pp1147-1161, January 2003

Kasdin, N. J., Belikov, R., Beall, J., Vanderbei, R. J., Littman, M. G., Carr, M., & Give'on, A. *Proc. of SPIE* Vol. 5905, pp. 0G1-0G9, 2005

Pueyo, L., Give'on, A., Carr, M., Littman, M., Kasdin, N. J., & Vanderbei, R. J., *Proc. of SPIE*, 5487(184), 2004

Redding, D., Green, J., Basinger, S., Cohen, D., Dumont, P., Hull, T., Moody, D., Trauger, J., Shaklan S., & Shi, F., *Proc. of SPIE*, 4854(40), 2002

Trauger, J., Brown, M., Brown, R. A., Burrows, A., Burrows, C., Ealey, M., Fischer, D., Ftaclas, C., Heap, S., Hull, T., Kasdin, J., Krist, J., Kuchner, M., Lunine, J., Marcy, G., Sahai, R., Spergel, D., Stapelfeldt, K., Traub, W., & Woodgate, B. *Proc. of AAS, DPS meeting 37*, 18.24, 2005

Vanderbei, R. J., Spergel, D. N., Kasdin, N. J., & Kuchner M., *Proc. of SPIE*, 5170(49), 2003

Wilhelmsen, Evans, J., Sommargren, G., Poyneer, L., Macintosh, B. A., Severson, S., Dillon, D., Sheinis, A. I., Palmer, D., Kasdin, N. J., & Olivier, S., *Proc. of SPIE* 5490(954), 2004

Direct Imaging of Exoplanets: Science & Techniques
Proceedings IAU Colloquium No. 200, 2005
C. Aime and F. Vakili, eds.
© 2006 International Astronomical Union
doi:10.1017/S1743921306009690

Free-standing Notch Filter Image Masks for Coronagraphy

J. Crepp[1], J. Ge[1], S. Miller[1], D. McDavitt[1], A. Vanden Heuvel[1], and M. Kuchner[2]

[1]Astronomy Department, University of Florida
Bryant Space Science Center, P.O. Box 112055
Gainesville, FL 32611-2055
contact email: jcrepp@astro.ufl.edu

[2]Exoplanets & Stellar Astrophysics Laboratory
NASA/Goddard Space Flight Center
Greenbelt, MD 20771, USA

Abstract. The Lyot coronagraph can achieve significant on-axis starlight suppression when equipped with a notch filter image mask. Notch filter masks are placed in the focal plane and have small intricate features. They must be manufactured using nanofabrication techniques, when designed for coronagraphs operating at visible or near infrared wavelengths. This process often involves depositing an opaque material onto a glass substrate whereby a pattern is then etched. The substrate however can be a source of internal reflections and chromatic aberrations that limit the coronagraph's dynamic range and broadband performance. We present theoretical and experimental work on a new series of image masks that do not require the support of a glass substrate. We show that by adding periodic bar structures of appropriate spacing, it may be possible to manufacture a Silicon binary notch filter mask that is "free-standing".

Keywords. extrasolar planets — high contrast imaging — coronagraphy — terrestrial planet finder — circumstellar matter.

1. Introduction

The detection of terrestrial exoplanets orbiting in the habitable zone of nearby stars will require an instrument with unprecedented wavefront stability and diffraction control. Additionally, the instrument must have a large collecting area, a broad bandwidth, and the ability to adaptively correct for imperfections in optical components that scatter light. At large contrast ratios, these capabilities correspond to tight tolerances. Subtle chromatic effects could easily limit the instrument's dynamic range and broadband performance.

One of the leading designs for the Terrestrial Planet Finder Coronagraph (TPF-C) is a Lyot coronagraph that is equipped with an 8th-order band-limited or notch filter image mask (Kuchner, Crepp, & Ge 2005). These masks perform well in theory and in practice. They have high off-axis throughput ($\sim$40%), a large search area, and a broad bandwidth (Kuchner & Traub 2002, Kuchner & Spergel 2003). They also provide resistance to low-order aberrations (Shaklan & Green 2005, Crepp *et al.* 2005), and have enabled the Lyot coronagraph to achieve the deepest contrast in the lab to date (Trauger *et al.* 2004).

Binary versions of these masks are manufactured by depositing an opaque material such as Chromium, Aluminum, or Gold onto a glass substrate; then a pattern is "etched out", and an anti-reflective coating is applied (Debes *et al.* 2004). The glass provides mechanical support, but is a source of unwanted internal reflections and chromatic aberrations. In this study, we aim to significantly reduce these effects in the image mask by eliminating the need for the glass all together.

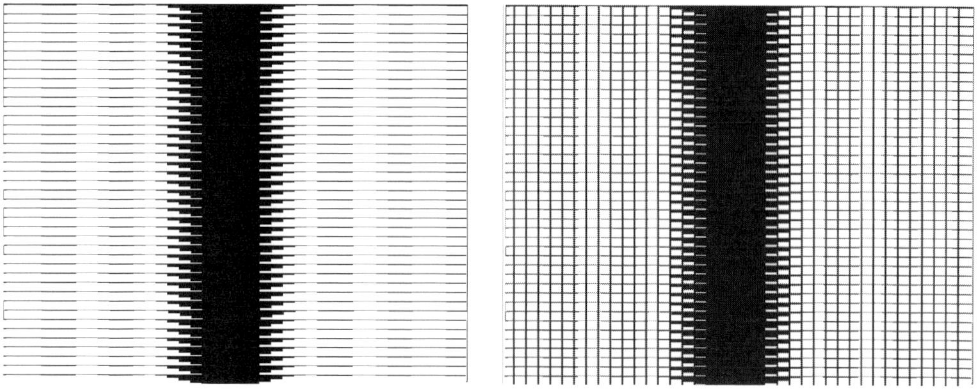

Figure 1. Simulated image of a sampled binary 8th-order notch filter image mask (left), and a free-standing mask equivalent (right). For a small penalty in off-axis attenuation, free-standing masks offer a way to circumvent the need for glass substrates.

Figure 1 displays a simulated image of a sampled binary 8th-order notch filter mask and a "free-standing" mask equivalent. The periodic bar structures in the free-standing mask provide strength in the absence of glass, and prevent the mask from collapsing under it's own weight, or from any external forces such as the harsh jostling of a shuttle lift-off. These structures clearly increase the off-axis attenuation of the coronagraph. However, depending on their breadth, the broadband performance of the masks may justify the loss in throughput. We intend to directly measure these quantities in the lab.

It is possible to model the performance of free-standing masks numerically, however a full electromagnetic treatment is required. The size of the smallest features are smaller than the wavelength of light, therefore it is necessary to include polarization, the finite thickness of mask, and other complications. Although such simulations are useful, the results must be confirmed experimentally. Moreover, it is not clear how narrow one can make the bar structures before they are no longer supportive. The contrast, wavelength dependence, and throughput of the mask are all functions of the mask thickness, width of the bars, and kind of material that is used, as well as the $f/\#$ of the system.

In the next section, we describe the basic theory for how binary free-standing masks diffract light to first order. Then we provide a summary of our nanofabrication techniques and progress towards constructing the first prototype.

2. Theory

Free-standing masks are notch filter masks that are multiplied by a dirac-comb filter of finite width. We use the notation of Kuchner & Spergel 2003 to describe how these masks interact with light in the image plane.

The equation describing a *linear* binary mask with vertical support bars is:

$$\hat{M}(x,y) = \sum_n \delta\left(y - n\frac{\lambda_{min}}{D}\right) * \hat{M}_{stripe}(x,y) \times \left\{\sum_m \delta(x - m\Delta x) * \left[1 - \Pi\left(\frac{x}{a}\frac{D}{\lambda_{min}}\right)\right]\right\} \tag{2.1}$$

where

$$\Pi\left(\frac{x}{a}\frac{D}{\lambda_{min}}\right) = \begin{cases} 1 & \text{for } |x| \leqslant \frac{a}{2}\frac{\lambda_{min}}{D} \\ 0 & \text{elsewhere} \end{cases} \tag{2.2}$$

and

$$\hat{M}_{stripe}(x,y) = \begin{cases} 1 & \text{where } y < \hat{M}_{notch}(x)\,\frac{\lambda_{min}}{D} \\ 0 & \text{elsewhere.} \end{cases} \tag{2.3}$$

The factor $\sum \delta(x - m\Delta x) * (1 - \Pi(xD/a\lambda_{min}))$ is a dirac-comb filter convolved with a top-hat. It represents a series of vertical bars, where Δx is their separation, and $0 \leqslant a \leqslant 1$ is the width of each individual bar, in units of λ_{min}/D. The function $\hat{M}_{notch}(x)$ is a one-dimensional notch filter function that can be either smooth or sampled.

Notch filter masks are designed such that only the low-spatial frequencies, $|u| \leqslant \frac{\epsilon}{2}\frac{D}{\lambda_{max}}$, diffract light inside the Lyot stop - this light interferes and identically cancels. Higher spatial frequencies $|u| \geqslant (1 - \frac{\epsilon}{2})\frac{D}{\lambda_{max}}$ diffract light outside the Lyot stop - this light is blocked by a stop in the subsequent pupil plane. In theory, the combination can yield infinite dynamic range for an unobstructed entrance aperture.

For free-standing masks, we can choose a spacing for the vertical stripes such that only one term diffracts light inside the Lyot stop and the rest diffract light outside. The Fourier transform of Equation 2.1 is

$$\begin{aligned} M(u,v) = {} & \sum_n \delta\left(v - n\frac{D}{\lambda_{min}}\right) \int \mathrm{sinc}\left(\pi v \frac{\lambda_{min}}{D}\hat{M}_{notch}(x)\right) \\ & \times\ \hat{M}_{notch}(x) e^{-i2\pi u x} dx \\ & *\ \frac{1}{\Delta x}\sum_m \delta\left(u - m\frac{1}{\Delta x}\right) \\ & \times\ \frac{\lambda_{min}}{D}(\delta(u) - a\ \mathrm{sinc}(\pi a u)), \end{aligned} \tag{2.4}$$

where m, n run over all integers. If $\Delta x \leqslant \lambda_{min}/D$, then only the $u = 0$ and $v = 0$ terms contribute, and at low-spatial-frequencies Equation 2.4 reduces to

$$\begin{aligned} M(u,v) &= (1 - a)\ \delta(u) * M_{notch}(u) \\ &= (1 - a)\ M_{notch}(u). \end{aligned} \tag{2.5}$$

Therefore, free-standing masks diffract light just like notch filter masks, with the exception that off-axis sources are attenuated by an additional factor of $(1 - a)^2$. The binary support bars are the mathematical equivalent of a neutral density filter.

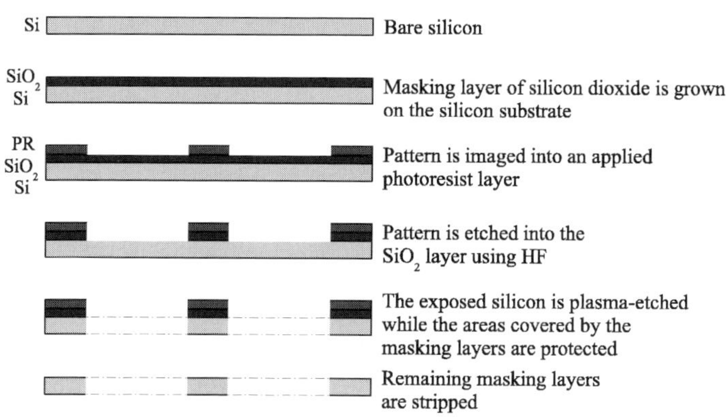

Figure 2. Edge-on view of the nanofabrication process for creating a free-standing mask.

It is also possible to add horizontal stripes. With a clever starting point, these stripes can be made to overlap or "hide" the smallest features in the mask. Of course, the off-axis throughput then goes as $\sim(1-a)^2(1-b)^2$, where b is the width of each horizontal stripe.

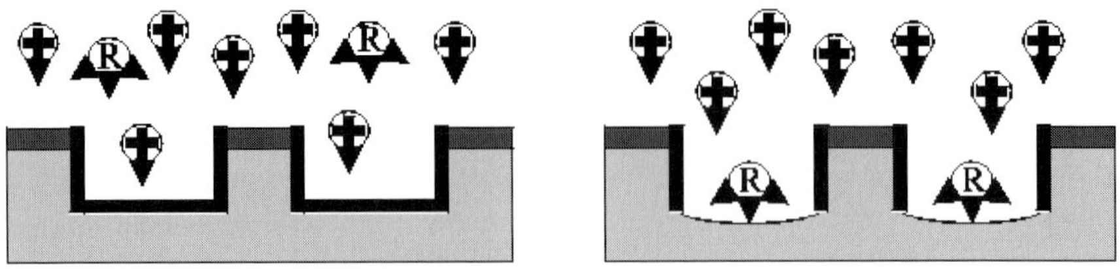

Figure 3. Edge-on view of the ASE process showing the ions selectively stripping the protecting Fluorocarbon layers along the trench bottoms, and the reactive species of the plasma etching away the exposed Silicon.

3. Nanofabrication: strategy & progress

We have chosen Silicon as the base structure for the masks. Silicon is opaque to light with $\lambda \lesssim 900$ nm and can be easily machined into microstructures with nanometer level surface roughness. Additionally, Silicon has a Young's Modulus of 131 GPa, which is greater than steel.

Preliminary patterns are being fabricated to determine the minimum allowable dimensions that can be"free standing", the maximum separation of these features, and what level of uniformity and roughness that can be achieved. Once these defining parameters are determined, we will fabricate a fully developed mask, similar to that shown in Figure 1. Our initial experiments are focused primarily on creating square-groove features with vertical sidewalls; although, it may be possible to direct unwanted diffracted light, created by the finite thickness of the masks, away from the optical axis with angled trenches. We will explore this option by experimenting with fast etch rates that compromise anisotropy.

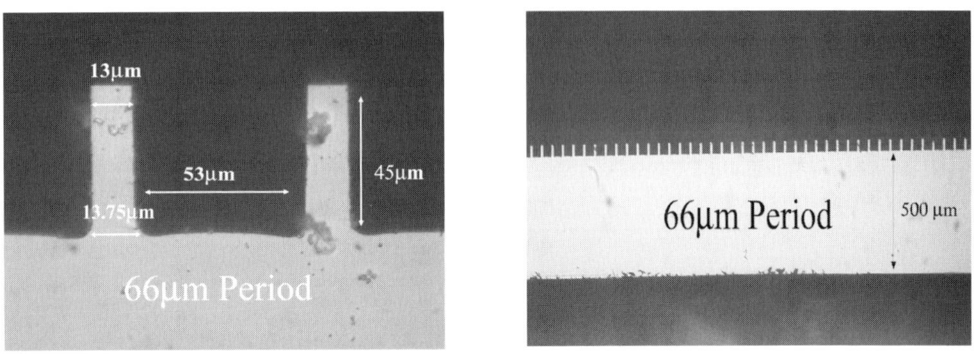

Figure 4. Microscope image of edge-on view of mask that has been partially etched. The goal is to etch all the way through the wafers and to maintain small structural support bars with minimal defects.

The process for fabricating free-standing masks involves the following steps and is illustrated in Figure 2. First, an etch-resistant SiO_2 layer is grown on the silicon substrate. Photolithography is then used to transfer the desired pattern from an electron-beam developed mask to the wafer surface with selected areas exposing the oxide layer. Currently we use contact photolithography to produce micron features. Electron-beam *direct* writing has a resolution on the order of 20 nm, and will also be used. The bare oxide is then wet etched with Hydrofluoric acid down to the underlying Silicon. Using the Advanced Silicon Etch (ASE) process with a Surface Technology Systems - Deep Reactive

Ion Etcher, plasma etches the pattern down through the Silicon wafer creating our free standing structures. The ASE process is detailed below. Finally, the remaining masking layer is removed and the substrate is cleaned.

The ASE process, based on the original "Bosch Process", anisotropically etches Silicon through a series of alternating passivation and etch steps, producing vertical sidewalls. The process is shown in Figure 3. The etch is free of the constraining geometries of silicon's crystallography that standard wet chemical etching is bound by. Each process cycle starts with a passivation step, where a C_4F_8-based plasma deposits a few Fluorocarbon polymer monolayers across surfaces exposed to the plasma. An isotropic etching plasma of SF_6 is then introduced. Ions bombard the surface of the wafer, selectively removing the polymer layers. The plasma, having an increased ion energy in the vertical direction, removes the Fluorocarbon from surfaces parallel to the wafer surface, leaving an exposed silicon surface at the base of trenches. The reactive Fluorine-based species of the plasma isotropically etches the bare Silicon while the remaining Fluorocarbon polymer protects the vertical sidewalls from lateral etching. The cycle is repeated to produce vertical sidewall profiles of $90^o \pm 2^o$ through a silicon wafer. Feature sizes can range from 1 to >500 μm and etch depths range from 10–800 μm while maintaining aspect ratios of up to 30.

Several error sources from the process have to be considered. These include: scallops, etch defects that form along the sidewalls created by the cycling of passivation and etching, "grass" that is observed as pillars of silicon at the bottom of trenches, microloading, the etch rate dependence on the aspect ratio, and finally RIE lag, where larger features locally decrease the number of etching species at the expense of narrower features and non-uniformity in the etch rate with respect to the location on the substrate.

4. Conclusions

We have discussed the potential utility of a free-standing notch filter image mask for coronagraphy, particularly for use with aggressive space missions such as TPF. We then derived the first order theory for how such masks diffract light: free-standing masks behave like notch filter masks with the exception that they attenuate off-axis sources by a small factor that is a function of the support bar widths. We then outlined our strategy and progress in making the first prototype mask using nanofabrication techniques.

"We acknowledge the support from NSF with grants NSF AST-0451407 and AST-0451408, NASA with grants NNG05G321G and NNG05GR41G, JPL TPF program, UCF-UF SRI program, UF and Penn State."

References

Crepp, J. R., Vanden Heuvel, A. D., Ge, J., Miller, S. P., & Kuchner, M. J. *Laboratory Testing of a Lyot Coronagraph Equipped with an 8th-order Notch Filter Image Mask*, submitted to ApJ

Debes, J. H., Ge, J., Kuchner, M. J., & Rogosky, M., *ApJ*, **608**, 1095–1099, 2004

Kuchner, M. J., Crepp, J. C., & Ge, J., *ApJ*, **628**, 466–473, 2005

Kuchner, M. J. & Spergel, D. N., *ApJ*, **594**, 617–626, 2003

Kuchner, M. J. & Traub, W. A., *ApJ*, **570**, 900–908, 2002

Shaklan, S. B. & Green, J. J., *ApJ*, **628**, 474–477, 2005

Trauger, J. *et al.*, Proc. SPIE, **5487**, 1330–1336, 2004

All photographs: Laurent Thareau [l.thareau@free.fr].

Direct Imaging of Exoplanets: Science & Techniques
Proceedings IAU Colloquium No. 200, 2005
C. Aime and F. Vakili, eds.

© 2006 International Astronomical Union
doi:10.1017/S1743921306009707

Combination of nulling interferometer, nulling coronagraph, and modified pupil method

J. Nishikawa and N. Murakami

MIRA Project, National Astronomical Observatory of Japan,
National Institute of Natural Science, Mitaka, Tokyo 181-8588, JAPAN
email: jun.nishikawa@nao.ac.jp, naoshi.murakami@nao.ac.jp

Abstract. We investigated two-stage combinations from three methods, nulling interferometer, nulling coronagraph, and modified pupil, calculating reduced intensity profile of a resolved central star and transmission for exo-planets. An achievable dynamic range can be derived by dividing the residual halo intensity of the star by the transmitted peak intensity of the planet. For observation parameters, here assumed are the wavelength of 600 nm in optical and the telescope diameter of 3 m. The combination of the nulling interferometer and the four-quadrant phase mask coronagraph showed the best performance reaching to 10^{-10} dynamic range at 100mas distance from the central star among five candidates of combination using nulling interferometer, achromatic interfero-coronagraph, four-quadrant phase mask, and shaped pupil method.

Keywords. instrumentation: interferometers, techniques: interferometric, techniques: image processing, telescopes, stars: imaging, planetary systems.

1. Introduction

Three methods are well known of achieving high dynamic range observations especially for the direct detection of extra-solar planets, nulling interferometer, nulling coronagraph, and modified pupil method. We investigate their combinations focusing on a light suppression for a resolved central star and a transmission for a planet detection area.

The nulling interferometer can achieve the perfect null at an on-axis point source but remains light for off-axis ray which dependence is a square of the angle for a two-phase nulling condition. The nulling interferometer can be considered as pre-optics producing weakened flat wave fronts, combined with any downstream single telescope method as seen in Nishikawa *et al.* (2005). Nulling coronagraphs (AIC: achromatic interfero coronagraph, e.g. Baudoz *et al.* 2000, FQPM: four quadrant phase mask coronagraph, e.g. Rouan *et al.* 2000) are also useful as pre-optics though they produce non-flat wave fronts. Modified pupil (apodization or SP: shaped pupil, e.g. Kasdin *et al.* 2003) can reduce halo intensity having an insensitive characteristics to the size of a resolved star, but in many cases the PSF (point spread function) core radius increases hiding close planets. The modified pupil method is used at the entrance of the optics and also at the Lyot stop. All of the methods reduce throughput for planets instead of the high dynamic range. In the present paper, we examined five cases of the combinations.

2. Calculation

We calculate the residual halo intensity of a resolved star and the transmission at its sorrounding area. In this paper, we assume observation parameters as follows, the wavelength λ of 600 nm, the telescope diameter D of 3 m, the baseline for NIB (nulling

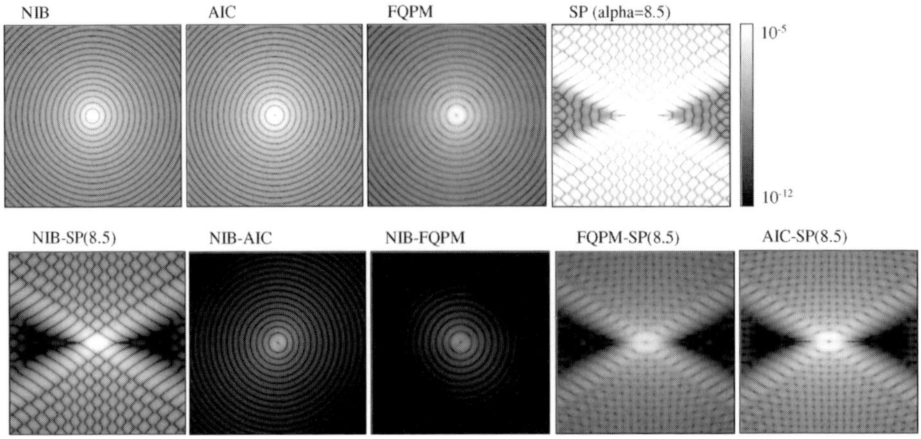

Figure 1. The central star images processed by high dynamic range methods of (upper) single way and (lower) combination of them. The central star diameter is assumed to be 0.02 λ/D, where the telescope diameter is 3m and the wavelength is 600nm. NIB: nulling interferometer in Bracewell's configuration with a baseline length of 3m, AIC: achromatic interfero coronagraph, FQPM: four quadrant phase mask coronagraph, SP: shaped aperture with $p(x) = I_0\left(\alpha\sqrt{1-x^2}\right)/I_0(\alpha)$ and $\alpha = 8.5$ used as an approximated prolate function, here I_0 is the zero-order modified Bessel function of the first kind. Lyot stop of 90% diameter is used for FQPM.

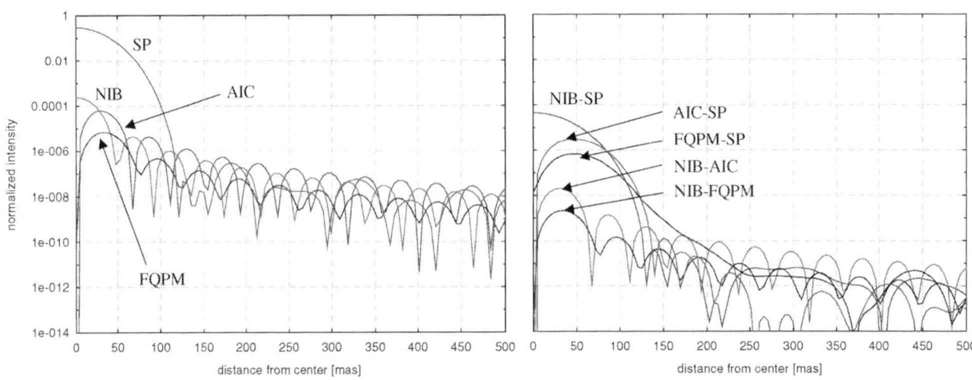

Figure 2. Radial profiles of the central star images along x-axis processed by high dynamic range methods of (left) single method and (right) combination of them, for the same parameters of Fig. 1. The central star diameter of 0.02 λ/D assumed here corresponds to 0.83 mas (equivalent to the Sun from 11.2 pc distance) and an Earth-like planet would be seen at 88mas. A combination of NIB-FQPM is the best reaching to 10^{-9} at 88mas, and also NIB-AIC. All of the combinations show reduced intensity of the central star lower than the original single method profiles and 10^{-10} farther than 2AU position.

interferometer of Bracewell's configuration) is 3 m, and the central star diameter is 0.02 λ/D (0.83 mas, equivalent to the Sun at 11.2 pc distance, then an Earth-like planet would be seen at $2.1\lambda/D$ or 88mas). The star images processed by high dynamic range methods in single way (NIB, AIC, FQPM, and SP) and combination of them (NIB-SP, NIB-AIC, NIB-FQPM, FQPM-SP, and AIC-SP) are calculated and shown in Fig. 1, and radial profiles of the images along x-axis are shown in Fig. 2. A shaped pupil is set by using

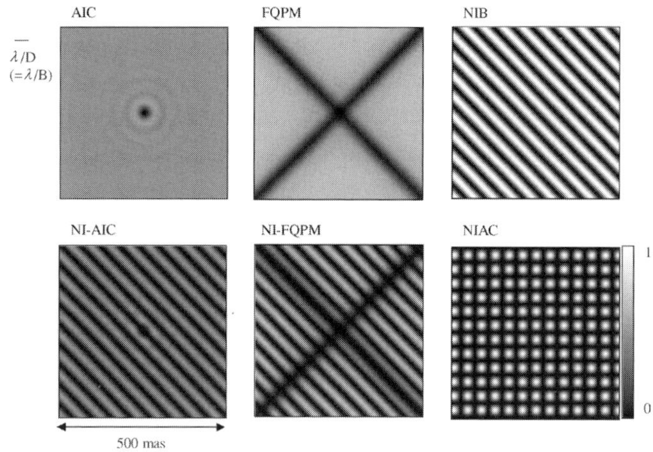

Figure 3. Transmission maps for the single and the combination methods. Maximum intensity of the images are used to estimate these maps.

an approximated prolate function as described in the figure caption. A combinations of NIB-FQPM is the best reaching to 10^{-10} at the Earth position, and NIB-AIC is also good. All of the combinations show reduced intensity of the central star lower than the original single method profiles and 10^{-10} farther than 2AU distance.

Residual intensity of dual-phase nulling methods has a square dependence on spatial size of a resolved star, and a forth-order law is seen for their double combination, which is similar to a four-aperture nulling interferometer of Angel's cross (NIAC: e.g., Dubovitsky & Lay 2004) or interferometric nulling coronagraph (Levine *et al.* 2003).

Possible combinations, configurations and parameters are not perfectly searched, here. The shaped pupil can be exchanged by a phase induced amplitude apodization (Guyon 2003). The two-telescope interferometer baseline can be reduced more using a Mach-Zehnder interferometer described as a visible nulling coronagraph (Levine *et al.* 2003). There is a possibility to achieve higher dynamic range than shown here even by a dual-stage combination when conditions are well-examined. We should note that, however, tip-tilt errors and wavefront errors are not considered here.

3. Transmission for planet and achievable dynamic range

Planet light is affected by the nulling or the apodization methods reducing the total energy and/or the peak intensity by diffusing the PSF. Transmission maps (here, maximum intensity of the images are used to estimate) for NIB, AIC and FQPM are shown in Fig. 3. as well as the combinations of NIB-AIC and NIB-FQPM, where that of NIAC is shown for comparison. The shaped pupil method has a constant value 0.29 of transmission in the field of view, which is the peak of the SP's profile in Fig. 2 (left). In Fig. 4, azimuthally averaged transmission profiles are plotted for the single and the combination methods. The transmission profiles are almost 0.2 for all of the combinations we considered under the present conditions.

The final achievable dynamic range would be derived by dividing the resolved star image profiles by the transmission profiles, so that multiplying a factor of about 5 is enough to consider the final result looking at the resolved star image profiles in Fig. 2 (right).

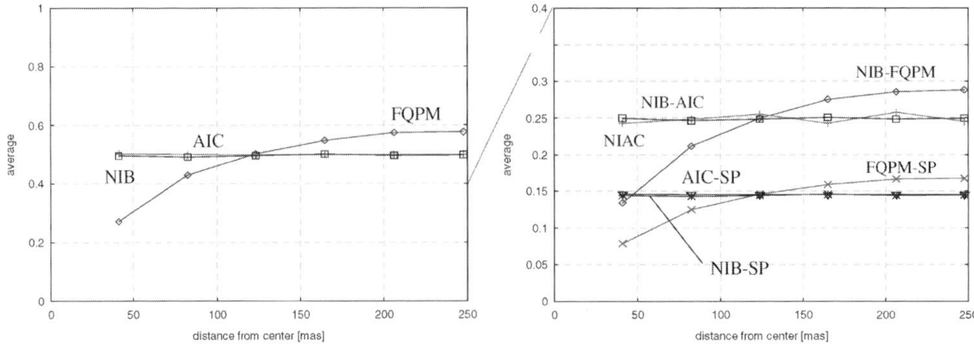

Figure 4. Radial profiles of azimuthally averaged transmission maps of Fig. 1 for the single and combination methods. Here a constant value of transmission for SP is used, the central peak value of PSF seen in Fig. 2. The resolved star image intensity profiles in Fig. 2 should be divided by these profiles to know achievable dynamic range. However, the transmission profiles are found to be about 0.2 for all of the combinations.

4. Conclusions

We investigated some combinations of high dynamic range methods. A two-stage combination of the nulling interferometer and the four-quadrant phase mask coronagraph showed the best performance among the candidates under the 600nm dual 3m-telescope conditions reaching to 10^{-10} dynamic range at 100mas distance from the central star whose diameter is 0.02 λ/D. There is a possibility to achieve higher dynamic range than shown here with other well-examined dual-stage combinations, when errors from tip-tilt and wavefront shape are not significant.

Acknowledgements

We would like to acknowledge the constant encouragements by Profs. M. Yoshizawa, N. Baba and M. Tamura.

References

Baudoz, P., Rabbia, Y., & Gay, J. 2000, *A&AS*, 141, 319

Dubovitsky, S. & Lay, O. P. 2003, *Proc. of SPIE* 5491, 284

Guyon, O. 2003, *A&A* 404, 379

Kasdin, J. N., Vanderbei, R. J., Spergel, D. N., & Littman, M. G. 2003, *ApJ*, 582, 1147

Levine, B. M., Shao, Michael, Liu, Duncan T., Wallace, James K. Lane, & Benjamin F. 2003, *Proc. of SPIE* 5170, 200

Nishikawa, J., Kotani, T., Murakami, N., Baba, N., Itho, Y., & Tamura, M. 2005, *A&A* 435, 379

Rouan, D., Riaud, P., Boccaletti, A., Clenet, Y., & Labeyrie, A. 2000, *Publ. Astron. Soc. Pacific*, 112, 1479

Direct Imaging of Exoplanets: Science & Techniques
Proceedings IAU Colloquium No. 200, 2005
C. Aime and F. Vakili, eds.
© 2006 International Astronomical Union
doi:10.1017/S1743921306009719

The Optical Telescope Assembly for the Terrestrial Planet Finder Coronagraph: Design and Analysis Results

R. G. Ohl[1], A. J. Martino[1], N. F. Martin[1]
and the TPFC team

[1]NASA/Goddard Space Flight Center, Greenbelt, MD 20771, USA
email: raymond.g.ohl@nasa.gov

Abstract. The Terrestrial Planet Finder Coronagraph (TPFC) is a National Aeronautics and Space Administration (NASA) exploration mission to directly detect and characterize terrestrial exoplanets at visible and near-infrared wavelengths. The TPFC mission is currently in a "pre-formulation" stage where requirements and designs are traded. TPFC must distinguish a planet that is more than 10 orders of magnitude fainter than its parent star at a separation of 62 mas ($\lambda = 600$ nm). Coronagraphic detection requires a large aperture telescope to resolve the exoplanet from its parent star, and great system (wavefront) stability during detection and characterization. This paper discusses the design considerations, trade studies and analysis leading to the current, "reference" design for the TPFC telescope. We present the salient features of the design and the most significant structural, thermal and optical analysis results. We also discuss the planned model validation and performance verification approach.

Keywords. telescopes, instrumentation: high angular resolution.

1. Introduction

Ford, *et al.* (2005) present an overview of the TPFC mission and baseline design ("flight baseline 1," FB1). The Jet Propulsion Laboratory, California Institute of Technology (JPL) manages the mission. NASA/Goddard Space Flight Center (GSFC) is responsible for the optical telescope assembly (OTA; Figure 1)

The primary mirror (PM) aperture is an 8×3.5 m ellipse. The 62 mas inner working angle drives the 8 m dimension and the perpendicular axis set by the shroud of the baseline launch vehicle (Boeing Delta IV-H). Diffraction produces an elliptical point spread function, so the highest spatial resolution is along the long dimension of the PM.

Shaklan, *et al.* (2005) describe the coronagraph instrument and requirements. The error budget contains static and dynamic terms. The dynamic requirements are driven by the speckle subtraction scheme, in which a stationary planet is separated from a speckle pattern that moves as the telescope is dithered 30° about its axis. The OTA wavefront must thus remain stable to $\sim$0.1 nm for certain Zernike components during the $\sim$8 hour integrations associated with dithering. The SM and M3 must remain aligned to the PM to better than of-order 10 nm in translations and 10 nrad in rotations during an integration. Content, *et al.* (2005) list static and dynamic WFE requirements for OTA component figure and alignment tolerances.

For planet finding, the capture range of the coronagraph deformable mirrors (DM) and the dynamic effects of "beam walk" drive the OTA static wavefront error (WFE) requirements for low- and mid-spatial frequency errors. A second instrument, the "general astrophysics instrument" (GAI), observes a wide field of view ($\sim 4 \times 4$ arcmin) around the target star. GAI science and higher-order effects on the coronagraph thus drive high

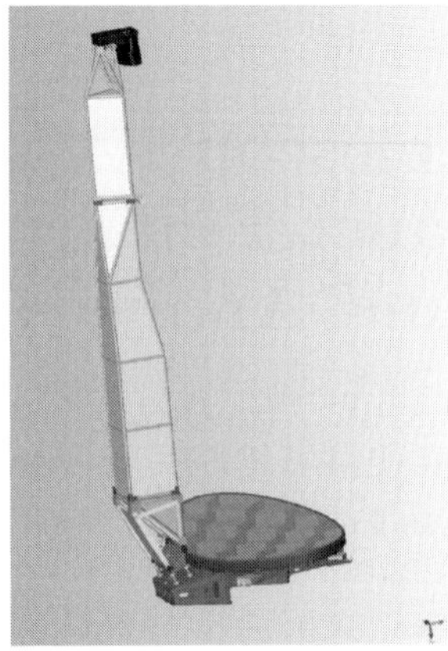

Figure 1. TPFC OTA (Ho/JPL, Engler/GSFC).

Table 1. OTA mirror prescriptions.

Name	Physical size (m)	Off-axis distance (m)	R(m)	$f/\#$	k
PM	8.0×3.5	2.3	26.75	3.82	−1.0019
SM	0.89×0.425	0.237	3.041	4.13	−1.47
M3	0.29×0.31		∞		

spatial frequency WFE tolerances — scattered light from the target star contributes to background on the GAI image plane. For TPFC, bands of WFE spatial frequency are defined relative to the correctable band limit of a high density, "fine" DM placed downstream at a pupil image of the PM. A power spectral density (PSD) specification is generated for each optic using a system performance model based on scalar diffraction theory. The integrated, static surface error requirement for the PM and secondary mirror (SM) is 5.4 nm RMS, while the requirement for the M3 fold mirror is 1.2 nm RMS. Both static and dynamic requirements have implications for mirror fabrication and subsystem- and OTA-level integration and test (I&T). We expect future work to somewhat relax the low-order WFE specification.

The mirrors are coated with protected silver. Wavefront amplitude variations also degrade performance, so reflectance uniformity over the PM must be better than ~0.5 % and is specified as a function of spatial frequency.

Driving mechanical requirements include launch loads, launch vehicle packaging and on-orbit vibration/jitter. Requirements and mitigation schemes are in flux, but we are keeping the minimum first resonant frequency of the payload above ~10 Hz.

The WFE stability requirements lead to challenging thermal stability tolerances, although large static thermal gradients are tolerable. For example, we require temperature stability of-order 1 mK on the PM during a science integration, while static gradients of-order 1 K are permitted.

2. Design and analysis

A schematic of the OTA FB1 concept is shown in Figure 1. The optical path starts at the 8×3.5 m elliptical, off-axis parabolic PM. The SM reflects the light toward a tertiary, flat, fold mirror (M3) that directs the beam into the coronagraph. A pick-off mirror with a through hole at the OTA focus sends the outer portion of the field to the GAI. The

PM is mounted semi-kinematically using a textbook, three-bipod flexure scheme to a metering structure (MS). The MS supports a set of heaters for maintaining the PM at ∼293 K. The SM is attached to a deployable tower. A laser "metrology truss" and SM hexapod mechanism actively correct for rigid body motions of the SM relative to the PM during a science integration(Shaklan, *et al.* 2004).

The results of several trade studies are critical to the FB1 design: monolithic vs. segmented PM aperture, Gregorian vs. Ritchey-Chrétien (RC) optical prescription, on-orbit WFE correction method, mirror blank material, "closed- back" vs. "open-back" mirror blank structural design and PM aperture shape (Content, *et al.* 2005).

The Lyot coronagraph design and high contrast requirement drive the architecture to an unobstructed, monolithic aperture. Very tight requirements on segment alignment stability and edge diffraction control rule out a segmented PM approach.

The FB1 OTA prescription is an off-axis RC design (Table 1). Compared to a Gregorian, the RC better accommodates packaging requirements for a given PM speed. The convex SM provides better polarization performance and more forgiving alignment sensitivity than a concave SM. However, it is more difficult to fabricate and test a meter-class, convex SM of this quality. Designs with one and three powered mirrors were also considered, but rejected (Howard, *et al.* 2005).

The OTA requires ∼5 μm of on-orbit WFE adjustment to correct for uncertainties in the prediction of PM ground-to-orbit surface error change due to gravity sag and compensating technology and modeling/test uncertainty. The FB1 design incorporates a "coarse DM" early in the coronagraph optical train. This DM has long-stroke actuators capable of correcting microns of low-order WFE, producing an output that is within the capture range of the later, finer pitch, low-stroke DMs. Future work will compare this concept with requirements for OTA manufacture, assembly and I&T. Actuated PM designs are under consideration and address mission requirements to different degrees.

The PM material trade is primarily driven by the thermal stability requirement (Section 1). The structure must be optimized for low-weight and high-stiffness to satisfy requirements from manufacture through launch. Such a lightweight mirror does not conduct heat well, so the thermal stability parameter (i.e., the ratio of conductivity to coefficient of thermal expansion, CTE) is a poor guide in this case. Heat flow is dominated by radiative, not conductive, coupling. This drives the PM to materials engineered for low-CTE and available in large size — i.e., ULE® and Zerodur®. Even for these materials, temperature changes must be limited to the ∼1 mK level over an integration. We considered many other factors like blank manufacturing capability, uniformity of CTE and material hardness for polishing, thermal hysteresis and radiation-induced dimensional change. We choose ULE® for the FB1 design, with some reservations.

We considered two structural designs for the PM blank: a "closed-back" or "sandwich" construction and an "open-back," triangular isogrid design (Content, *et al.* 2005). We examined their structural, acoustic and thermal performance. The open-back design is structurally inferior to the closed-back at this high-aspect ratio. Structural analysis showed first resonant frequencies of ∼25 Hz for the closed-back and ∼14 Hz for the open-back. Acoustic analysis showed higher launch loading for the open-back design. We built a highly detailed thermal model of a single, hexagonal piece of the PM. This analysis showed that the face sheet of an open-back mirror has better radiative communication to heater assemblies mounted around the mirror. Thermal analysis of the observatory using a lower-detail, integrated model showed small temperature changes on the PM during a dither maneuver. While thermal gradients in the open-back model are much smaller, the closed-back model performs with good margin against the <1 mK dynamic gradient requirement. The mechanical advantages drive FB1 to the closed-back design.

To analyze stray light performance, we built non-sequential, bi-directional reflectance distribution function models using the PSD specifications and a simple baffle design. Although invalid for very near angles, we predicted the point source transmission (PST) of the OTA and compared irradiance to the Zodiacal background. We compared the performance of the RC and Gregorian designs. For the Gregorian, a PM focus field stop shades the SM from illumination from the PM by sources outside the field. This makes microroughness less critical for mirrors after that stop. We found that the Gregorian's PST is less than a factor of a few lower than the RC PST for angles $\lesssim 3.5°$. We conclude that the different in irradiance does not alone warrant a change from the RC design.

3. Integration and test

The verification/validation approach is developing in parallel with the design (Smith, *et al.* 2005). Ideally, one would test the OTA under flight-like conditions on the ground to validate requirements and verify performance. However, contrast $<10^{-10}$ and stability $<10^{-11}$ may not be obtainable during testing. Therefore, we are planning a program that maximizes direct test and uses analysis to fill-in where testing is not possible or is prohibitively expensive. Test results feed back to the design early in the project so that we build an OTA that can be verified.

The modeled relationship between contrast and OTA output and component characteristics are the core of this approach (validated by sub-scale and flight hardware testing). Testing will be correlated with models starting at the lowest level of assembly or component test through integration, maintaining "traceability" from one step to the next.

Acknowledgements

We gratefully acknowledge the contributions of the JPL, GSFC and industry teams working on mission pre-formulation. Reference herein to any specific commercial product, process or service does not constitute or imply its endorsement by NASA or JPL.

References

Content, D., Ohl, R., Cafferty, T., Cohen, E., Egerman, R., Engler, C., Fantano, L., Feher, P., Green, J., Ho, T., Howard, J., Hunyadi, S., Irish, S., Jackson, C., Kissil, A., Kwack, E., Martin, N., Martino, A., Pattison, J., Peabody, S., Smith, A., & Weng, I. 2005, in: M. Kahan (ed.), *Optical Modeling and Performance Predictions II*, Proceedings of SPIE 5867, in press

Ford, V., Levine-West, M., Kissil, A., Kwack, E., Ho, T., Dumont, P., Lisman, D., Feher, P., & Cafferty, T. 2005, in: C. Aime & F. Vakili (eds.), *Direct Imaging of Exoplanets: Science and Techniques IAU Colloquium*, IAU Colloquium No. 200 (Cambridge University Press), this volume

Howard, J., Mouroulis, P., Thompson, A., Smith, A., Content, D., Ho, T., Jackson, C., Ohl, R., & Shaklan, S. 2005, in: P. Mouroulis, W. Smith & R. Johnson (eds.), *Current Developments in Lens Design and Optical Engineering VI*, Proceedings of SPIE 5874, in press

Shaklan, S., Marchen, L., Zhao, F., Peters, R., Ho, T., & Holmes, B. 2004, in: L. Peterson & R. Guyer (eds.), *Space Systems Engineering and Optical Alignment Mechanisms*, Proceedings of SPIE 5528, 22

Shaklan, S., Balasubramanian, K., Green, J., Hoppe, D., Lay, O., Lisman, D., & Mouroulis, P. 2005, in: C. Aime & F. Vakili (eds.), *Direct Imaging of Exoplanets: Science and Techniques IAU Colloquium*, IAU Colloquium No. 200 (Cambridge University Press), this volume

Smith, A., Blaurock, C., Krim, M., Levine, M., Liu, A., Martino, A., Ohl, R., & Pitman, J. 2005, in: D. Coulter (ed.), *Techniques and Instrumentation for Detection of Exoplanets II*, Proceedings of SPIE 5905, in press

Direct Imaging of Exoplanets: Science & Techniques
Proceedings IAU Colloquium No. 200, 2005
C. Aime and F. Vakili, eds.
© 2006 International Astronomical Union
doi:10.1017/S1743921306009720

Chromatic behavior of amplitude and phase compensators

Laurent Pueyo, Michael G. Littman, Jeremy Kasdin, Ruslan Belikov, and Amir Give'on

Department of Mechanical and Aerospace Engineering, Princeton University

Abstract. The achievable contrast level for space-based detection of exo-planets will be limited by the stability of the optics. As a consequence, active amplitude and phase compensation will be needed. High order aberrations can arise from both non linearity and Fresnel propagation, their wavelength dependence is studied here and it is concluded that λ independent errors are dominant. The behavior of a Michelson interferometer equipped with two deformable mirrors under chromatic light is then presented, and an achromatization method based on a dispersive element is suggested.

1. Reflectivity errors and Fresnel induced errors

1.1. *Contrast degradation due to amplitude errors.*

Visible light imaging of exo-planets will require both an instrument whose intrinsic design enables high dynamic range imaging and an extremely stable wavefront. In order to achieve successful planet detection the TPF mission will need a 10^{-10} contrast ratio in the zone of the image plane where planets photons are expected to land, called "dark zone". At this extreme level of starlight extinction, not only phase errors have to be corrected but amplitude fluctuations have to be taken into account 1. Maintaining a uniform illumination in spite of the uneven reflectivity of the primary mirror is the main problem justifying the need for an active amplitude compensator, but other amplitude errors can arise as a consequence of Fresnel propagation.

1.2. *Expansion of the pupil plane aberration*

In order to understand the wavelength behavior of Fresnel induced errors we will expand the aberration as a sum of λ^k, with $k \in \mathbb{Z}$. This approach relies on the propagation of spatially periodic aberrations, so we will adopt a Fourier representation for both the amplitude and the phase error. The phase $\phi(x)$ will be decomposed in harmonic series with Fourier coefficients $\{\frac{b_l}{\lambda}\}_{l\in\mathbb{Z}}$, the reflectivity $r(x)$ will be represented by the series $\{a_k\}_{k\in\mathbb{Z}}$. Expanding the phase exponential leads to a more explicit wavelength dependence of the aberration:

$$abb(x) = (1+r(x))e^{j\phi(x)} = 1+a_k e^{j2\pi\frac{k}{D}x} + j\frac{b_l}{\lambda}e^{j2\pi\frac{l}{D}x} + a_k\frac{b_l}{\lambda}e^{j2\pi\frac{m\ (k,l)}{D}x} + \left(\frac{b_l}{\lambda}\right)^2 e^{j2\pi\frac{l}{D}x} + \ldots \tag{1.1}$$

The next step is to evaluate the Fresnel integral for each one of these terms.

1.3. *Fresnel propagation of a Fourier component*

We choose not to take into account the $\frac{1}{\lambda z}$ factor in front of the propagation integral, and to consider the case of an infinite aperture. Then, if we isolate a given spatial frequency in the first order of the phase expansion, the field in the pupil plane after a distance z is:

$$E(\xi, z) = \int_{-\infty}^{+\infty} (1 + j\frac{b_l}{\lambda} e^{j2\pi \frac{l}{d} x}) e^{j\frac{\pi}{\lambda z}(x-\xi)^2} dx \tag{1.2}$$

The integration of the constant yields a uniform field.The second part has a closed form that can be written as:

$$E_2(\xi) = j\frac{b_l}{\lambda} e^{j2\pi \frac{l}{d} x} e^{j\pi \frac{l^2}{D^2} \lambda z} \tag{1.3}$$

This result exhibits a coupling between real and imaginary part of the aberration. This Fresnel factor justifies the existence of phase induced amplitude errors and amplitude induced phase errors. In order to generalize this result to a finite aperture we turned ourselves to numerical simulations: the case of a $D = 1$ m gaussian apodized pupil, under a $\lambda_0 = 600$ nm incident light was considered, the original aberration was phase only, pure imaginary. As shown on figure 1, after 10 m the periodic component of the field becomes both real and imaginary: we observe a phase induced amplitude error. After a propagation of a distance $z_T = \frac{2D^2}{\lambda l^2}$ the initial pupil is retrieved which concurs with the periodicity of the fresnel factor derived above.

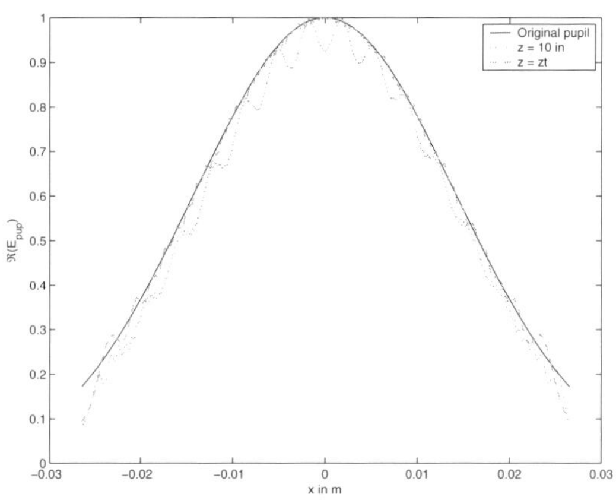

Figure 1. Illustration of phase induced amplitude errors: real part of the near field.

1.4. *Final expansion of the field*

Since $\pi \frac{l^2}{D^2} \lambda z$ is small, the final step in our expansion of the pupil plane field is to consider a Taylor series of this Fresnel factor. Then, every term in equation (1.1) is multiplied by a polynomial in λ. The final expression for the field at any distance z in the optical train can then be written as:

$$E_z(\xi) = \sum_n c_n(\lambda, z) e^{j2\pi \frac{n}{D} x} \tag{1.4}$$

Table (1) summarizes the contribution the wavelength dependence of the coefficients c_n and the contribution of the a_k and b_k in this final expansion. Note the wavelength independence of the phase induced amplitude term, whose spectral behavior is thus similar to reflectivity errors. Since, to the first order, amplitude errors are wavelength independent, a truly polychromatic amplitude actuator becomes necessary.

	Expansion of the aberration			
Fresnel Factor	i $\frac{b}{\lambda}$	a	$i\frac{ab}{\lambda}$	$\frac{b^2}{\lambda}$
1	$i\frac{b}{\lambda}$	a	$\frac{ab}{\lambda}$	$\frac{b^2}{\lambda}$
$i\frac{\pi z k^2}{D^2\lambda}$	$\frac{\pi z k^2}{D^2}b$	$i\frac{\pi z k^2\lambda}{D^2}a$	$\frac{\pi z k^2}{D^2}ab$	$\frac{\pi z k^2}{D^2\lambda^2}b^2$

Table 1. Wavelength dependence of each term in the expansion of a Fresnel Propagated aberration.

2. Chromatic leakage in a zero path length difference interferometer

2.1. *Michelson interferometer equipped with two deformable mirror*

In this section we consider a zero path length difference Michelson interferometer used as an amplitude compensator2. Two deformable mirrors provide two degrees of freedom, and using them in an interferometric configuration enables to convert their phase contribution into Path Length Difference and thus into amplitude variations. This can be written as:

$$\underline{E_{out}(x,y)} = \frac{E_{in}(x,y)}{2}\cos\left(\frac{\pi}{\lambda}(z_1(x,y) - z_2(x,y))\right)\exp\left(j\frac{\pi(z_1(x,y) + z_2(x,y))}{\lambda}\right) \quad (2.1)$$

Where E_{in} is the incident electrical field, E_{out} the outgoing one, z_1 and z_2 are the heights of each deformable mirror, and λ is the wavelength of the incoming light.

3. Chromatic leakage of a classical interferometer under polychromatic light

While phase correction is polychromatic, the amplitude compensation is only perfect for a central λ_0 and all other incident wavelengths will leak and degrade the contrast. As an example, if all the energy of a λ independent aberration is contained in one big speckle, a contraint of 10 orders of magnitude in contrast leads to the requirement of $\delta\lambda/\lambda = \frac{1}{20}$, with an amplitude error of 5×10^{-4}. While the dominant amplitude errors are wavelength independent, the expansion in (1) suggests to evaluate the influence of a zero path length difference interferometer on $1/\lambda^k$ errors. Figure 3 plots $< log(PSF) >$ as a function of the bandwidth for different wavelength dependences of the aberration. For $k = 2$ the correction is actually polychromatic. This result can be proved analytically by using a Taylor expansion of (2.1), which is justified since $d(x)/\lambda$ is expected to be small. The cancellation of a $E_{abb} = E_0(1+\epsilon cos(fx))$ amplitude ripple using an harmonic PLD,$d(x) = a_{corr}cos(f_{corr}x)$, leads to solve (3.1) for a_{corr}.

$$E_{out}(x) = (1 + \frac{\epsilon}{\lambda^2}cos(fx))\left(1 - \frac{\pi^2 a_{corr}^2}{2\lambda^2}cos^2(f_{corr}x)\right) \quad (3.1)$$

Hence, by choosing $a_{corr}^2 = \frac{4\epsilon}{\pi^2(1+\epsilon)}$ and $f_{corr} = f/2$, polychromatic correction can be achieved in the case of $1/\lambda^2$ amplitude errors.

4. Achromatization using a dispersive cell

The correction of λ independent amplitude errors remains the biggest concern since they are the largest source of luminosity non uniformity. In an effort to circumvent any chromatic leakage associated to them, we suggest in this section a concept of polychromatic amplitude control.

4.1. *Insertion of a dispersive gaseous cell*

Figure 4(b) illustrates the insertion of a gas cell in a one leg of a Michelson interferometer. The ides of tuning the a path length difference using dispersion and pressure changes has

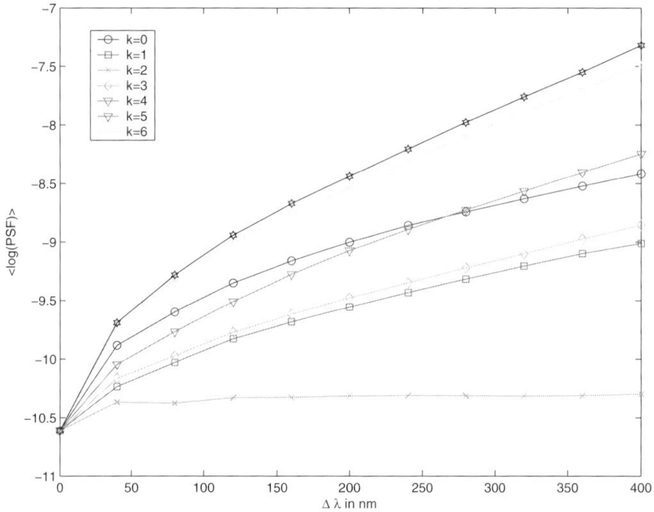

Figure 2. Chromatic leakage for several $1/\lambda^k$ dependence.

been used since the 50's, when Jacquinot 3 tuned a Fabry Perot spectroscope by enclosing it in a pressurized chamber. For our application we consider a cell with a gas having the linear dispersion $n(\lambda) = 1 + \alpha(P)\lambda$, where P is the pressure of the gas within the cell. Figure 3(a) illustrates the chromatic behavior of a classical Michelson interferometer, for which the response to a given Path Length Difference varies with wavelength. As stated previously, chromatic leakage is minimized if the amplitude compensator operates around Zero Path Length Difference. Using a dispersive gas enables to operate at constant ZPLD and introduces what is the equivalent of a polychromatic phase shift, as seen on figure 3(b). The necessary degree of freedom to control the amplitude, or to adjust the height of the intersection with the y axis, would come from a pressure adjustment in the gas cell.

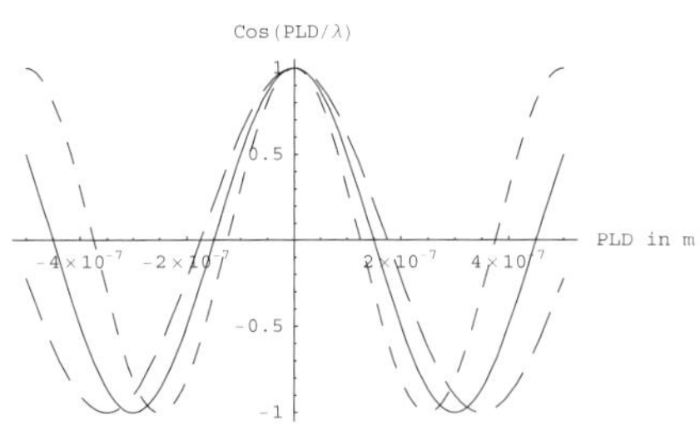

(a) Profile of the amplitude response of classical Michelson interferometer

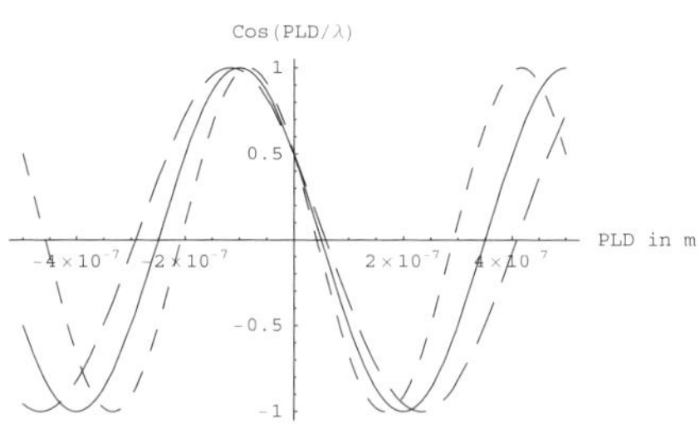

(b) Profile of the amplitude response of a Michelson interferometer with a gas cell inserted in one leg

Figure 3. Comparison of the response of the interferometer for three wavlength: 500 nm, short dashes, 600 nm, full line, 700 nm, long dashes.

A pixelated polychromatic amplitude control could thus be obtained by using two flat mirrors and stacking together several cells with independently controlled pressure. We are aware that such a system requires adressing technical issues that go far beyond the scope of this paper. Our objective here is merely to introduce a novel idea for polychromatic amplitude control, a critical component of the TPF mission, leaving aside technological feasibility for further studies.

4.2. *Proof of concept: linear dispersion of nitrogen*

The fundamental assumption underlying the achromatization is that the gas is linearly dispersive. This feature is confirmed experimentally for Nitrogen by Peck and Khanna 4. Figure 4(a) has been drawn for $P_0 = 1$atm and $T_0 = 288$K, the dotted line is the actual dispersion profile, the influence of the error introduced by the linearization remains under the critical value of the 10^{-5} requirement for the electrical field magnitude.

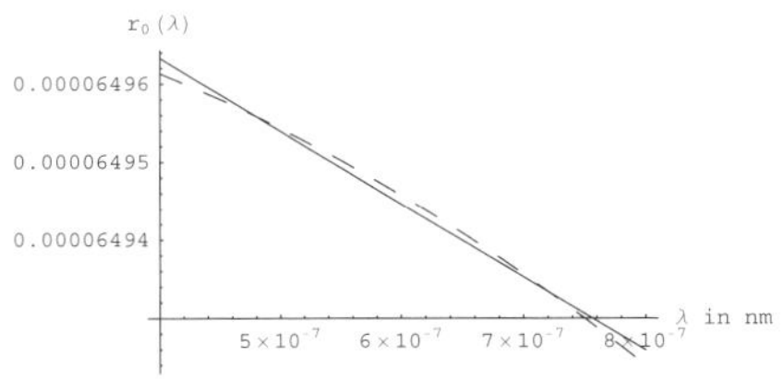

(a) $r_0(\lambda)$ in the visible, note the quasi linearity

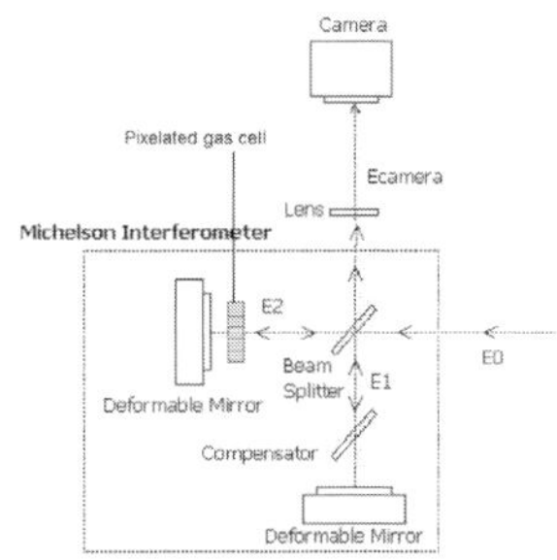

(b) Polychromatic Michelson interferometer with a gas cell

Figure 4. Concept of a dispersive element in an interferometer, and dispersion of Nitrogen in the visible.

5. Conclusion

The necessity to compensate for amplitude errors in coronographs dedicated to exoplanet detection comes from tight contrast requirements and Fresnel propagation of aberration has thus to be taken into account. We introduced a systematic approachto studying the chromatic effects of aberrations that is the first step to an eventual polychromatic correction. The interesting behavior of an interferometer a broadband corrector for $1/\lambda^2$ aberrations is proved, and chromatic leakage is quantified for other wavelength dependencies. Finally we proposed a concept to correct for λ independent amplitude errors and showed a short proof of its validity based on a literature review. Our next step is to carry out this idea experimentally.

References

Brown, R.A. & Burrows, C.J 1990, *Icarus* 87, 484

Littman, M.G., Carr, M., Leighton, J., Burke, E., Spergel, D.N., & Kasdin, N.J. (2002), *Proc SPIE* 4860, 182

Jacquinot, P. 1954, *JOSA* 44, 761

Peck, E. & Khanna, B.N. (1966) *JOSA* 56, 1059

All photographs: Laurent Thareau [`l.thareau@free.fr`].

Direct Imaging of Exoplanets: Science & Techniques
Proceedings IAU Colloquium No. 200, 2005
C. Aime and F. Vakili, eds.
© 2006 International Astronomical Union
doi:10.1017/S1743921306009732

The Effects of Instrumental Elliptical Polarization on Stellar Point Spread Function Fine Structure

Joseph C. Carson[1,2], Brian D. Kern[1], James B. Breckinridge[1], and John T. Trauger[1]

[1]Jet Propulsion Laboratory, California Institute of Technology, 4800 Oak Grove Dr., Pasadena, CA, 91101 USA
[2]California Institute of Technology, Pasadena, CA, USA

Abstract. We present procedures and preliminary results from a study on the effects of instrumental polarization on the fine structure of the stellar point spread function (PSF). These effects are important to understand because the the aberration caused by instrumental polarization on an otherwise diffraction-limited will likely have have severe consequences for extreme high contrast imaging systems such as NASA's planned Terrestrial Planet Finder (TPF) mission and the proposed NASA Eclipse mission. The report here, describing our efforts to examine these effects, includes two parts: 1) a numerical analysis of the effect of metallic reflection, with some polarization-specific retardation, on a spherical wavefront; 2) an experimental approach for observing this effect, along with some preliminary laboratory results. While the experimental phase of this study requires more fine-tuning to produce meaningful results, the numerical analysis indicates that the inclusion of polarization-specific phase effects (retardation) results in a point spread function (PSF) aberration more severe than the amplitude (reflectivity) effects previously recorded in the literature.

Keywords. polarization, instrumentation: high angular resolution, methods: analytical, methods: laboratory.

---

1. Introduction

Breckinridge & Oppenheimer (2004) published a numerical study on the effects of variations in reflectivity from an off-axis $F/1.5$ parabolic telescope primary mirror on the resulting point spread function (PSF), in the context of high-contrast imaging. That study investigated reflectivity variations across a curved mirror surface due to variations in the angle of incidence for light rays striking different parts of the mirror. Those reflectivity variations are a form of pupil apodization. The authors calculated that these reflectivity variations affected the PSF by a magnitude on the order of 10^{-5} of the peak intensity. They noted that this result was particularly important to high contrast imaging experiments such as TPF or Eclipse, which require supression of the on-axis PSF to factors of $\sim 10^{10}$. Hence, a 10^{-5} term must be investigated and understood.

Our study takes a similar starting point, regarding the reflection of light off a metallic mirror at a range of angles of incidence. We calculate both the amplitude and phase changes on reflection, for two orthogonal polarizations. In doing so, we conclude that the inclusion of polarization-specific phase effects (retardation) has a greater influence on the resulting PSFs than the amplitude effects (reflectivity) previously calculated. The expected effect should be large enough to be seen in relatively simple laboratory experiments.

The laboratory experiment proposed here does not involve reflection of a plane wave off of a powered metallic surface, but rather reflection of a spherical wave off a tilted flat mirror. The physics of these two configurations is quite similar, in that the angles of incidence of rays across either reflecting surface vary with position, giving rise to variations in reflected amplitude and phase that differ between polarizations. By measuring the resulting PSFs in orthogonal polarizations, the effect should be readily measurable. We are aware of no study that has verified any of these predictions in the laboratory, or examined how real-life optics (with the inclusion of dielectric coatings) may affect these predictions. In particular, we are investigating in-lab the effects of a protected silver coating used by ITT (Rochester, NY) for a number of telescope coatings, which is also a likely candidate for any future TPF or Eclipse mirror coating. The Jet Propulsion Laboratory (JPL) High Contrast Imaging Testbed (HCIT) group, of which J. Carson, B. Kern, and J. Trauger are members, is currently approaching in-laboratory high-contrast results where such polarization-induced aberrations may play a crucial role in the final achievable performance (Trauger *et al.* 2004). This fact provides a primary driver for the investigations described in this paper.

Section 2 describes our numerical predictions. Section 3 describes our experimental setup. Sections 4 presents our conclusions.

2. Numerical Predictions

The optical train used for the numerical analysis corresponding to this experiment was quite simple: a perfectly spherical converging wavefront reflects off a flat mirror, and propagates to a focus. For simplicity, we define "vertical" as the direction defined perpendicular to the propagation direction of the chief ray, lying in the plane of incidence of the chief ray with the mirror, and assume that the chief ray's angle of incidence is 45°. For each point in a grid in the input pupil, the angle of incidence and plane of incidence is calculated at the flat mirror. The complex reflection coefficients for p-polarized (in the plane of incidence) and s-polarized (perpendicular to the plane of incidence) light are copied from tabular data supplied by ITT from measurements on the coating in question. The corresponding Jones matrix for each point is tabulated, relating the amplitude and phase of output light in orthogonal linear polarizations to the amplitude and phase of input light in orthogonal linear polarizations.

If there were no retardation, an input containing only vertically polarized light would produce an output containing only vertically polarized light, and the resulting PSF could be analyzed along the lines of the Breckinridge & Oppenheimer study. With retardation, the output polarization state depends on the location in the input pupil, and will, in general, be elliptically polarized. By separating the output PSF, which was originally a vertically polarized input, into vertically and horizontally polarized components, the effect of retardation can be directly seen in the horizontally polarized PSF, without having to "subtract off" the unaberrated PSF or make additional differential measurements or theoretical assumptions. The difference between this analysis and traditional ellipsometry is that this analysis integrates the retardation over a number of angles and planes of incidence, and observes in the focal plane of the spherical wavefront.

For an input $F/3$ beam at 635 nm, using the protected silver coating from ITT, we predict that the "cross-polarized" PSF (the outputted horizontal polarization, given a 100% vertical input) has 1/260 the intensity of the orthogonally polarized PSF. The predicted PSFs in orthogonal polarizations, for a vertically polarized input, are shown in figures 1a and 1b.

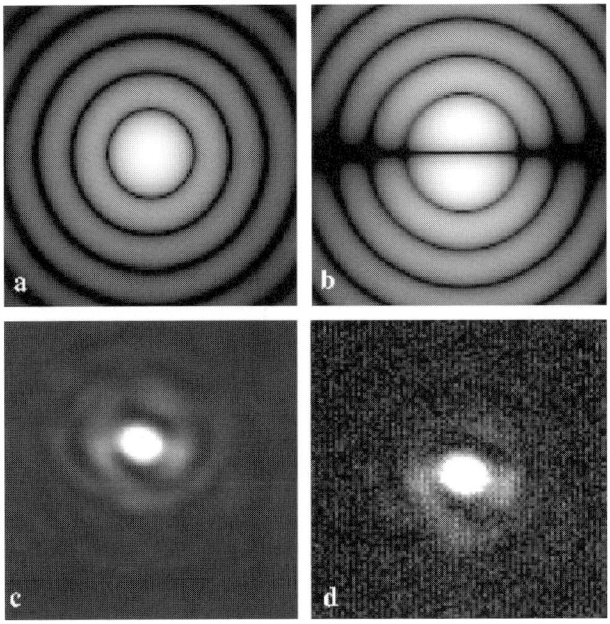

Figure 1. Analytic and experimental results for orthogonally polarized PSFs, given a vertically polarized input. Vertical polarization is on the left and horizontal on the right. Numerical predictions are above (logarithmically scaled) and experimental results below (linearly scaled). The intensity in *1b* is ~0.003 that of *1a*. While our experimental results (frames *1c* and *1d*) show a clear existence of cross-polarization, the noise levels are too high to confirm or verify the expected form of the PSF in *1b*.

3. The Experiment

Fig. 2 shows the setup of our laboratory experiment. We use a standard commercially available polarizing sheet to linearly polarize our incident light. A downstream calcite displacement polarizer (from Karl Lambrecht Corporation) and CCD camera (from Apogee) allow us to image the PSF in orthogonal polarizations simultaneously. A 635 nm laser ($\sim$3 mW) illuminating a 5 μm pinhole serves as our power source. Our iris, while controlling stray light, also gives us the option to change the $F/\#$ from our nominal $F/3$ beam. The lenses flanking the mirror flat were selected to provide a fast beam (which should enhance the cross-polarization effect and allow for a more easily observable aberrated PSF), and an imaged pinhole dimension significantly smaller ($<\lambda/[2\ D]$) than the diffraction limit.

As shown in Figure 1d, we currently do not observe the cross-polarized PSF's bifurcated features, predicted by our numerical simulations (see Figure 1b). The predicted fine-structure bifurcation PSF intensity level of 0.003 indicates that our current $\sim$1% laboratory intensity resolutions are too high to probe the desired PSF fine structure. Thus, we require a sensitivity improvement close to a factor of 10 to really observe the fine structure.

4. Conclusions

We have presented procedures and results from a numerical and laboratory experiments built to test for the effects of instrumental elliptical polarization on stellar point spread function fine structure. Our numerical experiments indicate that polarization-specific phase effects (retardation) have a larger effect on PSF aberrations than the amplitude (reflectivity) effects previously calculated by Breckinridge & Oppenheimer

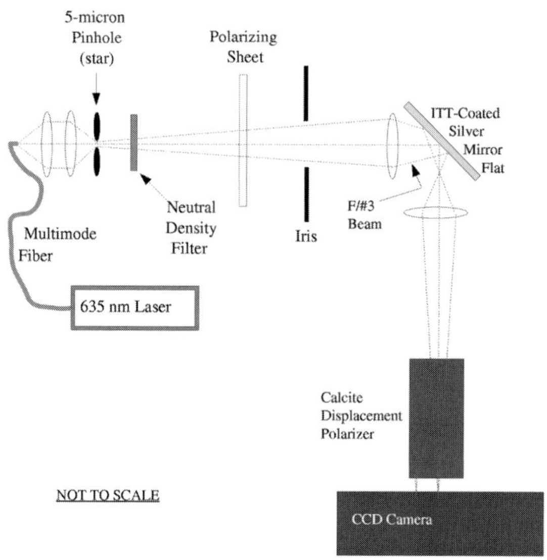

Figure 2. Our experimental setup. A 635-nm laser and 5-micron pinhole simulate our target source. The polarizing sheet ensures linearly polarized incident light for a straightforward analysis. The iris controls unwanted stray light and allows us to increase our $F/\#$, if desired. The lenses flanking the mirror flat ensure a fast beam (to exacerbate the cross-polarization effect and therefore provide an easily observable aberrated PSF) and an imaged pinhole significantly smaller ($<\lambda$ / (2 D)) than resolvable Fraunhofer diffraction pattern features. Our calcite displacement polarizer and CCD camera allow us to record the resultant PSF in orthogonal polarizations.

(2004). A comparison of the experimental results with the analytic results indicates that sensitivities in our laboratory measurements must improve by a factor of $\sim$10 before we may hope to observe the predicted deviations from the Fraunhofer PSF. Given the early stage of the project and the recognized opportunity to minimize certain noise sources through baffling and more rigorous calibrations, we remain guardedly optimistic that we may observe such an effect in the near future.

Acknowledgements

We would like to acknowledge the contribution of Albert Niessner (JPL) in helping to implement the software side of the experimental setup. We also thank Ben Oppenheimer (AMNH) for useful discussions. The research described in this paper was carried out at the Jet Propulsion Laboratory, California Institute of Technology, under a contract with the National Aeronautics and Space Administration.

References

Breckinridge, J. B. & Oppenheimer, B. R. 2004, *ApJ* 600, 1091

Trauger, J., Burrows, C., Gordon, B., Green, J., Lowman, A., Moody, D., Niessner, A., Shi, F., & Wilson, D. 2004, *SPIE* 5487, 1330

Direct Imaging of Exoplanets: Science & Techniques
Proceedings IAU Colloquium No. 200, 2005
C. Aime and F. Vakili, eds.
© 2006 International Astronomical Union
doi:10.1017/S1743921306009744

Interferometric apodization of rectangular aperture – Laboratory Experiments

Y. El Azhari[1,2], O. Azagrouze[1], F. Martin[3], R. Soummer[4] and C. Aime[3]

[1]Laboratoire de Physique des Hautes Énergies et Astrophysique, Département de Physique, Faculté des Sciences Semlalia, Université Cadi Ayyad, B.P. 2390, Marrakech, Morocco
email: elazhari@ucam.ac.ma

[2]Laboratoire d'Optique et Optoélectronique, Département de Physique, École Normale Supérieure, Marrakech, Morocco
email: youssef.elazhari@ensma.ac.ma

[3]Laboratoire Universitaire d'Astrophysique de Nice, UMR 6525 Université de Nice Sophia Antipolis, France
email: claude.aime@unice.fr

[4]American Museum of Natural History, 79th Street at Central Park West, New York, NY 10024 USA
email: rsoummer@amnh.org

Abstract. In this laboratory experiment, we study the possibility of producing an apodization of the pupil of a telescope using a classical Michelson interferometer. To simulate the star, we successively used a Laser source, a source of spectral light and a source of white light. Our goal is to study the performance of the assembly with polychromatic light. We present the results of experiments carried out with a rectangular aperture using a HeNe Laser and Na spectral light sources.

Keywords. instrumentation: interferometers; methods: analytical, laboratory; techniques: interferometric.

1. Introduction

The experimental work reported in this poster was made at the École Normale Supérieure of Marrakech, in collaboration with the University of Nice Sophia Antipolis. The goal is to realize an achromatic prolate apodized coronagraph (Aime (2005)), and one of the first steps is to study how the required pupil apodization can be produced by interferometry. Using a classical Michelson interferometer, it is only possible to obtain cosine apodizations, but Pueyo *et al.* (2004) have shown that this technique could be generalized to any apodizing function using deformable mirrors.

The principle of the experimental realization of the interferometric apodization was detailed by Aime *et al.* (2001), and first preliminary results were given by Soummer (2002). The basic principle is to use the Michelson interferometer to project fringes on the entrance aperture of the telescope, therefore producing a cosine transmission. We used a square aperture, and produced the apodization in a single direction.

2. Experimental Assembly

The assembly proposed to carry out the interferometric apodization is represented in figure 1. The Michelson interferometer allows to illuminate the lens L_3 by a coherent superposition of two plane waves of width D and wave vectors $\mathbf{k}_\pm = \frac{2\pi}{\lambda}(\pm\frac{\alpha}{2}\,\mathbf{x} + \mathbf{z})$. $\alpha = (\widehat{\mathrm{M}_1, \mathrm{M}'_2}) \ll 1$ is the angle between M_1 and M'_2. The amplitude of this wave at the entrance of L_3 is of the form $2\cos(\frac{\pi\alpha x}{\lambda})$. It equals zero at the edges of the aperture for $\alpha = \lambda/D$. If the quadratic phase term is ignored, the amplitude of the wave in the plane P_2 is given by the Fourier transform of ψ_{L_3} (Goodman (1996)):

$$\psi_{\mathrm{P}_2}(x) = \frac{\cos \pi X}{1 - 4X^2} \qquad \text{where} \qquad X = \frac{D\,x}{\lambda f_3} \tag{2.1}$$

This shows well that the distribution of amplitude obtained in the plane P_2 is similar to that corresponding to a slit of width D apodized by the function $\cos(\pi x/D)$.

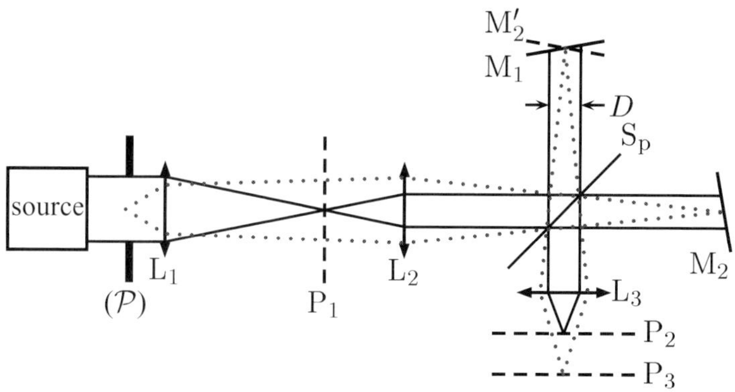

Figure 1. Experimental Assembly. The Michelson Interferometer is represented by the mirrors M_1 and $\mathrm{M}_2$and the beam splitter S_p. The lens L_1 represents the telescope of focal plane P_1. The lens L_2 allows to illuminate the interferometer by a plane wave and it forms an image of the entrance pupil $(\mathcal{P})$ of the telescope on the mirror M_2. The lens L_3 transports the final image of $(\mathcal{P})$ to the plane P_3 and the focal plane image of the telescope to the plane P_2. All optical surfaces have a flatness of $\lambda/20$ at 633 nm.

The source simulates a star which illuminates the entrance pupil $(\mathcal{P})$ of the telescope with a plane wave. The lens L_1 schematizes the telescope of focal plane P_1. The two arms of the Michelson have the same length. M'_2 is the image of M_2 by the beam splitter S_p; it forms a small angle α with M_1. The lens L_2 allows to illuminate the interferometer with a plane wave. It forms also an image of $(\mathcal{P})$ on the mirror M_2. The lens L_3 transports the final image of the pupil in the plane P_3. It also transports the focal plane of the telescope in the plane P_2. The interferometer is illuminated with a plane wave. It allows to obtain interferences between two plane waves. The fringes obtained are delocalized and have a period $i = \lambda/\alpha$.

3. Results and Discussion

The experimental tests were carried out successively with a HeNe laser and a sodium (Na) spectral source (figure 2).

3.1. *HeNe Laser Source*

The HeNe laser used has a power of 0.8 mW. It emits a red light of wavelength $\lambda = 632.8\,\mathrm{nm}$. The quasi-cylindrical beam, of small section ($\phi_0 \approx 2\,\mathrm{mm}$), is widened using the afocal system made up of $\mathrm{L}_{0_1}(+5\,\mathrm{mm})$ and $\mathrm{L}_{0_2}(+100\,\mathrm{mm})$ and represented in figure 2(a).

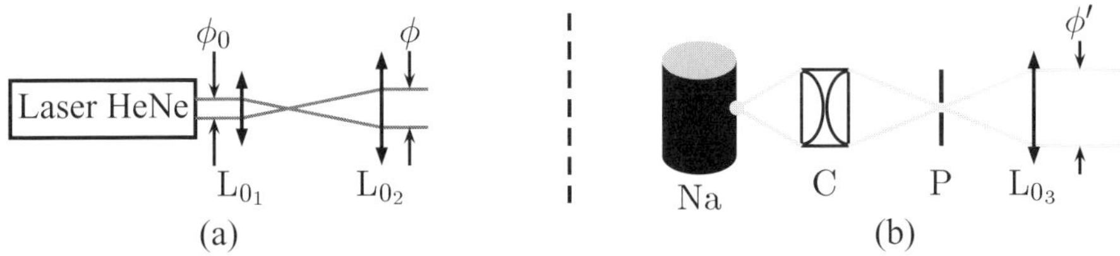

Figure 2. Sources used. (a): the afocal system formed by L_{0_1} (+5 mm) and L_{0_2} (+100 mm) lenses allows to enlarge the quasi-cylindrical beam emitted by the HeNe Laser. (b): the condenser C concentrates the beams coming from the Na Spectral Lamp on a pinhole P($\varnothing = 57\,\mu$m) placed at the first focus of the lens L_{0_3}.

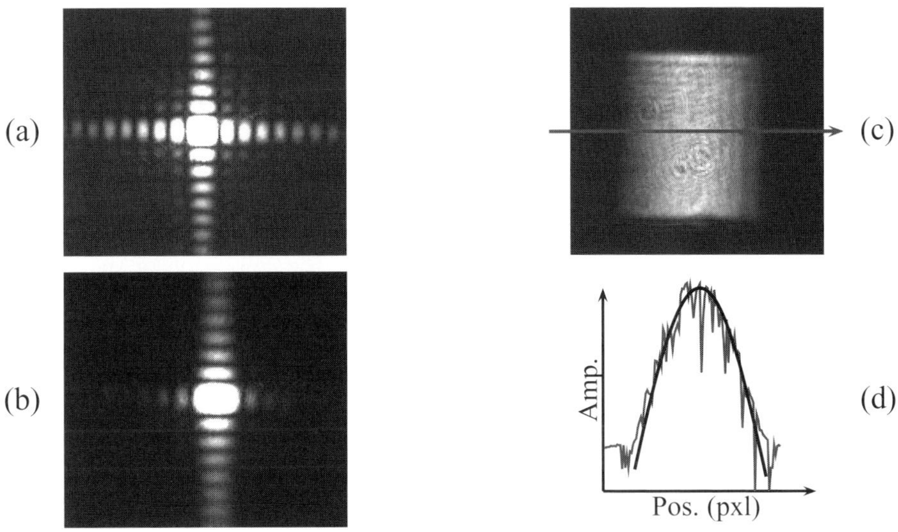

Figure 3. Example of results obtained with the Laser Source. (a): Non-apodized PSF obtained in the plane P_1. (b): Apodized PSF obtained in the plane P_2. (c)&(d) give respectively the distribution of intensity in the pupil plane P_3 and its cut parallel to the direction of apodization.

Figure 3 gives a typical example of results obtained with the laser source. The non-apodized PSF, obtained in the plane P_1, is represented in figure 3(a). It corresponds to a typical Fraunhoffer diffraction pattern of a rectangular aperture. In particular, the intensity of the first secondary maximum represents approximately 5.0% of that of the principal maximum. This experimental value is very close to the theoretical value which is 4.7%. Figure 3(b) gives the PSF obtained in the plane P_2 when the value of angle $\alpha = (\widehat{M_1, M_2'})$ is set to the optimal value $\alpha = \lambda/D$. The intensity of the 1st secondary maximum falls to 0.6% of that of the principal maximum whereas the theory envisages 0.5%. Figures 3(c) and (d) give respectively the distribution of intensity in the pupil plane P_3 and its cut parallel to the apodized direction. A fit of the intensity profile, thus obtained, with the function $\cos^2(\pi x/b\,D)$ gives $b = 1.14$.

3.2. *Na Spectral Source*

To simulate a star with a Na spectral source, we used the assembly represented in figure 2(b). The condenser C concentrates the beams coming from the spectral lamp on a pinhole P of a diameter $\varnothing = 57\,\mu$m. The pinhole P is placed at the first focus of the lens L_{0_3} which illuminates then the interferometer with a cylindrical beam of diameter ϕ'.

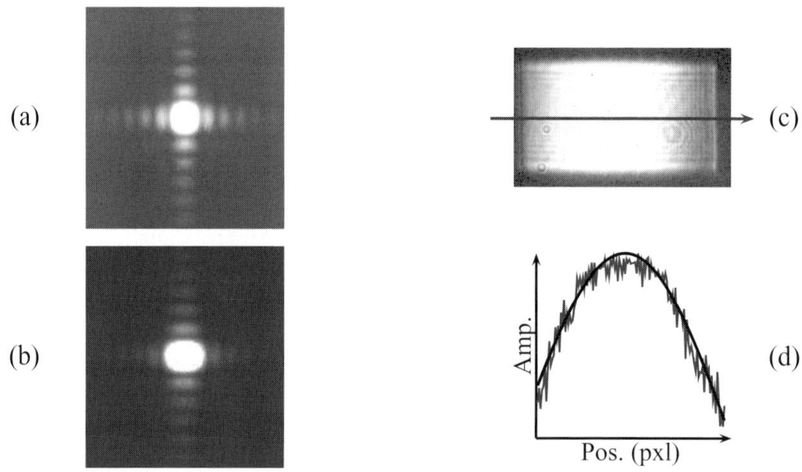

Figure 4. Example of results obtained with the Sodium Spectral Source. The comment is the same than the figure 3.

Figure 4 gives a typical example of results obtained with the light of Na spectral source. It shows the combined effects of the apodization and the resolution of the source (too large pinhole). In fact, it can be noted that despite the fact that the value $b = 1.10$ is close to the optimum one ($b = 1$), the apodization is less effective. This is due to the fact that the Na source is spatially incoherent. Indeed, in the non-apodized PSF, the intensity of the 1st secondary maximum in this case represents almost 8% of that of the principal maximum which is far from the theoretical value. In addition, the apodization disappears almost completely as soon as the diameter of the pinhole P exceeds 0.1 mm approximately. To obtain a better apodization with the Na source, it would have been necessary to use a pinhole with an even smaller diameter. The intensity is then likely to become too low.

4. Conclusion

We showed the experimental realization of the interferometric apodization of a rectangular pupil with a Michelson interferometer. For that we used a laser source then a sodium spectral source. The results obtained with the laser source show a good agreement with the theory. With the sodium source, the main limitation is due to the size of the pinhole, corresponding to a resolved source, for which the apodization effect is degraded.

The study was carried out in the case of only one dimension and has served as a precursor of a new experiment in which a Mach-Zehnder interferometer will be used. As described by Aime (2005), phase plates will be introduced to obtain a prolate apodization.

References

Aime, C., Soummer, R., & Ferrari, A. 2001, *A&A* 379, 697
Aime, C. 2005, *PASP* 117, 1012
Goodman, J.W., *Introduction to Fourier Optics, 2nd edition*, McGraw-Hill, San Francisco (1996)
Pueyo, L., Give´ on, A., Littman, M.G., Kasdin, J., & Vanderbei, R. 2004, SPIE 5490, 545
Soummer, R. 2002, PhD thesis, Sophia-Antipolis University, Nice

All photographs: Laurent Thareau [l.thareau@free.fr].

All photographs: Laurent Thareau [`l.thareau@free.fr`].

Direct Imaging of Exoplanets: Science & Techniques
Proceedings IAU Colloquium No. 200, 2005
C. Aime and F. Vakili, eds.
© 2006 International Astronomical Union
doi:10.1017/S1743921306009756

Three-dimensional common-path interferometer for AIC: nulling of polychromatic light

A. V. Tavrov[1], Yu. Otani[1], T. Kurokawa[1] and M. Takeda[2]

[1]Tokyo University of Agriculture and Technology,
2-24-16, Nakamachi, Koganei-shi, Tokyo, Japan
email: tavrov@cc.tuat.ac.jp

[2]The University of Electro-Communications,
1-5-1, Chofugaoka, Chofu, Tokyo, Japan
email: takeda@ice.uec.ac.jp

Abstract. Three-dimensional common-path interferometer is proposed to obtain achromatic nulling for the on-axial source; the off-axial source remains detectable. The 3D interferometer involves $\pm 90^\circ$ polarization rotations in each interferometer arm. That results in the achromatic 180° phase shift, so that the on-axial source interferes destructively. Depending on the source axial position, the light energy is split by different ratios between the Bright and the Nulled interferometer outputs. For the linearly polarized on-axial source, all the energy at nearly 100% is directed to the Bright port. For the off-axial source, the light is split by the ratio at nearly 50%/50%between the Bright and the Nulled ports. The common-path scheme compensates effectively the optical path difference (OPD) and it remains stable to mechanical vibrations. Theory, simulations and preliminary breadboard experiments are shown to be in reasonable agreement.

Keywords. Nulling interferometry, coronagraphy, common-path interferometry, achromatic nulling, mechanical stability.

---

1. Introduction

Achromatic photonic devices are in demand in variety of optical metrology applications from ultra-short pulse laser beams, to radar techniques and astronomical instrumentation Serabyn, Wallace, *et al.* (1999); Baudoz, Rabbia & Gay (2000); Baba, Murakami & Ishigaki (2001). A star emits in 10^7–10^{10} more light energy than the illuminating target planet depending on the wavelength band. Referring to the concept by Bracewell (1978), the long-base interferometry presents a high angular resolution technique in which the light from on-axial bright source is destructively interfering, when an off-axial low intensity target interferes constructively or remains not eliminated and therefore can be detected. To realize this principle, many schemes have been recently proposed, among which are phase masks, Rouan, *et al.* (2000), and pupil masks, delay lines arrays , Avoort, Mieremet, Pereira & Braat (2004), achromatic phase shifters, Rabbia, Gay, Bascou & Schneider (2001), and their sophisticated combinations Nishikawa, Kotani, Murakami, Baba, Itoh & Tamura (2005), just to mention a few. Among others, a fully achromatic π-shifter is a key device indispensable for the instrumentation of nulling interferometry and coronagraphy.

In the achromatic interfero-coronagraph, Baudoz, Rabbia & Gay (2000), in the Michelson interferometer, the both arms have to be adjusted accurately at the zero OPD, which

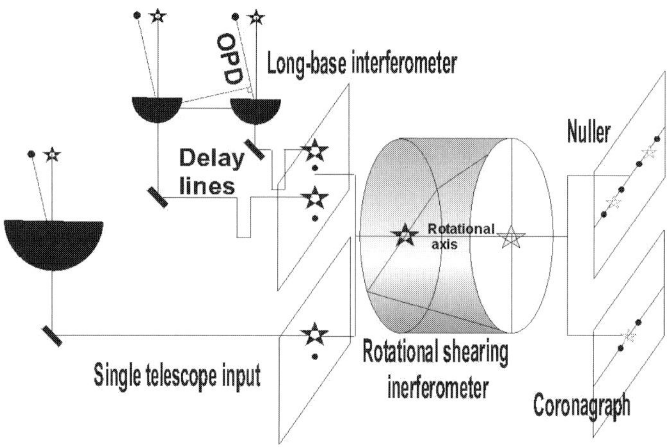

Figure 1. Nulling principle for long-base interferometry and coronagraphy.

is hard to realize with a mechanically unstable non-common-path interferometer. In this paper, a recently proposed common-path interferometer scheme, Tavrov, Tanaka, Shioda, Kurokawa & Takeda (2004), is analyzed, which effectively maintains the OPD between the interferometer arms and simultaneously adds an achromatic π-phase shift for destructive interference. For linearly polarized light, aside from the reflection losses on the mirrors, the interferometer throughput is enhanced then in the previous scheme, Tavrov, Bohr, Totzeck, Tiziani & Takeda (2002); for an on-axial source it can attain nearly a 100-to-0 percent power split ratio at the two exit ports of the interferometer, which are referred to Bright and Nulled ports, respectively. For an off-axial source, equal power comes out from the two ports with nearly 50-to-50 percent split ratio.

2. Combination of two beams with 180° RSI scheme

Let us explain, how, by means of the nulling interferometry, the light from an on-axial strong source is reduced via destructive interference, and how, by contrast, an off-axial faint source remains detectable. As shown in Figure 1, the OPD between two interferometer arms is equal to zero for on-axial source, denoted by star-mark. Its destructive interference occurs after the achromatic π- phase shifter by means of 180° RSI (rotational shearing interferometer), Serabyn, Wallace, *et al.* (1999); Scholl & Paez (1999). RSI means a differential interferometer with some rotation of the copied wavefront. A similar wavefront folding system was extensively studied by Itoh (1996) with a very small amount of the rotational shear. In Figure 1, shown by o-mark, the off-axial source introduces a non-zero optical path difference, its interference becomes no longer destructive and such an off-axial source can be detected. Both the stellar coronagraph and the long-base interferometer can use the proposed interferometer as an achromatic nulling beamcombiner.

3. Optical scheme

Conventional Sagnac interferometer, Hariharan (1995), maintains the OPD because of two interferometer arms are designed within the common path. For the nulling purpose, the Sagnac interferometer is extended in 3D, its optical scheme is shown in Figure 2.

This interferometer introduces $\pm 90°$ polarization rotations along the *forward* and the *backward* light paths. Three-dimensional mirror train consists of six plane mirrors

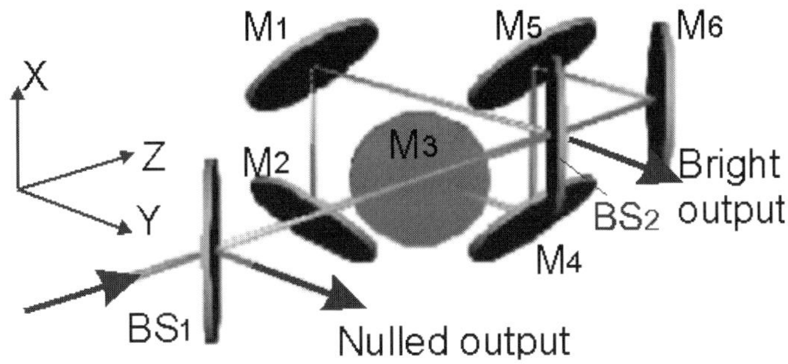

Figure 2. Out-of-plane common-path interferometer.

denoted consequently by $M_{1..6}$ and two beamsplitters: BS_1, BS_2. Principal rays are shown along the optical axis. Let us define the forward propagation following the route $BS_1 - BS_2 - M_1 - M_2 - M_3 - M_4 - M_5 - M_6 - BS_2 - BS_1$ that is the first interferometer arm. The backward path: $BS_1 - BS_2 - M_6 - M_5 - M_4 - M_3 - M_2 - M_1 - BS_2 - BS_1$ introduces the second interferometer arm. Local p- and s- planes of the mirrors $M_{1..6}$ and the beamsplitter BS_2 are oriented *out-of-plane*, Tavrov, Miyamoto, Kawabata, Takeda & Andreev (2000) and cause $\pm 90°$ polarization rotations in the first and the second interferometer arms correspondingly. In respect each other, the images in the arms become geometrically rotated to 180°. Electrical vectors follow the image rotation and it introduces the π-phase shift for destructive interference.

This is similar to the effect in the existing AIC scheme, Baudoz, Rabbia & Gay (2000), where one interferometer arm contains the plane mirror train and another arm contains the propagation through the focus so two images are superimposed, and flipped to 180° each to other.

To achieve a dark field, both interfering beams must have equal intensities, which is not realized in the common case by splitting and combining of the beams. An actual beam splitter works chromatically because it introduces different amounts of transmission **T** and reflection **R**, and the ratio depends on the wavelength. Following the reference, Serabyn, Wallace, *et al.* (1999), this problem is solved, making possible splitting and combining of beams with the equal **RT** and **TR** products.

Interference process always satisfies the energy conservation law, therefore in our scheme one BS_2 port reflects the light energy resulted by constructive interference and is denoted as the **"Bright output"**. Another BS_2 port transmits the rest light energy resulted by destructive interference to the **"Nulled output"**. The nulled output faced back to the incident light direction and the nulled result is observed via BS_1. To enhance the RSI transmittance for coronagraphy, the BS_1 is designed as the broadband polarization beamsplitter (PBS). Its transmitting mode is set matching with the incident linear polarization state on the input therefore nearly complete energy flux can be entered into the interferometer. Interferometer arms rotate the input polarization to $\pm 90°$ so the light polarization which is returned back to the BS1 matches now with the reflection mode of the PBS and again nearly complete energy flux can be directed to the Nulled output. Following the geometrical image rotation, the π-phase difference is added achromatically.

Ray tracing shows that an off-axial light component has another interference result: each tilted wave results in two images, which are spatially separated, as shown in Figure 1. They do not interfere because they are geometrically separated. By the closer axial position than the Airy diameter (close-sensing), Baudoz, Rabbia & Gay (2000), some

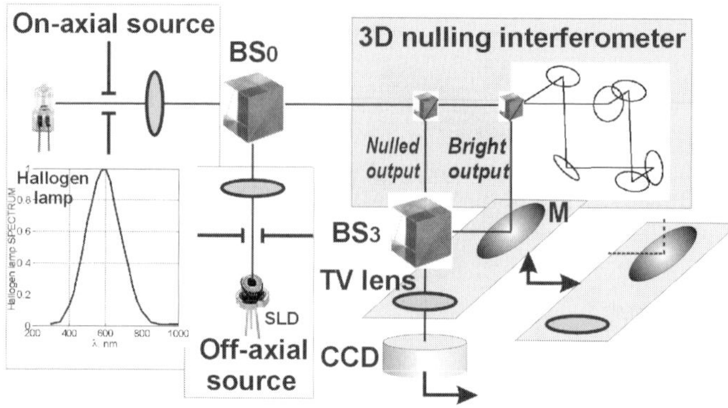

Figure 3. Experimental scheme.

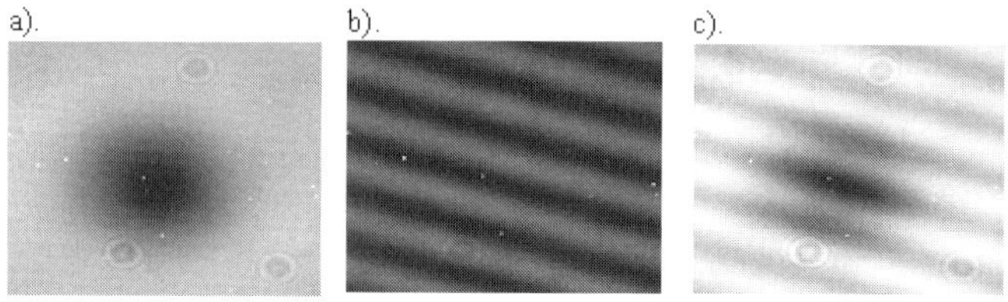

Figure 4. Pupil domain interference patterns: (a) *extended* on-axial source, (b) off-axial source, (c) superposition of the on-axial source and the off-axial source.

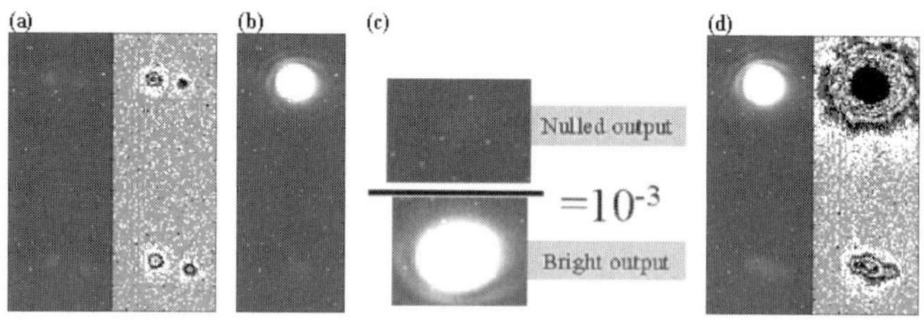

Figure 5. Achromatic nulling, the *Bright output* is above the *Nulled output*.(a) – Off-axial source is switched on. (b) – On-axial source is switched on. (c) – Intensity values were summed up in zooms, the nulling depth is about 10^{-3}. (d) – Both the on-axial and the off-axial sources are switched on.

of their fractions still interfere destructively, hence some non-interfering fractions cause the target light amount that is not nulled and can be detected. Due to the propagations forward and back, the proposed design compensates chromaticities of the beamsplitter and the mirrors. Achromatic phase shift between two interferometer arms is shown equal to 180° and the interference contrast of 100% is possible.

4. Experiment

In order to confirm an achromatic nulling property introduced by the proposed 3D common-path interferometer by simple experiments, we have constructed the bread-boarded demonstrator, its optical set-up is shown in the Figure 3. We made use of two separate light sources: the halogen lamp and the super-luminescent diode (SLD). The halogen lamp has emitted the spectral width over 300 nm, it was used to construct the on-axial source. The SLD emitting the light over 20 nm spectral width was used to construct the off-axial source. In order to obtain spatially coherent illumination we have constructed the spatial filters. Light from halogen lamp was transferred through the pinhole with the nearly 5 μm effective diameter, which was placed in the 6 mm focus of achromat lens. Second source shows the effective angular size about 4 arc minutes. SLD light was coupled into single-mode fiber and emitted by commercial fiber collimator. SLD source shows smaller angular size. The sources were aligned on-axially and off-axially and were combined by use of the beamsplitter BS_0; their lights can be clearly identified in the pupil space, Figure 4: (a) – central dark *ring-shaped* fringe was caused by the on-axial (extended) source; (b) – the fringes with more dense spacing are originated from the off-axial source tilted about 10 arcminutes; (c) -both the on-axial and the off-axial sources are superimposed, their fringes localizations have both the destructive interference on-axis. However the more dense fringes are caused by the off-axial source and we can detect its light on the background of the strong on-axial source, which is mostly suppressed due to destructive interference. Some residual energy remains non-nulled due to the finite size of the on-axial source and due to misalignments.

In order the CCD observes the image space, we converted the interference patterns from the pupil space to the image space attaching the TV lens, which focuses the light on the CCD. For demonstration purposes, we combined by the BS3 two interferometer output ports: the Nulled and the Bright within the same CCD frame, Figure 3. We aligned the Nulled and Bright images one above the other: the upper image area corresponds to the Bright output and in the bottom is the Nulled output.

In Figure 5 (a) – the only off-axial source is switched on and we observe its image in both areas: in the Nulled output and in the Bright output. Off-axial source is presented as the two pair of images, each with 0.25 fraction, located symmetrically from the RSI optical axis. We set low intensity for the off-axial light; shown on right, it is contrasted by another palette on the CCD thermal noise background. In (b), the only on-axial source is switched on, emitting relatively strong light intensity; anyway this intensity and the CCD exposure time have been optimized in order to omit the CCD saturation. We see the strong signal in the upper Bright area (shown saturated). In the bottom Nulled area we can still detect some amount of residual energy. In order to estimate the Nulling efficiency, in (c) we compare the Nulled and the Bright areas. Their pixel intensities have been summed up and their ratio is equal to coronagraphy contrast $\approx 10^{-3}$ was measured, which makes the reasonable nulling efficiency for the used sources size of several arcminutes. In (d), the both on-axial and off-axial sources are shown superimposed, here in the upper Bright area, the intensive on-axial component saturates the CCD and in Nulled area, the weak off-axial source can be visually separated. The CCD exposure time in Figure 5 (a), (b), (c) was set at 2 minutes, in Fig. 5 (d) was set at 4 minutes, that can demonstrate the mechanical stability of the system.

5. Conclusions

We proposed the three-dimensional (out-of-plane) common-path interferometer for the achromatic nulling. Both the computer simulations and the experiments show the reasonable agreement to demonstrate the achromatic nulling of the on-axial light. The nulling efficiency is affected if the interferometer mirrors are misaligned; anyway it is possible to obtain the 10^{-10} achromatic nulling for the on-axial source. We have constructed breadboard demonstrator and have obtained the nulled images with the white-light of the spectral width over 300 nm in visible range. For the used in our experiment, the *giant* source size of several arcminutes, we show experimentally the achromatic nulling depth around of 10^{-3}.

Acknowledgements

This work is a part of the 21st Century COE program on "Future Nano-Materials" conducted in the Tokyo University of Agriculture and Technology. We thank Y. Rabbia from the Observatoire de la Cote d'Azur and Ju. Nishikawa and M. Tamura from the National Astronomical Observatory Japan for helpful suggestions and discussions.

References

Serabyn, E., Wallace, J. K., Hardy, G. J., Schmidtlin, E. G. H., & Nguyen, H. T. 1999, "Deep nulling of visible laser light," *Appl. Optics.* 38, 712

Baudoz, P., Rabbia, Y., & Gay, J. 2000, "Achromatic interfero coronagraphy," *Astron. Astrophys. Suppl. Ser.* 141, 319–329

Baba, N., Murakami, N., & Ishigaki, T. 2001, "Nulling interferometry by use of geometric phase," *Opt. Lett.* 26, 1167

Bracewell, R. N. 1978, *Nature* 274, 780

Rouan, D., Riaud, P., Boccaletti, A., Clenet, Y., & Labeyrie, A. 2000, "The four-quadrant phase mask coronagraph," *Publications of the Astronomical Society of the Pacific.* 112, 1479–1486

C. van der Avoort, Mieremet, A., Pereira, S., & Braat, J. 2004, "Demonstration of nulling using delay line phase shifters," *In New Frontiers in Stellar Interferometry, ed. by W. Traub, Proc. SPIE* 5491, 816

Rabbia, Y., Gay, J., Bascou, E., & Schneider, J. L. 2001, *contract14398/00/NL/MV report European Space Research and Technology Centre, Noordwijk, Holland, rabbia@obs-azur.fr*

Nishikawa, Ju., Kotani, T., Murakami, N., Baba, N., Itoh, Y., & Tamura, M. 2005, "Combination of nulling interferometer and modified pupil for observations of exo-planets," *A&A* 379–384

Tavrov, A., Tanaka, Yo., Shioda, T., Kurokawa, T., & Takeda, M. 2004, "Achromatic coronagraph based on out-of-plane common-path nulling interferometer," *in New Frontiers in Stellar Interferometry, ed. Wesley A. Traub, Proc. SPIE* 5491, 824

Tavrov, A., Bohr, R., Totzeck, M., Tiziani, H., & Takeda, M. 2004, "Achromatic nulling interferometer based on a geometric spin-redirection phase," *Optics Letters* 27, 2070

Scholl, M. S. & Paez, G. 1999, "Cancellation of star light generated by a nearby star-planet systemupon detection with a rotationally-shearing interferometer," *Infrared Physics and Technology* 40, 357

Itoh, K. 1996, "Interferometric Multispectral Imaging," *In E. Wolf, Progress in Optics* XXXV, 145

Hariharan, P. 1995, "Interferometers," *Handbook of Optics 2, M. Bass, McGraw-Hill, New York*

Tavrov, A., Miyamoto, Yo., Kawabata, T., Takeda, M., & Andreev, V. 2000, "Generalized algorithm for the unified analysis and simultaneous evaluation of geometrical spin-redirection phase and Pancharatnam phase in a complex interferometric system," *JOSA A* 17, 1, 154

Direct Imaging of Exoplanets: Science & Techniques
Proceedings IAU Colloquium No. 200, 2005
C. Aime and F. Vakili, eds.
© 2006 International Astronomical Union
doi:10.1017/S1743921306009768

Study of Coronagraphic Techniques

Volker Tolls[1], Michael Aziz[2], Robert A. Gonsalves[3], Sylvain Korzennik[2], Antoine Labeyrie[4], Richard Lyon[5], Gary Melnick[1], Steve Somerstein[6], Gopal Vasudevan[6], and Robert Woodruff[7]

[1]Harvard-Smithsonian Center for Astrophysics, 60 Garden St., Cambridge, MA 02138, USA
email: vtolls@cfa.harvard.edu

[2]Harvard University, Cambridge, MA 02138, USA [3]Tufts University, 161 College Ave., Medford, MA 02155 [4]Observatoire de Haute-Provence, 04870 Saint-Michel l'Observatoire, France [5]NASA/Goddard Space Flight Center, Code 935, Greenbelt, MD 20771 [6]Lockheed-Martin Corporation, 3251 Hanover Street, Palo Alto, CA94304 [7]Lockheed-Martin Corporation, P.O. Box 179, Denver, CO 80201

Abstract. SAO has set up a testbed to study coronagraphic techniques, starting with Labeyrie's multi-step speckle reduction technique. This technique expands the general concept of a coronagraph by incorporating a speckle corrector (phase and/or amplitude) in combination with a second occulter for speckle light suppression. The correction function is derived applying the phase diversity method on images taken in focus and slightly out-of-focus. The occulter masks for the testbed will initially be produced lithographically. However, in a parallel program we are studying a new manufacturing method. This method utilizes focussed ion beams and will directly mill the mask shape into absorbing material deposited on a transparent substrate.

Keywords. instrumentation: adaptive optics, instrumentation: miscellaneous, methods: laboratory, techniques: miscellaneous, (stars:) planetary systems.

1. Introduction

Direct imaging of extra-solar planets is important for determining the properties of individual planets and to study multi-planet systems. Obtaining spectra of extra-solar planets enables us to constrain the composition of planetary atmospheres and surfaces, their climates, and their rotation periods. The techniques required to isolate and detect an extra-solar planet next to its host star are quite challenging and require significant improvement. The dominant program for imaging exo-planets is NASA Terrestrial Planet Finder (TPF). The current requirements for the TPF telescope system are very challenging in regards to manufacturing the primary mirror and then in controlling the telescope environment such that the changes are extremely small. Further challenging requirements are set for the star light suppression system which basically tries to correct the wavefront of the light coming from an observed host star to near perfection before it enters the coronagraph. In our ongoing program, we are studying an alternate approach, Labeyrie's multi-stage speckle reduction (MSSR) technique, Labeyrie (2002), which should lower the requirements on the telescope system and possibly on its metrology as well. In §2 we describe the MSSR technique and present simulations showing its performance. In order to demonstrate the MSSR technique in the lab, we have designed a testbed which is described in §3. Essential elements of the testbed are soft-edge occulter masks. These masks are typically manufactured using lithographic methods, e.g. optical lithographic methods and High Energy Beam Sensitive (HEBS) glass photomasks. Our program is

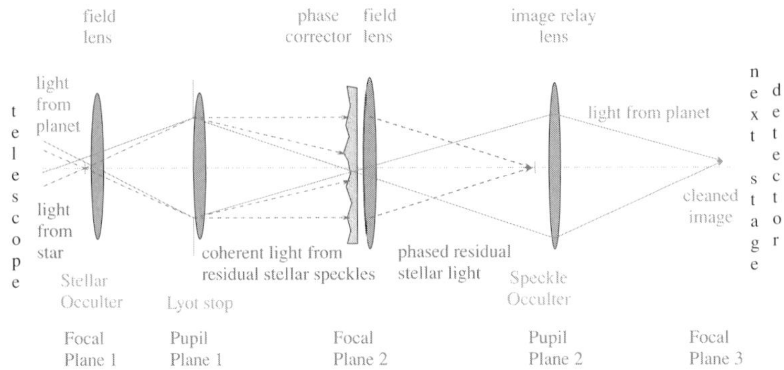

Figure 1. : Schematic setup for Labeyrie's multi-step speckle correction approach. The first part of this schematic shows a standard Lyot coronagraph (Focal Plane 1 and Pupil Plane 1) and a single correction step (phase corrector placed near Focal Plane 2, a second occulter near Pupil Plane 2), and a detection camera.

pursuing an alternate method utilizing an focused ion beam system to directly mill masks (§4).

2. Labeyrie's Multi-Step Speckle Correction Method

2.1. *Background*

An alternative approach has been suggested by Labeyrie. Figure 1 shows an optical layout of Labeyrie's MSSR approach. Instead of correcting the telescope wavefront before the coronagraph, Labeyrie suggests that measurements be made of "the phase of the speckle pattern" in the image plane of a Lyot coronagraph. The speckle pattern is the result of phase errors on the primary mirror and other optics in the system up to the coronagraph. An adaptive element (e.g., a deformable mirror) is then applied with a conjugate of the phase in the speckle pattern, setting the net phase to approximately zero. This results in the speckle having a plane wave component emanating from the coronagraph focal plane. A lens following the corrector re-images the telescope pupil. The plane wave of scattered light will be mostly focused into a small spot in the center of the re-imaged pupil plane. A second occulter blocks the central spot and the result is re-imaged to another focal plane. Measurement of the phase must be made with a field that does not contain a planet or any other faint object (e.g. by imaging another star). This correction will then improve the contrast between the faint object and the speckle background by as much as a factor of 10. Several additional stages of correction continue to improve the contrast ratio. Measurement and correction of the speckle phase need only be as accurate as 1/4 wave, greatly relaxing the requirements on the adaptive optics (AO) system. The pupil plane occulter can be fairly large (e.g. up to 10 % of the pupil diameter) without attenuating much of the signal of the off-axis planet or source. Imperfect correction of the focal plane phase causes the spot to spread out but it is still blocked by the stop. The Labeyrie approach will relax substantially the requirement that the wavefront accuracy of the large optics be better than 10^{-6}wave. Initial numerical simulations showed good results with a wavefront error of 1/1000 wave of the initial wavefront which has to be demonstrated in the proposed laboratory tests.

Figure 2. : Simulation of the Labeyrie multi-step method (for explanation see text).

2.2. *Simulations*

Before designing the testbed setup for the multi-step speckle correction technique, we simulated the approach using Fourier optics. The steps in our simulation are: (1) simulating the telescope, (2) simulating the coronagraph, (3) simulating the multistage speckle reduction, and (4) applying possible post-processing to the acquired images. The imperfections in the primary mirror are simulated using the specification of the power spectral distribution (PSD) for the TPF demonstration mirror. The resulting phase screen is then rescaled to the desired wavefront error. We assumed a wavefront error of $\lambda/1000$; however, this number can be changed to any wavefront error. The occulter in the coronagraph can be selected. The current choices are a sinc^2 or Gaussian profile absorbing mask. The mask parameters can be adjusted to optimize the system. Also, a rest-transmission can be specified to better simulate real occulters. We assumed a typical value of 10^{-10}. The phase corrector did correct the phase on the same pixel raster as the simulation was performed. In our testbed, the phase corrector will not have such a high resolution.

Figure 2 shows a typical result of our simulations. The resolution is 512×512 pixels showing a FOV of 25 λ/D. The left panel shows the ideal point spread function and the second panel the speckle background after the star light has been suppressed by a soft-edge occulter and the exit pupil has been reduced to 90%. The next three panels show the improvement of the speckle reduction after: 3 steps, 5 steps, and 7 steps. The simulated planets at (4, 0)*λ/D and at (-2,-2)*λ/D can be identified after 5 to 7 steps. The right panel shows a post-measurement data processing step. The image from step 7 was rotated by 180 degrees and subtracted from the original image. The speckles show a very high, but not perfect, rotational symmetry. In addition, the image had to be shifted by 1 pixel vertically and horizontally for the best line-up of the speckles. The planets still show up as circular objects with a point-symmetric, negative counterpart. Since the shape of the planets seems to be invariant when shifting the image and the "shadows" of the speckles (the negative component) move around the speckles, planets should be easily identifiable. The number of speckle reduction steps can be reduced, if both, the phase and the amplitude, can be corrected. In this case, only 1 to 2 steps might be required assuming sufficient resolution in the correcting components.

3. Testbed Design

In the initial design, SAO's coronagraphic testbed consists of the following elements: 1) the simulated source, 2) the Lyot coronagraph, 3) a single speckle reduction step, and 4) the camera. Initially, we are using a 1 mW He-Ne laser. The output beam is fed into a spatial filter with a 30 mm objective and a 10 μm pinhole. The transmitted light enters the coronagraph uncollimated. The coronagraph consists of a 2-inch diameter, spherical mirror with a surface accuracy of $\sim \lambda/1000$ at 633 nm at a distance of about 2 meters from the pinhole. The now converging beam passes through an apodizer with a cosine function and focuses onto the occulting mask. The final elements of the coronagraph are the Lyot stop followed by a 1-inch, 0.5-m radius spherical mirror.

The phase correcting element in our single step speckle corrector is a deformable mirror from Boston Micromachines Inc. This mirror has 140 elements, a 12-by-12 pixel array

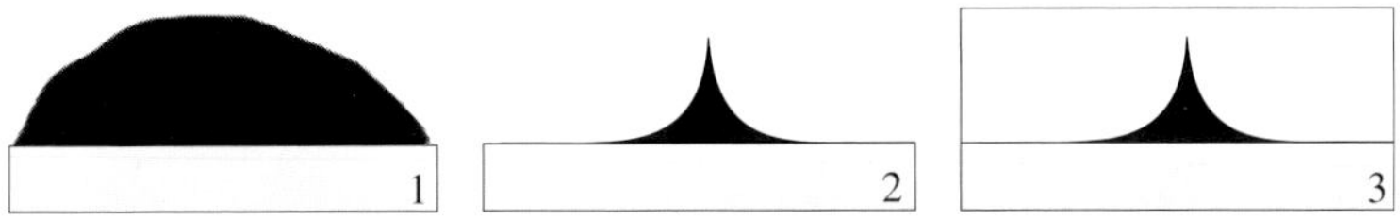

Figure 3. : Procedural steps in the manufacturing of ion beam milled occulter masks.

with the corner mirrors missing. The size of a single pixel is 275 × 275 μm (300 μm spacing), the total array size is 3.3 by 3.3 mm, and the stroke per element is 2 μm with 2 nm repeatability. The corrected speckle light is focused on a second occulter, effectively eliminating most of the speckle light. The final "cleaned" image is focused onto the detector, a Apogee ST-401 CCD camera with a Kodak full-frame CCD KAF-0401E. The camera is mounted on a z-direction translation stage. This enables us to use phase diversity to retrieve the phase and amplitude of the wave (see Gonsalves & Tolls 2005) and to calculate the correction function for the deformable mirror.

4. Ion Beam Milling of Occulter Masks

Our efforts include the study of a new method of mask manufacturing. The basic steps of this method are outlined in Figure 3. This effort uses the FEI Dual Beam Focused Ion Beam System (FIB) consisting of a fully digital Field Emission Scanning Electron Microscope integrated with Focused Ion Beam Technology at Harvard's Center for Imaging and Mesoscale Structures (CIMS). Currently we are studying how to control the ion beam in order to produce the desired shape. The build-in electron microscope enables us to tightly control the milling process. The challenge is to solve are finding the best milling patterns for removing the material without leaving marks on the mask or re-depositing removed material. In parallel we are studying how errors in the mask shape effect the mask properties and performance. Utilizing the beam propagation method for simulations, we can determine the mask behavior from theoretical shapes and wfrom measured shapes.

5. Mask Tester

For characterization of the manufactured occulter masks, we have designed a mask tester. The mask tester consists of a He-Ne laser source, collimating optics, x- and y-translation stage, pinhole, detector, and control electronics (a schematic of the setup can be found in Tolls *et al.* (2005)). The laser beam is focused with a focal spot size of about 1 μm. The location of the focal spot will be in the mask plane. The occulter will be mounted on an xy-translation stage whichcan be moved with 50 nm repeatability over a 1-inch range. The transmitted beam is then fed through a pinhole onto a pin diode detector. The test setup will be computer controlled and needs only an initial rough alignment such that the mask is completely within the scan range. After scanning, the mask profile can be retrieved using the maximum entropy method.

Acknowledgements

This work was supported by NASA through grant number NNG04GC57G and Smithsonian Astrophysical Observatory IRAD funding.

References

Gonsalves, R.A. & Tolls, V. *Proc. SPIE 5905*, in press
Labeyrie, A. *Proc. ESLAB 36 symposium*, ESTEC, ESA SP-514
Tolls, V. *et al.* *Proc. SPIE 5905*, in press
TPF Program *Technology Demonstration Mirror, Detailed Equipment Specification*, JPL

Direct Imaging of Exoplanets: Science & Techniques
Proceedings IAU Colloquium No. 200, 2005
C. Aime and F. Vakili, eds.
© 2006 International Astronomical Union
doi:10.1017/S174392130600977X

Adaptive Mask Coronagraph

P. Bourget[1,2], R. Vieira Martins[2], F. Colas[3], P. Assus[4] and A. Irbah[1]

[1]Service d'Aéronomie -CNRS- Verrières le Buisson, France.
email: pierre@aerov.jussieu.fr

[2]Observatório Nacional do Rio de Janeiro - MCT - Brazil.
[3]IMCCE, Paris.
[4]Observatoire de Nice -OCA.

Abstract. In this paper, the new coronagraph that will be mounted at the *Pic du Midi* Observatory (IMCCE - T1m telescope) is presented. To optimize the occulting process of a Lyot coronagraph, a compressed mercury (Hg) drop is used as an occulting disk and its size control offers an adaptation to the seeing conditions or to the Airy diameter fraction needed. In addition to the Hg-mask, a variation on the theme is proposed by a diameter and wavelength adaptive phase mask made of a gas bubble in immersion oil between two optical windows. The instrument concept offers a good versatility to test other mask type and pupil apodization techniques.

Keywords. Instrumentation: coronagraph, techniques: interferometry, planets and satellites: astrometry, stars: planetary systems:protoplanetary disks.

1. Introduction

The first Hg-mask coronagraph (Bourget *et al.* 2001) is a Lyot-type coronagraph projected for astrometric observations of faint satellites near Jovian planets and is in work since 2000 at the 1.6m telescope of the *Pico dos Dias* Observatory (LNA - Brazil). The instrument permitted to realize an astrometric survey of faint satellites. The results of systematic astrometric observations of the inner neptunian satellite Proteus (Vieira Martins *et al.* 2004) have been used for the determination of revised orbital elements (Jacobson *et al.* 2004). The Hg-mask has been developed to compensate the variations of the apparent diameter of the planets between the observing runs and to simplify the observation mode of the instrument (see figure 1). The new version of the instrument has been developed in 2004 at the *Observatório Nacional* (ON) of Rio de Janeiro in the framework of a cooperation program between CNPq (Brazil) and CNRS (France). It will be mounted in 2006 at the T1m telescope of the *Pic du Midi* Observatory (IMCCE). The instrument will be dedicated to observe faint satellites, double asteroids and protoplanetary disks. The integration of the instrument is now realized at the *Service d'Aéronomie* (CNRS).

In the theoretical case of a stellar image, the occulting disk of a classical Lyot coronagraph had to mask the central core and the first three or four rings in the Airy diffraction pattern to rise good contrast efficiency. If a direct imaging of extrasolar terrestrial planet (ETPs) is desired, the telescope must be able to resolve the orbit of the ETPs away of the fourth diffraction ring and this could only be possible with an extreme large telescope (ELT). Recent works on pupil apodization techniques (Guyon 2003; Aime *et al.* 2005) dedicated to the high dynamic range imaging and exoplanets detection show that it is possible to obtain a closest higher contrast. The pupil apodization induces a reduction of the point-spread function (PSF) diffraction rings energy. This process applied on a Lyot

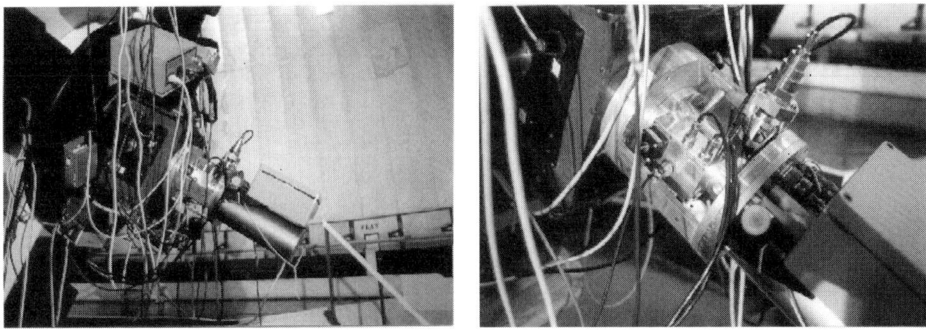

Figure 1. First generation of the Hg-Mask Coronagraph on the 1.6m telescope of the LNA-Brazil.

coronagraph does not limit the dynamic angular resolution efficiency to mask the three or four first rings of the PSF.

The phase mask proposed by Roddier & Roddier (1997), is much smaller than the Hg-mask. It covers only a fraction of the PSF and introduces a 180 degree phase shift on the incoming wave. Even if a pupil apodization (Aime *et al.* 2002; Guyon & Roddier 2002; Soummer *et al.* 2003) can reach a closest full coronagraphic extinction with the phase mask, the instrument is strongly dependent on the phase mask diameter to equalize the of opposite phases amplitudes. A bubble of gas in immersion oil gets a variable diameter phase mask able to be chromatically controlled. The properties of the immersion liquid superficial tension induce the circularity of the phase mask; the diameter control provides the optimization of the nulling efficiency. The main objective of the new instrument is the development of the adaptive mask concept by the means of "no-solid" masks.

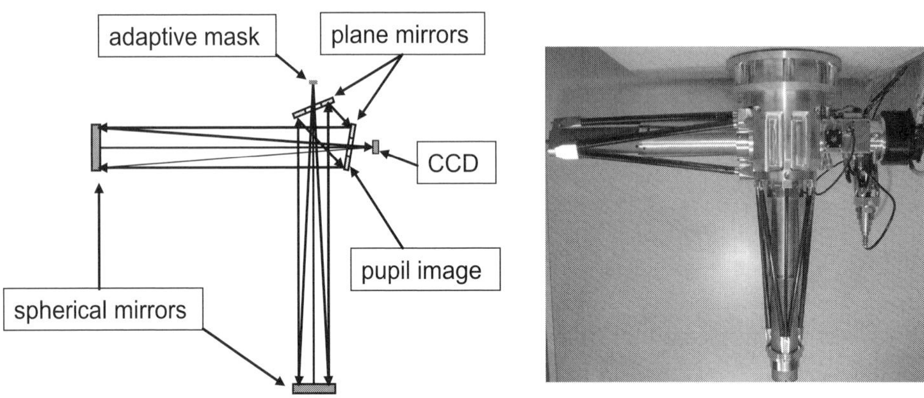

Figure 2. Optical layout of the Coronagraph and mechanical concept.

2. The instrument

The optical design of the instrument accepts an incoming beam of focal ratio near f/17. Instead of a big field instrument as the first version of the Hg-mask coronagraph dedicated to the astrometric observations of faint satellites (Bourget *et al.* 2004), the instrument presented in this work is a 9mm circular field. The optical design is an optical cavity of spherical mirrors. The secondary mirror occultation in the pupil image is used for the light entrance(see figure 2). The occulting disk unit accepts six different adaptive masks and offers an XY centering micrometric control. To avoid the mechanical flexion, a carbon-fiber structure is used in the geometrical extension of the instrument. As a first step of apodization, an iris diaphragm is placed on the pupil image plan near the second plane mirror. Both faces of the plane mirrors are polished to accept different kinds of reflection profiles for the amplitude apodization tests on the optical bench.

3. The adaptive masks

3.1. *The Hg-Mask*

The use of an ever finite numbers of fixed-diameter occulting disks makes difficult the Lyot coronagraphy. In fact, difficulties appear when changing the disk during observations since it has not often the good size and need to be centered with appropriate offsets. We proposed a solution to this problem by using mercury to realize variable diameter occulting disks. The mercury has often been used in astronomy due to its optical properties. We take into account the high density of the mercury by its superficial tension which allows to make a spherical drop with an optical quality. The drop is glued between optical windows by the van Der Waals forces and is used as the occulting disk (see figure 3). In addition, immersion oil is placed to avoid the mercury oxidation and the internal reflection between the windows. The whole system is used as the occulting disk. In a Lyot coronagraphic process the occulting mask acts on the scattered light as all the optical elements. The phase errors produced by the optical roughness of both (optics and mask) induce the scattered light ("Halo"). The minimum of the superficial tension potential function provides a toric free surface of mercury physically perfect (λ /100). The control of the distance between the optical windows acts on the tore diameter with a sub micron-precision. The three dimensional occulting process on the tore induces an apodization of the occultation. The diameter limits are 10 microns to 1.5 millimeters with different ranges of mercury drops compression. The surface of the tore is locally perfect and provides high diffraction quality of the occulting disk. On the optical bench or in the case of adaptive optics, the size control allows the adaptation of the mask to the Airy diameter fraction needed. In the case of a classical telescope, the process allows an adequation to the seeing conditions or to the apparent diameter of a resolved object.

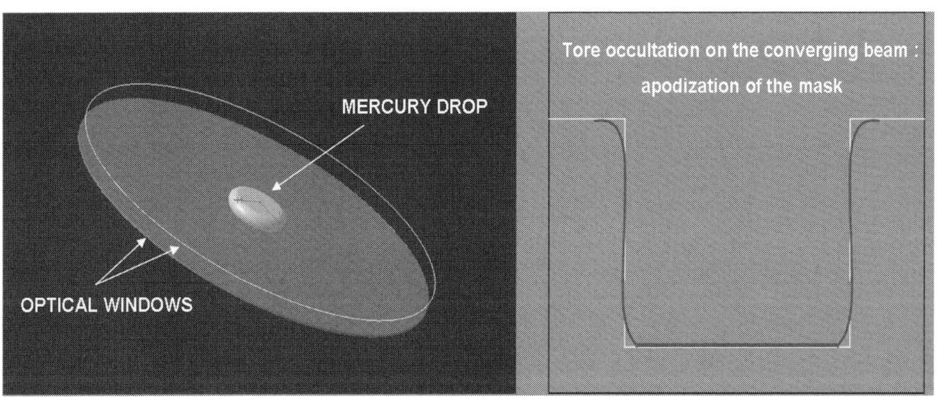

Figure 3. Hg Mask: The distance between the optical window and the pellicle film induces the size control of the toric free surface of mercury.

3.2. *The adaptive Phase-Mask*

A gas bubble in immersion oil between optical windows is used to obtain a phase retardation of 180 degree on the core of the PSF. The phase delay ϕ through the bubble is given by:

$$\phi = 2\pi \frac{e(n - n\prime)}{\lambda} \tag{3.1}$$

where n and $n\prime$ are respectively the refractive index of the liquid and the bubble, e the distance between the optical windows and λ the observation wavelength.

The distance parameter e allows to control both the chromatism and the diameter (see figure 4). Typical values for n and $n\prime$ are respectively about 1.5 and 1 leading to the delay control expression $e \approx k\lambda$ where k is an integer. The variation of the retardation is a linear function of e and the diameter is an $e^{-1/2}$ function. For an f/17 telescope (case of the 1m telescope of *Pic du Midi* Observatory), the Airy radius in the visible wavelength (0.5 microns) is about 10 microns. A bubble of 15 microns can be spread to obtain a 20 microns phase mask diameter with a 4 microns thickness. To minimize the total flux inside the Lyot stop, the optimum size of the phase mask is done by Roddier & Roddier (1997) as 0.43 times the radius of the first dark ring in the Airy pattern, which is 0.53 λ /D. These values are quite different in our case because of the internal reflections in the bubble and the transmission difference through the windows and the immersion oil. The adaptive phase mask allows a direct optimization of the nulling effect to the wavelength range of interest. The parameters to be controlled are e and the phase mask diameter. The original size of the bubble defines the range of the phase mask diameter. The non-linearity dependance between e and the phase mask diameter makes difficult the game of adjustment. To avoid this problem the bubble pressure (and refractive index $n\prime$) is another parameter which can be controlled even if depending on the tore diameter for a fixed thickness e. The coating used on the optical windows is another parameter to be computed to minimize the wavelength dependence of the delay. As a first step, the adaptive phase mask proposed by the instrument offers a thickness and bubble pressure control. The main technical challenge is the determination of the parameters able to reduce the wavelength dependence of the process preserving the diameter control to equalize the shifted phase amplitudes.

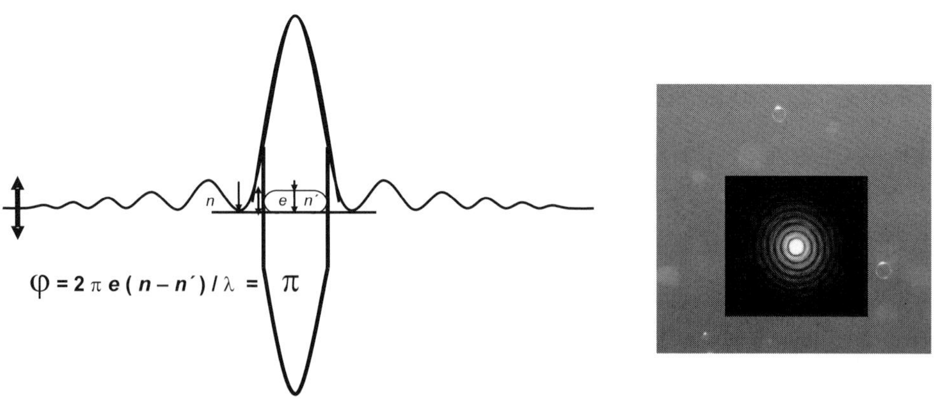

Figure 4. Phase Mask: The gas bubble acts as a phase mask on the core of the PSF, acting on the size control of the bubble we obtain a direct visibility of the nulling efficiency.

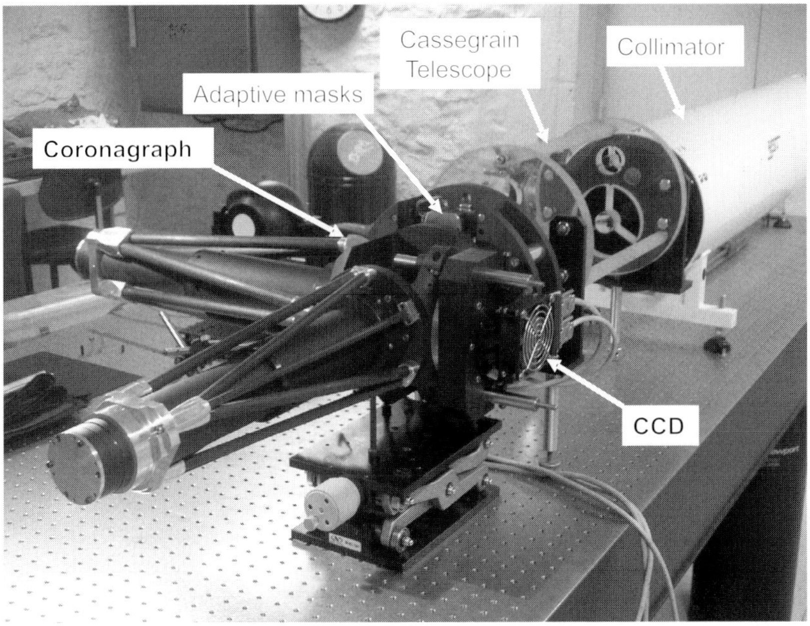

Figure 5. Coronagraph on the optical bench at the Service d'Aéronomie -CNRS. We use a collimator and a 12 cm Cassegrain telescope to create a PSF on the mask unit.

4. Conclusions

The instrument is dedicated to image double asteroids, faint satellites near Jovian planets and protoplanetary disks. The integration of the instrument on the optical bench at the *Service d'Aéronomie* (CNRS) figure 5 will permit the development of the adaptive phase mask. The instrument will be tested at the T1m telescope of the *Pic du Midi* Observatory (IMCCE), the optical qualities of the telescope and the 1m aperture combined with subarcsecond seeing conditions will allow the quantification of the extinction possibilities of the Adaptive Mask Coronagraph.

Acknowledgements

We would like to acknowledge the *Laboratório Nacional* de Astrofísica (LNA-Brazil) for the fellowship that permitted to conclude the first part of the project. The authors thank the CNPq-Brazil for partial support of this work.

References

Aime, C. 2005 *PASP*. 117, 1012
Aime, C., Soummer, R., & Ferrari, A. 2002, *A &A*. vol. 389, p. 334
Bourget, P., Veiga, C.H., & Vieira Martins, R. 2001, *PASP*. 113, 436
Bourget, P., Veiga, C.H., Vieira Martins, R., Assus, P., & Colas, F. 2004, *EAS Publication Series.*, Astronomy with High Constrast Imaging II, vol. 12, 205
Guyon, O. 2003, *A &A.*, vol. 404, p. 379
Guyon, O. & Roddier, F. 2002, *A &A*. vol. 391, p. 379
Jacobson, R.A. & Owen, W.M. Jr. 2004, *AJ*. 128, 1412
Roddier, F. & Roddier, C. 1997, *PASP*. 109, 815
Soummer, R., Aime, C., & Falloon, P.E. 2003, *A &A*. vol. 397, p. 1161
Vieira Martins, R., Veiga, C.H., Bourget, P., Andrei, A.H., & Descamps, P. 2004, *A &A.*, vol. 425, p. 1107

All photographs: Laurent Thareau [`l.thareau@free.fr`].

Direct Imaging of Exoplanets: Science & Techniques
Proceedings IAU Colloquium No. 200, 2005
C. Aime and F. Vakili, eds.
© 2006 International Astronomical Union
doi:10.1017/S1743921306009781

Cryogenic infrared optics for the SPICA coronagraph

Keigo Enya[1], Takao Nakagawa[1], Hirokazu Kataza[1], Hidehiro Kaneda[1], Yukari Yamashita Yui[1], Motohide Tamura[2], Lyu Abe[2], Yoshiyuki Obuchi[2], Takashi Miyata[3], Shigeyuki Sako[3], Takashi Onaka[4], Hidenori Takahashi[4] and the SPICA team

[1]Department of Infrared Astrophysics, Institute of Space and Astronautical Science, Japan Aerospace Exploration Agency, 3-1-1 Yoshinodai, Sagamihara, Kanagawa, 229-8510
[2]Optical and Infrared Astronomy Division, National Astronomical Observatory of Japan, 2-21-1 Osawa,Mitaka, Tokyo, 181-8588 Japan
[3]Institute of Astronomy, University of Tokyo, 2-21-1 Osawa, Mitaka, Tokyo, 181-8588 Japan
[4]Department of Astronomy, University of Tokyo, 7-3-1 Hongo, bunkyo, Tokyo, 113-0033 Japan
email: enya@ir.isas.jaxa.jp

Abstract. SPICA is a cooled, single large-mirror space-telescope, which is under discussion as an succsesor of the ASTRO-F mission. One of the most ambitious challenges of the SPICA mission is the direct observations of exoplanets with a coronagraph instrument. We report cryogenic infrared optics to realize high quality wavefronts for the SPICA coronagraph.

The SPICA satellite will be launched by an H-IIA rocket to Sun-Earth L2 Halo orbit early in the 2010s. The SPICA telescope is a Ritchey-Chretien optics with 3.5m diameter primary mirror, and cooled down to 4.5 K in orbit by radiation cooling and mechanical cryo-coolers. Main working wavelengths are 5–200 micron. Advantages of the SPICA coronagraph are the infrared wavelenths where the contrast between planets and central stars are smaller than the optical wavelengths, and that the cooled space telescope consists of monolithic mirrors.

Development of light-weight cooled telescope is one of the most important tasks to realize SPICA. At the present, sintered SiC and carbon fiber reinforced SiC (C/SiC) composite are candidate materials for the mirrors, truss, and optical bench. For these materials, estimations and improvements of basic property and surface roughness in cryogenic temperatures have been carried out. Deformation of trial product mirrors by cooling is also examined.

We are developing cryogenic deformable mirrors (DMs) because wave front accuracy of the SPICA telescope is 0.35 micron RMS, which is not enough for our coronagraphic instrument. For MEMS (Micro Electro Mechanical System) DM and some others, measurements of thermal deformation by cooling, electrical response, and heat generation are undergoing. Developments of a tip-tilt system for cryogenic usage started to cancel vibration caused by the cryo-coolers and other components and to realize a diffraction limit resolution. The first result of our binary mask coronagraph experiment is also shown.

Keywords. SPICA, infrared, telescope, cryogenic, tip-tilt, deformable mirror.

1. Outline of the SPICA mission

The SPICA (SPace Infrared telescope for Cosmology and Astrophysics) is the third Japanese infrared space-borne project (Nakagawa 2004, Onaka *et al.* 2005) following the Infrared Telescope in Space (IRTS) and ASTRO-F which will be launched in 2006 (Murakami 2004). SPICA is equipped a telescope with 3.5m aperture consists of a monolithic mirror. The whole telescope is cooled down to 4.5 K by radiation cooling and mechanical cryo-cooler. Required optical quality of the telescope is diffraction limited at

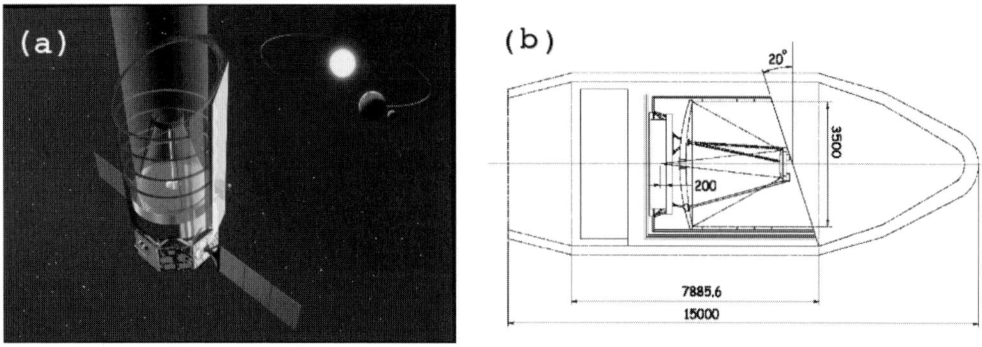

Figure 1. (a):Image of the SPICA satellite in orbit. (b):Illustrative view of the SPICA telescope in the H-IIA rocket fairing.

5 micron wavelength. Main observing wavelengths are 5–200 micron, therefore SPICA is complementary to James Webb Space Telescope (JWST)(Sabelhaus 2004). and Herschel Space Observatory (HSO)(Pilbratt 2004). SPICA will be launched into the Sun-Earth libration L_2 halo orbit by the H-IIA rocket early in the 2010s and operated for 5 years.

A coronagraph instrument will be equipped with other general purpose instrumetns on SPICA (Tamura *et al.* 2005, Abe *et al.* 2005). SPICA is a unique opprotunity because of its infrared observation with a large aperture and simple pupil provided by a monolithic mirror.

2. Develop of the telescope

2.1. *Design and material*

The SPICA telescope is a Ritchey-Chretien optics, consisting of two monolithic mirrors with 3.5m aperture and is cooled down to 4.5 K before operation.

At the present moment, sintered SiC (figure 2a, b) and carbon fiber reinforced SiC (C/SiC) composite (figure 2c, d) are candidate materials for the SPICA mirror, truss and optical bench (Enya *et al.* 2004). These materials have some advantages, especially their high stiffness to make light weight mirrors.

Sintered SiC is used in HSO and other missions, therefore this material is superior in their actual results. However, brittleness is a disadvantage of SiC. Improved C/SiC composite is superior in its fracture toughness. SiSiC slurry coating technique reduces surface roughness of the C/SiC composite mirror.

2.2. *ϕ700 mm C/SiC composite BBM mirror*

In the SPICA mission, one of the most important properties of the mirror is not to deform its figure in cooling down to 4.5K. We developed a BBM spherical mirror (Nakagawa 2004, Onaka *et al.* 2005) made of C/SiC composite with SiSiC slurry coating (figure 3a, b). The mirror consists of 6 segments to demonstrate feasibility of joint technology for fabrication of 3.5 m diameter mirror. Cryogenic test of the mirror was executed with cryo-chamber (figure 3c) that was the same one with tests of the ASTRO-F telescope (Kaneda *et al.* 2005) cooled by liquid Nitrogen and Helium. The surface figure was monitored by Zygo GPI-XP. The result are shown in (figure 3d, e). Significant deformation by cooling was confirmed.

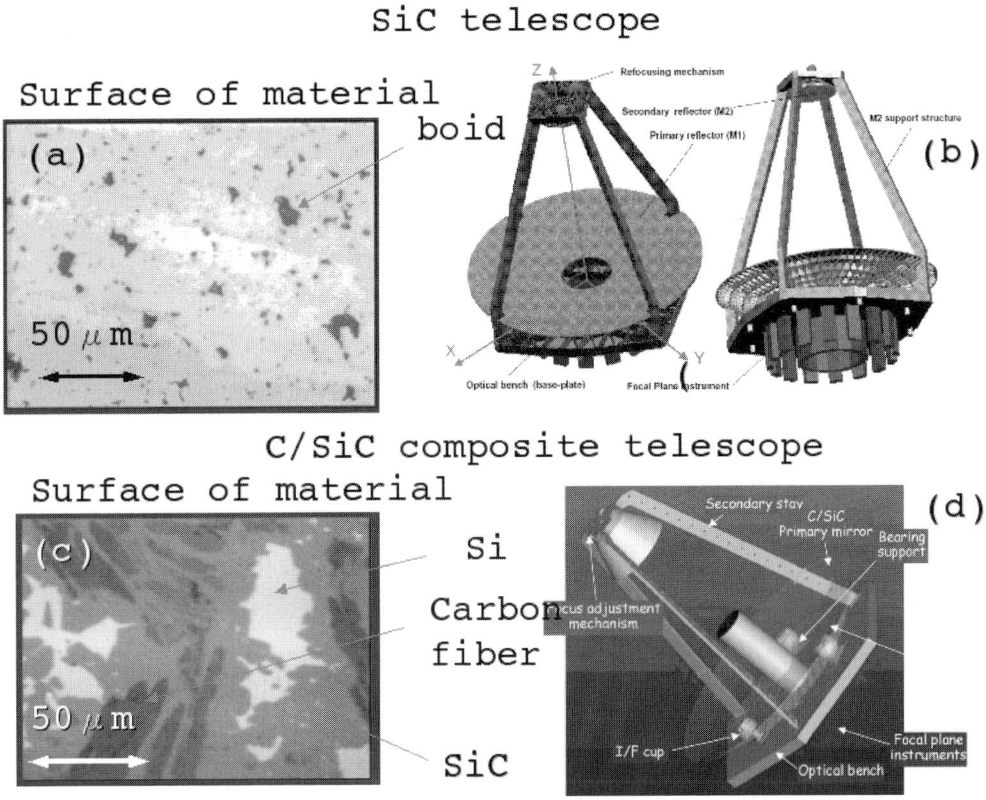

Figure 2. Designs of the SPICA telescope. Surface of polished Sintered SiC (a) and a telescope design with sintered SiC (b). Surface of polished bare C/SiC (c) composite and a telescope design with this material (d).

3. Cryogenic tip-tilt mirror system

3.1. *Why is it required?*

Pointing error in SPICA caused by vibration of cooler, compressor, and other components can be more than 1 arcsec, if any provisions are applied. On the other hand, pointing accuracy to be 0.03 arcsec is required for the coronagraph design. This gap is too large to be solved by improvement of pointing system. Therefore, a cryogenic tip-tilt mirror system is indispensable.

3.2. *Development and experiment*

Design of proto-type of a cryogenic tip-tilt mirror is shown in figure 4(a). Figure 4(b) shows the cryogenic chamber cooled down to 10 K with GM cryo-cooler. Motion of a cryogenic piezo actuator was confirmed in this chamber. The system also demonstrated mirror chopping in room temperature. A proto-type system is now under construction and examination.

It turned out that heat generation is a critical problem (10 times larger than specification of less than a few mW). We are going to test selection of actuators and adoption of support with a lever mechanism to reduce the heat generation.

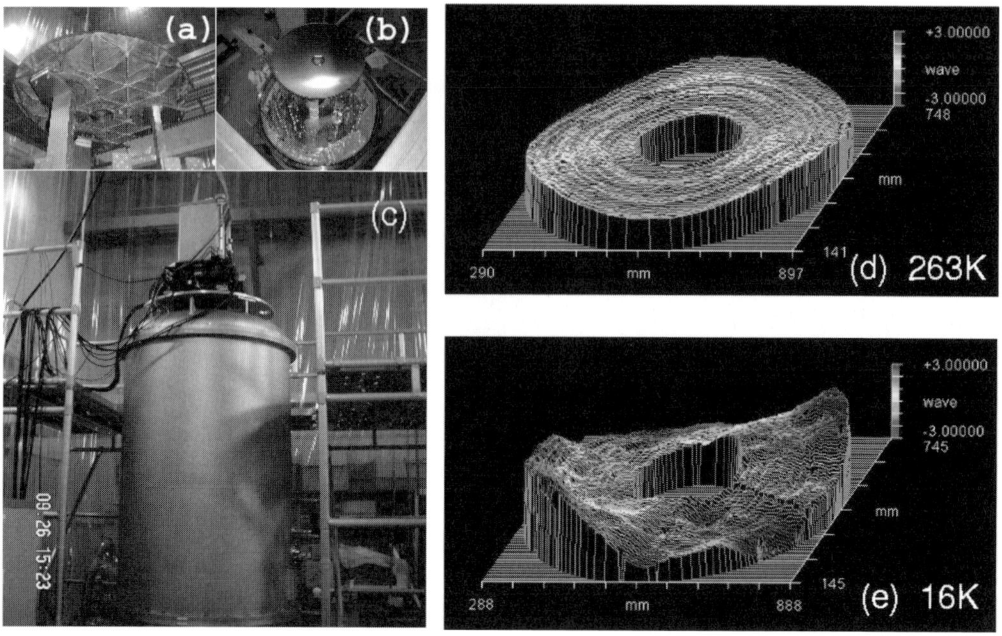

Figure 3. Cryogenic test of the $\phi 700$ mm C/SiC composite BBM mirror. The mirror (a), (b) and the cryogenic chamber for the test (c). (d) and (e) show thermal deformation of surface figure. Here figure of ambient temperature was subtracted and power was corrected. RMS of (d) and (e) are 0.40λ and 1.21λ respectively where $\lambda = 632.8$nm.

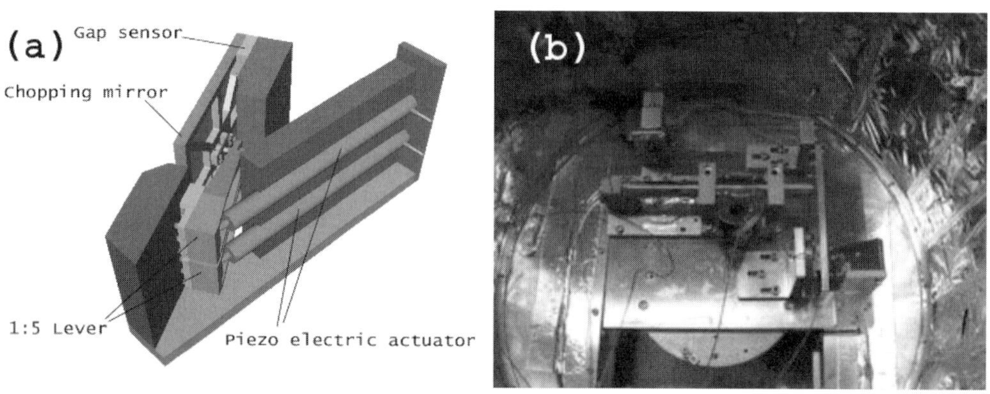

Figure 4. (a): Design of a proto-type of cryogenic tip-tilt mirror. (b): Piezo actuator set in a dewar with a 4K cryo-cooler. Normal motion was confirmed at cryogenic temperature, while its stroke was one order smaller than in ambient temperature.

4. Cryogenic deformable mirror

4.1. *Why is it required?*

Specification of wave front error of the SPICA telescope is 0.35 micron RMS, while the coronagraph requires ~10nm order RMS. This gap is too large to be solved by improvement of the telescope. Therefore, cryogenic deformable mirror (DM) is indispensable for the SPICA coronagraph.

The mirror of figure 5(a) is a MEMS DM with 2mm clear aperture and 32 channel actuators producted by BMC (Boston Micromachines Corporation).

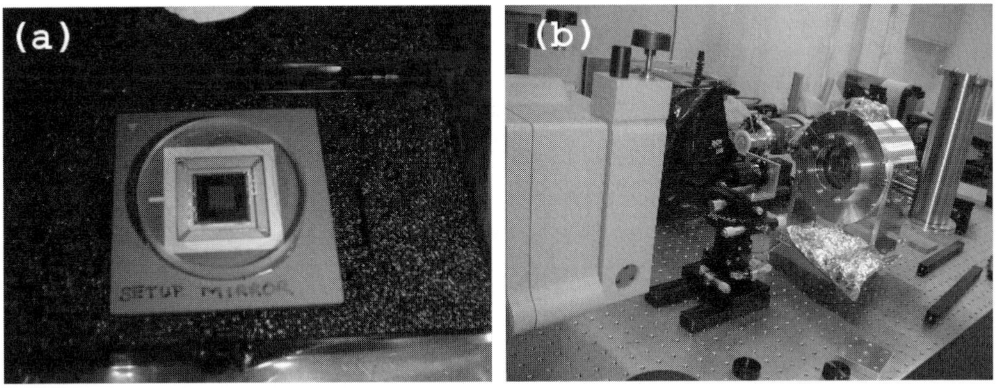

Figure 5. (a): 32 channel MEMS-deformable mirror producted by BMC (Boston Micromachines Corporation). (b): Set-up for cryogenic examination of DMs. This system consists of a ZYGO GPI-XP interferometer, cryo-chamber cooled by liquid nitrogen and zooming optics.

4.2. *Development and experiment*

Measurements of thermal deformation by cooling, electrical response, and heat generation are undergoing for some DMs (figure 5b). Zoomed surface figures of DMs in cryostat can be measured by Zygo GPI-XP. It turned out that deformation caused by difference of coefficients of thermal expansion between materials consisting of these devices is a critical problem. We are now developing a new configuration of substrate and adhesive in these devices to reduce the thermal deformation.

5. Coronagraph experiments

We think that binary shaped pupil mask is one of the most promising coronagraphic methods because of its achromatism, robustness against pointing error, and feasibility of mask fabrication (Tanaka *et al.* 2005a, 2005b). We started our experiments of coronagraph with visible light (He-Ne laser with wavelength to be 632.8nm) but without any wave front control and vacuum chamber. Binary mask pattern of aluminum thin coating (thickness is 100nm) was made on a glass substrate by nano-fabrication technology usually used for productions of semi-conductor devices (figure 6a, b). The mask was a kind of checkerboard mask (Vanderbei *et al.* 2005) designed to avoid obstruction by the secondary mirror and spiders.

The first result of our binary mask coronagraph experiments is shown in figure 6(c), (d). The profile of figure 6(c) is along an arrow shown in figure 6(d). At present, the achieved contrast is only 10^4, while the specification of contrast in this experiment is 10^6. Improvement of the experiment and examinations of other coronagraphic methods should be executed. Ultimately, we will develop an infrared and cryogenic coronagraph for the SPICA mission. The newest results obtained after the conference is shown in Abe *et al.* 2005.

Acknowledgements

We would like to acknowledge useful comments of J. Nishikawa, N. Murakami, K. Fujita and Y. Itoh. We are also grateful to support the mask fabrication by National Institute of Advanced Industrial Science and Technology. This work was supported by a Grant-in Aids for Science Research from the Japan Society for the Promotion of Science.

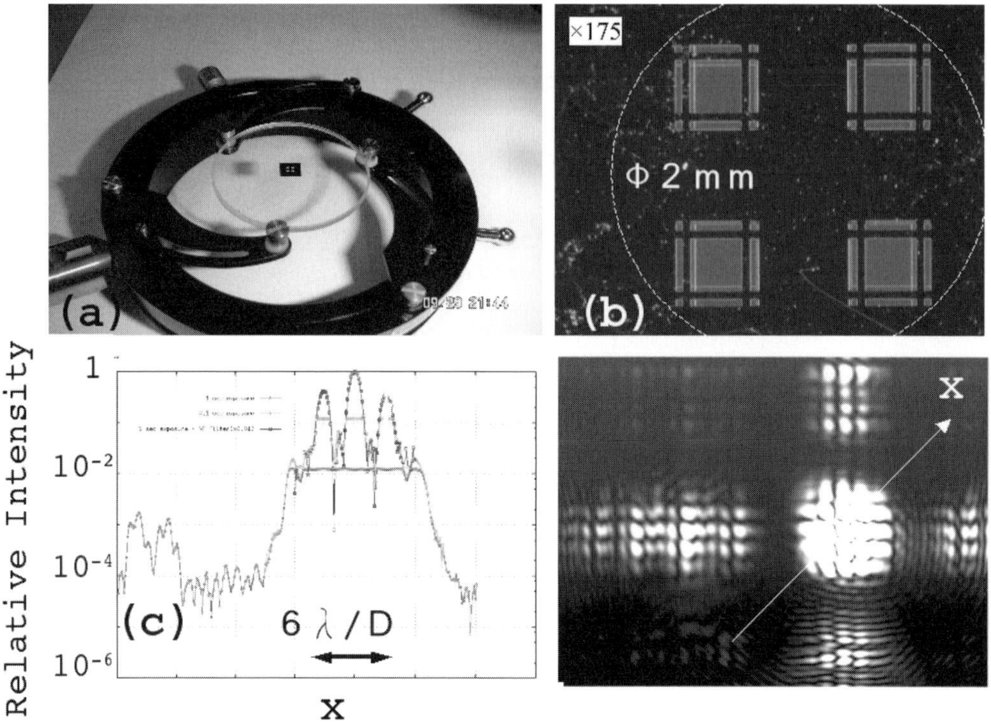

Figure 6. The first result of our coronagraph experiment made by nano-fabrication in visible light wavelength. (a, b): The mask consist of aluminum with 100nm thickness on a glass substrate. (c, d): Obtained profile and image of a point spread function respectively. Here *lambda* is wavelength of He-Ne laser(632.8nm) and D is diameter of the beam to be 2 mm.

References

Abe, L., Tamura, M., Enya, K.,Tanaka, S., Nakagawa, T., Kataza, H., Fujita,K., Nishikawa, J., & Murakami, N. 2005, *Proc. of IAUC200(this conference)*, submitted

Enya, K., Nakagawa, T., Kataza, H., Kaneda, H., Yamashita, Y. Y., Onaka, T., Oshima, T., & Ozaki, T. 2004, *Proc of SPIE* 5487, 1092

Kaneda, H., Onaka. T., Nakagawa, T., Enya, K., Murakami, H., Yamashiro, R., Ezaki, T., Numao, Y., & Sugiyama, Y. 2005, *Applied Optics*, submitted

Murakami, H. 2004, *Proc. of SPIE*5487, 330

Nakagawa, T. 2004, *Advances in Space Research* 34 , 645

Onaka, T. & Nakagawa, T. 2005, *Advances in Space Research*, in press

Pilbratt, G. T. 2004, *Proc. of SPIE* 5487 , 401

Sabelhaus, P. A. & Decker, J. 2004, *Proc. of SPIE* 5487, 550

Tanaka, S., Enya, K., Nakagawa, T., & Kataza, H. 2005, *PASJ*, submitted

Tanaka, S., Nakagawa, T., Enya, K., Kataza, H., Tamura, M., Abe, L., and the SPICA coronagraph team 2005, *Proc. of IAUC200(this conference)*, submitted

Tamura, M., Abe, L., HiCHAO team, SPICA coronagraph team & JTPF working group 2005, *Proc. of IAUC200(this conference)*, submitted

Vanderbei, R. J., Kasdin, N. J., & Spergel, D. N. 2004, *ApJ* 615, 555

Direct Imaging of Exoplanets: Science & Techniques
Proceedings IAU Colloquium No. 200, 2005
C. Aime and F. Vakili, eds.
© 2006 International Astronomical Union
doi:10.1017/S1743921306009793

Stellar coronagraphy: two mono-axial configurations for the Achromatic Interfero Coronagraph

J. Gay[1], J. P. Rivet[1], Y. Rabbia[2], G. Cassar[1], C. Buisset[2], F. Fressin[1] and T. Lepine[3]

[1]Observatoire de la Côte d'Azur, B.P. 4229, F-06304 Nice Cedex 4, France.
email: gay@obs-nice.fr

[2]Observatoire de la Côte d'Azur, Avenue Copernic, F-06130 Grasse, France.

[3]Ecole Supérieure d'Optique, LTSI, 10 rue Barrouin, F-42000 Saint-Etienne, France

Abstract. The standard AIC (Achromatic Interfero Coronagraph) has a "coudé" geometry (the output beam leaves at right angle from the input beam). Thus, some extra optical parts are required to fit such a device within the optical train between a telescope and its IR camera. To avoid this drawback, we present two mono-axial variants of the AIC.

Keywords. instrumentation: miscellaneous.

1. Basics on the standard AIC

The Achromatic Interfero-Coronagraph (AIC) has been introduced by Gay & Rabbia (1996) (see also Baudoz *et al.*, 2000) as a new concept of stellar coronagraphs. Instead of rejecting the light of a central star by a more or less smoothly varying complex transmission in the focal plane of a telescope, the AIC rejects the light by destructive interferences. The device works like a modified Michelson's interferometer (see Figure 1): the input light is divided by a beam splitter cube into two balanced beams leaving at right angle. The first one passes through the "passive" arm (thick solid lines on Figure 1), which involves only flat mirrors. The second one enters the "phase shifting" arm (thick dashed lines on Figure 1). The latter arm involves a cat's eye structure with a primary parabolic mirror and a small secondary spherical mirror, so that the light beam passes through a focus. Crossing a focal point is known to lead to an achromatic π phase shift for the amplitude (see Gouy, 1890 and Born & Wolf, 1999), and a central symmetry of the image. So, when both beams recombine, the light of a star lying on the axis (thus invariant by central symmetry) cancels because of the π phase shift, provided both arms of the interferometer have equal optical length. Cancellation does not occur for off-axis sources.

Being a Michelson's interferometer, the AIC has a "coudé" structure: the output and input beams are orthogonal. Thus, some light-, time- and space-consuming optical components need to be added in order to insert the AIC between an adaptative optics device and the camera of an existing telescope (see Figure 2 for a view of those extra components, in the CFHT implementation).

To avoid these bulky extra components, we propose two new optical combinations: the "CIAXE" and the "open-air CIAXE" or "CIAXA". Both are based upon the same principle of interferometric nulling as the standard AIC, but with a fully mono-axial and very compact design, so as to make them more "plug-and-play" (or "insert-and-see").

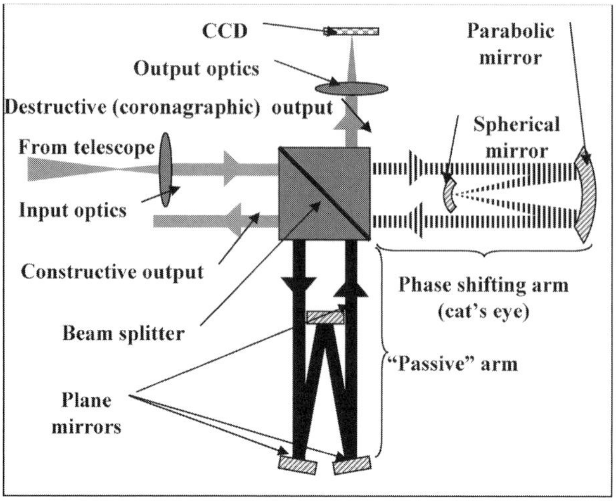

Figure 1. The configuration of the standard AIC.

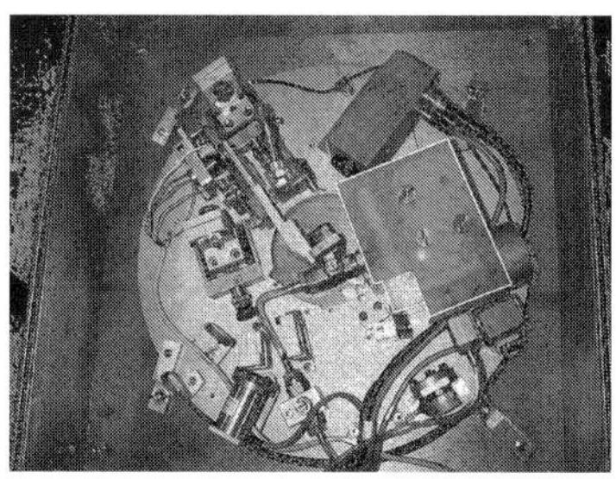

Figure 2. The standard AIC with extra optical components designed to fit it within the 3.6m telescope of CFH (Hawaii), between the adaptative optics PUEO and the IR camera KIR. **This is what CIAXE avoids!**

2. The CIAXE: a mono-axial version of the AIC

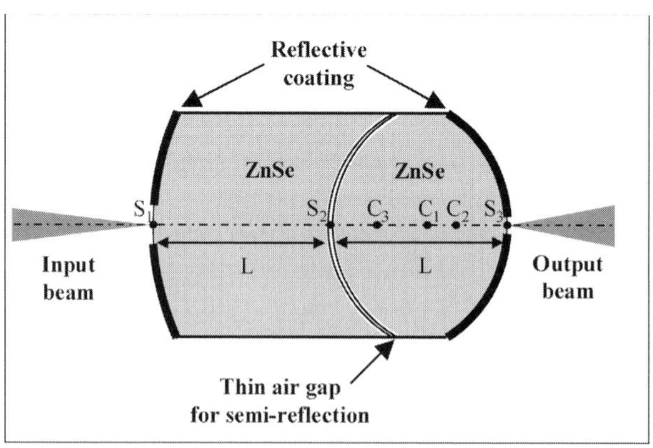

Figure 3. The configuration of the CIAXE.

The first and simplest of these two mono-axial alternatives to the AIC is sketched in Figure 3. It is based upon two equally thick lenses (made out of *ZnSe* for example), which are assembled together with a calibrated spacer ring, so as to leave between them a thin air gap. This air gap is chosen and possibly coated so as to act as a balanced beam splitter over the desired bandwidth. For optimal rejection, the thicknesses of both lenses have to be equal to the best possible precision (a difference of $20\,nm$ limits the rejection to 1000 in K band). The external surfaces have to be fully reflecting, except for two small circular zones surrounding the axis, so as to let the input and output beams enter and leave the CIAXE.

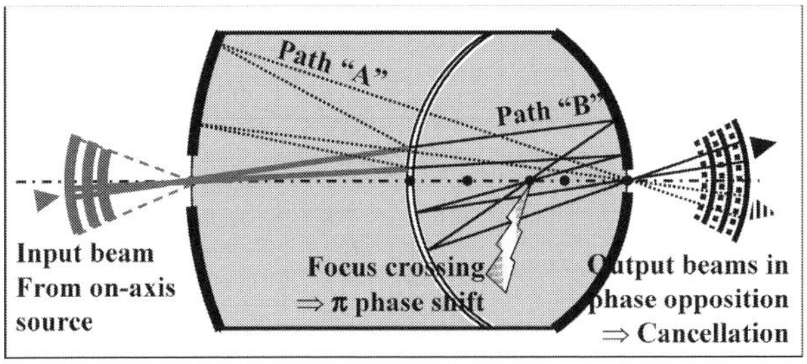

Figure 4. Ray tracing through the CIAXE for an on-axis source.

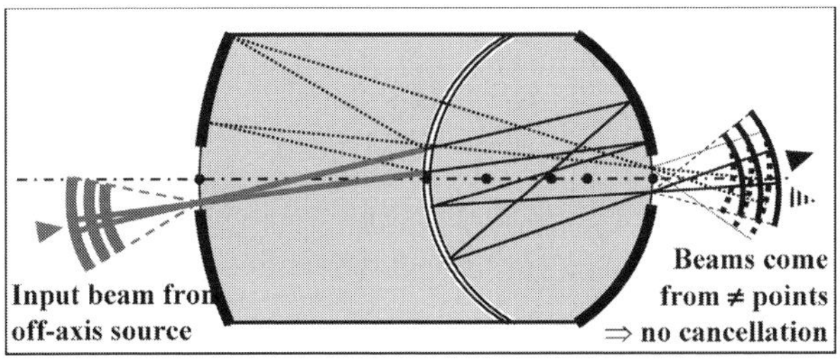

Figure 5. Ray tracing through the CIAXE for an off-axis source.

Figures 4 and 5 show the typical optical paths within the CIAXE, for a light ray coming from an on-axis source, or from an off-axis source. The constraints to be fulfilled by the three radii of curvature are (Gay *et al.*, 2005) :

- The image of S_1 must be S_3, through Paths A and B.
- The magnifications along Paths A and B must be opposite.
- The optical paths A and B must have equal length.
- Path B must cross a focus, and Path A must not.

Given the geometric thickness L, the solution is unique:

$$\begin{cases} R_1 = L\dfrac{4}{4-\sqrt{2}}, \\ R_2 = L\dfrac{1}{\sqrt{2}}, \\ R_3 = L\dfrac{-4}{4+\sqrt{2}}. \end{cases}$$

The light along Path A reaches the exit aperture around S_3 without crossing a focus. On the contrary, the light along Path B does cross a focus : it is achromatically phase-shifted of π. Thus, when recombining, the light from an on-axis source cancels (if the telescope pupil is axis-symmetric, and provided the optical paths be equal).

For an off-axis source, the lights along Paths A and B converge onto two different (symmetric) points of the exist aperture. Thus no cancellation occurs.

It can be shown (Gay *et al.*, 2005) that for an ideally working device, all the unwanted light of the central source is sent back to the entrance by successive reflections and transmissions on the beam-splitter gap.

3. The CIAXA, an "open-air" version of the CIAXE

As described above, the CIAXE has a drawback: the light travels in *Zinc Selenide*, which is a high refraction index medium (index above 2.4 in K band). Thus the wavelength is noticeably shorter than in the air. This leads to constraints on the optical surfaces accuracy which happen to be slightly above the capabilities of diamond turning techniques. Thus, an "open-air" version of the CIAXE, the "CIAXA" (for CIAXE + Air) has been designed, where the light essentially travels in air (see Figure 6).

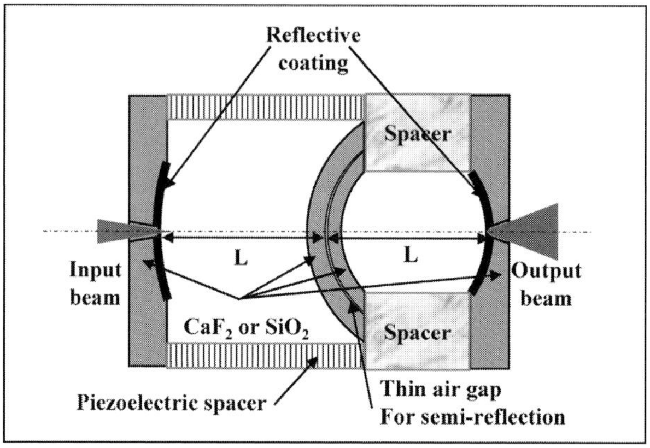

Figure 6. The "CIAXA", an open-air version of the CIAXE.

The piezoelectric spacers are needed to provide for an accurate fine tuning of the optical path difference. This is both an advantage (the accuracy less depends on machining efficiency), and a drawback (the device is more complex, and thus less autonomous).

4. Conclusion

The CIAXE and CIAXA work the same way as the standard AIC: they do stellar coronagraphy by destructive interferences between two balanced fractions of the input beam, one of which being achromatically π phase-shifted by focus crossing.

The specificity of the two variants under consideration is their easy integration in existing telescopes, due to their mono-axial structure. Some R&D effort is in progress to find machining procedures that meet the rather severe accuracy requirements of these components.

References

Gay, J. & Rabbia, Y. 1996, *C.R. Acad. Sci. Paris II* 322, 265

Baudoz, P., Rabbia, Y., & Gay, J. 2000, *Astr. Astroph. Suppl. Series* 141, 319

Gouy, L.G. 1890, *C.R. Acad. Sci. Paris* 110, 1251

Born, M. & Wolf, E. 1999, (Cambridge University Press), p. 499

Gay, J., Fressin, F., Rivet, J.P., Rabbia, Y., & Buisset, C. 2005, *C.R. Acad. Sci Paris II*, submitted.

Direct Imaging of Exoplanets: Science & Techniques
Proceedings IAU Colloquium No. 200, 2005
C. Aime and F. Vakili, eds.
© 2006 International Astronomical Union
doi:10.1017/S174392130600980X

An off-axis four-quadrant phase-mask coronagraph: concept and first results

E. Serabyn, E. E. Bloemhof, R. O. Gappinger, P. Haguenauer, B. Mennesson, M. Troy and J. K. Wallace

Jet Propulsion Laboratory, California Institute of Technology, Pasadena, CA 91109, USA
email: gene.serabyn@jpl.nasa.gov

Abstract. We have recently deployed a four-quadrant phase-mask coronagraph behind an unobscured, circular, off-axis section of the Palomar 200-inch telescope. To obtain very good wavefront correction across the 1.5 m subaperture, our relay optics reimage the sub-aperture onto the adaptive optics system's deformable mirror. This approach combines the advantages of low diffraction from obscuring elements, due to the off-axis aperture, with high wavefront correction, due to the magnification of the pupil, and high stellar rejection, due to the phase mask coronagraph. Our initial on-sky results include Strehl ratios exceeding 90%, peak stellar rejections of >100:1, and an improvement in contrast of 235:1 on a binary of separation $2\lambda/D$.

Keywords. coronagraphy.

1. Introduction

The goal of direct exoplanet detection requires the development of high-contrast imaging techniques such as coronagraphy. To detect faint planets in close proximity to bright stars, observations at a small inner working distance (IWD) and high contrast ratio are needed. Coronagraphic techniques show great promise in this regard, but coronagraphic observations on real telescopes tend to be limited by a number of factors such as a large IWD and wavefront errors which scatter large amounts of light. To image close to bright stars, three things are needed: a coronagraph to improve the intrinsic planet/star contrast ratio, a very good adaptive optics (AO) system to reduce scatter, and a means of reducing diffracted light from beam obscurations such as the secondary mirror. Our approach addresses and mitigates all three of these factors.

One obvious approach to better wavefront correction is to increase the number of elements in the AO system's deformable mirror (DM). However, another approach, which can be implemented quickly and cheaply, but which has not heretofore been pursued, is to reimage a smaller section of the primary mirror onto the DM, thereby reducng the effective spacing of the DM elements. Specifically, if the targeted subaperture is an unobscured quadrant of the primary mirror (between the spider support legs) in a Cassegrain telescope, a well-corrected, unobscured, off-axis telescope is generated, thus completely eliminating diffraction from vignetting elements, while also reducing scatter because of the closer actuator spacing. Note that while producing improved performance across a subaperture, no additional demands are placed on an already extant AO system. With e.g., the existing Palomar AO system (Troy *et al.* 2000), wavefront accuracies of under 100 nm and Strehl ratios >0.9 in the K band (2.2 μm) are expected.

In fact, at least one recent coronagraphic approach is ideally suited to an unvignetted, off-axis telescope configuration – the four-quadrant phase-mask (FQPM) coronagraph (Rouan it *et al.* 2000, Boccaletti *et al.* 2004)), which introduces phase shifts of π radians in two diametrically opposed quadrants of the focal plane (Fig. 1). Because there

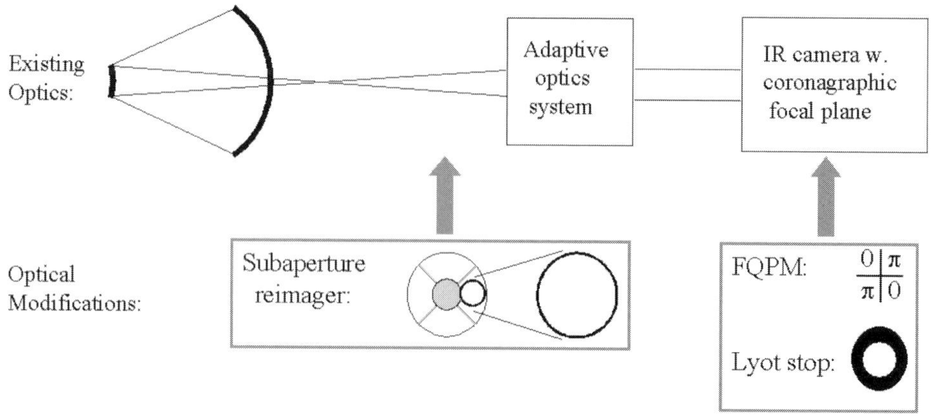

Figure 1. Optical schematic. Relay optics are inserted ahead of the existing AO bench to reimage the sub-pupil onto the DM, and the coronagraphic components are inserted into PHARO.

is no opaque starlight blocker in the FQPM approach (the star sits at the crosshairs of the phase plate), this approach is suited to observations very close to the optical axis (Riaud *et al.* 2001). However, the performance of a FQPM coronagraph is degraded by the presence of an on-axis secondary mirror (Lloyd *et al.* 2003). Thus, our proposed combination of an off-axis pupil with a FQPM coronagraph provides a solution for exploring the regions close to a bright star, by combining a high degree of wavefront control from the denser AO system, the total elimination of diffracted light from vignetting optical elements, and a small IWD from the phase-based coronagraph.

We therefore decided to test the performance of this off-axis FQPM approach on the sky. Fig. 1 shows the necessary components: a set of relay optics to magnify an off-axis sub-pupil onto the AO system's DM, a FQPM in the coronagraphic focal plane, and a clear, undersized Lyot stop in the pupil plane. The masks were fabricated and the optics tested at JPL over the past two years, after which the equipment was installed at Palomar (Haguenauer *et al.* 2005): the ambient-temperature relay optics module on the Palomar AO bench, and our phase and pupil masks in the cryogenic near-IR camera PHARO (Hayward *et al.* 2001). We carried out our first observations with this instrumentation on the Palomar 200-inch telescope on the nights of June 13–15, 2005.

2. Results

First we observed single stars, in order to determine the image quality obtainable on a well-corrected, unobscured sub-aperture. One of our "subaperture" images of the single star HD121107 is shown in Fig. 2, where an almost classical Airy pattern is evident. Indeed, as many as 7 Airy rings are seen quite distinctly, and bits of rings as far out as the 9th ring can be discerned. The Strehl ratio for our best images were in the 93% to 94% range, corresponding to a wavefront correction of ≈ 85 to 90 nm, significantly better than has previously been possible from the ground with the wider actuator spacings.

Next, we sent the light from this (and other) stars through the crosshairs of our FQPM coronagraph. One of our resultant images is shown in Fig. 3 (top center), where a peak rejection of 85 was obtained. The rejection depended somewhat on the seeing, as well as on our alignment onto the crosshairs, and in other images the rejection exceeded 100:1 slightly. However, this image is presented here because the symmetry of the residuals in this image allowed better stellar suppression after the next step – subtraction across

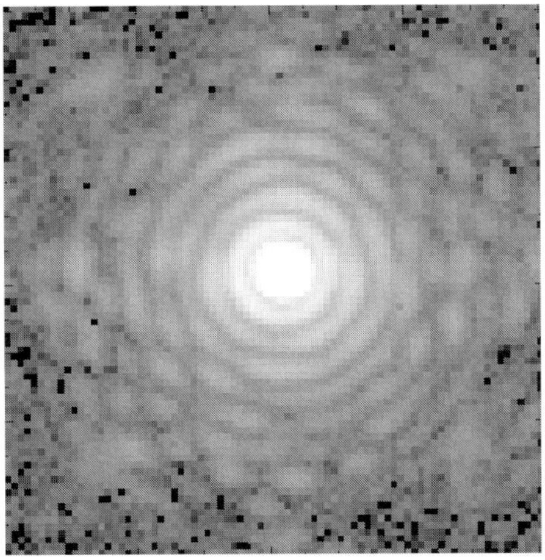

Figure 2. Image of the star HD121107 using our 1.5 m subaperture reimager on the Hale 200 inch telescope. The image is the sum of 20 individual integrations, each of 1.4 s duration. The best Strehl ratios obtained on such images were in the 93% to 94% range.

an image diagonal. In the top right-hand image, the central image has been subtracted across the diagonal of greatest symmetry, resulting in a final stellar supression of 220:1.

Finally, we observed binary stars, including HD148112. In the raw image of this binary shown in the bottom left panel of Fig. 3, the binary separation is seen to be $\approx 2\lambda/D$. The bottom center image shows the image of HD148112 transmitted by the FQPM coronagraph, and the lower right-hand image shows the subtraction across the diagonal of greatest symmetry in the coronagraphic residuals. While the dim companion is already visible in the raw (bottom left) image, the central and right hand images show the great contrast improvement which results from the FQPM coronagraph. The brighter star has a final peak suppression of 235:1, changing the binary contrast ratio from 23:1 to 1:10 in the opposite direction. In the final image, the dim companion completely dominates over the primary residuals, indicating that it should be possible to use this approach to detect much fainter companions even closer to bright stars.

3. Conclusions

Our first observing run with our off-axis FQPM coronagraph has allowed the demonstration of a well-corrected off-axis 1.5 m subaperture with Strehl ratios as high as 93% to 94%, and wavefront accuracies as good as 80 nm. Instantaneous peak stellar rejections of 100:1 were obtained, and the best (diagonal-subtracted) stellar rejection was 235:1. It is evident from the lower row of images in Fig. 3 that this approach allows high contrast observations very close to bright stars. We expect to obtain significant performance improvements beyond this level as we improve our instrumentation. Moreover, use of this technique on an off-axis section of a ten-meter class telescope should really bring its advantages to light, because an IWD of even $1\lambda/D$ with a 4 m off-axis sub-aperture is significantly closer to the axis than an IWD of e.g., $4\lambda/D$ using the full aperture.

Acknowledgements

This work was carried out at the Jet Propulsion Laboratory, California Institute of Technology, under contract with the National Aeronautics and Space Administration.

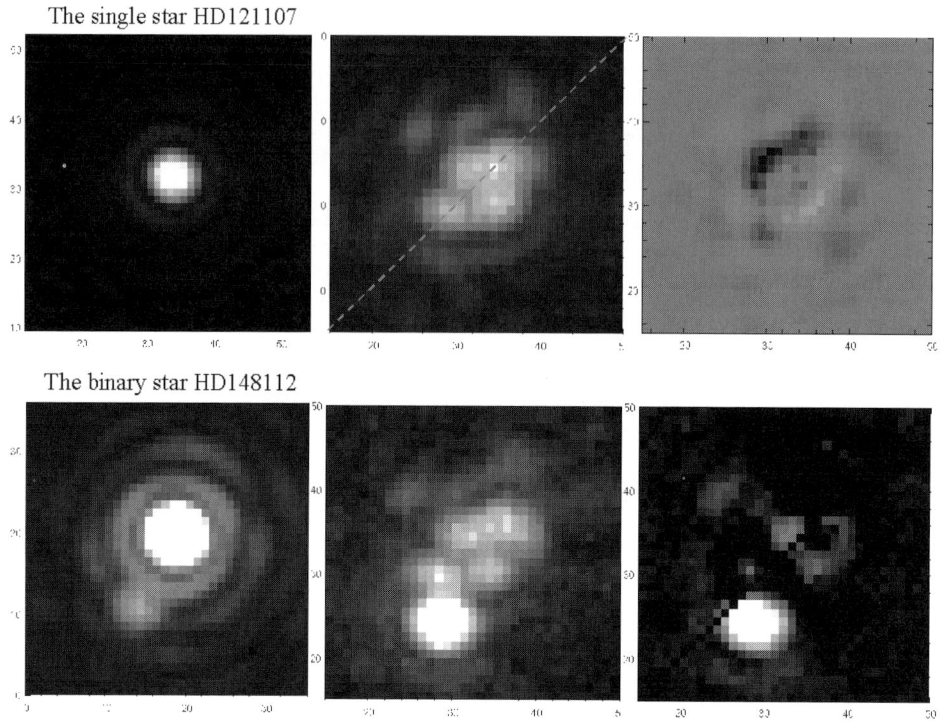

Figure 3. Top row: images of the single star HD121107. Top left: raw image, bypassing the coronagraph. Top center: raw image through the FQPM coronagraph. Top right: center image subtracted across the diagonal of greatest symmetry in the residuals. Bottom row: images of the binary star HD148112. Bottom left: raw image, bypassing the coronagraph. Bottom center: raw image through the FQPM coronagraph. Bottom right: center image subtracted across the diagonal of greatest symmetry in the residuals.

Observations at the Palomar Observstory were made as part of the continuing collaborative agreement between the Palomar Observatory and the Jet Propulsion Laboratory.

References

Boccaletti, A. *et al.* 2004, *PASP* 116, 1061
Haguenauer, P. *et al.* 2005, *Proc. SPIE* 5905, 59050S-1
Hayward, T.L. *et al.* 2001, *PASP* 113, 105
Lloyd, J.P. *et al.* 2003, *Proc. SPIE* 4860, 171
Riaud, P. *et al.* 2001, *PASP* 113, 1145
Rouan, D. *et al.* 2000, *PASP* 112, 1479
Troy, M. *et al.* 2000, *Proc. SPIE* 4007, 31

Direct Imaging of Exoplanets: Science & Techniques
Proceedings IAU Colloquium No. 200, 2005
C. Aime and F. Vakili, eds.
© 2006 International Astronomical Union
doi:10.1017/S1743921306009811

Shaped pupil coronagraphs for high-contrast imaging with SPICA

S. Tanaka[1], T. Nakagawa[1], H. Kataza[1], K. Enya[1], M. Tamura[2], L. Abe[2], and the SPICA coronagraph team

[1]Institute of Space and Astronautical Science, Japan Aerospace Exploration Agency, 3-1-1 Yoshinodai, Sagamihara, Kanagawa, 229-8510, Japan
email:stanka@ir.isas.jaxa.jp

[2]National Astronomical Observatory of Japan, 2-21-1 Osawa, Mitaka, Tokyo, 181-8588, Japan

Abstract. We present configurations of shaped pupil coronagraphs optimized for a realistic telescope to directly detect extrasolar giant planets in the mid-IR wavelengths. This study is linked to the development of the coronagraphic instrument aboard SPICA (*SPace Infrared telescope for Cosmology and Astrophysics*). We made a systematic assessment of the performance of the "checkerboard" and "concentric ring" masks, introducing a large central obstruction (C.O.) due to a secondary mirror and its related support spiders. With a small secondary mirror, we also propose a modification to the original symmetrical checkerboard apodization, which enables us to achieve a 10^{-7} contrast level at 4.0 λ/D. The transmission through the optimal binary masks exhibits abrupt increases and plateaus as the inner working angle (IWA) is increased. We attribute these properties of binary apodization function to the existence of threshold IWAs that allow large openings in the pupil.

Keywords. instrumentation: high angular resolution.

1. Introduction

It is planned to launch SPICA, a Japanese cooled infrared telescope with a 3.5 m-diameter primary mirror, in the early 2010s (Nakagawa 2004). One target of the SPICA mission is the direct imaging and spectroscopy of extrasolar giant planets in the mid-IR wavelengths (Abe *et al.* 2005; Enya *et al.* 2005; Tamura *et al.* 2005). The requirement on the SPICA coronagraphic instrument is to realize a contrast of 10^{-6} from an IWA of 3–5 λ/D to an OWA (outer working angle) of $\sim$10 λ/D during actual operation.

Shaped pupil techniques have important advantages such as the full achromaticity, the simplicity of optical system, and the ease of fabrication. However, few previous studies on binary pupils took C.O.s in the telescope pupil into account: most of them were for obstruction-free telescopes like an off-axis one. Hence, they cannot be applied directly to more realistic general-purpose telescopes, including SPICA, whose pupil functions are complicated due to the existence of a large secondary mirror and its related support spiders. The main objective of this paper is to find optimal binary apodization functions that maintain tight IWAs for a telescope pupil with a large C.O. We here require ideal performance of 10^{-7} contrast level. We also reveal the general properties of binary apodization function, from a careful analysis of the behavior of transmission through binary masks according to the values of IWA.

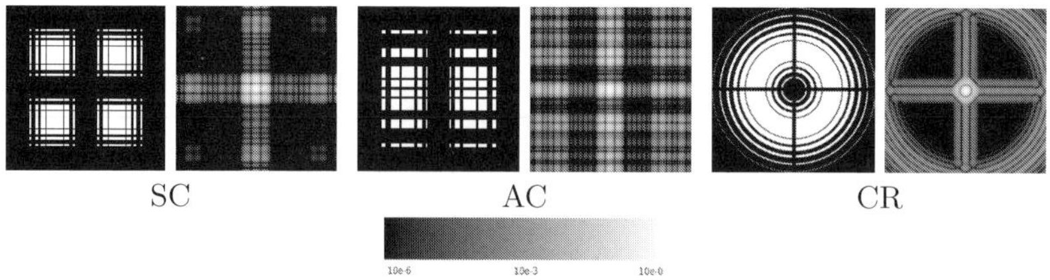

Figure 1. The optimal masks for $\alpha_s = 0.14$ and their associated PSFs. The side lengths of the mask images correspond to the pupil diameter. The PSF images show the central $40\ \lambda/D \times 40\ \lambda/D$ region in logarithmic scale. α_{spi} is 0.0228 for the AC and CR masks.

2. Masks for a pupil with a large C.O.

2.1. *Configuration*

A possible solution to overcome the obstacle presented by C.O.s is the use of either the concentric ring masks or the checkerboard masks, both proposed recently by Vanderbei *et al.* (2003, 2004). We considered three types of configurations based on those masks. The first one is the "symmetrical checkerboard" (SC) masks whose two aperture functions along x and y directions, $A_x(x)$ and $A_y(y)$, have the same large C.O. corresponding to the secondary mirror (with diameter $\alpha_s D$, D is the primary diameter). Secondly, we modified the constraint condition on $A_y(y)$ into a looser one: a C.O. corresponding to a spider of width $\alpha_{spi} D$, not to $\alpha_s D$. The benefit derived from this modification is that it allows us to shift the contrast requirement on $A_x(x)$ to $A_y(y)$, resulting in an improvement in the IWA. We call them the "asymmetrical checkerboard" (AC) masks. The last one is the concentric ring (CR) masks with cross-shaped spiders. The C.O. by the secondary mirror was taken into account in the optimization processes while the spiders were unavoidable.

We simulated large secondary mirrors with various diameters: $\alpha_s = 0.1$, 0.14, 0.2, and 0.24. The width of the spider arms is fixed at $\alpha_{spi} = 0.0228$. We required that the mask transmission, defined as the ratio of the total open area to a full circular pupil, be higher than 15 % and discarded the solutions that did not meet this criterion. Solutions whose IWAs are larger than 6.0 λ/D were excluded. Figure 1 shows examples of the optimal mask shapes we found and their PSFs.

2.2. *Comparison of the performance*

Table 1 shows that the AC masks have the tightest IWAs in almost all the cases. For $\alpha_s < 0.15$, we obtain an IWA of 4.0 λ/D with acceptable transmission. However, since the IWA of the checkerboard masks monotonically increases as the C.O. gets larger, the CR masks, whose performance is rather insensitive to the secondary mirror size, become a contender for $\alpha_s \geqslant 0.2$. The greater transmission of the CR masks, which is more than twice that of the checkerboard masks, could be the dominant factor that decides the supremacy in this regime. Figure 2 compares the total dark area obtained by the three configurations. In the cases of $\alpha_s \leqslant 0.14$, the AC masks have a significantly wider dark area inside the radius of $\sim$6.5 λ/D. However, the superiority of the AC masks to the CR masks is less obvious when $\alpha \geqslant 0.2$. In conclusion, the AC masks are promising due to the larger dark area for $\alpha_s \lesssim 0.15$. For larger C.O.s ($\alpha_s \gtrsim 0.15$), on the other hand, the CR masks provide the best compromise between search space and mask transmission.

Since the spider effects cannot be avoided with the CR masks, we are also interested in whether their performance is sensitive to the spider widths. When the spider arms are wider, the contrast gap between the cross-shaped diffraction halos and the PSF center decreases. Although the total dark area does not change much according to the arm

Table 1. Optimal solutions found for the symmetrical checkerboard (SC), asymmetrical checkerboard (AC), and concentric ring (CR) masks. The spider arm widths and the total contrast are fixed at $\alpha_{\rm spi} = 0.0228$ and 10^{-7}, respectively. The IWAs and OWAs are in λ/D unit.

$\alpha_{\rm s}$	Mask	IWA	OWA	Trans.	Mask	IWA	OWA	Trans.	Mask	IWA	OWA	Trans.
0.1	SC_1	$4\sqrt{2}$	$15\sqrt{2}$	23 %	AC_1	4.0	12.3	20 %	CR_1	4.7	17.0	48 %
0.14	SC_2	$4\sqrt{2}$	$13.5\sqrt{2}$	23 %	AC_2	4.0	11.4	18 %	CR_2	4.7	15.0	48 %
0.2	SC_3	No Solution Found			AC_3	5.0	11.4	15 %	CR_3	4.7	12.0	38 %
0.24	SC_4	No Solution Found			AC_4	5.4	11.4	17 %	CR_4	6.0	12.0	39 %

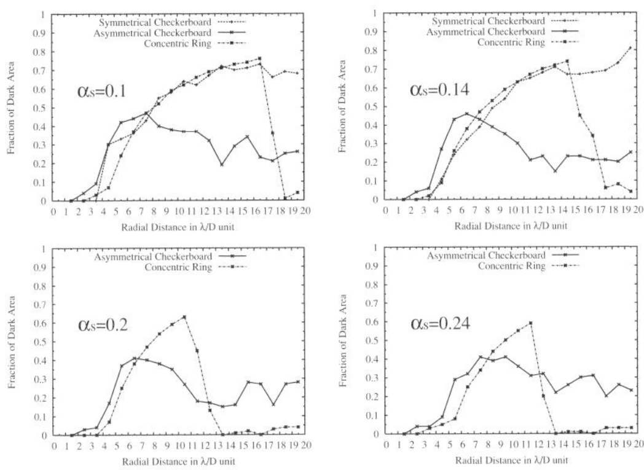

Figure 2. — The fraction of dark area (dimmer than the PSF core by 6 orders of magnitude or more) within a 1.0 λ/D-wide annuli each of which has the radius of 1.5, 2.5, . . . , and 19.5 λ/D.

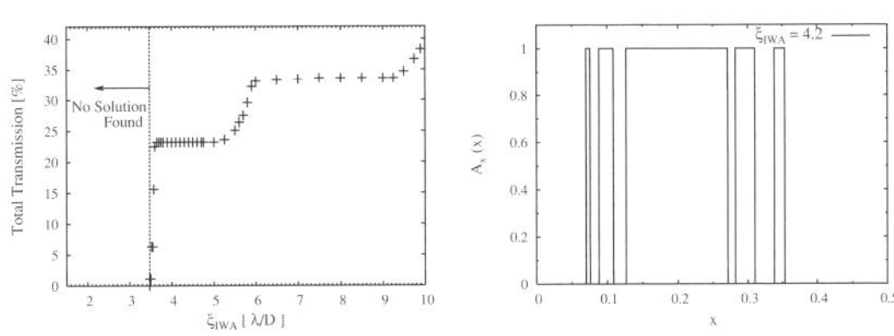

Figure 3. — *left*: Dependence of the mask transmission on the IWA (see text for the definition of the parameters). Under the current configuration, the maximum transmission available is 41 % . *right*: The cross-sectional cut of the optimal aperture functions, $A_x(x)$, for $\xi_{\rm IWA} = 4.2\ \lambda/D$.

width, therefore, it is still preferable to use narrower spider arms in view of total amount of light that gets into the main lobe of the PSF.

3. Behavior of Binary Aperture Functions

In this section, we discuss the relationship between the IWA and mask transmission. Although the following analysis can be extended to the AC and CR masks, we limit our discussion to SC masks with a 2-D OWA of $10\sqrt{2}\ \lambda/D$ (the corresponding 1-D OWA is $\xi_{\rm OWA} = 10\ \lambda/D$) and total contrast of 10^{-7}. We also impose a 14 % C.O. Figure 3 (*left*) shows that the transmission curve exhibits a stepwise behavior. There are two abrupt increases at 1-D IWA of $\xi_{\rm IWA} = 3.6\ \lambda/D$ and around $\xi_{\rm IWA} = 6.0\ \lambda/D$, and also two plateaus at transmission of 23 % and 34 %.

Fig. 3 also shows the cross-sectional cut of the aperture functions, $A_x(x)$, at $\xi_{\rm IWA} = 4.2\ \lambda/D$. From the Fourier transform of such multiple boxcar functions, we obtain the brightness contrast ratio realized at point ξ as follows:

$$C(\xi) = \left\{ \frac{\sum_j T_j(\xi) \cdot S_j(\xi)}{\sum_j W_j} \right\}^2 .$$

$T_j(\xi)$ refers to the trigonometric function term $\cos(2\pi\xi X_j)$ depending on $X_j \equiv (x_{2j} + x_{2j-1})/2$, the position of the midpoint of the j-th opening, and $S_j(\xi)$ denotes the sinc function term $W_j \sin(\pi\xi W_j)/(\pi\xi W_j)$ depending on $W_j \equiv x_{2j} - x_{2j-1}$, its width.

The denominator in the equation above, $\sum_j W_j$, represents the transmission of this apodization. While wider openings provide better transmission, the openings have to satisfy the requirements imposed by the values of the IWA, OWA, and contrast. We point out here the factors that constrain the widths of openings in the pupil. The first one is the value $|\sum_j T_j(\xi) S_j(\xi)|$, which must be very small in the dark region. In order to ensure this, all $T_j S_j$ should have roughly similar amplitudes inside the dark region under consideration. Any component which has significantly larger $|T_j S_j|$ in the dark region than those of the others is thus prohibited. The second factor is the extent of the dark region, or the distance from the IWA to the OWA. A large number of openings (i.e., a denser set of X_j) are needed to enlarge the dark region. The inner and outer openings are especially critical for the depth of the null. If the amplitude contributed by inner openings (with small X_j) is large (with large W_j), we need many other components with short T_j periods (large X_j) to compensate for it in the dark region. Therefore, it is necessary that the amplitudes contributed by the inner openings are small, and this is realized by narrower openings (small W_j). Components with short periods (large X_j) are also important because higher frequency components are used to finely tune the summation $\Sigma_j T_j S_j$, and thereby produce a deeper null. A wide opening (with large W_j) at the outermost region of the pupil causes a large inward shift of X_j, resulting in a loss of higher frequencies. Thus, in general, the openings near the inner and outer edges of the pupil become narrower than those in the intermediate region (see Fig. 3).

The saturation in the transmission at IWAs of $4 \sim 5\ \lambda/D$ is an important property which suggests that we ought to choose a solution just after the first transmission jump.

4. Summary

We discussed the dependence of coronagraphic performance of some binary masks, taking into account the effects from the secondary mirror and cross-shaped spiders. When the secondary mirror is rather small, we obtain a contrast level of 10^{-7} even at an IWA of 4.0 λ/D. The transmission curve of the optimal binary apodization function exhibits a rapid increase near a certain threshold value of the IWA and becomes saturated after that, indicating that for practical use we should choose the function at the onset of saturation.

References

Abe, L., *et al.* 2005, *this conference*

Enya, K., *et al.* 2005, *this conference*

Nakagawa, T. 2004, *Advances in Space Research*, 34, 645

Tamura, M., Abe, L., HiCIAO team, SPICA coronagraph team, & JTPF working group. 2005, *this conference*

Vanderbei, R. J., Spergel, D. N., & Kasdin, N. J. 2003, *ApJ*, 590, 593

Vanderbei, R. J., Kasdin, N. J., & Spergel, D. N. 2004, *ApJ*, 615, 555

Direct Imaging of Exoplanets: Science & Techniques
Proceedings IAU Colloquium No. 200, 2005
C. Aime and F. Vakili, eds.
© 2006 International Astronomical Union
doi:10.1017/S1743921306009823

Stellar coronagraphy: laboratory testing of the Achromatic Interfero Coronagraph

J. P. Rivet[1], J. Gay[1], J. Dalmau[1], G. Cassar[1] and Y. Rabbia[2]

[1]Observatoire de la Côte d'Azur, B.P. 4229, F-06304 Nice Cedex 4, France
email: rivet@obs-nice.fr

[2]Observatoire de la Côte d'Azur, Avenue Copernic, F-06130 Grasse, France

Abstract. We present new laboratory measurements of the intrinsic rejection performances in infrared ($1.9\,\mu$m to $2.5\,\mu$m) for a prototype of Achromatic Interfero Coronagraph (AIC). We first recall basics about the AIC, then describe the prototype under consideration. We give detailed explanations about the experimental setup and the procedures followed to measure the rejection rate. We end up with a discussion of the results obtained.

Keywords. instrumentation: miscellaneous.

1. Basics on the standard AIC

The Achromatic Interfero-Coronagraph (AIC) has been introduced by Gay & Rabbia (1996) (see also Baudoz *et al.*, 2000) as a new concept of stellar coronagraphs. Instead of rejecting the light of a central star by a more or less smoothly varying complex transmission in the focal plane of a telescope, the AIC rejects the light by destructive interferences. The device works like a modified Michelson's interferometer (see Figure 1): the input light is divided by a beam splitter cube into two balanced beams leaving at right angle. The first one passes through the "passive" arm (thick solid lines on Figure 1), which involves only flat mirrors. The second one enters the "phase shifting" arm (thick dashed lines on Figure 1). The latter arm involves a cat's eye structure with a primary parabolic mirror and a small secondary spherical mirror, so that the light beam passes through a focus. Crossing a focal point is known to lead to an achromatic π phase shift for the amplitude (see Gouy, 1890 and Born & Wolf, 1999), and to a central symmetry of the image. So, when both beams recombine, the light of a star lying on the axis (thus invariant by central symmetry) cancels because of the π phase shift, provided both arms of the interferometer have equal optical length. Cancellation does not occur for off-axis sources (see Figure 2).

The entrance pupil of the telescope has to be symmetric, and the odd-order aberrations (turbulence, optical parts) have to be as low as possible (Baudoz *et al.* (2000)). Extra input and output optics (sketched as lenses in Figure 1) are required to fit the AIC between a telescope and its IR detector.

2. The prototype under consideration

The prototype used for this experiment is a compact monolithic device based upon a bloc of fused silica with a cavity drilled inside for the phase shifting arm. The beam splitter cube, the parabolic mirror and the spacer containing the cavity for the passive arm are also made out of silica, and glued onto the main bloc (see Figure 3).

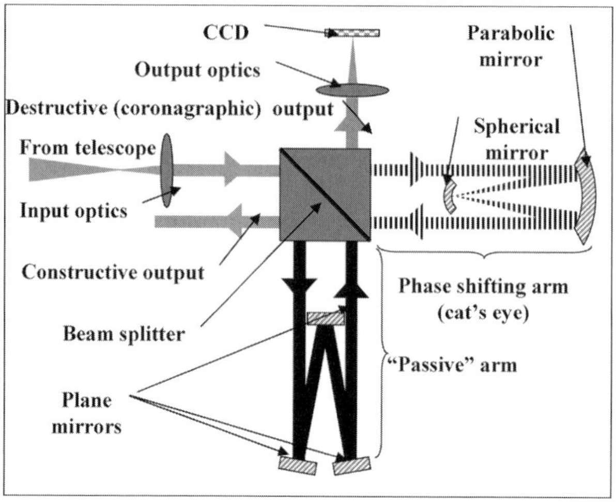

Figure 1. The configuration of the standard AIC.

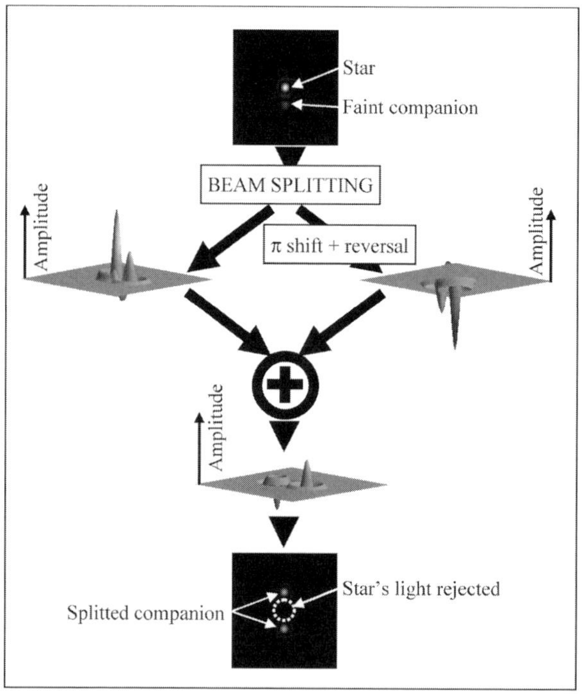

Figure 2. The principle of the AIC.

Since the equality of the geometrical lengths of both arms cannot be guaranteed to the required accuracy, the cavities are filled with dry nitrogen, and a suitable pressure difference is imposed, so as to compensate for the geometrical length difference ($\simeq 5\,\mu m$).

3. The experiment

This prototype, with a suitable input and output optics, has been inserted within the focal environment of the 3.6 m telescope at CFH (Hawaii), and tested on the sky. The

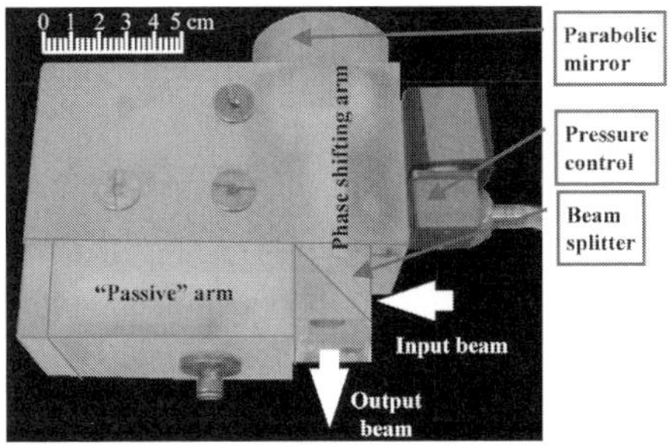

Figure 3. The prototype of AIC.

rejection performances were limited by the residual atmospheric turbulence, and possibly also by the quality of the optical components of the AIC (mainly the parabolic mirror).

To evaluate the intrinsic limitations of the AIC itself, a laboratory experiment has been conducted to measure the rejection rate of this prototype. The optical bench layout is sketched in Figure 4. The pinhole source diameter is $50\,\mu m$; the pupil stop diameter is $5.4\,mm$, and the focal length of the BK7 collimator is $106\,cm$ in the IR band under consideration ($1.9\,\mu m$ to $2.5\,\mu m$). The $ZnSe$ objective lens focuses the output beam on the receptor, a CEDIP Jade IR camera, with a Germanium filter.

Several parameters have to be very carefully fine tuned to achieve optimal rejection rates: The XY position of the pupil stop, the Z position of the collimator lens and of the objective lens, the nitrogen pressure difference, and the XY position of the source.

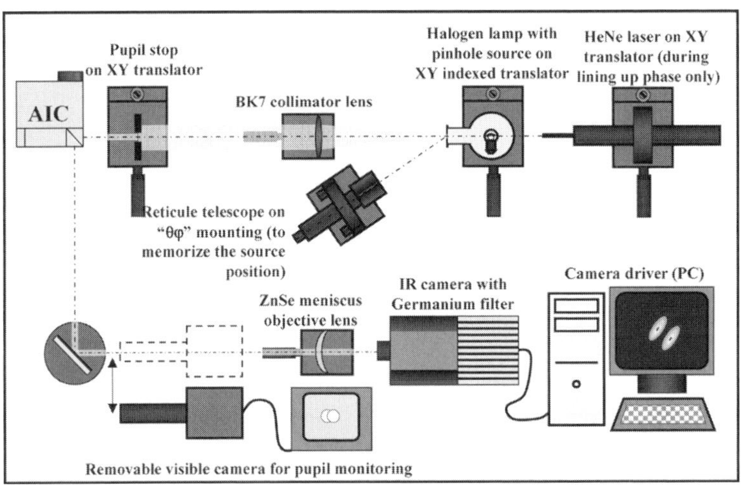

Figure 4. The experimental setup for rejection measurements.

4. The fine tuning procedures

The XY position of the pupil stop is tuned "manually", so as to make the two images of the pupil coincide on the visible camera monitor. For the other parameters, the "manual" tuning has revealed unsatisfactory, and the following more accurate procedure has been used:

- one first obtains a coarse grained value of the parameter.
- Then, one explores systematically several values around it.

• For each value, one obtains the IR intensity field by averaging over 64 suitably pre-processed snapshots, and one measures the peak intensity value at the center of the field.

• One fits the curve "value of the parameter versus peak intensity" with a parabola, and extract the optimal value of the parameter, corresponding to the minimum of the peak intensity.

After a complete fine tuning cycle for all the relevant parameters, the residual IR intensity can be measured by averaging over 512 snapshots.

5. The measured rejection rate

The coronagraphic rejection rate is the ratio of the input energy for an on-axis source, to the residual output energy. The input energy is evaluated by pushing the source far out of the axis. The most carefully processed data led to the following result:

$$\text{Measured rejection rate} \quad \simeq \quad 500 \text{ for } \lambda \in [1.9\,\mu m - 2.5\,\mu m]$$

6. The extinction profile

The normalized residual energy, when studied as a function of the source offset, yields the extinction profile. Figure 5 shows the theoretical extinction profile (solid curve), and the measured data (dots).

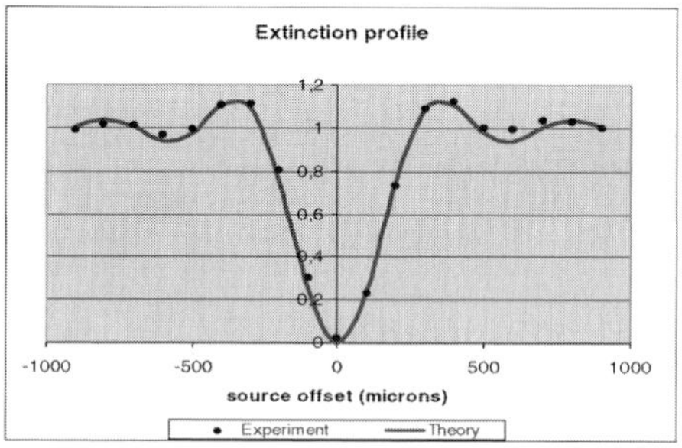

Figure 5. The rejection profile of the AIC (rejection versus distance from axis). Solid line: theoretical prediction; Dots: measured values.

7. Conclusion

Despite the poor quality of its parabolic mirror, the AIC prototype at hand has led to a measured rejection rate close to **500** over a relatively large bandwidth, even with a non-totally coherent source. Further on-sky tests at CFHT are now required to fully qualify this prototype for scientific applications.

References

Gay, J. & Rabbia, Y. 1996, *C.R. Acad. Sci. Paris II* 322, 265

Baudoz, P., Rabbia, Y., & Gay, J. 2000, *Astr. Astroph. Suppl. Series* 141, 319

Gouy, L.G. 1890, *C.R. Acad. Sci. Paris* 110, 1251

Born, M. & Wolf, E. 1999, (Cambridge University Press), p. 499

Direct Imaging of Exoplanets: Science & Techniques
Proceedings IAU Colloquium No. 200, 2005
C. Aime and F. Vakili, eds.
© 2006 International Astronomical Union
doi:10.1017/S1743921306009835

The Phase Induced Amplitude Apodization Coronagraph: an overview of simulations and laboratory effort.

Eugene A. Pluzhnik[1,6], Olivier Guyon[1], Stephen Ridgway[2], Robert Woodruff[3], Celia Blain[1], Frantz Martinache[4], and Raphael Galicher[5]

[1]Subaru Telescope, National Astronomical Observatory of Japan, 650 North A'ohoku Place, Hilo, HI 96720, USA, email: pea@naoj.org
[2]National Optical Astronomical Observatories
[3]Lockheed Martin Space Corporation
[4]Observatoire de Haute Provence
[5]Ecole Normale Supérieure, Paris, France
[6]Institute of Astronomy of Kharkov National University, Sumskaya 35, Kharkov 61022, Ukraine

Abstract. The high contrast (typically 10^{10}) and small angular separation between a planet and its parent star are the main challenges that need to be overcome to detect and characterize Earth-like planets around the nearest stars. Therefore, exoplanet imaging requires the use of a coronagraph, that ideally efficiently cancels the light from the star and has minimal influence on the planet image. The Phase Induced Amplitude Apodization Coronagraph relies on pupil apodization by geometrical remapping of the flux in the pupil plane. This method combines the advantages of classical pupil apodization with high throughput ($\approx$100%) and high angular resolution ($\approx\lambda/D$), and has some unique advantages over most coronagraphs, such as low chromaticity, low sensitivity to stellar angular size and to small pointing errors. As a result, planet detection time is about 50–100 times shorter in comparison with classical coronagraphic techniques (Martinache *et al.* 2005).

Both the advantages of the PIAAC and the main factors affecting the performance of the coronograph will be examined in our laboratory experiment in which high quality PIAA optics wilkl be combined with wavefront control to demonstrate achromatic high contrast imaging (10^6 or more) at small angular separation (less than $2\lambda/D$). We present here a description and current status of this experiment together with a short analyses of the main factors affecting the performance of the coronograph.

Keywords. direct exoplanet imaging, coronagraphy, apodization, pupil remapping, diffraction propagation.

1. Introduction

The main goal of the Terrestrial Planet Finder mission (NASA) is to find and characterize Earthlike planets around the nearest stars. However detection of such planets is difficult because of the high contrast (typically 10^{10}) and small angular separation (typically 50 mas) between the planet and its parent star. Therefore exoplanet imaging requires the use of a coronographic device. The Phase Induced Amplitude Apodization Coronagraph (PIAAC)(Guyon 2003) which relies on a loseless apodization technique, seems especially adapted to this task. Preliminary numerical and laboratory experiments with diamond turned lenses have confirmed the basic concept which this approach is based on (Galicher *et al.* 2005). However it was impossible to obtain high contrast with low-quality lenses. In the next phase of our laboratory experiment effort, high quality

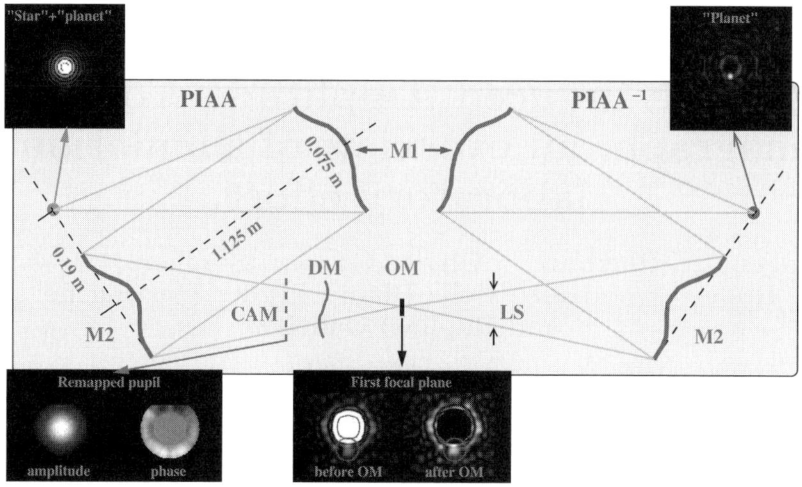

Figure 1. The optical layout of the PIAAC prototype. The maximal phase difference in the apodized pupil is 4°.2. The "planet" brightness is 10^{-4}. The "star-planet" distance is $2\lambda/D$.

mirrors will be combined with wavefront control to demonstrate achromatic high contrast imaging (10^6 or more) at small angular separation (less than $2\lambda/D$).

The goals of the experiment are to demonstrate/explore: (1) ability to polish the pupil remapping mirrors; (2) ability to correct off-axis abberations of the PIAAC with the second pupil remapping unit; (3) throughput, inner working angle and tip-tilt sensitivity of the PIAAC; (4) sensitivity to wavefront errors; (5) diffraction effects; (6) wavefront control with the remapped beam; (7) combination of the PIAAC with classical apodisation.

We present here a description and current status of our experiment together with analyses of the main factors affecting the performance of the coronograph, including diffraction effects, sensitivity to wavefront and pointing errors.

2. Optical layout

The experiment is designed to be operated in 3 modes: (1) without classical apodizer and Lyot mask (pure PIAA design); (2) without classical apodizer with Lyot mask (PIAA/Lyot design); (3) with classical apodizer and Lyot mask (PIAA/CPA design). The optical layout includes two identical Phase-Induced Amplitude Apodizers PIAA and PIAA^{-1}, (Fig. 1, Guyon *et al.* 2005), an occulting mask (OM) and a Lyot stop diaphragh (LS). The first PIAA apodizer, the occulting mask and the Lyot stop diaphragm produce a coronagraphic effect, while the second apodizer (PIAA^{-1}) corrects aberrations of off-axis sources introduced by the first apodizer. The source is collimated by the first PIAA mirror M1 and reimaged by the second PIAA mirror M2. The shape of the mirrors is a base off-axis paraboloid of revolution modified precisely to produce the designed output intensity profile. Each paraboloid has 1133 mm focal length and 75 mm diameter minumum clear aperture. The distance between the mirrors is 1125 mm and the centerline of the clear aperture is 190 mm off-axis. A deformable mirror (DM) is used to correct alignment and polishing errors. Additional apodization is obtained with the classical apodization mask CAM. Results of 3D raytracing simulations for amplitude /intensity and phase distributions in the main planes of the PIAAC are also presented in Fig. 1.

3. Apodized beam profile

Our goal is to achive a PSF contrast of 10^6 at $2\lambda/D$. Unfortunately, adopting a beam profile which exceed this requirement is dangerous because it increases the difficulty of

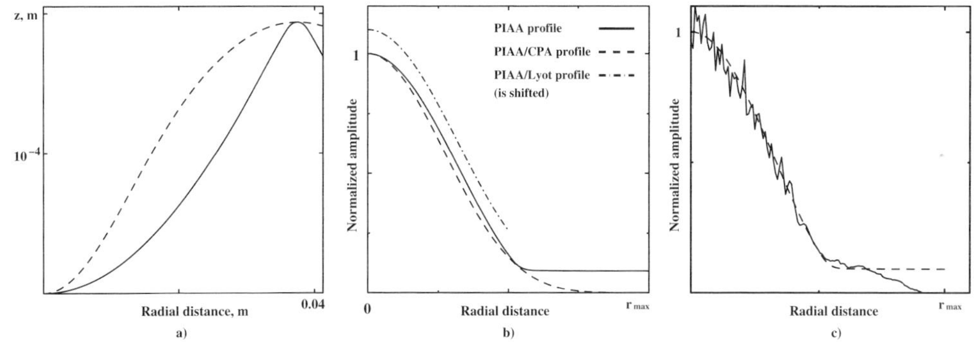

Figure 2. The M1 (*solid*) and M2 (*dashed*) profile terms (*a*), designed amplitude profiles (*b*) and a simulated amplitude profile (*c*) of the output beam.

polishing the optics. The most serious difficulty for the polishing of the optics is the sharp "bend" at the outside radius of the first mirror. This bend has a small radius of curvature and is therefore very challenging to polish, but it is essential because it distributes a small fraction of the entrance pupil's light into the broad wings of the output pupil.

In order to make the polishing of the optics easier we have chosen a hybrid design combining PIAAC and classical apodization. The classical apodization mask together with active wavefront control also can be used efficiently to correct the amplitude and phase errors produced with the first PIAA apodizer. This hybrid solution combines the advantages of both the PIAA and CPA, namely:(1) optics is easy to polish; (2) no significant loss in throughput; (3) no significant loss in angular resolution;

The chosen output beam profiles are presented in Fig. 2. The central part of the profile is a prolate function, which can be directly used in the PIAA/Lyot configuration, with maximal contrast of 10^{-6} at $2\lambda/D$. PIAA/CPA mode allows imaging with contrast up to 10^{-8} at $2\lambda/D$ (limited only by the wavefront errors).

4. Mirror shapes

The shape of both PIAA mirrors can be described by the equation:

$$Z_i = \textit{Off-axis\_paraboloid}_i(x,y) + \textit{Radial\_profile}_i(r) + \textit{Correcting\_term}_i(x,y) \tag{4.1}$$

where $\textit{Off-axis\_paraboloid}_i(x,y)$ is the base shape of i-th mirror, $\textit{Radial\_profile}_i(r)$ describes the modification of the base shape for on-axis system and $\textit{Correcting\_term}_i(x,y)$ is the corrective term for an off-axis system.

For any desired apodisation profile the remapping function $r_2 = f(r_1)$ can be determined such that the total flux within the radius r_1 of the input beam is equal to the total flux within the radius $f(r_1)$ of the output beam. This mapping provides us with a "correspondance table" between the radii at which geometrical light rays reflect on M1 and M2 mirrors for an on-axis source. For a given remapping function the radial profile terms are obtained by solving a differential equation (Guyon 2003)

$$\frac{dM_1}{dr_1} = \frac{dM_2}{dr_2} = \sqrt{1 + \frac{(M_2 - M_1)^2}{(r_2 - r_1)^2}} - \frac{M_2 - M_1}{r_2 - r_1} \tag{4.2}$$

Corrective terms can be calculated by using an iterative algorithm. Exact radial profiles are presented in Fig. 2 together with simulated apodization profile for designed shapes. Note that the size of the first mirror is 10% larger size than maximal size of the beam. It gives us some flexibility to decrease diffraction propagation effects and simplify an analytical presentation of radial profiles used for diamond turning of the mirrors.

Cycles per aperture	Maximum M2 surface RMS error (contrast 10^{-7})	Maximum output wavefront RMS error (contrast 10^{-6})
0–5	500 nm	100 nm
5–10	300 nm	100 nm
10–20	200 nm	50 nm
20–40	70 nm	35 nm
> 40	50 nm	25 nm

Table 1. The M2 surface quality and the output beam wavefront quality required to reach the 10¯7 and 10¯6 contrast respectively at $2\lambda/D$ (λ =600nm) with the use of wavefront control.

	Shift ($\pm\mu$m) (x,y)	z	Rotation ($\pm\mu$rad)	Inclination ($\pm\mu$rad)
M1	5	47	41	3.2
M2	14	60	52	4.5

Table 2. Maximal alignment errors for the PIAAC prototype.

5. Wavefront errors and alignment tolerance

Errors in the surface of the mirrors produce both amplitude and phase errors in the output beam. It was found through numerical simulations that for the goal of our experiment amplitude errors in the output beam are negligible in comparison with phase errors. The RMS phase error in the output beam compatible with the 10^{-6} contrast at $2\lambda/D$ is about $10^{-3}\lambda$ while a similar contrast degradation by amplitude errors would require 100 to 1000 times larger mirror surface errors. In Tables 1 we show phase errors tolerance for our experiment. Residual phase errors as well as alignment errors are corrected with the DM. In the aligned system the wavefront errors should not exeed 0.5 micron to be correctible with the DM. In Table 2 we show the alignment tolerance based on our raytracing simulations for 0.5μm wavefront errors. It has been identified through numerical simulations that these alignment errors can be easily corrected with the DM (32×32 elements) placed in the exit pupil. Amplitude errors can also be also corrected simultaneously with phase errors by the DM for half of the field of view. Our planed wavefront control scheme relies on focal plane wavefront sensing. The same focal plane array will be used for imaging and wavefront sensing. This approach provides optimal sensitivity for phase and amplitude sensing and offers optimal sensitivity to photon noise. We plan to use this technique to clear a "planet" search area out of speckles (details are given in (Guyon 2005)).

6. Diffraction effects

PIAA units have being designed and studied using geometrical optics approximations. However, differences between diffraction and geometrical propagation are not negligible at the 10^{-10} contrast level (Vanderbei 2005). Diffraction effects are introduced mainly by discontinuities or sharp transitions in the M1 mirror shape and can be easily observed on the M2 mirror surface. Diffraction is a source both amplitude and phase errors in output beam and can decrease coronagraphic performance. Since these diffraction effects are chromatic, they cannot be simultaneously corrected with the DM over a wide bandwidth.

There are two features of the output beam which can affect the high contrast PSF of an on-axis source (Fig. 3). The first one is high frequency oscillations at the edges of the output beam. The second is the peak/hole in the output beam center. It should be noted that these diffraction oscillations are mainly "sharp edges" effect. They decreases as the distanse from the edges increases and are smaller if boundaries are "smoother".

There are some possibilities to decrease diffractive oscillations (Pluzhnik *et al.* 2005):
(*a*) To use an oversized first mirror with an appropriate boundary profile (Fig. 4).

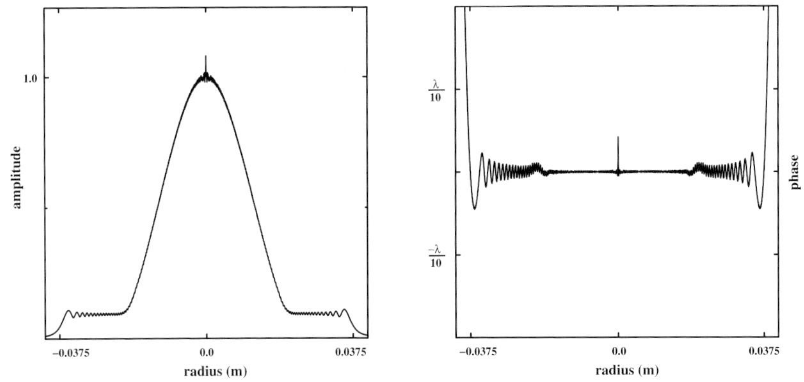

Figure 3. Diffraction effects in the PIAAC. The M1 mirror size is equal to the output beam size.

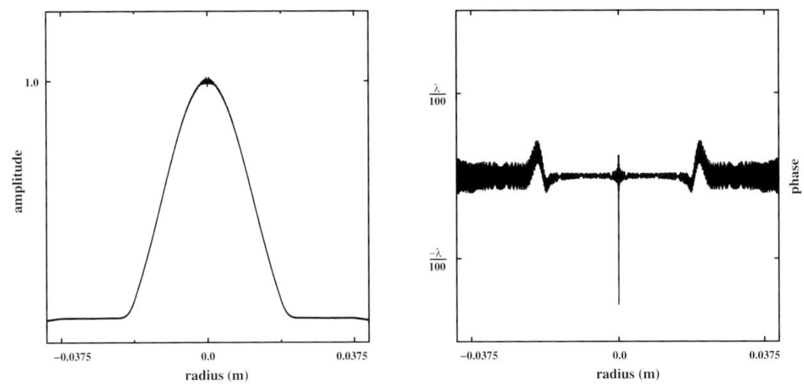

Figure 4. Diffraction effects for the PIAAC design with the "oversized" first mirror. The M1 mirror size is 10% larger than the output beam size.

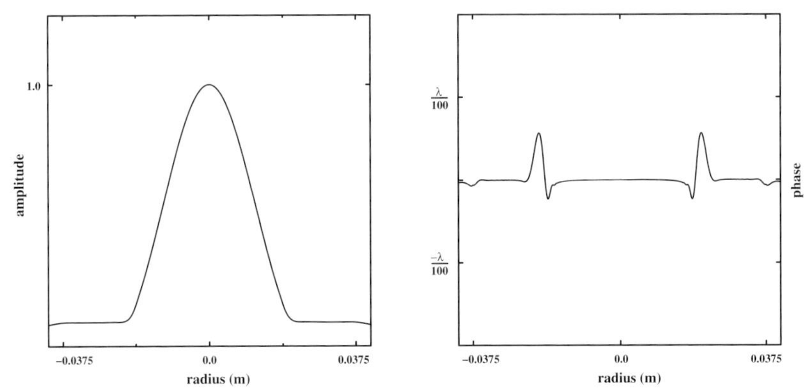

Figure 5. Diffraction effects for the "smoothed edge" PIAAC design. The sharp edge of the beam after the "oversized" M1 mirror is "smoothed" by using the 10% cosine taper window. The M1 mirror size is 10% larger than the output beam size.

(*b*) To smooth the first "oversized" mirror edge by using a classical apodization mask (Fig. 5).

It can be shown that it is possible to decrease diffraction effects to a level consistent with high-contrast (10^{-10}) imaging by using an appropriate optical design of the PIAA unit. Such a design keeps all advantages of the PIAA coronagraph (Pluzhnik *et al.* 2005): high throughput (80-90%, depending on the bandwith), small inner working angle (less then 2 λ/D), low chromaticity and relative insensitivity to pointing errors.

7. Current status

PIAA optics. Two set of PIAA mirrors are now beeing manufactured by AXSYS Technology, Inc. All mirrors have been diamond turned and verified by direct measurements of the surface profile. The overall shapes are in good agreement with our design with only 1λ RMS of deviation. There is no excess of error in any area of the mirrors: all surfaces were accurately produced by diamond turning. The beam apodization after this mirror is in good agreement with our numerical simulations. The optics are under fine polishing right now and will be finished before the end of 2005.

Light source. At the first stage of the experiment we are planing to use single point source, which will simulate an unresolved star. The $2\lambda/d$ for our optics set is approximately equal to 19 microns, so the source can be easily made using a 2 μm pinhole and a laser or a single mode fiber.

Classical apodization mask. A classical apodization mask is used to convert the "moderately" apodized beam into a fully apodized beam and, possibly, to correct slight errors of PIAA apodization. The mask only affects the outer parts of beam, where there is little light. As a result the overall transmition of the mask is high and it does not reduce the throughput and resolution of the system. The mask can be either continuous or binary, with no significiant difference of performance between them. Another mask (preapodizer) will be used to decrease diffraction effects in the PIAA optics.

Deformable mirror. The 1024 elements MEMS DM manufactured by Boston Micromachines will be used to correct alignment errors and cancel speckles in the "planet" search area. Electronics and control software for the DM control is developed with support of the Subaru telescope Adaptive Optics group.

8. Summary

The current status of the PIAA optics laboratory experiment under preparation at Subaru Telescope has been reviewed in this paper. We have given a short analyses of the main factors affecting the performance of the coronograph as well as some ideas how to decrease their influence.

Acknowledgements

This work was carried out under JPL contract numbers 1254445 and 1257767 for Development of Technologies for the Terrestrial Planet Finder Mission, with the support and hospitality of the National Astronomical Observatory of Japan.

References

Galicher, R., Guyon, O., Otsubo, M., Suto, H., & Ridgway, S. 2005, *PASP*, 117, 411
Guyon, O. 2003, *A&A*, 404, 379
Guyon, O., Pluzhnik, E.A., Galicher, R., Martinache, F., Ridgway, S., & Woodruff, R. 2005, *ApJ*, 622, 744
Guyon, O. 2005, *ApJ*, 629, 592
Martinache, F., Guyon, O., Pluzhnik, E.A., & Galicher, R. Accepted by *ApJ*
Pluzhnik, E.A., Guyon, O., Ridgway, S., Martinache, F., Woodruff, R., Blain, C., & Galicher, R. In preparation
Vanderbei, R.J. 2005, *astro-ph*/0506550

Direct Imaging of Exoplanets: Science & Techniques
Proceedings IAU Colloquium No. 200, 2005
C. Aime and F. Vakili, eds.
© 2006 International Astronomical Union
doi:10.1017/S1743921306009847

AO Assisted NIR 3D Spectroscopic Imaging – Can an ELT See the Earth at 10pc?

Markus Feldt[1], Alessandro Berton[1], Thomas Henning[1], Raffaele Gratton[2], Massimo Turatto[2], H. M. Schmid[3], M. Kasper[4] and J. L. Beuzit[5]

[1]Max Planck Institute for Astronomy, Königstuhl 17, D-69117 Heidelberg, Germany
email: mfeldt@mpia.de

[2]Osservatorio Astronomico di Padova, Vicolo dell'Osservatorio 5, I-35122 Padova, Italy,
email: gratton@astro.pd.it

[3]Institut für Astronomie, ETH Zentrum, CH-8092 Zürich, Switzerland
email: schmid@astro.phys.ethz.ch

[4]European Southern Observatory, Karl-Schwarzschild-Straße 2, D-85748 Garching, Germany
email: mkasper@eso.org

[5]Laboratoire d'Astrophysique, Observatoire de Grenoble, BP 53,
F-38041 Grenoble Cédex 9, France email: Jean-Luc.Beuzit@obs.ujf-grenoble.fr

Abstract. The European southern observatory ESO is currently undertaking the ambitious task of building a "Planet Finder" instrument for the VLT. The concept for this instrument includes a 3D spectroscopic imager assisted by a very powerful eXtreme AO system. In one of the two phase-A studies for this project, we have developed a simulation software for this kind of instrument which has now been extended and applied to the case of various ELTs. In this presentation, we give results of simulations and discuss achievable signal to noise ratios and prospects for detection and characterization from young gas giants down to terrestrial type planets. ELTs of various diameters are considered as well as a set of environmental conditions.

Keywords. instrumentation: spectrographs, methods: data analysis, techniques: high angular resolution, techniques: spectroscopic, (stars:) planetary systems.

1. Introduction

It is not long ago that astronomers believed it would always remain impossible to actually get an image of a planet that is not part of our own solar system. Nowadays, a number of project is dedicated to devise instruments that will achieve exactly that. ESO is planning to put a "Planet Finder" instrument on the VLT (Beuzit *et al.* 2005), and the GEMINI observatory has a similar project (Macintosh *et al.* 2004). In the context of the CHEOPS project, which was one of the two phase-A studies conducted in the context of the ESO Planet Finder, we devised the concept of using a 3D integral field spectrograph (IFS) as a sophisticated differential imaging device in order to suppress the residual speckle halo from a bright star and eventually to directly detect extra-solar planets (Feldt *et al.* 2004). As a result of this study, we found that such a device, supported by an eXtreme AO (XAO) system will, when placed on the VLT, be capable of detecting rather young (10 Myr) Jupiter-sized ($M = 5M_{\rm Jup}$) planets in orbit around stars up to 40 pc distant (Berton *et al.* 2005b). Older Planets would have to be slightly larger than Jupiter to be detected at close distances (~5 pc)

Not satisfied with such prospects, astronomers have already planned the next generation of such instruments for a new generation of "extremely large telescopes" (ELTs), the

Table 1. Values used in the simulation of the atmosphere.

No. of Layers	1
Fried radius r_0	0.17 m
Coherence time τ_0	3 ms
Wind speed v_{wind}	17.7 m/s
Outer scale L_0	25 m

most prominent example of which is ESO's "OverWhelmingly Large telescope (OWL)" with a primary mirror diameter of 100 m. The science case in the area of finding planets is for these telescopes and their instruments no longer limited to young giant planets, but explicitely includes the detection of Jupiter twins, and, more challenging and interesting, terrestrial planets (Verinaud *et al.* 2005). All science cases include not only direct imaging detections, but also characterization of the detected planets, e.g. by spectroscopy and/or polarimetry.

In order to evaluate the feasibility of such science cases and to estimate the observation times required to achieve these challenging goals, we applied our simulation code to the case of two ELTs hypothetically equipped with an extreme AO system and an integral field spectrograph basically identical to the one devised in the context of CHEOPS (Claudi *et al.* 2004).

2. Instrument and Simulation Code

A basic description of the IFS can be found in Claudi *et al.* 2004. In the context of CHEOPS, and still for "Planet Finder", the IFS consists of an array of 254×254 micro lenses in the focal plane covering a field-of-view of 3.″5×3.″5, followed by an Amici prism and a re-imaging optics that forms spectrally resolved ($R \sim 15$) micro-pupil images on the $2k\times2k$ detector. The system is placed behind a powerful XAO system with ~1600 actuators, running at loop frequencies between 1 kHz and 2 kHz. A description of the full instrument is given in Beuzit *et al.* (2005).

The simulation code is described briefly in Berton *et al.* (2005a) and in more detail in Berton *et al.* (2005b).

3. Simulation Set-up

The simulation approach chosen is a modular structure that propagates phase and amplitude arrays from one module to the next one, modifying it according to the physics involved. The discrete time step of the simulation is 1 ms, phase screens are generated and propagated in in a successive way in order to allow for the correlated nature of the speckle noise.

3.1. *Atmosphere*

The atmosphere is simulated by creating a continuous series of phase screens. the basic parameters used in this simulation are summarized in Tab. 1. For the generation of each phase screen, the predecessor is shifted a bit according to the wind speed and slightly modified using a Markov process to partially de-correlate the screen from the previous one. The amplitude is assumed to be flat, since it has been shown by Masciadri, Feldt and Hippler (2004) that scintillation can be neglected for intermediate Strehl ratios.

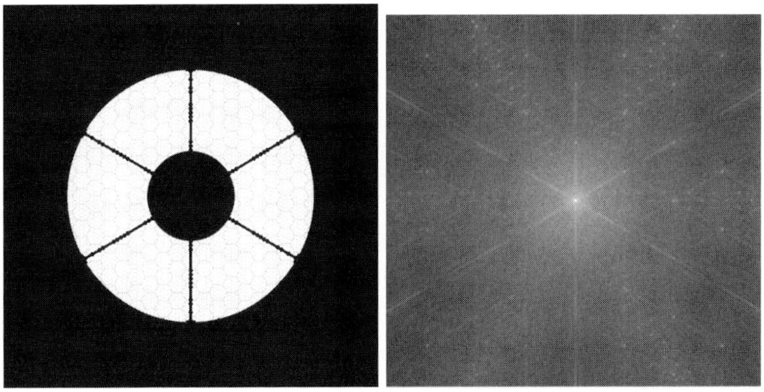

Figure 1. Shape of the OWL pupil and resulting PSF.

3.2. *Telescope*

The telescope is represented simply by its aperture which is multiplied onto the uniform input amplitude distribution. Two patterns were generated using the OWL Pupil Simulator by N. Yaitskova (priv. comm.). Two cases were simulated: A complete OWL design with a 100 m primary consisting of 1.6 m hexagonal segments, and a 26 m segmented secondary. Additionally, we simulated a scaled-down version with a primary diameter of 50 m. In this case the word "scaled-down" is to be taken literally, which will probably not match a real configuration since from an engineering point of view one would desire to keep the segment diameter at 1.6 m rather than using every component in half size.

There are no phase errors included that originate in the telescope whatsoever, neither in nor between the segments, no vibrations, and no missing segments. The amplitude map used for the simulations can be seen in Fig. 1 (For testing the reproduction quality, try to disentangle the small primary segments from the larger secondary segments) together with the resulting PSF (in the miraculous case where no atmosphere is present above OWL).

3.3. *Adaptive Optics*

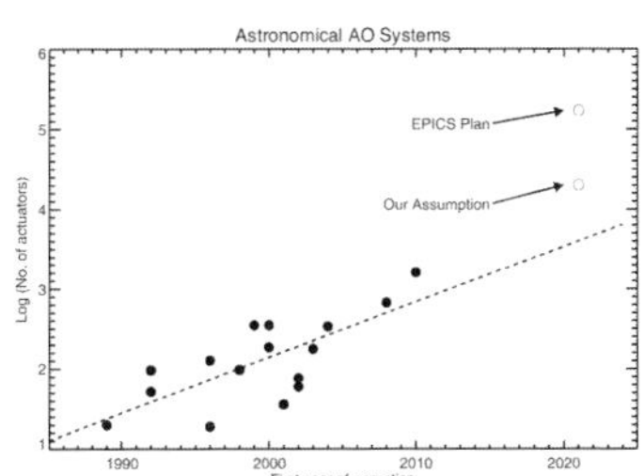

Figure 2. Growth of the number of actuators in astronomical AO systems over the years.

The AO system is simulated in a very simple way.: The atmospheric phase screens are down-sampled (using averaging) to the assumed sensor element resolution. This re-expanded version of the down-sampled phase screen is then subtracted from the atmospheric one arriving one AO loop time step (2 ms) later to provide the corrected phase map. This assumes an identical sampling for the wavefront sensor and the deformable mirror and automatically rules out all sources of noise (measurement noise in the WFS, aliasing, etc.). The two only errors included in this way are the temporal error and the fitting error. The two single parameters describing the system are the number of actuators (20,000) and the correction frequency (500 Hz).

It has been argued during the conference that the choice of the number of actuators is rather unsuitable, particularly since the resulting control radius is $r_c = \frac{\lambda}{2d} \approx 0\farcs14$ and thus most of the planets we are trying to detect are outside this radius in the (speckle)-noisy seeing halo that is not controlled by the XAO. We note however that our assumption is still quite optimistic (see also Fig. 2) and in a real system of our dimensions one would probably not notice the effect of a control radius-limited basin region in the PSF anyway.

3.4. *Integral Field Spectrograph*

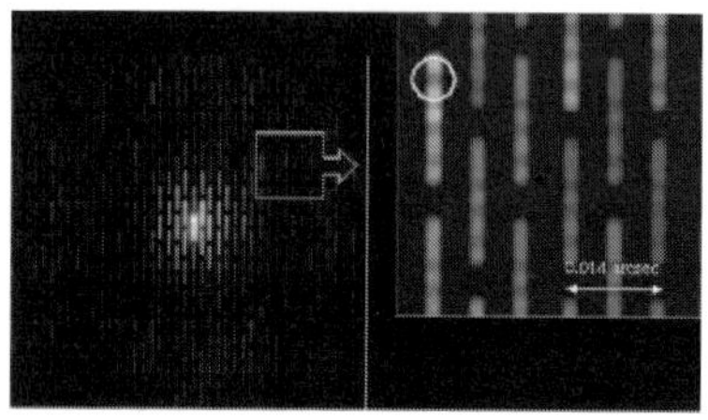

Figure 3. Spectra on the IFS detector.

The integral field spectrograph is the core module of our simulation code. This is precisely the reason why we do not discuss its details here. You can find an extended description in Berton *et al.* 2005a and a thorough explanation in Berton *et al.* 2005b. The output of the IFS module the detector image that shows the aligned spectra as seen in Fig. 3.

3.5. *Neglects*

Our simulations have been critisized for not starting systematically at the optimistic or pessimistic end of chain of effects that determines the final signal-to-noise ratio of a planetary detection.

We note that the primary goal of our simulations was to learn how the detectability of earth-like planets depends on the diameter of future giant telescopes under two key-assumptions. These key-assumptions are

(*a*) The AO system with which a system like the assumed one would be equipped is excellent and state of the art in about 2020, but it will very likely be far from perfect.

(*b*) The device for differential imaging will be a TIGRE-type IFS as suggested for the VLT Planet Finder.

Our results are to be seen in this context. Even though, simulations presented here do neglect a number of effects, the most important for which are:

- Flat fielding is not explicitly simulated, we simply multiply our pixel counts with a random number such that a residual error level of 10^{-4} is reproduced.
- The alignment of the spectra on the pixel grid is perfect. This will not be the case in a real system, but it will have to be achieved by sophisticated interpolation - which of course will cause additional errors. Investigation of this effect is ongoing.
- Interference between lenslets is not taken into account here as the micro pupil is calculated individually for each lenslet. We have however performed dedicated simulations with arrays of up to 40×40 lenslets that show the SNR reduction is only by a factor of ~0.98.
- Exploitation of more than one absorption feature is not simulated here, we present the simple subtraction of two wavelength. In Berton *et al.* 2005b we show a dedicated

simulation which makes use of two planetary absorption features in the J-band using up to 9 spectral elements provided by the IFS. This improves the detection SNR by a factor of 2.

In summary, the most important thing we have neglected in the present work is the strength of the IFS, so our results are to be seen on the pessimistic side with room for improvement by up to a factor of 2 in the detection SNR or a factor of 4 in the needed integration times presented in the following section.

4. Results

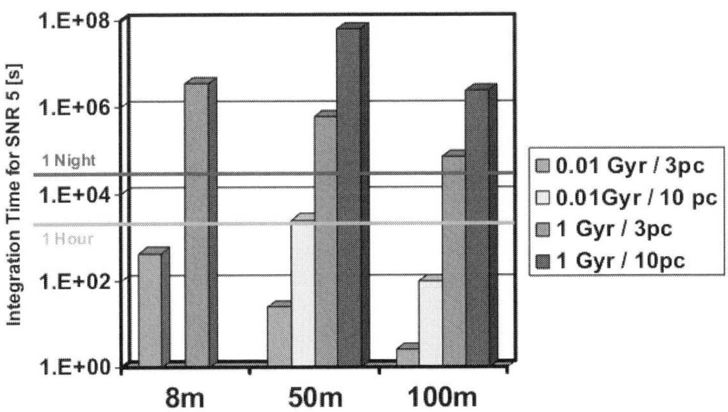

Figure 4. Integration times needed to detect a planet of 1 Jupiter mass in a 1 AU orbit around a G2 V star with a SNR of 5.

Fig. 4 gives the result of our simulations for Jupiter like planets, i.e. Planets of 1 Jupiter mass orbiting a G2 V star at 1 AU orbital radius. The times indicated are needed to achieve a detection SNR of 5. Note that these are standard cases for comparison, not actually existing planets. Note also that the numbers quoted for "8 m" were computed during the CHEOPS study. In CHEOPS, unlike in the present Planet Finder concept, it was not foreseen to observe closer than $0''.3$ from the star, which is why some planet classes are not shown for the 8 m case.

The single result we obtained so far for an exo-earth at 3 pc is as follows: We applied a spectral resolution of 150 (instead of 15 for the exo-Jupiters) in order to escape saturated absorption by both planet's atmospheres and rescaled the spectra accordingly in order to compensate the transmission by earth's atmosphere. In this way we could apply differential imaging to the exactly identical features that have just been compensated. Using this technique we calculate an observing time of 1000 hrs to detect the exo-earth in a 1 AU orbit around a G2 V star at 3 pc.

5. Discussion

It is clear that giant telescopes gain an enormous advantage over 8 m telescopes when trying to image planets. In some cases, integration times can become so short that completely new science fields can be opened - time series photometry or even spectroscopy – in fact even radial velocity measurements on exo-Jupiters in order to detect their satellites appear feasible in some cases. The detection of terrestrial planets however appears very difficult with an instrument suite as the one simulated here. However, by using the full advantage of an IFS – which gains a factor of 2 in SNR – and by possibly decreasing the spectral resolution by a factor of 10 and simply making use of more photons, the integration time needed might be reduced to levels that just appear achievable (of the order of 1 or a few nights – to find an exo-earth at 3 pc!!). We also note that the current EPICS study (Verinaud *et al.* 2005) suggests a much more sophisticated AO system plus a coronographic device which can potentially reduce the photon noise in the PSF wings. Such a device would - if the technology can indeed be developed - facilitate the detection of terrestrial planets out to distances of approximately 25 pc.

References

Berton, A., Feldt, M., Gratton, G., Henning, Th., Desidera, S., *et al.* 2005a, This proceedings

Berton, A., Gratton, R.G., Feldt, M., Henning, Th., Desidera, S., *et al.* 2005b, submitted to *Astronomy and Astrophysics*

Beuzit, J.-L., Mouillet, D., Moutou, C., Dohlen, K., Puget, P., *et al.* 2005, This Volume

Claudi, R. U., Turatto, M., Gratton, R., Antichi, J., Buson, S., *et al.* 2004, *Proceedings of the SPIE*, Vol. 5492, P. 1351

Feldt, M., Henning, Th., Hippler, S., Turatto, M., Neuhäuser, R. *et al.* 2004, *The CHEOPS Project: CHaracterizing Exo-planets by Opto-infrared Polarimetry and Spectroscopy*, in: "Exploring the Cosmic Frontier" eds. A. P. Lobanov *et al.*, ESO Astrophysics Symposia, 2005, in press

Macintosh, B. A., Bauman, B., Wilhelmsen Evans, J., Graham, J. R., Lockwood, C., *et al.* 2004, *Proceedings of the SPIE* vol. 5490, p. 359

Masciadri, E., Feldt, M., & Hippler, S. 2004, *Astrophysical Journal* 613, P. 572

Christophe Verinaud, Hubin Norbert, Kasper Markus, Antichi Jacopo, Beuzit Jean-Luc, *et al.* 2005, This proceedings

Discussion

CLOSE: Are you really claiming that at Strehl ratios of only 60% you see a cleared-out basin in the PSF centre?

We do, but this is only due to our rather perfect wavefront sensing and correction device – which is simply re-sampling the wavefront itself to the sensor and DM resolution. In this way, we of course have no aliasing whatsoever.

FRIEDLUND: Note that your 1 Myr Jupiter at 3 pc does of course not exist. Not many young stars are in the vicinity of the sun.

This is true, our calculations are so far limited to a few (fictitious) "standard" cases that were agreed upon in order to compare results of the two "Planet Finder" studies conducted by ESO.

Direct Imaging of Exoplanets: Science & Techniques
Proceedings IAU Colloquium No. 200, 2005
C. Aime and F. Vakili, eds.

© 2006 International Astronomical Union
doi:10.1017/S1743921306009859

Ground-based direct imaging of extra-solar planets supported by AO

Elena Masciadri [1], Kerstin Geissler[1], Stephan Kellner[1], Wolfgang Brandner[1], Thomas Henning[1], Reinhard Mundt[1], Laird Close[2], Beth Biller[2], and Alejandro Raga[3]

[1]Max-Planck Institut für Astronomie, Konigstuhl 17 – D69117, Heidelberg, Germany
email: masciadri@mpia.de

[2]Steward Observatory, Tucson, Arizona

[3]Ciencias Nucleares, UNAM, Mexico D.F., Mexico

Abstract. The detection of extra-solar planets made by direct imaging is an extremely challenging goal for astronomers due to the possibility to access the physical properties of planets and not only their existence. Using 8 m class telescopes joint with dedicated techniques (such as Simultaneous Differential Imaging aiming to suppress the speckle noise) it is possible, at present, to attain detection limit of 9–11 mag at 0.5 arcsec i.e. to access 3–10 M_J planets orbiting around young (100–200 Myr) nearby and late type stars. Searches for extra-solar planets carried out with the present technology are quite fundamental and critical not only for discovery of planets but also because it permits us to put constraints on theories of planets formation and migration. Besides, our understandings of the performances of sophisticated techniques such as the SDI is fundamental to plan new observational strategies, new generation instruments and telescopes. Speckles noise is, indeed, the main source of noise for observations in the NIR and visible and our ability in suppressing it is not so easily scaled at different parameters space. In this contribution I will present the main results that we obtained in on-going searches for planets carried out with NACO and NACO/SDI in the last years. A particular attention will be dedicated in comparing different observational strategies and in the employment of image processing techniques for recognizing, in an automatic way, planet features in deep images obtained with ground-based telescopes and AO facilities.

Keywords. planetary systems, image processing.

1. Introduction

Adaptive optics (AO) imaging techniques at 8–10 meter telescopes can detect young planets if they are sufficiently distant from their parent star. Since it is reasonable to assume a coevality between the parent star and the planet, we can retrieve from atmospherical models (Burrows *et al.* 1997, Baraffe *et al.* 2003) that planets having a mass in the range 3–10 M_J and orbiting around young late-type stars (10–200 Myr) have a typical brightness contrast with respect to the parent star of the order of 10^2–10^6, corresponding to a magnitude difference of ΔM of 5–15 mag. This means that a planet can be detected at a few tens of AUs from parent stars having distances of ∼50 pc.

According to recent statistical results (e.g. Marcy *et al.* 2003), most of known exoplanets orbit at distances smaller than 1 AU and the semi-major axes of the exo-planets found so far are not larger than ∼6 AU. Since these results are heavily biased towards short separations the derived major-axis distribution is certainly not representative. Not much is known at present about potential planets that could exist at distances larger than about 6 AU. Besides, there are models (for example the gravitational scattering model) that claim that in multi-planetary systems, due to gravitational interactions between the planets, it can happen that the lightest objects are ejected on hyperbolic trajectories and some planets can move on stable orbits having a distance of a few tens AU from the parent star. Such models were proposed by Weidenschilling & Marzari (1996) and Rasio & Ford (1996) and studied more recently by Papaloizou & Terquem (2001).

There are thus elements that indicate that such planets might exist at large distances from the central star. For this reason we planned and performed an extended survey aiming to search for massive exo-planets orbiting around nearby and young stars using one of the best available instruments at the present time for high-contrast AO imaging observations: NACO at the Very Large Telescope (VLT). Once the survey completed (Masciadri *et al.* 2005), we decided a follow-up of our project employing the SDI/NACO instrument. As explained in Section 3, such an instrument permits to improve the contrast reachable at sub arcsecond distances from the parent star. I report here the principal results obtained and I discuss the performances of the different observational strategies. Finally, I deal about a method that a few of us (Masciadri & Raga, 2004) developed for recognizing in an automatic way, planets features in deep images.

2. Deep direct imaging survey

We refer to Masciadri *et al.* (2005) for detailed informations on the target selection, observational strategy employed in the survey and data processing. Here we briefly summarize the principal informations. Around 30 targets (nearby stars) were observed during 3 runs. The targets are placed at a distance $d \leqslant 77$ pc (with 70 % of them in the first 32 pc), they have an age of 1–200 Myr and late spectral type (K and M). Firstly we observed each target in a narrow band filter for a few minutes in order to obtain a non-saturated point spread function (PSF). Secondly, we obtained deep (~20 minutes) K_s and H broadband observations of each target. Data were reduced using standard procedures to eliminate bad pixels, flat and background sky (IRAF and ECLIPSE). A dedicated pipeline has been employed to filter the low spatial frequencies in order to improve the detection limit at subarcsec distances (Masciadri *et al.* (2005)). Two different methods were used for the low spatial frequencies filtering.

We averaged the contrast ΔM obtained at 0.5 and 1 arcsec for all the observed targets having a total integration time within the range 20-30 minutes. We find, in the K_s band, a $\Delta M = 9$ mag at 0.5 arcesc and a $\Delta M = 11.5$ mag at 1 arcsec. We find, in H band, a $\Delta M = 9.2$ mag at 0.5 arcsec and a $\Delta M = 11.7$ mag at 1 arcsec. We recall that 0.5 arcsec corresponds, for our sample, to projected separations of 2-35 AU, 1 arcsec to projected separations of 4-70 AU. This illustrates nicely that NACO is a well performing instrument for high-contrast imaging observations.

No very promising massive-planet candidate was identified in our survey. Besides this we calculated the detection limit for 28 of these targets (Masciadri *et al.* 2005) and we retrieve the upper limit of the star/planet projected separation calculated for a potential 10 and 5 M_J planet (Masciadri *et al.* 2005, Table 4). By analysing these results from a statistical point of view, we can calculate (Masciadri *et al.* 2005, Fig.9) the cumulative distribution of the upper limit of the distance star/exo-planet for our sample of targets. The thin line represents the cumulative distribution obtained for $5M_J$ planets. We find that, with respect to our sample of targets, in 50% of cases (median value), there are not planets at distances larger than 14 AU and, in 100% of cases, there are no planets at distances larger than 65 AU. We underline that we are considering a $5M_J$ planet just as a reference, but the same calculation could be done for more and/or less massive planets in the range 3–10 M_J. In the case of more massive planets, the cumulative distribution should have a medium value (50%) associated to an angular separation smaller than 14 AU. In the case of less massive planets, the cumulative distribution should have a medium value (50 %) associated to a angular separation larger than 14 AU. To better appreciate the sensitivity with respect to the mass of the planets we also calculated the cumulative distribution for 10M_J planets (Fig.4, bold style line). We find that, in 50% of the cases (median value) there are no planets at distances larger than 8.5 AU and in 100% of the cases there are not planets at distances larger than 36 AU. The number given in Table 4 (Masciadri *et al.* 2005) and the cumulative distribution illustrate that it is relatively unlikely that a massive planet exist beyond a certain distance from the star. For example the fact that no 5M_J planet was found for 18 of our targets beyond 20 AU (see Table 4, Masciadri *et al.* 2005) means that the likelood for the existence of such planets beyond this distance is ~(1/18) % i.e. ~5.6%. This last estimation is useful for the reader to have an idea of how much the size of the sample affects the statistical results. We remind that we are using the term *probability* in a less conventional way. A precise (statistically speaking) definition of *probability* should need some positive detection that we obviously do not have.

3. Contrast at sub-arcsec distances: comparison of different techniques

In the survey carried out with NACO (Masciadri *et al.* 2005) no interesting candidates were detected but typical contrasts at 0.5 arcsec and 1 arcsec were calculated using the detection limit of 28 targets in H and K_s bands (see previous Section). The success of the search for extrasolar planets strictly depends on how efficient we are in increasing the reachable contrast at sub-arcsecond distances. We calculated, therefore, for deep direct images, the contrast vs. angular separation for different exposure times. Figure 1 (left) shows the result of this calculation. At distances larger than 1 arcsec, the larger is the exposure time, the larger is the reachable contrast. At sub-arcsec distances, the increasing of the exposure time does not supply a better contrast because we are fundamentally limited by the speckle noise. The Simultaneous Differential Technique (SDI) proposed by Racine *et al.* (1999) is a rather efficient method to reduce the speckle noise at small angular separations from the central star. Images taken simultaneously at slightly different wavelengths are subtracted from each other. Since the speckle noise of both images is nearly the same†, the speckle noise is eliminated in the resulting image providing one can sufficiently limit static aberrations. The differential technique finds a useful application in detecting cool methane-rich faint objects. Rosenthal *et al.* (1996) first proposed observations of the same star done simultaneously in two different wavelengths which are located *"on"* and *"off"* the methane (CH_4) absorption band. If one subtracts the image taken in the absorption band from the one taken outside the absorption band, planet/brown dwarf features appear in the continuum. The absorption at 1.62 μm, in the CH_4 band, is a clear signature of any object having temperature $T \leqslant 1300°$ K (giant exo-planets and T-type brown dwarfs). Supported by the preliminary results obtained by Close *et al.* (2004) with the commissioning data of the SDI/NACO instrument we decided to start a new survey in collaboration with the Close group (see Biller contribution, this Proceedings). In this contribution we intend to study the efficiency of the different observational strategies at sub-arcseconds distances from the target to estimate the improvement obtained with the

† We precise that speckles of images taken at slightly different wavelengths match if frames are rescaled opportunely by a λ_2/λ_1 factor.

Table 1. Mean contrast at 0.5 arcsec calculated with different observational techniques (deep direct imaging and SDI). [a]We remind to the reader that the SDI technique is employed in narrow band.

D (m)	Obs. Tech.	Filter	Δm (mag)	N°σ	Exp. Time (minutes)	N° Targets
8	Deep Imaging NACO	H	9.20	5σ	20-30	28
8	SDI/NACO[a]	1.57μm	8.61	5σ	10-20	15
8	SDI/NACO	H	10.31	5σ	10-20	15

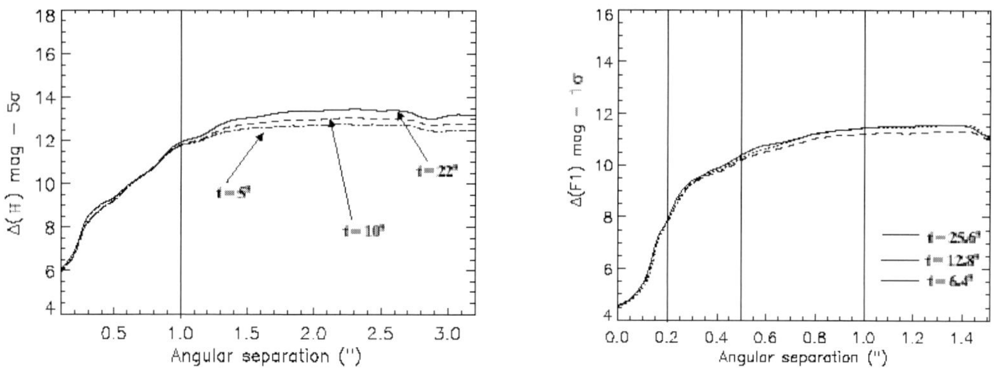

Figure 1. Left: detection limit vs. angular separation of one target of the survey done with deep direct imaging (Masciadri *et al.* 2005) obtained with integration times of 5, 10 and 22 minutes. Right: detection limit obtained with the SDI technique and a total exposure time of 6.4, 12.8 and 25.6 minutes.

SDI on sky on a 8 m class telescope. Table 1 summarize the mean contrast calculated at 0.5 arcsec employing the deep direct imaging and the SDI technique running with the same instrument (NACO). We want to precise that, employing the SDI technique, the calculation of the contrast refers to the narrow band, more precisley the narrow band at 1.57 μm (FWHM $= 0.024\mu$m). In order to compare the two observational strategies (Direct deep imaging and SDI) it is necessary to calculate the equivalent contrast in H broadband. This means that $\Delta m_H = \Delta m_{1.57\mu m} + \mathrm{A}$. The factor A depends on the spectral type of the planet-like object because the spectrum shape in proximity of the absorption band is different for different spectral type planet-candidates. In the case of a T6 type object we calculated that A = 1.7 mag. For spectral type later than T6, A is larger than 1.7 mag, for spectral type earlier than T6, A is smaller than 1.7 mag.

Looking at Table 1 we conclude that the **SDI technique permits to gain a factor of 1 mag in H band with respect to the Direct deep imaging at 0.5 arcsec**. We underline that this gain is a typical value, obtained averaging detection limit curves related to 15 targets and it is valid for the SDI applied to a 8 m telescope. Figure 1 (right) shows the detection limit at $F_1 = 1.57$ μm obtained for different exposure time: 6.4, 12.8 and 25.6 minutes. One can observe that, in the range 0.5–1 arcsec the gain is Δ $F_1 \sim 0.5$ mag in Δt = 15 minutes. In the range 0.2-0.5 arcsec, the gain Δ F_1 is $\sim$0 mag in the same Δt. This means that, at least in the actual version, the gain of the SDI is weakly depending on the total exposure time.

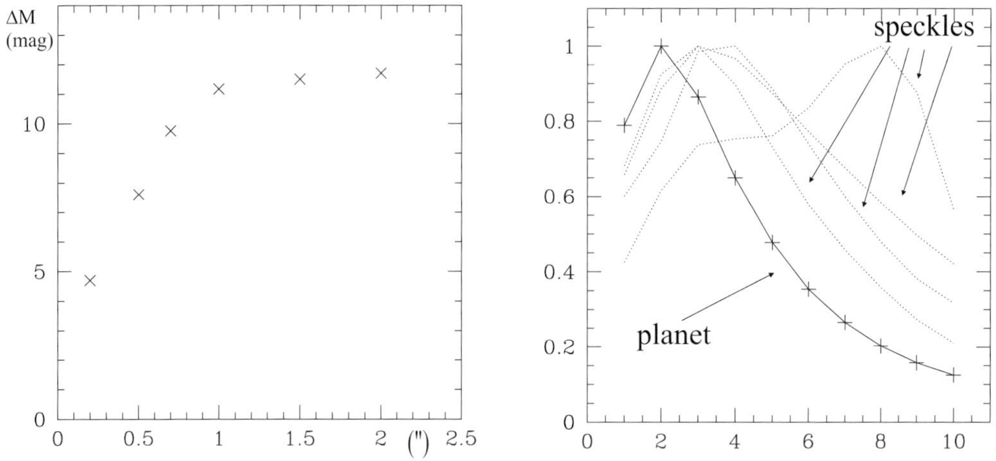

Figure 2. Left: ΔH (mag) between the central star and the template planet recognized by the automatic procedure vs. the angular separation (arcsec). Right: Example of a normalized spectrum of a planet (y-axis) vs. number of pixels (x-axis). This spectrum is used as template spectrum to recognize planets in a deep image. Dotted lines: spectra of a few speckles recognized by the procedure as local maxima and rejected because they did not follow the criteria of selection.

4. Automatic recognition of sub-stellar PSFs with wavelet technique

The fundamental problem for planets detection done with the direct imaging technique is that we have a huge contrast at very small angular separations. It would be therefore quite useful to develop a procedure, better if this is automatic, for recognition of features typical of sub-stellar PSFs. The method that we proposed (Masciadri & Raga 2005) is based on the wavelet technique. The detailed description of such a procedure can be found in Masciadri & Raga (2005). The principle is summarized here: a template planet, obtained scaling opportunely the no-saturated PSF of the star, is added in the deep image of the central star. The resulting deep image is filtered from the low spatial frequencies. The procedure characterizes the spectrum of the template planet (Fig.2-right, solid line) and then it looks for, in the deep image, features having the same spectra. These features have to be considered as potential sub-stellar or stellar -like PSF. A second epoch of observation is, therefore, necessary to discriminate between the two cases. This procedure was recently validated (Masciadri & Raga 2005) on a target (V343 Nor, m_k=6.3 mag, $d \sim 40$ pc, age ~12 Myr) of our NACO survey. Figure 2-right shows the contrast (ΔH) between the central star and the template planet versus the angular separation within the 0.2-2 arcsec range at which a planet is recognized by the automatic procedure. We note that, in the case of V343 Nor, a maximum $\Delta H = 4.7$ mag, associated to a planet, can be recognized in an automatic way at 0.2 arcsec with the present procedure. To give an idea to the reader of what we are talking about, we note that 4.7 mag is roughly the contrast in K band of 2M1207, the giant planet candidate recently discovered by Chauvin *et al.* (2004), with respect to its central star. Figure 2-right shows the spectra of a few "speckles" (dotted lines) recognized by the procedure as local intensity maxima that are rejected by the procedure because their spectra do not follow the criteria of selection (see *'tolerance criterium'* in Masciadri & Raga 2005). The spectrum shown in bold style (Fig.2-left) refers to the template planet placed in the deep image. One can observes that the procedure can discriminate the spectra of planets and speckles in an efficient way.

5. Conclusions

We can summarize the main results obtained in this study in the following way:

• $5M_J$ planets, having an age within 200 Myr, orbiting around stars placed at a distances $d \leqslant 70$ pc can be detected with NACO at distances smaller than 10 AU.

• We find a low probability (~10%) to find $5M_J$ planets at distances larger than 26 AU and $10M_J$ at distances larger than 16 AU around stars with properties defined above.

• The performances of different techniques (running at present on sky) at sub-arcsecond distances can be summarized in the following way. At 0.5 arcsec, the SDI technique permits to reach a typical contrast at 5σ, in [10-25] minutes, equal to ΔH = 10.31 mag. This is around 1 magnitude better than what obtained with the deep direct imaging (ΔH=9.2 mag). Results published recently (Boccaletti *et al.* 2004) indicate that the 4 quadrant phase mask (4QPM) coronograph implemented on NACO gives, at 0.5 arcseconds, a ΔH=9 mag at 5σ in K band. In order to do a fair comparison with the deep imaging and the SDI technique, results obtained in H band should be necessary. Considering that the SDI and the 4QPM are certainly techniques that will support the high-contrast future facilities of 8 m class telescopes and ELTs, we think that it is quite fundamental to verify on sky the performances of these techniques to be able to know where and how to improve such techniques and to scale performances at different parameters space.
At present we can state that the SDI can be considered a success giving access to an improvement of the contrast at sub-arcsecond distances of around 1 mag at 0.5 arcsecond with respect to the deep direct imaging observational strategy. On the other side, we have to note that, no planet candidates have been observed so far with the SDI technique that have not been also observed with the simpler deep direct imaging strategy. From an astrophysical point of view, it would be for sure interesting to be able to do that to optimize the scientific impact of such a promising technique.

• The wavelet technique is certainly a promising method for recognizing, in an automatic way, sub-stellar planet features in deep images.

References

Baraffe, I., Chabrier, G., Barman, T.S., Allard, F., & Hauschild, P.H. 2003, *A&A*, 402, 701

Boccaletti, A., Riaud, P., Baudoz, P., Baudrand, J., Rouan, D., Gratadour, D., Lacombe, F., & Lagrange, A.,-M. 2004, *PASP*, 116, 825

Burrows, A., Marley, M., Hubbard, W.B., Lunine, J.I., Guillot, T., Saumon, D., Freedman, R., Sudarsky, D., & Sharp, C. 1997 *ApJ*, 491, 856

Chauvin, G., Lagrange, A.M., Dumas, C., Zuckerman, B., Mouillet, D., Song, I., Beuzit, J.-L. & Lowrance, P. 2004, *A&AL*, 425, L29

Close, L.M., Lenzen, R., Biller, B., Brandner, W., & Hartung M., 2004, *ESO Workshop "Science with AO"*, 136

Marcy, G.W., Butler, R.P., Fischer, D.A., & Vogt, S.S. 2003 *ASP Conf.*, 294, 1

Masciadri, E., Mundt, R., Henning, Th., Alvarez, C., & Barrado y Navascues, D. 2005, *ApJ* 625, 1004

Masciadri, E. & Raga, A. 2004, *ApJL* 611, L137

Papaloizou J.C.B. & Terquem, C. 2001, *MNRAS*, 325, 221

Racine R., Walker G.A.H., Nadeau D., Doyon R., & Marois C. 1999, *PASP*, 111, 587

Rasio, F.A. & Ford, E.B. 1996, *Science*, 274, 954

Rosenthal, E.D., Gurwell, M.A., & Ho, P.T.P 1996, *Nature*, 384, 243

Weidenschilling, S. J. & Marzari, F. 1996, *Nature*, 384, 619

Direct Imaging of Exoplanets: Science & Techniques
Proceedings IAU Colloquium No. 200, 2005
C. Aime and F. Vakili, eds.
© 2006 International Astronomical Union
doi:10.1017/S1743921306009860

The EPICS project: Exoplanets detection with OWL

C. Verinaud[1], N. Hubin[1], M. Kasper[1], J. Antichi[2], P. Baudoz[4], J.-L. Beuzit[3], A. Boccaletti[4], A. Chalabaev[3], K. Dohlen[5], E. Fedrigo[1], C. Correia Da Silva[1], M. Feldt[6], T. Fusco[7], A. Gandorfer[8], R. Gratton[2], H. Kuntschner[1b], F. Kerber[1b], M. Le Louarn[1], R. Lenzen[6], E. Le Coarer[3], A. Longmore [9], D. Mouillet[10], R. Navarro[11], J. Paillet[12], P. Rabou[3], F. Rahoui[1], F. Selsis[12], H.M. Schmid[13], R. Soummer[14], D. Stam[15], C. Thalmann[13], J. Tinbergen[11], M. Turatto[2], and N. Yaitskova[1]

[1]European Southern Observatory,(ESO) Germany
[1b]Space Telescope European Coordinating Facility (ESO-STECF), Germany
[2]Istituto Nazionale di AstroFisica (INAF), Astronomical Observatory of Padova, Italy
[3]Laboratoire d'AstrOphysique de Grenoble (LAOG), France
[4]Laboratoire d'Etudes Spatiales et d'Instrumentation en Astrophysique (LESIA), France
[5]Laboratoire d'Astrophysique de Marseille (OAMP/LAM), France
[6]Max-Planck Institut fur Astronomie, Heidelberg (MPIA), Germany
[7]Office National d'Etudes et de Recherches Aérospatiales (ONERA), France
[8]Max-Planck Institut fur Sonnensystemforschung (MPS), Germany
[9]United Kingdom Astronomy Technology Centre (UKATC), UK
[10]Laboratoire d'Astrophysique de l'Observatoire de Midi-Pyrénées (LAOMP), France
[11]Netherlands Foundation for Research in Astronomy (ASTRON), Netherlands
[12]Ecole Normale Supérieure de Lyon (ENS), France
[13]Swiss Federal Institute of Technology of Zurich (ETHZ), Switzerland
[14]American Museum of Natural History, USA
[15]Astronomical Institute, University of Amsterdam, Netherlands

Abstract. This paper presents the status of the EPICS project, an Earth-like Planets Imaging Camera Spectrograph for OWL. We present the Top-Level-Requirements of the instrument and we describe the baseline of the Adaptive Optics system with optimized wave-front sensor. The expected performance in rejection of starlight in the near infrared and in the visible is given. The instruments concepts for detection and characterization of exo-planets will be briefly described. The Signal-to-Noise ratio estimation shows that Earth-like planets can be detected up to 20 pc in a reasonable amount of time. The extremely challenging requirements in terms of static residual errors and differential aberrations are discussed.

Keywords. planetary systems, adaptive optics, high angular resolution.

1. Introduction

This paper describes the status of the Earth-like Planets Imaging Camera Spectrograph (EPICS) for OWL. The EPICS project started in mid-april 2005, after the completion of the phase A studies of the Planet Finder project for the ESO Very Large Telescope (VLT), and is the result of the cooperation between the two competitive consortia. The core of the project is composed of an extreme Adaptive Optics (AO) system coupled with coronagraphs. Three instruments directly inspired from the VLT-PF conceptual design are briefly described and the expected starlight halo rejection is estimated. The Signal-to-Noise Ratio (SNR) for the detection of rocky planets and evolved gas giant planets is estimated. We emphasize that to achieve the 10^{-10} contrast needed for the primary science goal, the requirements in terms of instrumental static and differential aberrations will be extremely challenging.

2. Science case

One of the most ambitious science objectives of OWL is the detection and characterization of extra-solar systems in an advanced evolutionary stage, the understanding of the mechanisms of formation and evolution of planets, and possibly the discovery of other planets able to host life.

Rocky planets. The main goal of EPICS is the detection and characterization of rocky planets in the habitable zone (Kasting 1997). For characterization, we mean observation of a statistically significant sample of rocky planets, to understand the relation between rocky and giant planets, and to understand the main features of rocky planets physics. Based on the variety of atmosphere composition observed in the Solar system, it is expected that also the planets targets of EPICS will possess a wide range of atmospheric properties. Essentially, we consider at least four different classes of planets:

- Planets without atmosphere, or with a very thin atmosphere that does not produce significant features, like Mars. These are likely planets of either small mass and/or very close to the central star.
- Planets with atmospheres dominated by methane, like the giant planets in the Solar System. These are likely either massive and/or cold planets, able to maintain a substantial amount of H in their atmosphere
- Planets with atmospheres dominated by carbon dioxide, like Venus
- Planets with atmospheres dominated by oxygen, like the Earth

The three last classes of planets cited above have quite different spectra, both in the Visible and Near-IR, as well as in the Mid-IR. In the Visible-Near IR spectral range selected for EPICS, features due to H_2O, O_2, CH_4, and CO_2 are well represented. This should allow appropriate classification of the detected planets into the various classes defined above even from low resolution (survey type) spectra.

Evolved giant planets. EPICS will also permit a significant breakthrough in the detection and characterization of cold gas giant planets. The better contrast (the contrast of Jupiter at 5 AU is 10^{-9}) and larger separation, permits an easier detection, and opens the door to high resolution spectroscopy. In particular, radial velocity measurements and the analysis of atmospheric composition and dynamics of close-in giant planets will be possible. The contrast between a Jupiter mass planet at 0.5 AU and its star is around 10^{-7}, so roughly corresponding to the stellar AO residuals. For 10 pc distance from Earth, assuming a G2 star, its magnitude would be around 22.5 and the photon flux at resolution 50.000 would be about 0.5 photons per second and spectral bin (16% overall quantum efficiency). Therefore, a reasonably high SNR for the high resolution spectroscopy appears feasible in observing times of a couple of hours.

3. Top-Level Requirements

The detection of rocky planets with possibly Earth-like features is an extremely challenging goal for EPICS. The direct detection of exo-planets is made very difficult by the very high relative flux ratio from the star and planets orbiting it and their small angular separation. The Top-level requirements for the detection of Earth-like planets in the habitable zone can be expressed in terms of contrast requirement in function of the separation. We chose to restrict the targets for terrestrial planets to three spectral types,

G, K, and M, and investigated out to what distance one can find at least 100 stars of each type (see table below, Paranal latitudes $Z < 30$ deg.). This choice sets the limiting magnitude for adaptive optics as well as the minimum separation angle corresponding to the habitable zone.

Spectral type	Distance	V magnitude	Angular separation	Contrast
G	25 pc	7	40 mas	2.10^{-10}
K	20 pc	8	25 mas	8.10^{-10}
M	15 pc	10	15 mas	8.10^{-9}

4. Implementation concept

The core of the instrument is based on an extreme AO system of high performance that feeds three main instruments: a wave-length splitting Differential Imager, an Integral Field Spectrograph, a Differential Polarimeter. This choice reflects the concepts that have been developed for VLT-Planet Finder (Mouillet *et al.* 2004) (Gisler *et al.* 2004). Due to science goals that are significantly different from VLT-Planet Finder, the EPICS concept is based on somewhat different choices on the system level.

The EPICS concept should be compatible with the detection of both gas giants and rocky planets. Due to different locations of the spectral and polarimetric features of these two groups of planets, different channels split in the spectral domain are needed. Each scientific channel will be equipped with its own coronagraph.

- **The R band** [600–800 nm] is dedicated to the **Polarimetric Differential Imager** for detecting rocky planets and to the follow-up observations for the **detection of** O_2. The Differential Polarimeter for EPICS is directly based on the ZIMPOL concept (Gisler *et al.* 2004) proposed for the VLT Planet Finder. The main requirement is that the telescope polarisation remains low and relatively stable so that a suitable place can be found for the polarisation switch essential for calibration. Different possibilities are still under investigations.
- **The J band** [1100–1430 nm] will be equipped with a **Differential Imager** that will be sensitive to both CH_4 and H_2O absorption bands on a 4x4 arcsec field. The baseline is a dichroic based differential imager which main advantage is the high throughput and the possibility to implement 4 wave-lengths simultaneously. The most critical issue of this concept is the optical quality of the dichoics that should permit typically less than one nanometer differential aberration for the primary science goal requirements.
- **The H band** [1380–1800 nm] will be equipped with an **Integral Field Spectrograph**. The main features that can be detected in this band are CH_4 and CO_2 and H_2O. The EPICS IFS will operate on a 2x2 arcsec field with spectral resolutions per pixel from 15 to 30. Square and hexagonal shapes are studied in order to find a compromise between cross-coupling and size of the detector. A Fourier Transform Spectrograph is also being studied. This concept could have a better performance in terms of differential aberrations but has some other complications due to time dependent effects. One important advantage is that the spectral resolution can be adjusted from low resolution to very high resolution.
- **The I band** [800-1000 nm] is reserved for **wave-front sensing**. This band has been chosen because of the lesser scientific interest for planet detection. Moreover its location, spectrally speaking, between the visible and NIR instruments, is optimal with respect to important atmospheric chromatic limitations for extreme AO on ELTs.

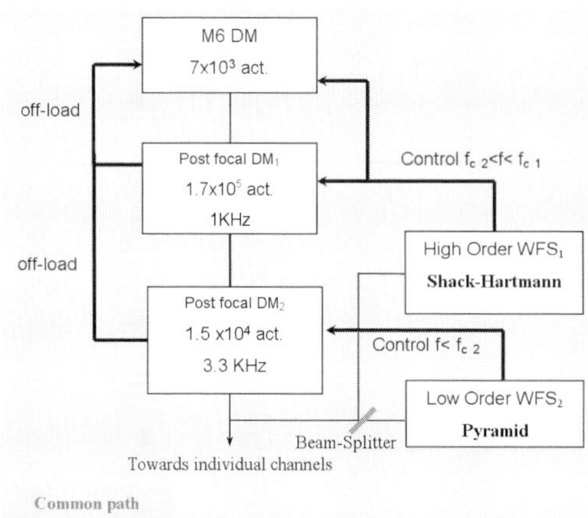

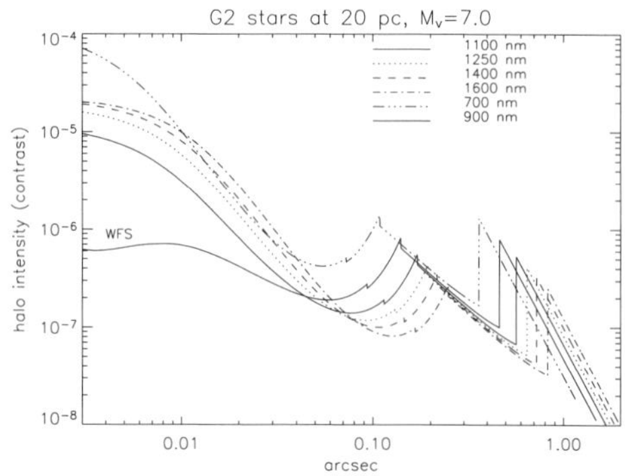

Figure 1. Left: EPICS XAO concept. Right: Halo intensity from AO residuals. Perfect coronagraph. Seeing:0.5 arcsec. Coherence time: 4 ms.

The EPICS ultimate contrast requirement is 4 orders of magnitude higher than the VLT-Planet Finder science goal of about 10^{-6} contrast at 0.1 arcsec. When scaling from a 10-m to a 100-m class telescope, the contrast naturally improves by a factor of 100 for a given rms value and power spectrum of the wave-front error. This means that the extreme AO system for EPICS should provide about 2 orders of magnitude better starlight halo rejection than a simply scaled version of the VLT Planet Finder system. This matter of fact calls for system specifications that are tremendously more stringent than for the VLT-PF project like a higher frame rate, low residual static errors and high sensitivity of the WFS for low to mid-spatial frequencies.

Adaptive optics. The EPICS XAO concept is displayed in Fig. 1. The OWL large adaptive mirror M6 is used for the correction of large amplitude and slow aberrations. The XAO system itself is based on a double stage. Two correctors in cascade each controlled by a WFS: a quad-cell Shack-Hartmann WFS controls a very high order adaptive mirror with 1.7×10^5 actuators (500x500 sub-apertures, $d_1 = 20\ cm$ inter-actuator distance as projected on OWL pupil) at 1 kHz. Using a fast Fourier reconstructor with modal control (Poyneer 2005), the correction is restricted only to the high spatial frequencies that are not corrected by the second stage. The latter is composed of a Pyramid WFS controlling a 1.5×10^4 adaptive mirror (150x150 sub-apertures, $d_2 = 65cm$) at 3 kHz. The main reason of this double stage comes from the computation time and CCD read-out time requirements and taking into account a projection at 10 years of future computing power availability. The advantages of this concept are the following:

• The Pyramid WFS is known to have a very low noise propagation for low spatial frequencies. Thus it favors a clean correction of the starlight halo at small angular separations (Verinaud 2005) where the most demanding contrast is required for rocky planets detection. A single stage with Pyramid sensor would have been ideal but revealed to be very demanding in terms of computing power required.

• The low number of sub-apertures of the Pyramid WFS (with respect to the SH WFS) permits a high frame rate for the read-out of the WFS CCD (300x300 pixels).

• Contrary to the Pyramid WFS, SH WFS-based systems can use fast algorithms for the reconstruction, allowing to control the high number of degrees of freedom of the first stage.

• The first stage SH-based system corrects only high spatial frequencies avoiding to propagate noise in the loop that would pollute the halo at small angular separations.

The corresponding theoretical Point Spread Functions for a perfect coronagraph at different wave-lengths are also represented in Fig. 1. One can notice three different regions: the innermost corrected by the Pyramid system, the intermediate corrected by the SH, and the outermost outside the corrected area. The main error sources affecting AO have been considered: servo-lag, photon noise, aliasing of pyramid sensor, static errors, anisoplanicity due to atmospheric differential refraction, chromatic seeing.

Coronagraphy. A special coronagraph has been designed for OWL (Yaitskova (2005)). It comprises a gaussian focal mask and several reticulated pupil stops that mimic the OWL segmentation gaps and follow their configuration during observation. The use of this system in double stage permits theoretically an almost perfect cancelation of the starlight. Micro-metric precision is needed for the alignment of the system. Apodized Lyot coronagraphs have also been considered (Soummer 2006).

5. Signal-to-noise ratio estimations

In this section the evaluation of the detection performance in terms of integration time needed to achieve a given SNR has been computed. The conditions of observations are :

• Differential detection, SNR=5, dual imaging either with DI or IFS and differential polarimetry (see also Schmid *et al.* (2006))

• Instrumental transmission 16%. Atmospheric transmission for altitude 4000-m: Transmission in H_2O band (40%) and in O_2 band (80 %).

• Seeing: 0.5 arcsec. $\tau_0 = 4\ ms$, AO residuals photon noise and speckle noise (perfect coronagraph).

The integration times needed for a 5σ detection are displayed in Fig. 2. H_2O (as in Earth-like planet) and CO_2 (10 % concentration) are detected in a couple of hours at 10 pc and in about one to two nights of observation at 15–20 pc. For the same amount of time Jupiter-like planets can be detected in CH_4 bands with a SNR of 50. Follow-up observations can be foreseen for O_2 detection in Earth-like planets (Fig. 2 bottom left) but require several tens of hours at 5–10 pc to several hundreds of hours at 20 pc. This amount of time, even if large, is probably not prohibitive regarding the philosophical impact such a detection could have. Planets without strong markers in CH_4, CO_2 and H_2O but with a high degree of polarisation could also be detected in the R band with the differential polarimeter (bottom right).

6. The challenge of systematic errors control

The instantaneous contrast delivered by the AO and the coronagraph will at best only be of about 5×10^{-8} at 0.1 arcsec separation (see Fig. 1). To attain the goal contrast of 10^{-10}, differential imaging and long exposures are needed. Differential imaging will permit to reject the residual atmospheric speckle noise but will inevitably introduce some new speckles that are due to systematic differential chromatic errors that can occur both before the coronagraph (because of atmospheric differential refraction) or after the coronagraph in the instrument itself (see Cavarroc 2005). Systematic errors for the future VLT-PF system are estimated to be of the order of several tens of nanometers. The requirements for EPICS will be of about a nanometer or less for some key optical elements. This kind of high precision control is comparable to what has already been achieved in the laboratory using very precise active mirrors (High Contrast Imaging Testbed (HCIT) experiment for the TPF-C mission, see http://planetquest.jpl.nasa.gov).

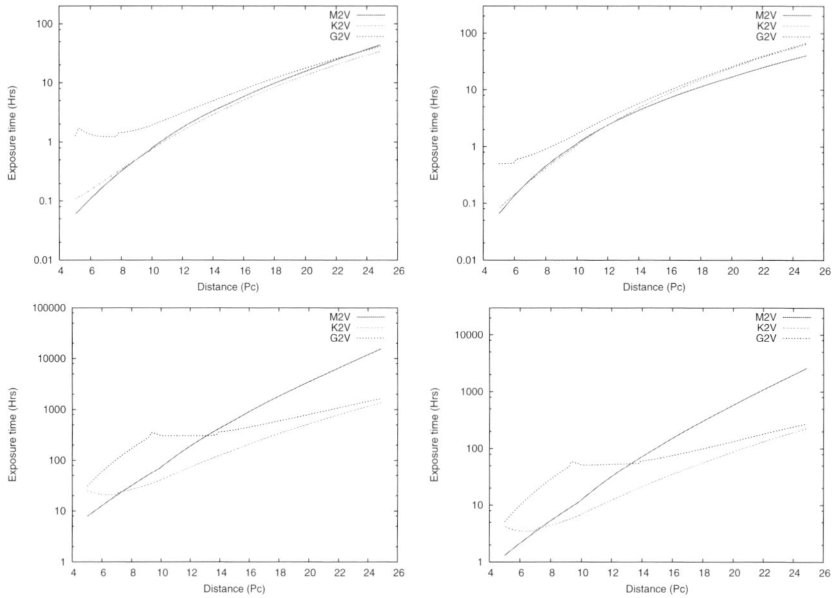

Figure 2. Integration times for a 5σ detection. Top-left: Earth-like planet, detection in water bands, R=15, $\lambda = 1400nm$. Top right: CO_2 rich (10%) Earth-sized planet,R=15, $\lambda = 1600\ nm$. Bottom left: detection of O_2 in Earth-like planet, R=150 $\lambda = 760\ nm$.Bottom right: Detection of Earth-like planet by differential polarimetry (15% polarization), 600-800 nm.

7. Conclusions

We have described a concept for a Planet Finder instrument for OWL. The performances show that, thanks to an optimised AO system and coronagraph, a high rejection of the starlight halo in J and H band can be obtained permitting to detect H_2O and CO_2 (if abundant >10%) in Rocky planets in about one or two observation nights up to 20 pc. Detection by polarimetry will also be possible, as well as O_2 in Earth-like planets, however in follow-up observation only. Detection of Jupiter-like planets in CH_4 bands in J and H band can be done with high SNR opening the door to high resolution spectroscopy. The requirements on systematic errors control are very challenging but reachable with existing high performance active correctors.

Acknowledgements

This activity is partly supported by the European Community (Framework Programme 6, ELT Design Study, contract No 011863)

References

Yaitskova, N. 2005 Proceedings of the SPIE, Volume 5905, pp. 292–302 (2005)
Schmid, *et al.* these proceedings
Kasting, J. F., Origins of Life and Evolution of the Biosphere, v. 27, Issue 1/3, p. 291–3107 (1997)
Gisler, D. *et al.*, Proceedings of SPIE Vol. 5492 (2004)
Mouillet, D. *et al.* ASP Conference Proceedings, Vol. 321, ISBN: 1-58381-183-4, 2004, p. 39 (2004)
Poyneer, L., Véran, J.-P., JOSA, Volume 22, Issue 8, pp. 1515–1526 (2005)
Soummer, R. *et al.*, these proceedings
Vérinaud, C., Le Louarn, M., Korkiakoski, V., Carbillet, M., MNRAS Letters, 357, 1 (2005)
Cavarroc, C. *et al.* in press A & A 2005

Direct Imaging of Exoplanets: Science & Techniques
Proceedings IAU Colloquium No. 200, 2005
C. Aime and F. Vakili, eds.
© 2006 International Astronomical Union
doi:10.1017/S1743921306009872

High order Adaptive Optics Requirements and feasibility for high contrast imaging

T. Fusco[1], G. Rousset[1], J.-L. Beuzit[2], D. Mouillet[3], and K. Dohlen[4]

[1]ONERA, B.P.72, 92322 Chatillon, France
[2]Observatoire de Grenoble, B.P. 53, 38041 Grenoble, France
[3]Observatoire Midi-Pyrénées, B.P. 826, 65008 Tarbes, France
[4]Observatoire de Marseille, B.P. 8, 13376 Marseille, France

Abstract. In the frame of the VLT Planet-Finder project, the phase A system study has demonstrated the feasibility of an extreme adaptive optics system aimed at the direct detection of extrasolar giant planets. The main results of this study are presented in this paper.

Keywords. high angular resolution, adaptive optics.

1. Introduction

We present here the results of the phase A study of the adaptive optics system for the VLT Planet Finder (VLT-PF) instrument (Beuzit *et al.* 2005). The main scientific objective of VLT-PF is the direct detection of photons coming from giant extrasolar planets (between 1 and 20 Jupiter masses). Any detection will then be followed by a first characterisation of the planet atmosphere (clouds, dust content, methane, water absorption...). In addition, the survey of an extended number of stars (typically a few hundreds) is mandatory in order to perform meaningful statistical studies. Such extremely challenging scientific objectives directly translate into a relatively complex high constrast instrument. Coronagraphic and smart imaging (differential imaging for instance) capabilities are essential to reach the high constrat (close to the optical axis) required for direct extrasolar planet detection. From the ground, the core of any high contrast instrument is an extreme adaptive optics (XAO) system correcting for the perturbation induced by the atmospheric turbulence as well as for the internal aberrations of the instrument itself.

Based on the specifications provided by the astronomers (Moutou *et al.* 2005) and on the constraints of the high contrast instrumentation, the XAO must fulfill the following three high level requirements: 1) ensure the measurement and correction of the turbulent phase perturbations, of the telescope and system common optics aberrations, of the non-common path aberrations (main AO loop); 2) ensure an extremely high stability (at low temporal frequency) of the optical axis at the level of the coronagraphic mask (Auxiliary Sensor [AS]); 3) ensure the measurement and the correction of any pupil motion (Pupil Motion Sensor [PMS]).In the following sections, we will mainly focus on the main AO loop design. The two other requirements (essential to ensure optimal coronagraphic extinction and, *in fine*, planet detection performance) will only be briefely addressed.

2. AO loop performance in the focal plane

The VLT-PF system aims at the detection of extremely faint sources (giant extrasolar planets) in the vinicity of bright stars. Such a challenging goal requires the use of a coronagraphic device to cancel out the flux coming from the star itself. Even if additionnal smart imaging techniques are added to reach the required contrast for extrasolar planet detection (typically 10^{-6} - 10^{-7} in contrast), the coronagraphic image is a basis for the AO optimisation. Moreover, unlike classical AO systems, residual variance or Strehl ratio are not sufficient anymore to optimize the system and to derive the pertinent trade-offs. They have to be replaced by a more accurate parameter which can provide information on the coronagraphic image shape in the focal plane. The purpose of the coronagraph is to remove the coherent light coming from the on-axis guide star (GS). Therefore one can analytically define a "perfect coronagraph" using the following equations:

$$I_{corona}(\rho) = \left\langle \left| \mathrm{FT}\left[P(r)A(r)e^{i\varphi_{res}(r)} - \sqrt{\exp\left[-\sigma^2_{\varphi+\log(A)}\right]}\, P(r) \right] \right|^2 \right\rangle \tag{2.1}$$

where $\langle . \rangle$ stands for a statistical average and with $A(r)$ the wavefront amplitude, $\varphi_{res}(r)$ the residual phase after AO correction, $P(r)$ the pupil function and $\sigma^2_{\varphi+\log(A)}$ the residual variance gathering phase and log-amplitude effects. When only a partial correction is performed, the coherent peak is removed and only the uncoherent light (the residual uncorrected speckles) remains. In that case, it is easy to show that, as a first approximation (first order expansion) the coronagraphic image is proportional to the residual phase power spectral density.

3. Balance of an error budget

The whole AO study is based on the balance of an error budget. A first approximation (first order expansion) of this coronagraphic shape is given by the residual phase power spectral density (PSD) as shown in Section 2. Therefore, the global error budget for the AO system can be summarized as follow:

$$\begin{aligned}\mathbf{PSD}_{res} = & \underbrace{\mathbf{PSD}_{scint} + \mathbf{PSD}_{diff} + \mathbf{PSD}_{refrac} + \mathbf{PSD}_{aniso}}_{\text{atmospheric limitation}} \\ & \underbrace{+\mathbf{PSD}_{fit} + \underbrace{\mathbf{PSD}_{temp} + \mathbf{PSD}_{alias} + \mathbf{PSD}_{noise}}_{\text{low order residual error}}}_{\text{AO loop residual error}} + \underbrace{\mathbf{PSD}_{calib} + \mathbf{PSD}_{aberr}}_{\text{calibration errors}}\end{aligned} \tag{3.1}$$

A complete description can be found in Fusco *et al.* 2005b. $\mathbf{PSD}_{res}$ is expressed in terms of residual variance per spatial frequency in order to highlight the range of spatial frequencies which is affected by each error item. The otimisation of this error budget will be performed, keeping in mind three main criteria:

- The corrected area, i.e. the focal plane area where the image contrast is significantly improved by the AO system. It mainly drives the choice of the number of actuators (the correction area is equal to λ_{im}/d in diameter, where d is the actuator spacing). Considering the typical targets which will be observed by VLT-PF and the imaging wavelengths (J, H and K bands), this area has to be larger than 0.8 arcsec in diameter.
- The detectivity level, i.e. the capability of the whole system to detect the planet signal. This level is affected by the AO loop errors (temporal, noise, aliasing...) which evolve rapidly with time and can be calibrated using differential imaging and a reference PSF. It can also be degraded by the telescope and the system high spatial frequencies and non-common path aberrations which slowly evolve with time and represent the ultimate limitation for the differential imaging and reference PSF subtraction techniques. The minimisation of the slowly variable defects implies the measurement and the correction

of non-common path aberrations (see Section 6) as well as the stabilisation of the optical beam during a whole observation sequence (see Section 7.2).

• The system sensitivity. This criterion is driven by the number of stars to be observed but highly depends on the detectivity level and the corrected area size. Indeed, the larger the corrected area the smaller the available flux per individual measurement zones (sub-aperture in the case of a SH device for instance). In addition, increasing the detectivity level implies a reduction in terms of temporal and noise errors, which leads to a faster system working on brighter guide stars for wavefront sensing. A first trade-off between scientific goals and system requirements has led to a limiting magnitude of 8 in H-band (corresponding to magnitudes 10-11 for visible band, depending on the GS types) for the system, implying that the detectivity capabilities have to remain optimum up to this magnitude.

In addition to this three scientific criteria, other constraints have to be taken into account during the instrument design: 1) use of well-proven technologies if possible; 2) new developments for critical issues only, with associated experimental validations; 3) the system has to be built on a tight schedule (5 years typically) with finite manpower and budget. These last three points are essential to minimize the risk factor during the instrument realisation.

4. Atmospheric limitations

The atmospheric limitations (Roddier & Roddier 1986) which gather all the errors due to propagation effects (scintillation, diffraction, anisoplanatism and differential refraction effects). Full correction of scintillation effects would only lead to an equivalent reduction of the phase variance smaller than 20 nm rms. In comparison of the small gain in performance, the system complexity is highly increased (two DM, measurement device, reconstruction process...). Considering the small expected gain, even if the scintillation is fully corrected (which is far from being obvious) it has been decided not to consider a scintillation corrector for the VLT Planet Finder instrument.

5. AO loop residual errors

The AO loop residual errors which gather all the "classical" AO loop errors and can be sub-divided in

• The fitting errors (high order modes not corrected by AO, i.e. frequencies higher than the AO spatial cut-off frequency). Considering the Planet Finder Top Level Specifications, the minimisation of the global error budget is not the only pertinent criterion, the spatial repartition of the errors has also to be taken into account. In particular, the detectivity performance is also linked to the capacity of the AO system to "clean up the PSF" (i.e. to be as close as possible to the diffraction pattern) in an area of more than 0.75" in radius from the optical axis. This corrected area (Corr) in the focal plane is directly linked to the inter-actuator distance ($d = D/(n_{act} - 1)$, where nact is the linear number of actuators in the telescope diameter D) with the following relation: corr $= \lambda_{im}/2d$. Decreasing d, that is increasing the number of actuators, has however consequences in term of the system limiting magnitude. Indeed, the larger the number of DM actuators, the larger the number of WFS sub-apertures and hence the smaller the number of available photons per sub-aperture and per frame.

• Low order modes (corrected by AO) gathering the aliasing, temporal and noise errors.

i Aliasing error: these effects are due to high spatial frequencies seen as low ones by the WFS device. The aliasing error is directly linked to the fitting error and in the specific case of SH WFS it corresponds to roughly 40% of the total fitting error variance. It dramatically increases the PSF residuals in its corrected area. A concept has been recently proposed (Poyneer & Macintosh 2004) to significantly reduce the aliasing effect. This device has been deeply studied and optimised with respect to the system and turbulence characteristics (spectral bandwidth, WFS sampling, turbulence ...). In addition, an experimental validation of the concept has been conducted using the ONERA AO bench (see Fusco *et al.* 2005a), in closed loop and with turbulence. A gain brought by a filtering device has been clearly demonstrated and it has been shown that the experimental results are in good agreement with the simulation, which validates the potentiality of the concept and its use in the VLT-PF design.
ii Temporal error: the AO loop temporal behaviour depends on each AO components (detector integration time, readout noise, real-time computations of commands from WFS data, numerical corrector, digital-analog converter, high voltage amplifier, DM actuator temporal response) and turbulence characteristics. The optimal way to deal with all these parameters is to design a Kalman filter based control algorithm (Petit *set al.* 2004). Nevertheless, practical implementation of such a control law is complex and requires more computing power, especially for high order systems. This leads to a significant increase of the RTC complexity. An hybrid solution has been considered to deal with this problem. An optimal modal gain integrator has been chosen to control high order modes while a Kalman filter has to be considered for tip-tilt modes to optimally correct turbulence and vibration effects
iii WFS measurement error: a comparison between SH and Pyramid WFS in the frame of the VLT-PF AO system has been performed. In both cases, performance, required calibrations and optimizations, as well as fundamental limitations have been identified and quantified. From the WFS performance, stability, complexity and risk evaluation, a spatially filtered SH WFS, combined with a new optimized slope estimation algorithm (Nicolle *et al.* 2004), has been chosen as a baseline for the VLT-PF AO system. It allows us to reach a limiting magnitude of around 10-11 (depending on GS type and CCD performance (Fusco *et al.* 2004))

6. Calibration error

The calibration errors (Fusco *et al.* 2003) that gather the AO loop mis-calibration (interaction matrix and reference slopes) and the mis-calibration of the non-common path aberrations (NCPA). Effects of mis-aligment on system performance have been studied leading to tight specifications on system stability (pupil conjugaison between DM and WFS, pupil motion ...). NCPA will be measured using a phase diversity approach Sauvage *et al.* 2005 and corrected in closed loop through a modification of the WFS references.

7. Auxiliary devices

7.1. *IR Tip-Tilt sensor*

The average image position (in other words the optical axis position) on the coronagraphic mask is a main specification for the VLT-PF system. The required accuracy for the mean image position is 0.5 mas or better. The global error for the average image position mainly depends on the differential refraction effect (between VIS and IR wavelengths) and the differential thermal or mechanical effects (between WFS and imaging paths). Considering the requirement, an open loop model of each differential evolution will not be accurate enough (considering all the possible parameters involved) to reach the absolute position performance. Therefore, to ensure that the specification will be fullfilled, an auxiliary IR tip-tilt sensor (AS) at the level of the coronagraphic mask has been proposed. This sensor will be coupled with a differential tip-tilt mirror (DTTM) located in an pupil plane in the WFS arms.

7.2. *Pupil motion sensor*

Pupil stability is a major issue to ensure the VLT-PF performance. The pupil has to remain motionless during the whole observation process. When located after the Nasmyth focus of the telescope, this stability requirement implies a pupil de-rotator and a pupil re-centering device. It has been shown that a pupil shift of 1% of the pupil diameter or a pupil rotation of 1 degree will reduce by a factor of 1.5 to 2 (for typical conditions at the VLT) the detection capability of a coronagraphic + differential imaging system. This

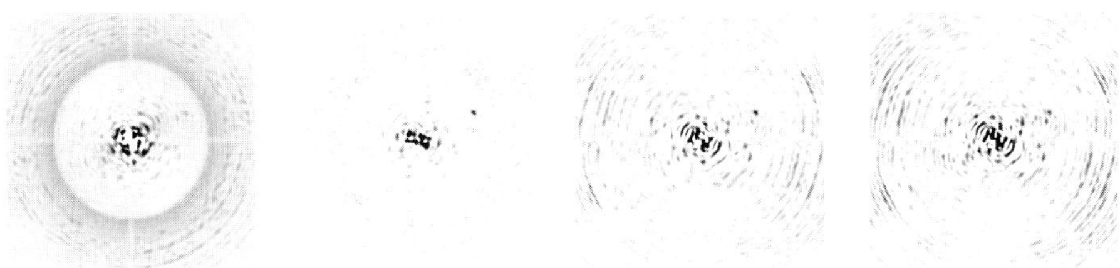

differential image — motionless pupil — pupil motion = 0.6% — pupil motion = 1.2%

Differential image + calibration on reference star

$I(\lambda_1) - I(\lambda_2)$ — $[I(\lambda_1) - I(\lambda_2)] - [Ref(\lambda_1) - Ref(\lambda_2)]$

Figure 1. [Left] Differential coronagraphic (4-quadrant) image ($\lambda_1 = 1.56\mu m$, $\lambda_2 = 1.59\mu m$), [Right] differential coronagraphic image + reference subtraction: pupil shift between object and reference star = 0, 0.6 and 1.2 % of the full pupil. The companion ($\Delta_m = 15$, separation = 0.6 arcsec) is really distinguishable from residual fixed speckles for a fixed pupil.

had led to impose a pupil stability in translation better than 0.2 (goal 0.1 %) of the full VLT pupil. This performance is achieved using a pupil tip-tilt mirror located close to the entrance focal plane of the VLT-PF. This mirror is controlled by a pupil motion sensor (PMS). The PMS directly uses the SH-WFS data to measure pupil motion. Because pupil motion is rather slow, a measurement-correction process has to be performed typically every minute which ensures a good SNR on the PMS data.

8. Global system design

A global trade-off from all the points mentioned above (combined with optical design, technological aspect, cost and risk issues) leads to the following AO system main characteristics:

- A 41x41 actuator DM of 180 mm diameter, located in a pupil plane with an inter-actuator stroke $>\pm 1\mu m$ (mechanical), a maximum stroke $>\pm 3.5\mu m$ (mechanical), plus a 2-axis TTM with a ± 0.5 mas resolution.
- 40x40 Shack-Hartmann WFS, with a spectral range between 0.45 and 0.95 μm, 6x6 pixels per sub-aperture (Shannon sampling @ 0.65 μm), a focal plane filtering device with variable size (from λ/d to $3\lambda/d$ at $0.7\mu m$) and a temporal sampling frequency 1kHz (goal 1.5 kHz). The foreseen detector is a 256x256 pixels Electron Multiplication CCD detector with a read-out-noise $<1e^-$ and a 1.4 excess photon noise Fusco *et al.* 2004.

- Mixed numerical control law with a Kalman filter law for Tip-Tilt control and an Optimal Modal Gain Integrator law for DM control. The global AO loop delay has to be lower than 1 ms (goal 666 μs).
- Non-common path aberrations: off-line measurements and on-line compensation using a phase diversity algorithm.
- Auxiliary IR tip-tilt sensor and pupil sensor to measure and xcorrect for optical axis and pupil displacement.

End-to-end simulations have been performed with this conceptual design. Results are summarized in Table 1. Using these AO simulations in a complete systeme model (in-

Seeing conditions	0.65"		0.85		1.05"	
Spatial repartition	Low freq.	High freq.	Low freq.	High freq.	Low freq.	High freq.
Mag 9	46	51	54	67	67	81
Mag 11	64	51	73	67	90	81

Table 1. AO error budget (in nm rms) for various seeing and GS flux conditions. In each case, both low spatial frequency error (i.e corrected by AO) and hig spatial frequencies error (fitting error) are given.

cluding AO, coronagraphy, differential imaging and detectivity process) it has been shown that these values satisfy the VLT-PF high level requirements alowing direct detection of hot Jovian-like planets. In conclusion, an AO for Planet Finder represent a large step forward both in terms of system components and calibration procedures, nevertheless a complete analysis (with a detailed error budget) has shown that a AO system fulfilling all the requirement mandatory for the direct detection of hot Jupiter like planet is feasible in a reasonable time scale (5 years) with tried and tested technologies.

References

J.-L. Beuzit *et al.* "A planet Finder instrument for the VLT ", This conf., 2005

C. Moutou *et al.* "Science case for the next generation VLT AO instrument Planet Finder", This conf., 2005

T. Fusco *et al.* "Conceptual design of an extreme AO dedicated to extra-solar planet detection by the VLT-planet Finder intrument", **5903**, Proc. Soc. Photo-Opt. Instrum. Eng., 2005

F. Roddier and C. Roddier, "Noao infrared adaptive optics program ii: modeling atmospheric effects in adaptive optics systems for astronomical telescopes," in *Advanced Technology Optical Telescopes III*, **628**, Proc. Soc. Photo-Opt. Instrum. Eng. (Washington), 1986

C. Petit *et al.* "Kalman filter based control loop for adaptive optics," in *Advancements in Adaptive Optics*, **5490**, SPIE, 2004

L. A. Poyneer and B. Macintosh, "Spatially filtered wave-front sensor for high-order adaptive optics," *J. Opt. Soc. Am. A* **21**(5), pp. 810–819, 2004

T. Fusco, M. Nicolle, G. Rousset, V. Michau, J.-L. Beuzit, and D. Mouillet, "Optimisation of Shack-Hartmann-based wavefront sensor for XAO system," in *Advancements in Adaptive Optics*, **5490**, Proc. Soc. Photo-Opt. Instrum. Eng., 2004

T. Fusco *et al.*, "Closed-loop experimental validation of the filtered Shack-Hartmann," *Opt. Lett.* **29**, pp. 1255–1257, Jun. 2005

M. Nicolle *et al.* "Improvement of shack-hartmann wavefront sensor measurement for extreme adaptive optics," *Opt. Lett.* **29**, pp. 2743–2745, Dec. 2004

T. Fusco, G. Rousset, and A. Blanc, "Calibration of AO system. Application to NAOS-CONICA," in *Science with Adaptive Optics*, W. Brandner and M. Kasper, eds., Springer-Verlag, 2004. Sept. 2003, Garching, Germany

J.-F. Sauvage *et al.* "Fine calibration and pre-compensation of non-common path aberrations for high performance AO system," in *This conference*, 2005

Direct Imaging of Exoplanets: Science & Techniques
Proceedings IAU Colloquium No. 200, 2005
C. Aime and F. Vakili, eds.
© 2006 International Astronomical Union
doi:10.1017/S1743921306009884

Analysis of ground-based differential imager performance

A. Boccaletti[1], D. Mouillet[2], T. Fusco[3], P. Baudoz[1] C. Cavarroc[1], J.-L. Beuzit[4], C. Moutou[5] and K. Dohlen[5]

[1]LESIA, Observatoire de Meudon, 5 pl. J. Janssen, 92195 Meudon, France
email: anthony.boccaletti@obspm.fr

[2]Observatoire Midi-Pyrénées, 57 av. d'Azereix, 65008 Tarbes, France
[3]ONERA, 29 avenue de la Division Leclerc, 92320 Chatillon, France
[4]LAOG, Observatoire de Grenoble, 38041 Grenoble, France
[5]LAM, Traverse du Siphon, 13376 Marseille, France

Abstract. In the context of extrasolar planet direct detection, we evaluated the performance of differential imaging with ground-based telescopes. This study was carried out in the framework of the VLT-Planet Finder project and is further extended to the case of Extremely Large Telescopes. Our analysis is providing critical specifications for future instruments mostly in terms of phase aberrations but also regarding alignments of the instrument optics or offset pointing on the coronagraph. It is found that Planet Finder projects on 8m class telescopes can be successful at detecting Extrasolar Giant Planets providing phase aberrations, alignments and pointing are accurately controlled. The situation is more pessimistic for the detection of terrestrial planets with Extremely Large Telescopes for which phase aberrations must be lowered at a very challenging level.

Keywords. instrumentation: high angular resolution, stars: low-mass, brown dwarfs, planetary systems.

---

1. Future ground-based instruments to search for exoplanets

Many techniques and projects were proposed in the past few years to image directly extrasolar planets from the ground. Specific instruments like Adaptive Optics (AO) systems and coronagraphs were developed in this way. However, it is expected that even an extreme AO system coupled with a high rejection coronagraph will not be sufficient to attenuate the diffracted light form the on-axis star down to the level of gazeous planets like Jupiter. For illustration, Fig. 1 shows the 5σ detectivity radial profile considering a 40×40 actuators AO system on an 8m telescope and an achromatic phase mask coronagraph. The symbols on the plot correspond to different planet masses assuming an old M0V star at 10pc observed in the H band. The contrast at 0.5" (corresponding to a 5AU orbit at 10pc) is only 6.10^{-5}. In that case, the detection of planets (with a mass lower than $13M_J$) is definitely precluded. The situation would be more favorable for younger stars but not sufficient to explore the very low masses at close angular separations.

The radial contrast presented in Fig. 1 arises from a residual diffraction pattern made with atmospheric speckles left uncorrected by the AO system plus static speckles mostly originating from the non common path. It is worth to remind here that a coronagraph has the ability to suppress the coherent light only and then random interferences like speckles are escaping its action. Nevertheless, several smart techniques of speckle calibrations were already proposed. The most developed so far is probably the spectral differential imaging (Marois *et al.* 2000) first implemented inside TRIDENT at CFHT (Marois *et al.*

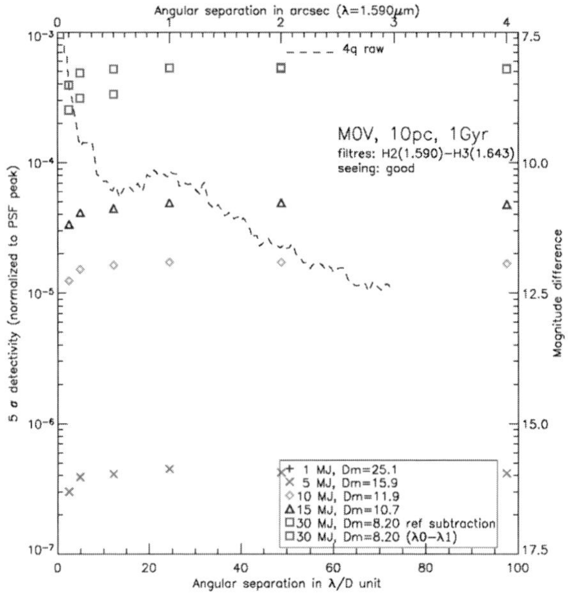

Figure 1. Raw contrast achieved with an 8m telescope, a 40×40 actuators AO system and an achromatic phase mask coronagraph. The expected intensities of planets are over-plotted for several masses and several angular separations. The contrast at 0.5" is only 6.10^{-5} which is not allowing the detection of planetary masses. A speckle pattern calibration unit is definitely needed to improve the contrast.

2005) and more recently at the VLT (Lenzen *et al.* 2004). The basic idea is to record simultaneously two images in two different but spectrally close filters on the same detector. If phase aberrations are small, the images are identical once rescaled in intensity and spatially matched (speckle pattern has a radial dependence with the wavelength). But they actually differs since the instrument has always chromatic aberrations and because the 2-beam separations implies 2 different optics which means differential phase aberrations. The question is: how much the detection level can be lowered by this technique when realistic assumptions are made on the whole system (atmosphere, telescope, instrument)?

In this paper, we present some results of numerical simulations we performed to assess the performance of differential technique. Section 2 describes the principle of the simulation we performed in the case of the VLT Planet Finder project (Beuzit *et al.* 2005) and gives the list of parameters we have considered. Section 3 shows some examples of sensitivity analysis for some particular parameters. Finally, a similar study is carried out for the particular case of Extremely Large Telescopes.

2. Modeling differential imaging

The current design of the VLT Planet Finder includes an extreme AO system (40×40 actuators) to achieve large Strehl ratios, several coronagraphic masks to block out the on-axis starlight, an appropriate stop to cover the residual diffraction inside the pupil and a dual band imager to calibrate the speckle pattern in real time according to the technique of Spectral Differential Imaging (Marois *et al.* 2000).

To be realistic the simulation must include a thorough model of the atmosphere, of the telescope, and of the instrument. The simulation is performed with a sequence of 3

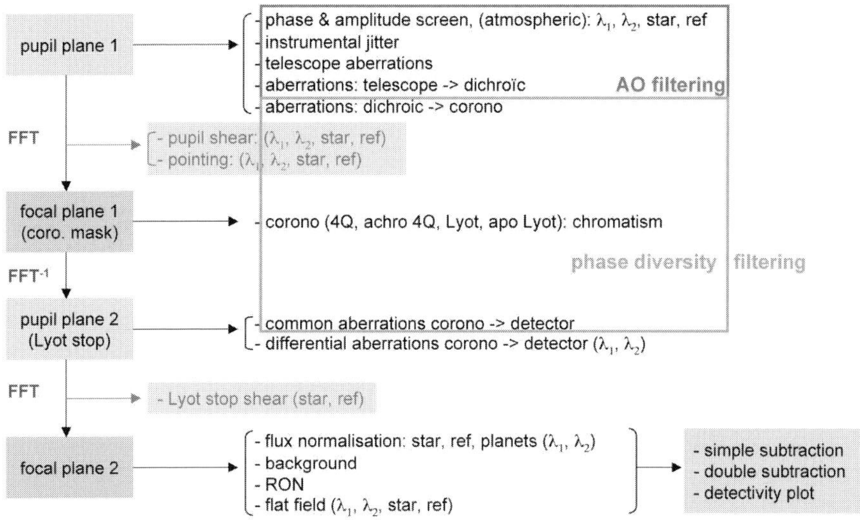

Figure 2. Flow chart of the numerical simulation presenting the main steps at the left and a more detailed structure at the right. The simulation is divided in 2 codes, one for calculating the diffraction images and the second one for including photometry. As outputs, the simulation provides detectivity plots at 5σ as a function of the angular separation.

Fourier transforms starting from the first pupil plane and up to the detector plane. For the telescope, we considered the actual phase maps as provided by ESO (for M1, M2 and M3 telescope mirrors). For the instrument, we took into account the number of optical surfaces we have in our optical design on which we added a f^{-2} power spectrum density (PSD) phase defects. Upstream the dichroic of the AO, the phase maps (including the atmosphere and the telescope) were filtered by the PSD of the AO system (below the cutoff frequency). Similarly, phase maps upstream the coronagraph were filtered by the PSD of the phase diversity algorithm (36 Zernike polynomials are assumed perfectly corrected). The differential optics in front of the detectors were not filtered. When relevant, we also considered the chromatism of the coronagraph (for phase masks).

In addition, we included differential offset pointing (on the coronagraph) and differential misalignment (of the telescope pupil with respect to the instrument pupil). The term differential refers to a chromatic and temporal variation. In other words, the pointing will be slightly different in the 2 simultaneous spectral channels but also when observing sequentially a reference star as required to calibrate the differential optics. We also accounted for possible misalignment of the Lyot stop of the coronagraph.

A second numerical code models the photon noise, the detector noise (readout and flat field) and the background noise. Images are normalized in intensity using numerical models for several ages and masses of the planets. Then, we provided several types of data: raw coronagraphic images; single subtraction after proper rescaling in size and intensity following the principle of spectral differential imaging ($I_s(\lambda_1) - I_s(\lambda_2)$ where I_s and I_r denotes the image of the star and of the reference); and double subtraction obtained by combining the single subtraction and the calibration on a reference star ($[I_s(\lambda_1) - I_s(\lambda_2)] - [I_r(\lambda_1) - I_r(\lambda_2)]$). Single subtraction provides calibration of the common aberrations (both atmospheric and instrumental) but double subtraction has also the advantage to provide calibration of the differential aberrations.

The flow chart of the simulations is shown in Fig. 2.

parameters	values
pupil alignment error in translation	$< 0.2\%$
pupil alignment error in rotation	$< 0.1^\circ$
image alignment and stability on coronagraph	< 0.5 mas
image defocus on coronagraph	< 4 nm rms
coronagraphic phase mask chromatism	$< 10^{-2}$ rd
Lyot stop alignment	$< 0.2\%$
wavefront errors non common to simultaneous images	< 10 nm rms
total wavefront errors downstream Lyot stop	< 90 nm rms
$\lambda/\Delta\lambda$ between 2 spectral channels	> 10
Flat Field residual errors	$< 10^{-3}$

Table 1. System defects included in the VLT-PF end-to-end simulations.

3. Sensitivity analysis

Prior to performance evaluation we carried out a sensitivity analysis of the parameters modeled in the simulation and draw the error budget. Table 1 shows the error budget where each parameters were constrained by simulations while keeping in mind the issue of feasibility. Figures 3 and 4 give some examples of sensitivity analysis.

Given those values, we calculated the detectivity for several test cases. Our set of 240 cases covers various observing conditions (turbulence conditions), various science targets (stellar types and ages, planets from 1 to 30 Jupiter masses), and various observing modes (5 filter pairs ranging from $1.08\mu m$ to $2.25\mu m$. Planet spectra are taken from Allard *et al.* (2001) models. The signal to noise ratio is calculated on radial profiles (azimuthally averaged) obtained on the coronagraphic image, on the single subtraction and on the double subtraction. A complete set of results makes possible an extensive discussion on the science case of the instrument, and the optimal observing strategy (Moutou *et al.* 2005) . For and old M star (1Gyr) in the solar vicinity a giant planet of about 7 times the mass of Jupiter can be detected. The situation is more favorable for young stars for which 1 M_J planets are detectable in orbit as close as 0.2" (8AU at 40pc) since they are relatively bright and feature strong methane absorption. Regarding nearby old stars (3pc), the detection is also improved to 1 M_J for irradiated planets (separation = 1 AU) according to the model of Sudarsky *et al.* (2003).

4. The particular case of Extremely Large Telescopes: terrestrial planets

Terrestrial planets have became the top priority targets of Extremely Large Telescopes. The feasibility of this program is overwhelming the single astrophysical interest and therefore needs to be evaluated thoroughly. The star/planet contrast is about 10^{-10} at a projected separation of 0.1" for a system placed at 10pc. Therefore, compared to the performance delivered by VLT Planet Finder a gain of 4 orders of magnitude is needed. For the same reason as above a differential technique is needed to improve the detectability.

Our team developed a simple analytical model which is including the atmospheric phase residual left uncorrected by the AO system, the static common aberrations in the instrument and the differential (non common) aberrations (Cavarroc *et al.* 2005). For sake of generality we did not assume any particular type of differential technique. So we just considered two images recorded simultaneously one showing the planet and the other not. It is shown that the fundamental limit (independently of the integration time) is given by the amount of static aberrations and that the effects of common and

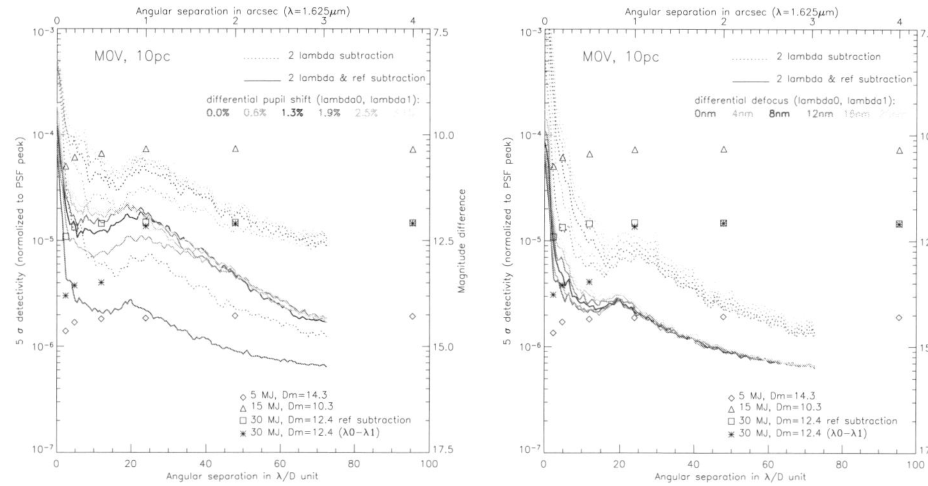

Figure 3. Sensitivity analysis of the chromatic pupil shear between the telescope and the instrument (left) and of the chromatic defocus (right). Dotted lines give the radial contrast at 5σ for a single subtraction and solid lines are for double subtraction. The pupil shear is given in percent of the pupil diameter and the defocus is expressed in nanometers rms. Color codes are detailed on the plots.

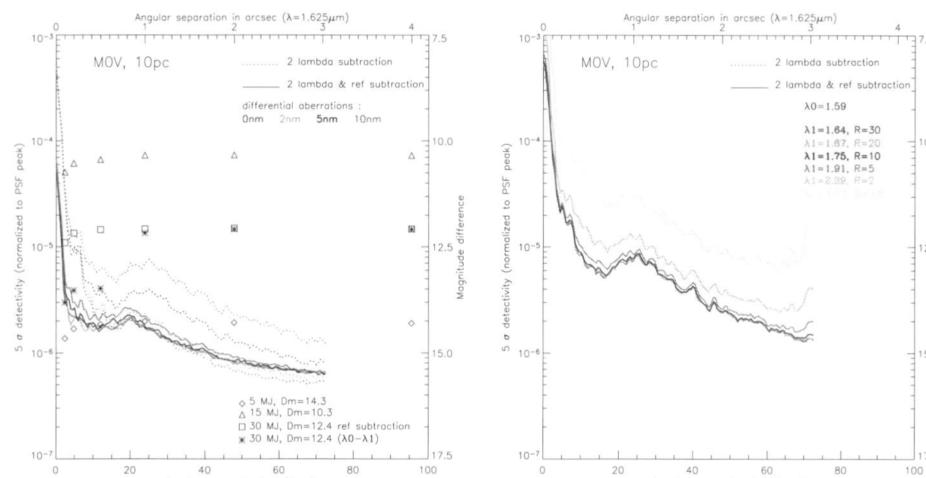

Figure 4. Identical to Fig. 3 for the case of differential aberrations (left) and spectral separation between filters (right).

differential optics are correlated. The amplitudes of aberrations is extremely challenging to allow planet detection. For a 100m telescope common and differential optics must be controlled respectively at 20nm rms and 0.01nm rms to achieve a planet/star contrast of 10^{-10}. Moreover, our assumptions on the AO system performance lead to much longer integration time (a few hundreds of hours for a 100m telescope) than what is previously announced in the literature. It is obvious that simulated performance of ELTs cannot be realistic if static aberrations are neglected. The other important result is that if we manage to lower the static aberrations to the aforesaid level we still need a telescope diameter of 100m to achieve the detection. Two results of simulation are shown in Fig. 5

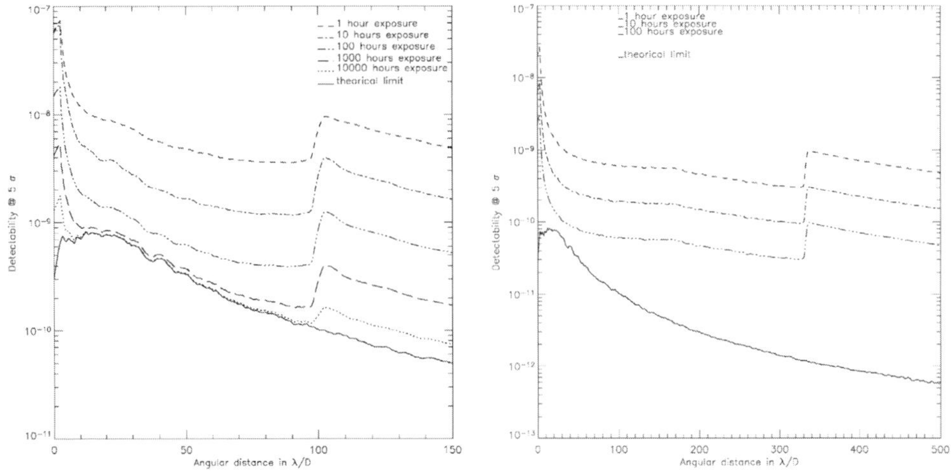

Figure 5. Results of simulation obtained by Cavarroc *et al.* (2005) using a simple analytical model of differential imager in ELTs. The left plot stands for a 30m telescope and differential aberrations of 0.1nm rms and the right plots is for a 100m telescope having reduced the differential aberrations to 0.01nm rms. In both cases the common aberrations are set to 20nm rms.

for respectively a 30m telescope with differential aberrations of 0.1nm rms and a 100m telescope with differential aberrations 0.01nm rms.

5. Conclusion

Differential techniques are becoming available on current AO systems and are providing first results on young star associations. In the near future, extreme AO systems on 8m ground-based telescopes will be equipped with spectral differential imaging units. Extrasolar Giant Planets are expected to be discovered and characterized using such instruments since the phase aberrations requirements for such programs are within the current capabilities of differential techniques and providing the alignments and pointing inside the coronagraph are accurately controlled. For the longer term, ELTs will have the very challenging objective to image terrestrial planets from the ground but the performance of differential techniques will have to be pushed at a limit which is several orders of magnitude above the current state of the art. More developments and modeling will be mandatory in this area.

References

Allard, F., Hauschildt, P.H., Alexander, D.R., *et al.* 2001, ApJ, 556, 357

Beuzit *et al.* 2005, this proceeding

Cavarroc, C., Boccaletti, A., Baudoz, P., Fusco, T., & Rouan, D. 2005, to appear in A&A, astroph/0509713

Lenzen, R., Close, L., Brandner, W., *et al.* 2004, in Proc. SPIE, vol. 5492, 970

Marois, C., Doyon, R., Racine, R., & Nadeau, D. 2000, PASP, 112, 91

Marois, C., Doyon, R., Nadeau, D., Racine, R. & *et al.* 2005, to appear in ApJ, astroph/0504652

Moutou, *et al.* 2005, this proceeding

Sudarsky, D., Burrows, A., & Hubeny, I. 2003, ApJ, 588, 1121

Direct Imaging of Exoplanets: Science & Techniques
Proceedings IAU Colloquium No. 200, 2005
C. Aime and F. Vakili, eds.
© 2006 International Astronomical Union
doi:10.1017/S1743921306009896

Calibration of Residual Speckle Pattern in a Coronagraph

Michael Shao[1], Joseph J. Green[1], Benjamin Lane[2], James Kent Wallace[1], Bruce Martin Levine[1], Rocco Samuele[3], Shanti Rao[1], and Edouard Schmidtlin[1]

[1]Jet Propulsion Laboratory, California Institute of Technology 4800 Oak Grove Drive, Pasadena, CA 91109, USA, email: Michael.Shao@jpl.nasa.gov

[2]Center for Space Research, Department of Physics, Massachusetts Institute of Technology. Cambridge, MA 02139, USA

[3]Northrop Grumman Space Technology, One Space Park, Redondo Beach, CA 90278, USA

Abstract. Ground and space based coronagraphs have been proposed to suppress the light of the star so a planet nearby can be imaged. But even when starlight has been suppressed by 10^{10}, the residual starlight is as bright as the planet, and must be subtracted to $2 * 10^{-11}$ for a 5 sigma detection of the planet. For a ground based AO coronagraph, the problem is even more severe. Typically suppression of starlight to 10^{-5} of the star is possible and the residual speckle pattern must not have any "bumps" that mimic a planet at $10^{-7} - 10^{-8}$. This paper describes a speckle calibration approach that measures the electric field of the light after it exits the coronagraph, in order to estimate the speckle pattern in the image plane. This technique makes use of the coherence of star light or rather the incoherence of starlight to planet light, and has very significant advantages compared to other techniques.

For a space based coronagraph, an alternative approach is to rotate the telescope /coronagraph and subtract two images. The calibration interferometer described here has the advantage that the temporal stability of the system can be relaxed by several orders of magnitude. For a ground based AO coronagraph system this approach has none of the serious limitations of the techniques based on the radial expansion of the speckle pattern with wavelength and enables ground based AO coronagraphs to approach the photon limit rather than the atmospheric limit. The calibration interferometer is being built for a NASA sounding rocket experiment by BU, JPL, MIT, and GSFC (PICTURE) with a 50cm telescope and a nulling coronagraph to be launched in 2007. It is also part of a design study for an extreme AO coronagraph for the Gemini Telescope, and a conceptual study of an extreme AO coronagraph for the TMT.

Keywords. coronagraph, direct detection of planets, nulling interferometer.

1. Introduction

Removal of the residual speckle pattern after a coronagraph is essential for direct detection of planets whose flux is at or below the residual speckle level. If the average flux after a coronagraph is say 10^{-10} and the speckles amplitudes have a Gaussian distribution with a variance of $\approx 10^{-10}$, then in a typical image with a 30*30 airy spot field of view, there will be $\approx$45 speckles brighter than $2 * 10^{-10}$ and $\approx$3 speckles brighter than $3 * 10^{-10}$. A clear detection of a planet at 10^{-10} means the fluctuation in the background has to be significantly smaller than 10^{-10}, typically $\approx 2 * 10^{-11}$.

Many techniques have been proposed to "subtract" the speckle pattern. One is based on assuming the speckle pattern is stable while the telescope is rotated about the line of sight. The main drawback of this approach is that it requires extreme stability. If we

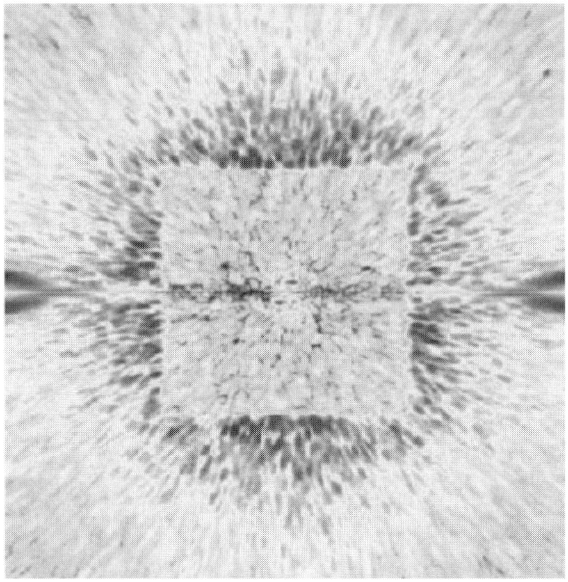

Figure 1. TPF-C simulation of residual speckles in white light.

want a speckle pattern of intensity $\approx 10^{-10}$ to be stable to $\approx 10^{-11}$ for 2–4 hrs, we are asking for the wavefront of the telescope to be stable to a few picometers for 2–4 hrs. For ground base extreme AO coronagraphs this approach doesn't work at all. A 2nd approach that has been proposed is to use the radial smearing of the speckle pattern with wavelength. This approach works if the imperfections in the wavefront are due to phase and amplitude errors in the pupil. But for a coronagraph looking at Earths, deformable errors will remove most of the errors at the telescope pupil and residual errors that form the 10^{-10} speckles will in general not originate at the pupil, and will not smear radially with wavelength.

The TPF-C project did a simulation of their coronagraph, shown in figure 1. The main feature is the dark hole. Outside the darkhole, these speckles are cause by wavefront errors at spatial frequencies that can not be corrected by the deformable mirror. These originate mostly from errors in the telescope optics (at a pupil) and they show the characteristic radial smearing with wavelength. Inside the dark hole, almost all the wavefront errors at a pupil have been removed and those speckles some times do and other times don't show radial smearing with wavelength.

A third technique for removing the residual speckle pattern has been called differential imaging. This technique relies of the target (e.g. planet) having a strong spectral feature, such as the methane band in the near IR. If we don't change wavelengths very much, the speckle features should be identical, and the difference between an "in band" and out of band image can show the presence of the planet. This technique has two drawbacks, one is it can't be used for Earthlike planets (which may not have a strong methane feature) and the images are limited to rather narrow bandwidths limiting SNR.

The speckle subtraction approach described here has none of the drawbacks of the above approaches. It doesn't require extreme wavefront stability (picometers) over hours only over fractions of a second. It can be used over a moderate bandwidth (10%–20%) and not just a <1% bandpass. It can be used for space based or ground based coronagraphs. The concept is based on the coherence of the speckle pattern with starlight. The planet light is incoherent with the starlight.

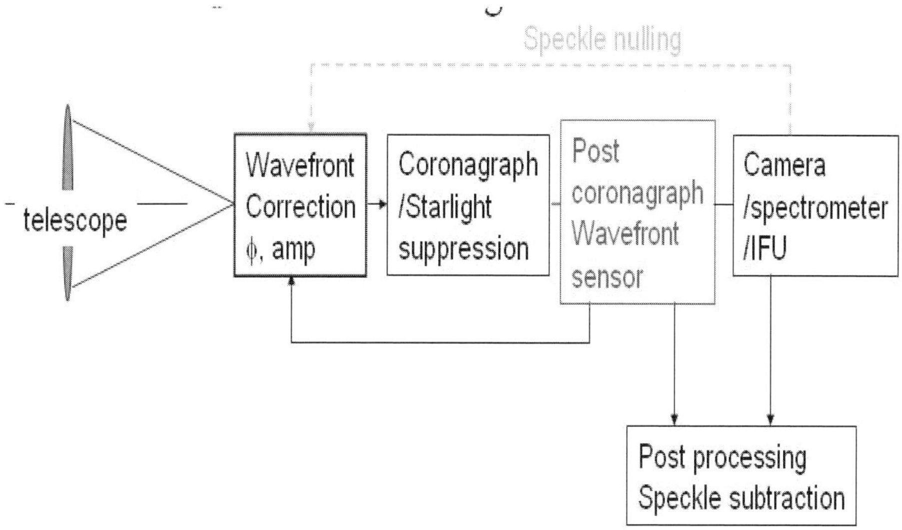

Figure 2. A block diagram of coronagraphic system.

2. Post Coronagraph Wavefront Sensor/Speckle calibration system

A block diagram of the coronagraphic system is show in figure 2. A post coronagraph wavefront sensor (PCWS) is placed between the coronagraph and the science camera/spectrometer. This wavefront sensor measures the electric field of the light that has not been suppressed by the coronagraph, typically because the amplitude and phase errors were not perfectly corrected. The PCWS is used in real time to control the deformable mirror(s) to correct phase and amplitude. The more important use is in post processing. The E-field of the light after the coronagraph in the pupil can be Fourier transformed and squared to yield the speckle pattern of the scattered light in the science camera.

The PCWS can be implemented as a machzender interferometer shown in figure 3 below. If the coronagraph is a "nulling interferometer", the "bright" output port of the nuller is spatially filtered (with a pinhole) and interfered with ≈50% of the light from the "dark" output port with a phase shifting type machzender interferometer. If the coronagraph suppresses all but $\approx 10^{-7}$ of the light, and that light is spread over ≈1000 airy spots/pixels in the science focal plane, the average scattered starlight is $\approx 10^{-10}$ of the original star in each airyspot/pixel. In the pupil plane however, the average flux is $0.5 * 10^{-7}$ in the "dark" arm of the machzender interferometer and ≈1.0 in the bright arm.

3. Post Coronagraph Wavefront sensor SNR

This is not the first time a machzender interferometer has been suggested for use behind a coronagraph. By using the full starlight flux in the "reference" arm, there is a significant SNR advantage over alternative post coronagraph wavefront sensing schemes. We are trying to in effect measure the flux of speckles that are 10^{-10} the brightness of the star. The measurements of these speckles have added noise, from CCD dark current, CCD read noise, and for a space coronagraph, local and exo-zodi contamination. But none of these additional noise sources are "coherent" with the starlight in the reference

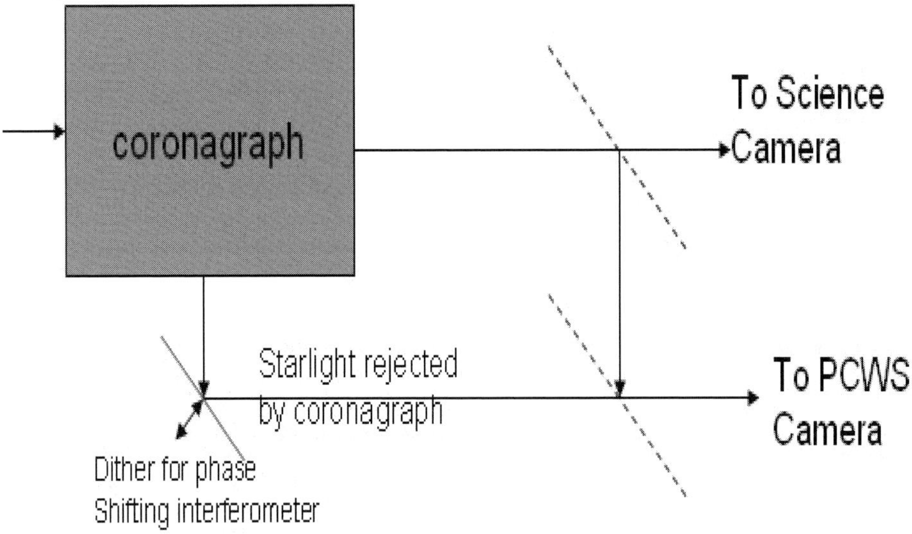

Figure 3. A post coronagraph wavefront sensor (PCWS).

arm of the interferometer. Taking the intensity of the two interfering beams as roughly 1.000 and $0.5 * 10^{-7}$, the fringe amplitude is $\approx 1.0 * fringeVis$ or $4.47 * 10^{-4}$. If the reference beam were attenuated to $5 * 10^{-8}$ to match the "dark" arm, the fringe amplitude would be 10^{-7}. At 10^{-7} CCD dark current, local/exo-zodi would dominate the SNR. The "unbalanced" machzender in a TPF-C mission would reduce the integration time needed for wavefront sensing by about a factor of 100 (from ≈2 hrs to ≈1 minute to measure). The measurement of the post coronagraph wavefront is now limited by photon flux at the dark output of the coronagraph, rather than other larger noise sources such as local/exo-zodi or detector noise.

4. Summary

The PCWS also provides a method for measuring the speckle pattern in a ground based extreme AO system, in a way that is limited only by photon statistics instead of speckle statistics. This makes ground based detection of "normal" Jupiters potentially possible, if they are 1–1.5 AU from the parent stars with contrast of $\approx 10^{-8}$ or better.

In space this approach will be demonstrated in the PICTURE project, a sounding rocket experiment designed to look for a possible planet around E. Eri and scheduled for launch in early 2007.

This concept is also used in a design study for the TPF-C mission, one of 5 funded studies of instrument concepts for the TPF coronagraph mission.

Direct Imaging of Exoplanets: Science & Techniques
Proceedings IAU Colloquium No. 200, 2005
C. Aime and F. Vakili, eds.
© 2006 International Astronomical Union
doi:10.1017/S1743921306009902

Spectroscopy with the Terrestrial Planet Finder Coronagraph

Sara R. Heap[1] **and the CorSpec Team**†

[1]Exploration of the Universe Division, Goddard Space Flight Center,
Greenbelt MD, USA
email: Sally.Heap@NASA.gov

Abstract. We describe the rationale and requirements for an integral-field spectrograph aboard NASA's Terrestrial Planet Finder Coronagraph.

Keywords. space vehicles: instruments, (stars:) planetary systems, (stars:) planetary systems: formation, (stars:) planetary systems: protoplanetary disks, instrumentation: spectrographs, instrumentation: detectors, techniques: image processing, planets and satellites: general, astrobiology.

1. Introduction

Ten years from now, we should be familiar with planetary systems around nearby stars. Radial-velocity monitors will have identified those systems having giant planets in Jupiter-like orbits. Corot and Kepler will have discovered rocky planets transiting their parent stars, and NASA's Space Interferometry Mission will have discovered systems having massive terrestrial planets orbiting nearby stars. The prime targets for follow-up characterization are planets orbiting bright, nearby stars found by Doppler techniques and SIM interferometry. These techniques, however, are indirect: their evidence for planets comes from measurements of the orbital motions of the host stars. NASA is therefore planning a mission called the Terrestrial Planet Finder Coronagraph (TPF-C) that will enable us to detect and analyze light from an earth-like planet itself. Its light can tell us about the physical conditions on the planet, whether the planet is habitable, and perhaps even if the planet has life. TPF-C's prime instrument for analyzing light is a spectrograph that will be used to simultaneously characterize all planets, gas and dust within the coronagraphic field.

2. Instrument Concept

NASA recently selected us to study a coronagraphic spectrograph (CorSpec for short) for TPF-C. We envision a spectrograph with four optical channels operating simultaneously. Together, the four channels cover the spectral range, 0.45–1.10 microns. Each channel is fed by a starlight suppression system optimized for the wavelength region of that channel. Each channel has its own integral field unit (IFU) sampling a 140 × 140-spaxel field, a set of dispersers (prisms and grating), and a photon-counting detector system optimized for that passband, which is about 22% wide.

† P. Bordé, C. Bowers, B. Dean, D. Ebbets, D. Fabrycky, J. Kasdin, S. Kilston, M. Kuchner, D. Lindler, V. Meadows, C. Noecker, B. Rauscher, S. Seager, W. Sparks, D. Spergel, A. M. Smith, W. Traub, J. Trauger, B. Woodgate

Each of the four starlight suppression systems should produce a high-quality coronagraphic image in the central wavelength region of the passband, but the flanking wavelength regions may be corrupted by bright speckles produced by phase and amplitude wavefront errors. We therefore envision a wavefront-sensing system that uses the short- and long-wavelength flanks of all four optical channels to detect changes in the wavefront, to diagnose the wavefront, and to send back instructions for correcting the wavefront in each of the four starlight suppression systems.

The wavelength of transition between adjacent channels can be adjusted by selection of the appropriate dichroic beamsplitter that intercepts light from the telescope before entering the starlight suppression systems. We expect that two exposures, each employing its own set of dichroic beamsplitters, will be sufficient to cover the full spectral range of CorSpec without gaps.

3. Scientific Objectives

One of the primary objectives of TPF-C is to characterize ... terrestrial planets spectroscopically, searching for absorption caused by O_2, O_3, H_2O, and possibly CO_2 and CH_4. The ability to measure Rayleigh scattering and photosynthetic pigments is also desired. Such information may provide evidence of habitability and of life itself. By definition, a potentially habitable planet is a terrestrial planet lying within the Habitable Zone, the annulus around a star where H_2O would be in its liquid state (Des Marais *et al.* 2001). The inner and outer radii of the HZ are approximately $0.7\sqrt{L_{\rm bol}}$ and $1.5\sqrt{L_{\rm bol}}$ AU.

The science objectives combined with the intrinsic and observed properties of the target stars dictate the performance requirements of CorSpec. We therefore compiled a plausible TPF-C target list composed of sun-like stars having no companion within 8". Most are metal-rich, which makes them among the most likely to have planets (Valenti & Fischer 2005; Santos *et al.* 2005; Fischer & Valenti 2005), although there is a healthy diversity in metallicity. For example, we should get a good clue to the influence on metallicity on the formation of planets by comparing the disk around τ Ceti ([Fe/H] $= -0.52$) with that of ϵ Eri ([Fe/H] $= -0.03$). Although the estimated ages have very large uncertainties, most appear to be older than a billion years, so the chances of life having started are higher (if the Earth is a good example), and the background presented by their exo-zodiacal disks should be fainter (Habing *et al.* 2001).

Figure 1-Left plots the separation of Earth-like planets at the inner edge of the Habitable Zone (IHZ) at quadrature (circular orbits assumed) vs. the visual magnitudes. It shows that the HZ of all the targets lie outside TPF-C's Inner Working Angle for Detection, $\mathrm{IWA_D}$=62 mas, at 6000 Å (The IWA scales with wavelength.) Except for α Cen, their angular separations from the star are less than 360 mas at the outer edge of the habitability zone (OHZ). All have planet-star brightness ratios in the range from 0.8×10^{-10} (F stars) to 17×10^{-10} (K stars) at the IHZ. At the OHZ, however, some planet-star brightness ratios are as low as 0.2×10^{-10}, well below TPF-C's nominal capabilities of 1×10^{-10}. The faintest planets have V=30.3 at the IHZ and V=31.9 at the OHZ.

Not only are Earth-like planets faint sources, but they are seen close to stars that can be 10 billion (10^{10}) or more times brighter. Figure 1-Right illustrates the stellar-glare challenge for a prime TPF-C target, 47 UMa, which is known to have two outer giant planets in near-circular orbits (Fischer *et al.* 2002). If a potentially habitable planet has only half the surface area of the Earth ($M = 0.35 M_{\oplus}$) and lies in the outer edge of the Habitable Zone, CorSpec may have to characterize planets with planet-star brightness

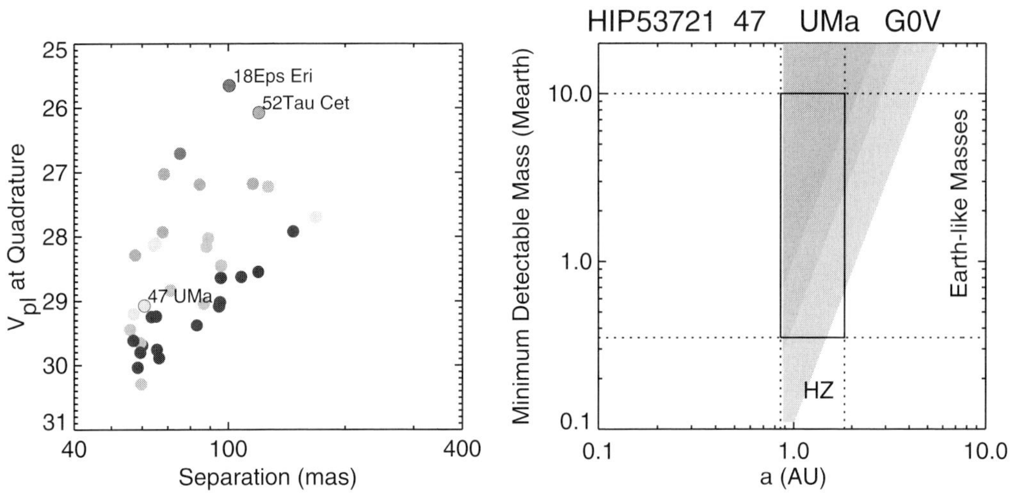

Figure 1. Properties of target stars. Left: Visual magnitude and angular separation of Earth-like planets at the IHZ as seen at quadrature. Right: Coverage of the HZ for three different contrasts, $L_p/L_s = 2, 1$, and 0.4×10^{-10} (dark to light shades).

ratios less than $L_p/L_s = 0.4 \times 10^{-10}$. This brightness ratio is lower than the nominal capabilities of TPF-C, 1×10^{-10}.

4. Performance Requirements

Sensitivity. The faintness of terrestrial planets puts severe demands on the CorSpec detectors. The main problem is read noise from the CCD detector ($3e^-$), which – because it must be read out every 1000 seconds to limit the build-up of cosmic-ray hits – is usually much stronger than the planet flux. Only with a photon-counting detector can we obtain spectra of planets orbiting all of TPF-C's core stars.

Efficiency. Even with a photon-counting detector, a typical exposure take days to obtain with the standard starlight suppression system; and as noted above, two exposures are required to cover the full spectral range without gaps. We are therefore working to design special masks for spectroscopy that would further shorten the time. The standard system employs masks suited for search and discovery but only probes to 62 mas at 6000 Å and is proportionately larger at longer wavelengths. The spectroscopic masks would cover only a small portion of the coronagraphic field but would probe in to half that separation, so that we could observe a planet at gibbous phase (rather than at maximum elongation) when it would be a full magnitude brighter. A 2.5 X increase in brightness translates to a 5 X reduction in exposure time needed to reach the same signal to noise ratio. Counteracting this gain is the loss in overall throughput by the spectroscopic masks, so the actual gain will be less. However, independent of the throughput question, is the gain in terms of contrast: with the spectroscopic masks, the brightness of a planet is increased by a factor of 2.5 X with respect to the speckle background.

Spatial Multiplexing. Since the exposure time for an observation takes days or even weeks, and the wavefront-corrected spectral band may be only 0.1μ wide, it is imperative to maximize CorSpec's efficiency. An integral field spectrograph (IFS) can provide that high degree of multiplexing efficiency by simultaneously obtaining a spectrum of each image element in the coronagraphic field – and do so without having had to acquire the target in some slit or orient the telescope at a specific roll angle.

CorSpec will need to Nyquist-sample $\geqslant$90% of the entire Habitable Zone around "core stars" at the TPF-C angular resolution, to disperse the light of each sample for a resolving power of 70, and to record the 2×10^4 spectra with just one detector per channel. The table below gives the implied optical parameters of the IFU, evaluated at 0.5 and 0.8 microns.

Optical Parameters of the CorSpec IFU

Parameter	0.5μ	0.8μ
Inner Working Angle (mas)	52	83
Outer Working Angle (mas)	620	990
Angular resolution (mas)	18.4	29.5
Angular sample size (mas)	9.2	14.7
Samples across the FOV	134	134

Wavefront Sensing. The TPF-C Starlight Suppression System (SSS) is currently thought to be operated in a "set and forget" mode. It is therefore limited by changes in the telescope wavefront during an observation. However, we cannot expect the Starlight Suppression System to remain stable over the days needed to obtain a spectrum of a terrestrial planet. We therefore plan to use CorSpec to monitor the wavefront while an observation is being made, diagnose changes in wavefront, and send instructions for correction back to the Starlight Suppression System.

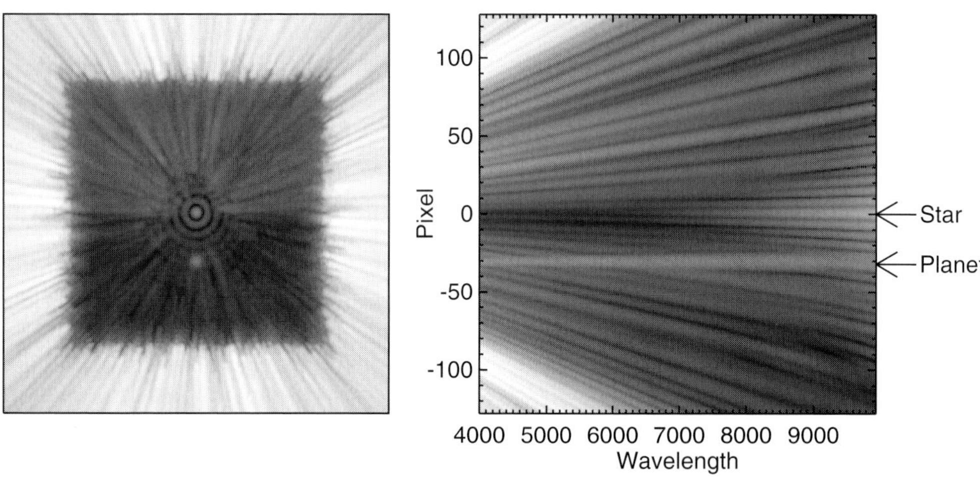

Figure 2. Simulated cross-cuts of a TPF-C datacube. Left: A wide-band image reconstructed from a swath of monochromatic images in the datacube. Right: A cross-cut at x_p showing the divergence of the background "speckles" with increasing wavelength. Note the edge of the dark hole at short wavelengths, and stellar leakage around the mask at long wavelengths.

5. CorSpec Data

The datacube (x, y, λ) produced by an IFS observation is rich in information. Not only can the datacube be manipulated to yield the background-subtracted spectrum of a planet that can be examined for biomarkers (Des Marais *et al.* 2001), but it can be reformatted into a wide-band image for planet detection (Fig. 2-Left). It can also be reformatted into a set of narrow-band images to show the distribution and chemistry of gas in disks during the late stages of planet formation (Brandeker *et al.* 2004). Quantities of gas as small as one Earth mass may be crucial to forming terrestrial planets by causing them to migrate, damping their eccentricities, altering their formation mechanisms, and supplying them with proto-atmospheres and essential volatiles.

A crosscut of the datacube at a specific x or y produces a pseudo-long-slit spectrum showing how artifacts of wavefront errors ("speckles") change in position and brightness with wavelength (Figure 2-Right). Such information will be invaluable for modeling and removing the speckle background. It will also be invaluable to the wavefront-correction system on-board TPF-C.

In summary, an IFU is the ideal device for coronagraphic spectroscopy, and a coronagraphic spectrograph in space is an ideal application for an IFU. The main design challenge is how to pack about 20,000 spectra onto one $2K \times 4K$ photon-counting CCD without serious cross-talk between adjacent spectra.

References

Brandeker, A., Liseau, R., Olofsson, G., & Fridlund, M. 2004, *A&A* 413, 681
Des Marais, D. *et al.* 2001, *JPL Publication* 01-008
Fischer, D. *et al.* 2002, *ApJ* 564, 1028
Fischer, D. & Valenti, J. 2005, *ApJ*, 622, 1102
Habing, H. *et al.* 2001, *A&A* 365, 545
Marcy, G. & Butler, P. 1998, *ARAA* 36, 56
Santos, N. *et al.* 2005, *A&A* 415, 1153
Valenti, J. & Fischer, D. 2005, *ApJS*, 159, 141

All photographs: Laurent Thareau [`l.thareau@free.fr`].

Direct Imaging of Exoplanets: Science & Techniques
Proceedings IAU Colloquium No. 200, 2005
C. Aime and F. Vakili, eds.
© 2006 International Astronomical Union
doi:10.1017/S1743921306009914

Speckle nulling with space-based coronagraphs

Pascal J. Bordé[1]†, and Wesley A. Traub[2]

[1]Harvard-Smithsonian Center for Astrophysics, 60 Garden Street, Cambridge, MA 02138, USA
email: pborde@cfa.harvard.edu

[2]Jet Propulsion Laboratory, M/S 301-451, 4800 Oak Grove Drive, Pasadena, CA 91109, USA
email: wtraub@jpl.nasa.gov

Abstract. Following the tracks of Malbet, Yu, & Shao (1995) on dark hole algorithms, we present analytical methods to measure and correct the speckle noise behind an ideal coronagraph. We show that, in a low aberration regime, wavefront sensing can be accomplished with only three images, the next image being fully corrected (no iterative process needed). The only hardware required is the coronagraph deformable mirror and an imaging detector in the focal plane, thus there are no non-common path errors to correct. Our first method, *speckle field nulling*, is a fast FFT-based algorithm requiring the deformable mirror influence functions to have identical shapes. Our second method, *speckle energy minimization* is more general and based on matrix inversion. Numerical simulations show that these methods can improve the contrast by several orders of magnitude.

Keywords. Instrumentation: adaptive optics – techniques: high angular resolution – planetary systems.

1. Introduction

By *speckle nulling*, we refer to any technique whose goal is the reduction of speckle noise down to a level compatible with exoplanet detection. In the focal plane, speckle noise appears as a background made of a collection of star-like spots, the speckles. Speckles doubly hamper planet detection: they worsen the photon noise by increasing the background, and they introduce confusion because speckles and planets look alike.

Speckle noise arises from the scattering of star light by optical defects acting either as phase gratings (e.g. polishing errors), or as amplitude gratings (e.g. inhomogeneous reflectivity). Speckle nulling is achieved by creating destructive interferences at the locations of speckles thanks to a deformable mirror (DM for short; see Trauger, Burrows, Gordon, *et al.* 2004). In terms of wave propagation, optical defects cause wavefront aberrations that the DM compensate for.

Speckle nulling belongs to the broader topic of wavefront sensing and control. As explained by Malbet, Yu, & Shao (1995) in their pioneering work, it differs from conventional adaptive optics (AO) in the sense that its goal is not to maximize the Strehl ratio (i.e. minimize the standard deviation of phase errors), but to minimize the light scattered in a given zone of the image plane, dubbed *dark hole*, where planet detection becomes possible. Moreover, conventional AO measures the wavefront with a separate sensing channel, most of the time at a wavelength shorter than the wavelength used for science operation. For high-contrast imaging, it is highly desirable to measure the wavefront directly with the science image in order to avoid non-common path errors.

† Michelson Postdoctoral Fellow; present address: Michelson Science Center, California Institute of Technology, 770 S Wilson Avenue, MS 100-22, Pasadena, CA 91125.

In their paper, Malbet, Yu, & Shao (1995) discuss only wavefront control assuming prior knowledge of the wavefront thanks to some phase retrieval techniques. We provide here a full approach to wavefront sensing and control. However, we narrow the discussion to the case where aberrations are small, i.e. the problem remains linear. Non-linearity could be dealt with by iterating (see Give'on, Kasdin, & Vanderbei 2005) or other methods could be used in the first place to get to the wavefront quality needed here.

This paper is organized as follows: in §2 we expose the wavefront sensing and control theory, in §3 we present one- and two-dimensional simulations with white and colored speckle noise, and we conclude in §4.

2. Speckle nulling theory

We present here a simplified theory in which diffraction by the pupil is not included. For a more complete exposition of the theory, see Bordé & Traub (2005).

2.1. *Wavefront sensing*

Let $E(u,v) = e^{i\phi(u,v)} \approx 1+i\phi(u,v)$ be the normalized electric field due to the star alone in a pupil plane before the coronagraph. Here, ϕ is a complex function accounting for small departures from a plane wave, i.e. phase and amplitude aberrations. In order to assess the performance of speckle nulling alone, we assume an ideal coronagraph which perfectly removes the star: in a reimaged pupil after the coronagraph, the field is $E'(u,v) = i\phi(u,v)$. On the detector, the image-plane field is then $\widehat{E'}(\alpha,\beta) = i\widehat{\phi}(\alpha,\beta)$, where hats denote Fourier transforms. Because the detector measures the field intensity, $I(\alpha,\beta) = |\widehat{\phi}(\alpha,\beta)|^2$, complex aberrations cannot be obtained from a single image. Fortunately, in an AO system, the DM provides a natural means to acquire the missing information by modifying the wavefront phase, which leads to a field intensity variation (this is a form of *phase diversity*).

It is easy to show that complex aberrations can be measured from a set of three images, numbered here 0, 1, and 2. Let ψ be the phase change introduced by the DM. Although this is not mandatory, we assume for the sake of simplicity that the DM is originally flat ($\psi_0 = 0$). The intensity takes the successive values

$$\left\{\begin{array}{l} I_0 = |\widehat{\phi}|^2 \\ I_1 = |\widehat{\phi} + \widehat{\psi}_1|^2 \\ I_2 = |\widehat{\phi} + \widehat{\psi}_2|^2 \end{array}\right. \iff \left\{\begin{array}{l} (\widehat{\psi}_1)^* (\widehat{\phi}) + (\widehat{\psi}_1)(\widehat{\phi})^* = I_1 - I_0 - |\widehat{\psi}_1|^2 \\ (\widehat{\psi}_2)^* (\widehat{\phi}) + (\widehat{\psi}_2)(\widehat{\phi})^* = I_2 - I_0 - |\widehat{\psi}_2|^2. \end{array}\right. \tag{2.1}$$

Provided $\Delta \equiv (\widehat{\psi}_1)^* \widehat{\psi}_2 - \widehat{\psi}_1 (\widehat{\psi}_2)^* \neq 0$, we derive for the complex aberrations

$$\widehat{\phi} = \frac{\widehat{\psi}_2 \left(I_1 - I_0 - |\widehat{\psi}_1|^2\right) - \widehat{\psi}_1 \left(I_2 - I_0 - |\widehat{\psi}_2|^2\right)}{\Delta}. \tag{2.2}$$

The condition $\Delta \neq 0$ reflects that (2.1) should not contain twice the same equation, i.e. the DM should be driven so that, between images 1 and 2, the intensity is modified in a measurable way. In practice, ψ_1 is computed to modify the intensity in every point by an amount comparable to the intensity in that point. Then, ψ_2 is computed to maximize Δ in every point, so as to maximize the measurement signal-to-noise ratio.

2.2. *Wavefront control*

With the knowledge of the complex aberrations, speckles would instantly go away if we could make ψ equal to $-\phi$. However, (*a*) the DM cannot take an arbitrary shape as it is made of a facesheet supported by a finite number of actuators (see Trauger, Moody, Gordon, *et al.* 2003 for a detailed description), and

(*b*) the DM introduces a real, not complex, phase change.

Point (*a*) implies that aberrations can only be corrected up to a maximum spatial frequency set by the density of actuators (with respect to the wavelength), which means that the dark hole cannot be arbitrarily large. Although point (*b*) seems to imply that only phase aberrations could be corrected, it really means – as we will show later on – that phase and amplitude aberrations can both be corrected, but at the cost of half the dark hole area.

We model the DM shape as the linear superposition of two-dimensional influence functions. Each of these functions describes the DM surface deformation in response to the actuation of a given actuator. The phase change introduced by a $N \times N$-element DM is then

$$\psi(u, v) = \sum_{k=0}^{N} \sum_{l=0}^{N} a_{kl}\, f_{kl}(u, v), \tag{2.3}$$

where the a_{kl} (the actuator strokes) are the unknowns in the wavefront control problem.

2.2.1. *Speckle field nulling*

As a first approach to the a_{kl} calculation, let us examine under what conditions $\widehat{E'}$ could be made zero in the simplified case where all influence functions have an identical shape, i.e. $f_{kl}(u, v) = f(u - k\frac{d}{\lambda}, v - l\frac{d}{\lambda})$, where d is the actuator spacing. We have

$$\widehat{\phi}(\alpha, \beta) + \widehat{\psi}(\alpha, \beta) = 0 \iff \sum_{k=0}^{N-1} \sum_{l=0}^{N-1} a_{kl}\, e^{-i\frac{2\pi d}{\lambda}(k\alpha + l\beta)} = -\frac{\widehat{\phi}(\alpha, \beta)}{\hat{f}(\alpha, \beta)}. \tag{2.4}$$

We recognize in the left-hand size of (2.4) a truncated Fourier series. By setting the strokes equal to the Fourier coefficients of $-\widehat{\phi}/\hat{f}$, we minimize the total energy of $\widehat{E'}/\hat{f}$ in the dark hole. This is close, but not quite the same thing as minimizing the total speckle energy in the dark hole, which would be the total energy of $\widehat{E'}$ (more on that in §2.2.2). With the Fourier solution to (2.4), we learn that:

(*a*) because the Fourier series is truncated (as actuators are in limited number), there can be no exact solution to (2.4) and no infinitely deep dark hole, unless the truncated part of the series happen to be equal to zero, i.e. aberrations are band-limited;

(*b*) because Fourier series are periodic, the dark hole will have an extension limited to that period. Precisely, it will be a centered square with a maximum size of $\pm\frac{\lambda}{2d} = \pm\frac{N}{2}\frac{\lambda}{D}$, where D is the pupil's diameter. In other words, a dark hole created with a $N \times N$-DM will be at most $N \times N$ element of resolution wide, which is consistent with the sampling theorem.

The solution by Fourier series is attractive because it can be coded with FFTs and is computationally very fast. However, it has two drawbacks: (1) it cannot handle influence function variations across the DM, and (2) it does not exactly minimize the speckle field energy in the dark hole which would seem desirable.

2.2.2. *Speckle energy minimization*

As a second approach, let us now compute the a_{kl} as the result of the minimization of the total speckle energy in the dark hole, defined as

$$\mathcal{E} \equiv <\widehat{\phi} + \widehat{\psi}, \widehat{\phi} + \widehat{\psi}> \equiv \iint_{\mathcal{H}} (\widehat{\phi} + \widehat{\psi})(\widehat{\phi} + \widehat{\psi})^{*}. \tag{2.5}$$

Given that $\partial\widehat{\psi}/\partial a_{kl} = \hat{f}_{kl}$, the total energy is minimized when, $\forall(k,l)$,

$$\frac{\partial \mathcal{E}}{\partial a_{kl}} = 0 \quad \Longleftrightarrow \quad \sum_{n=0}^{N-1}\sum_{m=0}^{N-1} a_{nm}\,\Re\left(< \hat{f}_{nm}, \hat{f}_{kl} >\right) = -\Re\left(< \widehat{\phi}, \hat{f}_{kl} >\right), \tag{2.6}$$

where $\Re$ stands for the real part.

We recognize in (2.6) a linear system that can be solved in matrix format:

$$FA = \Phi \Leftrightarrow A = F^{-1}\Phi, \tag{2.7}$$

where F is a $N^2 \times N^2$-matrix, and A and Φ are N^2-vectors.

It happens that if influence functions have the property of being separable, that is to say if $f_{kl}(u,v) = g_k(u)g_l(v)$, then (2.7) can be rewritten as

$$G\,A\,G = \Phi \Leftrightarrow A = G^{-1}\,\Phi\,G^{-1}, \tag{2.8}$$

where G, A and Φ are now $N \times N$-matrices. This has the advantage of requiring much less computing power. Box-shape functions and bidimensional Gaussians are two examples of fully separable functions. If instrumental influence functions are not fully separable but the wavefront is corrected as if they were, a degradation of the dark hole performance is to be expected (see §3.2 for an example).

When phase and amplitude aberrations are simultaneously present, ϕ is complex, and the careless use of any of the two algorithms would lead to complex values for the a_{kl}. The key to correct amplitude aberrations with real actuator strokes is to treat them as phase aberrations. This can be done by deriving from $\widehat{\phi}$ a function with Hermitian symmetry and therefore a real Fourier transform. However, that Hermitian function can be made equal to $\widehat{\phi}$ only in half the dark hole, therefore half the dark hole area is sacrificed.

3. Simulations

3.1. *White speckle noise*

To start with, we simulate our speckle nulling algorithms in one-dimension with white speckle noise (Fig. 1). The standard deviation of phase aberrations is set to $\lambda/1000$ so that the linear approximation of §2.1 would remain valid. Refer to Bordé & Traub (2005) for simulations with amplitude aberrations as well. The DM features 64 actuators and top-hat influence functions. As shown in the bottom panel of Fig. 1, energy minimization yields a dark hole slightly deeper than field nulling: 1.4×10^{-11} vs. 5.8×10^{-11} (intensities are scaled with respect to the peak of the star Airy function). This is always the case by typically a factor of two. It is possible to obtain deeper dark holes with energy minimization but at the cost of smaller angular extensions. For instance, if one accepts to reduce the dark hole size from 64 to 44 resolution elements, the dark hole floor drops down to 2.7×10^{-15}.

3.2. *Colored speckle noise*

Figure 2 is the result of a two-dimensional simulation including realistic phase aberrations from the surface map of an 8.2-m mirror. In this map, the power spectral density of surface errors decreases as the third power of the spatial frequency above 0.35 m^{-1}. The DM features 64×64 actuators, and influence functions are as measured on the High-Contrast Imaging Testbed (HCIT; Trauger, Burrows, Gordon, *et al.* 2004) at NASA's Jet Propulsion Laboratory. These simulations does not include quantum or read-out noise and assume a perfect DM. With field nulling (as shown in Fig. 2), the average dark hole floor is 5.9×10^{-12}. Energy minimization assuming separable influence functions yields

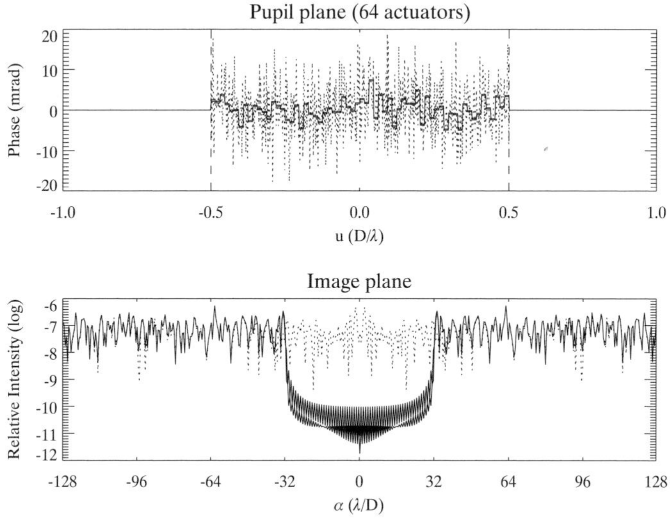

Figure 1. One-dimensional speckle nulling simulation. Top: Phase-aberrated wavefront (dotted line) and negative of DM shape (thick solid line). Bottom: Uncorrected coronagraphic image (dotted line), and corrected coronagraphic image with speckle field nulling (thin solid line) and with speckle energy minimization (thick solid line).

a higher average floor of 7.1×10^{-11}, because of a 5 % mismatch between the real HCIT functions and their separable approximations.

4. Conclusion

In this paper, we presented a full theory for wavefront sensing and control using science camera images (no separate wavefront sensing channel), assuming that phase and amplitude aberrations remain small. Sensing is performed with only three images, and relies on the deformable mirror to introduce the needed phase diversity.

Since our theory is general, it can adapted to most coronagraph designs. A good example is the application to shaped pupil coronagraphs by Give'on, Kasdin, & Vanderbei (2005). We are currently working on the adaptation to coronagraphs with image-plane band-limited masks (Kuchner & Traub 2002) to start experimenting with the HCIT. In parallel, we plan to develop a polychromatic theory to accommodate the spectral bandwidth necessary for realistic planet detection and spectroscopy.

Acknowledgements

We acknowledge many helpful discussions with Chris Burrows, John Trauger, Joe Green, Stuart Shaklan, Amir Give'on, Anthony Boccaletti, and Pierre Baudoz. This work was performed in part under contract 1256791 with the Jet Propulsion Laboratory (JPL), funded by NASA through the Michelson Fellowship Program, and in part under contract 1260535 from JPL. JPL is managed for NASA by the California Institute of Technology. This research has made use of NASA's Astrophysics Data System.

References

Bordé, P. & Traub, W. 2005, *ApJ*, accepted (astro-ph/0510597)
Give'on, A., Kasdin, N. J., & Vanderbei, R. 2005, these proceedings

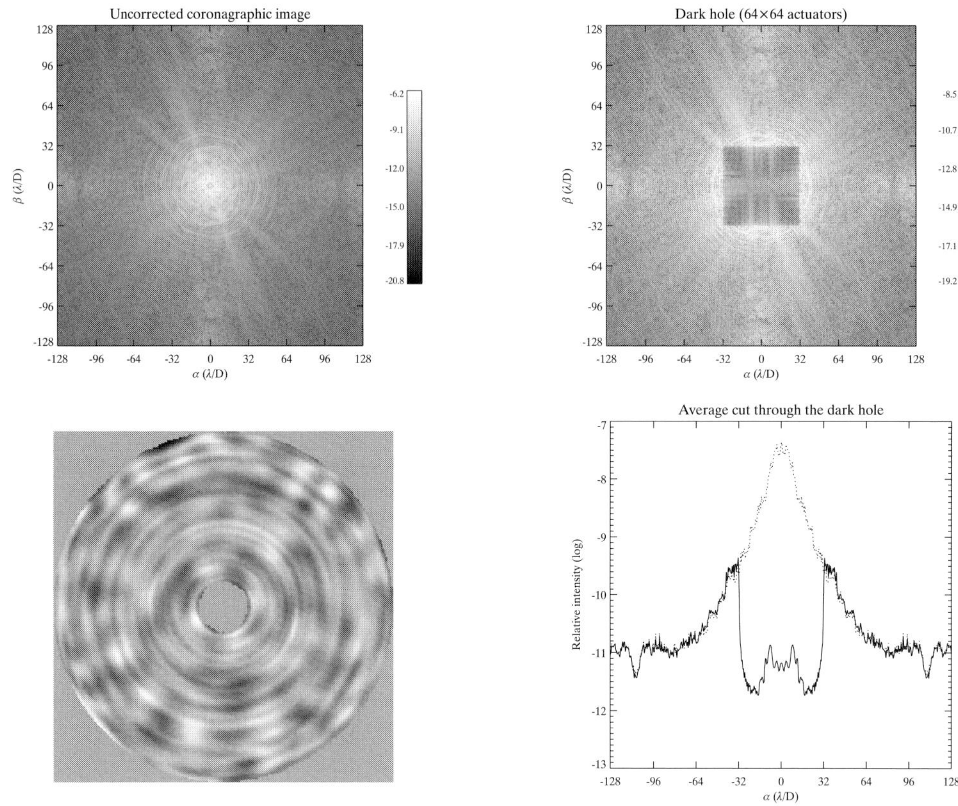

Figure 2. Two-dimensional speckle nulling simulation. Bottom left: phase map of the first 8.2-m primary mirrors of ESO's Very Large Telescope. Top left: uncorrected coronagraphic image obtained with this phase map scaled to a standard deviation of $\lambda/1000$. Top right: corrected image obtained with the field nulling algorithm (§2.2.1). Bottom right: average cut through the dark hole.

Kuchner, M. & Traub, W. 2002, *ApJ*, 570, 900
Malbet, F., Yu, J., & Shao, M. 1995, *PASP*, 107, 386
Trauger, J., Moody, D., Gordon, B., Gürsel, Y., Ealey, M., & Bagwell, R. 2003, *SPIE*, 4854, 1
Trauger, J., Burrows, C., Gordon, B., Green, J., Lowman, A., Moody, D., Niessner, A., Shi, F., & Wilson, D. 2004, *SPIE*, 5487, 1330

Discussion

TRAUGER: Your model assumes that coronagraph optics downstream from the DM do not introduce additional phase and amplitude errors. However, these errors will necessarily exist and affect the image. Could you comment on that?

BORDÉ: We have indeed assumed an optically perfect coronagraph. I guess that the solution to this problem will be to iterate the algorithm, so that unknown errors will be taken care of in the process. In addition, I should say that the effect of the coronagraph is not only to remove the star PSF as in our simplistic model, but is also to modify the DM influence functions in a way that depends on their positions in the pupil. This can easily be (and will be) incorporated in the energy minimization algorithm, but cannot be dealt with in the field nulling algorithm.

Direct Imaging of Exoplanets: Science & Techniques
Proceedings IAU Colloquium No. 200, 2005
C. Aime and F. Vakili, eds.
© 2006 International Astronomical Union
doi:10.1017/S1743921306009926

Closed-loop Wavefront Correction for High Contrast Imaging: The "peak-a-boo" Algorithm

Amir Give'on[1]†, N. Jeremy Kasdin[1], and Robert J. Vanderbei[2]

[1]Department of Mechanical and Aerospace Engineering,
Princeton University, Princeton, NJ 08544

[2]Department of Operational Research and Financial Engineering,
Princeton University, Princeton, NJ 08544

Abstract. High contrast imaging from space must overcome photon noise of the diffracted star light and scattered light from optical components defects. The very high contrast required (up to 10^{-10} for terrestrial planets) puts severe requirements on the wavefront control system, as the achievable contrast is limited by the quality of the wavefront. In this paper we present a closed loop correction method for the shaped pupil coronagraph to minimize the energy in a pre-defined region in the image where terrestrial planets would be found. The reconstruction part uses three intensity measurements in the image plane with a pinhole added to the shaped pupil for diversity. This method has been shown in simulations to be effective in the presence of photon noise, deformable mirror modeling errors and overcoming defective actuators in the deformable mirror.

Keywords. instrumentation: adaptive optics.

1. Introduction

Recently, the problem of high-contrast imaging (typically using a coronagraph) has become of increasing interest. Here, we refer to the problem of imaging a very dim point source very near a much brighter point or extended source (star). There are numerous missions planned for both ground and space to search for planets and other objects close to nearby stars [Beichman (1999)]. Our group at Princeton University has been pioneering the Shaped Pupil Coronagraph (SPC) [Kasdin (2003)]. This imaging system uses a binary apodization mask (shaped pupil) to shape the point spread function (PSF) such that a predefined region in the image plane will have the high contrast needed. The very high contrast required (up to 10^{-10} for terrestrial planets) puts severe requirements on the wavefront control system, as the achievable contrast is limited by the quality of the wavefront.

The performance of conventional AO systems typically has been limited by the accuracy of the wavefront phase estimation (known as phase reconstruction) and the ability of the deformable mirror (DM) to achieve the arbitrary shapes required [Tyson (2000)]. However, as improvements have been made in both these areas, the ability to achieve the deep nulls required for high contrast imaging has still been stalled [Poyneer (2004)]. We showed in [Give'on (2005)] that this is due to what we refer to as frequency folding. We demonstrated that even with a perfect estimation of the wavefront and a DM limited

† Michelson graduate fellow, Michelson Science Center, California Institute of Technology, 770 S Wilson Avenue, MS 100-22, Pasadena, CA 91125

only by the highest spatial frequency it can achieve, conventional phase conjugation is limited by intermodulation products of its high frequency content.

In this paper we present a closed loop correction method for the shaped pupil coronagraph to minimize the energy in a pre-defined region in the image where terrestrial planets would be found. The correction part of the method is based on [Borde' (2005)]. The innovative reconstruction part uses three intensity measurements in the image plane with a pinhole added to the shaped pupil for diversity.

2. Wavefront Reconstruction – The "peak-a-boo" Algorithm

The electric field at the image plane, in the presence of aberrations in the system is a complex valued function. The measured quantity by the CCD is proportional to the intensity of the field where the complex field information is lost. The wavefront reconstruction problem is to recover of the real and imaginary parts of the complex field at the image plane.

Given a shaped pupil described by $A(x,y)$, The electric field in the image plane for the aberrated system operating in monochromatic light with wavelength λ and a focal length f, is given by:

$$\begin{aligned} E(\xi,\eta) &= \frac{1}{\lambda f}\int_{-\infty}^{\infty}\int_{-\infty}^{\infty} A(x,y)e^{\alpha(x,y)+\frac{2\pi}{\lambda}\beta(x,y)}e^{-i\frac{2\pi}{\lambda f}(x\xi+y\eta)}dxdy \\ &= \frac{1}{\lambda f}\int_{-\infty}^{\infty}\int_{-\infty}^{\infty} A(x,y)\left[1+\Phi(x,y)\right]e^{-i\frac{2\pi}{\lambda f}(x\xi+y\eta)}dxdy \end{aligned} \tag{2.1}$$

where we define

$$\Phi(x,y) = e^{\alpha(x,y)+\frac{2\pi}{\lambda}\beta(x,y)} - 1 \tag{2.2}$$

which can be written as:

$$\begin{aligned} E(\xi,\eta) &= E_0(\xi,\eta) + E_\Phi(\xi,\eta) \\ &= E_0(\xi,\eta) + E_\Phi^R(\xi,\eta) + iE_\Phi^I(\xi,\eta) \end{aligned} \tag{2.3}$$

where $E_\Phi^R(\xi,\eta)$ and $E_\Phi^I(\xi,\eta)$ are the real and imaginary parts of $E_\Phi(\xi,\eta)$, respectively, and

$$\begin{aligned} E_0(\xi,\eta) &= \frac{1}{\lambda f}\int_{-\infty}^{\infty}\int_{-\infty}^{\infty} A(x,y)e^{-i\frac{2\pi}{\lambda f}(x\xi+y\eta)}dxdy \\ E_\Phi(\xi,\eta) &= \frac{1}{\lambda f}\int_{-\infty}^{\infty}\int_{-\infty}^{\infty} A(x,y)\Phi(x,y)e^{-i\frac{2\pi}{\lambda f}(x\xi+y\eta)}dxdy \end{aligned} \tag{2.4}$$

Since $E_0(\xi,\eta)$ is theoretically known (since we know $A(x,y)$), the reconstruction problem is to recover $E_\Phi^R(\xi,\eta)$ and $E_\Phi^I(\xi,\eta)$.

The diversity for image plane measurements suggested here is a pinhole in the center of the shaped pupil and using the actuators on the DM that are aligned with the pinhole location to change the phase across it. Let the complex valued function $h_k(x,y)$ ($k = 1,2,3$) denote the electric field due to the pinhole and the phase difference across it induced by the DM, then the corresponding electric field at the image plane in each case, is given by:

$$\begin{aligned} E_k(\xi,\eta) &= \frac{1}{\lambda f}\int_{-\infty}^{\infty}\int_{-\infty}^{\infty} \left\{A(x,y)\left[1+\Phi(x,y)\right]+h_k(x,y)\right\}e^{-i\frac{2\pi}{\lambda f}(x\xi+y\eta)}dxdy \\ &= E_0(\xi,\eta) + E_\Phi^R(\xi,\eta) + iE_\Phi^I(\xi,\eta) + \Delta_k(\xi,\eta) \end{aligned} \tag{2.5}$$

where,

$$\Delta_k(\xi,\eta) = \frac{1}{\lambda f}\int_{-\infty}^{\infty}\int_{-\infty}^{\infty} h_k(x,y)e^{-i\frac{2\pi}{\lambda f}(x\xi+y\eta)}dxdy \tag{2.6}$$

and the three PSF's are given by:

$$I_k = E_k^* E_k \tag{2.7}$$

then, if we let

$$\begin{aligned}
G_{12} &= I_2 - I_1 - \Delta_2^*\Delta_2 + \Delta_1^*\Delta_1 - 2\Re\{E_0\Delta_2\} + 2\Re\{E_0\Delta_1\} \\
G_{13} &= I_3 - I_1 - \Delta_3^*\Delta_3 + \Delta_1^*\Delta_1 - 2\Re\{E_0\Delta_3\} + 2\Re\{E_0\Delta_1\} \\
A_1 &= 2(\Re\{\Delta_2\} - \Re\{\Delta_1\}) \\
A_2 &= 2(\Im\{\Delta_2\} - \Im\{\Delta_1\}) \\
A_3 &= 2(\Re\{\Delta_3\} - \Re\{\Delta_1\}) \\
A_4 &= 2(\Im\{\Delta_3\} - \Im\{\Delta_1\})
\end{aligned} \tag{2.8}$$

then the real and imaginary parts of $E_\Phi(\xi,\eta)$ are given by:

$$E_\Phi^R = \frac{A_4G_{12} - A_2G_{13}}{A_1A_4 - A_2A_3}, \quad E_\Phi^I = \frac{A_1G_{13} - A_3G_{12}}{A_1A_4 - A_2A_3} \tag{2.9}$$

3. Wavefront Correction

Let the DM be modeled with $N \times N$ actuators on a square grid and its surface, $\phi(x,y)$, following the super position principle,

$$\phi(x,y) = \sum_{k=0}^{N-1}\sum_{l=0}^{N-1} a_{k,l} f_{k,l}(x,y) \tag{3.1}$$

where $a_{k,l}$ is the height of the actuator on the kth row and l^{th} column and $f_{k,l}(x,y)$ is the DM's distinct influence function, centered at the actuator's position.

Suppose we add this DM in the pupil plane to the expression in Eq. 2.2, assuming a first order approximation for its effect and neglecting any cross terms between with DM and the aberration. Furthermore, assuming the star light has been suppressed at the high contrast region (either using a coronagraph or a shaped pupil), then the electric field in the high contrast region at the image plane is given by

$$\widehat{E_{\mathcal{H}}}(\xi,\eta) = \widehat{A\Phi}(\xi,\eta) + i\frac{2\pi}{\lambda}\widehat{A\phi}(\xi,\eta) \tag{3.2}$$

where $\mathcal{H}$ denotes the high contrast region and the hat represents the Fourier Transform. Following the correction method described in [Borde' (2005)] which minimizes the total intensity of light in the high contrast region, the condition for a stationary point with respect to the heights of the actuators is given by:

$$\sum_{m=0}^{N-1}\sum_{n=0}^{N-1} a_{m,n}\Re\left\{\left\langle \widehat{Af}_{m,n}, \widehat{Af}_{k,l}\right\rangle\right\} = \frac{\lambda}{2\pi}\Im\left\{\left\langle \widehat{A\Phi}, \widehat{Af}_{k,l}\right\rangle\right\} \tag{3.3}$$

for all $k,l = 1,2,\ldots,N$. This condition can be put in a matrix notation and solved for the heights of the actuators. The resulting DM configuration will minimize the total intensity of light assuming a first order approximation for the DM. In order to take into account the non-linearities in the system, we may close the loop and iterate. The following section shows the simulation of this closed loop system.

4. Simulation

Figure 1 shows the pupil used in these simulations. The image on the left is the shaped pupil with its ideal PSF in the center (All PSF images are given in log scale. The axes are in units of λ/D). In order to simulate the correction method described in this paper, system aberrations were added to the model as described in Eq. 2.2. The phase aberration was simulated with $RMS = \lambda/100$ and a $1/f^{3/2}$ power law (f in this case is the spatial frequency). The amplitude aberration was simulated with $RMS = 0.005$ and a $1/f^2$ power law. A pinhole with a diameter of $5 * 10^{-4}m$ was added at the center of the shaped

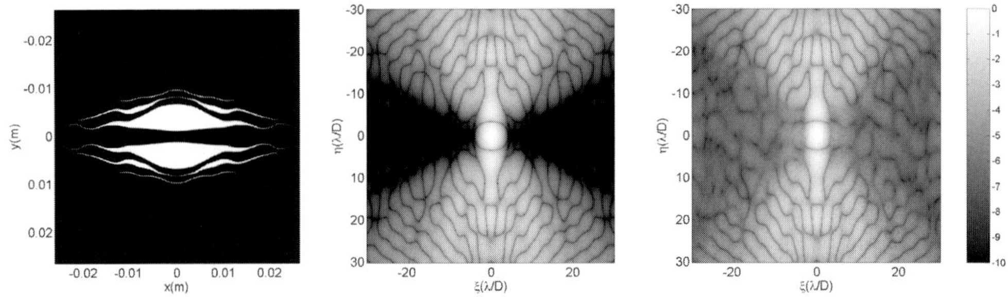

Figure 1. The shaped pupil (left) used in the simulations here and its corresponding ideal PSF (center) and the PSF of the system with aberrations (right). As seen in the figure on the right the aberrations in the system effect the contrast in the image.

pupil, as described in Eq. 2.6. The DM actuators, aligned with the pinhole were used to apply three different phase differences across the pinhole. The phase differences used in this simulation were 0, $\lambda/3$ and $2\lambda/3$. The following figure shows the PSF's of the system, simulating this diversity:

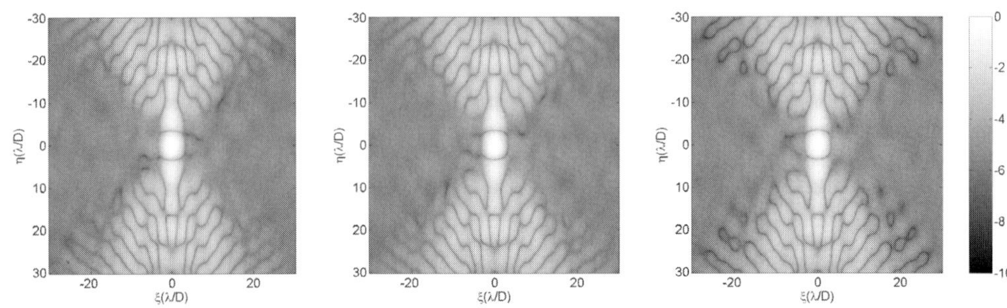

Figure 2. The three different images taken for the reconstruction part. As seen in the images, the effect of the pinhole is to add more light everywhere in the image plane. This is the reason this diversity performs well.

Simulations of this method has been successful in reconstructing the real and imaginary parts of $E_\Phi(\xi, \eta)$. The error in reconstruction (its absolute valued squared) is of the order of 10^{-25} is the dark hole region.

Figure 3 shows the PSF at the end of each iteration for the closed loop system. "Iteration 0" refers to the PSF of the system before the correction is applied. The DM was modeled with 52×52 actuators and the correction was on one side of the PSF (in order to correct for both amplitude and phase), up to 25 λ/D. As seen in the figure, in this ideal case, the correction converges after the third iteration.

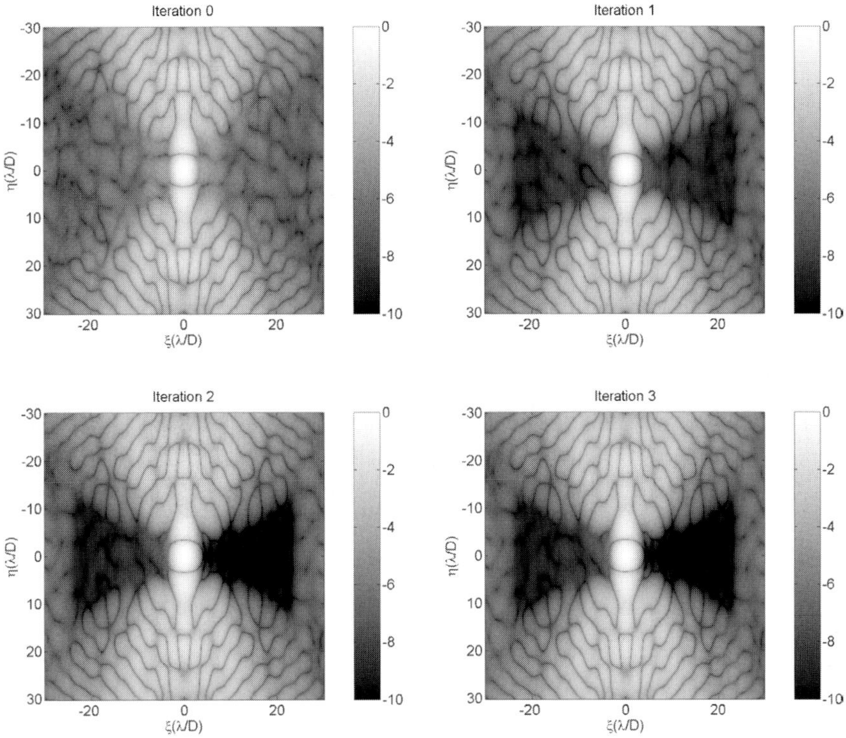

Figure 3.

5. Algorithm Robustness

The simulation shown above describes the ideal case. This section shows the simulation results of the closed loop method described in this paper in non-ideal cases. The first case is when the measured PSF follows a Poisson statistics for the arrival of photons at the image plane. The second case is when the shape of the DM is not exactly a sum of influence functions as assumed in the correction part (see Eq. 3.1) but an average between this model and a spline interpolation between the actuators heights. The third case is when some of the actuators fail and the DM surface at the location of the failed actuators is pinned to the substrate. Figure 4 compares the convergence of these three non-ideal cases to one another and to the ideal case.

The curve labeled "Ideal" describes the performance of the closed loop system when all the actuators are active, the DM model is exact and the measurement of the three images is deterministic. The curve labeled "Dead actuators" refers to the case when two actuators that are aligned with the open area of the shaped pupil are pinned to the substrate of the DM. The curve labeled "DM modeling error" refers to the case when the surface of the DM is a spline interpolation between the actuators' heights, rather than a sum of influence functions. The curve labeled "Photon noise" refers to the case where the PSF is the expected value of photons arriving at each CCD pixel. The measurement of the PSF is Poisson distributed. The curve labeled "All" refers to the combination case of all three modification to the model of the system.

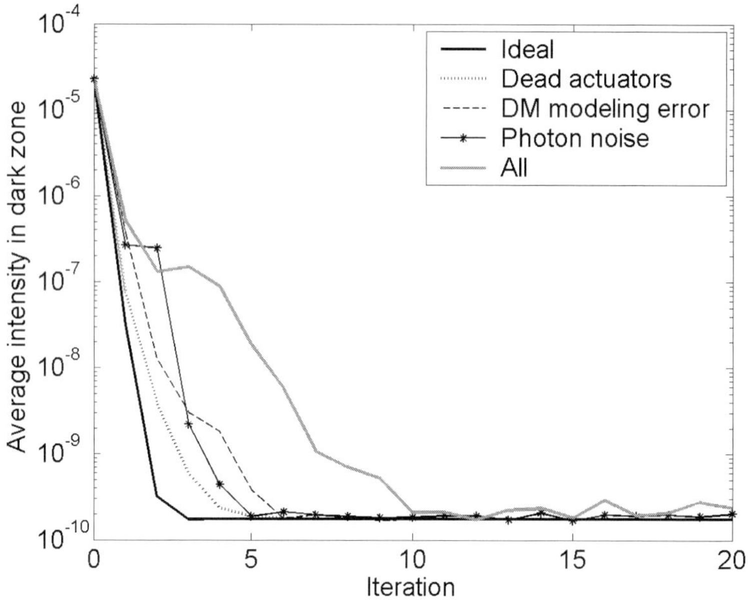

Figure 4. A plot comparing the convergence of the closed loop correction method in the presence of photon noise, deformable mirror modeling errors and defective actuators in the deformable mirror. The closed loop system was simulated in each case independently and then in the case of all modifications in effect at the same time.

As seen in the plots, the three model modifications, in the distinct cases, caused the closed loop system to converge in several more iterations. In the case of photon noise, ringing occurs in the convergence, which is expected from the stochastic nature of the system.

6. Summary and Future Work

This paper presents an innovative reconstruction method as part of a closed loop wavefront correction system for shaped pupil coronagraphs. This method has been shown in simulations to be effective in the presence of photon noise, deformable mirror modeling errors and overcoming defective actuators in the deformable mirror. We hope to test this reconstruction method in the lab in the near future and ultimately close the loop around it with DMs.

References

Beichman, C. A., Woolf, N. J., & Lindensmith, C. A., "The Terrestrial Planet Finder", volume 99-3, Pasadena: JPL Publication, 1999.

Borde', P. J. & Traub, W. A., "High-contrast imaging from space: speckle nulling in a low aberration regime", Accepted to ApJ, 2005, (astro-ph/0510597).

Give'on, A., Kasdin, N. J., Vanderbei, R. J., & Avitzour, Y., "On representing and correcting wavefront errors in high-contrast imaging systems", Accepted to JOSA A, Oct. 2005.

Kasdin, N. J., Vanderbei, R.J., Spergel, D. N., & Littman, M. G., "Extrasolar Planet Finding via Optimal Apodized and Shaped Pupil Coronagraphs", ApJ, 582: 1147–1161, 2003.

Poyneer, L. A. & Macintosh, B., "Spatially filtered wave-front sensor for high-order adaptive optics", JOSA A, 21(5):810–819, May 2004.

Tyson, R. K., "Introduction to Adaptive Optics", SPIE Press, 2000.

Direct Imaging of Exoplanets: Science & Techniques
Proceedings IAU Colloquium No. 200, 2005
C. Aime and F. Vakili, eds.
© 2006 International Astronomical Union
doi:10.1017/S1743921306009938

Maximum *a posteriori* planet detection and characterization with a nulling interferometer

Eric Thiébaut[1] and Laurent Mugnier[2]

[1]Jean-Marie Mariotti Center/Centre de Recherches Astronomiques de Lyon
9, avenue Charles André; F-69561 Saint Genis Laval Cedex, France
email: thiebaut@obs.univ-lyon1.fr

[2]Jean-Marie Mariotti Center/Office National d'Etudes et de Recherches Aérospatiales
B.P. 72, 92322 Châtillon Cedex, France
email: Laurent.Mugnier@onera.fr

Abstract. Nulling interferometers such as Darwin or TPF will require a rather sophisticated data processing in order to perform a reliable planet detection and characterization. We propose a Bayesian method, which follows the maximum *a posteriori* (MAP) approach, to solve this problem. Our method accounts for the noise statistics and optimally combines the data from a nulling interferometer at all observed wavelengths to perform reliable planet detection. The problem to be solved is however multi-modal. We show how, in practice, the global optimum of the MAP criterion can be found by our method; the latter also provides the most likely spectral energy distributions of all planets. Additionally, we show that a proper regularization allows us to achieve an improved robustness of the detection and could lead to shorter observation times.

Keywords. techniques: interferometric, methods: data analysis, methods: numerical, methods: statistical, planetary systems.

1. Introduction

Nulling interferometers make use of destructive interferences to get rid of the stellar light on the line of sight and strongly reduce the contrast between the light received from a star and from its surrounding planets Bracewell (1978). It is expected that this feature, possibly combined with internal and external modulations, will allow us to detect direct evidences of exo-solar planet emission. A nulling interferometer however gives data of very low signal-to-noise ratio and which are far from a direct image of the observed scene. The planet detection and their characterization by their spectral signatures therefore require a tailor-made data processing. In this paper, we propose a new method based on the maximum *a posteriori* (MAP) approach, in order to achieve an optimal planet detection and characterization. First, the assumed data model for a nulling interferometer is established. Then we explain how the MAP approach can be applied to this kind of data. We also show how to effectively obtain the globally optimal solution: that is the most likely planet positions and spectral energy distributions. Finally, we demonstrate the capabilities of our algorithm by the processing of realistic simulated data.

2. Data Model

A general model of the output from a nulling interferometer writes Mennesson (1999):

$$A_{t,\lambda} = \iint R_{t,\lambda}(\boldsymbol{\theta})\, I_\lambda(\boldsymbol{\theta})\, \mathrm{d}\boldsymbol{\theta}, \tag{2.1}$$

where $A_{t,\lambda}$ is the recorded amplitude at time t and for the effective wavelength λ, $R_{t,\lambda}(\boldsymbol{\theta})$ is the response of the instrument, *i.e.*, the transmission map as a function of angular coordinates $\boldsymbol{\theta}$, and $I_\lambda(\boldsymbol{\theta})$ is the specific brightness distribution of the observed object (or scene). In this model, we assume that the scene does not vary during the whole data acquisition. Doing otherwise would make the inversion prohibitively complicated and probably not robust at least for the detection stage.

Spatial nulling interferometers such as Darwin will use both external modulation (*i.e.*, rotation of the interferometric array with time) and internal modulation (*i.e.*, modification of the transmission map *via* phase shifts in the beam combination), so one should bear in mind that index t in Eq. (2.1) actually codes for both the position in time and for the configuration of the phase shifts of the array. Historically, the concept of internal modulation was imagined by Mariotti in 1997 Mennesson, Léger & Ollivier (2005).

Because the data model of Eq. (2.1) is linear, any linear combination of the data gives *synthetic* data that follow the same model, with a *synthetic* transmission map that is the corresponding linear combination of transmission maps. When using internal modulation it is possible to find coefficients for such a linear combination that make the synthetic transmission map an odd function with respect to the angular direction $\boldsymbol{\theta}$ Absil (2001). Additionally, it is trivial to show from Eq. (2.1) that the data corresponding to an odd R and an even object's brightness distribution is zero. This is of paramount importance as all components of the observed scenes except the planets should be even within a good approximation. Indeed, the star and the exo-zodiacal light should be symmetrical, and the local zodiacal dust as well as the instrumental thermal emissions are at a constant level across the field of view. This means that we can, in our data model, retain only the contribution of the planets, and search only for their parameters in the inversion. We can write the specific object's brightness distribution as:

$$I_\lambda(\boldsymbol{\theta}) = \sum_{i=1}^{N_{\mathrm{src}}} F_{i,\lambda}\,\delta(\boldsymbol{\theta} - \boldsymbol{\theta}_i) + I_\lambda^{\mathrm{sym}}(\boldsymbol{\theta})\,, \tag{2.2}$$

where N_{src} is the number of planets and $F_{i,\lambda}$ is the spectral energy distribution (SED) of the i-th planet and $I_\lambda^{\mathrm{sym}}(\boldsymbol{\theta})$ accounts for all the symmetrically distributed sources. After proper linear combination of the data, $I_\lambda^{\mathrm{sym}}(\boldsymbol{\theta})$ does not contribute to the expected value of the signal; it is however the main source of noise. Combining the data model of Eq. (2.1) with the object's model of Eq. (2.2) and accounting for modulation yields:

$$A_{t,\lambda} = \sum_{i=1}^{N_{\mathrm{src}}} R_{t,\lambda}(\boldsymbol{\theta}_i) F_{i,\lambda} + n_{t,\lambda}, \tag{2.3}$$

where $n_{t,\lambda}$ is the detection noise.

3. Detection Algorithm

The problem at hand is to estimate the positions $\boldsymbol{\theta}_i$ and the SED's $F_{i,\lambda}$ of the planets, assuming that their number N_{src} is known. Let us denote this set of parameters by $(\boldsymbol{\theta}, \mathbf{F}) \triangleq \{\boldsymbol{\theta}_i, F_{i,\lambda}; i = 1, \ldots, N_{\mathrm{src}}; \lambda_{\min} \leqslant \lambda \leqslant \lambda_{\max}\}$.

A well-known approach for the detection task is correlation. One could for instance consider correlating the data recorded at a given wavelength with the noiseless data model obtained for a point-source at all possible locations Angel & Woolf (1997), Mennesson & Mariotti (1997). The location yielding the highest correlation would be a probable position for a planet. Yet, let alone the fact that such a scheme is not statistically appropriate for an instrument that is not shift-invariant, it neither makes use of statistical information on the noise, nor provide a way to optimally combine the data at several wavelengths.

The maximum *a posteriori* (MAP) approach has none of these shortcomings; it defines the solution $(\hat{\boldsymbol{\theta}}, \hat{\mathbf{F}})$ as the set of parameters that has maximum probability given the data:

$$(\hat{\boldsymbol{\theta}}, \hat{\mathbf{F}}) = \arg\max_{(\boldsymbol{\theta},\mathbf{F})} \Pr(\boldsymbol{\theta}, \mathbf{F}|\mathbf{A}) \,, \tag{3.1}$$

where $\mathbf{A}$ denotes the data set. By Bayes' theorem and since the probability $\Pr(\mathbf{A})$ of the data alone does not depend on the sought parameters, the MAP solution also solves:

$$(\hat{\boldsymbol{\theta}}, \hat{\mathbf{F}}) = \arg\max_{(\boldsymbol{\theta},\mathbf{F})} \Pr(\mathbf{A}|\boldsymbol{\theta}, \mathbf{F}) \Pr(\boldsymbol{\theta}, \mathbf{F}) = \arg\min_{(\boldsymbol{\theta},\mathbf{F})} J_{\text{MAP}}(\boldsymbol{\theta}, \mathbf{F}) \,, \tag{3.2}$$

where the so-called maximum *a posteriori* cost function (or criterion) is:

$$J_{\text{MAP}}(\boldsymbol{\theta}, \mathbf{F}) = J_{\text{data}}(\boldsymbol{\theta}, \mathbf{F}) + J_{\text{prior}}(\boldsymbol{\theta}, \mathbf{F}) \,, \tag{3.3}$$

and where $J_{\text{data}}(\boldsymbol{\theta}, \mathbf{F}) \propto -\log \Pr(\mathbf{A}|\boldsymbol{\theta}, \mathbf{F})$ is the neg-log-likelihood of the data given the model and $J_{\text{prior}}(\boldsymbol{\theta}, \mathbf{F}) \propto -\log \Pr(\boldsymbol{\theta}, \mathbf{F})$ is the negative logarithm of the *a priori* probability of the model. Assuming independent Gaussian noise, J_{data} writes:

$$J_{\text{data}}(\boldsymbol{\theta}, \mathbf{F}) \triangleq \sum_{t,\lambda} \frac{1}{\sigma_{t,\lambda}^2} \left(A_{t,\lambda} - \sum_{i=1}^{N_{\text{src}}} R_{t,\lambda}(\boldsymbol{\theta}_i)\, F_{i,\lambda} \right)^2 \tag{3.4}$$

where the noise variance $\sigma_{t,\lambda}^2$ can be estimated from the data. Minimizing the term J_{data} enforces agreement of the parameters with the data; and minimizing J_{prior} enforces agreement of the parameters with the available *a priori* information. In inverse problem jargon, J_{prior} is called a regularization term. Since a properly sampled planet spectrum should be a smooth function of the wavelength, we define J_{prior} so as to estimate the roughness of the SED's:

$$J_{\text{prior}}(\boldsymbol{\theta}, \mathbf{F}) \triangleq \sum_{i=1}^{N_{\text{src}}} \mu_i \sum_{\lambda=\lambda_{\min}}^{\lambda_{\max}} \left(\frac{\partial^m F_{i,\lambda}}{\partial \lambda^m} \right)^2 \tag{3.5}$$

where the mth-order spectral derivative is computed by finite differences and were μ_i are hyper-parameters allowing us to tune the relative weight of regularization.

Finding the minimum of J_{MAP} with respect to $(\boldsymbol{\theta}, \mathbf{F})$ yields the optimal solution. However, such a criterion is multi-modal, hence its deepest minimum cannot be found by a descent optimization algorithm: *global optimization* is required. Nevertheless, since J_{data} and J_{prior} are convex with respect to the subset of parameters $\mathbf{F}$, the most likely SED's can be obtained by solving:

$$\frac{\partial J_{\text{MAP}}(\boldsymbol{\theta}, \mathbf{F})}{\partial F_{i,\lambda}} = 0 \,; \forall i, \forall \lambda \,, \tag{3.6}$$

for given planet positions. Moreover, because both criteria J_{data} of Eq. (3.4) and J_{prior} of Eq. (3.5) are quadratic with respect to the SED's, J_{MAP} is also quadratic and Eq. (3.6) is a set of linear equations. Its solution, denoted in the following by $\hat{\mathbf{F}}(\boldsymbol{\theta})$, is analytical and depends on the data and on the considered planet positions.

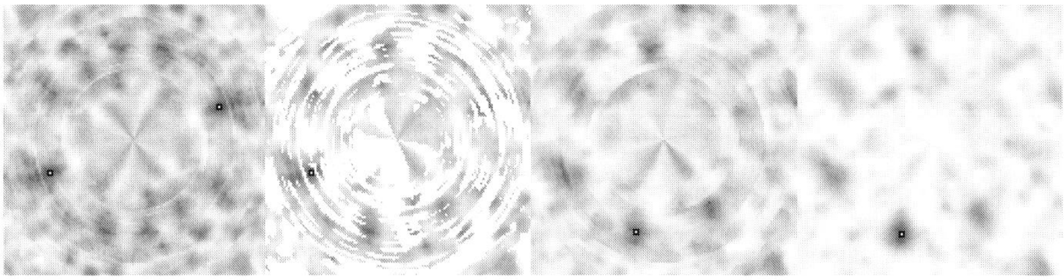

Figure 1. Detection of a faint planet by various criteria. From left to right: unconstrained $J^\dagger_{\text{data}}$, $J^\dagger_{\text{data}}$ with negative SED's removed, $J^\dagger_{\text{data}}$ with positivity and, $J^\dagger_{\text{MAP}}$ with positivity. Darkest values indicate most likely planet positions. The global minima of the criteria are highlighted as a white dot surrounded by a black border.

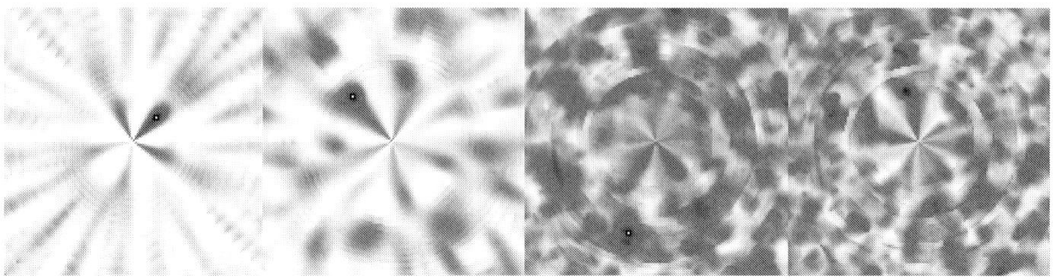

Figure 2. Multiple planet detection steps at low signal-to-noise ratio (RSB ∼ 2). From left to right: criterion mapped for the 1st, 2nd, 3rd and 4th candidate planet.

If we replace the SED's by $\hat{\mathbf{F}}(\boldsymbol{\theta})$ in the cost function $J_{\text{MAP}}(\boldsymbol{\theta}, \mathbf{F})$ of Eq. (3.2) we obtain a *partially optimized* cost function with many less parameters, which only depends (explicitly) on the assumed planet positions:

$$J^\dagger_{\text{MAP}}(\boldsymbol{\theta}) \triangleq J_{\text{MAP}}(\boldsymbol{\theta}, \hat{\mathbf{F}}(\boldsymbol{\theta})). \tag{3.7}$$

It must be noted that this new cost function $J^\dagger_{\text{MAP}}(\boldsymbol{\theta})$ is nothing but the original MAP cost function $J_{\text{MAP}}(\boldsymbol{\theta}, \mathbf{F})$ that has been optimized on a subset of its parameters (namely, on the SED's). Consequently, the former has exactly the same minima as the latter and is significantly simpler to optimize.

Simply by setting $\mu_i = 0, \forall i$, it is possible to turn regularization off to obtain a solution which does not depend on any prior. In this case, $J_{\text{MAP}} = J_{\text{data}}$ and hence $J^\dagger_{\text{MAP}} = J^\dagger_{\text{data}}$. To demonstrate the improvement brought by regularization, we must compare the results obtained by optimizing the unregularized criterion $J^\dagger_{\text{data}}$ and those obtained with $J^\dagger_{\text{MAP}}$. In the following and for sake of simplicity, we denote by $J(\boldsymbol{\theta}, \mathbf{F})$ the actual criterion (either J_{MAP} or J_{data}) and $J^\dagger(\boldsymbol{\theta})$ the corresponding partially optimized criterion.

The minimum of $J^\dagger$ provides the most likely set of planet positions. As expected and shown by Fig. 1, $J^\dagger$ is still multi-modal with respect to $\boldsymbol{\theta}$. To find the global minimum of $J^\dagger$, the most obvious approach is to sample the field of view and map the cost function $J^\dagger$ onto the grid of planet positions. However if the grid has N_{grid} cells, this requires the solving of Eq. (3.6) and the computation of $J^\dagger$ for ${N_{\text{grid}}}^{N_{\text{src}}}$ cases. Such a global search is therefore limited to a modest number of planets on a reasonably small grid. In the single planet case (see Fig. 1), $J^\dagger$ is a 2-D pseudo-image of the field of view where local minima corresponds to the most likely locations of the planet.

Obviously, the first pseudo-image shown by Fig. 1 is even, which constitutes a sign ambiguity on the planet position. This ambiguity is due to the fact that the considered synthetic transmission map R is odd so that for any $(\boldsymbol{\theta}, \mathbf{F})$, $(-\boldsymbol{\theta}, -\mathbf{F})$ has exactly the same likelihood. Simply selecting the SED that is mostly positive removes the ambiguity. The

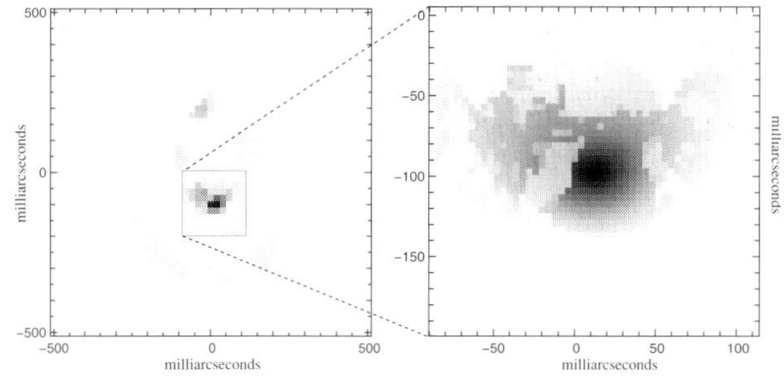

Figure 3. Planet detection map for the *blind-test*. The darkest parts of the maps indicate the minima of the criterion, hence where the planet is most likely located.

fact that the SED of each planet must actually be positive at all wavelengths suggests an enhancement to the SED estimation, which consists in minimizing Eq. (3.2) or Eq. (3.4) with respect to the SED's under a positivity constraint:

$$\hat{\mathbf{F}}(\boldsymbol{\theta}) = \arg\min_{\mathbf{F}} J(\boldsymbol{\theta}, \mathbf{F}) \quad \text{s.t.} \quad F_{i,\lambda} \geqslant 0\,; \forall i, \forall \lambda\,. \tag{3.8}$$

The solution is no longer analytical but improves the planet detection, as seen by the second and third pseudo-images in Fig. 1. Many numerical algorithms have been devised to solve Eq. (3.8) when J is quadratic Lawson & Hanson (1974), Bro & Jong (1997), Stark & Parker (1995). In our simulations, we solved Eq. (3.8) with VMLM-B Thiébaut (2002), which is suitable even for non-quadratic criteria.

4. Simulation Results

We first checked our algorithm with a scene made of a central star, an exo-zodiacal cloud and 1 or 3 planets. We simulated data for 4 different rotation angles of a Robin-Laurance interferometer with 15 spectral channels. In the case of a single planet, Fig. 1 clearly demonstrates the improvements brought by non-negativity and regularization: not only do these constraints help the method remove ambiguities (there are fewer local minima of similar depth in the criterion map) but they also yield the correct position, which would have been completely wrong otherwise. In order to be able to detect more than one planet we iteratively applied our single planet detection scheme to a 3 planet scene as follows: for each new planet, we optimize $J^{\dagger}$ with respect to all SED's but we only allow the new planet to move; this yields a 2-D pseudo-image, whose global minimum is the most likely position of the new planet. Figure 2 shows that the 3 first planets are readily detected by our algorithm, whereas the (non-existing) 4th one is not.

In order to test our algorithm when the components of the observed scene are really unknown, we applied our method to nulling data simulated by someone else. The data consists in 1000 exposures of a rotating linear nulling interferometer with 3 telescopes and 2 internal modulation states (the so-called L-TTN configuration). There are 27 wavelength channels from $6.7\,\mu$m to $20\,\mu$m, the duration of each exposure is 100 s and we estimated the signal-to-noise ratio to be SNR $\simeq 1.5$ per datum. Figure 3 shows the regularized criterion $J^{\dagger}_{\text{MAP}}$ computed for this data set onto a $1'' \times 1''$ field of view around the central star. It clearly exhibits a likely point-like source at $(0''\!.012, -0''\!.097)$ from the central star. After unveiling of the simulation parameters, a planet was indeed simulated at 1 AU away from a star at a distance of 10 pc. To check the robustness of our algorithm, we repeated the processing for smaller and smaller data subsets. Figure 4 shows the result of the detection with 100 %, 50 %, 10 %, 5 % and 1 % of the original data, *i.e.*, with

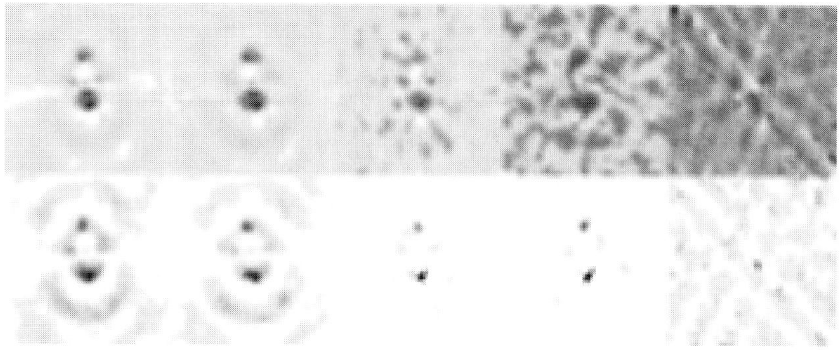

Figure 4. Effects of reducing the total observing time on planet detection. Top row: planet detection by unregularized criterion $J^\dagger_{\rm data}$. Bottom row: planet detection with spectral regularization criterion $J^\dagger_{\rm MAP}$. From left to right: 1000, 500, 100, 50 and 10 exposures. Strict positivity constraint was used in all cases.

1000, 500, 100, 50 and only 10 exposures of 100 s each. From this figure it is clear that spectral regularization improves the sensitivity of the detection. Indeed, with spectral regularization, the detection is still reliable with only ~50 exposures; whereas at least ~100 exposures are needed without the regularization.

5. Conclusions

We have shown how a maximum *a posteriori* approach can lead to a planet detection algorithm suitable for nulling interferometry data. Unlike the correlation method used in earlier studies, our approach optimally combines all the available pieces of information. Thanks to simulated data, we have demonstrated that non-negativity and spectral regularization (enforcement that the spectral energy distribution of a planet should be somewhat smooth) are critical features to disentangle the numerous ambiguities in the detection from such kind of data. The most likely SED's of the detected planets are a by-product of our algorithm and can be used to perform planet characterization. We have also shown how the detection scheme can be modified to account for several planets.

Acknowledgements

This work has been supported by ESA under contract *Reconstruction of Exo Solar System Properties* (RESSP). The scene for our multiple planet detection tests was computed by A. Belu (LUAN). The data for our blind-test were kindly provided by R. den Hartog (ESA).

References

Absil, O. 2001, PhD thesis, Université de Liège, Belgique
Angel, J. R. P. & Woolf, N. J. 1997, *Ap. J.* **475**, 373–379
Bracewell, R. N. 1978, *Nature* **274**, 780–781
Bro, R. & Jong, S.D. 1997, *Journal of Chemometrics* **11**, 393–401
Lawson, C. L. & Hanson, R. J. 1974, *Solving Least Squares Problems*, Prentice-Hall
Mennesson, B. & Mariotti, J.-M. 1997, *Icarus* **128**, 202–212
Mennesson, B. 1999, PhD thesis, Université de Paris VII, France
Mennesson, B., Léger, A. & Ollivier, M. 2005, *Icarus* **178**, 570–588
Stark P.B. & Parker R.L. 1995, *Comp. Stat.* **10**, 129–141
Thiébaut, E. 2002, in *Proc. SPIE* **4847**, 174–183

Direct Imaging of Exoplanets: Science & Techniques
Proceedings IAU Colloquium No. 200, 2005
C. Aime and F. Vakili, eds.
© 2006 International Astronomical Union
doi:10.1017/S174392130600994X

The Self-Coherent Camera: a new tool for planet detection

P. Baudoz, A. Boccaletti, J. Baudrand, and D. Rouan

LESIA, Observatoire de Paris-Meudon, France
email: pierre.baudoz@obspm.fr

Abstract. The performance of high contrast imaging systems is very often limited by the presence of static speckles in the point-spread function of the central source. Several techniques have already been proposed to discriminate a faint companion from these residual speckles. These techniques used different criteria to separate a speckle from a companion: polarization, spectral information or coherence. Here, we propose a new imaging device, the Self-Coherent Camera (SCC), that is based on the lack of coherence between the stellar light and the planet that is searched for. This SCC is a simple instrument that allows us to reach the fundamental limitation of the photon noise by calibrating the speckles in the recorded images. After the description of the general problem of discriminating speckles from planets, we will explain the principle of the SCC. Then, we will analyze the different limitations of this technique as well as the performance that can be reached with current telescopes.

Keywords. speckles, coherence.

1. Introduction

Direct imaging of exoplanets requires extremely high dynamical range, high spatial resolution and high sensitivity simultaneously. Different imaging techniques have been proposed to reach such a challenging goal. For example, coronagraphs (Rouan *et al.* 2000), apodized (Soummer *et al.* 2003) or shaped pupil (Kasdin *et al.* 2003) could help reaching the high dynamical range mandatory for direct imaging of exoplanets. However, an essential limitation of these systems is the presence of speckles in the Point Spread Function (PSF). Indeed, while Adaptive Optics (AO) improve greatly the quality of the wavefront of ground based image, there are still residual fast speckles in AO images. These residual atmospheric effects should be averaged quickly with time but, because of defects introduced in the optical system, speckles are still detected on images recorded on ground-based instruments. Slow drifts of the phase errors of optical elements make it difficult to calibrate these speckles that are not fully static. The same kind of drift occurs for space-based telescope where the same kind of speckles are observed. Thus, when attempting to detect a faint companion on long exposure images, one finds that "quasi-static" speckles can be a dominant source of error (Marois *et al.* 2005). The best way to overcome this difficulty is to calibrate these "quasi-static" speckles. Since they are slowly drifting, it is mandatory to find a way to discriminate the speckles of the star from a faint companion during the exposure. This is the purpose of differential imaging techniques. Several criteria have been proposed so far to discriminate the speckles from the planets: spectrophotometry (Racine *et al.* 1999, Marois *et al.* 2005), polarimetry (Seager *et al.* 2000, Baba & Murakami 2003), and coherence (Codona & Angel 2004, Guyon 2004, Labeyrie 2004). While both concepts based on spectrophotometry and polarimetry depends on the physical properties of the planets, the coherence is a robust criterion when no physical information is available from the planets that could be observed. The techniques based on coherence split the beam in two parts, one used as a coherent reference for the other. The beams are recombined on the same optical axis as in a Michelson interferometer. The recorded interferences are used to control the phase errors with a deformable mirror in Codona & Angel (2004) or an active hologram in Labeyrie (2004) or to calibrate the speckles using a modulated optical path difference (OPD) in Guyon (2004). Here, we suggest another way to calibrate the speckles using coherence as a criterion. The technique we propose rely on a simple Fizeau recombination of the science beam and the reference beam. It enables to discriminate speckles from a planet without using fast OPD modulation or deformable mirrors. After presenting the proposed concept and a possible implementation, we estimate the signal to noise ratio reachable with this technique and show simulations of observations with SCC on a 8 meter telescope.

2. The Self-Coherent Camera concept

The concept of the Self-Coherent Camera (SCC) is described in figure 1 (left). We propose to split the light coming from the telescope into two beams. One of the beam is spatially filtered using a pinhole or an optical fiber. The two beams are recombined in the focal plane in a Fizeau scheme. To do so, the two pupil beams are optically brought at the same plane right before a focusing lens. The distance between the fully coherent pupil and the corrugated pupil is large enough so that their autocorrelation are never superimposing. A possible set-up is shown in figure 1 (right).

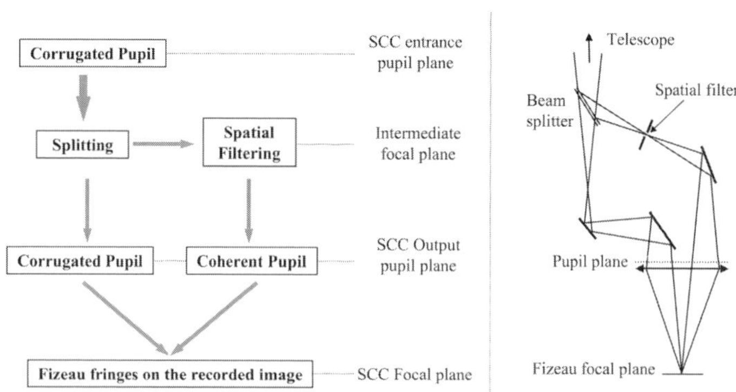

Figure 1. Left: Principle of the Self-Coherent Camera. Right: Possible set-up for the SCC.

The Fizeau image in the focal plane will show fringes pinning the intensity distribution of the stellar flux. Since the flux of the companion is fully removed from the reference beam by spatial filtering, the intensity of the companion will be unaltered by the reference beam. This is what can be seen in figure 2 where the point spread function of the stellar flux interfere with the reference beam creating a fringe pattern over all the diffraction rings and speckles. A companion located at a distance of 2 λ/D and in the upper right quadrant does not show fringes because it is not coherent with the reference beam.

3. Formalism

Assuming the electromagnetic field in the entrance pupil plane is described by $\Psi(\boldsymbol{\xi}) = \widehat{\sqrt{I(\boldsymbol{x})}}$ with $\sqrt{I(\boldsymbol{x})}$ describing the complex amplitude in the focal plane, we can write the field of both the star A and its companion B in the pupil plane:

$\Psi_A(\boldsymbol{\xi}) + \Psi_B(\boldsymbol{\xi}) = \widehat{\sqrt{I_A(\boldsymbol{x})}} + \widehat{\sqrt{I_B(\boldsymbol{x})}}$

The coherent pupil of the reference beam can be described in the pupil plane by: $\Psi_R(\boldsymbol{\xi}) = \widehat{\sqrt{I_R(\boldsymbol{x})}}$

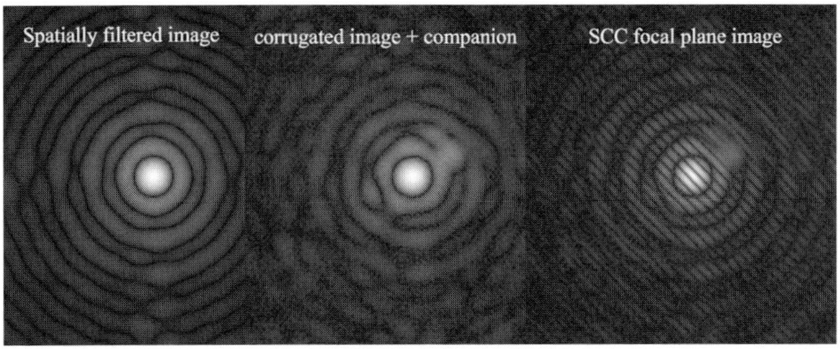

Figure 2. Left: reference beam. Middle: corrugated beam with a main star and a companion located at a distance of 2 λ/D and in the upper right quadrant. Right: SCC final image plane where the companion does not show Fizeau fringes. The companion is only 20 times fainter that the star. The simulation hypothesis are the same than the one of figure 5.

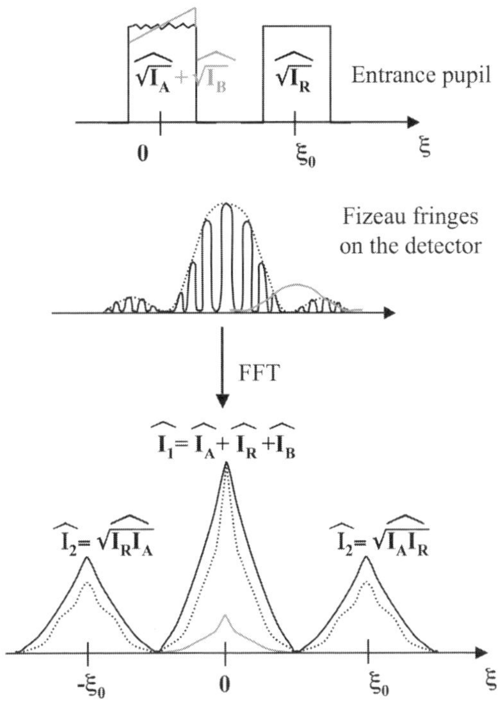

Figure 3. Principle of the SCC. Top: Electromagnetic field distribution in the SCC output pupil plane for a corrugated star (black) and its companion (grey). Middle: Fizeau fringes recorded on the detector for both objects. Bottom: Numerical Fourier Transform of the detected images showing autocorrelations and intercorrelations between the reference beam and the stellar beam.

Assuming that the vector $\boldsymbol{\xi}_0$ describes the distance between the coherent pupil and the corrugated pupil (supposed to be centred on zero). The field at the SCC pupil output plane is given by:

$$\Psi(\boldsymbol{\xi}) = \Psi_A(\boldsymbol{\xi}) + \Psi_B(\boldsymbol{\xi}) + \Psi_R(\boldsymbol{\xi}) * \delta(\boldsymbol{\xi} - \boldsymbol{\xi}_0) \tag{3.1}$$

where $*$ is the convolution symbol. The image recorded at the focal plane following this pupil plane is:

$$I(\boldsymbol{x}) = \mid \widehat{\Psi(\boldsymbol{\xi})} \mid^2 = I_A(\boldsymbol{x}) + I_B(\boldsymbol{x}) + I_R(\boldsymbol{x}) + 2\sqrt{I_A(\boldsymbol{x}) I_R(\boldsymbol{x})} \cdot \cos(2\pi \boldsymbol{x} \boldsymbol{\xi}_0) \tag{3.2}$$

Equation 3.2 assumes that there is no optical path difference between both pupils when recombined. If an optical path difference is introduced, it will introduce a phase in the cosinus. Following equation 3.2, the image will look like a classical Fizeau image with fringes pinning the intensity distribution of the main star. However, the image of the companion does not show any fringes because it is not coherent with the reference beam. Since the intensity of the main star is coded with fringes while the image of the companion is not, it looks clearly that one can discriminate the image of the companion from the stellar flux. A good approach is to use the numerical FFT of the image to separate the three areas limited by the autocorrelation of the pupil function. The centred area where appears the autocorrelation of the two different image of the pupil and two lateral terms where appears the correlation between the fields Ψ_A and Ψ_R (Fig. 3). The separation between the two image of the pupil $\boldsymbol{\xi}_0$ is large enough so that the different terms are not superimposed and can be numerically separated.

Assuming that we describe the centred area by $\widehat{I_1(\boldsymbol{x})} = \widehat{I_A(\boldsymbol{x})} + \widehat{I_B(\boldsymbol{x})} + \widehat{I_R(\boldsymbol{x})}$ and the two other area by $\widehat{I_2(\boldsymbol{x})} = \widehat{\sqrt{I_A(\boldsymbol{x}) I_R(\boldsymbol{x})}}$, we can write the intensity of the companion using the following formula:

$$I_B(\boldsymbol{x}) = I_1(\boldsymbol{x}) - I_R(\boldsymbol{x}) - \frac{I_2(\boldsymbol{x})^2}{I_R(\boldsymbol{x})} \tag{3.3}$$

To detect the companion $I_B(\boldsymbol{x})$, one needs to record separately $I_R(\boldsymbol{x})$. It can be done because $I_R(\boldsymbol{x})$ is a spatially filtered beam that can be very stable over time.

This equation is valid for an exposure shorter than the coherent time of the atmospheric speckles. Indeed for a long exposure (N times longer than the coherent time), the intensity I_1 and I_2 can be written:

$$I_1(\boldsymbol{x}) = \sum_{i=0}^{N} [I_A^i(\boldsymbol{x}) + I_B^i(\boldsymbol{x})] + N.I_R(\boldsymbol{x})$$

and

$$I_2(\boldsymbol{x}) = I_R(\boldsymbol{x}) \sum_{i=0}^{N} [\sqrt{I_A^i(\boldsymbol{x})}]$$

where $I_A^i(\boldsymbol{x})$ and $I_A^i(\boldsymbol{x})$ varies with atmospheric turbulence (corrected or not).

Obviously equation 3.3 is not any more valid for long exposure because

$$\sum_{i=0}^{N} I_A^i(\boldsymbol{x}) \neq \left[\sum_{i=0}^{N} \sqrt{I_A^i(\boldsymbol{x})}\right]^2$$

For the same reason, equation 3.3 can only be applied to monochromatic beam. However, if observing on small angular distances where the SCC is the most efficient, the bandwidth can be increased up to a reasonable value (R=10). Another way to avoid this issue is to use a chromatic corrector as proposed by Wynne (1979).

The division that appears in equation 3.3 could also be an issue because it increase the noise in the low photon area of $I_R(\boldsymbol{x})$. However, the exposure time for I_R can be much larger than the coherent time because it is a stable beam. So the division will increase the noise at the dark ring positions but can be reduced using a statistical analysis of the recorded images.

4. Simulation

To evaluate the capabilities of the SCC, we first simulated the instrument for a perfect PSF and compared it with direct imaging photon noise. Since the SCC can only record short exposure times, the number of photon per image will be limited and an estimation

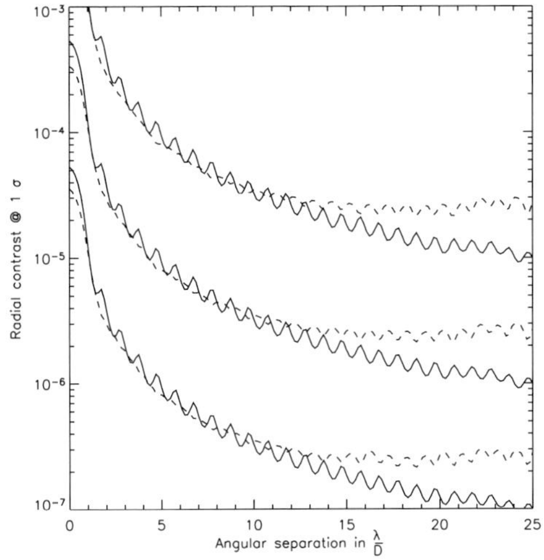

Figure 4. Averaged radial contrast at 1 σ with direct imaging (solid) and using SCC (dashed) for 1,100 ,10000 images. Hypothesis: Perfect PSF, $m_H = 5$, $\Phi_{tel} = 8m$, $\lambda_0 = 1.6\mu m$, $\Delta\lambda = 0.3\mu m$, exp. time = 5 ms, total transmission= 0.2. Chromaticity of the PSF is not simulated. RON=0.

of the photon noise effect is needed. To take into account the photon noise, we assumed a star with a magnitude $m_H = 5$ ($\lambda_0 = 1.6\mu m$, $\Delta\lambda = 0.3\mu m$) observed with an 8m-telescope (total transmission of 0.2). The exposure time is set to 5 ms which corresponds roughly to the coherent time in the H band. The Readout Noise (RON) is equal to zero. No chromaticity of the PSF is simulated for this example. The reference beam $I_R(\boldsymbol{x})$ is recorded separately in a single exposure with exposure time of 5 ms , 0.5s and 50s respectively for the 1, 100, 10000 simulated SCC images. This obviously helps decreasing the photon noise introduced by $I_R(\boldsymbol{x})$ in the equation 3.3 for long exposure. Figure 4 shows the averaged radial contrast at 1 σ for the SCC and for direct imaging photon noise. Both observation techniques give the same limiting contrast up to 13 λ/D where the SCC techniques hits its limits because the average photon level per pixel becomes lower than one on a single image.

As described in the introduction, the limiting noise when dealing with real AO images is not the photon noise, neither the atmospheric speckle noise that is averaged very quickly but the slow drift of "quasi-static" speckles. To show that the SCC is independent of this noise, we simulated AO images with the same observed star than above introducing phase defects in the wavefronts. The reference beam shows a 10 nm rms typical optical defects (spatial frequency in f^{-2}) and 10000 AO-corrected wavefronts have been simulated. These wavefronts are simulated from an analytical expression of the power spectral density (PSD). This PSD approximates the low frequency by a law in f^{-2} up to the corrected spatial frequency. The level of this part of the PSD is arbitrary chosen to reach a realistic Strehl ratio of about 0.8 in H band. The frequencies above this limit are classical atmospheric law with an outer scale of 20m and a Fried parameter $r_0 = 25cm$. The AO system is supposed to correct 40 by 40 subapertures in the pupil. A Strehl ratio of 0.8 corresponds to a phase aberration of 120 nm rms in the pupil. We add to all the simulated wavefront a fixed phase aberration with the same PSD but 12 times less aberrated (10 nm). As above, the reference beam is recorded separately in a single exposure corresponding to the number of simulated wavefront used when resolving equation 3.3. An example of the reference image, the corrugated image, and the interference between both of them is shown in Figure 2. In Figure 5, the averaged radial contrast at 1 σ of the SCC is compared to the theoretical photon noise and to the "quasi-static"

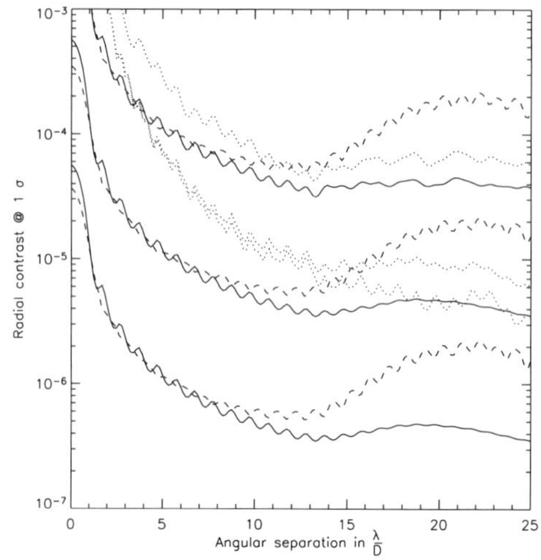

Figure 5. Averaged radial contrast at 1 σ for direct imaging with fixed wavefront errors (dotted), for SCC imaging(dashed) and for theoretical photon noise (solid). The simulations for 1,100, and 10000 images are shown. Photon noise with direct imaging and using SCC. Corrugated PSF. Hypothesis: $r_0 = 25cm$, AO =40x40, $Sr = 0.8$ in H band, $m_H = 5$, $\Phi_{tel} = 8m$, $\lambda_0 = 1.6\mu m$, $\Delta\lambda = 0.3\mu m$, exp. time = 5 ms, total transmission= 0.2. Chromaticity of the PSF is not simulated. RON=0. Fixed wavefront error =10nm, Reference beam wavefront error= 10nm.

speckle noise. The SCC is again limited only by the photon noise up to 13 λ/D for the same reason than above. For the "quasi-static" speckle noise, the contrast is limited by the fixed speckle level. The level being 12 times lower than the AO residual aberration, integrating more than 150 images does not improve the contrast.

5. Conclusion

We described a new imaging device that relies on coherence to differentiate a speckle from a true companion. This Self-Coherent Camera (SCC) uses Fizeau fringes in the focal plane to encode the speckles of the star. A simple image processing technique is used to compute from each recorded image the signal of a true companion which is not encoded by fringes. Simulation of a perfect case showed that the photon noise limit can be reached if the flux per pixel is larger than one photon. The SCC is then especially efficient for small angular separation. A more realistic simulation showed that at these small angular separation, the SCC improves the contrast that can be detected even if the images are limited by slowly drifting speckles, which is the case for real AO images.

References

Baba, N. & Murakami, N. 2003, 115, 1363

Codona, J. & Angel, R. 2004, ApJ, 604, L117

Guyon, O. 2004, ApJ, 615, 562

Kasdin, N. J., Vanderbei, R. J., Spergel, D. N., & Littman, M. G. 2003, ApJ, 582, 1147

Labeyrie, A. 2004, Proc. "Astronomy with High Contrast Imaging II", EAS Pub. Series, 12, 3

Marois, C., Doyon, R., Nadeau, D., Racine, R., Riopel, M. Vallée, P., & Lafrenière, D. 2005, PASP, 117, 745

Racine, R., Walker, G., Nadeau, D., Doyon, R., & Marois, C. 1999, PASP, 111, 587

Rouan, D., Riaud, P., Boccaletti, A., Clénet, Y., & Labeyrie, A. 2000, PASP 112, 1479

Seager, S., Whitney, B., & Sasselov, S. 2000, ApJ, 540, 504

Soummer, R., Aime, C., & Falloon, P. E. 2003, A&A, 403, 369

Wynne C. 1979, Opt. Comm., 28, 21

Direct Imaging of Exoplanets: Science & Techniques
Proceedings IAU Colloquium No. 200, 2005
C. Aime and F. Vakili, eds.
© 2006 International Astronomical Union
doi:10.1017/S1743921306009951

High Contrast Adaptive Optics for Exoplanet Detections: performance limits and optimal wavefront sensing strategy

Olivier Guyon

Subaru Telescope, National Astronomical Observatory of Japan 659 N. A'ohoku Pl. Hilo, HI 96720, USA
email: guyon@naoj.org

Abstract. The effects of photon noise, aliasing, wavefront chromaticity and scintillation on the PSF contrast achievable by ground-based adaptive optics (AO) are evaluated for different wavefront sensing schemes. I show that "classical" AO (sensing in the visible, imaging in the near-IR) is limited to about 10^5 PSF contrast in the central arcsecond because of scintillation chromaticity.

This comparative study shows that a focal-plane based wavefront sensor (WFS), combining wavefront sensing and scientific imaging on the same detector is optimal for high contrast imaging. This approach combines high WFS sensitivity, immunity to aliasing and non common path errors and optical design simplicity. Its theoretical performance is compared to commonly used WFSs, illustrating the advantages of this technique.

I show that such a system can be efficiently used as a second stage after a low-order AO system. Control and data reduction algorithms are presented, as well as possible optical designs incorporating a coronagraph. A laboratory demonstration of this technique is currently being done at Subaru Telescope.

Keywords. adaptive optics, speckle, wavefront sensor.

1. Introduction

Adaptive optics and coronagraphy are the key enabling technologies for the direct imaging of extrasolar planets. Recent developpements in coronagraphic techniques theoretically allow $\approx 10^{10}$ PSF contrast level a few λ/d from the optical axis. This opens the possibility of direct detection of extrasolar terrestrial planets from space (NASA TPF-C mission).

Atmospheric turbulence, even partially corrected by an AO system, prevents ground-based imaging system from delivering PSFs with such high contrast. In most cases, the achieved PSF contrast limit is therefore set by the AO system performance rather than the coronagraph itself.

In this paper, upper limits on how faint the wings of an Adaptive Optics PSF can be made are computed. In this work, it is assumed that a coronagraph cancels the "static" diffraction (Airy rings). The PSF surface brightness normalized the PSF peak surface brightness is used as a quantifier, and thereafter noted **PSF contrast**. Actual detection limits for faint companions are not derived in this work, but the PSF contrast is clearly an important factor towards successfull detection:

- **Photon noise.** Improving the PSF contrast reduces the bright "background" superimposed on a faint companion, and therefore the photon noise that comes with it.
- **Speckle noise.** Reducing the PSF wings brightness also reduces the speckle noise (Bloemhof *et al.* 2001, Bloemhof 2003, Aime & Soummer 2004).

• **Calibration.** Detection of a faint companion requires accurate calibration of the PSF wings which needs to be subtracted. This calibration is more difficult if the PSF wings are brighter.

2. PSF contrast limit

2.1. *The main terms*

The PSF surface brightness is decomposed in this work in 6 terms, which are not a priory correlated, and can therefore be added together:

• C_0: PSF contrast limit imposed by OPD aberrations in uncorrected atmospheric turbulence.

• C_1: PSF contrast limit imposed by amplitude aberrations in uncorrected atmospheric turbulence (scintillation).

• C_2: PSF contrast limit imposed by residual OPD aberrations after AO correction.

• C_3: PSF contrast limit imposed by residual amplitude aberrations after AO correction of OPD and amplitude.

• C_4: PSF contrast limit imposed by the differential OPD between the WFS and imaging wavelengths. This term is caused by the chromaticity of Fresnel propagation.

• C_5: PSF contrast limit imposed by the differential scintillation between the WFS and imaging wavelengths. This term is caused by the chromaticity of Fresnel propagation.

• C_6: PSF contrast limit imposed by the differential OPD between the WFS and imaging wavelengths. This term is caused by the chromaticity of the refraction index of air.

With an AO system correcting the wavefront phase (but without scintillation correction), the PSF surface brightness is $C = C_1 + C_2 + C_4 + C_6$. If phase and amplitude are corrected, the PSF surface brightness becomes $C = C_2 + C_3 + C_4 + C_5 + C_6$.

Analytical expressions for each of these terms are derived in Guyon (2005) for observations at the zenith, and results shown in this paper are derived from these expressions. The relative importance of these terms is shown in figure 1 in a particular example (which details can be found in Guyon (2005)).

It should be noted that terms C_2 and C_3 are function of the **wavefront sensor efficiency**, which is defined and computed for several WFSs in Guyon (2005). This topic will be discussed in more detail in §3.

2.2. *Wavelength choices*

In a well-optimized AO system, able to correct both the phase and amplitude of the incoming wavefront, the residual PSF surface brightness due to chromatic effects (C_4, C_5 and C_6) dominates the contribution due to photon noise and time lag (C_2 and C_3) if $\lambda_{WFS} \neq \lambda_{PSF}$. In the example given in figure 1, where $\lambda_{WFS} = 0.55\mu m$ and $\lambda_{PSF} = 1.6\mu m$, this difference exceeds 3 orders of magnitudes within the central arcsecond.

Unless the chromatic effects can be calibrated (which appears to be feasible for C_6 – Fusco T., private communication), high PSF contrast within the central arcsecond therefore requires $\lambda_{WFS} \approx \lambda_{PSF}$. It is especially interesting to notice that in the example given in figure 1, the contribution of scintillation to the PSF surface brightness is quite close to the contribution due to scintillation chromaticity.

3. Wavefront sensor choice

The results shown in figure 1 assume a wavefront sensor (WFS) with optimal efficiency, which makes optimal use of every detected photon to reconstruct the incoming

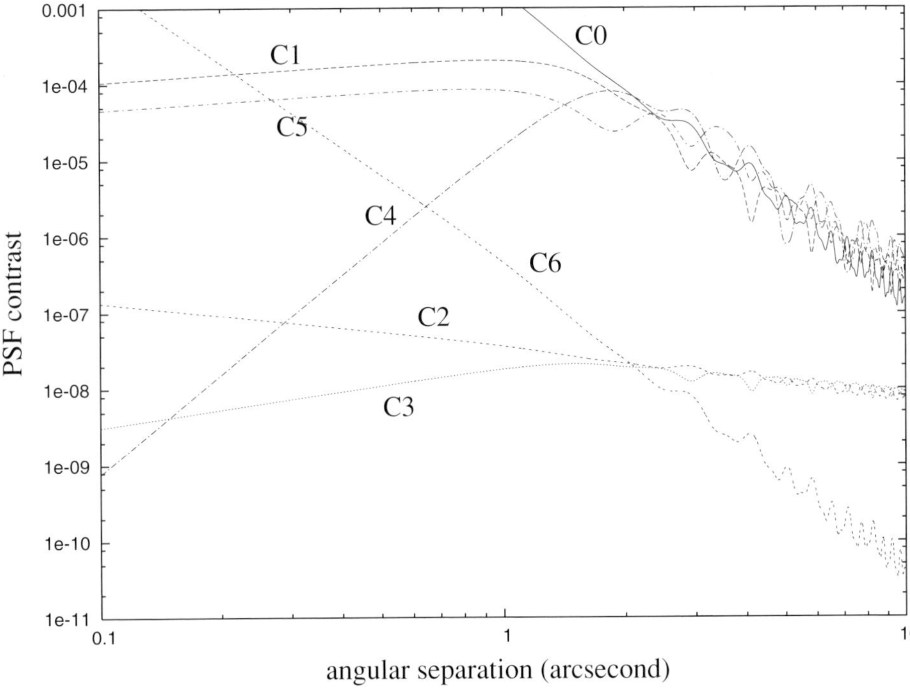

Figure 1. Main sources of PSF contrast limit as a function of angular separation for an 8m telescope. C0 shows what would be the PSF contrast if the atmospheric turbulence were uncorrected. C1 is the atmospheric scintillation. The residual phase (C2) and amplitude (C3) errors after correction if an "ideal" WFS are shown here for a mv=5 guide star. Chromatic effects C4/C5 (chromaticity of OPD and amplitude due to diffractive propagation through the atmosphere) and C6 (chromaticity of the air refraction index) are shown for $\lambda_{WFS} = 0.55\mu m$ and $\lambda_{PSF} = 1.6\mu m$. Exact details of this simulation and analytical expressions for these terms can be found in Guyon (2005).

wavefront's phase and amplitude. As detailed in Guyon (2005), the **efficiency** of a WFS can be quantified by the parameters β_p and β_a for respectively pupil phase and pupil amplitude measurements. In most WFSs, β_p and β_a are a function of the spatial frequency in the pupil plane, or, equvalently, the angular distance to the PSF center.

As can be seen in figure 2, WFSs fall into 2 categories:

• **WFSs sensitive to "noise propagation":** Shack-Hartmann (SHWFS), Curvature (CWFS) and Modulated Pyramid (MPYRWFS). In these WFSs, the efficiency is poor at small angular separation, especially if the number of sensing elements is high.

• **WFSs insensitive to "noise propagation":** Pupil Plane Mach Zehnder (PPMZWFS), Focal Plane (FPWFS, a type of phase diversity measurement), Zernike (ZWFS) and Fixed Pyramid (FPYRWFS). These WFSs maintain a very good efficency at all spatial frequencies simultaneously.

WFSs insensitive to noise propagation appear theoretically superior for high contrast imaging, but offer limited dynamical range, poor linearity, and in some cases limited spectral bandwidth due to chromatic effects. They are therefore adapted to the high-Strehl regime where linearity can be maintained (residual WF errors less than 1 radian at the WFS wavelength).

Although WFSs which are sensitive to noise propagation offer poor efficency, they are usually very linear and are therefore more robust if the residual WF errors are large (which is usually the case in current AO systems). It should be noted that with the proper

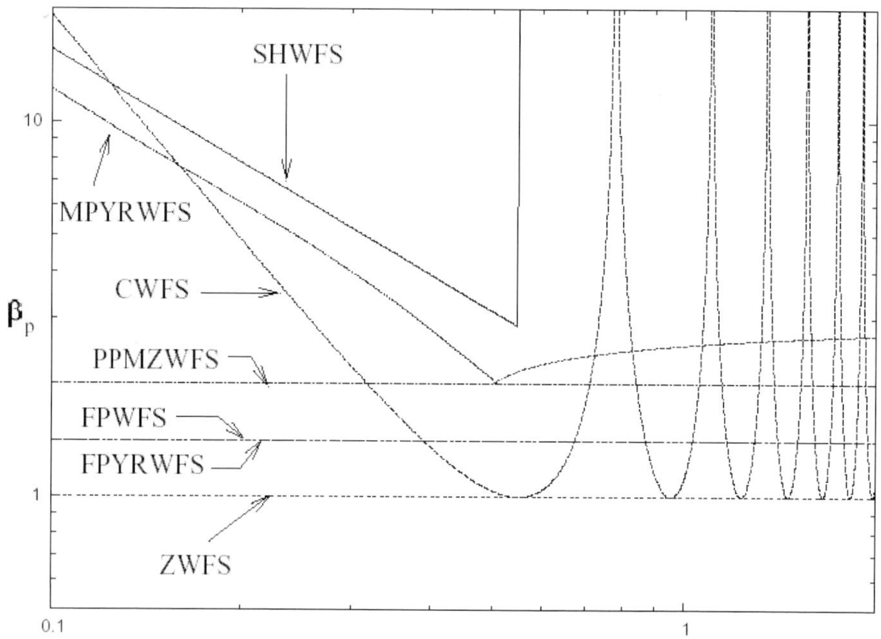

Figure 2. Efficiency of different WFSs for the measurement of phase aberrations in the pupil plane. The efficiency β_p is optimal when equal to 1. The number of photons required to reach a given measurement accuracy is proportionnal to ${\beta_p}^2$. The efficiency is shown as a function of angular separation for a 8m telescope (with $\lambda_{WFS} = 0.55\mu m$ and $\lambda_{PSF} = 1.6\mu m$). Exact details of this simulation and analytical expressions for these terms can be found in Guyon (2005).

choice of actuator sampling in the pupil plane, and by avoiding the use of a coronagraph which is too sensitive to low odrer aberrations, such WFSs can still successfully be used to efficiently "clean" the PSF at a given angular distance. This is illustrated in figure 2 where these WFSs were tuned to deliver optimal efficiency at a 0.5" angular separation.

4. Conclusion

4.1. *Wavefront sensing & control strategies, coronagraphy*

To approach the PSF contrast limit theoretically given by photon noise (terms C_2 and C_3 for $\beta = 1$), a high performance WFS should be used at the imaging wavelength. A phase diversity technique (referred to Focal Plane Wavefront Sensor, or FPWFS, in this work), using the same focal plane detector array for wavefront sensing and scientific imaging, seems ideal for this application. Its main advantages are (1) excellent efficiency at all spatial frequencies (2) low aliasing (3) absence of non-common path errors. The DM, upstream of the coronagraph, introduces the diversity necessary for wavefront reconstruction. The wavefront should preferably be "cleaned" by a first stage AO system prior to the coronagraph/ FPWFS unit. This scheme is shown in figure 3.

Ideally, the second stage AO system needs to be fast (bandwidth >10kHz on bright stars), requiring a frame rate of about 100 kHz: as soon as a few photons per focal plane speckle are detected, a command should be sent to the fast DM (Angel 2003). A fast low readout noise detector is therefore essential for this application. An equally challenging aspect of this system is the required computation speed for the second stage AO system.

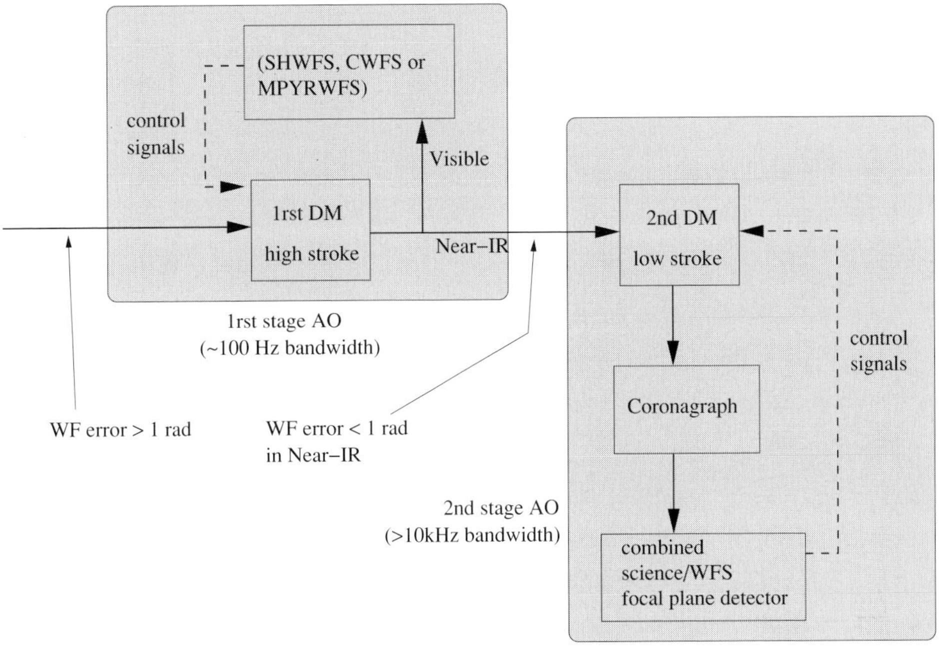

Figure 3. Schematic representation of a 2 stages AO system + coronagraph optimized for high contrast imaging. The first stage AO system can be a conventionnal Shack-Hartmann or Curvature based system, while the second faster AO system is a focal plane WFS. In this example, the science wavelength is "Near-IR", but the same scheme could be used for visible imaging (in which case light might have to be share with the first stage AO system WFS).

4.2. *Imaging extrasolar terrestrial planets from the ground ?*

An Earth-sized planet is about 10^{-10} the brightness of its central star, and a Jupiter-like planet is only about 10 times brighter. The angular separation for these targets is less than 1". Even the optimistic PSF contrast limit derived in this work are quite far from this level. One way to improve the PSF contrast is to increase the telescope diameter: this concentrates the planet's light in a tighter core (size λ/D) which better stands out of the PSF wings. As shown in figure 4, under favorable conditions, a 10^8 contrast is theoretically achievable at 0.2" with a 60m diameter telescope and a high quality coronagraph which suppress the Airy pattern by 3 orders of magnitude. The residual speckles then need to be calibrated to a 10% accuracy to recover a Jupiter-like planet: this calibration requirement is not particularly challenging, especially if most of the speckles average into a smooth background in long exposures.

With the same assumptions, direct detection of an Earth-like planet would require 1% speckle calibration accuracy, which, although challenging, is quite concievable. In this case, the required number of planetary photon for a SNR=7 detection is 5000: with a 60m diameter telescope, the detection would not require prohibitive exposure times.

References

Angel, J.R.P. 2003, *ASP Conf. Ser, 294, 543-556, "Scientific Frontiers in Research on Extrasolar Planets", eds. S. Seager and D. Deming, Washington D.C. 2003*

Bloemhof, E.E., Dekany, R.G., Troy, M., & Oppenheimer, B.R. 2001, *ApJ*, 558, L71

Bloemhof, E.E. 2003 *ApJ*, 582, L59

Guyon, O. 2005, *ApJ* 629, 592

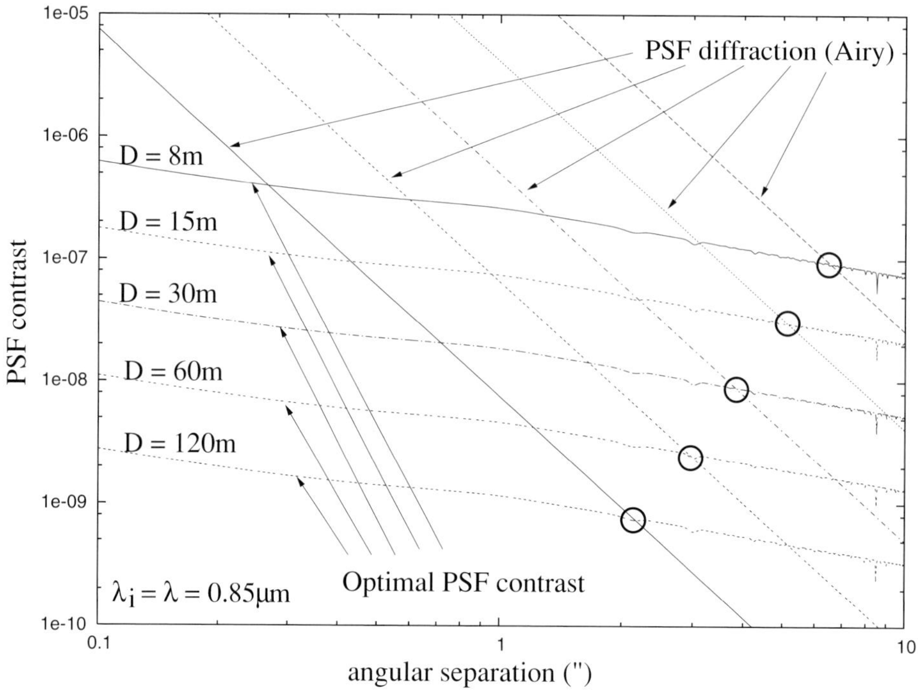

Figure 4. PSF contrast with an optimal WFS ($\beta = 1$) and Airy pattern diffraction for 8m, 15m, 30m, 60m and 120m diameter telescopes. The black circles mark the points where the theoretical PSF contrast achievable with a perfect coronagraph equals the Airy pattern diffraction contrast level.

Discussion

Fusco: Another important effect not mentionned in your work is the chromatic shear of the wavefront due to atmospheric dispersion. This effect can easily dominate others when $\lambda_{WFS} \neq \lambda_{PSF}$ and observations are not made at Zenith.

Fusco: It should be possible to correct for the term C_6 (chromaticity of the refraction index of air) since it is a well known scaling constant.

Direct Imaging of Exoplanets: Science & Techniques
Proceedings IAU Colloquium No. 200, 2005
C. Aime and F. Vakili, eds.
© 2006 International Astronomical Union
doi:10.1017/S1743921306009963

Data processing for exoplanet detection using direct imaging

A. Ferrari[1], M. Carbillet[1], E. Serradel[1], C. Aime[1] and R. Soummer[2]

[1]LUAN, Universié de Nice Sophia-Antipolis,
06108 Nice cedex 2, France – email: {ferrari,marcel,serradel,aime}@unice.fr

[2]Department of Astrophysics, American Museum of Natural History,
79th Street at Central Park West, New York, NY 10024, USA – email: soummer@amnh.org

Abstract. This communication is devoted to data processing of images obtained using an extreme adaptive optics (AO) system and a coronagraph. Specific attention is given to the following degrading factors: the residuals of atmospheric turbulence after AO correction and the "side" effects of the coronagraph. Relying on a statistical modeling of the measurements a test based on short exposure images is proposed. This processing, which generalizes the dark-speckle technique, takes into account the "local" variance of the complex amplitude residuals and the deterministic response of the system (i.e. without atmospheric turbulence).

Keywords. instrumentation: high angular resolution, methods: data analysis, techniques: high angular resolution, techniques: image processing, stars: planetary systems.

---

1. Assumptions and data model

Let us start considering a high-flux regime. We adopt in the sequel the model derived in Aime & Soummer (2004). The complex amplitude resulting from the propagation of light through the atmosphere, to the telescope, and then through an adaptive optics (AO) system and a coronagraph can be decomposed into two terms in the focal plane. The first term is the deterministic response of the coronagraph, we denote it $c(\vec{x})$ where $\vec{x}$ represents the position in the focal plane. The second term comes from the atmospheric turbulence residuals not entirely corrected by the AO system and propagated through the coronagraph, $\Psi_s(\vec{x})$. The complex value $\Psi_s(\vec{x})$ is assumed to be circular Gaussian distributed: $\Psi_s(\vec{x}) \sim \mathcal{N}_c(0, \sigma^2(|\vec{x}|))$. As a consequence, the instantaneous intensity $i(\vec{x})$ is written $|\Psi_s(\vec{x})+c(|\vec{x}|)|^2$. Moreover, a planet in $\vec{x}$ will basically shift $i(\vec{x})$ by a deterministic quantity $m(\vec{x})$.

The Gaussian distribution assumption for the complex amplitude is well established in a single position $\vec{x}$. However, the extension of this hypothesis to a vector containing the complex amplitudes at different positions of the focal plane is not natural, mostly due to the complexity of the transformations operated in the AO system. For this reason we will limit the study to a monodimensional statistical model. Note that, besides the fact that a multidimensional model will introduce extra unknown parameters (the spatial correlation of the complex amplitude), it leads to results which are not analytical (Ferrari *et al.* (2004), Tourneret *et al.* (2005)). To simplify the notations the position $\vec{x}$ will be omitted.

According to the previous assumptions, when $m = 0$ i is proportional to a random variable distributed as a noncentral χ^2_2 distribution (Johnson *et al.* (1995)), also denoted as a Rice distribution in statistical optics. Hence, the probability distribution of i is:

$$p(i;m,c,\sigma) = \frac{1}{\sigma^2}\exp\left(-\frac{i-m+c^2}{\sigma^2}\right) I_0\left(\frac{2c\sqrt{i-m}}{\sigma^2}\right) H(i-m) \tag{1.1}$$

where I_0 is the order 0 first type modified Bessel function. We recall the first two moments of this distribution:

$$\mathsf{E}[i] = \sigma^2 + c^2 + m, \; \mathsf{var}[i] = \sigma^4 + 2c^2\sigma^2. \tag{1.2}$$

Let us now consider the low-flux regime, and in particular the processing of M short exposure images. The number of detected photons in a pixel is distributed as:

$$\Pr(N=n;m,c,\sigma) = \int_0^{+\infty} \frac{i^n}{n!}\exp(-i)p(i;m,c,\sigma)di. \tag{1.3}$$

When $m=0$, this writes:

$$\Pr(N=n;0,c,\sigma) = \frac{\exp\left(-\frac{c^2}{\sigma^2}\right)}{1+\sigma^2}\left(\frac{\sigma^2}{1+\sigma^2}\right)^n {}_1F_1\left(n+1;1;\frac{c^2}{\sigma^2+\sigma^4}\right), \tag{1.4}$$

where ${}_1F_1(a;b;z)$ is a Kummer function which reduces here to a polynomial:

$${}_1F_1(n+1;1;z) = \exp(z)\sum_{k=0}^{n}\frac{n!}{k!k!(n-k)!}z^k. \tag{1.5}$$

Instead, when $m \neq 0$, Eq. (1.3) directly leads to:

$$\Pr(N=n;m,c,\sigma) = \exp(-m)\sum_{k=0}^{n}\frac{m^{n-k}}{(n-k)!}\Pr(N=k;0,c,\sigma) \tag{1.6}$$

2. Test derivation

When all the parameters (here c, σ and m) are known, the optimal test is the Neyman Pearson Detector (NPD) which consists in comparing the Likelihood Ratio (LR) of the data under both hypothesis (here $m=0$ vs. $m \neq 0$) to a threshold. This detector maximizes the probability detection for a fixed Probability of False Alarm (PFA), which can be tuned by the threshold value. When the parameters are unknown a suboptimal technique is the Generalized Likelihood Ratio detector where the unknown parameters are replaced by their maximum likelihood estimators under both hypothesis (see for example Bickel & Doksum (2001)).

The estimation of $m(\vec{x})$ from $\Pr(N=n;m,c,\sigma)$ is complex. However m being assumed small we retain the Locally Most Powerful (LMP) test. This one relying on a Taylor expansion of the LR is optimal for m close to 0. The test decides: $m(\vec{x}) \neq 0$ if $T_{LMP}(\vec{x}) > \xi$ where:

$$T_{LMP}(\vec{x}) = \sum_{k=1}^{M}\left.\frac{\partial \log\left(\Pr(N=n_k;m,c,\sigma)\right)}{\partial m}\right|_{m=0} \tag{2.1}$$

$$= \sum_{k=0}^{\infty}\left.\frac{\partial \log\left(\Pr(N=k;m,c,\sigma)\right)}{\partial m}\right|_{m=0} N_k = -M + \sum_{k=1}^{\infty}\beta(k;c,\sigma)N_k \tag{2.2}$$

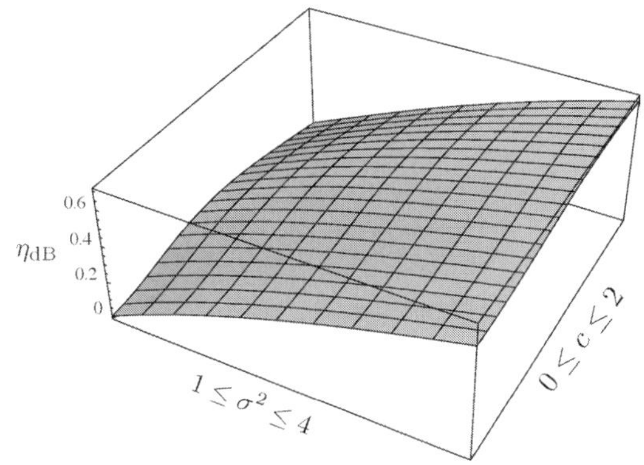

Figure 1. Comparison between the SNR of the long exposure image and the LMP test.

where N_k is the number of times where k photons are detected in $\vec{x}$ and:

$$\beta(n; c, \sigma) = \left(1 + \frac{1}{\sigma^2}\right) \frac{{}_1F_1\left(n; 1; \frac{c^2}{\sigma^2+\sigma^4}\right)}{{}_1F_1\left(n+1; 1; \frac{c^2}{\sigma^2+\sigma^4}\right)} \tag{2.3}$$

Let us note that the well-known Dark Speckle (DS) technique, Boccaletti *et al.* (2000), is equivalent to the test statistic (2.2) were $\forall k > 0$, $\beta(k; c, \sigma^2) = 1$.

3. Test performances

Once the test statistic as been defined, an important point is the derivation of its distribution in order to relate the threshold to the PFA. The N_k follows a joint multinomial distribution with index M and cell probability given by Eq. (1.4) or (1.6). This distribution was used by Ferrari *et al.* (2005) to prove that $\mathrm{SNR}_{\mathrm{NPD}} - \mathrm{SNR}_{\mathrm{DS}} = O(c^2)$ for the model under scope.

However, due to the complexity of the exact distribution, we will focus on the standard asymptotic (M large) distribution of the LMP test statistic (see for example Bickel & Doksum (2001)). When M is large we have:

$$S \stackrel{def}{=} \frac{T_{LMP}}{MI(0)} \stackrel{a}{\sim} \begin{cases} H_0 : \mathcal{N}(0,1) \\ H_1 : \mathcal{N}(m\sqrt{MI(0)}, 1) \end{cases} \tag{3.1}$$

where $I(0)$ is the Fisher information for m when $m \to 0$. A straightforward calculus shows that:

$$I(0) = -\lim_{m \to 0} \mathsf{E}\left[\frac{\partial^2 \Pr(N = n-1; 0, c, \sigma)}{\partial^2 m}\right] = \sum_{n=1}^{\infty} \frac{\Pr(N = n-1; 0, c, \sigma)^2}{\Pr(N = n; 0, c, \sigma)} - 1 \tag{3.2}$$

Using (3.1) the test simplifies to:

$$\text{decide } m = 0 \text{ if } S < \sqrt{MI(0)} z_{1-\mathrm{PFA}} \tag{3.3}$$

where z_α is the α quantile of the standard normal distribution.

At this point, let us note that the SNR of the LMP can be compared to the SNR of the long exposure image, computed using Eqs. (1.2), thanks to the ratio:

$$\eta = \frac{\mathrm{SNR}_{\mathrm{LMP}}}{\text{SNR of long exposure image}} \propto I(0)(\sigma^4 + 2c^2\sigma^2), \tag{3.4}$$

exoplanet	
intensity ratio wrt star	$5\,10^{-5}$
separation angle wrt star	0.5 arcsec
atmosphere+telescope	
Fried parameter r_0 (at 500 nm)	16 cm
number of turbulent layers	5
layers velocities	8—24 m/s
wave-front outer-scale $\mathcal{L}_0$	20 m
telescope diameter	8.22 m
AO system	
sensor type	Shack-Hartmann
sensor configuration	40×40
number of modes	744
time-filter type	pure integration
coronagraph	
coronagraph type	Lyot
mask diameter	$2\lambda/D$
Lyot stop diameter	$0.78\,D$
imaging device	
wavelength band	K (2.2 μm)
nb. of planet phot./ms	≈1
short exposure	1 ms
decorellation interval	5 ms
long exposure	800 ms, $M = 800$

Table 1. Main parameters of the data simulation.

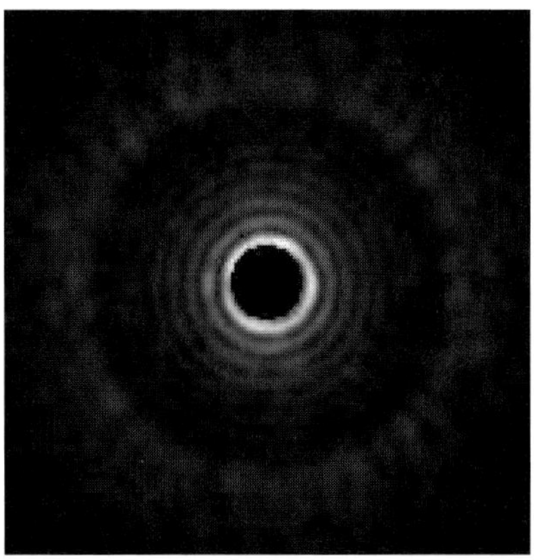

Figure 2. Long exposure masked image.

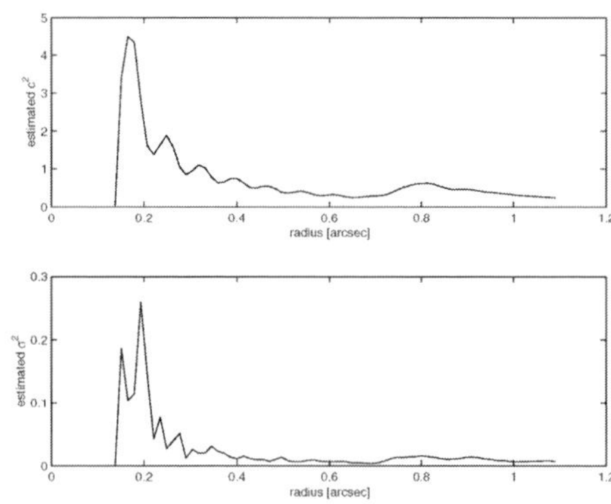

Figure 3. Estimated $c^2(|\vec{x}|)$ and $\sigma^2(|\vec{x}|)$.

A plot of η as a function of c and σ^2 is given in Fig. 1. It shows that the benefit of the proposed detection procedure increases principally with the variance of the turbulence residual. Note that this gain is relevant only in the photon noise limited case, in particular the read-out noise is not considered.

4. Simulations

Table 1 contains the main physical parameters of the performed numerical simulation, made with the Software Package `CAOS` (Carbillet *et al.* (2005)), while Fig. 2 shows the resulting long-exposure image, obtained integrating the data cube short exposures.

Computation of the test statistic $S(\vec{x})$ requires the knowledge of $c^2(|\vec{x}|)$ and $\sigma^2(|\vec{x}|)$. The first could be obtained by a calibration procedure. Analytical expressions of the second have been recently proposed by Fusco & Conan (2004) and Chelli (2005). However, the solution retained here was to estimate $c^2(|\vec{x}|)$ and $\sigma^2(|\vec{x}|)$ directly from the data. These quantities are computed solving Eq. (1.2) where the first and second order moments of $i(\vec{x})$ are estimated by sample averaging on rings of radius $|\vec{x}|$.

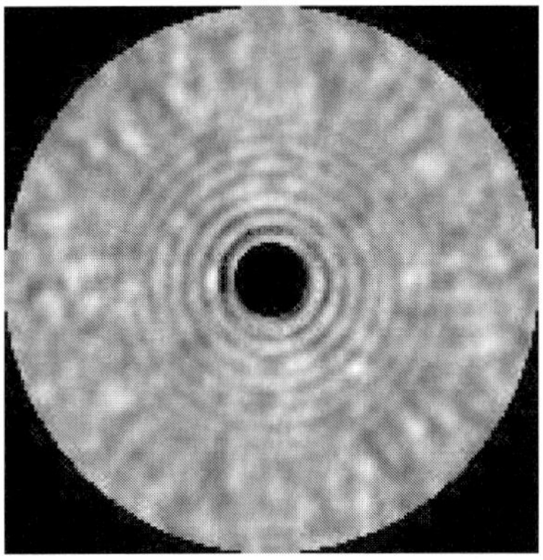

Figure 4. Test statistic $S(\vec{x})$ of the LMP test.

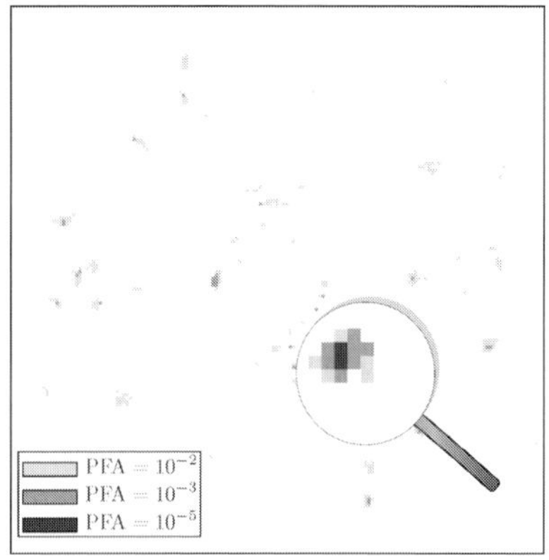

Figure 5. Result of the LMP test.

The result is given in Fig. 3. Figure 4 shows the test statistic $S(\vec{x})$ and Fig. 5 the result of the test for different values of the PFA. The lens is at the location of the simulated planet. In this simulation the planet is the unique event detected with the lower PFA.

5. Conclusions

We have derived a test based on short exposure images for the direct detection of faint unresolved objects in post-AO imaging data. A detailed simulation of the atmospheric and instrumental effects (8-m class telescope, 40×40 elements AO, Lyot coronagraph, K-band imaging) was performed. Our test was applied to a faint companion ($5\,10^{-5}$ relative intensity) at 0.5 arcsec from the main bright star, leading to promising results. Further developments include modeling of additional noise contributions and estimation procedure of the variance of the AO residual.

References

Aime, C. & Soummer, R. 2004, *EUSIPCO* (Vienna)

Bickel, P.J. & Doksum, K.A. 2001, *Mathematical Statistics* (Prentice Hall)

Boccaletti, A., Moutou, C., & Abe, L. 2000, *Astron. Astrophys. Suppl. Ser.*, 141, 157

Carbillet, M., Vérinaud, C., Femenía, B., *et al.* 2005, *MNRAS* 356 (4), 1263

Chelli, A. 2005, *Astron. Astrophys.* 441 (3), 1205

Ferrari, A., Letac, G., & Tourneret, J.-Y. 2004, *EUSIPCO* (Vienna)

Ferrari, A., Tourneret, J.-Y., Serradel, E., & Aime, C. 2005, *GRETSI* (Louvain-la-Neuve)

Fusco, T. & Conan, J.-M. 2004, *J. Opt. Soc. Am. A*, 21 (7), 1277

Tourneret, J.-Y., Ferrari, A., & Letac, G. 2005, *Statistical Signal Processing Conference* (Bordeaux)

Johnson, N. L., Kotz, S., & Balakrishnan, N. 1995, *Continuous Univariate Distributions, Vol. 2* (John Wiley)

Photographs (this page only): Thomas Aime.

Direct Imaging of Exoplanets: Science & Techniques
Proceedings IAU Colloquium No. 200, 2005
C. Aime and F. Vakili, eds.
© 2006 International Astronomical Union
doi:10.1017/S1743921306009975

Suppressing Speckle Noise for Simultaneous Differential Extrasolar Planet Imaging (SDI) at the VLT and MMT

Beth A. Biller[1], **Laird M. Close**[1], **Rainer Lenzen**[2], **Wolfgang Brandner**[2], **Donald McCarthy**[1], **Eric Nielsen**[1], **Stephan Kellner**[2], and **Markus Hartung**[3]

[1]Steward Observatory, University of Arizona, Tucson, AZ 85721
email: bbiller@as.arizona.edu

[2]Max-Planck-Institut für Astronomie, Königstuhl 17, 69117 Heidelberg, Germany
[3]European Southern Observatory, Alonso de Cordova 3107, Santiago 19, Chile

Abstract. We discuss the instrumental and data reduction techniques used to suppress speckle noise with the Simultaneous Differential Imager (SDI) implemented at the VLT and the MMT. SDI uses a quad filter to take images simultaneously at 3 wavelengths surrounding the 1.62 μm methane bandhead found in the spectrum of cool brown dwarfs and gas giants. By performing a difference of images in these filters, speckle noise from the primary can be significantly attenuated, resulting in photon noise limited data. Non-trivial data reduction tools are necessary to pipeline the simultaneous differential imaging. Here we discuss a custom algorithm implemented in IDL to perform this reduction. The script performs basic data reduction tasks but also precisely aligns images taken in each of the filters using a custom shift and subtract routine. In our survey of nearby young stars at the VLT and MMT (see Biller *et al.*, this conference), we achieved H band contrasts >25000 (5σ ΔF1(1.575 μm) >10.0 mag, ΔH > 11.5 mag for a T6 spectral type object) at a separation of 0.5" from the primary star. We believe that our SDI images are among the highest contrast astronomical images ever made from ground or space for methane rich companions.

Keywords. instrumentation: adaptive optics, methods: data analysis, techniques: image processing, (stars:) planetary systems.

1. Introduction

Adaptive optics (AO) systems that are "photon noise limited" can detect an object up to 10^5 times fainter its primary at separations >1" – sufficient to detect a young, self-luminous giant extrasolar planet. However, the contrast problem isn't the only obstacle in directly detecting extrasolar planets – Beyond the contrast limit problem, all AO systems suffer from a limiting "speckle noise" floor (Racine *et al.* 1999). Within 1" of the primary star, the field is filled with speckles left over from instrumental features and residual atmospheric turbulence after adaptive optics correction. These speckles vary as a function of time and color. For photon noise limited data, the signal to noise S/N is proportional to $t^{0.5}$, where t is the exposure time. For speckle-noise limited data, the S/N does not increase with time past a specific speckle-noise floor (limiting contrasts to $\sim 10^3$ at 0.5"). This speckle-noise floor is considerably above the photon noise limit and makes planet detection very difficult. Interestingly, space telescopes such as HST also suffer from a similar limiting speckle-noise floor due to imperfect optics and "breathing" (Schneider *et al.* 2003). Direct detection of extrasolar giant planets requires special instrumentation to suppress this speckle noise floor and produce photon noise limited images.

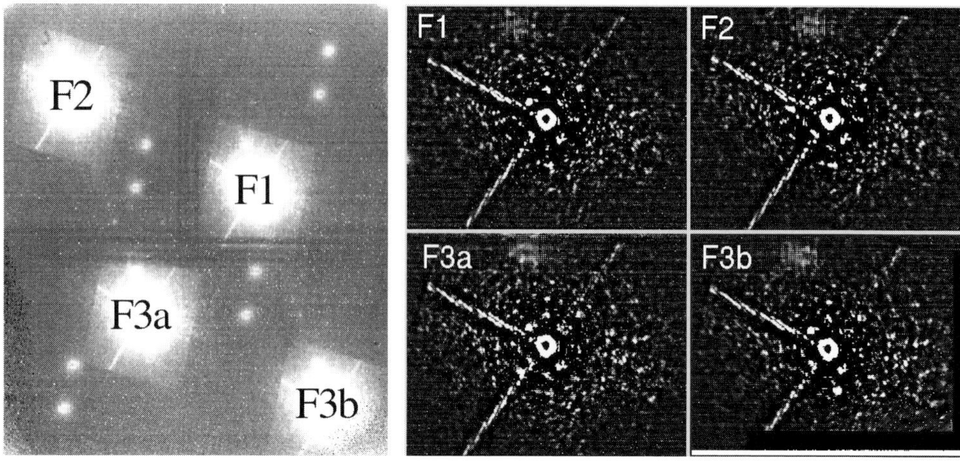

Figure 1. Left: Two minutes of raw SDI data from NACO SDI's 1024×1024 Aladdin array in the CONICA AO camera (Lenzen *et al.* 2004). Right: Same dataset, slightly processed. Apertures have been selected around each filter image. In order to reveal the speckle pattern, a heavily smoothed image was subtracted from the raw images (unsharp masking). The resulting speckle patterns are very similar between the 4 simultaneous images which means that an effective subtraction of speckles can be obtained between the filters.

Simultaneous Differential Imaging is an instrumental method which can be used to calibrate and remove the "speckle noise" in AO images, while also isolating the planetary light from the starlight. This method was pioneered by Racine *et al.* (1999), Marois *et al.* (2000), Marois *et al.* (2002) and Marois *et al.* (2005). It exploits the fact that all cool (T_{eff} <1200 K) extra-solar giant planets are thought to have strong CH_4 (methane) absorption redwards of 1.62 μm in the H band infrared atmospheric window (Burrows *et al.* 2001, Burrows *et al.* 2003). Our SDI device obtains four images of a star simultaneously through three slightly different narrowband filters (sampling both inside and outside of the CH_4 features). These images are then differenced. This subtracts out the halo and speckles from the bright star to reveal any massive extrasolar planets orbiting that star. Since a massive planetary companion will be brightest in the F1 1.575 μm filter and absorbed in the rest, while the star is bright in all three, a difference can be chosen which subtracts out the star's light and reveals the light from the companion. Thus, SDI also helps eliminate the large contrast difference between the star and substellar companions.

2. The SDI AO Cameras at the VLT and MMT

An SDI imaging scheme is currently implemented at the 6.5m MMT (using the MMT AO adaptive secondary mirror and the ARIES AO camera – McCarthy *et al.* 1998) and at the ESO VLT (using the 8m UT4 and the NAOS-CONICA (NACO) AO system) by a group headed by L. Close and R. Lenzen (Close *et al.* 2005, Lenzen *et al.* 2004, Lenzen *et al.* 2005). Both devices are fully commissioned and available for observing.

The SDI technique requires some specialized optics consisting of a cryogenic custom double calcite Wollaston device and a focal plane quad CH_4 filter. Our custom Wollaston splits the beam into four identical beams while minimizing non-common path errors. The *differential* non-common path errors are less than 10 nm RMS per Zernike mode (Lenzen *et al.* 2004). Each of the four beams is fed through one of the filters on the quad filter. Filter wavelengths were chosen on and off the methane absorption feature at 1.62 μm and were spaced closely (every 0.025 μm) in order to limit residuals due to speckle and

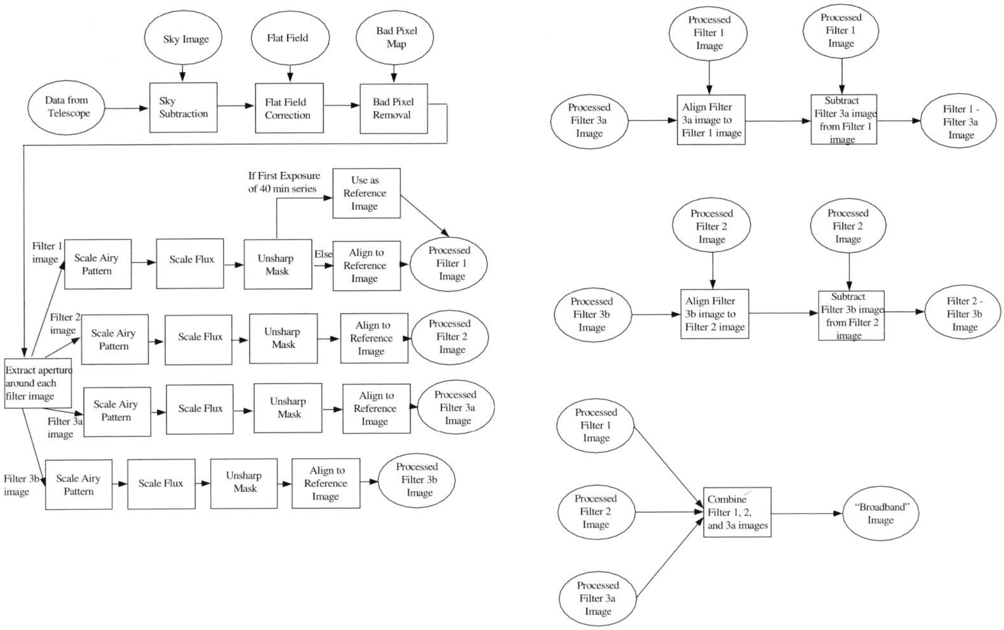

Figure 2. Pipeline Block Diagrams.

calcite chromatism. We used four filters F1, F2, F3a, and F3b with central wavelengths F1=1.575 μm, F2=1.600 μm, and F3a=F3b=1.625 μm. The filters are approximately 0.025 μm in bandwidth. A cold 5"×5" focal plane mask has been implemented as a field stop for the VLT device. No coronagraph is currently used, since the Strehl ratios (~20–30%) are too low to increase the contrasts significantly. The SDI camera has a platescale of $0.017''$/pix.

3. Observational Technique and Data Reductions

A raw dataset from NACO SDI is shown in Fig. 1. The inner 0.2" diameter core is saturated in each image to increase signal in the halo. After unsharp masking, we find that the speckle patterns in each of the separate filters are nearly identical. (See Fig. 1). This bodes well for our ability to attenuate speckle noise.

To distinguish between faint planets and any residual speckles, we observe each object at a variety of position angles. Instrumental and telescope "super speckles" (Racine *et al.* 1999) should not rotate with a change of rotator angle; however, a real planet should appear to rotate by the change in rotator angle. The data is reduced using a custom IDL script. A pipeline block diagram for this IDL script is presented in Fig. 2. Alignments are performed using a custom shift and subtract algorithm. We calculate 2 differences (and one non-differenced combination) which are sensitive to substellar companions of spectral types "T" ($T_{eff} < 1200$ K), "Y" ($T_{eff} \leqslant 600$ K), and "L" ($T_{eff} > 1200$ K). Data taken at different position angles are subtracted (e.g. 20 minutes of data at 0 degrees minus 20 minutes of data at 33 degrees) in order to further attenuate speckle noise.

A fully reduced dataset from the VLT SDI device as well as the same dataset reduced in a standard AO manner is presented in Fig. 3. Simulated planets were inserted into

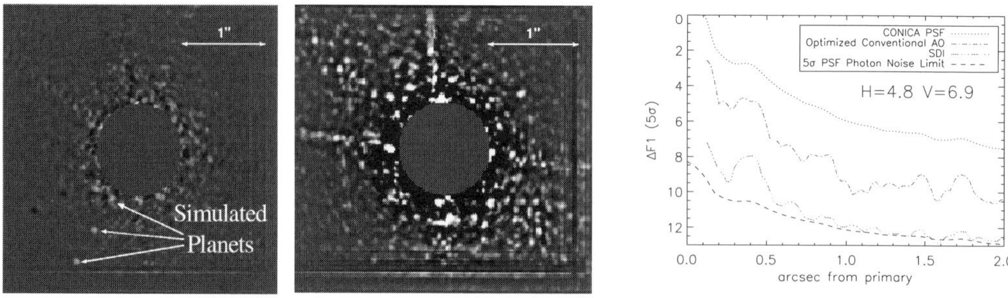

Figure 3. Left: A complete reduced dataset (40 minutes of data at a series of rotator angles – 0°, 33°, 33°, 0°) from the VLT SDI device. Simulated planets have been added at separations of 0.55, 0.85, and 1.35" from the primary, with $\Delta F1 = 10$ mag (attenuation in magnitudes in the F1 1.575 μm filter) fainter than the primary. These planets are scaled from unsaturated images of the example star taken right before the example dataset (and have fluxes in each filter appropriate for a T6 object). Past 0.7", the simulated planets are detected with S/N > 10. Center: Standard AO data reduction of the same dataset. Images have been coadded, flat-fielded, sky-subtracted, and unsharp-masked. Simulated planets have been added with the same properties and at the same separations as before. None of the simulated planets are clearly detected in the standard AO reduction. Additionally, numerous bright super speckles remain in the field. Right: ΔF1 (5σ attenuation in magnitudes in the 1.575 μm F1 filter) vs. Separation for 40 minutes of VLT SDI data for a 70 Myr K1V star at 15 pc. The top curve is the AO PSF. The next curve is the "classical AO PSF" unsharp masked. The third curve down is 40 minutes of SDI 0°-33° data. The last curve is the theoretical contrast limit due to photon-noise. At star-companion separations >0.5", we are photon-noise limited and achieve star to planet H band contrasts >25000 (5σ ΔF1(1.575 μm) >10.0 mag, ΔH > 11.5 mag for a T6 spectral type object) at a separation of 0.5" from the primary star.

Table 1. Properties of Example SDI Survey Stars and Comparison Stars.

Case	Spectral Type	Age	Distance	H	V	ΔF1[1]	ΔH[1]	Exposure Time
A	K2V	30 Myr	45.5 pc	7.1	9.1	10.5	12	40 min
B	K1V	70 Myr	15 pc	4.8	6.9	10.5	12	40 min
C	M3V	30 Myr	24 pc	7.1	12.2	10	11.5	40 min
10 K-M stars[2]	K-M	0-1 Gyr	10-50 pc	6.4-8.7	8-12	8.61	10.31	10-25 min
Gl 86[3]	K1V	10 Gyr	10.9 pc	4.2	6.2	12.8	14.3	80 min

[1] 5σ at 0.5" [2] Masciadri *et al.* IAUC 200 [3] Mugrauer & Neuhäuser 2005.

the dataset pre-reduction. In the SDI reduction, simulated planets (with ΔF1=10 – 5σ attenuation in magnitudes in the 1.575 μm F1 filter) are detected with S/N > 10 past 0.7". In comparison, none of the simulated planets are clearly detected in the standard AO data reduction and numerous bright super speckles remain in the field. A plot of ΔF1 (5σ) vs. separation from the primary is presented in Fig. 3. For this dataset, we achieved star to planet H band contrasts (5σ) >25000 (5σ ΔF1(1.575 μm) >10.0 mag, ΔH > 11.5 mag for a T6 spectral type object) at a separation of 0.5" from the primary star – approaching the photon-noise limit in 40 minutes of data. ΔF1(1.575 μm) and ΔH (for a methane object) for 3 of our survey stars (see Biller *et al.* this conference) as well as for two other comparison objects are shown in Table 1 – it is clear that the achievable contrast varies according to the magnitude of the object and total exposure time. We believe these are among the highest contrast astronomical images taken to date of methane rich companions.

Acknowledgements

BAB acknowledges support through the NASA GSRP program. LMC acknowledges support through NSF CAREER and NASA Origins grants.

References

Burrows, A., Hubbard, W.B., Lunine, J., & Liebert, J. 2001, *Reviews of Modern Physics* 73, 719

Burrows, A., Sudarsky, D., & Lunine, J. 2003, *ApJ* 596, 587

Close, L.M., Lenzen, R., Biller, B.A., Brandner, W., & Hartung, M. 2005, *Science with AO ESO Workshop September 16-19 2003*, 136

Lenzen, R., Close, L., Brandner, W., Hartung, M., & Biller, B. 2004, *Proc. SPIE* 5492

Lenzen, R., Close, L., Brandner, W., Biller, B., & Hartung, M. 2005, *Science with AO ESO Workshop September 16-19 2003*, 46

Masciadri, E. *et al.* 2006, IAUC 200 proceedings

Marois, C., Doyon, R., Racine, R., & Nadeau, D. 2000, *Proceedings of the Astronomical Society of the Pacific* 112, 91

Marois, C., Doyon, R., Racine, R., & Nadeau, D. 2002, *Proc. SPIE* 4008, 788

Marois, C., Doyon, R., Nadeau, D., Racine, R., Riopel, M., Vallée, P., & Lafrenière, D 2005, *Proceedings of the Astronomical Society of the Pacific*, 117, 745

McCarthy, D., Burge, J., Angel, R., Ge, J., Sarlot, R., Fitz-Patrick, B., & Hinz, J. 1998, *Proc. SPIE* 3354, 750

Mugrauer, M. & Neuhäuser, R. 2005, *MNRAS*, 361, L15

Racine, R., Walker, G.A.H., Nadeau, D., Doyon, R., & Marois, C. 1999, *Proceedings of the Astronomical Society of the Pacific* 111, 587

Schneider, G., Becklin, E., Close, L., Figer, D., Lloyd, J., Macintosh, B., Hines, D., Max, C., Potter, D., Rieke, M., Scoville, N., Thompson, R., Weinberg, A., & Windhorst, R. 2003, *solicited by STScI in preparation for HST Cycle 12*

All photographs: Laurent Thareau [`l.thareau@free.fr`].

Direct Imaging of Exoplanets: Science & Techniques
Proceedings IAU Colloquium No. 200, 2005
C. Aime and F. Vakili, eds.
© 2006 International Astronomical Union
doi:10.1017/S1743921306009987

Terrestrial Planet Finder Coronagraph Instrument Design

S. B. Shaklan, K. Balasubramanian, D. Ceperly, J. J. Green, D. J. Hoppe, O. P. Lay, P. D. Lisman and P. Z. Mouroulis

Jet Propulsion Laboratory, California Institute of Technology, 4800 Oak Grove Dr., Pasadena, CA 91009
email: stuart.b.shaklan@jpl.nasa.gov

Abstract. We describe the baseline TPF-C instrument design as of October 2005.

1. Introduction

For the past 2 years, NASA has invested substantial resources to study the design and performance of the Terrestrial Planet Finder Coronagraph (TPF-C). The work, led by the Jet Propulsion Laboratory with collaboration from Goddard Space Flight Center and several university and commercial entities, encompasses the starlight suppression system, observatory design, performance modeling, materials characterization, primary mirror studies, and a significant technology development effort including a high-contrast imaging testbed that has achieved 1e-9 contrast in a laboratory experiment (Trauger *et al.* (2005)). TPF-C papers covering the full range of these topics appear in the proceedings of the 2005 SPIE meeting (conferences 5905, 5867, and 5895) and are too numerous to list in this brief manuscript.

2. Requirements

A group of U.S.-based scientists is presently preparing the high-level science requirements. The notional requirements call for the capability to detect ~30 terrestrial planets in the habitable zone (HZ) assuming all stars have one such planet (and likewise, ~3 planets if the probability of occurrence of a terrestrial planet around any star is 10%).

The science requirements lead to a challenging set of performance requirements. The reflected visible/NIR light from a half-illuminated earth-like planet at 1 AU from a solar-type star is $\sim 10^{10}$ times fainter than the starlight. For a star at a distance of 10 pc, the planet appears 100 mas from the star. Studies of observing completeness (Brown (2004), Brown (2005)) and our own mission completeness studies find that to meet the science requirements for terrestrial planet detection, TPF-C should achieve the performance requirements listed in Table 1.

Table 1: TPF-C Coronagraph Contrast Error Budget Requirements

Parameter	Requirement	Comment
Static Contrast	$6\text{x}10^{-11}$	Coherent terms
Contrast Stability	$2\text{x}10^{-11}$	Thermal + Jitter
Instrument Stray Light	$1.5\text{x}10^{-11}$	Incoherent light
Inner Working Angle	$4\ \lambda/D_{long}$	57 mas at λ=550 nm, D_{long} = 8 m
Outer Working Angle	$48\ \lambda/D_{short}$	1.5 arcsec at λ=550 nm, D_{short}=3.5 m

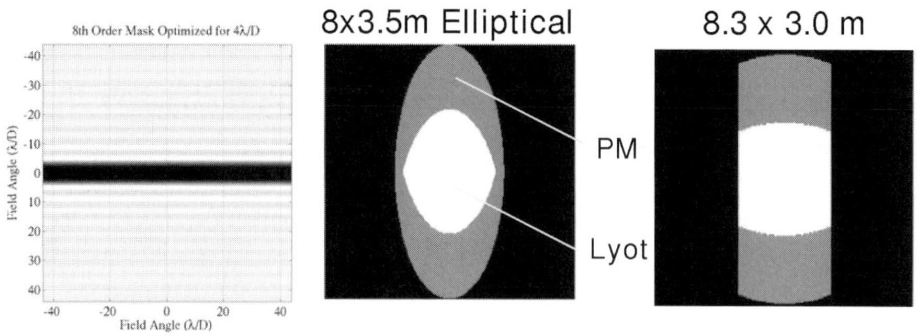

Figure 1. Left: the transmission function of the 8th-order coronagraph mask. Middle: the original 8x3.5 m elliptical paerture and Lyot stop for operation at $4\lambda/D$. Right: the 8.3 x 3 m 'bandaid' design and corresponding Lyot stop.

"Contrast" refers to the level of light scattered by the instrument, relative the incident star light. The contrast stability is the parameter that determines the limiting signal-to-noise ratio (SNR). At a level of 2 x 10^{-11}, a planet having contrast 10^{-10} (mag = 25) can be detected with an instrument-limited SNR = 5. The inner working angle (IWA) is the closest point to the star where a planet can be detected. Extensive modeling (e.g. Green & Shaklan (2003)) has shown that for a Lyot coronagraph, it is impractical to work within 3-4 /D ($\sim$ the third Airy ring) because as one removes diffraction at smaller working angles, the Lyot aperture is reduced while aberration sensitivity increases. Our inner working angle of 4 λ/D represents a compromise between the required resolution ($\sim$60 mas), the largest aperture that can fit in an existing launch shroud, and the engineering requirements at the IWA. The outer working angle (OWA) is fundamentally limited by the number of actuators in the deformable mirror (DM). An OWA of 1.5 arcsec is achieved by spanning the short axis of the primary mirror with 96 actuators.

3. Instrument Design

3.1. *Apertures*

The aforementioned models show that monolithic apertures are far superior to segmented apertures in a high-contrast coronagraph. Segments lead to significant throughput losses at the segment edges, but more importantly relative piston motions of the segments scatters light across the coronagraph 'dark hole' and must be controlled to picometer levels. Monolithic apertures are required but are limited in their dimension by the launch vehicle shroud. Engineers at Ball Aerospace developed the concept of an elliptical aperture coronagraph (Noecker *et al.* (2003)). The long axis delivers the required resolution while the total area provides adequate sensitivity. We have recently evolved this approach from its original elliptical shape to a nearly rectangular aperture (it has the shape of a common bandage), as shown in Fig. 1

The advantages of the 'band aid' aperture are two-fold. First, in the presence of a linear occulter mask, the rectangular shape improves Lyot stop efficiency. For the case shown, it improves from 34% (elliptical) to 44% (rectangular). Second, the shape allows one to employ a so-called 'bar-code' shaped-pupil mask (Vanderbei *et al.* (2004)). These masks are straightforward to build and model, and can be made broad-band and polarization-independent.

Table 2. Coronagraph Trades

Property	Occulter / Lyot	Shaped Pupil	PIAA	Visible Nuller
Practical IWA	3.0-3.5 λ/D	3.5-4.5 λ/D	2.0-3.0 λ/D ?	3.0-3.5 λ/D
Stop Efficiency	40-60%	25%	~100%	40-60%
Immediate Discovery Space	>70% Constrained by occulter	>70% Constrained by limited diffraction shaping	> 90% Constrained by field dependent aberrations	<25% Constrained by fringe pattern and fiber attenuation
Minimum Primary Mirror Dimensions	8x3.5 m	8x3.5 m	< 6x3.5 m	8x3.5 m
SM Despace Stability	20 nm	20 nm	1 nm ?	1 nm
Implementation Complexity	Low	Low	Moderate	High
Critical Issue	Complex errors induced at a focal plane may limit broadband utility	Interaction between high-angle diffraction and high-frequency speckles	Fabrication of optical components. Diffraction?	Implementation complexity

3.2. *Coronagraphs*

Several coronagraph forms have been considered, including band-limited Lyot coronagraphs (Kuchner & Traub (2002)), shaped pupils (Kasdin *et al.* (2003)), remapped pupils (Guyon *et al.* (2005)), a visible nulling instrument (Mennesson *et al.* (2003)), and a vortex-mask coronagraph (Palacios (2005)). Four-quadrant phase masks are not considered because their 2nd-order dependence on wave front tilt does not sufficiently suppress starlight. Vortex masks appear to have good aberration rejection and high throughput but have not been modeled to the same level of detail as band-limited masks. The baseline design includes accommodation for Lyot coronagraphs and shaped-pupils. Both can be implemented using a filter wheel approach to select different masks for different purposes (e.g. improved discovery space vs. deeper contrast over a restricted space).

Table 2 outlines the trade-offs between the various concepts. PIAA (pupil remapping) is the most promising but least mature concept. PIAA offers 100% throughput and 100% discovery space for planets outside the 2nd Airy ring. However fabrication of components and potential diffraction limitations (Vanderbei (2005)) must be explored before this approach is deemed viable. Laboratory tests in a high-dynamic range coronagraph are underway.. The baseline Lyot coronagraph is a linear 8th-order mask (Kuchner *et al.* (2005), Shaklan & Green (2005)). The 8th-order null is effective at filtering low-order aberrations that will result when the primary mirror sees thermal gradients and when the secondary mirror moves relative to the primary. Compared to a 4th-order null (e.g. the visible nuller), the 8th-order mask is 1–2 orders of magnitude less sensitive to changes in aberration content at the IWA. It also offers excellent discovery space – a planet can be detected with nearly equal efficiency beyond the IWA except for a central strip running orthogonal to the mask oscillations. The mask will likely be fabricated in HEBS glass (Trauger *et al.* (2005)) as this continuous-tone approach is the least polarization sensitive. We are currently addressing achromatization of the mask densification process. Binary mask implementations (e.g. notch-filter masks, (Kuchner & Spergel (2003)) have severe polarization and chromaticity issues (Lay *et al.* (2005)). The bar-code shaped-pupil mask shares low aberration sensitivity and large discovery space with the 8th-order mask. It has comparable throughput at the IWA but worse throughput for planets at $> \sim 6\ \lambda/D$. It is the simplest form of coronagraph to implement.

3.3. *Wave Front Control (WFC)*

Our WFC system employs two high-performance Xinetics DMs (Ealey & Trauger (2004)) arranged in a Michelson interferometer configuration. This allows independent control of both phase and amplitude over the dark hole, but the wave front correction has wavelength dependence that does precludes fulll correction of reflectivity and phase-induced amplitude errors over a broad spectral band. We are also considering ways to implement sequential DMs to help conjugate far-from-pupil optics. This approach may also eliminate the need for the Michelson beamsplitter and wedges which are likely to introduce additional chromatic and polarization issues. The DM format is 96 x 96 actuators on a 1 mm pitch. Thermal stabilization of the DMs is crucial – sub-Angstrom wave front stability is required during the extent of our 'set-and-forget' observing scenario (Shaklan *et al.* (2005)).

3.4. *Other Aspects*

The optical design includes an afocal cylindrical telescope that reimages the short axis of the telescope across the full width of the DM (thus forming a square image of the primary aperture on the DM). This increase the outer working angle, significantly expanding the detection space of detection of Jovian planets. The design also includes two orthogonal polarization channels. This was originally implemented for three reasons. First, the slightly different aberration content of each polarization leaked around the 4th order mask resulting in unacceptable static contrast. Second, binary masks are not simultaneously effective in both polarizations. These issues are remedied through the use of 8th order masks and by employing HEBS masks rather than binary ones. Third, the secondary channel offers full redundancy in case of DM or detector failure. We are currently considering eliminating the second channel and readdressing the redundancy issues.

Acknowledgements

The work described here has been performed at the Jet Propulsion Laboratory, California Institute of Technology, under a contract with the National Aeronautics and Space Administration.

References

Brown, R.A. 2004 *ApJ* 607, 1003
Brown, R.A. 2005 *ApJ* 624, 1010
Green, J.J., & Shaklan, S.B. 2003 *Proc. SPIE* 5170, 38
Guyon, O. *et al.* 2005 *ApJ* 622, 744
Ealey, M.A & Trauger, J.T. 2004 *Proc. SPIE* 5166, 172
Kasdin, N.J. *et al.* 2003 *ApJ* 582, 1147
Kuchner, M.J. & Traub, W.A. 2002 *ApJ* 570, 900
Kuchner, M.J. & Spergel, D.N. 2003 *ApJ* 594, 617
Kuchner, M.J. Crepp J. & Ge J. 2005 *ApJ* 628, 466
Lay, O.P., *et al.* 2005 *Proc. SPIE* 5905 (in press)
Mennesson, B. *et al.* 2003 *Proc. SPIE* 4860, 32
Noecker, N.C, Woodruff, R.A., & Burrows C. 2003 *Proc. SPIE* 4860, 72
Palacios, D. 2005 *Proc. SPIE* 5905 (in press)
Shaklan, S.B. & Green, J.J. 2005 *ApJ* 628, 474
Shaklan, S.B. *et al.* 2005 textitProc. SPIE 5905 (in press)
Trauger, J.T. *et al.* 2005 *Proc. SPIE* 5905 (in press)
Vanderbei, R.J., Kasdin, N.J, & Spergel, D.N. 2004 *ApJ* 615, 555.
Vanderbei, R.J. 2005 *ApJ* (submitted)

Direct Imaging of Exoplanets: Science & Techniques
Proceedings IAU Colloquium No. 200, 2005
C. Aime and F. Vakili, eds.
© 2006 International Astronomical Union
doi:10.1017/S1743921306009999

Speckle statistics in direct and coronagraphic imaging

Rémi Soummer[1] †, Claude Aime[2], André Ferrari[2], Anand Sivaramakrishnan[1], Laurent Jolissaint[3], James Lloyd[4], Ben R. Oppenheimer[1], Russell Makidon[5], and Marcel Carbillet[2]

[1]American Museum of Natural History, 79th St at Central Park West, New York, USA
email: rsoummer@amnh.org, anand@amnh.org, bro@amnh.org

[2]Laboratoire Universitaire d'Astrophysique de Nice, Parc Valrose, Nice, France
email: Claude.Aime@unice.fr, Andre.Ferrari@unice.fr

[3]Herzberg Institute of Astrophysics, 5071 West Saanich Road, Victoria, B.C. V9E 2E7, Canada
email: laurent.jolissaint@nrc-cnrc.gc.ca

[4]Department of Astronomy, Cornell University, Ithaca, NY 14853, USA
email: jpl@astro.cornell.edu

[5]Space Telescope Science Institute, 3700 San Martin Drive, Baltimore, USA
email: makidon@stsci.edu

Abstract. In this communication, we study the statistical properties of the light intensity in direct and coronagraphic images, in the context of ground-based Extreme Adaptive Optics observations. The same approach can also be used for space observations with different scales. We show that a coronagraph only affects the perfect part of the wave and leaves the uncorrected part of the wavefront almost unaffected. This statistical model can explain the 'speckle pinning' effect (presence of speckles at the position of the diffraction rings), as an amplification of the speckle noise. This statistical approach can be verified on real adaptive optics data.

Keywords. instrumentation: high angular resolution, instrumentation: adaptive optics.

---

1. Introduction

Direct imaging of a faint companion around a bright star is a difficult task, where the contrast ratio and the angular separation are the relevant observable parameters. The problem consists of detecting a faint source on top of a bright background mainly caused by the star diffracted light. In the case of ground-based observations with adaptive optics, the residual uncorrected aberrations produce random intensity fluctuations of the background, and lead to the presence of speckles in the field. In direct non-coronagraphic images, these speckles mainly appear at the position of the diffraction rings of the star. This phenomenon, also known as 'speckle pinning', Bloemhof et al. (2001), can be explained using a statistical model showing that the speckle fluctuation is amplified at the position of the diffraction rings.

2. Statistical properties of residual speckles in AO- corrected coronagraphic images

In this section, we study the statistical properties of speckles after a coronagraph, for ground based observations using Extreme Adaptive Optics (ExAO). This presentation extends that of Aime & Soummer (2004a), and Aime & Soummer (2004b).

† Michelson Fellow

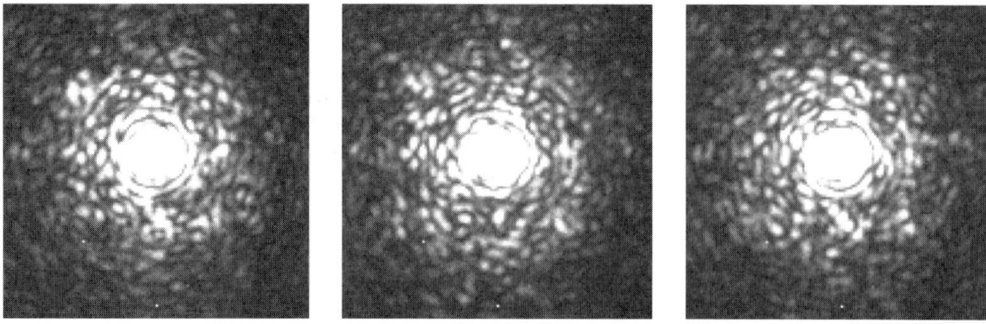

Figure 1. Direct (non-coronagraphic) PSFs from Palomar with AO. The narrow band, short exposures show that the speckles are preferentially located at the position of the PSF rings (speckle pinning). The fluctuations of these speckles is the main source of noise in high contrast imaging.

We consider an instrument with an ExAO system and a generic coronagraph to accomodate any type of focal plane mask designs. We denote by the subscripts 1,2,3,4 the four successive coronagraphic planes: the first plane corresponds to the entrance aperture (possibly apodized), the second plane is the focal plane where a coronagraphic mask is applied, the third plane corresponds to a relay pupil plane where a diaphragm is applied (the Lyot Stop), and finally the fourth plane corresponds to the final focal plane.

Following the notations of Aime & Soummer (2004a), at any instant, we can write the wavefront complex amplitude at the entrance pupil as the coherent sum of two terms, a deterministic term A corresponding to a perfect plane wave, and a random term $a(\mathbf{r})$ corresponding to the uncorrected part of the wavefront. This term $a(\mathbf{r})$ can also include either phase or amplitude errors and has zero mean:

$$\Psi_1(\mathbf{r}) = [A + a(\mathbf{r})]\ P(\mathbf{r}), \tag{2.1}$$

where the function $P(\mathbf{r})$ describes the aperture transmission, and $\mathbf{r} = (x, y)$ is the coordinate vector, used in both pupil and field.

In the first focal plane, a coronagraphic mask is applied at the center of the image of the star image. Writing the mask transmission as $1 - M(\mathbf{r})$, allows to accomodate any type of mask coronagraphs. For example, a classical hard-edged Lyot coronagraph, is described using a top-hat function for M. The complex amplitude of the wave in the focal plane is given by a scaled Fourier Transform (FT) of this pupil amplitude Goodman (1996):

$$\Psi_2(\mathbf{r}) = \frac{1}{\imath\lambda f}\ \mathcal{F}[\Psi_1(\mathbf{r})]\,(1 - M(\mathbf{r}))\,, \tag{2.2}$$

where the symbol $\widehat{}$ denotes the FT, f the telescope focal length, λ the monochromatic wavelength, and the notation $\mathcal{F}$ the scaled FT:

$$\mathcal{F}[\Psi_1(\mathbf{r})] = \widehat{\Psi}_1\left(\frac{\mathbf{r}}{\lambda f}\right). \tag{2.3}$$

For clarity, we will omit in the following the scaling factor $1/\imath\lambda f$ that appears between each of the planes, as a proper re-scaling leads to the same result, and we have:

$$\Psi_2(\mathbf{r}) = (A\,\mathcal{F}[P(\mathbf{r})] + \mathcal{F}[a(\mathbf{r})\,P(\mathbf{r})])\ (1 - M(\mathbf{r}))\,, \tag{2.4}$$

In the next pupil plane, the complex amplitude before the Lyot stop $P'(\mathbf{r})$ is:

$$\Psi_3(\mathbf{r}) = A\,P(\mathbf{r}) - \frac{1}{\lambda^2 f^2}\,A\,P(\mathbf{r}) * \mathcal{F}[M(\mathbf{r})] \\ + a(\mathbf{r})P(\mathbf{r}) - \frac{1}{\lambda^2 f^2}\,a(\mathbf{r})P(\mathbf{r}) * \mathcal{F}[M(\mathbf{r})] \tag{2.5}$$

The first line is proportional to the Lyot stop complex amplitude for a perfect coronagraph Soummer *et al.* (2003). The factor $1/\lambda^2 f^2$ is due to the two-dimension change of variable in the convolution product. In the case of an APLC, the Lyot stop is identical to the pupil. In either case, we have: $P(\mathbf{r})P'(\mathbf{r}) = P'(\mathbf{r})$.

In the final focal plane, with the notations $S(\mathbf{r}) = \mathcal{F}[a(\mathbf{r})\,P'(\mathbf{r})]$, we obtain:

$$\Psi_4(\mathbf{r}) = A\,\Psi_{coro}(\mathbf{r}) + S(\mathbf{r}) - \frac{1}{\lambda^2 f^2}(S(\mathbf{r})M(\mathbf{r})) * \mathcal{F}[P'(\mathbf{r})], \tag{2.6}$$

where Ψ_{coro} denotes the focal wave amplitude of the coronagraph in the perfect case.

The convolution term of Eq.2.6 can be neglected outside the mask area: the extension of the term $S(\mathbf{r})M(\mathbf{r})$ is limited to the mask size and the equivalent width of $\mathcal{F}[P'(\mathbf{r})]$ is one resolution element. Therefore, the spatial extension of their convolution product is limited approximately to the mask size, which is illustrated in Fig. 2.

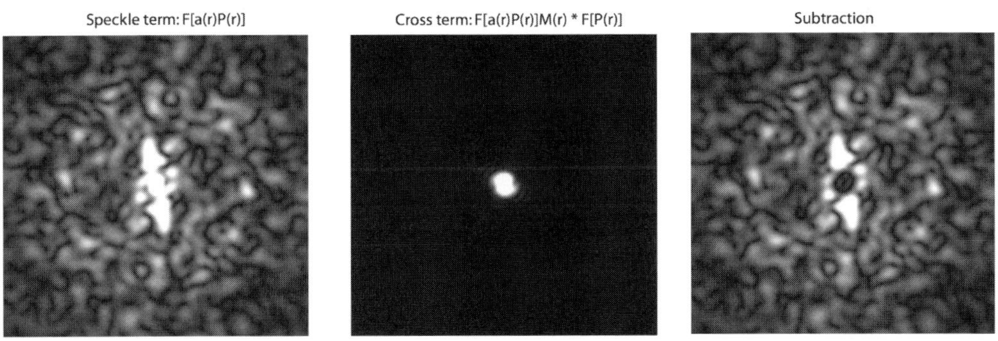

Figure 2. illustration corresponding to Eq.2.6. Left is the speckle term $S(\mathbf{r}) = \mathcal{F}[a(\mathbf{r})\,P(\mathbf{r})]$ (modulus). Center: the cross term $S(\mathbf{r})M(\mathbf{r}) * \mathcal{F}[P'(\mathbf{r})]$. This illustrates that the cross-terms can be neglected outside the mask area. Right: Modulus of the subtraction of the two amplitudes: this shows there is no modification outside the central region corresponding to the mask size.

We obtain a similar expression to the case without coronagraph with $C(\mathbf{r}) = A\,\mathcal{F}[P(\mathbf{r})]$ Aime & Soummer (2004a):

$$\Psi_4(\mathbf{r}) = \tilde{C}(\mathbf{r}) + S(\mathbf{r}). \tag{2.7}$$

The wave amplitude after a coronagraph appears as a sum of a deterministic term $\tilde{C}(\mathbf{r})$, and a random term $S(\mathbf{r})$, at each position in the focal plane, outside the mask area. The term $\tilde{C}(\mathbf{r})$ corresponds to the focal amplitude after the coronagraph. Static aberrations (deterministic) can also be included in this term. The second component is a random term, associated with the speckles, identical to that of Aime & Soummer (2004a): $S(\mathbf{r}) = \mathcal{F}[a(\mathbf{r})\,P(\mathbf{r})]$. The coronagraph has a negligible effect of the speckle part, as illustrated and Fig. 2.

In the case of coronagraphic images as well as in direct images Aime & Soummer (2004a), the statistics of the focal plane complex amplitude follows a non-central gaussian statistics. This problem is formally equivalent to the study of laser speckles over a coherent background in the context of holography. The probability density function

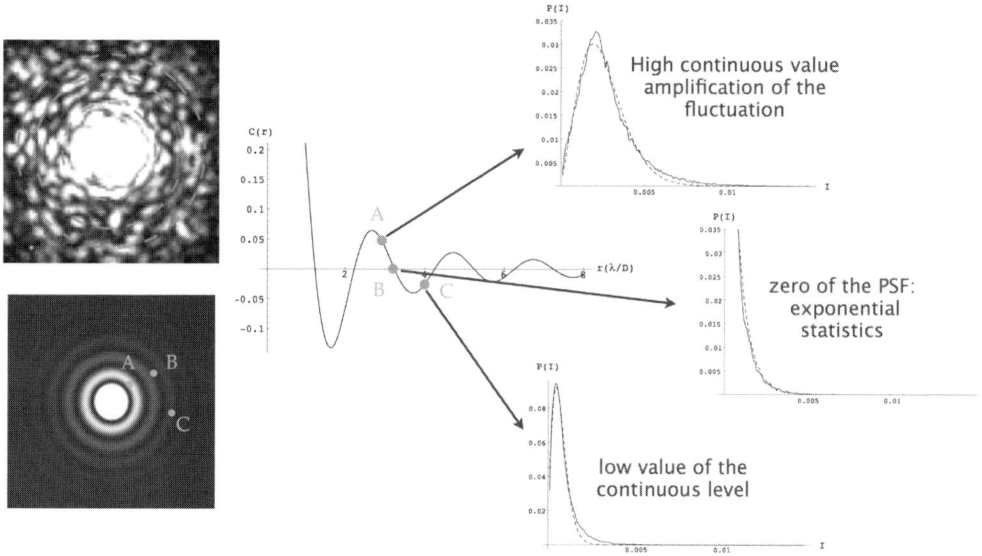

Figure 3. Probability Density Function of the light intensity at 3 different positions in the focal plane, corresponding to different background field amplitude C(r) (or intensity levels Ic. The hard lines correspond to the result of a numerical simulation using 3000 instantaneous PSF (PAOLA Software, Jolissaint 2004). The dotted curves correspond to the theoretical model fitted to the numerical simulations.. The width of the distribution clearly increases with an increase in the level of the constant intensity background: this explains speckle pinning: speckle fluctuations are amplified by the constant background corresponding to the perfect part of the wave.

(PDF) of $\Psi_2(x, y)$ was given by Goodman (1975):

$$\mathcal{P}(\xi, \eta) = \frac{1}{\pi < |S(x,y)|^2 >} \exp\left(\frac{-(\xi - C(x,y))^2 + \eta^2}{< |S(x,y)|^2 >}\right), \tag{2.8}$$

where ξ and η denote the real and imaginary part of $\Psi_2(x, y)$ at the position (x, y).

The corresponding PDF for the intensity, known as a *modified Rician density*, was given by Goodman (1975) and also used by Cagigal and Canales (1998), Cagigal and Canales (2000):

$$\mathcal{P}_I(I) = \frac{1}{I_s} \exp\left(-\frac{I + I_c}{I_s}\right) \mathrm{I}_0\left(\frac{2\sqrt{I}\sqrt{I_c}}{I_s}\right), \tag{2.9}$$

where I_0 denotes the zero-order modified Bessel function of the first kind. This PDF is also valid in the case of coronagraphic images, I_c being the coronagraphic response to a perfect wave. A numerical simulation is shown in Fig. 3 to illustrate this Rician distribution. The intensity statistics are generated from the simulated PSFs, at a given radius. PDFs examples are given for three different radii, one at the top of an Airy ring (strong pinning effect), one at a PSF zero (no speckle pinning) and one at an intermediate position.

Speckle pinning can be easily explained from the analysis of these intensity PDFs: speckle intensity and fluctuations are amplified by the coherent part of the wave amplitude $C(r)$, or intensity I_c. This can be seen directly on the PDFs in Fig. 3, where the widths increase with I_c. Depending on the amplitude of the Airy pattern at successive rings, the intensity I_c is alternatively large and small and the variance of the speckles is amplified accordingly. At the zeroes of the PSF, no amplification occurs and the statistics is equivalent to that of a fully developed speckle pattern (exponential statistics). In the

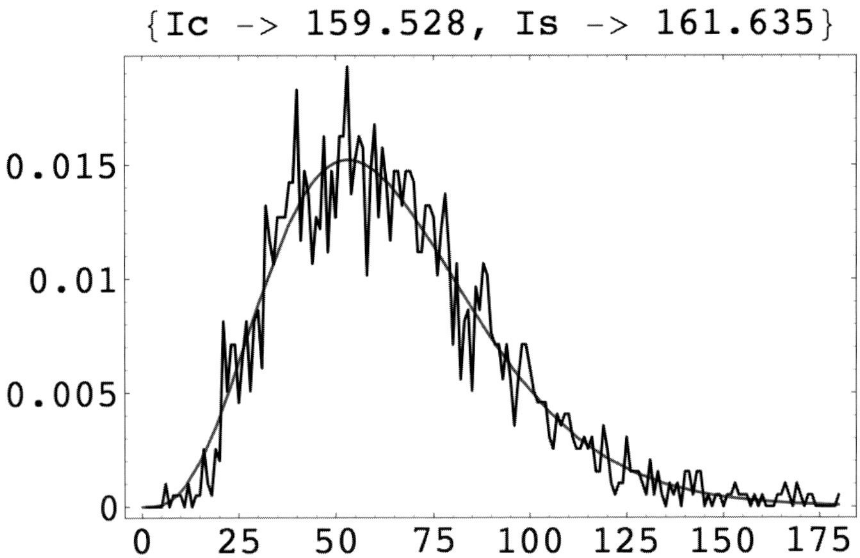

Figure 4. This is the comparison between the empirical PDF and the statistical model, for a given pixel in the field. The model used here corresponds to the integration of 4 independent speckle realizations. The empirical PDF at this location has been obtained using 1500 short exposure images in the K-band with the Palomar AO system.

case of a perfect coronagraph, speckle pinning is totally suppressed and all the residual speckles are similar to laser speckles with exponential statistics.

These statistics has been verified on direct images (non-coronagraphic) by Fitzgerald & Graham (2005) using AO data at the Lick Observatory. Using data from Palomar, we also find good agreement between the statistical model and the data. The data consists of non-coronagraphic images, obtained with the Palomar AO system and narrow band filters in the K-band. The exposure times of 120ms integrate several realizations of speckles, assuming a typical lifetime of 30-40 ms in K-band. It is possible to generalize the rician model to obtain the PDF for the sum of M independant speckle realizations. The corresponding statistics for such integrated speckles is also known as a non-central chi square distribution of order M:

$$\mathcal{P}(I) = \frac{M}{I_s}\left(\frac{I}{I_c}\right)^{\frac{M-1}{2}} e^{-\frac{M\,(I+I_c)}{I_s}} I_{M-1}\left(\frac{2\sqrt{I}\sqrt{I_c}M}{I_s}\right) \tag{2.10}$$

Using a Kolmogorov-Smirnov test, it is possible to study the compatibility between the model and the data as a function of the speckle lifetime. We find the best agreement between model and data for $M = 4$ (sum of 4 independent realizations in each exposure), which is compatible with the expected lifetime in K-band. We give an example in Fig. 4 where a good agreement is obtained between data and model.

A direct application of this statistical model is to provide the variance of the speckle noise which can be used for further simulations to predict the performance of instruments with coronagraphs.

Acknowledgements

Rémi Soummer is supported by a Michelson Fellowship, under contract with the Jet Propulsion Laboratory funded by NASA. JPL is managed for NASA by the California Institute of Technology.

References

Aime, C. & Soummer, R., 2004, *ApJ*, 612, L85
Aime, C. & Soummer, R., 2004, *EAS Publications Series*, 12, 89
Aime, C., Soummer, R., & Ferrari, A., 2002, *A&A*, 389, 334
Bloemhof, E. E., Dekany, R. G., Troy, M., & Oppenheimer, B. R., 2001, *ApJ* 558, L71
Bloemhof, E. E., 2003, *ApJ* 582, L59
Cagigal, M. P. & Canales, V. F., 1998, *Optics Letters* 23, 1072
Cagigal, M. P. & Canales, V. F., 2000, *JOSA* 17, 1312
Fitzgerald, M. & Graham, J., 2005, *ApJ preprint doi:10.1086/'498339'*
Goodman, J., 1975, in *topics in applied physics: laser speckle and related phenomena, Dainty Ed.*, springer verlag berlin
Goodman, J., 1996, *Introduction to Fourier Optics*, Mac Graw Hill
Soummer, R., Aime, C., & Falloon, P. E., 2003, *A&A*, 397, 1161

Direct Imaging of Exoplanets: Science & Techniques
Proceedings IAU Colloquium No. 200, 2005
C. Aime and F. Vakili, eds.
© 2006 International Astronomical Union
doi:10.1017/S1743921306000019

Earth-like planet detection with Extremely Large Telescopes: fundamental limitations

Cavarroc Céline[1], **Boccaletti Anthony**[1], **Baudoz Pierre**[1], **Fusco Thierry**[2] **and Rouan Daniel**[1]

[1]Observatoire de Paris-Meudon, [2]ONERA

Abstract. Direct detection and characterization of Earth-like planets from the ground is a very challenging issue. Among the projects, the Extremely Large Telescopes are very promising to improve the angular resolution and to increase the total number of collected photons. We studied this type of instruments in a very optimistic case to evaluate what level of aberrations limits fundamentally the detection. For that purpose, we considered a perfect coronagraph coupled with an extreme adaptive optics device. Even with a Strehl ratio of more than 96%, it only provides a contrast of $10^{-6} - 10^{-7}$ at $30\lambda/D$. A calibration system downstream the coronagraph is therefore mandatory to reach the contrast of 10^{-10} between a terrestrial planet and its star in the near infra-red. We modelized a very general system taking into account dynamic aberrations left uncorrected by the adaptive optics system, static aberrations of the system and differential static aberrations due to the calibration channel. Numerical simulations demonstrate that the static aberrations are becoming very limitative and must not be neglected. Indeed, to achieve a contrast of 10^{-10}, with common aberrations of 5 nm on a 100 meter telescope, the differential aberrations must be controlled at the level of 200 picometers. We also compare this speckle noise to the limitation due to the photon noise.

Keywords. instrumentation: high angular resolution, instrumentation: adaptive optics, planetary systems.

1. Introduction

After the discovery of more than 150 extra-solar planetary systems for ten years by indirect methods, direct imaging and characterization of Earth-like planets has became the next challenge. Extremely Large Telescopes (ELT) are promising tools to reach the required performance. Several concepts of ELT exist around the world with sizes from 30 to 100 m. This new generation of telescopes will require complete studies to evaluate their performances. The present study takes place in this context.

The detection of Earth-like planet in the near infra-red imposes to reach a contrast of 10^{-10} at close angular distance to a star. For that purpose, we consider a coronagraph associated with a extreme adaptive optics system. With this system, even very performing, we may not expect a contrast larger than 10^{-6}–10^{-7} at $30\lambda/D$. To enhance it, we consider a simultaneous calibration system with one channel which shows the planet whereas the other does not.

Section 2 presents the system and the hypotheses we made. Then, we demonstrate (section 3) the performance which can be reached from this system and what is limitative. Finally, in section 4, we analyze the limitations due the photon noise and the performance of the wavefront sensor.

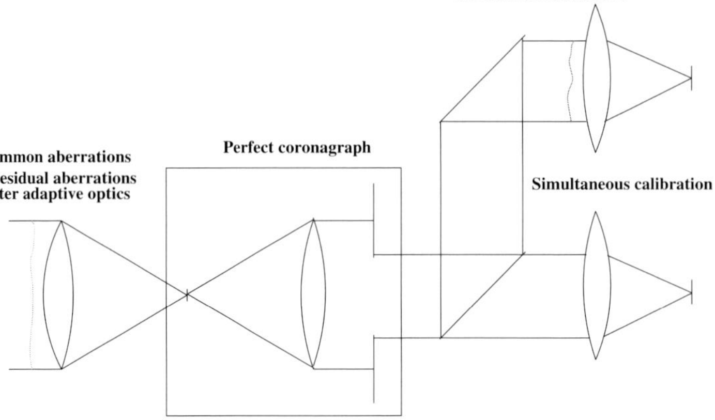

Figure 1. Generic concept of the studied system with from left to right the telescope pupil, the focal plane, the relayed optics with the Lyot stop and differential optics. The intensity distribution at a given plane is defined by the Fourier transform of the upstream plane. Phase aberrations are included in the telescope pupil and in the pupil after the beamsplitter.

2. System

We assume that the instrument is a one hundred meter telescope composed with a classical coronagraphic system and a calibration channel downstream the coronagraph (Fig. 1). This calibration is performed simultaneously to reduce the speckle noise. Besides we consider the perfect case in which the planet is shown in one channel and not in the other. We do not assume any particular system of calibration (several types exist using properties of the planet: its spectral properties Racine (1999), its polarization or its coherence with respect to the star).

The aim of this paper is to study only the effects of the aberrations on the performance of the system and the limitations they impose. So, we assume a monolithic entrance pupil (without segmentation) and a perfect coronagraph, i.e. a coronagraph which removes perfectly all the coherent light. It is simulated by subtracting a perfect pupil to the actuall pupil.

Meanwhile, the system is not perfect: we assume an adaptive optics (AO) system which provides a Strehl ratio of about 98%. The atmospheric phase screens are generated from analytical expressions Rigaut (1998). We represent in Fig. 2 their power spectrum densities (PSD). We consider a Shack-Hartmann wavefront sensor (SHS) with 200 actuators across the pupil diameter (the choice has been made from an extrapolation from VLT-PF). The seeing is 0.75 arcsecond at $lambda = 0.5\mu m$. The imaging wavelength is 1.63 μm. We assume a 5th magnitude star and a loop frequency of 2.5 kHz. The PSD includes fitting errors, the servo-lag and the photon noise of the wavefront sensor. The aliasign is filtered in the wavefront sensor.

We also consider static aberrations of the instrument: common aberrations upstream the coronagraph and differential aberrations due to the difference of optical path between the two channels of the calibration. The classical optical aberrations are represented by a PSD with a slope in f^{-2} with f the spatial frequency Duparré (2002). We consider this type of PSD but with an improvement: we estimate that low frequencies can be corrected (by a better polishing or by a correction -phase diversity for instance) under a given cut-off frequency. The two PSDs are represented in Fig. 2.

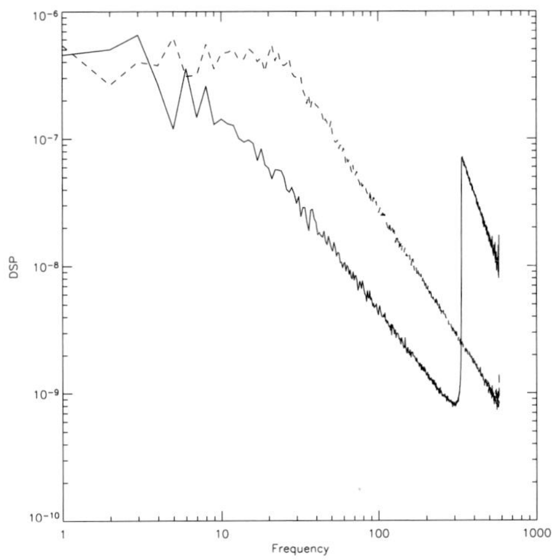

Figure 2. One run of the PSD used in our simulations: the solid line corresponds to the PSD of the residual aberrations after the AO system whereas the dashed line corresponds to the static aberrations (common and differential).

3. Speckle noise

First of all, we will only study the speckle noise and the limitation it imposes on the detectability level. This detectability corresponds to the angular standard deviation.

3.1. *Gain brought by the differential imaging*

We compare the initial detectability level to the one of the image after subtraction for respectively a short and a long exposures. This long exposure is made by adding a few hundreds of images (about 500) obtained from decorrelated phase screens while static aberrations remain unchanged.

Fig. 3 compares these detectability levels at 5 σ assuming a static common path aberration of 5 nm and a differential aberration of 0.2 nm.

The instantaneous detectability is impressed by the AO correction: for a short exposure it limits the detectability, even after subtraction. But by subtracting, we almost gain a factor 1000 for a short exposure.

We note, in Fig. 3, that the final level of detectability for a long exposure is only set by the static aberrations (both common and differential). Indeed, the residual aberrations after AO average azimuthally and do not influence the standard deviation anymore. A more detailed study of this limitation is made in Cavarroc (2005). With common static aberrations of 5 nm and differential ones of 0.2 nm the limit of detectability is about 10^{-10} at 0.1" which is enough to detect Earth-like planets but which are also very challenging values.

3.2. *Influence of static aberrations*

Because the level is set by static aberrations, the shape of the PSF is crucial: if we choose a cut-off frequency lower (or a pure slope in f^{-2}) with the same error budget, the final level will be lower at high frequencies but higher in the region of interest.

Fig. 4 shows the influence of the amplitude of the different types of static aberrations for an angular separation of $30\lambda/D$ (0.1" for a diameter of 100 m). The effect of the common and differential aberrations is clearly different: the detectability varies linearly with the differential aberrations and quadratically with common aberrations.

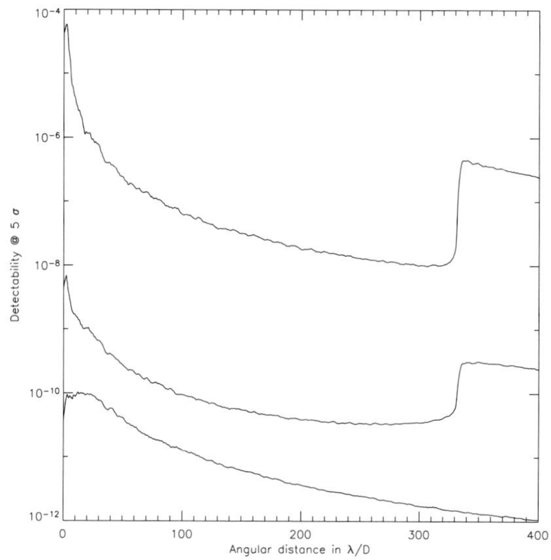

Figure 3. Detectability at 5 σ. The upper curve corresponds to the detectability for a short exposure in a direct imaging mode, the middle curve for a short exposure after subtraction and the lower for a long exposure after subtraction.

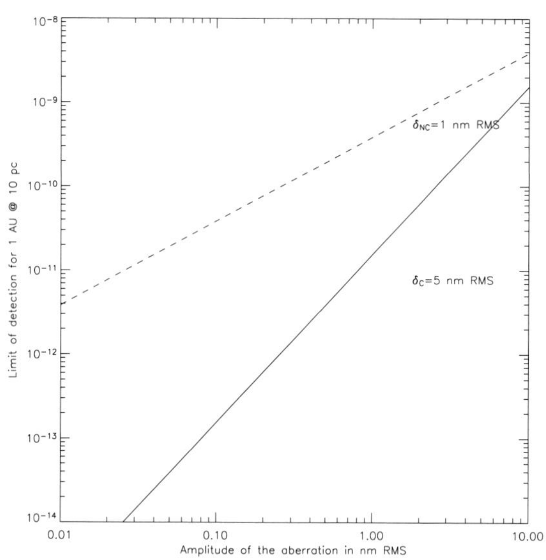

Figure 4. Detectability at 5 σ for an angular separation of $30\lambda/D$ corresponding to 1 AU at 10 pc for a long exposure. The solid line stands for the impact of the common static aberration when fixing the differential aberrations and conversely for the dashed line.

Thus, to reach the desired contrast of 10^{-10} at 0.1", for a common aberration of 5 nm, we have to control differential aberrations to 0.2 nm, which is the assumption we used in this study, and, conversely, for differential aberrations of 1 nm the common aberrations must be reduced under 0.3 nm RMS.

Common aberrations, i.e. aberrations upstream the coronagraph, are dominant and must be reduced as far as possible.

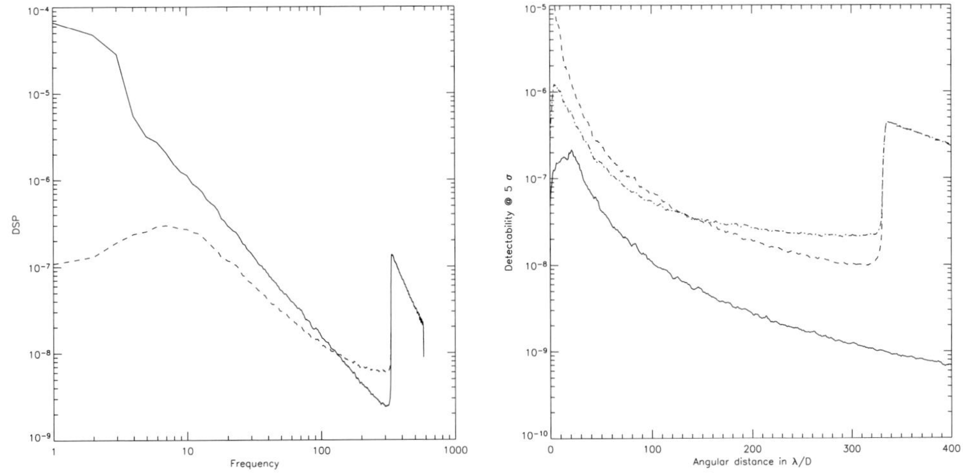

Figure 5. Left figure compares smoothed PSD of AO defects for a SHS (solid line) and a pyramidal sensor (dashed line). Right one compares the detectability level at 5 σ of the images: the solid line corresponds to the level impressed only by static aberrations, the dashed line to the level obtained with a Shack-Hartmann sensor and the dash-dotted line to the level obtained with a pyramidal sensor.

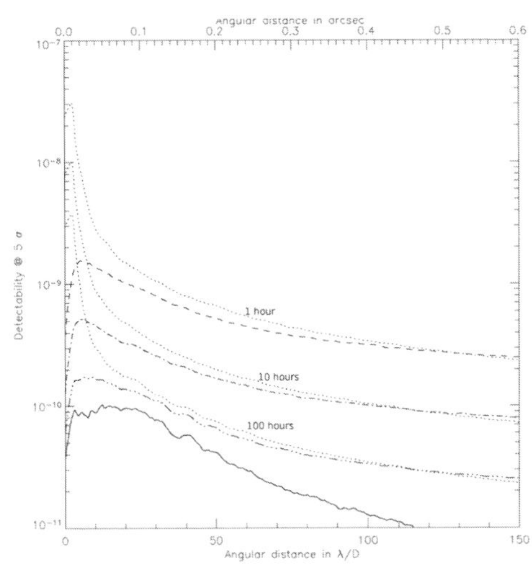

Figure 6. Radial profile of the detectability at 5 σ for different exposure times. The lower curve corresponds to the theoretical limit for an infinitely long exposure. Dotted line corresponds to the level of detectability for a Shack-Hartmann sensor whereas other lines stand for a pyramidal wavefront sensor.

4. Photon noise

Until now, the whole study was led assuming a Shack Hartmann sensor and considering only the speckle noise. This section is dedicated to the study of the limitation of detectability due to photon noise and the interest of using a pyramidal sensor. It is indeed important to know if the level of the contrast set by static aberrations is achievable in a reasonable amount of time. For that purpose, we assume a G2V star of magnitude 5 located at 10 pc, considering a spectral resolution of $\lambda/\Delta\lambda = 5$ and a total transmission of 5%.

Figure 5 compares the Shack-Hartmann and the pyramidal sensors with the same parameters: until $150\lambda/D$, the latter is more interesting than the former because it better corrects these low frequencies.

Figure 6 gives the limit of detectability for several exposure times from 1 to 100 hours for a system with a Shack-Hartmann sensor or with a pyramidal sensor and compares these levels to the limit set by static aberrations. By increasing the exposure time, the detectability converges towards the limit. With a telescope of 100 m, we reach $1.5.10^{-10}$ in one hundred of hours at 0.1" from the center for both systems.

Another relevant point is that the photon noise depends on the level impressed by the adaptive optics. Indeed, it is included in each direct image before subtraction and, as said in section 3, these images are dominated by the residual of AO.

Thus, beyond static aberrations, the detectability capacity of the ELTs relies on the performance of the AO system. Close to the star, the pyramidal sensor is more performing than the Shack-Hartmann sensor but in the region of interest, this difference becomes low and allow us to gain less than a factor of 2 on the photon noise.

5. Conclusion

For this study, we made simple assumptions: the coronagraph is perfect, the pupil is not segmented and has no central obstruction; we chose a quite simple model for the AO system and we took into account static aberrations upstream the coronagraph and downstream to account for the calibration imperfection. In spite of the simplicity, the results are important: to be able to detect an Earth-like planet in the neighborhood (0.1") of a Sun-like star, we have to constrain the static aberrations very strongly, limiting common path aberrations to 5 nm and differential aberrations to 0.2 nm. Another fundamental limitation comes from the exposure time: the aberrations which limit the detectability in the case of direct imaging, namely residual aberrations after AO, will determine the level of the photon noise. Thus reaching 10^{-10} with our hypotheses requires a few hundreds of hours.

The detection of Earth-like planets, even with a telescope of 100 m, is a great challenge and a thorough analysis of the whole system is necessary to determine precisely the capability of detection of such telescopes.

References

Cavarroc, C., Boccaletti, A., Baudoz, P., Fusco, T. & Rouan, D. 2005, *accepted by A&A*

Duparré, A., Ferre-Borrull, J., Gliech, S., Notni, G., Steinert, J., & Bennett, J. M. 2002, *AO* 41, 154

Racine, R., Walker, G. A. H., Nadeau, D., Doyon, R., & Marois, C. 1999, *PASP* 111, 587

Rigaut, F., Salmon, D., Arsenault, R., Thomas, J., Lai, O., Rouan, D., Véran, J. P., Gigan, P., Crampton, D., Fletcher, J. M., Stilburn, J., Boyer, C., & Jagourel, P. 1998, *PASP* 110, 152

Direct Imaging of Exoplanets: Science & Techniques
Proceedings IAU Colloquium No. 200, 2005
C. Aime and F. Vakili, eds.

© 2006 International Astronomical Union
doi:10.1017/S1743921306000020

The potential for imaging low-mass companions to solar-type stars: Utilising radial-velocity data from the AAPS, Keck & Lick projects

J. S. Jenkins[1], H. R. A. Jones[1], C. G. Tinney[2], C. McCarthy[3], B. D. Carter[4], A. J. Penny[5] & D. J. Pinfield[1]

[1]University of Hertfordshire, College Lane, Hatfield, Hertfordshire, AL10 9AB

[2]Anglo-Australian Observatory, PO Box 296, Epping, NSW 1710, Australia

[3]Department of Physics and Astronomy, San Francisco State University, San Francisco, CA 94132

[4]Faculty of Sciences, University of Southern Queensland, Toowoomba, 4350, Australia

[5]Harvard-Smithsonian Center for Astrophysics, 60 Garden St, MA 02138,USA

Abstract. We are about to embark on a project to directly image low-mass companions orbiting within 1.5" of their host stars. Our long-term aim is to reach down into the planetary-mass regime. In order to start this project we have time on the VLT NAOS/CONICA (NACO) facility and in the near future we plan to make use of the Near-Infrared Coronographic Imager (NICI) instrument mounted on Gemini-South. Our targets are F-M type dwarfs, and have been selected from the Anglo-Australian Planet Search (AAPS), Keck and Lick search lists. These surveys have been periodically monitoring over 1200 stars between them for the past ~8 years with precisions typically around 3ms^{-1}. Our target list comprises objects which show significant evidence for companions on long-period orbits. Having *a priori* knowledge of companion characteristics allows us the advantage of prioritising our target list. Given the relatively high incidence of multiple planet systems found so far our search may also serve to provide useful constraints on any large separation, massive companions in our target systems.

Keywords. instrumentation: adaptive optics, instrumentation: high angular resolution, stars: low-mass, brown dwarfs; stars: planetary systems.

1. Introduction

The detection of over 150 planets orbiting sun-like stars has revolutionised our knowledge of our local neighbourhood and our position in it. Yet planets are not the sole close companions to solar-type stars. For instance, Duquennoy, Mayor & Halbwachs (1991) discussed stellar multiplicity in a series of papers, focusing primarily on spectroscopic binaries (SB) found using the CORAVEL spectrometers. The bridge in mass-space between giant planets and SBs is huge, and it is here where cool objects such as Brown Dwarfs (BD) reside.

Radial-velocity (RV) surveys have revealed very few BDs orbiting solar-type stars (leading to the phrase 'brown dwarf desert' being coined to describe this paucity e.g. Marcy & Butler (2000)) though beyond ~4AU there are few RV data points. This arises because the RV method requires *at least* one full orbit to reliably constrain a companion's characteristics, which is a very time consuming process. Direct imaging techniques can be used in conjunction with RV projects to probe the extremities of these systems, without the need to wait for a long period of time, admittedly at the cost of performing a very technically challenging observation. Liu *et al.* (2002) used the Gemini-North and Keck Adaptive Optics (AO) systems to obtain three epochs of images of the BD companion to HR 7672, which had been indicated by the stellar RV signature. The flux ratio was found to be 8.6 magnitudes at a separation of 0.79". This level of contrast at this separation was pushing the limits of the instruments used. However the introduction of Simultaneous Differential Imaging (SDI) has allowed higher contrasts at smaller separations for the coolest stellar companions, e.g. $\Delta H \sim 13$ at $\sim$0.5" from Mugrauer & Neuhauser (2005).

2. Target Selection

2.1. *Angular Separation*

Our target list is drawn from the AAPS, Keck and Lick planet search projects. The lowest mass companions, containing all our planetary-mass objects, have been taken from the compendium of Butler *et al.* (2006). The stars are FGKM-type stars, bright ($H < 7.7$) and are accessible to the VLT and Gemini-South facilities (dec $< +40°$). The other important characteristic these stars share is a long term radial-velocity profile consistent with an orbiting low-mass companion. These companions are inferred to have a wide range of masses ($1 \leqslant M \sin i \leqslant 725\ M_J$), all masses below $1M_J$ were fixed to 1 to fit in with the model limits. They also have accurate distances ($<$ 110pc), which were extracted from the Hipparcos (ESA, 1997) catalogue. It must be noted that unlike the Hipparcos data, the derived orbital properties are not well constrained, mostly because the companion orbits are much longer than the monitoring baselines of the surveys, and in some cases the companion properties have been derived with no inflection in the RV curve. The fit to these objects produce semimajor axis lower limits, which will most likely be much larger than the values we have derived. Also the eccentricity of most of the BDs and SBs are fixed to zero, causing more separation ambiguity.

2.2. *Contrast Ratio*

We found JHK magnitudes for the primary stars in the 2MASS catalogue (http://irsa.-ipac.caltech.edu/) and using the Hipparcos distances we were able to generate accurate absolute JHK magnitudes. For the absolute magnitudes of the companions however, we used the non-grey evolutionary tracks of Burrows *et al.* (1997) and the NextGen models of Allard *et al.* (1997). We assumed ages from Valenti & Fischer (2005) and minimum masses from our Keplerian fits. We split the companions into two groups depending on which model we could use to generate robust magnitudes. The split was made on the basis of companion effective temperatures, with all the objects using the Burrows *et al.* models having an $T_{eff} < 2000$K ($M \sin i < 85M_J$) and all other objects using the NextGen models. A custom spline fitting procedure obtained the bolometric luminosities for all our companions. The model split between the cooler and hotter companions is clearly apparent. We have 156 companions below the strong methane absorption boundary (1200K) (shown by the dot-dashed vertical line on the plot), apart from their too low luminosity we expect them to be good SDI targets. There are also 23 companions with effective temperatures above 1200K, 22 of which have $T_{eff} > 2000$K and are represented by triangles in Fig. 2.

A major problem faced when attempting to image close-in companions to bright stars, is the contrast difference. To work out the contrasts between the stars and their companions we had to obtain the JHK magnitudes of the companions, both from internal and reflected flux. To accomplish this we used the median colours and bolometric corrections (BC) from Leggett *et al.* (2002) (BC_T=2.06; BC_L=3.25, the subscripts denote

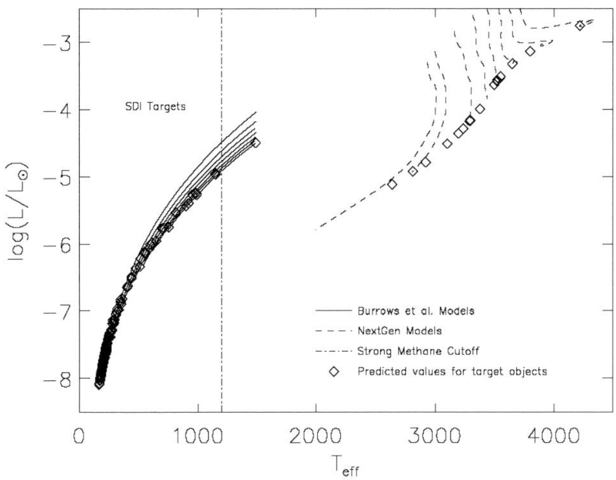

Figure 1. The solid lines are Burrows *et al.* (1997) models for different masses and ages. The dashed lines represent a range of masses and ages for the NextGen models of Allard *et al.* (1997), which is populated by our high-mass targets. The dot-dashed line marks the turn-off of strong methane absorption in cool dwarfs and provides a useful boundary for SDI targets. The discontinuity between the models is the LT boundary region, which is not very well modelled, thus the Burrows *et al.* models were truncated.

spectral type). This correction gave the absolute K magnitudes and, using the colours ($H\text{-}K_T$=-0.04; $H\text{-}K_L$=0.70), we were able to generate their expected absolute H-band magnitudes. For all planetary-mass companions we assumed Jupiter characteristics. We then simplistically simulated the expected reflected flux in the H-band and added this component to the internal flux. This was done by using simple geometry and assuming Jupiter's radius and albedo. The total absolute magnitudes of the companions were then subtracted from the absolute magnitudes derived for the primaries giving the contrast ΔH. Fig. 2 shows the contrasts and separations of all the objects we propose to image. The objects represented by filled circles are all companions with strong methane absorption. The bulk of the objects lie within 0.3" of the primaries due to the bias introduced by the short RV baseline. The majority of these lower-mass objects (M sin $i < 15M_J$) also have high contrasts (ΔH>10), putting them below the NACO-SDI threshold. However, two objects have larger separations >0.3", approaching the separations of the already discovered objects of McCaughrean *et al.* (2003), Chauvin *et al.* (2004) and Neuhauser *et al.* (2005) and, combined with a ΔH<13.5, they could be amenable to SDI imaging. The NACO-SDI (Mugrauer & Neuhauser (2005)) and NICI (Gemini (2005)) sensitivities are highlighted on Fig. 2 by dotted and dashed lines respectively. Once the masses and semimajor axes are more precisely defined, the companion magnitudes and separations will most likely increase giving lower contrasts and more viable targets. This has been highlighted on the plot by arrows on one of the objects near to current detection thresholds. The arrows represent the direction in which all objects are expected to move once inclination and eccentricity effects are considered and more RV data points acquired. Another major source of uncertainty is age. For example, the (1-sigma) age uncertainty of the solid circle companion with arrows plotted is ±2.1 Gyr, giving rise to a ±2.1 mag error in ΔH with the primary. All companions with $T_{\rm eff} > 2000$K (triangles in Fig. 2) have H-band magnitudes less than 15, allowing direct imaging using normal AO

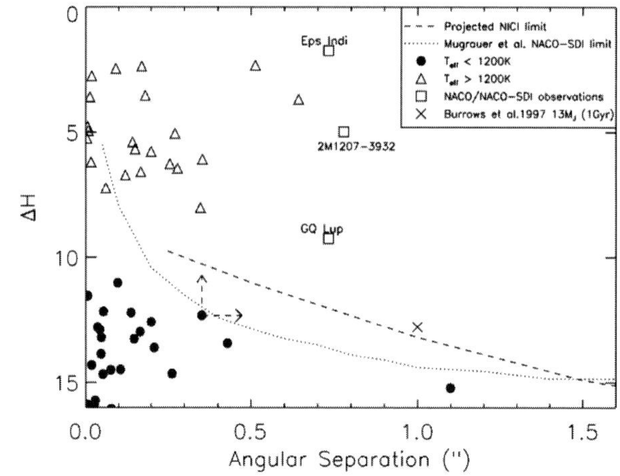

Figure 2. Triangles represent targets with $T_{eff} > 1200K$, filled circles are targets $< 1200K$ (amenable to SDI) and the squares are companions to young stars already found by NACO & SDI. For comparison, the dotted line shows the NACO-SDI sensitivity (Mugrauer & Neuhauser (2005)) and the dashed line shows the NICI sensitivity (Gemini (2005)). The arrows show the most probable shift direction for one of the objects once better RV parameters have been derived and inclination and eccentricity effects considered. The single cross represents a $13M_J$ object at 5AU orbiting a G5V star with an age of 1 Gyr and a distance of 5pc.

techniques. Four of these objects have separations larger than 0.35" and ΔH less than 8, making excellent targets. All planetary-mass companions are off the Fig. 2 plot scale. The cross in Fig. 2 shows the position of a $13M_J$ planet in a Jupiter-like orbit as a companion to a 1 Gyr, Sun-like star at 5pc. This type of object resides extremely close to the plotted instrument thresholds, highlighting just how difficult it is to obtain a direct image of any planetary-mass object with the current suite of instruments available. However, RV studies have revealed a high number of multiple planet systems, therefore imaging planet-host stars may provide useful constraints on any longer period, massive companions not yet revealed in the RV dataset.

References

Allard, F., *et al.* 1997, *ARA&A* 35, 137
Blundell, J., *et al.*, *MNRAS* submitted
Burrows, A., *et al.* 1997, *SPIE* 491, 856
Butler, R.P., *et al.* 2006, *ApJ* in prep
Chauvin, G., *et al.* 2004, *A&A* 425, L29
Duquennoy, A., Mayor, M., & Halbwachs, J.L. 1991, *A&A* 88, 281
Gemini 2005, http://www.gemini.edu/sciops/instruments/nici/niciCampaign.html
Leggett, S.K., *et al.* 2002, *ApJ* 564, 452
Liu, M.C., *et al* 2002, *ApJ* 571, 519
Marcy, G.W. & Butler, R.P. 2000, *PASP* 112, 137
McCaughrean, M.J., *et al.* 2003, *A&A* 413, 1029
Mugrauer, M. & Neuhauser, R. 2005, *MNRAS* 361, L15
Neuhauser, R., *et al.* 2005, *A&A* 435, L13
Valenti, J. & Fischer, D. 2005, *ApJS* 159, 141

Direct Imaging of Exoplanets: Science & Techniques
Proceedings IAU Colloquium No. 200, 2005
C. Aime and F. Vakili, eds.
© 2006 International Astronomical Union
doi:10.1017/S1743921306000032

Advantage of a predictive control law for extreme adaptive optics imaging

B. Le Roux[1] and M. Carbillet[2]

[1] INAF-Osservatorio Astrofisico di Arcetri, Largo E. Fermi, I-50125 Firenze
email: leroux@arcetri.astro.it

[2]LUAN-UMR 6525, Université de Nice-Sophia Antipolis, Parc Valrose, F-06108 Nice cedex 2
email: marcel.carbillet@unice.fr

Abstract. We study the application of a predictive control law based on a Kalman filter for an extreme adaptive optics system. In particular, we discuss the minimization of temporal error and show the evolution of prediction errors with the order of the model. We also discuss the choice of the optimal temporal frequency as a function of the control law and the level of noise. Finally, the gain expected with respect to a non-predictive law is presented.

Keywords. atmospheric effects; instrumentation: adaptive optics; stars: imaging.

1. Introduction

Imaging faint astrophysical objects near bright stars, like exoplanets, is a real challenge for modern astrophysical instrumentation. Recently, a planet has been imaged by Chauvin *et al.* (2005), using the NACO instrument: the faint companion was one hundred times fainter than the brown dwarf. In this case the atmospherical correction provided by the adaptive optics [OA] system NAOS was enough to image the planet, but in general an instrument dedicated to exoplanets imaging should be able to image objects 10^5 to 10^{10} times fainter than the star. This means, among others, that a very powerfull AO system is needed. This is called an eXtreme AO [XAO] system, characterized by a resulting Strehl ratio in the imaging band as extreme as $90-95\,\%$. This challenge has recently motivated the publication of a number of ideas to improve AO, for example in wave-front sensing [WFS] (Nicolle *et al.* (2004), Vérinaud *et al.* (2005)).

We propose here to improve again the AO performance by considering a predictive control law, applying the Kalman-filter-based optimal control law (Le Roux *et al.* (2004)) to the XAO case. We first remind the principle of this optimal approach in Sec. 2. In Sec. 3 we then discuss some of its limits and the way to apply it to XAO, showing how using a high order autoregressive [AR] model to describe the turbulence evolution would decrease the temporal error. We also discuss the trade-off to be found between the temporal error and the WFS error due to the noise, leading to the notion of optimal temporal frequency. Finally we show in Sec. 4 how decreasing the temporal frequency and using a predictive control law permit to increase the performance of an XAO system, with respect to simply use an integrator control law with a high temporal frequency.

2. State-space formalism and optimal control for AO

The optimal control law used here (Le Roux *et al.* (2004)) is based on a Kalman filter in a state-space formalism, built on a state-space model recalled hereafter.

Let us first consider the chronology of events in an AO system: CCD integration for WFS, read-out of CCD, computation and application of new voltages for the deformable mirror. Assuming the same time T for the CCD integration and for the global delay, there is a $2T$ time-lag between the measurement and the application of the correction, time-lag to which is due the so-called temporal error.

The Kalman filter is an estimator of the turbulent phase that is based on a physical representation of the system turbulence$\oplus$AO, the so-called state model. This model should sum up the behaviour of the system and the chronology described before which can be read in three fundamental equations:

- the definition of the residual phase: $\phi^{\rm res} = \phi^{\rm tur} - \phi^{\rm cor} = \phi^{\rm tur} - \mathbf{Nu}$, where $\mathbf{N}$ is the matrix that links the voltages basis to the turbulent modes basis, $\mathbf{u}$ are the voltages, $\phi^{\rm tur}$, $\phi^{\rm res}$ and $\phi^{\rm cor}$ are denoting the turbulent, residual and correction phase;
- the definition of the measurement: $\mathbf{Y}_n = \mathcal{D}\phi^{\rm res}_{n-1} + \mathbf{w}_n$, where $\mathcal{D}$ is the interaction matrix, and $\mathbf{w}_n$ the measurement noise at instant nT;
- the equation of evolution of the turbulence written as: $\phi^{\rm tur}_{n+1} = \mathcal{F}[\phi^{\rm tur}_n, \phi^{\rm tur}_{n-1}, \dots, \phi^{\rm tur}_{n-p+1}] + \nu_n$, where ν is a white noise and $\mathcal{F}$ is being choosen as an AR model of order p (AR_p).

In Le Roux *et al.* (2004), and for the sake of simplicity, the evolution of the turbulence was described by the AR_1 model: $\phi^{\rm tur}_{n+1} = \mathbf{A}\phi^{\rm tur}_n + \nu_n$. It is then possible to describe the system by choosing the state vector: $\mathbf{X}_n = \begin{pmatrix} \phi^{\rm tur}_{n+1} & \phi^{\rm tur}_n & \phi^{\rm tur}_{n-1} & \mathbf{u}_{n-1} & \mathbf{u}_{n-2} \end{pmatrix}^T$. The so-called state-space model is then:

$$\mathbf{X}_{n+1} = \begin{pmatrix} \mathbf{A} & 0 & 0 & 0 & 0 \\ \mathbf{Id} & 0 & 0 & 0 & 0 \\ 0 & \mathbf{Id} & 0 & 0 & 0 \\ 0 & 0 & 0 & 0 & 0 \\ 0 & 0 & 0 & \mathbf{Id} & 0 \end{pmatrix} \mathbf{X}_n + \begin{pmatrix} 0 \\ 0 \\ 0 \\ \mathbf{Id} \\ 0 \end{pmatrix} \mathbf{u}_n + \begin{pmatrix} \nu_n \\ 0 \\ 0 \\ 0 \\ 0 \end{pmatrix} \tag{2.1}$$

$$\mathbf{Y}_n = \mathcal{D}\begin{pmatrix} 0 & 0 & \mathbf{Id} & 0 & -\mathbf{N} \end{pmatrix} \mathbf{X}_n + \mathbf{w}_n. \tag{2.2}$$

The enhancement brought by the Kalman approach with respect to a more classical approach, the Optimized Modal Gain Integrator [OMGI] in classical AO (Gendron & Léna (1994)), was presented in Le Roux *et al.* (2004), where an improvement essentially due in that case to the predictive ability of the optimal control law was obtained. The use of a temporal modeling of the turbulence allows to predict its evolution during the $2T$ delay between WFS and correction and then reduces the temporal delay. The AR_1 model used to characterize the temporal evolution of the atmospherical turbulence was enough to improve the performance of generic AO and MCAO systems.

But the temporal decorrelation of such a model is exponential while the turbulence one is slower. Figure 1 presents the behaviour of the temporal decorrelation of a Taylor-Kolmogorov turbulence and two AR1 models. Increasing the order of the model would allow to better model the temporal evolution of the turbulence and therefore to better predict its behavior.

3. Control for XAO: optimal prediction and noise filtering

3.1. *Prediction and noise filtering*

Current projects of XAO systems usually consider high sampling frequencies, higher than 1 kHz, and WFS able to sense very high spatial frequencies, as a Shack-Hartmann [SH] with many sub-apertures (e.g. 40×40 for an 8-m telescope). The first consequence of such a configuration is a very low signal for each sub-aperture (i.e. a high photon noise). This should be considered as a consequence of the use of an integrator control law.

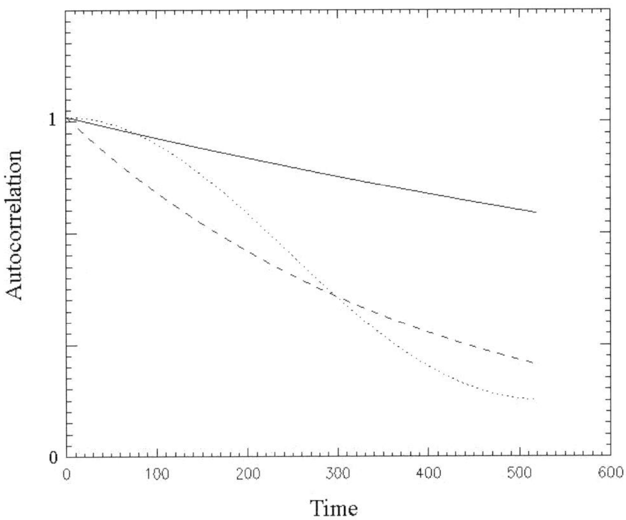

Figure 1. General form of the temporal decorrelation for the atmospheric turbulence [dotted line] and for two AR1 models [continuous line and dashed line].

The choice of the temporal frequency f_{temp} being a trade-off between decreasing the temporal error by increasing f_{temp} and decreasing the photon noise by decreasing f_{temp}, there is an optimal value of f_{temp} to be found, minimizing the sum of the two error terms. In the case of a non-predictive control law, as the integrator one, the only way to decrease the temporal error is to highly increase f_{temp}. In the other hand a predictive control law allows to decrease the temporal error by using an efficient temporal model to predict the turbulent evolution. The resulting optimal f_{temp} is then lower, allowing to simultaneously reduce the temporal error, thanks to prediction, and the photon noise.

We propose here to compare the performance of a system based on an integrator control law with a high f_{temp} and a system based on a predictive control law, here based on a Kalman filter, with a lower f_{temp}.

There are many questions to be answered. First of all, the order of the AR model necessary to do such a prediction is a crucial parameter that decides its practical feasibility. It has to be determined, as well as f_{temp} in the predictive case, for a given photon noise. Finally, we should express the gain brought by this predictive approach.

3.2. *Choice of the state-space model for XAO*

The choice of the state-space model for XAO depends essentially upon the choice of the order of the AR model used. In order to address the influence of the AR model order on the prediction error we have simulated a turbulence made up of 7 layers moving in random angles with a wind speed of 12 m/s and following a Kolmogorov spectrum with a Fried parameter of 12.5 cm (at 500 nm), over an 8-m telescope. The turbulent phase arriving on the telescope aperture is decomposed onto the first 30 radial orders of the Zernike modes basis for 5000 iterations, at a rate of 500 Hz (the temporal frequency has been choosen here as an example).

For each Zernike mode, we fit an AR_1, AR_2, ..., AR_{11} on 500 iterations. We then use these AR models to predict the evolution of the turbulence in the next 500 iterations. The computed resulting global prediction error is plotted on Fig. 2: we obtain that it decreases continously with the order of the AR model. Note that from an AR_1 to an AR_{11} model a huge gain in variance is obtained, but this gain is almost already acheived with a 4th order model, which then represents the best order in this case.

Now let us remark that in reality, the optimal order of the model depends upon f_{temp}. The ability of the model to represent the turbulence evolution and hence to predict it

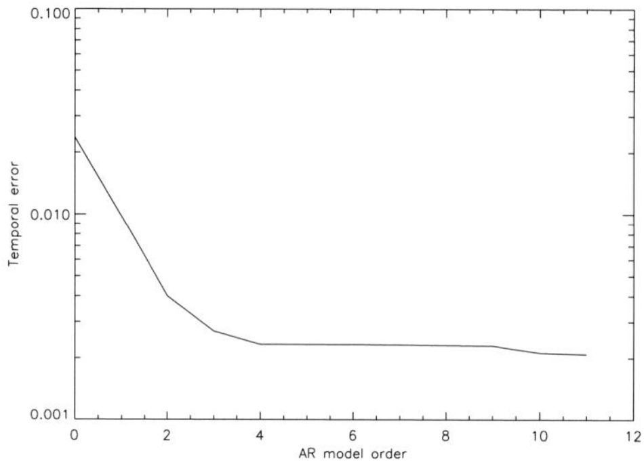

Figure 2. AR model prediction error vs. order (order 0 means no prediction).

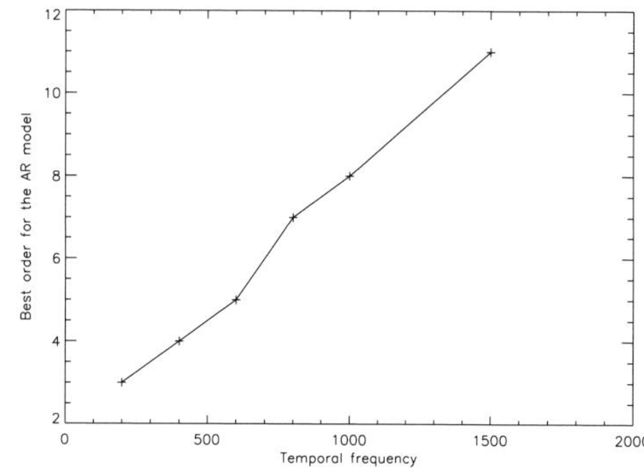

Figure 3. Best order of the AR model vs. temporal frequency.

can be linked to the temporal range on which it is written, meaning the order of the model times the temporal frequency. As an example, 4 iterations at 500 Hz correspond to the same temporal range than 8 iterations at 1.5 kHz. Of course this argument is totally qualitative and needs to be quantified.

We reproduced the same steps as before with the same parameters but for a range of f_{temp} between 100 Hz and 1 kHz, in order to find the best order for the AR model. The best f_{temp} is the frequency from which there is no relevant gain in increasing again the frequency. Figure 3 presents the best model order p_{best} vs. f_{samp}. The expected behaviour of p_{best} increasing with f_{ech} is clearly shown.

3.3. *Choice of the temporal frequency and gain obtained*

In a second step, it is possible to quantify the improvement brought by a predictive approach with respect to the integrator control law and to determine the corresponding f_{temp}. We have hence simulated again the same Kolmogorov turbulence. The temporal frequency explored a range from 100 to 2000 Hz, and for each frequency we computed the same way as before the parameters of the AR$_p$ model, with p optimized, as in Fig. 3.

Such a control law is based on the same type of model that the one presented in Sec. sec:rappels but with more turbulent phase vectors in the state vector, in order to be able to write the full AR model. To use an AR$_5$ model, the state vector must be $\mathbf{X}_n = \begin{pmatrix} \phi^{\mathrm{tur}}_{n+1} & \phi^{\mathrm{tur}}_{n} & \phi^{\mathrm{tur}}_{n-1} & \phi^{\mathrm{tur}}_{n-2} & \phi^{\mathrm{tur}}_{n-3} & \mathbf{u}_{n-1} & \mathbf{u}_{n-2} \end{pmatrix}^T$, in order to be able to write $\phi^{\mathrm{tur}}_{n+2}$ (within $\mathbf{X}_{n+1}$) as a function of $\phi^{\mathrm{tur}}_{n+1}$, ϕ^{tur}_{n}, $\phi^{\mathrm{tur}}_{n-1}$, $\phi^{\mathrm{tur}}_{n-2}$ and $\phi^{\mathrm{tur}}_{n-3}$ (within $\mathbf{X}_n$). Of course this means that all vectors and matrices are much bigger than with an AR$_1$ model. If we consider in the same time that the number of measurements (given for example by a 40×40 SH) can be already large and that recent propositions to increase the sensitivity of WFS in XAO (Le Roux *et al.* (2005)) could increase even more this number, one should understand that an XAO system represents a huge number of parameters to deal with. A full study must be done to understand how the computing time can be decreased, probably by using sparse matrix properties.

4. Discussion

Figure 4 shows, as a function of the temporal frequency, the temporal errors corresponding to the predictive case [full line] and the non-predictive case [dashed line]. The photon noise as a function of the temporal frequency is also plotted [dotted line], as well as the sum of the temporal errors and of the noise [dashed-dotted and dashed-triple-dotted

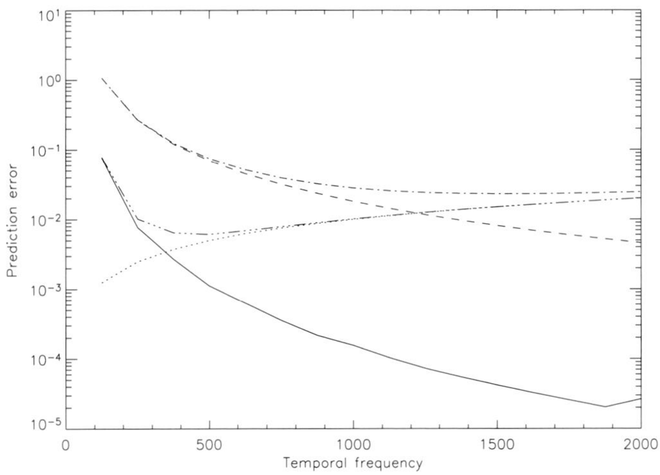

Figure 4. Temporal errors in the integrator case and the predictive control case, and WFS error due to photon noise.

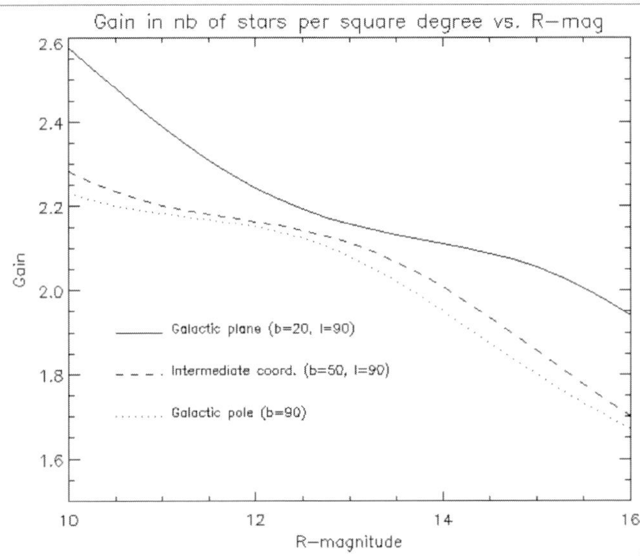

Figure 5. Evolution of the number of candidate stars for a predictive approach with respect to a non-predictive approach.

lines] in order to be able to compare the relevant parameters, which are the sum of both error terms. In the predictive case, an optimum is obtained at a temporal frequency $f_{\rm pred}$ which is about half the optimal frequency obtained in the non-predictive case. In terms of variance of error, choosing $f_{\rm pred}$ and a predictive control reduces the variance by a factor 2.

It should be indeed noticed that the goal of this first study is the understanding of the prediction aspect of the Kalman-based filter. It does not take into account parameters like for example a difference of noise propagation between predictive and non-predictive control. A full AO simulation would be necessary to confirm the gain of the predictive approach in a realistic case. We are currently implementing an optimal control based on a high-order AR-model within the Software Package `CAOS` (Carbillet *et al.* (2005)) in order to simulate a full AO system working with a control law and quantitatively compare it to an integrator control law.

For what concerns the number of candidate stars for exoplanet search, the previous result has the following implication. A gain of two in variance of residual phase implies a gain of one mag. in limiting magnitude. Figure 5 shows the resulting gain in number of observable stars by square degree vs. the magnitude and for three typical positions in the sky, based on the Bachall & Soneira model of distribution of stars (Bahcall & Soneira (1984)). This gain in number of observable stars is in average in the field larger than 2.

References

Chauvin, G., Lagrange, A.-M., Dumas, C., *et al.*, 2005, *Astron. Astrophys.*, 438 (2), L25
Nicolle, M., Fusco, T., Rousset, G., & Michau, V., 2004, *Opt. Letters*, 29 (23), 2743
Vérinaud, C., Le Louarn, M., Korkiakoski, V., & Carbillet, M., 2005, *MNRAS*, 357 (1), L26
Le Roux, B., Conan, J.-M., Kulcsár, C., *et al.*, 2004, *J. Opt. Soc. Am. A*, 21(7), 1261
Gendron, É. & Léna, P., 1994, *Astron. Astrophys.*, 291 (1), 337
Carbillet, M., Vérinaud, C., Femenía, B., *et al.*, 2005, *MNRAS*, 356 (4), 1263
Le Roux, B., Coyne, J., & Ragazzoni, R., 2005, *App. Opt.*, 44 (2), 171
Bahcall, J. N. & Soneira, R. M., 1984, *Ap. J. Suppl. Ser.*, 55, 67

Direct Imaging of Exoplanets: Science & Techniques
Proceedings IAU Colloquium No. 200, 2005
C. Aime and F. Vakili, eds.
© 2006 International Astronomical Union
doi:10.1017/S1743921306000044

The Lyot Project: Understanding the AEOS Adaptive Optics PSF

Russell B. Makidon[1,5], A. Sivaramakrishnan[2,5], R. Soummer[2,5], B. R. Oppenheimer[2], L. C. Roberts, Jr.[3], J. R. Graham[4,5], and M. D. Perrin[4,5]

[1]Space Telescope Science Institute, 3700 San Martin Drive, MD 21218, USA
[2]Department of Astrophysics, American Museum of Natural History, 79th Street at Central Park West, New York NY 10024, USA
[3]The Boeing Company, 535 Lipoa Pkwy, Suite 200, Kihei, HI 96753, USA
[4]Astronomy Department, University of California, Berkelely CA 94000, USA
[5]NSF Center for Adaptive Optics

Abstract. Adaptive optics (AO) systems have significantly improved astronomical imaging capabilities over the last decade, and are revolutionizing the kinds of science possible with 4-5 m class ground-based telescopes. A thorough understanding of AO system performance at the telescope can enable new frontiers of science as observations push AO systems to their performance limits. We look at the understanding we have gained from recent Lyot Project images at the Advanced Electro-Optical System (AEOS) 3.6 m telescope to show how progress made in improving WFR can be measured directly in improved science images. We describe how wave front errors affect the AO point-spread function (PSF), and model details of AEOS AO to simulate a PSF which matches the actual AO PSF in the astronomical H-band. Finally, we estimate the impact of improvements to wave front reconstruction techniques on diffraction-limited coronagraphy with the Lyot Project near-infrared coronagraph.

Keywords. instrumentation: adaptive optics, planetary systems: protoplanetary disks.

1. Introduction

The 3.6 m AEOS telescope, part of the U.S. Air Force Research Laboratory's Maui Space Surveillance System (MSSS), is arguably the best telescope on which to develop the nascent field of Extreme Adaptive Optics (ExAO). This is a result both of the telescope's location at a prime astronomical site possessing good seeing and, more importantly, the AEOS adaptive optics system's 941-actuator deformable mirror, with 35 actuators spanning the diameter of the DM. This provides the highest actuator density available to any civilian astronomical observing program, with a projected actuator spacing of order 0.11 m per actuator at the primary mirror, and currently produces images with Strehl ratios (SRs) of typically ∼0.15 in the I-band for objects brighter than 6th magnitude and Strehl ratios as high as ∼0.25 observed during upper-quartile seeing conditions (see Roberts & Neyman (2002), Makidon *et al.* (2005)).

The Lyot Project near-infrared coronagraph, at coudé focus, delivers H-band SRs in the ∼0.65–0.83 range and occasionally higher. This near-IR performance agrees in the main with predictions based on results from simple AO simulations anchored in Palomar and AEOS AO data (see Makidon *et al.* (2005), Sivaramakrishnan *et al.* (2001)). However, dynamic range with the coronagraph still falls short of predictions based on numerical modeling.

In order to make best use of the advantages offered by AEOS for the Lyot Project coronagraph, we first needed to understand the characteristics of the AO system. For the Lyot Project, we have found that one of the best ways to understand the AEOS AO system – which is a facility system not associated with the Lyot Project coronagraph – is through modeling the PSF itself.

2. Modeling AEOS Adaptive Optics

A detailed simulation of an AO system including all aspects of wave front sensing, reconstruction, and correction requires substantial computational effort. For some purposes it is sufficient to model the action of an AO system as a transfer function or spatial frequency filter acting on the incident wave front. Following the method presented in Sivaramakrishnan *et al.* (2001), we assume we use a high-pass filter with a power law distribution in spatial frequency space to mimic the action of the AEOS AO system,

$$F(k) = \begin{cases} (k/k_{AO})^n, & \text{for } k < k_{AO} \\ 1, & \text{otherwise.} \end{cases} \tag{2.1}$$

In Makidon *et al.* (2003), we used a 'parabolic' filter (e.g. $n = 2$) to model this AO system. However, further refinement of our simulations based on comparision with early AEOS I-band imaging data have shown that an $n = 0.9$ power law distribution best fits the AEOS system. We discuss this in further detail in Makidon *et al.* (2005).

We simulate a set of H-band images by generating thirty-seven monochromatic images across the bandpass, from $\lambda = 1.45$ to $1.81\ \mu m$. We assume a single-layer, Kolmogorov-spectrum phase screen to represent atmospheric turbulence, and use the $n = 0.9$ power law filter in spatial frequency space to represent correction of the incoming wave front by the AEOS AO system. We assume we are in the Fraunhofer regime throughout our simulations, where the image plane is the Fourier Transform of the pupil plane and vice versa.

Our simulations produced model PSFs which matched many of the features of the AEOS AO direct imaging PSF. However, details of the coronagraphic images suggest other sources of static phase aberration remained uncorrected in the optical train.

3. The Lyot Project: First Light

First light with the Lyot Project coronagraph at the AEOS 3.63 m telescope and AO system was in March 2004 (see Fig. 1). While the coronagraph performed at an extremely high level, as predicted from laboratory tests and simulation, the presence of strong and persistent residual speckles led us to consider elements in the AEOS AO optical train as potential sources of phase aberrations in the PSF.

One of the challenges in analyzing coronagraphic image data from this or any other high-Strehl ratio system is to understand the sources of residual speckles, and how to mitigate the effects of these speckles on coronagraphic imaging. For AEOS, we are working to understand the speckles produced by phase aberrations in the telescope and adaptive optics system upstream of the coronagraph optics, as well as issues related to scintillation and pupil illumination.

4. Modeling Phase Errors in AEOS Optics

In an attempt to understand the residual speckle structure observed in coronagraphic images acquired with the AEOS AO system, we considered sources of phase error in the telescope and AO optical train, including pinned actuators on the AO systems deformable mirror (DM). As has been reported by Gullapalli *et al.* (2000), there are a number of

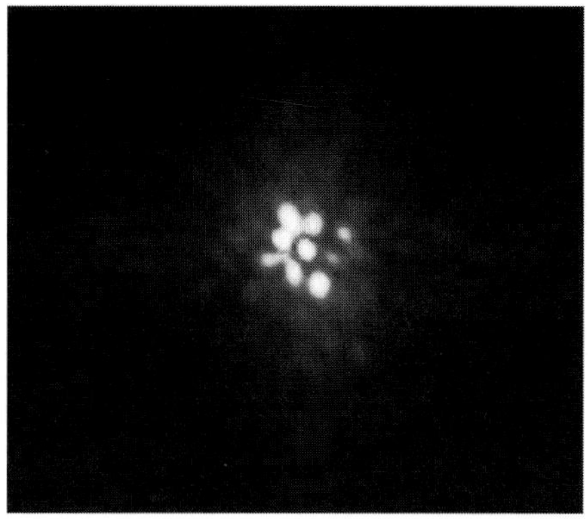

Figure 1. Coronagraphic image of a nearby star observed with the Lyot Project coronagraph at the AEOS 3.63 m telescope in March 2004. This image was acquired with the Kermit Infrared Camera Perrin *et al.* (2004) behind the Lyot Projects 0.35″ Focal Plane Mask in the H-band ($4\lambda/D$ diameter) and optimized Lyot stop used to suppress light from central star. Residual speckles dominate the noise in this image (see Racine *et al.* (1999)). Of note is the presence of symmetric and anti-symmetric speckle structure, as predicted from the first- and second-order terms of the PSF expansion of Perrin *et al.* (2003).

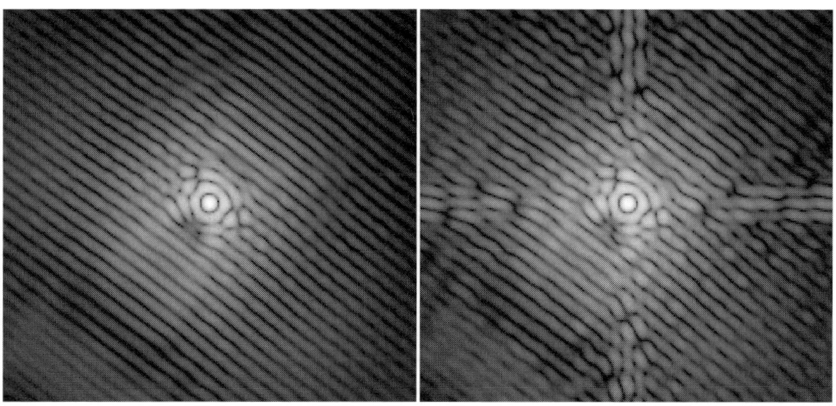

Figure 2. Simulated monochromatic coronagraphic PSFs at $\lambda = 1.63$ μm resolution assuming the Lyot Project coronagraphs 445 μm-diameter Focal Plane Mask and Lyot Stop. The PSF at left assumes five Gaussian phase aberrations of varied depths in the pupil plane representing pinned DM actuators. Residual speckle structure close to the PSF core dominates the image. The PSF at right includes the same five Gaussian phase aberrations in the pupil plane as well as a secondary mirror support structure that is misaligned with respect to the Lyot Mask. These PSFs are shown at $\lambda/4D$ resolution on a logarithmic grayscale, with the upper limit of the grayscale at right 1/100th that at left. Noise sources due to Poisson statistics or detector effects are absent from these simulations.

actuators on the AEOS DM that are pinned at levels below the nominal zero-level of the DM. Here, we model a DM with five pinned actuators, and represent each pinned actuator as a phase aberration in the form of a Gaussian with FWHM of 1/34th of the pupil diameter, with the a depth of this phase aberration dependent on the physical characteristics of the DM. We then propagate this phase aberration through our simulated AO system and the Lyot Project coronagraph optics. In these simulations, we assume a focal plane mask (FPM) of diameter 445 μm in an $f/70$ beam followed by an optimally-undersized Lyot stop matched to this FPM (see Figure 2 – left).

In addition, we considered the situation where, in addition to phase errors induced by the DM, we see phase aberrations due to the incorrect masking the AEOS secondary mirror (SM) support structure in the Lyot pupil (*e.g.,* as in the case of misalignment). We show this for the monochromatic case in Fig. 2 (right). As in the previous example, we see residual speckle close to the PSF core dominating the coronagraphic image. However, the presence of unmasked secondary supports in the Lyot pupil show a significant contribution to the coronagraphic image from a radius of $\sim 12\lambda/D$ from the PSF core, translating to 1.05″ at the Kermit Infrared Camera with the Lyot Project coronagraph.

5. Conclusions

As ground-based and space-based coronagraphic imaging systems push to higher and higher contrasts, the control of – and understanding of – phase aberrations induced by intervening optics becomes more and more important. Knowledge of how those phase aberrations create structure in the final image plane will be necessary for correct interpretation of the observations. Indeed, evern for very high-Strehl imaging systems such as the Advanced Camera for Surveys on the *Hubble Space Telescope*, it is the presence of residual phase aberrations of the optics which ultimately limits the performance of HST as a high-dynamic range instrument (Krist 2004, *private communication*).

Acknowledgements

The authors are grateful to D. T. Gavel, G. Smith, and P. E. Hodge for helpful discussions, and to the the staff of the Maui Space Surveillance System for their assistance in taking the Lyot Project data. R.S. and M.D.P. are supported by NASA Michelson Postdoctoral and Graduate Fellowships, respectively, under contract to the Jet Propulsion Laboratory (JPL) funded by NASA. The JPL is managed for NASA by the California Institute of Technology. The research presented here was supported by the STScI Director's Discretionary Research Fund, NSF grants AST-0088316 and AST-0215793, and AFRL/DE through Contract Number F29601-00-D-0204. Our work has also been supported by the National Science Foundation Science and Technology Center for Adaptive Optics, managed by the University of California at Santa Cruz under cooperative agreement No. AST – 9876783. This research made use of the SIMBAD database, operated at CDS, Strasbourg, France.

References

Gullapalli, S. N., Abreu, R., Rappoport, W. M., Zmek, W. P., & Pringle, R. 2000, *Proc. SPIE* 3931, 285-299

Makidon, R. B., Sivaramakrishnan, A., Roberts, L. C., Oppenheimer, B. R., & Graham, J. R. *Proc. SPIE* 4860, 315-323

Makidon, R. B., Sivaramakrishnan, A., Perrin, M. D., Roberts, L. C., Jr., Oppenheimer, B. R., Soummer, R., & Graham, J. R. 2005, *PASP* 117, 831

Perrin, M. D., Sivaramakrishnan, A., Makidon, R. B., Oppenheimer, B. R., & Graham, J. R. 2003, *ApJ* 596, 702

Perrin, M. D., Graham, J. R., Trumpis, M., Kuhn, J., Whitman, K., Coulter, R., Lloyd, J. P., & Roberts, L. C., Jr. 2004, *2003 AMOS Tech. Conf.*

Racine, R., Walker, G. A. H., Nadeau, D., Doyon, R., & Marois, C. 1999, *PASP* 111, 587

Roberts, L. C., Jr. & Neyman, C. R. 2002, *PASP* 114, 1280

Sivaramakrishnan, A., Koresko, C. D., Makidon, R. B., Berkefeld, T., & Kuchner, M. J. 2001, *ApJ* 552, 397

Direct Imaging of Exoplanets: Science & Techniques
Proceedings IAU Colloquium No. 200, 2005
C. Aime and F. Vakili, eds.
© 2006 International Astronomical Union
doi:10.1017/S1743921306000056

Real-time and post-facto correction for differential aberrations in direct planet imaging by adaptive optics

J.-F. Sauvage, T. Fusco, G. Rousset, and C. Petit

Département d'Optique Théorique et Appliquée, ONERA BP 72, 92322 Chtillon, France
email: jean-francois.sauvage@onera.fr

Abstract. We present experimental results of a new procedure of measurement and pre-compensation of the non-common path aberrations in Adaptive Optics. A significant Strehl ratio increase (from 70 to 95.5 % in R band) is demonstrated.

Keywords. Adaptive optics, Wave-front sensing, Phase diversity.

1. Introduction

The correction of non-common path aberrations (NCPA) is one of the critical issues to achieve ultimate adaptive optics (AO) performance. These aberrations have to be measured by a dedicated tool in the imaging camera. An efficient way to obtain such a correction is to use phase diversity (PD) [1, 2, 3, 4]. They are then pre-compensated by a modification of the Wavefront Sensor (WFS) references in the AO loop to obtain the wavefront error at the scientific detector. It is clear that such a process imposes linear WFS and thus is well adapted to Shack Hartmann (SH) WFS. This approach has been successfully applied on NAOS-CONICA [3, 5, 6] and has led to a significant improvement of the global system performance. Even if a real improvement can be seen on pre-compensated images, a significant amount of aberrations are still not corrected. Several effects may explain such a limitation. Firstly, phase diversity measurements are not perfect (due to noise effects, detector transfert function and algorithm approximations [3]). An optimized Phase Diversity algorithm is proposed in section 2. In addition, the pre-compensation of these measured aberrations is only partial due to uncertainties on AO models (WFS, Deformable Mirror (DM)). In order to avoid these model uncertainty effects (which are probably the more important limitation of the approach), a pseudo-closed loop scheme is proposed in section 3 for the measurement and the correction of NCPA. These new procedure is experimentally validated on the ONERA AO bench.

2. NCPA measurements

Phase Diversity [1] is a focal plane wavefront sensor able to estimate the pupil phase thanks to focal plane images of an artificial point source. In order to remove undetermination in the phase estimation, two focal plane images differing from a known aberration are needed. In our case, the known aberration is the defocus because of its simplicity to implement. A modification of the SH WFS reference slopes allows us to introduce or not the defocus on the DM by closing the loop on an artificial calibration source.

A PD algorithm [3] is used to measure NCPA on focused and defocused images i_f and i_d. We propose here a new optimized algorithm for the phase estimation. It is based on a MAP approach, by maximizing the probability of obtaining simultaneously the object o and the phase ϕ knowing the images i_f and i_d. This probability can be rewritten as shown in equation 2.1 in order to introduce the *a priori* on the object and the phase $f(o)$ and $f(\phi)$:

$$f(i_f, i_d, o, \phi) = f(i_f, i_d | o, \phi) f(o) f(\phi) \tag{2.1}$$

The *a priori* knowledge on object is taken equal to 1 (meaning taht we assume the object to be totally unknown), but the *a priori* on the phase will take into account the phase spectrum in n^{-2} and constraint the noise error on high order Zernike modes. In practice, the algorithm minimizes the criterium $J(o, \phi)$ defined in equation 2.2 as the opposite of logarithm of the probability defined in 2.1 :

$$\begin{aligned} J(o, \phi) &= -\ln\left(f(i_f, i_d | o, \phi)\right) \qquad (2.2) \\ &= -\ln\left(f(i_f, i_d | o, \phi)\right) - \ln\left(f(\phi)\right) \\ &= \left\| \frac{i_f(\vec{r}) - h_f(\phi, \vec{r}) * o(\vec{r})}{\sigma_f(\vec{r})} \right\|^2 + \left\| \frac{i_d(\vec{r}) - h_d(\phi, \phi_d, \vec{r}) * o(\vec{r})}{\sigma_d(\vec{r})} \right\|^2 + \phi^t R_\phi^{-1} \phi \end{aligned}$$

The expression of this criterium makes appear the noise standard deviation $\sigma_f(\vec{r})$ and $\sigma_d(\vec{r})$ in the focused and defocused images, with $\vec{r}$ the position vector in the image plane. This statistics allows us to deal with a non uniform gaussian noise, varying from one pixel to another , and modeling with a good approximation both photon and electronics noises. $h_f(\phi, \vec{r})$ and $h_d(\phi, \phi_d, \vec{r})$ are respectively the focused and defocused PSF of the intrument. h_f only depends on the aberrant phase ϕ and on the position vector, while h_d also depends on the known defocus phase ϕ_d.

The last term in equation 2.2 is the *a priori* knowledge on the phase : a typical spatial spectrum and phase variance, via the covariance matrix R_ϕ. The two different optimizations proposed here (non-uniform noise model and phase regularisation) are tested in section 3.

3. NCPA compensation

NCPA pre-compensation requires to compute the WFS slopes corresponding to the measured aberrations. These slopes are then substracted to te SH reference slopes, the new references being used in the closed loop operation. This slope modification requires a precise knowledge of the system parameters and a good model of the WFS itself. Any error on the WFS model directly affects the slope computation from measured NCPA and thus limits the performance of the global pre-compensation process. As an example, an error of 10% on the WFS pixel scale directly translates into an uncorrected error of 10% of the amplitude of the NCPA. These model errors are probably today the main limitation for the NAOS-CONICA system [5, 6].

3.1. *Pseudo-closed loop process*

3.1.1. *Principle of the Pseudo- closed loop process*

A first way to reduce these errors is of course to perform accurate calibrations of the WFS. Nevertheless, uncertainties on system parameters as well as remaining errors on calibrations will always degrade the ultimate performance of the pre-compensation process. In order to overcome this problem and to reach the ultimate performance in the NPCA pre-compensation process, a new approach is proposed. We called it the "pseudo closed loop" process.

The idea is to overcome model uncertainties using a multiple step process for the NCPA estimation and pre-compensation. First we perform a WFS calibration and we derive the initial WFS model. The pseudo closed loop approach can be summarized as follows:

1) first measurement of NCPA with PD
2) modification of reference slopes to account for NCPA using the WFS model
3) measurement of residual NCPA after the first pre-compensation
4) addition of the new slopes deduced from the residual NCPA to the current slopes
5) points 3) and 4) are performed until convergence. With such an approach, the only impact of the model uncertainties is to reduce the speed of convergence.

3.1.2. *First experimental validation*

The pseudo-closed loop process has been tested on the ONERA AO bench using the initial PD algorithm [3]. This laboratory bench is composed of a 9x9 actuator Deformable Mirror (DM) and a 8x8 sub-aperture SH-WFS. The imaging camera is a 512x512 Princeton camera with a 4e-/pixel/frame readout noise. A unresolved fibered laser diode source working at 630 nm, is used at the system entrance focal plane for calibration and closed-loop test.

The accuracy of the NCPA compensation is quantified thanks to Strehl ratio estimation in the focused images. The Strehl ratio is a widely used quality estimator for optical system, and is a way to quantify the residual phase in the pupil plane of the instrument. This Strehl ratio may be estimated in two different ways : firstly by a measurement SR_{im} in the focal plane image, and secondly by computing SR_{Zern} by using the PD measured coefficient (see equation 3.1).

$$SR_{Zern} = \exp(-\sum_{i=4}^{i=N} a_i^2) \tag{3.1}$$

with a_i the measured Zernike coefficients, and N the number of Zernike polynomials. Piston ($i = 1$) has no effect on image formation, Tip and Tilt ($i = 2, 3$) are interpreted as a movement of the observed object. The results presented in figure 1 show both the diminution of the measured Zernike coefficients with the number of iteration of the pseudo-closed loop process, and the increase of Strehl ratio with iteration number.

In our case, 75 Zernike coefficients are estimated (from a_4 to a_{78}) but only 33 are corrected (from a_4 to a_{36}) by the AO system because of the relatively small number of subapertures on the SHWFS. The large number of estimated coefficients also avoids aliasing effects coming from the high order Zernike modes estimated.

We plot on the figure 1 SR_{Im}, SR_{Zern} and the ratio between these two quantities, $SRim/SRzern$. The difference between SR_{Im} and SR_{zern} can be explained by the uncorrected high order coefficients (from a_{42} to a_∞) and the SR measurement bias due to mis-knowledge of the system (exact over-sampling factor, exact fibre size and shape). This ratio $SRim/SRzern$ is roughly constant whatever the iteration number is and its average value can be estimated to 98% which corresponds to a 15 nm rms phase error. We plot in figure 2 the PSF images obtained on the imaging camera without NCPA pre-compensation and after 1, 2 and 3 iterations of pseudo-closed loop NCPA corrections. It clearly shows the high order aberrations (we only compensate for 33 Zernike modes here). It is interesting to notice that the PSF structures far from the optical axis (due to the high order non-compensated aberrations) roughly remain the same whatever the number of iterations is, as expected (they remain uncorrected). The correction of low order aberrations allows us to clean the centre of the PSF. It tends to a perfect Airy pattern with two complete Airy rings and a third clearly visible.

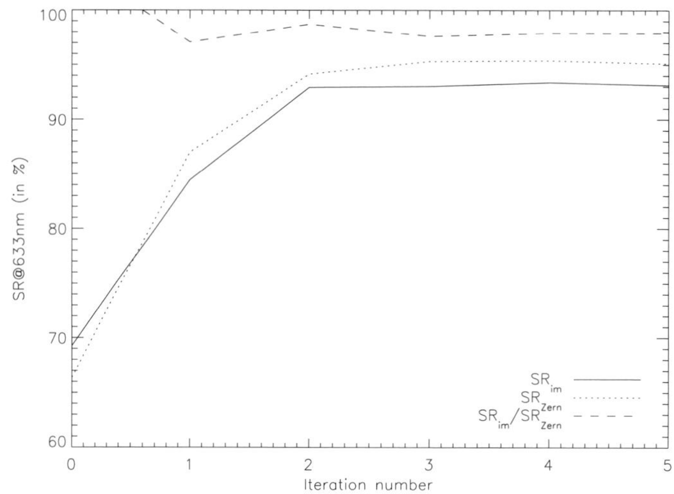

Figure 1. Experimental results of non-common path aberrations pre-compensation using phase diversity measurements and reference slope modifications. SR_{im} is measured on the focal plane PSF recorded at 633 nm. In addition, SR_{zern} computed from the measured NCPA is plotted in dotted line. The ratio between SR_{im} and SR_{zern} is plotted in dashed line.

Figure 2. PSF corresponding to the 4 first points of Fig 1, represented in invert logarithmic scale. The PSF on the left is the PSF obtained without any pre-compensation and used to perform the first PD measurement (iteration 0). The 3 other PSF correspond to iteration 1, 2 and 3 respectively. $\lambda = 632nm$.

3.2. *Tests of optimized algorithms*

Until now, the algorithm used to perform phase diversity was the classical one : uniform noise model, and no regularisation term [3]. We are going now to test the optimized algorithm with the non-uniform noise model and the phase regularisation with the experimental pseudo-loop procedure.

In order to test these algorithms, two different regimes of Signal to Noise Ratio (SNR) have been used. The SNR is defined as the ratio of the maximum of the focused image on the noise standard deviation.

- high SNR regime as in section 3, $SNR = 3 \times 10^4$ in order to test influence of photon noise on the non-uniform algorithm. This SNR regime ensures us that the predominant noise is the photon noise.
- low SNR regime, $SNR = 10^2$ (obtained with smallest exposure time while acquiring the pair of images) in order to test the gain brought by non-uniform noise and by regularisation.

Table 1 gathers the different level of Strehl Ratio after 5 iterations obtained with the different algorithm.

	Classical algorithm	Non uniform algorithm	Regularised algorithm	Regularised and non uniform algorithm
Saturation level for high SNR	93.5%	95.5%	93.5%	95.5%
Saturation level for low SNR	72.1%	81.2%	91.9%	92.3%

Table 1. SR values obtained with the different algorithms : classical, non-uniform, and regularised for high and low SNR regimes. Accuracy of SR measurement is estimated to 0.2% rms.

3.2.1. *Gain brought by non-uniform noise model*

The gain brought by non uniform noise model is revealed even for high SNR conditions. As in photon noise regime the noise variance evolves from one pixel to another, a non uniform algorithm is usefull to be more accurate in the Zernike coefficient measurements.

As written in table 1, the level reached with non uniform noise model at high SNR regime (95.5%) is higher than with the classical uniform algorithm (93.5%) at the same SNR regime. The use of non uniform model at low SNR regime brings a substantial gain of 9%.

At very high SNR the Phase diversity measurement is anyway very good even with the classical algorithm. The gain is therefore only 2% in Strehl Ratio, but such a gain at high SR is mandatory for the condition required by exoplanet detection.

3.2.2. *Gain brought by regularized algorithm*

The gain brought by the use of the regularisation term is revealed for low SNR regime. At low SNR, the classical algorithm without regularisation is unable to estimate correctly the phase and therefore the compensation is not efficient. The SR value remains at 72.1%. But with the use of regularisation, the SR value is at 91.9%, the NCPA have been estimated with almost the same accuracy than at high SNR regime with the classical algorithm.

4. Conclusion

The proposed new procedure for the measurements and the pre-compensation of the NCPA allows us to achieve a high performance (SR of 95.5%) on our own AO bench. This at the level of the requirement for extrem AO in order to detect exoplanets from the ground.

References

[1] R. A. Gonsalves, "Phase retrieval and diversity in adaptive optics," *Optical Engineering* **21**(5), pp. 829–832, 1982

[2] R. G. Paxman, T. J. Schulz, & J. R. Fienup, "Joint estimation of object and aberrations by using phase diversity," *Journal of the Optical Society of America A* **9**(7), pp. 1072–1085, 1992

[3] A. Blanc, T. Fusco, M. Hartung, L. M. Mugnier, & G. Rousset, "Calibration of NAOS and CONICA static aberrations. Application of the phase diversity technique," *Astron. Astrophys.* **399**, pp. 373–383, 2003

[4] A. Blanc, L. M. Mugnier, & J. Idier, "Marginal estimation of aberrations and image restoration by use of phase diversity," *J. Opt. Soc. Am. A* **20**(6), pp. 1035–1045, 2003

[5] G. Rousset, F. Lacombe, P. Puget, N. Hubin, E. Gendron, J.-M. Conan, P. Kern, P.-Y. Madec, D. Rabaud, D. Mouillet, A.-M. Lagrange, & F. Rigaut, "Design of the Nasmyth Adaptive Optics System (NAOS) of the VLT," in *Astronomical Telescopes & Instrumentation*, D. Bonaccini and R. K. Tyson, eds., **3353**, Proc. Soc. Photo-Opt. Instrum. Eng., (Kona, Hawaii), Mar. 1998

[6] M. Hartung, A. Blanc, T. Fusco, F. Lacombe, L. M. Mugnier, G. Rousset, & R. Lenzen, "Calibration of NAOS and CONICA static aberrations. Experimental results," *Astron. Astrophys.* **399**, pp. 385–394, 2003

Direct Imaging of Exoplanets: Science & Techniques
Proceedings IAU Colloquium No. 200, 2005
C. Aime and F. Vakili, eds.
© 2006 International Astronomical Union
doi:10.1017/S1743921306000068

Scintillation and pupil illumination in AO coronagraphy

Anand Sivaramakrishnan[1,6], B. R. Oppenheimer[1], M. D. Perrin[2,6], L. C. Roberts, Jr[4]., R. B. Makidon[3,6], R. Soummer[1,6] A. P. Digby[1], L. W. Bradford[4], M. A. Skinner[4], N. H. Turner[5], and T. A. ten Brummelaar[5]

[1]Department of Astronomy, American Museum of Natural History, 79th Street at Central Park West, New York NY 10024, USA
[2]Astronomy Department, University of California, Berkelely CA 94000, USA
[3]Space Telescope Science Institute, 3700 San Martin Drive, MD 21218, USA
[4]The Boeing Company, 535 Lipoa Pkwy, Suite 200, Kihei, HI 96753, USA
[5]Center for High Angular Resolution Astronomy, Georgia State University Mt. Wilson CA 91023, USA
[6]NSF Center for Adaptive Optics

Abstract. The Lyot Project near-infrared *JHK* coronagraph achieved first light on the Advanced Electro-Optical System (AEOS) in March 2004. Optical pupil plane imaging at video rates from this coronagraph provides data on atmospheric scintillation and quasi-static pupil intensity variations. We examine the effect of these variations on coronagraphic performance. Early simulations suggested Strehl ratio reductions of the order of 2–3% due to residual uncorrected phase aberrations in *H*-band. We find that static or quasi-static pupil illumination non-uniformity in *I*-band reduces Strehl by ~2%. A lower bound on the effects of dynamic illumination variation over the pupil is also ~2% in *I*-band. Some of the static intensity variations in the pupil are due to pinned deformable mirror (DM) actuators. We simulate the effects a pinned actuator has on the coronagraph. The resultant speckles in simulated coronagraphic images show similarities to some Lyot Project PSFs. This highlights the importance of knowledge of the pupil in next-generation extreme AO coronagraphs in order to realize the predicted photometric dynamic range of their images.

Keywords. instrumentation: adaptive optics, planetary systems: protoplanetary disks.

1. Introduction

The 3.6 m AEOS adaptive optics (AO) system's 941-element deformable mirror (DM) provides higher order wave front correction than any other astronomical AO system today. It produces images with Strehl ratios (SRs) of ~0.15 in the *I*-band for objects brighter than 6th magnitude, with SR's as high as ~0.25 during the best seeing conditions (Roberts & Neyman 2002). The Lyot Project near-infrared coronagraph (Oppenheimer *et al.* 2004), at its coudé focus, delivers *H*-band SRs in the ~0.65–0.83 range (Perrin *et al.* 2004, Makidon *et al.* 2005), and occasionally higher. This performance agrees in the main with predictions based on results from simple AO simulations anchored in Palomar and AEOS AO data (Sivaramakrishnan *et al.* 2001, Makidon *et al.* 2005), but dynamic range with the coronagraph still falls short of predictions based on numerical modelling. Incorporating knowledge of pinned actuators on the deformable mirror (DM) into simulations brings modelling and coronagraphic data into closer agreement.

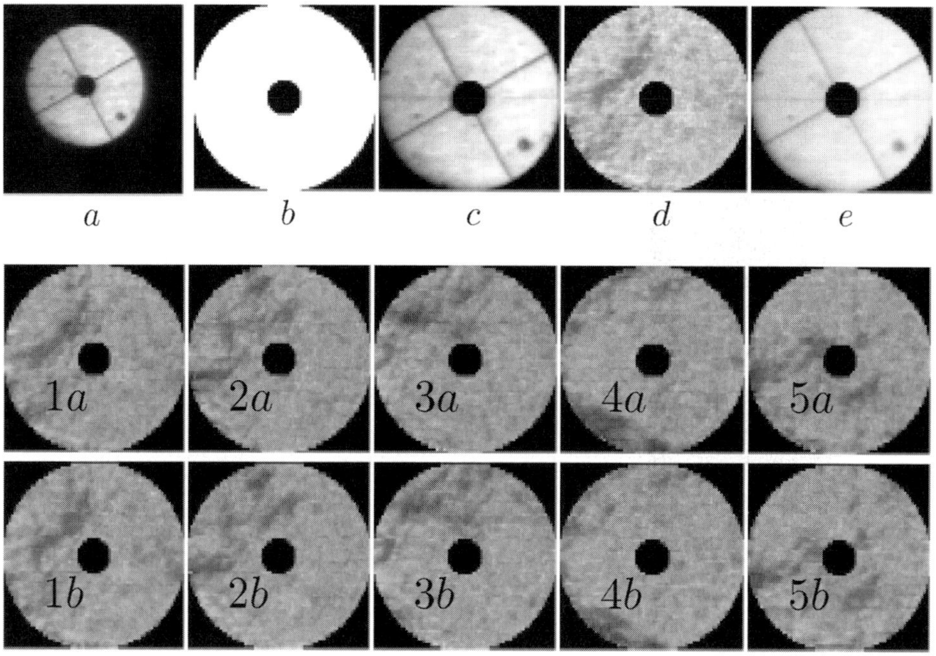

Figure 1. Raw and processed video rate CCD images of the pupil (see text).

2. Observations, calibration and theory

The Lyot Project's pupil-viewing mode uses a video rate CCD to record the intensity distribution of light in the pupil (Bradford *et al.* 2005). These images were taken in a non-coronagraphic mode, with the focal plane stop (which at H is a $4\lambda/D$ diameter hole in a mirror) moved out of the beam. While the SR is reduced slightly because of pinned DM actuators on the DM, the localized nature of the resulting aberration produces image plane speckles which are not suppressed by a coronagraph. This decreases the dynamic range of the coronagraph. Normally an interlaced video-rate CCD camera observes light that is rejected by a Lyot mask (at a reimaged pupil). Using a flat mirror instead of this mask, we collected full pupil images, in 176×180 pixel frames with a 9 μm pixel pitch. The illuminated pupil is approximately 116 pixels, so each pixel is 3.12 cm across on the pupil, or 3.64 pixels per AO system subaperture (11.3 cm). The CCD is read using 12-bit digitization. The AO system was not actively correcting atmospheric aberrations during this experiment. We present data from one observation of a $V = 2.9$ star, HR 7235 (spectral class A0V, with a parallax of 39.18 mas), on 8 June 2004. The rising star was at a zenith distance of 13° and an azimuth of 122° at the start of observations. We obtained 2048 frames at 30 Hz in a little over a minute elapsed time.

2.1. *Calibration*

When the CCD is evenly illuminated it shows no detectable features or pixel to pixel variations that are larger than $1{:}10^3$. We therefore used the data frames without flat-fielding the detector response. Bias subtraction was done with un-illuminated regions of sets of 512 frames; the median of 16 dark columns on either end of the detector was subtracted from each full frame. After bias subtraction, mean and median frames of the stack were computed. To compensate for pupil wander (a single CCD frame is shown in Fig. 1*a*, on a different scale than all the other pupil images in this paper), the images

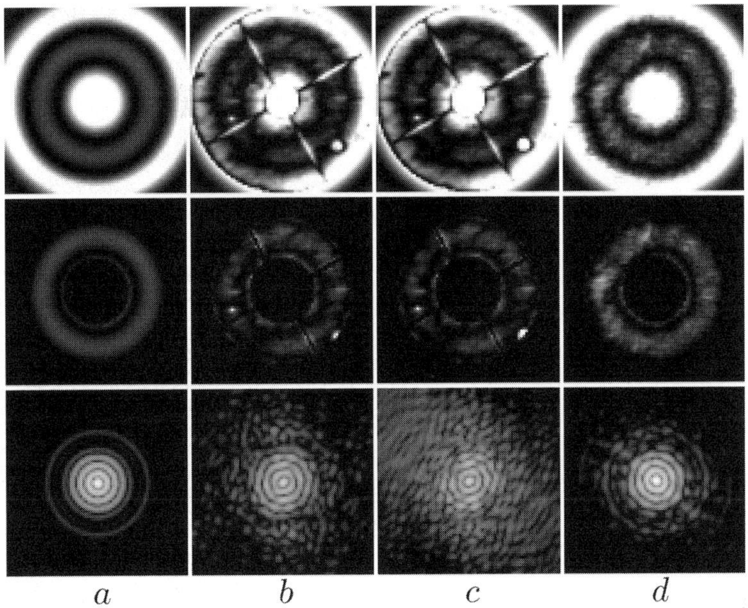

Figure 2. Lyot pupil and coronagraphic image intensities (see text).

were registered to the nearest pixel by cross-correlation with a numerical pupil mask representing the centered pupil, with outside diameter 116 pixels, inside diameter 22 pixels (Fig. 1*b*).

Secondary support spiders are narrower than our projected pixel scale, and are not precisely in a pupil plane. They partially vignette our pupil image. This can be calibrated out by "illumination-flat-fielding". The normalized mean of stacked, registered, images is the "illumination flat" (Fig. 1*c*). Each bias-subtracted data frame in the stack was divided by this flat (e.g. Fig. 1*d*) — these video rate frames measure the atmospheric scintillation. The square root of the flat field (Fig. 1*e*) is the quasi-static pupil illumination function $A_o(\mathbf{x})$ of the coronagraph. The scintillation frames were then de-interlaced by creating two sub-frames (labelled *a* and *b*) from each parent frame. Five pairs of sub-frames from successive parent frames are shown in Fig. 1 (bottom). Inter-frame timing between the *a* and *b* sub-frames is tightly controlled by CCD camera hardware. Operating system characteristics prevented reliable timing between succesive full frames of the CCD.

2.2. *Non-uniform Pupil Illumination Effects in Coronagraphy*

We describe pupil illumination non-uniformity with the real function

$$A(\mathbf{x}, t) = A_o(\mathbf{x})\, f(t)\, (1 + Z(\mathbf{x}; t)) \tag{2.1}$$

where t represents time, and $\mathbf{x}$ the pupil plane position vector. The real function $Z(\mathbf{x}; t)$ is constrained to possess a zero mean over the clear aperture at any instant. $A(\mathbf{x}, t)$ is the modulus of the complex-valued pupil field strength. We do not concern ourselves with the photometric ramifications of scintillation (e.g. Dravins *et al.* 1998), since direct detection and, if possible, spectroscopy are our focus. We subsume the temporal variation of the net power entering the telescope into the function $f(t)^2(1 + Z(\mathbf{x}; t)^2)$. The two-dimensional spatial frequency content of instantaneous 'realizations' of $Z(\mathbf{x}; t)$ cause scintillation to be a concern for ground-based near-IR coronagraphy. In the weak scintillation regime in the H-band, $Z^2 \ll 1$. We use observed I-band pupil intensities as input to our models.

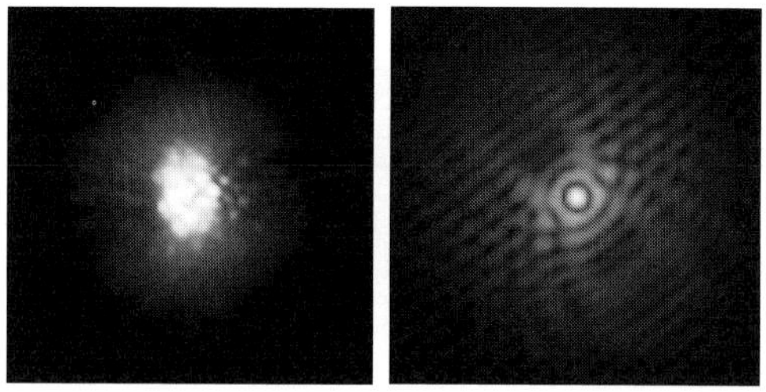

Figure 3. Vega (8 s, H-band) taken with the Lyot Project coronagraph (left). The pixel scale is 13.5 mas (the image is $3''.2$ across). Elongation of the PSF is due to pinned actuators on the AO system's DM. The phase aberration induces $\sim$10% reduction in Strehl ratio. A simulated H-band coronagraphic image with five pinned actuators on the DM (right).

3. Conclusion

In Fig. 2 we show the Lyot pupil intensity before (top row) and after (middle row) the Lyot stop, and the coronagraphic PSF (log scale, bottom row) simulated in the H-band. We simulate uniform pupil illumination (column a), the observed quasi-static intensity without (b) and with (c) phase errors from a pinned actuator 1.25 microns below the DM surface, and only atmospheric scintillation sub-frame (d). The pinned actuator's phase errors contribute the most speckle noise to the image by far. Fig. 3 shows real data (left) and a simulated image with five pinned actuators on the DM (right). The morphology of the actual coronagraphic PSF's core and inner regions is reproduced qualitatively by this model.

Acknowledgements

This research was supported by the U.S. Air Force Office of Scientific Research through AFRL/DE (Contract Number F29601-00-D-0204), the U.S. National Science Foundation (NSF) through grants No. AST-0088316 and AST-0215793, the NSF Science and Technology Center for Adaptive Optics, managed by the University of California at Santa Cruz under cooperative agreement AST-9876783, the Cordelia Corporation, Hillary Lipsitz, an anonymous supporter, and Space Telescope Science Institute's Research Programs Office and Director's Discretionary Research Fund. Soummer and Digby are supported by NASA Michelson Postdoctoral fellowships, and Perrin by a NASA Michelson Graduate Fellowship, under contract to the Jet Propulsion Laboratory (JPL) funded by NASA. The JPL is managed for NASA by the California Institute of Technology.

References

Bradford, L. W., Roberts, L. C., Skinner, M. A., Digby, A. P., Oppenheimer, B. R., Perrin, M. D., & Turner, N. H. 2005, *AMOS Tech. Conf.*

Dravins, D., Lindegren, L., Mezey, E., & Young, A. T. 1998, *PASP*, Vol. 110, 610

Lyot, B. 1939, *MNRAS*, Vol. 99, 580

Makidon, R. B., Sivaramakrishnan, A., Perrin, M. D., Roberts, L. C., Jr., Oppenheimer, B. R., Soummer, R., & Graham, J. R. 2005, *PASP* Vol. 117, 831

Oppenheimer, B. R. *et al.* 2004, *Proc. SPIE* Vol. 5490

Perrin, M. D., Graham, J. R., Trumpis, M., Kuhn, J., Whitman, K., Coulter, R., Lloyd, J. P., & Roberts, L. C., Jr. 2004, *2003 AMOS Tech. Conf.*

Roberts, L. C., Jr. & Neyman, C. R. 2002, *PASP* Vol. 114, 1280

Sivaramakrishnan, A., Koresko, C. D., Makidon, R. B., Berkefeld, T., & Kuchner, M. J. 2001, *ApJ* Vol. 552, 397

Direct Imaging of Exoplanets: Science & Techniques
Proceedings IAU Colloquium No. 200, 2005
C. Aime and F. Vakili, eds.

© 2006 International Astronomical Union
doi:10.1017/S174392130600007X

High-Order Band-Limited Coronagraphs in Adaptive Optics Systems

A. Vanden Heuvel, Justin Crepp and Jian Ge

Astronomy Department, University of Florida
211 Bryant Space Science Center, P.O. Box 112055
Gainesville, FL 32611-2055
email: avh@astro.ufl.edu, jcrepp@astro.ufl.edu, jge@astro.ufl.edu

Abstract. In principle, band-limited coronagraphic image masks are capable of removing all on-axis starlight providing infinite dynamic range for the direct imaging of exoplanets. In practice, optical aberrations left uncorrected by the adaptive optics system and practical limitations on mask alignment prevent band-limited masks from reaching contrast levels necessary for exoplanet imaging. So-called, "high-order" band-limited masks reduce the coronagraph's sensitivity to low order aberrations, and have been measured to reach contrast levels of 10^{-7} at $4\,\lambda/D$ without the aid of a deformable mirror. In this work, we simulate the performance of high-order band-limited coronagraphic image masks in a variety of adaptive optics systems. We compare their performance with that of hard-edge stops, soft-edge Gaussian masks, and traditional (4th-order) band-limited coronagraphs.

1. Introduction

Despite recent advances in adaptive optics (AO) technology, the full potential of even the most basic coronagraphic image masks is not yet achievable on state-of-the-art AO telescopes. Many designs for sophisticated image masks capable of very high contrast levels have been proposed in recent years, but the full potential of these masks can only be reached when wavefront errors are pushed very near to zero. A new class of band-limited masks, dubbed "high-order" masks, have been shown analytically and experimentally to be less sensitive to low-order optical aberrations that may be left uncorrected by an adaptive optics system (Kuchner, Crepp, & Ge 2005, Shaklan & Green 2005, Crepp *et al.* 2005). We simulate the performance of these and other more traditional coronagraphic image masks in a variety of AO correction levels. Along the way, we characterize the performance degeneracy among various coronagraph designs when wavefront errors dominate, we identify threshold levels of AO correction where it becomes observationally advantageous to switch to more sophisticated masks, and we explore the importance of wavefront correction levels in determining which masks respond best to small misalignments (tip/tilt errors).

2. High-Order Band-Limited Masks

Band-limited coronagraphic image masks (Kuchner & Traub 2002) have mask transmission functions that are composed of a limited number of low spatial frequencies. This forces all the unwanted on-axis starlight to be diffracted to a finite region near the edges of the Lyot plane. These traditional band-limited masks have intensity transmission profiles that increase as the fourth power of the distance from the optical axis. So-called "high-order" band-limited masks (Kuchner, Crepp, & Ge 2005) have transmission profiles that grow as the eighth, twelfth, or higher power of the distance from the optical axis while still diffracting all of the starlight to a narrow region around the edge of the pupil. The presence of a wider central null reduces the mask's sensitivity to low-order optical aberrations (Shaklan & Green 2005). In particular, Eighth-order masks have been shown experimentally to be less sensitive to low-order aberrations than traditional band-limited coronagraphs (BLCs) with only a small trade-off in Lyot stop throughput (Crepp *et al.* 2005).

2.1. *Coronagraph Performance Degeneracy*

We simulate the corongraphic performance of hard, soft (Gaussian), 4th-, 8th-, and 12th-order band-limited coronagraphs in a variety of AO correction levels in order to identify key rms wavefront error thresholds where moving to more sophisticated masks becomes observationally advantageous.

When wavefront correction lies below a certain threshold, $S \sim 0.82$ (Sivaramakrishnan *et al.* 2001), there is no substantial advantage (in terms of managing diffracted light) in using *any* type of coronagraphic image mask. Slightly above this threshold, we find that *all* of the varieties of image masks we examined perform equally well at rejecting starlight (i.e. there is a performance degeneracy among the masks). There is, therefore, no clear preference between hard masks and more sophisticated masks until a new threshold is reached (Figure 1a). Above this level of AO correction ($S \sim 0.9$), masks that are more sophisticated than a traditional hard-edge stop begin to reach improved levels of stellar flux rejection. A performance degeneracy remains among these sophisticated masks until still higher levels of AO correction can be achieved. It is not until very high wavefront correction levels that the degeneracy between the masks is clearly broken (Figure 1c-d). Even then, however, for certain non-optimal choices of Lyot stop size, different masks will achieve the same stellar flux rejection. For example, in Figure 1d, a 4th-order band-limited mask provides approximately the same stellar attenuation as a Gaussian image mask, but the band-limited mask can reach this level of attenuation with a larger Lyot stop size (i.e. with more throughput). These results suggest that for managing diffracted light, there is no advantage in switching to masks more sophisticated than hard stops until AO correction levels reach rms wavefront errors of $\sim\lambda/18$. In addition, once rms wavefront errors reach $\sim\lambda/271$ one would clearly choose 4th-order band-limited masks over Gaussian masks. Below this level however, either mask will manage diffracted light equally well. It would appear from this study that there is no advantage to using high-order band-limited masks, but of course there are other considerations beyond 'diffraction management' which influence one's choice of image mask. Next we examine one such consideration: mask alignment sensitivity.

3. Tip/Tilt Errors

A mask's sensitivity to tip/tilt alignment errors can greatly impact the overhead time required for conducting high-contrast observations. It is this consideration that led to the invention of high-order band-limited masks. We simulate coronagraphic performance as a function of mask mis-alignment to determine each mask's sensitivity to tip/tilt errors and to gain insight into reducing observing overhead time. Again, we simulate the coronagraphs in various levels of AO correction since different masks will excel in different conditions. In a recent study, Lloyd and Sivaramakrishnan (2005) find the surprising result that with otherwise perfect wavefronts, soft-edge Gaussian image masks are less sensitive to small mask alignment errors than traditional hard-edge masks. In our study we identify this same effect; moreover, we recognize that this result changes with changing wavefront correction. In Figure 2a, we can see that for AO correction levels of $S \sim 0.89$,

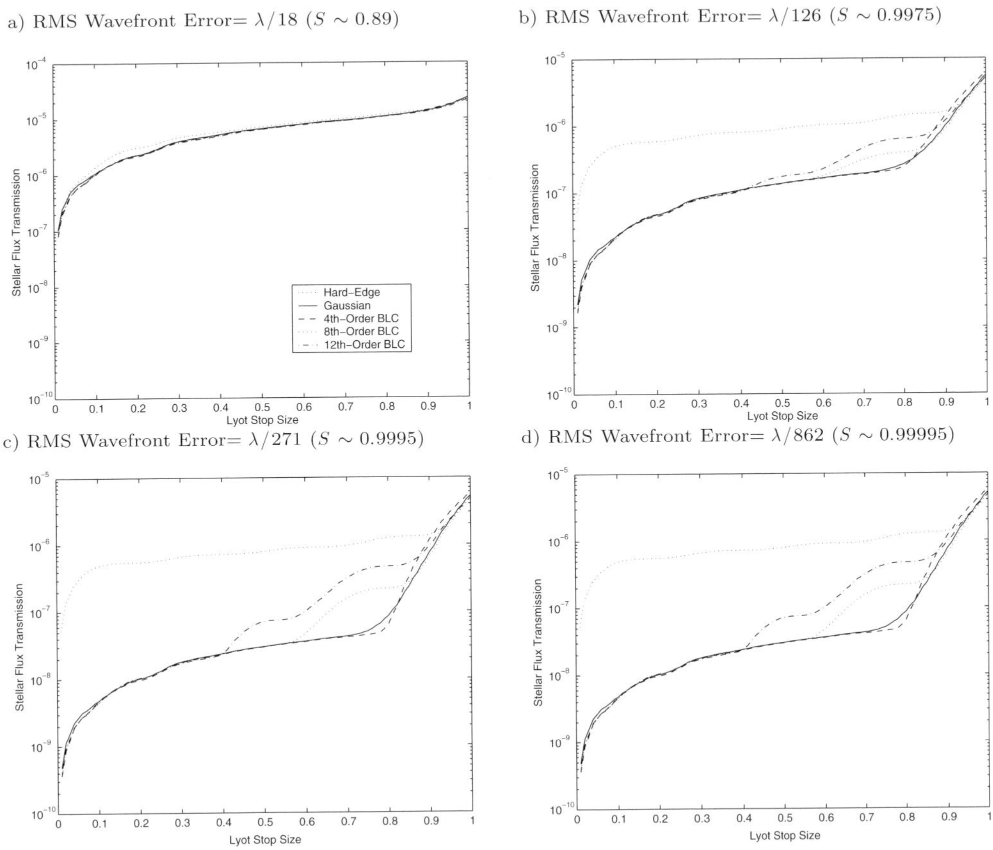

Figure 1. Breaking the mask degeneracy. Each figure displays the coronagraphic performance, characterized by the total stellar flux tranmission, versus the Lyot stop size (in percentage of the aperture). The stellar flux transmission is defined as the integrated stellar flux after the Lyot stop divided by the total stellar flux incident on the aperture normalized by the mask transmission profile. Figure (a) shows the level of AO correction where the performance degeneracy between traditional hard-edge masks and more sophisticated masks is just being broken. For wavefront correction worse than this threshold, each image mask performs equally well. Figure (b) demonstrates the reduced performance of the high-order (8th and 12th) band-limited masks for over-sized Lyot stops. Figure (c) shows the AO correction level at which the BLC-Gaussian degeneracy is broken. Figure (d) illustrates more clearly that at the sufficient level of AO correction, a traditional band-limited mask with an optimized Lyot stop can provide better contrast than a Gaussian image mask. Notice however, that for Lyot stop sizes below a certain threshold (~0.7), BLC and Gaussian masks continue to perform equally well. As the AO-correction level continues to increase, this threshold Lyot stop size pushes toward zero. So, for RMS wavefront error better than $\lambda/271$, 4th-order band-limited image masks can always provide contrast levels (at least) equivalent to a Gaussian mask, but at a higher throughput.

the hard-edge mask is indeed the least sensitive to mask alignment errors as one would expect. As wavefront correction improves, light diffracted by the hard-edge mask into the center of the Lyot plane begins to dominate and the high-order band-limited masks become the least sensitive to tip/tilt errors. In this regime (Figure 2b-d) 4th-order masks (including Gaussian and traditional band-limited masks), while providing better contrast than hard-edge masks, are clearly the most sensitive to mask alignment errors, and high-order masks should be preferentially chosen to reduce overhead time and improve observing efficiency.

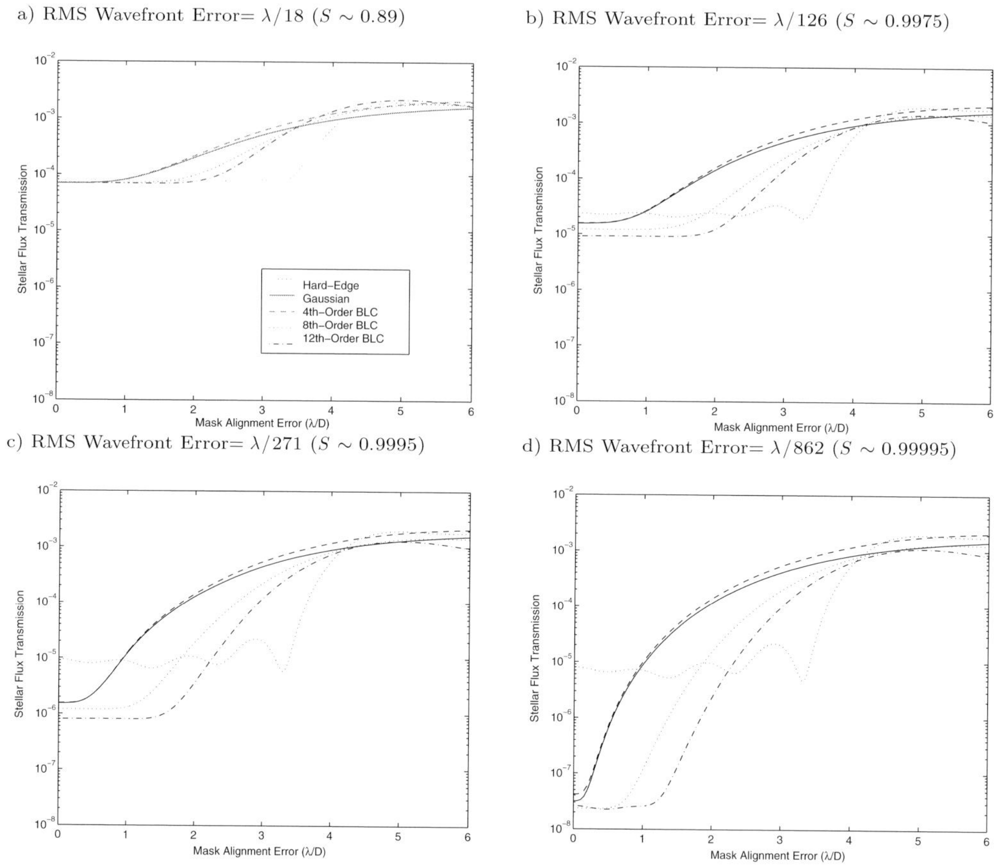

Figure 2. Tip/tilt mask alignment sensitivity for each mask in a variety of AO correction levels.

References

Crepp, J., Vanden Heuvel, A., Ge, J., Miller, S., & Kuchner, M.J. Oct. 2005, Submitted to ApJ

Kuchner, M.J. & Traub, W.A. 2002, ApJ, 570, 900

Kuchner, M.J., Crepp, J., & Ge, J. 2005, ApJ, 628, 466

Lloyd, J.P. & Sivaramakrishnan, A. 2005, ApJ, 621, 1153

Shaklan, S.B. & Green, J.J. 2005, ApJ, 628, 474

Sivaramakrishnan, A., Koresko, C.D., Makidon, R.B., Berkefeld,T., & Kuchner, M.J. 2001, ApJ, 552, 397

Direct Imaging of Exoplanets: Science & Techniques
Proceedings IAU Colloquium No. 200, 2005
A.C. Editor, B.D. Editor and C.E. Editor, eds.
© 2006 International Astronomical Union
doi:10.1017/S1743921306000111

Point Process Algorithm: A New Bayesian Approach for TPF-I Planet Signal Extraction

T. Velusamy, K. A. Marsh and B. Ware

Jet Propulsion Laboratory, 4800 Oak Grove Drive, Pasadena, California 91109, U.S.A.

email: velu@jpl.nasa.gov; Kenneth.A.Marsh@jpl.nasa.gov; Brent.Ware@jpl.nasa.gov

Abstract. TPF-I capability for planetary signal extraction, including both detection and spectral characterization, can be optimized by taking proper account of instrumental characteristics and astrophysical prior information. We have developed the Point Process Algorithm (PPA), a Bayesian technique for extracting planetary signals using the sine/cosine chopped outputs of a dual nulling interferometer. It is so-called because it represents the system being observed as a set of points in a suitably defined state space, thus providing a natural way of incorporating our prior knowledge of the compact nature of the targets of interest. It can also incorporate the spatial covariance of the exozodi as prior information which could help mitigate against false detections. Data at multiple wavelengths are used simultaneously, taking into account possible spectral variations of the planetary signals. Input parameters include the sigma of measurement noise and the *a priori* probability of the presence of a planet. The output can be represented as an image of the intensity distribution on the sky, optimized for the detection of point sources. Previous approaches by others to the problem of planet detection for TPF-I have relied on the potentially non-robust identification of peaks in a "dirty" image, usually a correlation map. Tests with synthetic data suggest that the PPA provides greater sensitivity to fainter sources than does the standard approach (correlation map + CLEAN), and will be a useful tool for optimizing the design of TPF-I.

Keywords. methods: data analysis, techniques: interferometric, stars: planetary systems.

1. Introduction

A key component of the proposed Terrestrial Planet Finder mission is a nulling interferometer (TPF-I) for the wavelength range 7–15 μm, currently envisaged as a free-flying four-element dual-Bracewell array. The outputs of two nulling pairs are combined coherently to produce a modulated signal as the interferometer is rotated about the line of sight to the star (Beichman & Velusamy 1999). This signal may be "sine-chopped" by differencing two phase-shifted versions ($\pi/2$ & $3\pi/2$) to maximize the information content for planetary signals (Velusamy & Beichman 2001).

Various inversion procedures exist for extracting the fluxes and locations of possible planets (see, for example, Angel & Woolf 1997; Velusamy & Marsh 2004); the one most widely used involves making a correlation map (referred to as a "dirty image") followed by deconvolution with the CLEAN algorithm (Draper *et al.* 2005). A key step in that procedure is an iterative search for peaks in the dirty image, which can be non-robust when noise bumps fall on "side lobes" of the responses to other sources.

In this paper we propose a Bayesian technique for planet detection which avoids the noise-vulnerable peak-finding step. It can process data at many wavelengths simultaneously, and can incorporate prior information such as the spatial covariance of the exozodi. Since the planetary signals are represented as a random set of points in a suitably-defined state space, we refer to the resulting algorithm as a "Point Process Algorithm" (PPA) following the terminology of Richardson & Marsh (1992).

2. Measurement Model

The starting point of our approach is a measurement model which relates a data vector, **d** (whose components are the complete set of samples of the sine-chop signal at all wavelengths of observation), to the intensity distribution, $I(x, y)$, on the sky. The latter is modeled as the superposition of a set of point sources of unknown number, fluxes and positions, upon an extended background whose intensity at position (x_j, y_j) is denoted by ζ_j. The distribution of point sources is represented as a set of occupation numbers in a 3-dimensional state space whose axes are flux and x, y position. The state space is divided into a regularly-sampled grid of cells, such that the cell labeled (f_k, x_j, y_j) represents the kth possible flux value, f_k, at the jth spatial position, (x_j, y_j). If a point source of that particular flux density is present at that particular position, then the occupation number of that cell, Γ_{jk}, will be equal to 1; otherwise, it will equal zero. This representation provides a convenient way of incorporating our prior knowledge of spatial dilution.

The measurement model must include a spectral model which relates the planetary fluxes at the different wavelengths. There are several possibilities, one of which is to treat the planets as black bodies at the local radiation temperature. If the orbital inclination and position angle of the orbital tilt axis are known, the planet temperature, $T_j \equiv T(x_j, y_j)$, can then be determined at each (x_j, y_j) from knowledge of the stellar luminosity and radial distance. Our measurement model can then be written as:

$$d_i = \sum_{j,k} f_k \left[B(\lambda_i, T_j)/B(\lambda_0, T_j)\right] H_i(x_j, y_j) \Gamma_{jk} + \sum_j H_i(x_j, y_j) \zeta_j(\lambda_i) + \nu_i \qquad (2.1)$$

where λ_i represents the wavelength corresponding to the ith measurement, $B(\lambda, T)$ represents the Planck function, and the f_k represent possible values of planetary flux at some suitably-defined reference wavelength, λ_0; $H_i(x_j, y_j)$ represents the response of the measurement system to a source of unit flux at (x_j, y_j), and ν_i is the measurement noise. We can rewrite (2.1) in more compact notation as:

$$\mathbf{d} = \mathbf{A\Gamma} + \mu \qquad (2.2)$$

where μ is a stochastic quantity representing the combined effects of ν_i and $\zeta_j(\lambda_i)$, and $\mathbf{\Gamma}$ is the state vector whose components are Γ_{jk}; $\mathbf{A}$ is a matrix whose elements are derived from values of $H_i(x_j, y_j)$ and other quantities as indicated by (2.1).

We assume ν_i is an uncorrelated zero-mean Gaussian random process (GRP). The exozodi distribution, $\zeta_j(\lambda_i)$ is also assumed to be a GRP, but with covariance C_ζ corresponding to the spatial correlation properties. If an estimate of the exozodi distribution is available from lower-resolution (possibly ground-based) observations, its contribution can be subtracted *a priori* from **d**, so that $\zeta_j(\lambda_i)$ may be regarded as zero-mean. The state vector, $\mathbf{\Gamma}$, is also regarded as a stochastic process whereby each cell is statistically uncorrelated, and the *a priori* probability of cell occupancy is a small number, P_1.

3. Estimation Procedure

The central operation is to estimate the state vector, $\mathbf{\Gamma}$, given the observed data. Specifically, we will obtain the expectation value of this quantity, defined by:

$$\rho\,(x_j, y_j, f_k\,|\mathbf{d}) \equiv E\,\Gamma_{jk} \tag{3.1}$$

where E is the expectation operator. We refer to $\rho\,(x_j, y_j, f_k\,|\mathbf{d})$ as a density, since it represents the average local density of occupied cells in the state space of position and flux.

Estimating the density in state space given a set of observed data is a generic problem in statistical mechanics, and has been applied to acoustical imaging (Richardson & Marsh 1987) and target tracking (Richardson & Marsh 1992). The solution procedure involves the solution of a hierarchy of integro-differential equations which fortunately can be truncated, to a good approximation, at the first member. We then obtain:

$$\frac{\partial \rho}{\partial t} + \phi_1 \rho = 0 \tag{3.2}$$

where t is a dimensionless progress variable representing the degree of conditioning on the data, and ϕ_1 is a conditioning factor which incorporates the data, $\mathbf{d}$, and the system response matrix, $\mathbf{A}$.

Equation (3.2) is to be integrated from $t = 0$ to a terminal value, t_f, defined by a stopping rule based on consistency with the observed data, corresponding to a reduced chi squared value (χ^2_ν) of unity.

The initial condition on ρ for the numerical solution of (3.2) is $\rho\,(t = 0) = \rho_0$, where ρ_0 is the *a priori* density equal to the constant value P_1; its integral over all state space represents the *a priori* expectation number of planets present.

Our corresponding estimate of the source intensity distribution is then:

$$\hat{I}\,(x_j, y_j) = \sum_k f_k \rho\,(x_j, y_j, f_k) \tag{3.3}$$

Estimates of the planet fluxes themselves may be obtained from the integrated value around each peak in this image; the uncertainties correspond to the standard *a posteriori* variances of maximum likelihood estimates.

4. Tests with Synthetic Data

We have tested the PPA using a set of 15 test cases designed to replicate realistic observing scenarios for TPF-I in the wavelength range 7–15 μm. Each case involved 0–5 planets at radial distances of 0.4–5.25 AU from a solar-type star at a distance of 10 or 15 pc, superposed on a 1 Zodi dust distribution. The 15 cases involved a total of 37 planets, 24 of which were of 1 Earth flux, and 2 of which were $<$ 1 Earth flux. The orbital inclination was 60° except for two face-on cases. All planets were at Earth temperature (260 K). Radial distances and signal-to-noise ratios were distributed as shown in Fig. 1.

The measurement configuration was the "X array" consisting of four 4-m detectors at the vertices of a rectangle of width 12 m (corresponding to the nulling baseline length) and a length of either 36 m or 72 m. Note that we have adopted 12m for nulling in lieu of the nominal design value of 20m to improve the sensitivity at the short wavelengths. The measurements consisted of the simulated sine chop signals through a full rotation of the array, at wavelengths of 7.44, 8.50, 9.92, 11.90 & 14.90 μm, and a total integration time of 1 day. The data generation included the effects of stellar leakage (due to incomplete

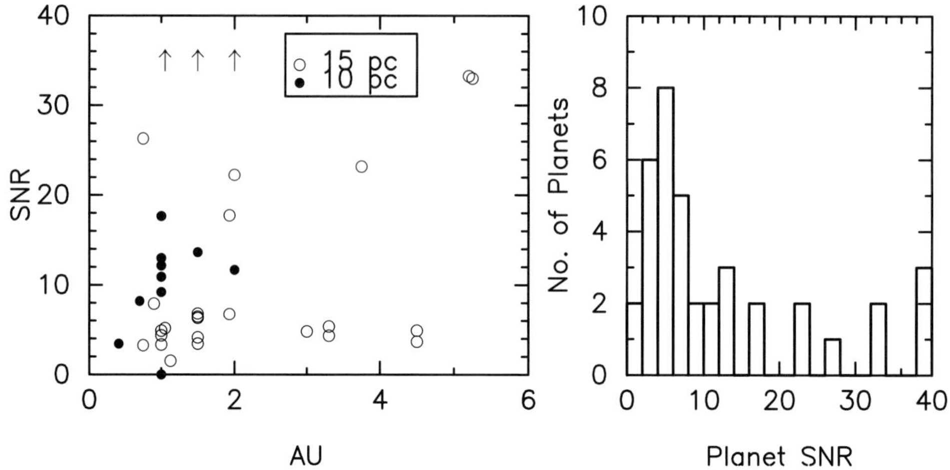

Figure 1. The distribution of planet parameters in the test cases, with respect to radial distance from the star and corresponding ideal SNR for an isolated planet. The distances to the stars are indicated.

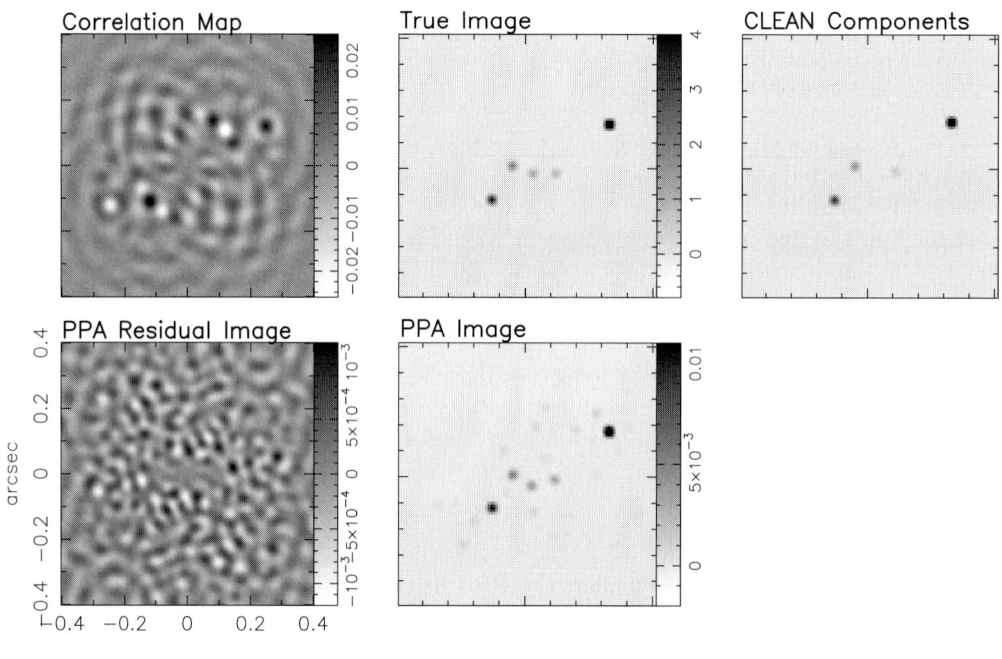

Figure 2. Images obtained by PPA and Clean methods for a test case involving 5 planets with the X-36m array. Similar images were obtained for all 15 test cases.

nulling of the stellar photosphere) and thermal emission from a symmetrical exozodiacal dust cloud. The noise model included the effects of Poisson noise from all components.

The data were inverted using the PPA, and also by the standard technique of "correlation map + CLEAN" for comparison purposes. The spatial covariance of the exozodi was ignored (i.e., C_ζ was set at zero) since symmetrical sources cancel in the sine chop signal. Full account was, however, taken of the Poisson noise contribution of this component. Figure 2 shows the results for a case involving 5 assumed planets.

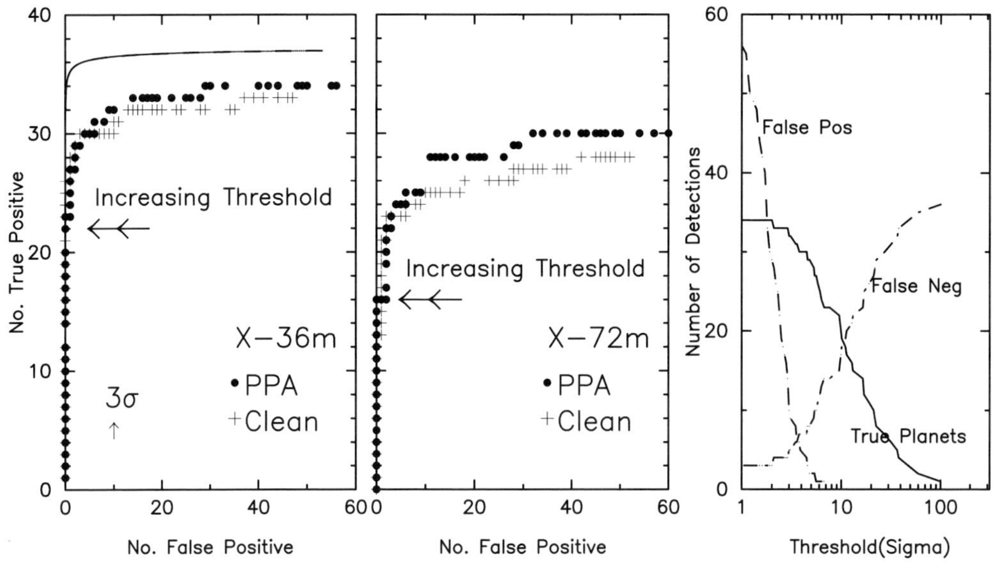

Figure 3. Planet detection statistics. The left two panels show the operating characteristic curves for PPA and CLEAN. The solid line on the left hand plot represents the idealized theoretical curve for an isolated planet. The right panel shows the detection performance of PPA as a function of threshold in sigmas.

These inversions yielded images for all 15 test cases (as exemplified by Figure 2) from which we extracted estimates of point source locations, fluxes, and corresponding uncertainties. We identified the six brightest peaks in each image, then applied a SNR threshold and counted the detections. Comparison with the true (assumed) planet positions then led to the number of true detections and false alarms as a function of the detection threshold in sigmas. Figure 3 shows a plot of true positives vs. false negatives, known as the operating characteristic curve of the detection system, for both PPA and CLEAN. It is apparent that for a large range of abscissa values, the PPA detected more planets than CLEAN, particularly with the 72 m array. The PPA performance as a function of detection threshold in sigmas is shown in Figure 3.

5. Discussion

The PPA is theoretically a near-optimal approach, and the present results support this. For the 72 m array in particular, the PPA has effectively increased the sensitivity of the array for planet detection over standard techniques exemplified by CLEAN.

The results also indicate that the 72 m array would be less sensitive than the 36 m array for this ensemble of planets, presumably due to the smaller relative range of spatial frequencies involved. By combining the data from both configurations (with same total integration time), we found that the added spatial frequency coverage provided by the 72 m array did not offset the reduced signal to noise of the 36 m data.

The detection statistics from Figure 3 indicates that:

- At the 5.8σ level we detect 26 of the 37 planets, but obtain 11 false positives; the 26 detections include 15 of the 24 Earth flux planets.
- A threshold of 5σ would facilitate detection of more than half of the Earths in a TPF survey, with only one false positive.
- Lowering the thresold to 4σ would facilitate detection of 19 of the 24 Earths in the sample, but with a penalty of 5 false positives.

Future work will emphasize the effects of exozodi, including the effects of spatial inhomogeneities which may masquerade as planets in the interferometer signals, and their mitigation via appropriate incorporation of *a priori* estimates and spatial covariance. For the latter purpose we will make use of recent observations of debris disks using the Spitzer Space Telescope (Beichman *et al.* 2005).

Acknowledgements

This work was performed by the Jet Propulsion Laboratory, California Institute of Technology, under contract with the National Aeronautics and Space Administration.

References

Angel, J. R. P. & Woolf, N. J. 1997, *ApJ* 475, 373

Beichman, C. A. *et al.* 2005, *ApJ* 622, 1160

Beichman, C. A. & Velusamy, T. 1999 in: S. C. Unwin & R. V. Stachnick (eds.), *Optical & IR Interferometry from Ground and Space* (ASP Conference Series) Vol. 194, pp 405–419

Draper, D. W. *et al.* 2005, *AJ*, submitted

Richardson, J. M. & Marsh, K. A. 1987, *Acoustical Imaging* 15, 615

Richardson, J. M. & Marsh, K. A. 1992, in: C. R. Smith *et al.* (eds.), *Maximum Entropy and Bayesian Methods* (Kluwer Academic Publishers, Netherlands), pp 213–220

Velusamy, T. & Beichman, C. A. 2001, *IEEE Aerospace Conference Proceedings* Vol 4, p 2013

Velusamy, T. & Marsh, K. A. 2004, *Proceedings of 2nd TPF/Darwin International Conference (Dust Disks and the Formation, Detection and Evolution of Habitable Planets)* San Diego, July 28–29, 2004

Discussion

KALTENEGGER: Do you find that the use of two baselines improves the results such that it justifies the double integration time needed then?

VELUSAMY: We find that for a given total integration time, it is better to put all of it into the 36 m configuration rather than splitting it between 36 m and 72 m, i.e. the added (u, v) coverage provided by the 72 m data did not quite compensate for the reduction in signal to noise of the 36 m data.

KALTENEGGER: Do your calculations take the properties of the instrument over a full rotation into account, i.e., do you extract the signal over a full rotation of the instrument, or over a fraction of a full rotation, or do you assume no rotation for your calculation?

VELUSAMY: The inversion makes simultaneous use of the data at all rotation angles of the array, from 0° to 360° in increments of typically 1°, as in our simulations. However the algorithm can work on any given data set with partial or complete rotation.

Direct Imaging of Exoplanets: Science & Techniques
Proceedings IAU Colloquium No. 200, 2005
Claude Aime and Farrokh Vakili, eds.
© 2006 International Astronomical Union
doi:10.1017/S1743921306000081

Concluding Remarks, or Pilgrim's Progress: The Road Ahead

Wesley A. Traub[1,2]†

[1]Jet Propulsion Laboratory, M/S 301-451, 4800 Oak Grove Dr., Pasadena, CA 91109, USA
email: wtraub@jpl.nasa.gov

[2]Harvard-Smithsonian Center for Astrophysics, 60 Garden St., Cambridge, MA 02138, USA

Abstract. This has been a wonderful conference, in a beautiful setting, with many of the world's planet-seekers on hand to present their ideas, at a time when we stand technically almost ready to embark on designing space missions to search for Earth-like planets and signs of life on them. In these remarks, I wish to simply list the 'big questions' that face us today, and offer some personal comments on how we might address these questions.

For a bit of inspiration from the past, I will close with a parallel that might be drawn between the journey that we all wish to embark upon, to search for extrasolar planets and to characterize them, and the one described by John Bunyan in his 1678 allegory, Pilgrim's Progress.

A remarkable aspect of this conference has been the agreement on our common goal, to detect and characterize extrasolar planets. While it is true that various of us suscribe to different theories of how planetary systems are formed, or how to best build the telescopes that can investigate these planetary systems, nevertheless we do all believe that the next great frontier in astrophysics is the study of extrasolar planets, and in particular Earth-like planets.

The topic of this conference can be phrased in terms of the following 'big questions'. Are there Earth-like planets around nearby stars? Are these planets habitable? Do they show signs of life?

These questions generate many sub-questions that we need to answer. In the following I list the questions as I see them today. I will phrase them in terms that can be easily understood by most of the public. Doubtless you will have other questions, but this is a start.

Keywords. stars:planetary systems, techniques: high angular resolution

1. What does a 'normal' planetary system look like?

Until 1995 most astronomers thought that other planetary systems probably looked like the Solar System. Since then radial-velocity (RV) data have shown that giant planets orbiting very close to their parent stars are clearly the (observationally-biased) norm. Models continue to predict that planetary systems are probably formed more along traditional lines (i.e., roughly similar to the Solar System), and that migration is required to produce the RV-observed systems. Improved RV data will permit us to see more planets in existing systems, if they are there, but we are not likely to reach the terrestrial planet range with this technique. We need Kepler and SIM to tell us more, but these too are slightly biased toward large inner planets. Ultimately we need images from TPF-C and TPF-I to tell us what the systems look like, although each of these has its own biases as well. Regarding zodiacal dust, recall that we are at present quite uncertain about whether the median planetary system contains a lot of dust, or a little. By itself the level of dust will tell us about processes in the system, but in addition we are hoping that a relatively low level of dust is present because this will keep the background noise low, and make it easier for TPF-C or TPF-I to find and characterize the planet.

† Present location: Jet Propulsion Laboratory

2. What are the characteristics of its terrestrials, giants, and disk?

In addition to knowing that an object simply exists, we also need to know the physical and chemical characteristics of the terrestrial planets, giant planets, and zodiacal dust (and gas) in a planetary system. For this we first need to obtain colors and spectroscopy, following the traditional astronomical path of measurement priorities. Planets are likely to be even more diverse than stars, so it is not obvious how color, for example, tells us about a planet, but by the twin routes of modeling and observation in the Solar System, we can expect to learn how to characterize a planet using color. Of course, color is discussed before spectroscopy because the targets are very faint, and the first information we can expect to have is existence, followed by color, and after much more integration time, spectroscopy. The advantage of spectroscopy is that it directly relates to atomic and molecular species (as color does not). We can add to this list the properties of surfaces, both solid and cloudy, as something that spectroscopy will tell us about. Finally recall that the color of the thermal-infrared continuum will in principle tell us about the effective temperature, however complications of clouds and spectral bands will make this interpretation problematic unless we also understand something about the spectrum.

3. What is the frequency of occurrence of 'normal' systems?

Asking about the probability of finding a 'normal' stellar system is a broader version of the question, what is the probability of finding an Earth-like planet in the habitable zone? From Kepler and SIM we will get distribution functions, biased toward the several-Earth size and mass range, but with minimum detectable values sufficiently close to the one-Earth level that we will be able to extrapolate to that level with reasonable confidence. However we will need to image systems with TPF-C and TPF-I to know with certainty if the planets in the one-Earth range are present, and if present, what types of planets they are. Thus even though we expect that Kepler and SIM will tell us the distribution function of planets in the mass range down to about an Earth, we will also need to spectroscopically characterize an Earth-size object to find out if it is Venus-like, Mars-like, or truly Earth-like.

4. Do Earth-like planets exist nearby?

Kepler will tell us about the likely abundance of Earth-size planets around its target stars, but these stars are several hundred parsecs distant, which is too far to be able to characterize them using either TPF-C or TPF-I. Therefore we will also need SIM to make a close examination of nearby planets, where it will be most sensitive, to tell us if Earth-mass planets are in the Solar neighborhood (roughly less than 30 pc). TPF-C and TPF-I can also search this same set of nearby stars, but at the cost of spending time searching instead of characterizing, and at the cost of not having dynamical masses for the objects that they will see in common with SIM.

5. Does life exist on these planets?

The search for life in the universe, or even around nearby stars, has progressed over the past 10 years from being a topic that was essentially the exclusive domain of supermarket tabloids, to a topic that professional scientists can now discuss in public without fear of being laughed at by their colleagues. My personal view is that the goal of astrophysics is to explain the evolution of the *entire* Universe, starting with the Big Bang, followed by galaxy formation, star formation, planet formation, and the development of life on a planet. To defend the last point, recall that major issues in planet formation include the following: the initial presence or delivery of water to a planet, the degree of bombardment suffered by a planet, the influence of radiation, the possible delivery of complex molecules, the origin and protective role of planetary magnetic fields, the influence of massive planets in the same system, etc. All these elements, and more, can have a major influence on whether life can start and flourish on a planet, and all of these are in the proper domain of astrophysics.

We have within our grasp the technical means to detect the likely presence of life on an extrasolar planet. I emphasize 'likely' because we do not have a handy 'life molecule' whose signature will tell us unequivocally that it was produced by living organisms. (However a laser beacon from a planet would do quite nicely!) What we do have is spectral signatures of molecules whose presence strongly indicates the presence of life. (a) The strongest of these is nitrous oxide, which can only be produced (in quantity) by life; unfortunately it is present only in a blended region of the infrared spectrum overlapped by water and methane, and also carbon dioxide. However missions beyond TPF-I, with greater collecting area and resolving power, could unambiguously identify this species. (b) The presence of a large amount of molecular oxygen is a good indicator of life, because only life can produce this species in abundance. Photochemistry can generate small amounts, but other chemical pathways consume it before long. Only the sustained production by life can overcome the loss mechanisms, and produce a high abundance. Ozone, produced photochemically from oxygen, is a good secondary indicator of oxygen. (c) The simultaneous presence of a reduced species with an oxidizing species is a good indicator of a system that is not in local thermodynamic equilibrium, and in particular, life. Thus, for example, if we can see both methane and oxygen on the same planet, we will have a good indication of life. (d) Habitability is a precondition for life, and is indicated by the combination of surface temperature and water vapor, together pointing to liquid water on the surface. (e) The 'red edge' of vegetation on land indicates the presence of photosynthesis in the leaves of land plants on Earth. It is conceivable that a different process operates on another planet, for example a 3-photon one in place of the Earth's 2-photon process, which might move the red edge farther to the red. (f) Other indicators of environments conducive to life might include the presence of a substantial, cloud-free atmosphere (by Rayleigh scattering), active weather systems (from random variations of cloud reflectance or emission), and rotation period (from periodic variations of reflectance or emission).

6. What information can we learn about a planet?

All of the above indicators of life on a planet can be detected spectroscopically in the visible, infrared, or both. In addition we can get mass from SIM, and size from the occasional instance of a transit. We get effective temperature from the thermal infrared, and effective area as well. From mass and area we get density, hence the type of planet (ice, gas, rock). From visible brightness and area we get albedo, hence the type of reflecting material (in broad terms, gas, cloud, rock, or a combination). From mass and radius we get surface gravity. Combined with temperature we get the lapse rate. Rayleigh scattering gives the total air mass above a reflecting surface. Spectroscopy tells us the degree of greenhouse warming, and the altitude of the effective temperature, from which we can extrapolate (using the lapse rate) to get surface temperature. The temperature and spectrum will tell us if liquid water is present on the surface, and therefore if any initially large amount of carbon dioxide might have been weathered into rock. Mass, gravity, and density will suggest if plate tectonics might be active, and therefore if the sequestered carbon dioxide may have been pushed further below the surface. The variation of reflectivity will indicate possible land-ocean partitioning, and thus the fraction of land and water.

7. How do we prepare ourselves, scientifically, to learn about expected, and unexpected, characteristics of extrasolar planets?

From a combination of observation and modeling we can explore possible types of planet systems, but if we base our instruments on TPF-C and TPF-I solely on these examples, we may miss out on an unexpected color or spectral signature. Recall that the first Mars surface sampler sent back an apparently positive sign of life, but this was a mistake caused by a trace amount of Martian hydrogen peroxide that had not been anticipated. It is possible that we might be quite satisfied with discovering Earth-like planets and life around other stars, however if we do this with blinders we might miss something else. To guard against missing the unexpected, the only solution is to use as broad a spectral range as is feasible, and to use as high a spectral resolving power as possible. (Needless to say the instrument should be optically efficient too, otherwise we cannot utilize the increased resolving power.) Perhaps the best solution is to use a general-purpose spectrograph, as opposed to a potentially more efficient spectral filter bank.

8. Is our experience with the Earth and the Solar System sufficient preparation?

This question presumes that we will do all that we can to understand what an Earth-twin, and a Solar-System twin would look like, if placed around a nearby star. To this we must add theoretical modeling, to see what a different set of planets would look like, and evolutionary modeling to see what the Solar-System planets looked like during the past 4.5 billion years, and in the future.

9. How do we detect these planets and disks?

After much debate and study, the US has settled on the dual approach of both a visible coronagraph and an infrared interferometer, and this combination has some very strong merits. However we should still remain alert to other methods. We should also remain open to using other observatories to probe planetary systems, some outstanding examples of which are using HST for transits (unfortunately no longer possible) and spectroscopy, Spitzer for infrared transits and dust emission, and ground-based interferometers for zodiacal light measurements. In the future, JWST in space and extreme adaptive optics on the ground are also of interest.

10. How do we solve the 'physics problem' of detection and characterization?

The 'physics problem' is the question of how build an optical system that will reduce the star's diffracted and scattered light to a low enough level that we can see the direct light from a planet without significant interference. At present we have a wonderful array of potential techniques to do both of these, as discussed in many papers during this conference. It is remarkable that most of these methods cannot be found in optics textbooks, because they have been newly invented specifically for planet-finding. We still need more such inventions. However today we have in hand techniques that will allow us to begin designing TPF-C and TPF-I, and to refine the design as we go along. We need to continue to carry out laboratory demonstrations, and to work on theoretical methods of improving our current designs.

11. Are precursor missions needed?

This is a difficult question. The pro argument says that a precursor mission (e.g., a small version of the ultimately-planned TPF-C or TPF-I) is valuable because it provides technological development and experience at a reduced cost, thereby reducing risk in the follow-on full-scale mission. The con side says that there is a political risk in flying a precursor mission because it may turn out that people will mentally convert it to the final mission, and therefore never fund the latter. One way out of this is to do the equivalent of a precursor mission under a different heading, and at a strictly capped cost, so that the small mission cannot easily be viewed as being the same as the final one.

12. Is international cooperation necessary?

Cooperation is good in principle (to eliminate duplication, and waste of resources), but can be difficult in practice (culture clashes, trade and information restrictions). My view is that we should work hard to develop international cooperation to help ourselves search for and characterize extrasolar planets, because this is a project for which there is strong interest by the people of all nations, but which will be expensive enough that sharing the cost is a prudent step. We already have in place official agreements to initiate joint work between Europe and the US on the TPF-I/Darwin concept. We should consider expanding this agreement to cover the TPF-C project, and to include other interested nations, such as Japan.

13. Can we open this new branch of astrophysics, and also provide benefits to other branches?

Realizing that TPF-C and TPF-I are potentially expensive projects, and that funding is always limited, it would be wise for us to design these missions to serve as broad a segment of the astrophysics community as is possible. Given that these missions have much greater power in areas such as collecting area, angular resolution, wavelength coverage, stability, and other key aspects, it seems logical that more traditional areas of astrophysics could benefit from shared use of mission time. We should actively explore these areas of cooperation, and engage our colleagues in the process from the start.

14. How do we prepare for the public response to an Earth-like planet? Or to life? Or to no terrestrial planets?

We all believe that it would be a historically major event to discover a planet that looks like the Earth, and perhaps has life on it. It would be major because people have wondered about such questions throughout the ages. We are highly privileged to live in a time when this question can be answered, and even more privileged to be the scientists who can make these discoveries. Even the statement from us that we have searched high and low for Earth-like planets but found nothing, would be a major discovery. We should think about how we want to approach these topics, especially in our dealings with the public, first to be careful to explain that this is good science, and second to not over-promise results. We should also think about what we would say if a reporter calls up and asks what we think about such a discovery. It is possible that the public will see such a discovery as not being important to their daily lives. But I suspect that instead it will be seen as a major turning point, analogous to the time when Copernicus moved the center of the universe from the Earth to the Sun.

15. An allegorical afterword.

This year we had the exhilarating experience of being on the verge of a new start for TPF, closely followed by the dispiriting experience of having our funding drastically reduced. This was not the first life-death episode for TPF, nor is it likely to be the last, judging from the history of other missions. The need for governments to spend their resources on other projects will always be with us, just like floods, famine, hurricanes, and wars will always be with us (and in part just because of these things). We must have the confidence of our convictions that the search for extrasolar planets, and their characterization, is a long-term project of major scientific, and societal, importance. Therefore we must weather these changes, and be prepared for the opportunity when better times return.

These thoughts bring to mind the allegory 'Pilgrim's Progress', written in 1678 by John Bunyan in England. Bunyan was thinking of religion in this book, but the images can easily be applied to our labors as well. Specifically, Bunyan envisioned the journey of a person, the eponymous pilgrim, from a land of troubles, down a long road filled with tests of pilgrim's resolution, to the promised land beyond. Our journey from this year to the one in which we have an operating mission, is similar. Pilgrim first passed through the Slough of Despond (that's us today, of course). He then encountered Mr. Worldly Wiseman, the Evangelist, Mr. Obstinate, Mr. Legality, Messrs. Passsion, Patience, Despair, Simple, Sloth, Presumption, Mistrust, and Timorous. Pilgrim climbs the Hill of Difficulty, enters the valley of Humiliation, the Valley of the Shadow of Death, and many more such trials. He finally climbs the Delectable Mountains, and enters the Coelestial City. It was easy for Bunyan's readers to apply the allegory to their lives, and just as easy for us to apply it to our enterprise today. Pilgrim learns the lesson that dedication and steadfastness were the keys to achieving his goal, and so they can be for us.

References

Bunyan, J. 1678 *The Pilgrim's Progress*, Barnes and Nobel Classics, New York, 2005

Author Index

Object Index